Digital Computer Electronics

Third Edition

Albert Paul Malvino, Ph.D.
Jerald A. Brown

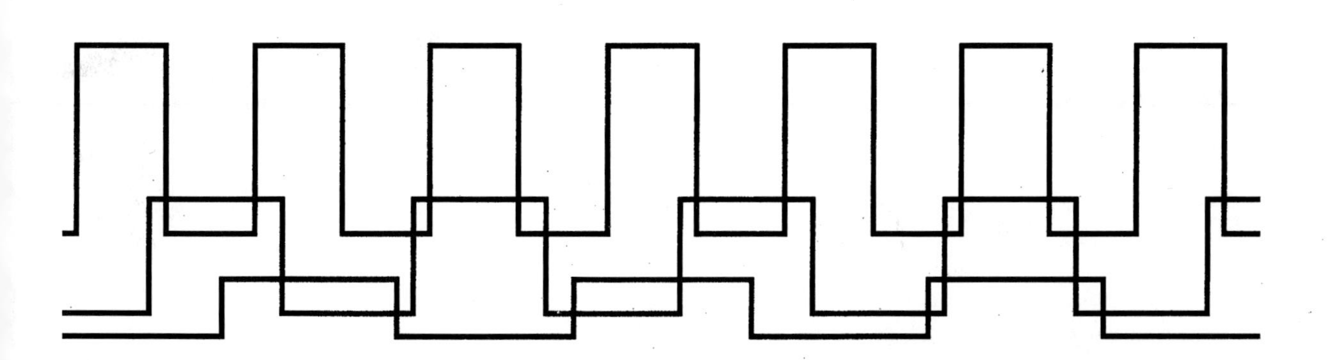

GLENCOE

Macmillan/McGraw-Hill

New York, New York Columbus, Ohio Mission Hills, California Peoria, Illinois

This textbook was prepared with the assistance of Publishing Advisory Service.

LSI circuit photo: Manfred Kage/Peter Arnold Inc.

To my wife, Joanna, who encourages me to write.
And to my daughters, Joanna, Antonia, Lucinda,
Patricia, and Miriam, who keep me young.
—A.P.M.

. . . to my wife Vickie
dearest friend
fellow adventurer
love of my life
—J.A.B.

Library of Congress Cataloging-in-Publication Data

Malvino, Albert Paul.
Digital computer electronics / Albert Paul Malvino, Jerald A.
Brown. — 3rd ed.
p. cm.
Includes index.
ISBN 0-02-800594-5 (hardcover)
1. Electronic digital computers. 2. Microcomputers. 3. Intel
8085 (Microprocessor) I. Brown, Jerald A. II. Title.
TK7888.3.M337 1993
621.39'16—dc20 92-5895
CIP

Digital Computer Electronics, Third Edition

ISBN 0-02-800594-5

Printed in the United States of America

2 3 4 5 6 7 8 9 A-KP 99 98 97 96 95 94 93 92

Contents

PART 4

Microprocessor Instruction Set Tables 379

Preface

Textbooks on microprocessors are sometimes hard to understand. This text attempts to present the various aspects of microprocessors in ways that are understandable and interesting. The only prerequisite to using this textbook is an understanding of diodes and transistors.

A unique aspect of this text is its wide range. Whether you are interested in the student-constructed SAP (simple-as-possible) microprocessor, the 6502, the 6800/6808, the 8080/8085/Z80, or the 8086/8088, this textbook can meet your needs.

The text is divided into four parts. These parts can be used in different ways to meet the needs of a wide variety of students, classrooms, and instructors.

Part 1, Digital Principles, is composed of Chapters 1 to 9. Featured topics include number systems, gates, boolean algebra, flip-flops, registers, counters, and memory. This information prepares the student for the microprocessor sections which follow.

Part 2, which consists of Chapters 10 to 12, presents the SAP (simple-as-possible) microprocessor. The student constructs this processor using digital components. The SAP processor contains the most common microprocessor functions. It features an instruction set which is a subset of that of the Intel 8085—leading naturally to a study of that microprocessor.

Part 3, Programming Popular Microprocessors (Chapters 13 to 23), simultaneously treats the MOS/Rockwell 6502, the Motorola 6800/6808, the Intel 8080/8085 and Zilog Z80, and the 16-bit Intel 8086/8088. Each chapter is divided into two sections. The first section presents new concepts; second section applies the new concepts to each microprocessor family. Discussion, programming examples, and problems are provided. The potential for comparative study is excellent.

This part of the text takes a strong programming approach to the study of microprocessors. Study is centered around the microprocessor's instruction set and programming model. The 8-bit examples and homework problems can be performed by using either hand assembly or cross-assemblers. The 16-bit 8086/8088 examples and problems can be performed by using either an assembler or the DOS DEBUG utility.

Part 4 is devoted to the presentation of the instruction sets of each microprocessor family in table form. Several tables are provided for each microprocessor family, permitting instructions to be looked up alphabetically, by op code, or by functional category, with varying levels of detail. The same functional categories are correspondingly used in the chapters in Part 3. This coordination between parts makes the learning process easier and more enjoyable.

Additional reference tables are provided in the appendixes. Answers to odd-numbered problems for Chapters 1 to 16 follow the appendixes.

A correlated laboratory manual, *Experiments for Digital Computer Electronics* by Michael A. Miller, is available for use with this textbook. It contains experiments for every part of the text. It also includes programming problems for each of the featured microprocessors.

A teacher's manual is available which contains answers to all of the problems and programs for every microprocessor. In addition, a diskette (MS-DOS 360K 5¼-inch diskette) containing cross-assemblers is included in the teacher's manual.

Special thanks to Brian Mackin for being such a patient and supportive editor. To Olive Collen for her editorial work. To Michael Miller for his work on the lab manual. And to Thomas Anderson of Speech Technologies Inc. for the use of his cross-assemblers. Thanks also to reviewers Lawrence Fryda, Illinois State University; Malachi McGinnis, ITT Technical Institute, Garland Texas; and Benjamin Suntag.

Albert Paul Malvino
Jerald A. Brown

A man of true science uses but few hard words,
and those only when none other will answer his purpose;
whereas the smatterer in science thinks that
by mouthing hard words he understands hard things.

Herman Melville

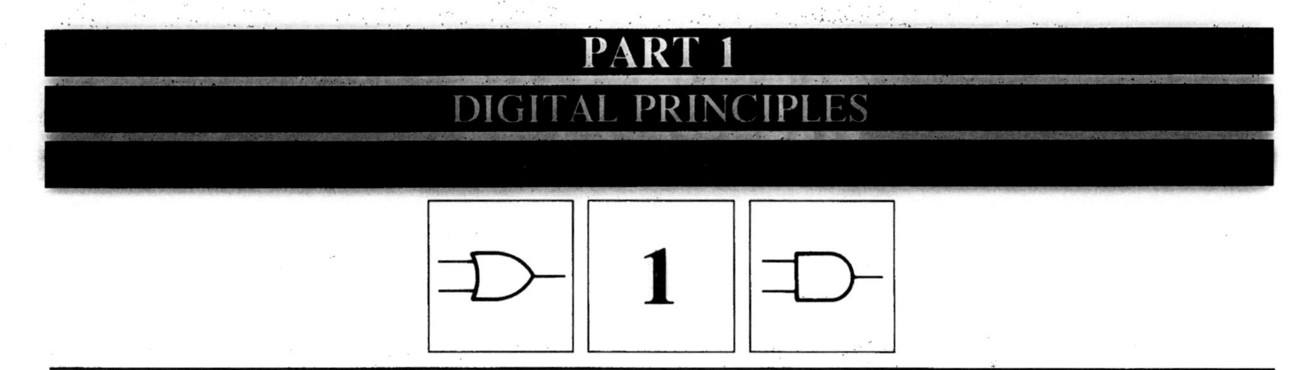

NUMBER SYSTEMS AND CODES

Modern computers don't work with decimal numbers. Instead, they process *binary numbers,* groups of 0s and 1s. Why binary numbers? Because electronic devices are most reliable when designed for two-state (binary) operation. This chapter discusses binary numbers and other concepts needed to understand computer operation.

1-1 DECIMAL ODOMETER

René Descartes (1596–1650) said that the way to learn a new subject is to go from the known to the unknown, from the simple to the complex. Let's try it.

The Known

Everyone has seen an odometer (miles indicator) in action. When a car is new, its odometer starts with

00000

After 1 mile the reading becomes

00001

Successive miles produce 00002, 00003, and so on, up to

00009

A familiar thing happens at the end of the tenth mile. When the units wheel turns from 9 back to 0, a tab on this wheel forces the tens wheel to advance by 1. This is why the numbers change to

00010

Reset-and-Carry

The units wheel has reset to 0 and sent a carry to the tens wheel. Let's call this familiar action *reset-and-carry.*

The other wheels also reset and carry. After 999 miles the odometer shows

00999

What does the next mile do? The units wheel resets and carries, the tens wheel resets and carries, the hundreds wheel resets and carries, and the thousands wheel advances by 1, to get

01000

Digits and Strings

The numbers on each odometer wheel are called *digits.* The decimal number system uses ten digits, 0 through 9. In a decimal odometer, each time the units wheel runs out of digits, it resets to 0 and sends a carry to the tens wheel. When the tens wheel runs out of digits, it resets to 0 and sends a carry to the hundreds wheel. And so on with the remaining wheels.

One more point. A *string* is a group of characters (either letters or digits) written one after another. For instance, 734 is a string of 7, 3, and 4. Similarly, 2C8A is a string of 2, C, 8, and A.

1-2 BINARY ODOMETER

Binary means two. The binary number system uses only two digits, 0 and 1. All other digits (2 through 9) are thrown away. In other words, binary numbers are strings of 0s and 1s.

An Unusual Odometer

Visualize an odometer whose wheels have only two digits, 0 and 1. When each wheel turns, it displays 0, then 1, then back to 0, and the cycle repeats. Because each wheel has only two digits, we call this device a *binary odometer*.

In a car a binary odometer starts with

0000 (zero)

After 1 mile, it indicates

0001 (one)

The next mile forces the units wheel to reset and carry; so the numbers change to

0010 (two)

The third mile results in

0011 (three)

What happens after 4 miles? The units wheel resets and carries, the second wheel resets and carries, and the third wheel advances by 1. This gives

0100 (four)

Successive miles produce

0101 (five)
0110 (six)
0111 (seven)

After 8 miles, the units wheel resets and carries, the second wheel resets and carries, the third wheel resets and carries, and the fourth wheel advances by 1. The result is

1000 (eight)

The ninth mile gives

1001 (nine)

and the tenth mile produces

1010 (ten)

(Try working out a few more readings on your own.)

You should have the idea by now. Each mile advances the units wheel by 1. Whenever the units wheel runs out of digits, it resets and carries. Whenever the second wheel runs out of digits, it resets and carries. And so for the other wheels.

Binary Numbers

A binary odometer displays binary numbers, strings of 0s and 1s. The number 0001 stands for 1, 0010 for 2, 0011 for 3, and so forth. Binary numbers are long when large amounts are involved. For instance, 101010 represents decimal 42. As another example, 111100001111 stands for decimal 3,855.

Computer circuits are like binary odometers; they count and work with binary numbers. Therefore, you have to learn to count with binary numbers, to convert them to decimal numbers, and to do binary arithmetic. Then you will be ready to understand how computers operate.

A final point. When a decimal odometer shows 0036, we can drop the leading 0s and read the number as 36. Similarly, when a binary odometer indicates 0011, we can drop the leading 0s and read the number as 11. With the leading 0s omitted, the binary numbers are 0, 1, 10, 11, 100, 101, and so on. To avoid confusion with decimal numbers, read the binary numbers like this: zero, one, one-zero, one-one, one-zero-zero, one-zero-one, etc.

1-3 NUMBER CODES

People used to count with pebbles. The numbers 1, 2, 3 looked like ●, ●●, ●●●. Larger numbers were worse: seven appeared as ●●●●●●●.

Codes

From the earliest times, people have been creating codes that allow us to think, calculate, and communicate. The decimal numbers are an example of a code (see Table 1-1). It's an old idea now, but at the time it was as revolutionary; 1 stands for ●, 2 for ●●, 3 for ●●●, and so forth.

Table 1-1 also shows the binary code. 1 stands for ●, 10 for ●●, 11 for ●●●, and so on. A binary number and a decimal number are equivalent if each represents the same amount of pebbles. Binary 10 and decimal 2 are equivalent because each represents ●●. Binary 101 and decimal 5 are equivalent because each stands for ●●●●●.

TABLE 1-1. NUMBER CODES

Decimal	Pebbles	Binary
0	None	0
1	●	1
2	●●	10
3	●●●	11
4	●●●●	100
5	●●●●●	101
6	●●●●●●	110
7	●●●●●●●	111
8	●●●●●●●●	1000
9	●●●●●●●●●	1001

Equivalence is the common ground between us and computers; it tells us when we're talking about the same thing. If a computer comes up with a binary answer of 101, equivalence means that the decimal answer is 5. As a start to understanding computers, memorize the binary-decimal equivalences of Table 1-1.

EXAMPLE 1-1

Figure 1-1*a* shows four light-emitting diodes (LEDs). A dark circle means that the LED is off; a light circle means it's on. To read the display, use this code:

(a) (b)

Fig. 1-1 LED display of binary numbers.

LED	Binary
Off	0
On	1

What binary number does Fig. 1-1*a* indicate? Fig. 1-1*b*?

SOLUTION

Figure 1-1*a* shows off-off-on-on. This stands for binary 0011, equivalent to decimal 3.

Figure 1-1*b* is off-on-off-on, decoded as binary 0101 and equivalent to decimal 5.

EXAMPLE 1-2

A binary odometer has four wheels. What are the successive binary numbers?

SOLUTION

As previously discussed, the first eight binary numbers are 0000, 0001, 0010, 0011, 0100, 0101, 0110, and 0111. On the next count, the three wheels on the right reset and carry; the fourth wheel advances by one. So the next eight numbers are 1000, 1001, 1010, 1011, 1100, 1101, 1110, and 1111. The final reading of 1111 is equivalent to decimal 15. The next mile resets all wheels to 0, and the cycle repeats.

Being able to count in binary from 0000 to 1111 is essential for understanding the operation of computers.

TABLE 1-2. BINARY-TO-DECIMAL EQUIVALENCES

Decimal	Binary	Decimal	Binary
0	0000	8	1000
1	0001	9	1001
2	0010	10	1010
3	0011	11	1011
4	0100	12	1100
5	0101	13	1101
6	0110	14	1110
7	0111	15	1111

Therefore, you should memorize the equivalences of Table 1-2.

1-4 WHY BINARY NUMBERS ARE USED

The word "computer" is misleading because it suggests a machine that can solve only numerical problems. But a computer is more than an automatic adding machine. It can play games, translate languages, draw pictures, and so on. To suggest this broad range of application, a computer is often referred to as a *data processor*.

Program and Data

Data means names, numbers, facts, anything needed to work out a problem. Data goes into a computer, where it is processed or manipulated to get new information. Before it goes into a computer, however, the data must be coded in binary form. The reason was given earlier: a computer's circuits can respond only to binary numbers.

Besides the data, someone has to work out a *program*, a list of instructions telling the computer what to do. These instructions spell out each and every step in the data processing. Like the data, the program must be coded in binary form before it goes into the computer.

So the two things we must input to a computer are the program and the data. These are stored inside the computer before the processing begins. Once the computer run starts, each instruction is executed and the data is processed.

Hardware and Software

The electronic, magnetic, and mechanical devices of a computer are known as *hardware*. Programs are called *software*. Without software, a computer is a pile of "dumb" metal.

An analogy may help. A phonograph is like hardware and records are like software. The phonograph is useless without records. Furthermore, the music you get depends on the record you play. A similar idea applies to computers. A computer is the hardware and programs are the software. The computer is useless without programs. The program stored in the computer determines what the computer will do; change the program and the computer processes the data in a different way.

Transistors

Computers use *integrated circuits* (ICs) with thousands of transistors, either bipolar or MOS. The parameters (β_{dc}, I_{CO}, g_m, etc.) can vary more than 50 percent with temperature change and from one transistor to the next. Yet these computer ICs work remarkably well despite the transistor variations. How is it possible?

The answer is *two-state* design, using only two points on the load line of each transistor. For instance, the common two-state design is the cutoff-saturation approach; each transistor is forced to operate at either cutoff or saturation. When a transistor is cut off or saturated, parameter variations have almost no effect. Because of this, it's possible to design reliable two-state circuits that are almost independent of temperature change and transistor variations.

Transistor Register

Here's an example of two-state design. Figure 1-2 shows a transistor register. (A *register* is a string of devices that store data.) The transistors on the left are cut off because the input base voltages are 0 V. The dark shading symbolizes the cutoff condition. The two transistors on the right have base drives of 5 V.

The transistors operate at either saturation or cutoff. A base voltage of 0 V forces each transistor to cut off, while a base voltage of 5 V drives it into saturation. Because of this two-state action, each transistor stays in a given state until the base voltage switches it to the opposite state.

Another Code

Two-state operation is universal in digital electronics. By deliberate design, all input and output voltages are either low or high. Here's how binary numbers come in: low voltage represents binary 0, and high voltage stands for binary 1. In other words, we use this code:

Voltage	Binary
Low	0
High	1

For instance, the base voltages of Fig. 1-2 are low-low-high-high, or binary 0011. The collector voltages are high-high-low-low, or binary 1100. By changing the base voltages we can store any binary number from 0000 to 1111 (decimal 0 to 15).

Bit

Bit is an abbreviation for binary digit. A binary number like 1100 has 4 bits; 110011 has 6 bits; and 11001100 has 8 bits. Figure 1-2 is a 4-bit register. To store larger binary numbers, it needs more transistors. Add two transistors and you get a 6-bit register. With four more transistors, you'd have an 8-bit register.

Nonsaturated Circuits

Don't get the idea that all two-state circuits switch between cutoff and saturation. When a bipolar transistor is heavily saturated, extra carriers are stored in the base region. If the base voltage suddenly switches from high to low, the transistor cannot come out of saturation until these extra carriers have a chance to leave the base region. The time it takes for these carriers to leave is called the *saturation delay time* t_d. Typically, t_d is in nanoseconds.

In most applications the saturation delay time is too short to matter. But some applications require the fastest possible

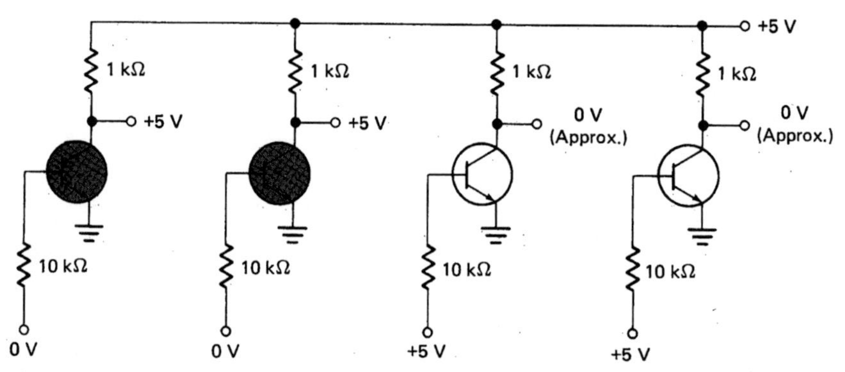

Fig. 1-2 Transistor register.

switching time. To get this maximum speed, designers have come up with circuits that switch from cutoff (or near cutoff) to a higher point on the load line (but short of saturation). These nonsaturated circuits rely on clamping diodes or heavy negative feedback to overcome transistor variations.

Remember this: whether saturated or nonsaturated circuits are used, the transistors switch between distinct points on the load line. This means that all input and output voltages are easily recognized as low or high, binary 0 or binary 1.

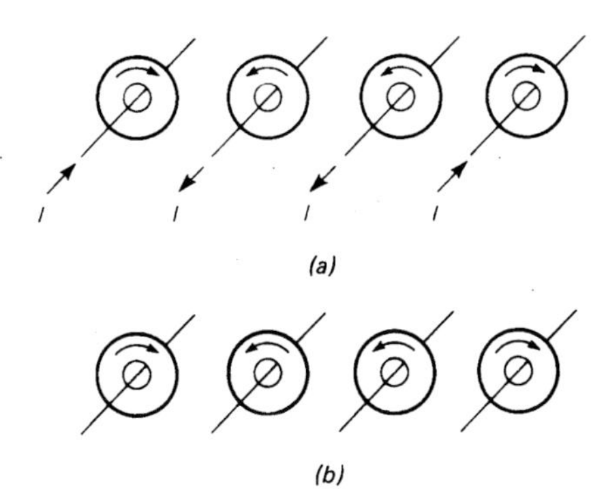

Fig. 1-3 Core register.

Magnetic Cores

Early digital computers used magnetic cores to store data. Figure 1-3*a* shows a 4-bit core register. With the right-hand rule, you can see that conventional current into a wire produces a clockwise flux; reversing the current gives a counterclockwise flux. (The same result is obtained if electron-flow is assumed and the left-hand rule is used.)

The cores have rectangular hysteresis loops; this means that flux remains in a core even though the magnetizing current is removed (see Fig. 1-3*b*). This is why a core register can store binary data indefinitely. For instance, let's use the following code:

Flux	Binary
Counterclockwise	0
Clockwise	1

Then, the core register of Fig. 1-3*b* stores binary 1001, equivalent to decimal 9. By changing the magnetizing currents in Fig. 1-3*a* we can change the stored data.

To store larger binary numbers, add more cores. Two cores added to Fig. 1-3*a* result in a 6-bit register; four more cores give an 8-bit register.

The *memory* is one of the main parts of a computer. Some memories contain thousands of core registers. These registers store the program and data needed to run the computer.

Other Two-State Examples

The simplest example of a two-state device is the on-off switch. When this switch is closed, it represents binary 1; when it's open, it stands for binary 0.

Punched cards are another example of the two-state concept. A hole in a card stands for binary 1, the absence of a hole for binary 0. Using a prearranged code, a card-punch machine with a keyboard can produce a stack of cards containing the program and data needed to run a computer.

Magnetic tape can also store binary numbers. Tape recorders magnetize some points on the tape (binary 1), while leaving other points unmagnetized (binary 0). By a prearranged code, a row of points represents either a coded instruction or data. In this way, a reel of tape can store thousands of binary instructions and data for later use in a computer.

Even the lights on the control panel of a large computer are binary; a light that's on stands for binary 1, and one that's off stands for binary 0. In a 16-bit computer, for instance, a row of 16 lights allows the operator to see the binary contents in different computer registers. The operator can then monitor the overall operation and, when necessary, troubleshoot.

In summary, switches, transistors, cores, cards, tape, lights, and almost all other devices used with computers are based on two-state operation. This is why we are forced to use binary numbers when analyzing computer action.

EXAMPLE 1-3

Figure 1-4 shows a strip of magnetic tape. The black circles are magnetized points and the white circles unmagnetized points. What binary number does each horizontal row represent?

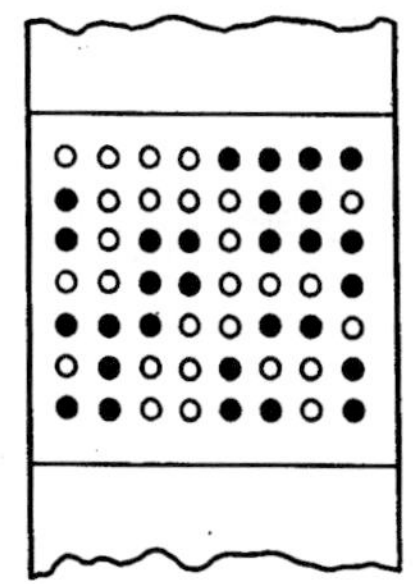

Fig. 1-4 Binary numbers on magnetic tape.

SOLUTION

The tape stores these binary numbers:

Row 1	00001111	Row 5	11100110
Row 2	10000110	Row 6	01001001
Row 3	10110111	Row 7	11001101
Row 4	00110001		

(*Note:* these binary numbers may represent either coded instructions or data.)

A string of 8 bits is called a *byte*. In this example, the magnetic tape stores 7 bytes. The first byte (row 1) is 00001111. The second byte (row 2) is 10000110. The third byte is 10110111. And so on.

A byte is the basic unit of data in computers. Most computers process data in strings of 8 bits or some multiple (16, 24, 32, and so on). Likewise, the memory stores data in strings of 8 bits or some multiple of 8 bits.

1-5 BINARY-TO-DECIMAL CONVERSION

You already know how to count to 15 using binary numbers. The next thing to learn is how to convert larger binary numbers to their decimal equivalents.

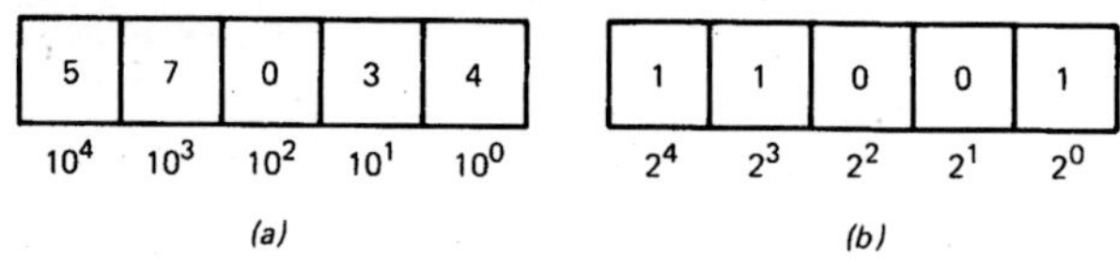

Fig. 1-5 (*a*) Decimal weights; (*b*) binary weights.

Decimal Weights

The decimal number system is an example of positional notation; each digit position has a *weight* or value. With decimal numbers the weights are units, tens, hundreds, thousands, and so on. The sum of all digits multiplied by their weights gives the total amount being represented.

For instance, Fig. 1-5*a* illustrates a decimal odometer. Below each digit is its weight. The digit on the right has a weight of 10^0 (units), the second digit has a weight of 10^1 (tens), the third digit a weight of 10^2 (hundreds), and so forth. The sum of all units multiplied by their weights is

$$(5 \times 10^4) + (7 \times 10^3) + (0 \times 10^2) + (3 \times 10^1) + (4 \times 10^0) = 50{,}000 + 7000 + 0 + 30 + 4 = 57{,}034$$

Binary Weights

Positional notation is also used with binary numbers because each digit position has a weight. Since only two digits are used, the weights are powers of 2 instead of 10. As shown in the binary odometer of Fig. 1-5*b*, these weights are 2^0 (units), 2^1 (twos), 2^2 (fours), 2^3 (eights), and 2^4 (sixteens). If longer binary numbers are involved, the weights continue in ascending powers of 2.

The decimal equivalent of a binary number equals the sum of all binary digits multiplied by their weights. For instance, the binary reading of Fig. 1-5*b* has a decimal equivalent of

$$(1 \times 2^4) + (1 \times 2^3) + (0 \times 2^2) + (0 \times 2^1) + (1 \times 2^0) = 16 + 8 + 0 + 0 + 1 = 25$$

Binary 11001 is therefore equivalent to decimal 25.

As another example, the byte 11001100 converts to decimal as follows:

$$(1 \times 2^7) + (1 \times 2^6) + (0 \times 2^5) + (0 \times 2^4) + (1 \times 2^3) + (1 \times 2^2) + (0 \times 2^1) + (0 \times 2^0) = 128 + 64 + 0 + 0 + 8 + 4 + 0 + 0 = 204$$

So, binary 11001100 is equivalent to decimal 204.

Fast and Easy Conversion

Here's a streamlined way to convert a binary number to its decimal equivalent:

1. Write the binary number.
2. Write the weights 1, 2, 4, 8, . . . , under the binary digits.
3. Cross out any weight under a 0.
4. Add the remaining weights.

For instance, binary 1101 converts to decimal as follows:

1.	1 1 0 1	(Write binary number)
2.	8 4 2 1	(Write weights)
3.	8 4 $\not{2}$ 1	(Cross out weights)
4.	8 + 4 + 0 + 1 = 13	(Add weights)

You can compress the steps even further:

1 1 0 1	(Step 1)
8 4 $\not{2}$ 1 → 13	(Steps 2 to 4)

As another example, here's the conversion of binary 1110101 in compressed form:

1 1 1 0 1 0 1
64 32 16 $\not{8}$ 4 $\not{2}$ 1 → 117

Base or Radix

The *base* or *radix* of a number system equals the number of digits it has. Decimal numbers have a base of 10 because digits 0 through 9 are used. Binary numbers have a base of 2 because only the digits 0 and 1 are used. (In terms of an odometer, the base or radix is the number of digits on each wheel.)

A subscript attached to a number indicates the base of the number. 100_2 means binary 100. On the other hand, 100_{10} stands for decimal 100. Subscripts help clarify equations where binary and decimal numbers are mixed. For instance, the last two examples of binary-to-decimal conversion can be written like this:

$$1101_2 = 13_{10}$$
$$1110101_2 = 117_{10}$$

In this book we will use subscripts when necessary for clarity.

1-6 MICROPROCESSORS

What is inside a computer? What is a microprocessor? What is a microcomputer?

Computer

The five main sections of a computer are input, memory, arithmetic and logic, control, and output. Here is a brief description of each.

Input This consists of all the circuits needed to get programs and data into the computer. In some computers the input section includes a typewriter keyboard that converts letters and numbers into strings of binary data.

Memory This stores the program and data before the computer run begins. It also can store partial solutions during a computer run, similar to the way we use a scratchpad while working out a problem.

Control This is the computer's center of gravity, analogous to the conscious part of the mind. The control section directs the operation of all other sections. Like the conductor of an orchestra, it tells the other sections what to do and when to do it.

Arithmetic and logic This is the number-crunching section of the machine. It can also make logical decisions. With control telling it what to do and with memory feeding it data, the arithmetic-logic unit (ALU) grinds out answers to number and logic problems.

Output This passes answers and other processed data to the outside world. The output section usually includes a video display to allow the user to see the processed data.

Microprocessor

The control section and the ALU are often combined physically into a single unit called the *central processing unit* (CPU). Furthermore, it's convenient to combine the input and output sections into a single unit called the *input-output* (I/O) *unit*. In earlier computers, the CPU, memory, and I/O unit filled an entire room.

With the advent of integrated circuits, the CPU, memory, and I/O unit have shrunk dramatically. Nowadays the CPU can be fabricated on a single semiconductor chip called a *microprocessor*. In other words, a microprocessor is nothing more than a CPU on a chip.

Likewise, the I/O circuits and memory can be fabricated on chips. In this way, the computer circuits that once filled a room now fit on a few chips.

Microcomputer

As the name implies, a *microcomputer* is a small computer. More specifically, a microcomputer is a computer that uses a microprocessor for its CPU. The typical microcomputer has three kinds of chips: microprocessor (usually one chip), memory (several chips), and I/O (one or more chips).

If a small memory is acceptable, a manufacturer can fabricate all computer circuits on a single chip. For instance, the 8048 from Intel Corporation is a one-chip microcomputer with an 8-bit CPU, 1,088 bytes of memory, and 27 I/O lines.

Powers of 2

Microprocessor design started with 4-bit devices, then evolved to 8- and 16-bit devices. In our later discussions of microprocessors, powers of 2 keep coming up because of the binary nature of computers. For this reason, you should study Table 1-3. It lists the powers of 2 encountered in microcomputer analysis. As shown, the abbreviation K stands for 1,024 (approximately 1,000).† Therefore, 1K means 1,024, 2K stands for 2,048, 4K for 4,096, and so on.

Most personal microcomputers have 640K (or greater) memories that can store 655,360 bytes (or more).

TABLE 1-3. POWERS OF 2

Powers of 2	Decimal equivalent	Abbreviation
2^0	1	
2^1	2	
2^2	4	
2^3	8	
2^4	16	
2^5	32	
2^6	64	
2^7	128	
2^8	256	
2^9	512	
2^{10}	1,024	1K
2^{11}	2,048	2K
2^{12}	4,096	4K
2^{13}	8,192	8K
2^{14}	16,384	16K
2^{15}	32,768	32K
2^{16}	65,536	64K

† The abbreviations 1K, 2K, and so on, became established before K- for *kilo-* was in common use. Retaining the capital K serves as a useful reminder that K only approximates 1,000.

1-7 DECIMAL-TO-BINARY CONVERSION

Next, you need to know how to convert from decimal to binary. After you know how it's done, you will be able to understand how circuits can be built to convert decimal numbers into binary numbers.

Double-Dabble

Double-dabble is a way of converting any decimal number to its binary equivalent. It requires successive division by 2, writing down each quotient and its remainder. The remainders are the binary equivalent of the decimal number. The only way to understand the method is to go through an example, step by step.

Here is how to convert decimal 13 to its binary equivalent. Step 1. Divide 13 by 2, writing your work like this:

```
    6  1 → (first remainder)
2 ) 13
```

The quotient is 6 with a remainder of 1.
Step 2. Divide 6 by 2 to get

```
    3  0 → (second remainder)
2 ) 6  1
2 ) 13
```

This division gives 3 with a remainder of 0.
Step 3. Again you divide by 2:

```
    1  1 → (third remainder)
2 ) 3  0
2 ) 6  1
2 ) 13
```

Here you get a quotient of 1 and a remainder of 1.
Step 4. One more division by 2 gives

```
           Read
           down
    0  1    |
2 ) 1  1    |
2 ) 3  0    |
2 ) 6  1    |
2 ) 13      ↓
```

In this final division, 2 does not divide into 1; therefore, the quotient is 0 with a remainder of 1.

Whenever you arrive at a quotient of 0 with a remainder of 1, the conversion is finished. The remainders when read downward give the binary equivalent. In this example, binary 1101 is equivalent to decimal 13.

Double-dabble works with any decimal number. Progressively divide by 2, writing each quotient and its remainder. When you reach a quotient of 0 and a remainder of 1, you are finished; the remainders read downward are the binary equivalent of the decimal number.

Streamlined Double-Dabble

There's no need to keep writing down 2 before each division because you're always dividing by 2. From now on, here's how to show the conversion of decimal 13 to its binary equivalent:

```
     0  1  |
    ) 1  1 |
    ) 3  0 |
    ) 6  1 ↓
2 ) 13
```

EXAMPLE 1-4

Convert decimal 23 to binary.

SOLUTION

The first step in the conversion looks like this:

```
    11  1
2 ) 23
```

After all divisions, the finished work looks like this:

```
     0  1  |
    ) 1  0 |
    ) 2  1 |
    ) 5  1 |
   ) 11  1 |
2 ) 23     ↓
```

This says that binary 10111 is equivalent to decimal 23.

1-8 HEXADECIMAL NUMBERS

Hexadecimal numbers are extensively used in microprocessor work. To begin with, they are much shorter than binary numbers. This makes them easy to write and remember. Furthermore, you can mentally convert them to binary form whenever necessary.

An Unusual Odometer

Hexadecimal means 16. The hexadecimal number system has a base or radix of 16. This means that it uses 16 digits to represent all numbers. The digits are 0 through 9, and A through F as follows: 0, 1, 2, 3, 4, 5, 6, 7, 8, 9, A, B, C, D, E, and F. Hexadecimal numbers are strings of these digits like 8A5, 4CF7, and EC58.

An easy way to understand hexadecimal numbers is to visualize a hexadecimal odometer. Each wheel has 16 digits on its circumference. As it turns, it displays 0 through 9 as before. But then, instead of resetting, it goes on to display A, B, C, D, E, and F.

The idea of reset and carry applies to a hexadecimal odometer. When a wheel turns from F back to 0, it forces the next higher wheel to advance by 1. In other words, when a wheel runs out of hexadecimal digits, it resets and carries.

If used in a car, a hexadecimal odometer would count as follows. When the car is new, the odometer shows all 0s:

0000 (zero)

The next 9 miles produce readings of

0001 (one)
0002 (two)
0003 (three)
0004 (four)
0005 (five)
0006 (six)
0007 (seven)
0008 (eight)
0009 (nine)

The next 6 miles give

000A (ten)
000B (eleven)
000C (twelve)
000D (thirteen)
000E (fourteen)
000F (fifteen)

At this point the least significant wheel has run out of digits. Therefore, the next mile forces a reset-and-carry to get

0010 (sixteen)

The next 15 miles produce these readings: 0011, 0012, 0013, 0014, 0015, 0016, 0017, 0018, 0019, 001A, 001B, 001C, 001D, 001E, and 001F. Once again, the least significant wheel has run out of digits. So, the next mile results in a reset-and-carry:

0020 (thirty-two)

Subsequent readings are 0021, 0022, 0023, 0024, 0025, 0026, 0027, 0028, 0029, 002A, 002B, 002C, 002D, 002E, and 002F.

You should have the idea by now. Each mile advances the least significant wheel by 1. When this wheel runs out of hexadecimal digits, it resets and carries. And so on for the other wheels. For instance, if the odometer reading is

835F

the next reading is 8360. As another example, given

5FFF

the next hexadecimal number is 6000.

Equivalences

Table 1-4 shows the equivalences between hexadecimal, binary, and decimal digits. Memorize this table. It's essential that you be able to convert instantly from one system to another.

TABLE 1-4. EQUIVALENCES

Hexadecimal	Binary	Decimal
0	0000	0
1	0001	1
2	0010	2
3	0011	3
4	0100	4
5	0101	5
6	0110	6
7	0111	7
8	1000	8
9	1001	9
A	1010	10
B	1011	11
C	1100	12
D	1101	13
E	1110	14
F	1111	15

1-9 HEXADECIMAL-BINARY CONVERSIONS

After you know the equivalences of Table 1-4, you can mentally convert any hexadecimal string to its binary equivalent and vice versa.

Hexadecimal to Binary

To convert a hexadecimal number to a binary number, convert each hexadecimal digit to its 4-bit equivalent, using Table 1-4. For instance, here's how 9AF converts to binary:

$$\begin{array}{ccc} 9 & A & F \\ \downarrow & \downarrow & \downarrow \\ 1001 & 1010 & 1111 \end{array}$$

As another example, C5E2 converts like this:

$$\begin{array}{cccc} C & 5 & E & 2 \\ \downarrow & \downarrow & \downarrow & \downarrow \\ 1100 & 0101 & 1110 & 0010 \end{array}$$

Incidentally, for easy reading it's common practice to leave a space between the 4-bit strings. For example, instead of writing

$$C5E2_{16} = 1100010111100010_2$$

we can write

$$C5E2_{16} = 1100\ 0101\ 1110\ 0010_2$$

Binary to Hexadecimal

To convert in the opposite direction, from binary to hexadecimal, you again use Table 1-4. Here are two examples. The byte 1000 1100 converts as follows:

$$\begin{array}{cc} 1000 & 1100 \\ \downarrow & \downarrow \\ 8 & C \end{array}$$

The 16-bit number 1110 1000 1101 0110 converts like this:

$$\begin{array}{cccc} 1110 & 1000 & 1101 & 0110 \\ \downarrow & \downarrow & \downarrow & \downarrow \\ E & 8 & D & 6 \end{array}$$

In both these conversions, we start with a binary number and wind up with the equivalent hexadecimal number.

EXAMPLE 1-5

Solve the following equation for x:

$$x_{16} = 1111\ 1111\ 1111\ 1111_2$$

SOLUTION

This is the same as asking for the hexadecimal equivalent of binary 1111 1111 1111 1111. Since hexadecimal F is equivalent to 1111, x = FFFF. Therefore,

$$FFFF_{16} = 1111\ 1111\ 1111\ 1111_2$$

EXAMPLE 1-6

As mentioned earlier, the memory contains thousands of registers (core or semiconductor) that store the program and data needed for a computer run. These memory registers are known as *memory locations*. A typical microcomputer may have up to 65,536 memory locations, each storing 1 byte.

Suppose the first 16 memory locations contain these bytes:

0011 1100
1100 1101
0101 0111
0010 1000
1111 0001
0010 1010
1101 0100
0100 0000
0111 0111
1100 0011
1000 0100
0010 1000
0010 0001
0011 1010
0011 1110
0001 1111

Convert these bytes to their hexadecimal equivalents.

SOLUTION

Here are the stored bytes and their hexadecimal equivalents:

Memory Contents	Hex Equivalents
0011 1100	3C
1100 1101	CD
0101 0111	57
0010 1000	28
1111 0001	F1

0010 1010	2A
1101 0100	D4
0100 0000	40
0111 0111	77
1100 0011	C3
1000 0100	84
0010 1000	28
0010 0001	21
0011 1010	3A
0011 1110	3E
0001 1111	1F

What's the point of this example? When talking about the contents of a computer memory, we can use either binary numbers or hexadecimal numbers. For instance, we can say that the first memory location contains 0011 1100, or we can say that it contains 3C. Either string gives the same information. But notice how much easier it is to say, write, and think 3C than it is to say, write, and think 0011 1100. In other words, hexadecimal strings are much easier for people to work with. This is why everybody working with microprocessors uses hexadecimal notation to represent particular bytes.

What we have just done is known as *chunking*, replacing longer strings of data with shorter ones. At the first memory location we chunk the digits 0011 1100 into 3C. At the second memory location we chunk the digits 1100 1101 into CD, and so on.

EXAMPLE 1-7

The typical microcomputer has a typewriter keyboard that allows you to enter programs and data; a video screen displays answers and other information.

Suppose the video screen of a microcomputer displays the hexadecimal contents of the first eight memory locations as

A7
28
C3
19
5A
4D
2C
F8

What are the binary contents of the memory locations?

SOLUTION

Convert from hexadecimal to binary to get

1010 0111
0010 1000
1100 0011
0001 1001
0101 1010
0100 1101
0010 1100
1111 1000

The first memory location stores the byte 1010 0111, the second memory location stores the byte 0010 1000, and so on.

This example emphasizes a widespread industrial practice. Microcomputers are programmed to display chunked data, often hexadecimal. The user is expected to know hexadecimal-binary conversions. In other words, a computer manufacturer assumes that you know that A7 represents 1010 0111, 28 stands for 0010 1000, and so on.

One more point. Notice that each memory location in this example stores 1 byte. This is typical of first-generation microcomputers because they use 8-bit microprocessors.

1-10 HEXADECIMAL-TO-DECIMAL CONVERSION

You often need to convert a hexadecimal number to its decimal equivalent. This section discusses methods for doing it.

Hexadecimal to Binary to Decimal

One way to convert from hexadecimal to decimal is the two-step method of converting from hexadecimal to binary and then from binary to decimal. For instance, here's how to convert hexadecimal 3C to its decimal equivalent.

Step 1. Convert 3C to its binary equivalent:

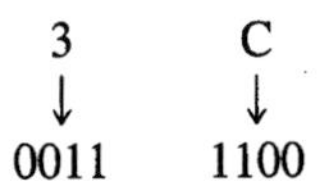

Step 2. Convert 0011 1100 to its decimal equivalent:

0 0 1 1 1 1 0 0
~~128~~ ~~64~~ 32 16 8 4 ~~2~~ ~~1~~ → 60

Therefore, decimal 60 is equivalent to hexadecimal 3C. As an equation,

$$3C_{16} = 0011\ 1100_2 = 60_{10}$$

Positional-Notation Method

Positional notation is also used with hexadecimal numbers because each digit position has a weight. Since 16 digits are used, the weights are the powers of 16. As shown in

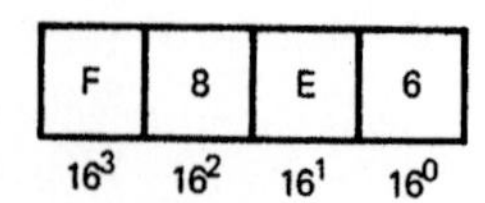

Fig. 1-6 Hexadecimal weights.

the hexadecimal odometer of Fig. 1-6, the weights are 16^0, 16^1, 16^2, and 16^3. If longer hexadecimal numbers are involved, the weights continue in ascending powers of 16.

The decimal equivalent of a hexadecimal string equals the sum of all hexadecimal digits multiplied by their weights. (In processing hexadecimal digits A through F, use 10 through 15.) For instance, the hexadecimal reading of Fig. 1-6 has a decimal equivalent of

$$\begin{aligned}&(F \times 16^3) + (8 \times 16^2) + (E \times 16^1) + (6 \times 16^0)\\&\quad = (15 \times 16^3) + (8 \times 16^2) + (14 \times 16^1) + (6 \times 16^0)\\&\quad = 61{,}440 + 2{,}048 + 224 + 6\\&\quad = 63{,}718\end{aligned}$$

In other words,

$$F8E6_{16} = 63{,}718_{10}$$

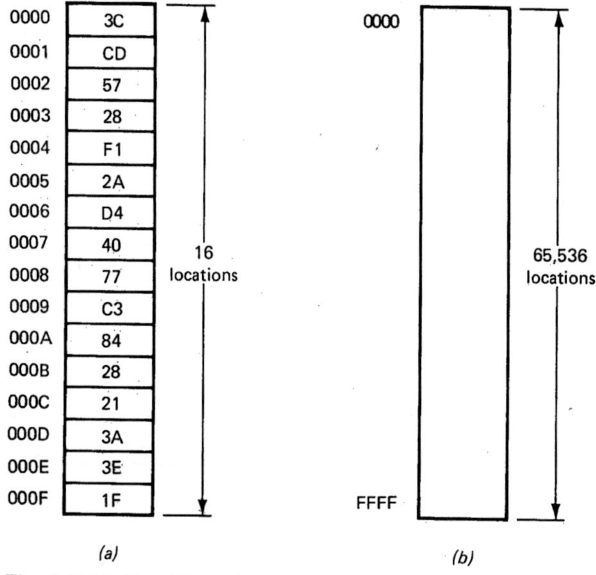

Fig. 1-7 (*a*) First 16 words in memory; (*b*) 64K memory.

Memory Locations and Addresses

If a certain microcomputer has 64K memory, meaning 65,536 memory locations, each is able to store 1 byte. The different memory locations are identified by hexadecimal numbers called *addresses*. For instance, Fig. 1-7*a* shows the first 16 memory locations; their addresses are from 0000 to 000F.

The address of a memory location is different from its stored contents, just as a house address is different from the people living in the house. Figure 1-7*a* emphasizes the point. At address 0000 the stored contents are 3C (equivalent to 0011 1100). At address 0001 the stored contents are CD, at address 0002 the stored contents are 57, and so on.

Figure 1-7*b* shows how to visualize a 64K memory. The first address is 0000, and the last is FFFF.

Table of Binary-Hexadecimal-Decimal Equivalents

A 64K memory has 65,536 hexadecimal addresses from 0000 to FFFF. The equivalent binary addresses are from

$$0000\ 0000\ 0000\ 0000$$

to

$$1111\ 1111\ 1111\ 1111$$

The first 8 bits are called the *upper byte* (UB); the second 8 bits are the *lower byte* (LB). If you have to do a lot of binary-hexadecimal-decimal conversions, use the table of equivalents in Appendix 2, which shows all the values for a 64K memory.

Appendix 2 has four headings: binary, hexadecimal, UB decimal, and LB decimal. Given a 16-bit address, you convert the upper byte to its decimal equivalent (UB decimal), the lower byte to its decimal equivalent (LB decimal), and then add the two decimal equivalents. For instance, suppose you want to convert

$$1101\ 0111\ 1010\ 0010$$

to its decimal equivalent. The upper byte is 1101 0111, or hexadecimal D7; the lower byte is 1010 0010, or A2. Using Appendix 2, find D7 and its UB decimal equivalent

$$D7 \rightarrow 55{,}040$$

Next, find A2 and its LB decimal equivalent

$$A2 \rightarrow 162$$

Add the UB and LB decimal equivalents to get

$$55{,}040 + 162 = 55{,}202$$

This is the decimal equivalent of hexadecimal D7A2 or binary 1101 0111 1010 0010.

Once familiar with Appendix 2, you will find it enormously helpful. It is faster, more accurate, and less tiring than other methods. The only calculation required is adding the UB and LB decimal, easily done mentally, with pencil and paper, or if necessary, on a calculator. Furthermore, if you are interested in converting only the lower byte, no calculation is required, as shown in the next example.

EXAMPLE 1-8

Convert hexadecimal 7E to its decimal equivalent.

SOLUTION

When converting only a single byte, all you are dealing with is the lower byte. With Appendix 2, look up 7E and its LB decimal equivalent to get

$$7E \rightarrow 126$$

In other words, Appendix 2 can be used to convert single bytes to their decimal equivalents (LB decimal) or double bytes to their decimal equivalents (UB decimal + LB decimal).

1-11 DECIMAL-TO-HEXADECIMAL CONVERSION

One way to perform decimal-to-hexadecimal conversion is to go from decimal to binary then to hexadecimal. Another way is *hex-dabble*. The idea is to divide successively by 16, writing down the remainders. (Hex-dabble is like double-dabble except that 16 is used for the divisor instead of 2.)

Here's an example of how to convert decimal 2,479 into hexadecimal form. The first division is

```
        154   15  F
16 ) 2,479
```

The next step is

```
          9   10  A
       ) 154  15  F
16 ) 2,479
```

The final step is

```
                       Read
                       down
        0    9  9       |
      ) 9   10  A       |
    ) 154   15  F       |
16 ) 2,479              v
```

Notice how similar hex-dabble is to double-dabble. Also, remainders greater than 9 have to be changed to hexadecimal digits (10 becomes A, 15 becomes F, etc.).

If you prefer, use Appendix 2 to look up the decimal-hexadecimal equivalents. The next two examples show how.

EXAMPLE 1-9

Convert decimal 141 to hexadecimal.

SOLUTION

Whenever the decimal number is between 0 and 255, all you have to do is look up the decimal number and its hexadecimal equivalent. With Appendix 2, you can see at a glance that

$$8D \leftarrow 141$$

EXAMPLE 1-10

Convert decimal 36,020 to its hexadecimal equivalent.

SOLUTION

If the decimal number is between 256 and 65,535, you need to proceed as follows. First, locate the largest UB decimal that is less than 36,020. In Appendix 2, the largest UB decimal is

$$\text{UB decimal} = 35{,}840$$

which has a hexadecimal equivalent of

$$8C \leftarrow 35{,}840$$

This is the upper byte.

Next, subtract the UB decimal from the original decimal number:

$$36{,}020 - 35{,}840 = 180$$

The difference 180 has a hexadecimal equivalent

$$B4 \leftarrow 180$$

This is the lower byte.

By combining the upper and lower bytes, we get the complete answer: 8CB4. This is the hexadecimal equivalent of 36,020.

After a little practice, you will find Appendix 2 to be one of the fastest methods of decimal-hexadecimal conversion.

1-12 BCD NUMBERS

A *nibble* is a string of 4 bits. *Binary-coded-decimal* (BCD) numbers express each decimal digit as a nibble. For instance, decimal 2,945 converts to a BCD number as follows:

2	9	4	5
↓	↓	↓	↓
0010	1001	0100	0101

As you see, each decimal digit is coded as a nibble.

Here's another example: $9{,}863_{10}$ converts like this:

9	8	6	3
↓	↓	↓	↓
1001	1000	0110	0011

Therefore, 1001 1000 0110 0011 is the BCD equivalent of $9{,}863_{10}$.

The reverse conversion is similar. For instance, 0010 1000 0111 0100 converts as follows:

0010	1000	0111	0100
↓	↓	↓	↓
2	8	7	4

Applications

BCD numbers are useful wherever decimal information is transferred into or out of a digital system. The circuits inside pocket calculators, for example, can process BCD numbers because you enter decimal numbers through the keyboard and see decimal answers on the LED or liquid-crystal display. Other examples of BCD systems are electronic counters, digital voltmeters, and digital clocks; their circuits can work with BCD numbers.

BCD Computers

BCD numbers have limited value in computers. A few early computers processed BCD numbers but were slower and more complicated than binary computers. As previously mentioned, a computer is more than a number cruncher because it must handle names and other nonnumeric data. In other words, a modern computer must be able to process *alphanumerics* (alphabet letters, numbers, and other symbols). This why modern computers have CPUs that process binary numbers rather than BCD numbers.

Comparison of Number Systems

Table 1-5 shows the four number systems we have discussed. Each number system uses strings of digits to represent quantity. Above 9, equivalent strings appear different. For instance, decimal string 128, hexadecimal string 80, binary string 1000 0000, and BCD string 0001 0010 1000 are equivalent because they represent the same number of pebbles.

Machines have to use long strings of binary or BCD numbers, but people prefer to chunk the data in either decimal or hexadecimal form. As long as we know how to convert from one number system to the next, we can always get back to the ultimate meaning, which is the number of pebbles being represented.

TABLE 1-5. NUMBER SYSTEMS

Decimal	Hexadecimal	Binary	BCD
0	0	0000 0000	0000 0000 0000
1	1	0000 0001	0000 0000 0001
2	2	0000 0010	0000 0000 0010
3	3	0000 0011	0000 0000 0011
4	4	0000 0100	0000 0000 0100
5	5	0000 0101	0000 0000 0101
6	6	0000 0110	0000 0000 0110
7	7	0000 0111	0000 0000 0111
8	8	0000 1000	0000 0000 1000
9	9	0000 1001	0000 0000 1001
10	A	0000 1010	0000 0001 0000
11	B	0000 1011	0000 0001 0001
12	C	0000 1100	0000 0001 0010
13	D	0000 1101	0000 0001 0011
14	E	0000 1110	0000 0001 0100
15	F	0000 1111	0000 0001 0101
16	10	0001 0000	0000 0001 0110
32	20	0010 0000	0000 0011 0010
64	40	0100 0000	0000 0110 0100
128	80	1000 0000	0001 0010 1000
255	FF	1111 1111	0010 0101 0101

1-13 THE ASCII CODE

To get information into and out of a computer, we need to use numbers, letters, and other symbols. This implies some kind of alphanumeric code for the I/O unit of a computer. At one time, every manufacturer had a different code, which led to all kinds of confusion. Eventually, industry settled on an input-output code known as the *American Standard Code for Information Interchange* (abbreviated ASCII). This code allows manufacturers to standardize I/O hardware such as keyboards, printers, video displays, and so on.

The ASCII (pronounced *ask'-ee*) code is a 7-bit code whose format (arrangement) is

$$X_6X_5X_4X_3X_2X_1X_0$$

where each X is a 0 or a 1. For instance, the letter A is coded as

1000001

Sometimes, a space is inserted for easier reading:

100 0001

TABLE 1-6. THE ASCII CODE

$X_3X_2X_1X_0$	$X_6X_5X_4$					
	010	**011**	**100**	**101**	**110**	**111**
0000	SP	0	@	P		p
0001	!	1	A	Q	a	q
0010	"	2	B	R	b	r
0011	#	3	C	S	c	s
0100	$	4	D	T	d	t
0101	%	5	E	U	e	u
0110	&	6	F	V	f	v
0111	’	7	G	W	g	w
1000	(	8	H	X	h	x
1001	)	9	I	Y	i	y
1010	*	:	J	Z	j	z
1011	+	;	K		k	
1100	,	<	L		l	
1101	−	=	M		m	
1110	•	>	N		n	
1111	/	?	O		o	

Table 1-6 shows the ASCII code. Read the table the same as a graph. For instance, the letter A has an $X_6X_5X_4$ of 100 and an $X_3X_2X_1X_0$ of 0001. Therefore, its ASCII code is

100 0001 (A)

Table 1-6 includes the ASCII code for lowercase letters. The letter a is coded as

110 0001 (a)

More examples are

110 0010 (b)
110 0011 (c)
110 0100 (d)

and so on.

Also look at the punctuation and mathematical symbols. Some examples are

010 0100 ($)
010 1011 (+)
011 1101 (=)

In Table 1-6, SP stands for space (blank). Hitting the space bar of an ASCII keyboard sends this into a microcomputer:

010 0000 (space)

EXAMPLE 1-11

With an ASCII keyboard, each keystroke produces the ASCII equivalent of the designated character. Suppose you type

PRINT X

What is the output of an ASCII keyboard?

SOLUTION

P (101 0000), R (101 0010), I (100 1001), N (100 1110), T (101 0100), space (010 0000), X (101 1000).

GLOSSARY

address Each memory location has an address, analogous to a house address. Using addresses, we can tell the computer where desired data is stored.
alphanumeric Letters, numbers, and other symbols.
base The number of digits (basic symbols) in a number system. Decimal has a base of 10, binary a base of 2, and hexadecimal a base of 16. Also called the radix.
bit An abbreviation for binary digit.
byte A string of 8 bits. The byte is the basic unit of binary information. Most computers process data with a length of 8 bits or some multiple of 8 bits.
central processing unit The control section and the arithmetic-logic section. Abbreviated CPU.
chip An integrated circuit.
chunking Replacing a longer string by a shorter one.
data Names, numbers, and any other information needed to solve a problem.
digital Pertains to anything in the form of digits, for example, digital data.
hardware The electronic, magnetic, and mechanical devices used in a computer.
hexadecimal A number system with a base of 16. Hexadecimal numbers are used in microprocessor work.
input-output Abbreviated I/O. The input and output sections of a computer are often lumped into one unit known as the I/O unit.
microcomputer A computer that uses a microprocessor for its central processing unit (CPU).
microprocessor A CPU on a chip. It contains the control and arithmetic-logic sections. Sometimes abbreviated MPU (microprocessor unit).
nibble A string of 4 bits. Half of a byte.
program A sequence of instructions that tells the computer how to process the data. Also known as software.
register A group of electronic, magnetic, or mechanical devices that store digital data.
software Programs.
string A group of digits or other symbols.

SELF-TESTING REVIEW

Read each of the following and provide the missing words. Answers appear at the beginning of the next question.

1. Binary means ______. Binary numbers have a base of 2. The digits used in a binary number system are ______ and ______.
2. (*two; 0, 1*) Names, numbers, and other information needed to solve a problem are called ______. The ______ is a sequence of instructions that tells the computer how to process the data.
3. (*data, program*) Computer ICs work reliably because they are based on ______ design. When a transistor is cut off or saturated, transistor ______ have almost no effect.
4. (*two-state, variations*) A ______ is a group of devices that store digital data. ______ is an abbreviation for binary digit. A byte is a string of ______ bits.
5. (*register, Bit, 8*) The control and arithmetic-logic sections are called the ______ (CPU). A microprocessor is a CPU on a chip. A microcomputer is a computer that uses a ______ for its CPU.
6. (*central processing unit, microprocessor*) The abbreviation K indicates units of approximately 1,000 or precisely 1,024. Therefore, 1K means 1,024, 2K means 2,048, 4K means ______, and 64K means ______.
7. (*4,096, 65,536*) The hexadecimal number system is widely used in analyzing and programming ______. The hexadecimal digits are 0 to 9 and A to ______. The main advantage of hexadecimal numbers is the ease of conversion from hexadecimal to ______ and vice versa.
8. (*microprocessors, F, binary*) A typical microcomputer may have up to 65,536 registers in its memory. Each of these registers, usually called a ______, stores 1 byte. Such a memory is specified as a 64-kilobyte memory, or simply a ______ memory.
9. (*memory location, 64K*) Binary-coded-decimal (BCD) numbers express each decimal digit as a ______. BCD numbers are useful whenever ______ information is transferred into or out of a digital system. Equipment using BCD numbers includes pocket calculators, electronic counters, and digital voltmeters.
10. (*nibble, decimal*) The ASCII code is a 7-bit code for ______ (letters, numbers, and other symbols).
11. (*alphanumerics*) With the typical microcomputer, you enter the program and data with typewriter keyboard that converts each character into ASCII code.

PROBLEMS

1-1. How many bytes are there in each of these numbers?

a. 1100 0101
b. 1011 1001 0110 1110
c. 1111 1011 0111 0100 1010

1-2. What are the equivalent decimal numbers for each of the following binary numbers: 10, 110, 111, 1011, 1100, and 1110?

1-3. What is the base for each of these numbers?

a. 348_{10}
b. $1100\ 0101_2$
c. 2312_5
d. $F4C3_{16}$

1-4. Write the equation

$$2 + 2 = 4$$

using binary numbers.

1-5. What is the decimal equivalent of 2^{10}? What does 4K represent? Express 8,192 in K units.

1-6. A 4-bit register has output voltages of high-low-high-low. What is the binary number stored in the register? The decimal equivalent?

Fig. 1-8 An 8-bit LED display.

1-7. Figure 1-8 shows an 8-bit LED display. A light circle means that a LED is ON (binary 1) and a dark circle means a LED is OFF (binary 0). What is the binary number being displayed? The decimal equivalent?

1-8. Convert the following binary numbers to decimal numbers:

a. 00111
b. 11001
c. 10110
d. 11110

1-9. Solve the following equation for x:

$$x_{10} = 11001001_2$$

1-10. An 8-bit transistor register has this output:

low-high-low-high-low-high-low-high

What is the equivalent decimal number being stored?

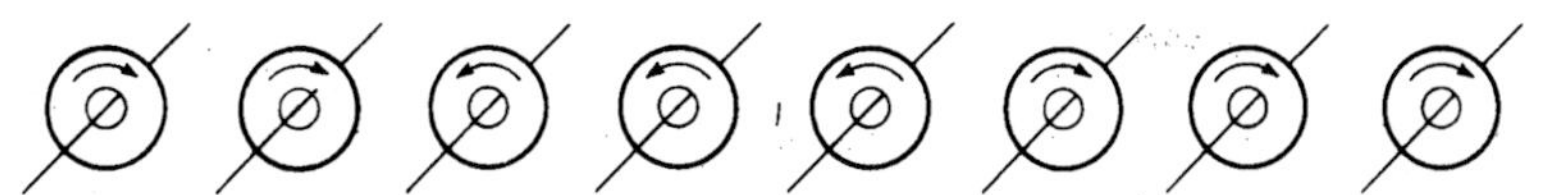

Fig. 1-9 An 8-bit core register.

1-11. In Fig. 1-9 clockwise flux stands for binary 1 and counterclockwise flux for binary 0. What is the binary number stored in the 8-bit core register? Convert this byte to an equivalent decimal number.

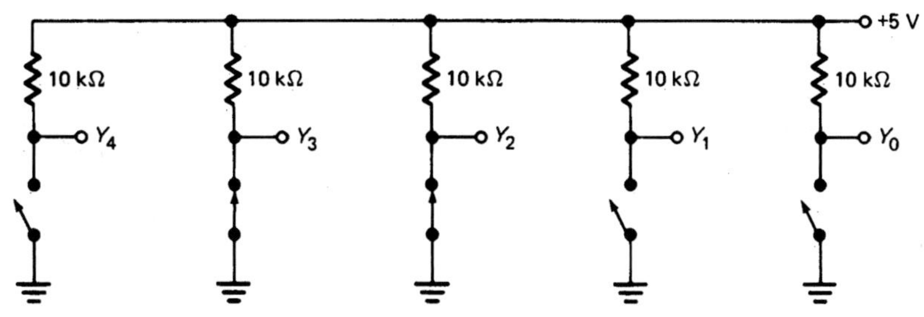

Fig. 1-10 A 5-bit switch register.

1-12. Figure 1-10 shows a 5-bit switch register. By opening and closing the switches you can set up different binary numbers. As usual, high output voltage stands for binary 1 and low output voltage for binary 0. What is the binary number stored in the switch register? The equivalent decimal number?

1-13. Convert decimal 56 to its binary equivalent.

1-14. Convert 72_{10} to a binary number.

1-15. An 8-bit transistor register stores decimal 150. What is the binary output of the register?

1-16. How would you set the switches of Fig. 1-10 to get a decimal output of 27?

1-17. A hexadecimal odometer displays F52A. What are the next six readings?

1-18. The reading on a hexadecimal odometer is 27FF. What is the next reading? Miles later, you see a reading of 8AFC. What are the next six readings?

1-19. Convert each of the following hexadecimal numbers to binary:
a. FF
b. ABC
c. CD42
d. F329

1-20. Convert each of these binary numbers to an equivalent hexadecimal number:
a. 1110 1000
b. 1100 1011
c. 1010 1111 0110
d. 1000 1011 1101 0110

1-21. Here is a program written for the 8085 microprocessor:

Address	Hex Contents
2000	3E
2001	0E
2002	D3
2003	20
2004	76

Convert the hex contents to equivalent binary numbers.

1-22. Convert each of these hexadecimal numbers to its decimal equivalent:
a. FF
b. A4
c. 9B
d. 3C

1-23. Convert the following hexadecimal numbers to their decimal equivalents:
a. 0FFF
b. 3FFF
c. 7FE4
d. B3D8

1-24. A microcomputer has memory locations from 0000 to 0FFF. Each memory location stores 1 byte. In decimal, how many bytes can the microcomputer store in its memory? How many kilobytes is this?

1-25. Suppose a microcomputer has memory locations from 0000 to 3FFF, each storing 1 byte. How many bytes can the memory store? Express this in kilobytes.

1-26. A microcomputer has a 32K memory. How many bytes does this represent? If 0000 stands for the first memory location, what is the hexadecimal notation for the last memory location?

1-27. If a microcomputer has a 64K memory, what are the hexadecimal notations for the first and last memory locations?

1-28. Convert the following decimal numbers to hexadecimal:

a. 4,095
b. 16,383
c. 32,767
d. 65,535

1-29. Convert each of the following decimal numbers to hexadecimal numbers:

a. 238
b. 7,547
c. 15,359
d. 47,285

1-30. How many nibbles are there in each of the following:

a. 1000 0111
b. 1001 0000 0100 0011
c. 0101 1001 0111 0010 0110 0110

1-31. If the numbers in Prob. 1-30 are BCD numbers, what are the equivalent decimal numbers?

1-32. What is the ASCII code for each of the following:

a. 7
b. W
c. f
d. y

1-33. Suppose you type LIST with an ASCII keyboard. What is the binary output as you strike each letter?

1-34. For each of the following rows, provide the missing numbers in the bases indicated.

	Base 2	Base 10	Base 16
a.	0100 0001		
b.		200	
c.			3CD
d.		125	
e.	1101 1110 1111		
f.			FFFF
g.		2,000	

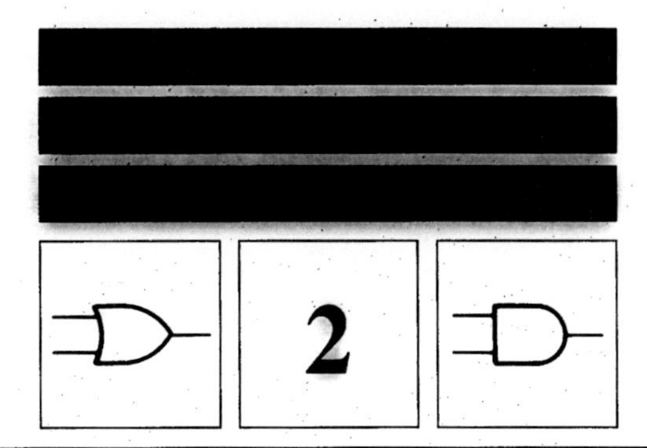

GATES

For centuries mathematicians felt there was a connection between mathematics and logic, but no one before George Boole could find this missing link. In 1854 he invented symbolic logic, known today as *boolean algebra*. Each variable in boolean algebra has either of two values: true or false. The original purpose of this two-state algebra was to solve logic problems.

Boolean algebra had no practical application until 1938, when Claude Shannon used it to analyze telephone switching circuits. He let the variables represent closed and open relays. In other words, Shannon came up with a new application for boolean algebra. Because of Shannon's work, engineers realized that boolean algebra could be applied to computer electronics.

This chapter introduces the *gate*, a circuit with one or more input signals but only one output signal. Gates are digital (two-state) circuits because the input and output signals are either low or high voltages. Gates are often called *logic circuits* because they can be analyzed with boolean algebra.

2-1 INVERTERS

An *inverter* is a gate with only one input signal and one output signal; the output state is always the opposite of the input state.

Transistor Inverter

Figure 2-1 shows a transistor inverter. This common-emitter amplifier switches between cutoff and saturation. When V_{IN} is low (approximately 0 V), the transistor cuts off and V_{OUT} is high. On the other hand, a high V_{IN} saturates the transistor, forcing V_{OUT} to go low.

Table 2-1 summarizes the operation. A low input produces a high output, and a high input results in a low output. Table 2-2 gives the same information in binary form; binary 0 stands for low voltage and binary 1 for high voltage.

An inverter is also called a NOT gate because the output is not the same as the input. The output is sometimes called the *complement* (opposite) of the input.

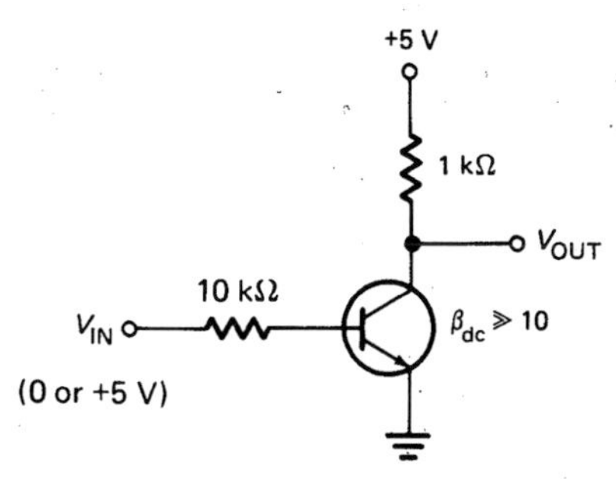

Fig. 2-1 Example of inverter design.

TABLE 2-1

V_{IN}	V_{OUT}
Low	High
High	Low

TABLE 2-2

V_{IN}	V_{OUT}
0	1
1	0

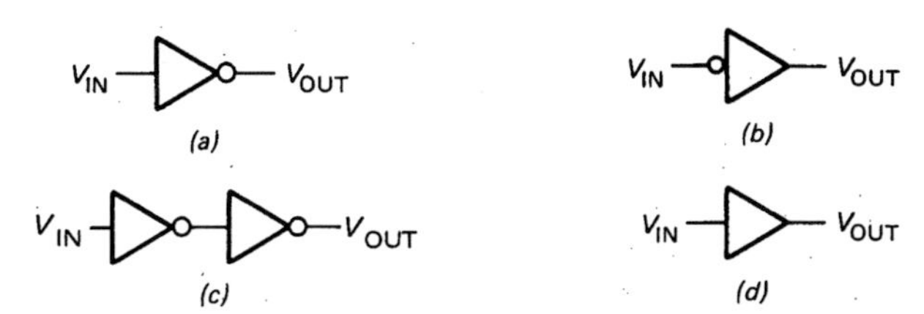

Fig. 2-2 Logic symbols: (*a*) inverter; (*b*) another inverter symbol; (*c*) double inverter; (*d*) buffer.

Inverter Symbol

Figure 2-2*a* is the symbol for an inverter of any design. Sometimes a schematic diagram will use the alternative symbol shown in Fig. 2-2*b*; the bubble (small circle) is on the input side. Whenever you see either of these symbols, remember that the output is the complement of the input.

Noninverter Symbol

If you cascade two inverters (Fig. 2-2*c*), you get a noninverting amplifier. Figure 2-2*d* is the symbol for a noninverting amplifier. Regardless of the circuit design, the action is always the same: a low input voltage produces a low output voltage, and a high input voltage results in a high output voltage.

The main use of noninverting amplifier is buffering (isolating) two other circuits. More will be said about buffers in a later chapter.

EXAMPLE 2-1

Fig. 2-3 Example 2-1.

Figure 2-3*a* has an output, *A* to *F*, of 100101. Show how to complement each bit.

SOLUTION

Easy. Use an inverter on each signal line (Fig. 2-3*b*). The final output is now 011010.

A *hex inverter* is a commercially available IC containing six separate inverters. Given a 6-bit register like Fig. 2-3*a*, we can connect a hex inverter to complement each bit as shown in Fig. 2-3*b*.

One more point. In Fig. 2-3*a* the bits may represent a coded instruction, number, letter, etc. To convey this variety of meaning, a string of bits is often called a binary word or simply a *word*. In Fig. 2-3*b* the word 100101 is complemented to get the word 011010.

2-2 OR GATES

The OR gate has two or more input signals but only one output signal. If any input signal is high, the output signal is high.

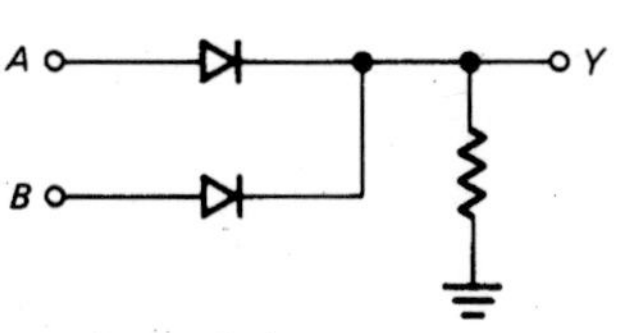

Fig. 2-4 A 2-input diode OR gate.

Diode OR Gate

Figure 2-4 shows one way to build an OR gate. If both inputs are low, the output is low. If either input is high, the diode with the high input conducts and the output is high. Because of the two inputs, we call this circuit a 2-input OR gate.

Table 2-3 summarizes the action; binary 0 stands for low voltage and binary 1 for high voltage. Notice that one or more high inputs produce a high output; this is why the circuit is called an OR gate.

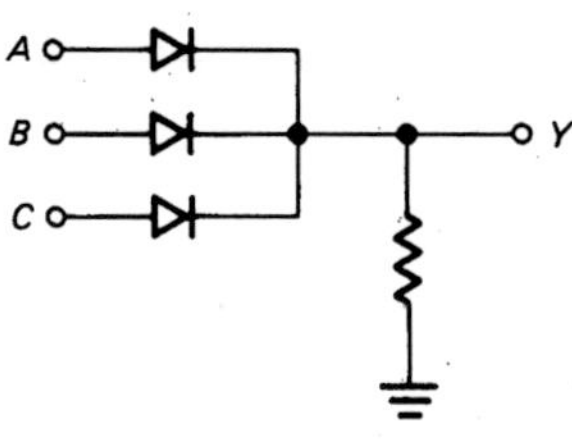

Fig. 2-5 A 3-input diode OR gate.

More than Two Inputs

Figure 2-5 shows a 3-input OR gate. If all inputs are low, all diodes are off and the output is low. If 1 or more inputs are high, the output is high.

Table 2-4 summarizes the action. A table like this is called a *truth table;* it lists all the input possibilities and the corresponding outputs. When constructing a truth table, always list the input words in a binary progression as shown (000, 001, 010, . . . , 111); this guarantees that all input possibilities will be accounted for.

An OR gate can have as many inputs as desired; add one diode for each additional input. Six diodes result in a 6-

TABLE 2-3. TWO INPUT OR GATE

A	*B*	*Y*
0	0	0
0	1	1
1	0	1
1	1	1

TABLE 2-4. THREE-INPUT OR GATE

A	*B*	*C*	*Y*
0	0	0	0
0	0	1	1
0	1	0	1
0	1	1	1
1	0	0	1
1	0	1	1
1	1	0	1
1	1	1	1

input OR gate, nine diodes in a 9-input OR gate. No matter how many inputs, the action of any OR gate is summarized like this: one or more high inputs produce a high output.

Bipolar transistors and MOSFETs can also be used to build OR gates. But no matter what devices are used, OR gates always produce a high output when one or more inputs are high. Figure 2-6 shows the logic symbols for 2-, 3-, and 4-input OR gates.

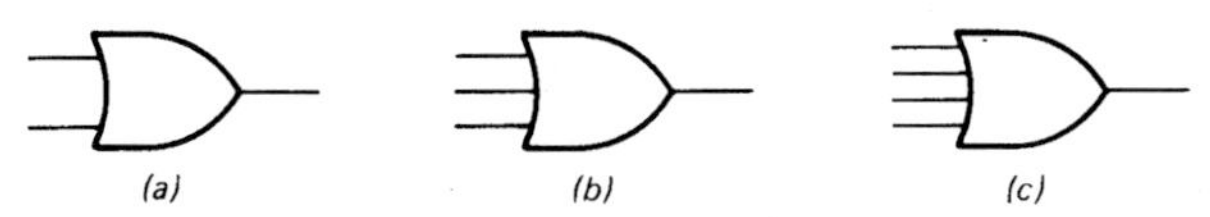

Fig. 2-6 OR-gate symbols.

EXAMPLE 2-2

Show the truth table of a 4-input OR gate.

SOLUTION

Let Y stand for the output bit and A, B, C, D for input bits. Then the truth table has input words of 0000, 0001, 0010, . . . , 1111, as shown in Table 2-5. As expected, output Y is 0 for input word 0000; Y is 1 for all other input words.

As a check, the number of input words in a truth table always equals 2^n, where n is the number of input bits. A 2-input OR gate has a truth table with 2^2 or 4 input words; a 3-input OR gate has 2^3 or 8 input words; and a 4-input OR gate has 2^4 or 16 input words.

TABLE 2-5. FOUR-INPUT OR GATE

A	B	C	D	Y
0	0	0	0	0
0	0	0	1	1
0	0	1	0	1
0	0	1	1	1
0	1	0	0	1
0	1	0	1	1
0	1	1	0	1
0	1	1	1	1
1	0	0	0	1
1	0	0	1	1
1	0	1	0	1
1	0	1	1	1
1	1	0	0	1
1	1	0	1	1
1	1	1	0	1
1	1	1	1	1

EXAMPLE 2-3

How many inputs words are in the truth table of an 8-input OR gate? Which input words produce a high output?

SOLUTION

The input words are 0000 0000, 0000 0001, . . . , 1111 1111. With the formula of the preceding example, the total number of input words is $2^n = 2^8 = 256$.

In any OR gate, 1 or more high inputs produce a high output. Therefore, the input word of 0000 0000 results in a low output; all other input words produce a high output.

EXAMPLE 2-4

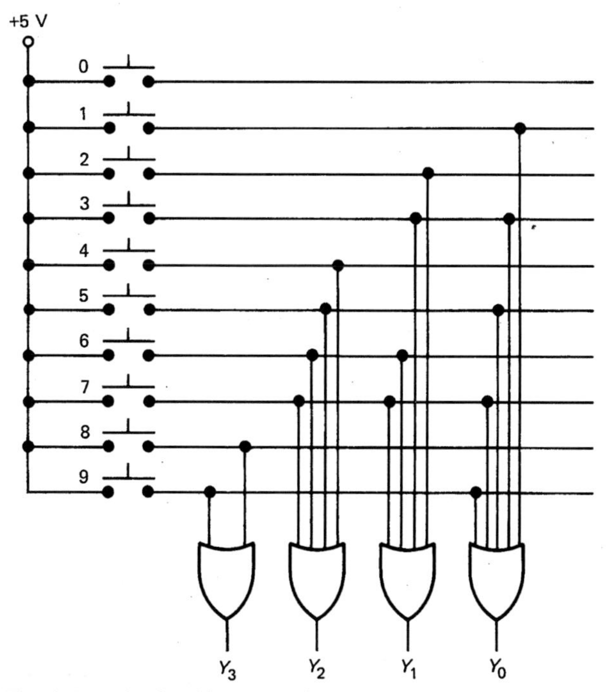

Fig. 2-7 Decimal-to-binary encoder.

The switches of Fig. 2-7 are push-button switches like those of a pocket calculator. The bits out of the OR gates form a 4-bit word, designated $Y_3Y_2Y_1Y_0$. What does the circuit do?

SOLUTION

Figure 2-7 is a decimal-to-binary *encoder,* a circuit that converts decimal to binary. For instance, when push button 3 is pressed, the Y_1 and Y_0 OR gates have high inputs; therefore, the output word is

$$Y_3Y_2Y_1Y_0 = 0011$$

If button 5 is keyed, the Y_2 and Y_0 OR gates have high inputs and the output word becomes

$$Y_3Y_2Y_1Y_0 = 0101$$

When switch 9 is pressed,

$$Y_3Y_2Y_1Y_0 = 1001$$

Check the other input switches to convince yourself that the output word always equals the binary equivalent of the switch being pressed.

2-3 AND GATES

The AND gate has two or more input signals but only one output signal. All inputs must be high to get a high output.

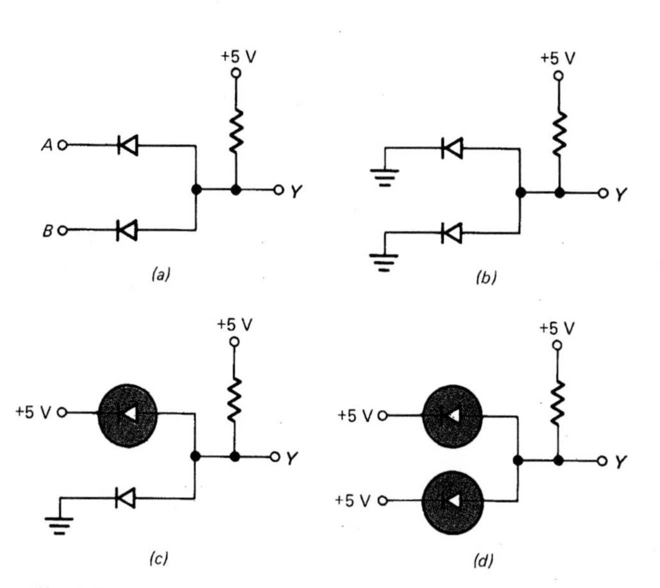

Fig. 2-8 A 2-input AND gate. (*a*) circuit; (*b*) both inputs low; (*c*) 1 low input, 1 high; (*d*) both inputs high.

Diode AND Gate

Figure 2-8*a* shows one way to build an AND gate. In this circuit the inputs can be either low (ground) or high (+5 V). When both inputs are low (Fig. 2-8*b*), both diodes conduct and pull the output down to a low voltage. If one of the inputs is low and the other high (Fig. 2-8*c*), the diode with the low input conducts and this pulls the output down to a low voltage. The diode with the high input, on the other hand, is reverse-biased or cut off, symbolized by the dark shading in Fig. 2-8*c*.

When both inputs are high (Fig. 2-8*d*), both diodes are cut off. Since there is no current in the resistor, the supply voltage pulls the output up to a high voltage (+5 V).

TABLE 2-6. TWO-INPUT AND GATE

A	*B*	*Y*
0	0	0
0	1	0
1	0	0
1	1	1

Table 2-6 summarizes the action. As usual, binary zero stands for low voltage and binary 1 for high voltage. As you see, *A and B* must be high to get a high output; this is why the circuit is called an AND gate.

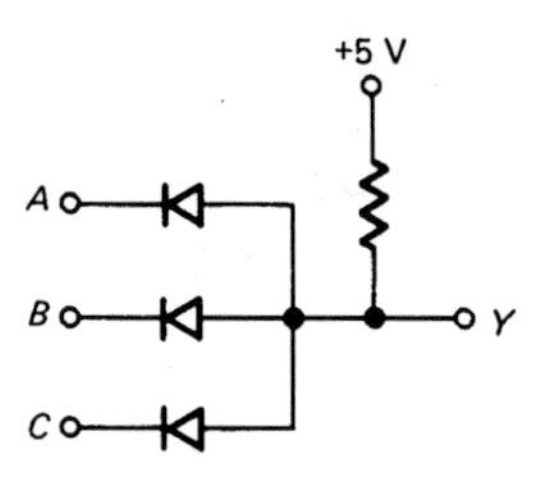

Fig. 2-9 A 3-input AND gate.

More than Two Inputs

Figure 2-9 is a 3-input AND gate. If all inputs are low, all diodes conduct and pull the output down to a low voltage. Even one conducting diode will pull the output down to a low voltage; therefore, the only way to get a high output is to have all inputs high. When all inputs are high, all diodes are nonconducting and the supply voltage pulls the output up to a high voltage.

Table 2-7 summarizes the 3-input AND gate. The output is 0 for all input words except 111. That is, all inputs must be high to get a high output.

AND gates can have as many inputs as desired; add one diode for each additional input. Eight diodes, for instance, result in an 8-input AND gate; sixteen diodes in a 16-input

TABLE 2-7. THREE-INPUT AND GATE

A	*B*	*C*	*Y*
0	0	0	0
0	0	1	0
0	1	0	0
0	1	1	0
1	0	0	0
1	0	1	0
1	1	0	0
1	1	1	1

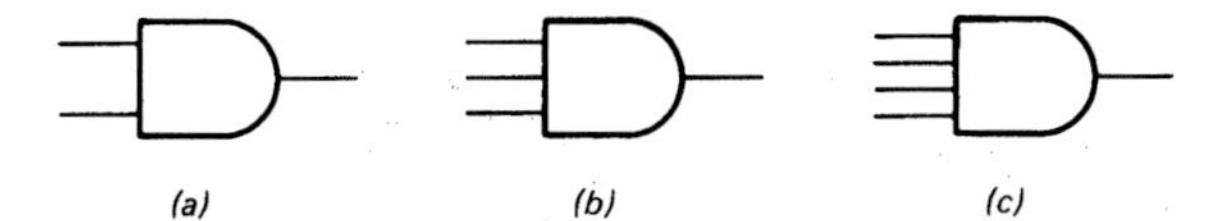

Fig. 2-10 AND-gate symbols.

AND gate. No matter how many inputs an AND gate has, the action can be summarized like this: All inputs must be high to get a high output.

Figure 2-10 shows the logic symbols for 2-, 3-, and 4-input AND gates.

EXAMPLE 2-5

Describe the truth table of an 8-input AND gate.

SOLUTION

The input words are from 0000 0000 to 1111 1111, following the binary progression. The total number of input words is

$$2^n = 2^8 = 256$$

The first 255 input words produce a 0 output. Only the last word, 1111 1111, results in a 1 output. This is because all inputs must be high to get a high output.

EXAMPLE 2-6

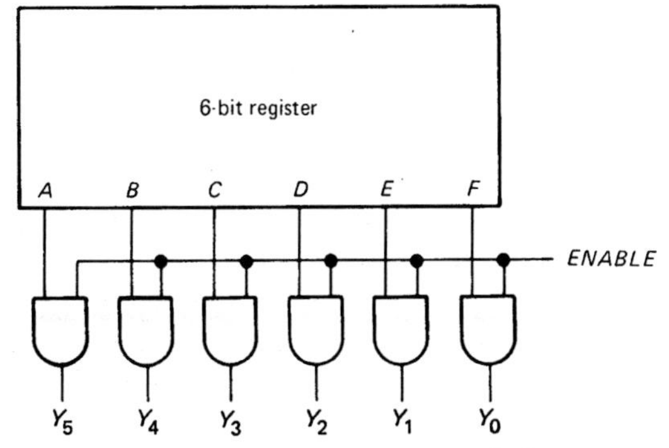

Fig. 2-11 Using AND gates to block or transmit data.

The 6-bit register of Fig. 2-11 stores the word ABCDEF. The *ENABLE* input can be low or high. What does the circuit do?

SOLUTION

One use of AND gates is to transmit data when certain conditions are satisfied. In Fig. 2-11 a low *ENABLE* blocks the register contents from the final output, but a high *ENABLE* transmits the register contents.

For instance, when

$$ENABLE = 0$$

each AND gate has a low *ENABLE* input. No matter what the register contents, the output of each AND gate must be low. Therefore, the final word is

$$Y_5Y_4Y_3Y_2Y_1Y_0 = 000000$$

As you see, a low *ENABLE* blocks the register contents from the final output.

On the other hand, when

$$ENABLE = 1$$

the output of each AND gate depends on the data inputs (*A*, *B*, *C*, . . .); a low data input results in a low output, and a high data input in a high output. For example, if ABCDEF = 100100, a high *ENABLE* gives

$$Y_5Y_4Y_3Y_2Y_1Y_0 = 100100$$

In general, a high *ENABLE* transmits the register contents to the final output to get

$$Y_5Y_4Y_3Y_2Y_1Y_0 = ABCDEF$$

2-4 BOOLEAN ALGEBRA

As mentioned earlier, Boole invented two-state algebra to solve logic problems. This new algebra had no practical use until Shannon applied it to telephone switching circuits. Today boolean algebra is the backbone of computer circuit analysis and design.

Inversion Sign

In boolean algebra a variable can be either a 0 or a 1. For digital circuits, this means that a signal voltage can be either low or high. Figure 2-12 is an example of a digital circuit because the input and output voltages are either low or high. Furthermore, because of the inversion, *Y* is always the complement of *A*.

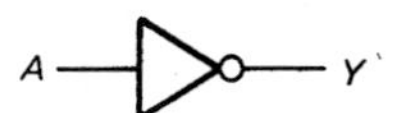

Fig. 2-12 Inverter.

A word equation for Fig. 2-12 is

$$Y = \text{NOT } A \tag{2-1}$$

If A is 0,

$$Y = \text{NOT } 0 = 1$$

On the other hand, if A is 1,

$$Y = \text{NOT } 1 = 0$$

In boolean algebra, the overbar stands for the NOT operation. This means that Eq. 2-1 can be written

$$Y = \overline{A} \qquad (2\text{-}2)$$

Read this as "Y equals NOT A" or "Y equals the complement of A." Equation 2-2 is the standard way to write the output of an inverter.

Using the equation is easy. Given the value of A, substitute and solve for Y. For instance, if A is 0,

$$Y = \overline{A} = \overline{0} = 1$$

because NOT 0 is 1. On the other hand, if A is 1,

$$Y = \overline{A} = \overline{1} = 0$$

because NOT 1 is 0.

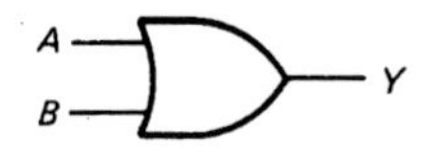

Fig. 2-13 OR gate.

OR Sign

A word equation for Fig. 2-13 is

$$Y = A \text{ OR } B \qquad (2\text{-}3)$$

Given the inputs, you can solve for the output. For instance, if $A = 0$ and $B = 0$,

$$Y = 0 \text{ OR } 0 = 0$$

because 0 comes out of an OR gate when both inputs are 0s.

As another example, if $A = 0$ and $B = 1$,

$$Y = 0 \text{ OR } 1 = 1$$

because 1 comes out of an OR gate when either input is 1. Similarly, if $A = 1$ and $B = 0$,

$$Y = 1 \text{ OR } 0 = 1$$

If $A = 1$ and $B = 1$,

$$Y = 1 \text{ OR } 1 = 1$$

In boolean algebra the + sign stands for the OR operation. In other words, Eq. 2-3 can be written

$$Y = A + B \qquad (2\text{-}4)$$

Read this as "Y equals A OR B." Equation 2-4 is the standard way to write the output of an OR gate.

Given the inputs, you can substitute and solve for the output. For instance, if $A = 0$ and $B = 0$,

$$Y = A + B = 0 + 0 = 0$$

If $A = 0$ and $B = 1$,

$$Y = A + B = 0 + 1 = 1$$

because 0 ORed with 1 results in 1. If $A = 1$ and $B = 0$,

$$Y = A + B = 1 + 0 = 1$$

If both inputs are high,

$$Y = A + B = 1 + 1 = 1$$

because 1 ORed with 1 gives 1.

Don't let the new meaning of the + sign bother you. There's nothing unusual about symbols having more than one meaning. For instance, "pot" may mean a cooking utensil, a flower container, the money wagered in a card game, a derivative of *cannabis sativa* and so forth; the intended meaning is clear from the sentence it's used in. Similarly, the + sign may stand for ordinary addition or OR addition; the intended meaning comes across in the way it's used. If we're talking about decimal numbers, + means ordinary addition, but when the discussion is about logic circuits, + stands for OR addition.

Fig. 2-14 AND gate.

AND Sign

A word equation for Fig. 2-14 is

$$Y = A \text{ AND } B \qquad (2\text{-}5)$$

In boolean algebra the multiplication sign stands for the AND operation. Therefore, Eq. 2-5 can be written

$$Y = A \cdot B$$

or simply

$$Y = AB \qquad (2\text{-}6)$$

Read this as "Y equals A AND B." Equation 2-6 is the standard way to write the output of an AND gate.

Given the inputs, you can substitute and solve for the output. For instance, if both inputs are low,

$$Y = AB = 0 \cdot 0 = 0$$

because 0 ANDed with 0 gives 0. If A is low and B is high,

$$Y = AB = 0 \cdot 1 = 0$$

because 0 comes out of an AND gate if any input is 0. If A is 1 and B is 0,

$$Y = AB = 1 \cdot 0 = 0$$

When both inputs are high,

$$Y = AB = 1 \cdot 1 = 1$$

because 1 ANDed with 1 gives 1.

Decision-Making Elements

The inverter, OR gate, and AND gate are often called *decision-making elements* because they can recognize some input words while disregarding others. A gate recognizes a word when its output is high; it disregards a word when its output is low. For example, the AND gate disregards all words with one or more 0s; it recognizes only the word whose bits are all 1s.

Notation

In later equations we need to distinguish between bits that are ANDed and bits that are part of a binary word. To do this we will use italic (slanted) letters (A, B, Y, etc.) for ANDed bits and roman (upright) letters (A, B, Y, etc.) for bits that form a word.

For example, $Y_3Y_2Y_1Y_0$ stands for the logical product (ANDing) of Y_3, Y_2, Y_1, and Y_0. If $Y_3 = 1$, $Y_2 = 0$, $Y_1 = 0$, and $Y_0 = 1$, the product $Y_3Y_2Y_1Y_0$ will reduce as follows:

$$Y_3Y_2Y_1Y_0 = 1 \cdot 0 \cdot 0 \cdot 1 = 0$$

In this case, the italic letters represent bits that are being ANDed.

On the other hand, $\mathrm{Y_3Y_2Y_1Y_0}$ is our notation for a 4-bit word. With the Y values just given, we can write

$$\mathrm{Y_3Y_2Y_1Y_0} = 1001$$

In this equation, we are not dealing with bits that are ANDed; instead, we are dealing with bits that are part of a word.

The distinction between italic and roman notation will become clearer when we get to computer analysis.

Positive and Negative Logic

A final point. *Positive logic* means that 1 stands for the more positive of the two voltage levels. *Negative logic* means that 1 stands for the more negative of the two voltage levels. For instance, if the two voltage levels are 0 and −5 V, positive logic would have 1 stand for 0 V and 0 for −5 V, whereas negative logic would have 1 stand for −5 V and 0 for 0 V.

Ordinarily, people use positive logic with positive supply voltages and negative logic with negative supply voltages. Throughout this book, we will be using positive logic.

EXAMPLE 2-7

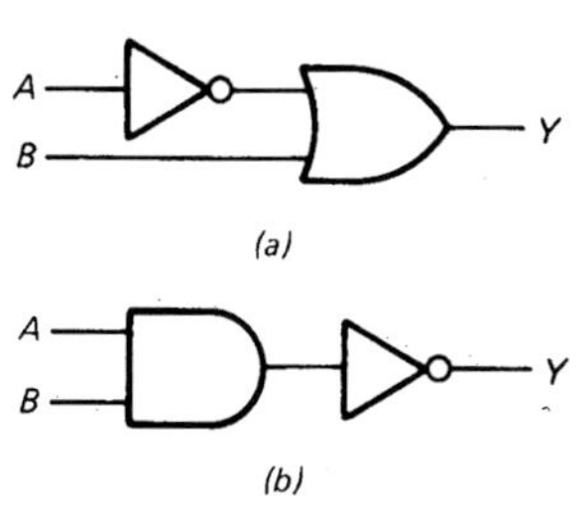

Fig. 2-15 Logic circuits.

What is the boolean equation for Fig. 2-15*a*? The output if both inputs are high?

SOLUTION

A is inverted before it reaches the OR gate; therefore, the upper input to the OR gate is $\overline{A}$. The final output is

$$Y = \overline{A} + B$$

This is the boolean equation for Fig. 2-15*a*.

To find the output when both inputs are high, either of two approaches can be used. First, you can substitute directly into the foregoing equation and solve for Y

$$Y = \overline{A} + B = \overline{1} + 1 = 0 + 1 = 1$$

Alternatively, you can analyze the operation of Fig. 2-15*a* like this. If both inputs are high, the inputs to the OR gate are 0 and 1. Now, 0 ORed with 1 gives 1. Therefore, the final output is high.

EXAMPLE 2-8

What is the boolean equation for Fig. 2-15*b*? If both inputs are high, what is the output?

SOLUTION

The AND gate forms the logical product AB, which is inverted to get

$$Y = \overline{AB}$$

Read this as "Y equals NOT AB" or "Y equals the complement of AB."

If both inputs are high, direct substitution into the equation gives

$$Y = \overline{AB} = \overline{1 \cdot 1} = \overline{1} = 0$$

Note the order of operations: the ANDing is done first, then the inversion.

Instead of using the equation, you can analyze Fig. 2-15*b* as follows. If both inputs are high, the AND gate has a high output. Therefore, the final output is low.

EXAMPLE 2-9

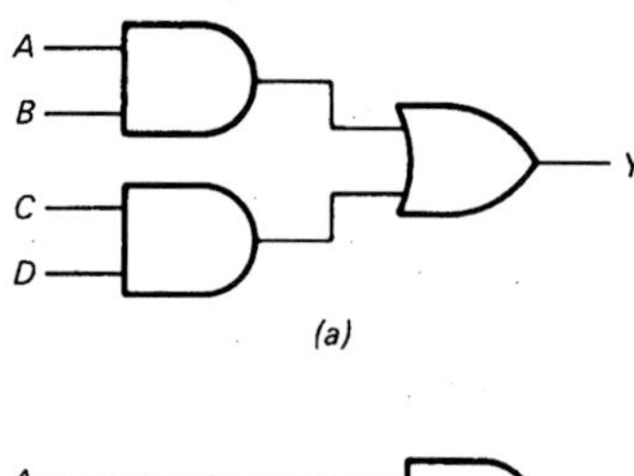

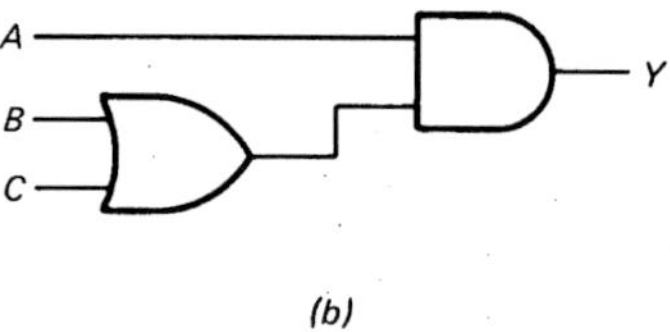

Fig. 2-16 Logic circuits.

What is the boolean equation for Fig. 2-16*a*? The truth table? Which input words does the circuit recognize?

SOLUTION

The upper AND gate forms the logical product AB, and the lower AND gate gives CD. ORing these products results in

$$Y = AB + CD$$

Read this as "Y equals AB OR CD."

Next, look at Fig. 2-16*a*. The final output is high if the OR gate has one or more high inputs. This happens when AB is 1, CD is 1, or both are 1s. In turn, AB is 1 when

$$A = 1 \quad \text{and} \quad B = 1$$

TABLE 2-8. TRUTH TABLE FOR $Y = AB + CD$

A	B	C	D	Y
0	0	0	0	0
0	0	0	1	0
0	0	1	0	0
0	0	1	1	1
0	1	0	0	0
0	1	0	1	0
0	1	1	0	0
0	1	1	1	1
1	0	0	0	0
1	0	0	1	0
1	0	1	0	0
1	0	1	1	1
1	1	0	0	1
1	1	0	1	1
1	1	1	0	1
1	1	1	1	1

CD is 1 when

$$C = 1 \quad \text{and} \quad D = 1$$

Both products are 1s when

$$A = 1 \quad B = 1 \quad C = 1 \quad \text{and} \quad D = 1$$

Therefore, the final output is high when A and B are 1s, when C and D are 1s, or when all inputs are 1s.

Table 2-8 summarizes the foregoing analysis. From this it's clear that the circuit recognizes these input words: 0011, 0111, 1011, 1100, 1101, 1110, and 1111.

EXAMPLE 2-10

Write the boolean equation for Fig. 2-16*b*. If all inputs are high, what is the output?

SOLUTION

The OR gate forms the logical sum $B + C$. This sum is ANDed with A to get

$$Y = A(B + C)$$

(Parentheses indicate ANDing.)

One way to find the output when all inputs are high is to substitute and solve as follows:

$$Y = A(B + C) = 1(1 + 1) = 1(1) = 1$$

Alternatively, you can analyze Fig. 2-16*b* like this. If all inputs are high, the OR gate has a high output; therefore, both inputs to the AND gate are high. Since all high inputs to an AND gate result in a high output, the final output is high.

EXAMPLE 2-11

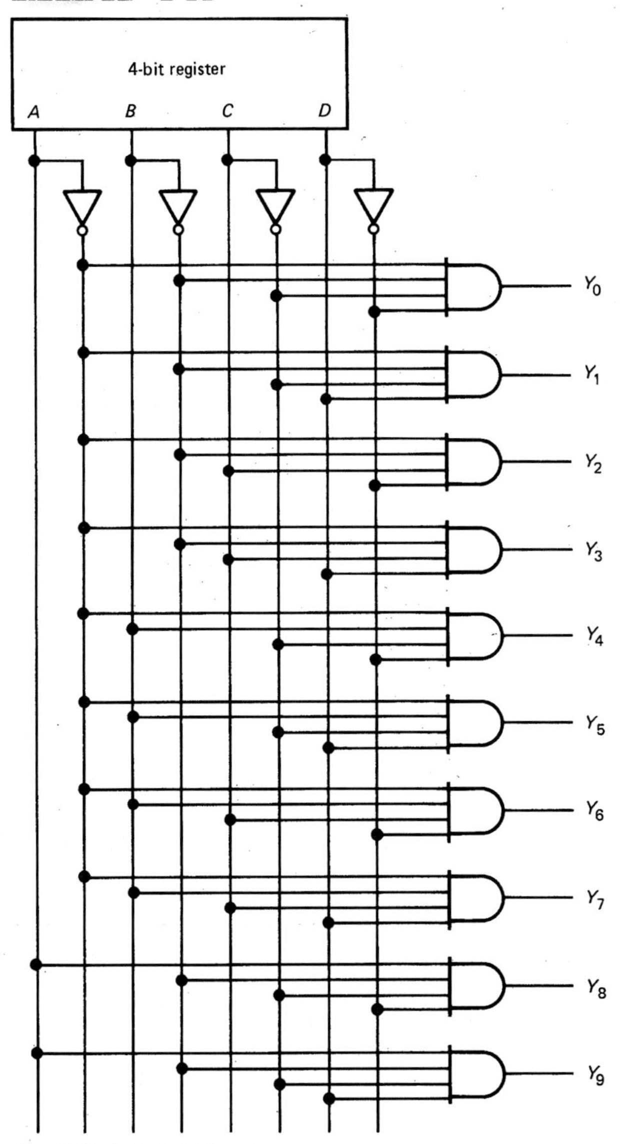

Fig. 2-17 A 1-of-10 decoder.

What is the boolean equation for each *Y* output in Fig. 2-17?

SOLUTION

Each AND gate forms the logical product of its input signals. The inputs to the top AND gate are $\overline{A}$, $\overline{B}$, $\overline{C}$ and $\overline{D}$; therefore,

$$Y_0 = \overline{A}\,\overline{B}\,\overline{C}\,\overline{D}$$

The inputs to the next AND gate are $\overline{A}$, $\overline{B}$, $\overline{C}$ and D; this means that

$$Y_1 = \overline{A}\,\overline{B}\,\overline{C}D$$

Analyzing the remaining gates gives

$$Y_2 = \overline{A}\,\overline{B}C\overline{D}$$
$$Y_3 = \overline{A}\,\overline{B}CD$$
$$Y_4 = \overline{A}B\overline{C}\,\overline{D}$$
$$Y_5 = \overline{A}B\overline{C}D$$
$$Y_6 = \overline{A}BC\overline{D}$$
$$Y_7 = \overline{A}BCD$$
$$Y_8 = A\overline{B}\,\overline{C}\,\overline{D}$$
$$Y_9 = A\overline{B}\,\overline{C}D$$

EXAMPLE 2-12

What does the circuit of Fig. 2-17 do?

SOLUTION

This is a binary-to-decimal decoder, a circuit that converts from binary to decimal. For instance, when the register contents are 0011, the Y_3 AND gate has all high inputs; therefore, Y_3 is high. Furthermore, register contents of 0011 mean that all other AND gates have at least one low input. As a result, all other AND gates have low outputs. (Analyze the circuit to convince yourself.)

If the register contents change to 0100, only the Y_4 AND gate has all high inputs; therefore, only Y_4 is high. If the register contents change to 0111, Y_7 is the only high output.

In general, the subscript of the high output equals the decimal equivalent of the binary number stored in the register. This is why the circuit is called a *binary-to-decimal* decoder.

The circuit of this example is also called a 4-line–to–10-line decoder because there are 4 input lines and 10 output lines. Another name for it is a 1-of-10 decoder because only 1 of 10 output lines has a high voltage.

GLOSSARY

AND *gate* A logic circuit whose output is high only when all inputs are high.

boolean algebra Originally known as symbolic logic, this modern algebra uses the set of numbers 0 and 1. The operations OR, AND, and NOT are sometimes called *union, intersection,* and *inversion*. Boolean algebra is ideally suited to digital circuit analysis.

complement The output of an inverter.

gate A logic circuit with one or more input signals but only one output signal.

inverter A gate with only 1 input and 1 output. The output is always the complement of the input. Also known as a NOT gate.

logic circuit A circuit whose input and output signals are two-state, either low or high voltages. The basic logic circuits are OR, AND, and NOT gates.

OR gate A logic circuit with 2 or more inputs and only 1 output; 1 or more high inputs produce a high output.

truth table A table that shows all input and output possibilities for a logic circuit. The input words are listed in binary progression.

word A string of bits that represent a coded instruction or data.

SELF-TESTING REVIEW

Read each of the following and provide the missing words. Answers appear at the beginning of the next question.

1. A gate is a logic circuit with one or more input signals but only ________ output signal. These signals are either ________ or high.

2. (*one, low*) An inverter is a gate with only ________ input; the output is always in the opposite state from the input. An inverter is also called a ________ gate. Sometimes the output is referred to as the complement of the input.

3. (*1,* NOT) The OR gate has two or more input signals. If any input is ________, the output is high. The number of input words in a truth table always equals ________, where *n* is the number of input bits.

4. (*high,* 2^n) The ________ gate has two or more input signals. All inputs must be high to get a high output.

5. (AND) In boolean algebra, the overbar stands for the NOT operation, the plus sign stands for the ________ operation, and the times sign for the ________ operation.

6. (OR, AND) The inverter, OR gate, and AND gate are called decision-making elements because they can recognize some input ________ while disregarding others. A gate recognizes a word when its output is ________.

7. (*words, high*) A binary-to-decimal decoder is also called a 4-line–to–10-line decoder because it has 4 input lines and 10 output lines. Another name for it is the 1-of-10 decoder because only 1 of its 10 output lines is high at a time.

PROBLEMS

2-1. How many inputs signals can a gate have? How many output signals?

2-2. If you cascade seven inverters, does the overall circuit act like an inverter or noninverter?

2-3. Double inversion occurs when two inverters are cascaded. Does such a connection act like an inverter or noninverter?

2-4. The contents of the 6-bit register in Fig. 2-3*b* change to 101010. What is the decimal equivalent of the register contents? The decimal equivalent out of the hex inverter?

2-5. An OR gate has 6 inputs. How many input words are in its truth table? What is the only input word that produces a 0 output?

2-6. Figure 2-18 shows a hexadecimal encoder, a circuit that converts hexadecimal to binary. Pressing each push-button switch results in a different output word $Y_3Y_2Y_1Y_0$. Starting with switch 0, what are the output words? (NOTE: The new symbol in Fig. 2-18 is another way to draw an OR gate.

2-7. In Fig. 2-18 what switches would you press to produce

0011 1001 1100 1111

(Work from left to right.)

2-8. What is the 4-bit output in Fig. 2-18 when switch A is pressed? Switch 4? Switch E? Switch 6?

2-9. An AND gate has 7 inputs. How many input words are in its truth table? What is the only input word that produces a 1 output?

2-10. Visualize the register contents of Fig. 2-19 as the word $A_7A_6 \cdots A_0$, and the final output as the word $Y_7Y_6 \cdots Y_0$. What is the output word for each of the following conditions:

a. $A_7A_6 \cdots A_0$ = 1100 1010, *ENABLE* = 0.
b. $A_7A_6 \cdots A_0$ = 0101 1101, *ENABLE* = 1.
c. $A_7A_6 \cdots A_0$ = 1111 0000, *ENABLE* = 1.
d. $A_7A_6 \cdots A_0$ = 1010 1010, *ENABLE* = 0.

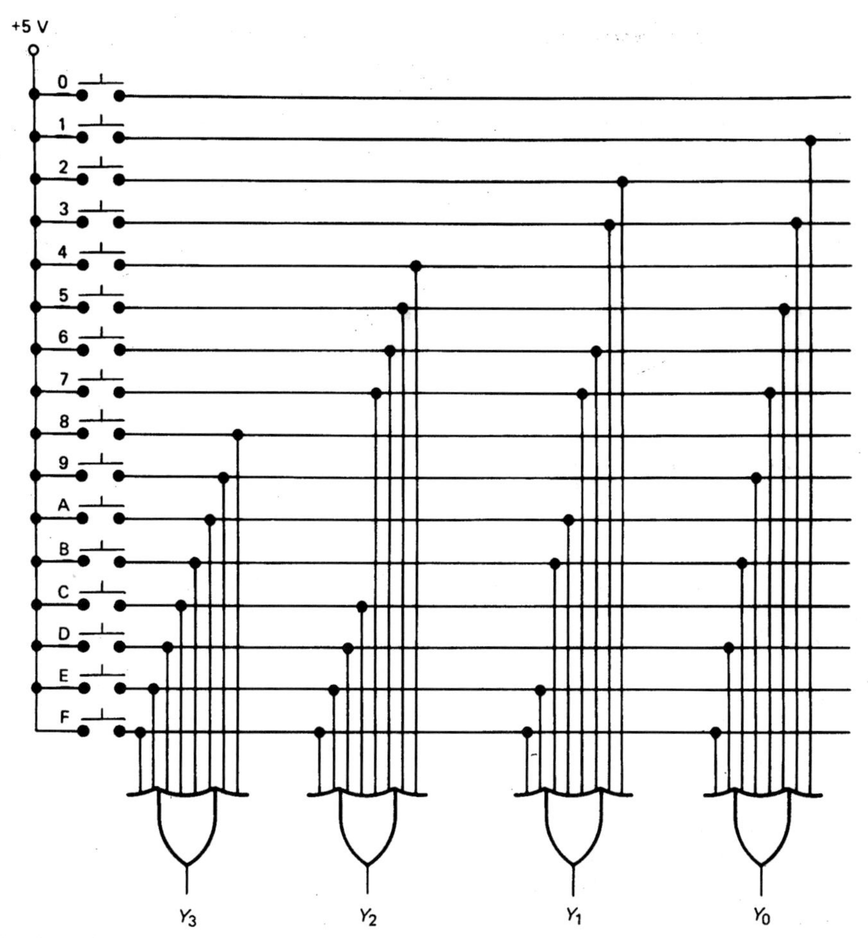

Fig. 2-18 Hexadecimal encoder.

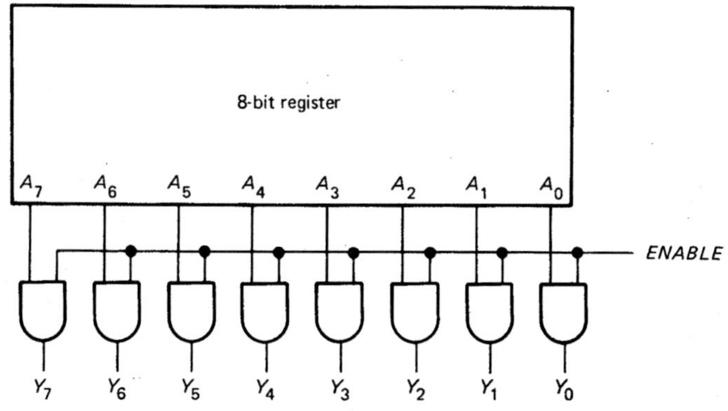

Fig. 2-19

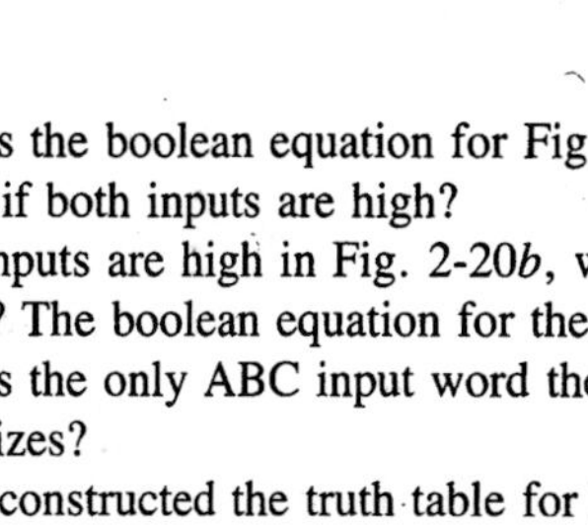

A
B
Y
(a)

A
B
C
Y
(b)

Fig. 2-20

2-11. The 8-bit register of Fig. 2-19 stores 59_{10}. What is the decimal equivalent of the final output word if *ENABLE* = 0? If *ENABLE* = 1?

2-12. Answer these questions:

a. What input words does a 6-input OR gate recognize? What word does it disregard?

b. What input word does an 8-input AND gate recognize? What words does it disregard?

2-13. What is the boolean equation for Fig. 2-20*a*? The output if both inputs are high?

2-14. If all inputs are high in Fig. 2-20*b*, what is the output? The boolean equation for the circuit? What is the only ABC input word the circuit recognizes?

2-15. If you constructed the truth table for Fig. 2-20*b*, how many input words would it contain?

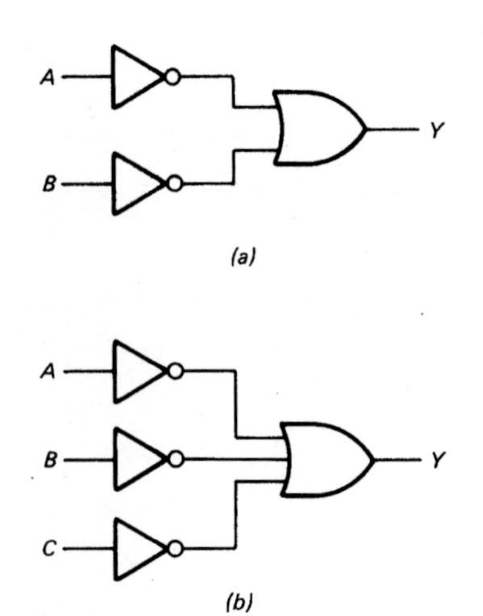

Fig. 2-21

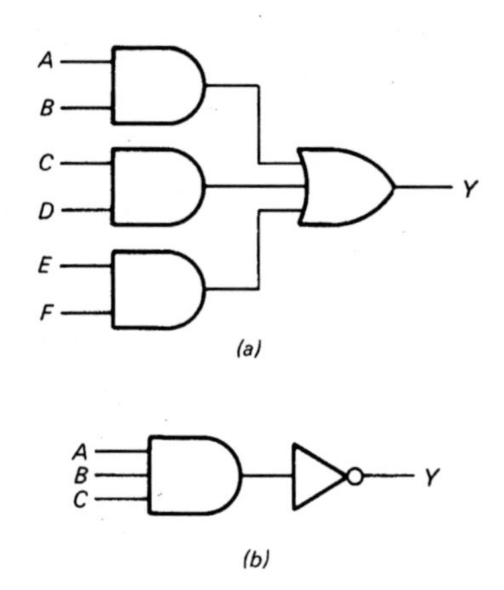

Fig. 2-22

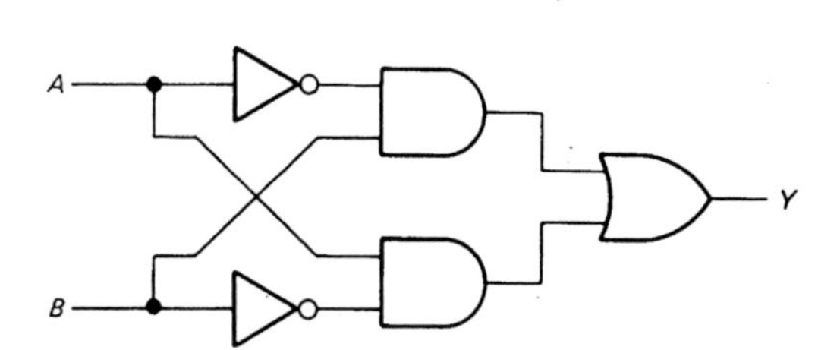

Fig. 2-23

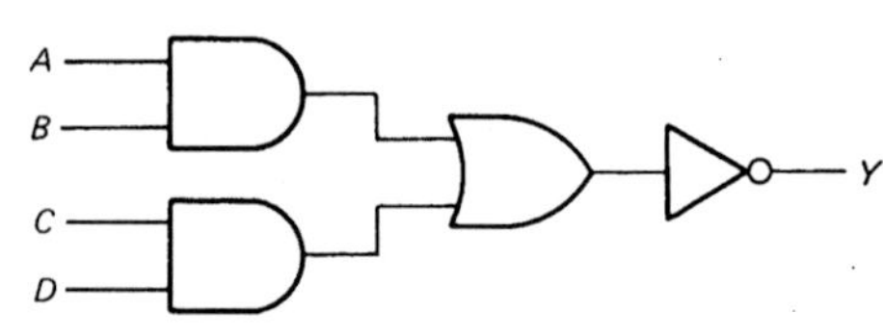

Fig. 2-24

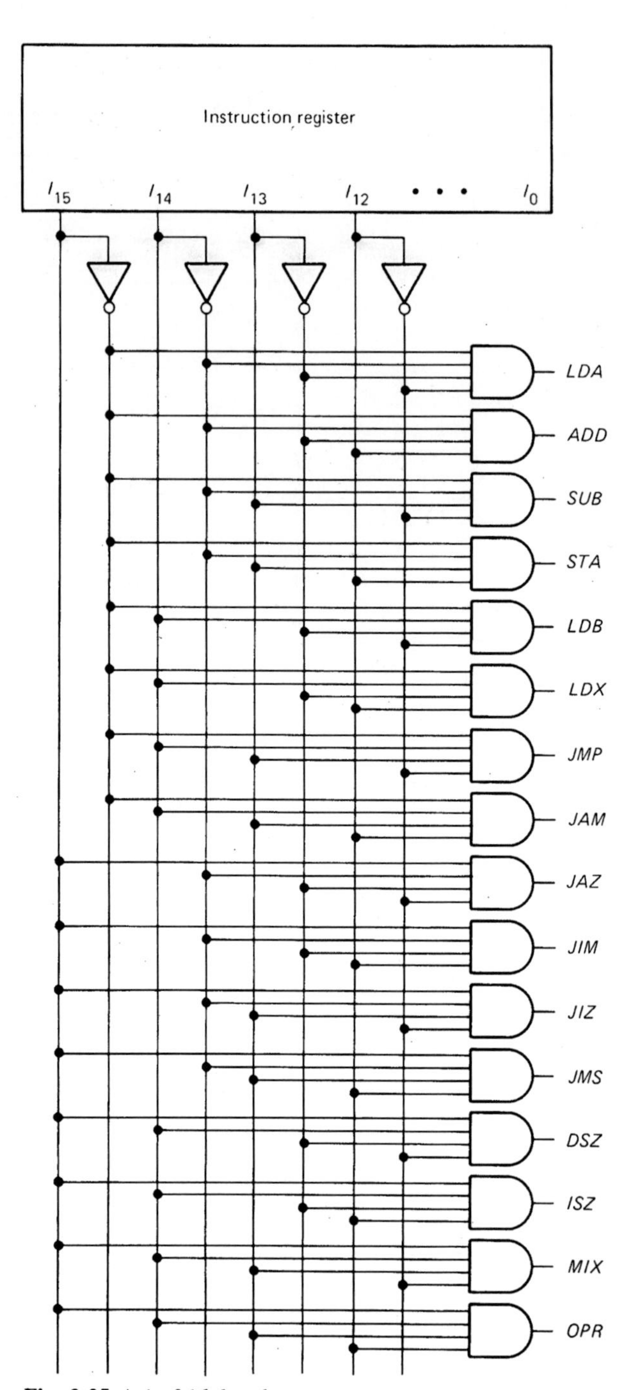

Fig. 2-25 A 1-of-16 decoder.

2-16. What is the boolean equation for Fig. 2-21*a*? The output if both inputs are high?

2-17. If all inputs are high in Fig. 2-21*b*, what is the output? What is the boolean equation of the circuit? What ABC input words does the circuit recognize? What is the only word it disregards?

2-18. What is the boolean equation for Fig. 2-22*a*? The output if all inputs are 1s? If you were to construct the truth table, how many input words would it have?

2-19. Write the boolean equation for Fig. 2-22*b*. If all inputs are 1s, what is the output?

2-20. If both inputs are high in Fig. 2-23, what is the output? What is the boolean equation for the circuit? Describe the truth table.

2-21. What is the boolean equation for Fig. 2-24? How many ABCD input words are in the truth table? Which input words does the circuit recognize?

2-22. Because of the historical connection between boolean algebra and logic, some people use the words "true" and "false" instead of "high" and "low" when discussing logic circuits. For instance, here's how an AND gate can be described. If any input is false, the output is false; if all inputs are true, the output is true.

a. If both inputs are false in Fig. 2-23, what is the output?
b. What is the output in Fig. 2-23 if one input is false and the other true?
c. In Fig. 2-23 what is the output if all inputs are true?

2-23. Figure 2-25 shows a 1-of-16 decoder. The signals coming out of the decoder are labeled *LDA*, *ADD*, *SUB*, and so on. The word formed by the 4 leftmost register bits is called the **OP CODE.** As an equation,

$$\textbf{OP CODE} = I_{15}I_{14}I_{13}I_{12}$$

a. If *LDA* is high, what does **OP CODE** equal?
b. If *ADD* is high, what does it equal?
c. When **OP CODE** = 1001, which of the output signals is high?
d. Which output signal is high if **OP CODE** = 1111?

2-24. In Fig. 2-25, list the **OP CODE** words and the corresponding high output signals. (Start with 0000 and proceed in binary to 1111.)

2-25. In the following equations the equals sign means "is equivalent to." Classify each of the following as positive or negative logic:

a. 0 = 0 V and 1 = +5 V.
b. 0 = +5 V and 1 = 0 V.
c. 0 = −5 V and 1 = 0 V.
d. 0 = 0 V and 1 = −5 V.

2-26. In Fig. 2-25 four output lines come from the decoder. Is it possible to add more op codes without increasing the number of output lines?

2-27. How many output lines from the decoder would be needed to have 256 op codes?

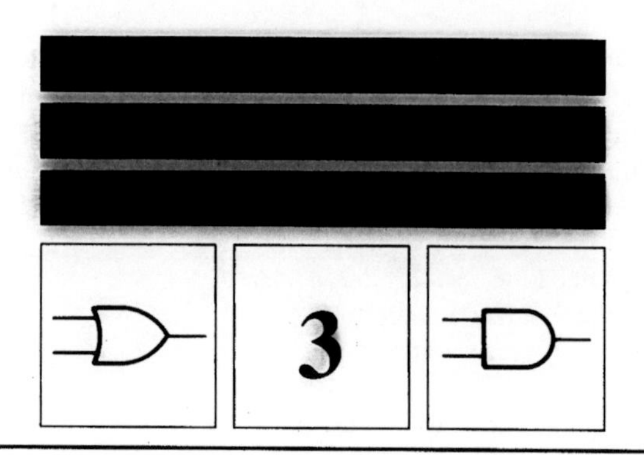

MORE LOGIC GATES

This chapter introduces NOR and NAND gates, devices that are widely used in industry. You will also learn about De Morgan's theorems; they help you to rearrange and simplify logic circuits.

3-1 NOR GATES

The NOR gate has two or more input signals but only one output signal. All inputs must be low to get a high output. In other words, the NOR gate recognizes only the input word whose bits are all 0s.

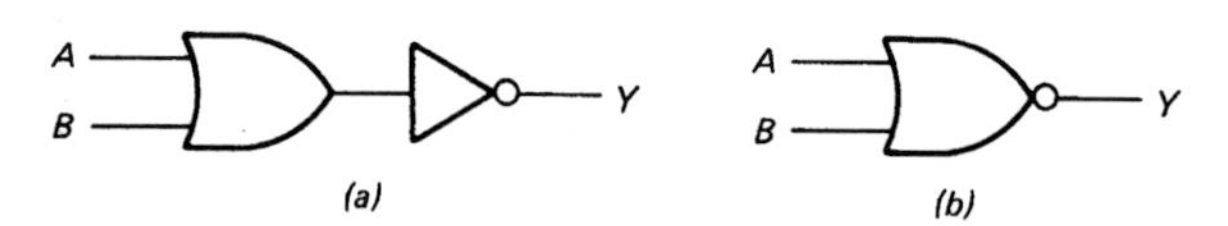

Fig. 3-1 NOR gate: (*a*) logical meaning; (*b*) standard symbol.

Two-Input Gate

Figure 3-1*a* shows the logical structure of a NOR gate, which is an OR gate followed by an inverter. Therefore, the final output is NOT the OR of the inputs. Originally called a NOT-OR gate, the circuit is now referred to as a NOR gate.

Figure 3-1*b* is the standard symbol for a NOR gate. Notice that the inverter triangle has been deleted and the small circle or bubble moved to the OR-gate output. The bubble is a reminder of the inversion that follows the ORing.

With Fig. 3-1*a* and *b* the following ideas are clear. If both inputs are low, the final output is high. If one input is low and the other high, the output is low. And if both inputs are high, the output is low.

Table 3-1 summarizes the circuit action. As you see, the NOR gate recognizes only the input word whose bits are all 0s. In other words, all inputs must be low to get a high output.

TABLE 3-1. TWO-INPUT NOR GATE

A	B	$\overline{A+B}$
0	0	1
0	1	0
1	0	0
1	1	0

Incidentally, the boolean equation for a 2-input NOR gate is

$$Y = \overline{A + B} \tag{3-1}$$

Read this as "Y equals NOT A OR B." If you use this equation, remember that the ORing is done first, then the inversion.

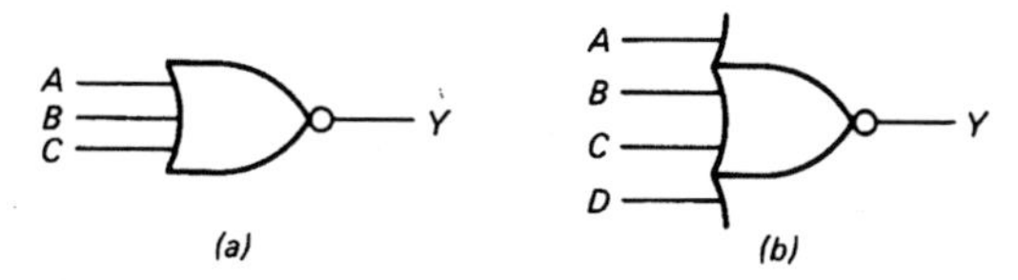

Fig. 3-2 NOR gates: (*a*) 3-input; (*b*) 4-input.

Three-Input Gate

Regardless of how many inputs a NOR gate has, it is still logically equivalent to an OR gate followed by an inverter. For instance, Fig. 3-2*a* shows a 3-input NOR gate. The 3 inputs are ORed, and the result is inverted. Therefore, the boolean equation is

$$Y = \overline{A + B + C} \tag{3-2}$$

The analysis of Fig. 3-2*a* goes like this. If all inputs are low, the result of ORing is low; therefore, the final output

TABLE 3-2. THREE-INPUT NOR GATE

A	B	C	$\overline{A+B+C}$
0	0	0	1
0	0	1	0
0	1	0	0
0	1	1	0
1	0	0	0
1	0	1	0
1	1	0	0
1	1	1	0

is high. If one or more inputs are high, the result of ORing is high; so the final output is low.

Table 3-2 summarizes the action of a 3-input NOR gate. As you see, the circuit recognizes only the input word whose bits are 0s. In other words, all inputs must be low to get a high output.

Four-Input Gate

Figure 3-2*b* is the symbol for a 4-input NOR gate. The inputs are ORed, and the result is inverted. For this reason, the boolean equation is

$$Y = \overline{A + B + C + D} \tag{3-3}$$

The corresponding truth table has input words from 0000 to 1111. Word 0000 gives a 1 output; all other words produce a 0 output. (For practice, you should construct the truth table of the 4-input NOR gate.)

3-2 DE MORGAN'S FIRST THEOREM

Most mathematicians ignored boolean algbebra when it first appeared; some even ridiculed it. But Augustus De Morgan saw that it offered profound insights. He was the first to acclaim Boole's great achievement.

Always a warm and likable man, De Morgan himself had paved the way for boolean algebra by discovering two important theorems. This section introduces the first theorem.

The First Theorem

Figure 3-3*a* is a 2-input NOR gate, analyzed earlier. As you recall, the boolean equation is

$$Y = \overline{A + B}$$

and Table 3-3 is the truth table.

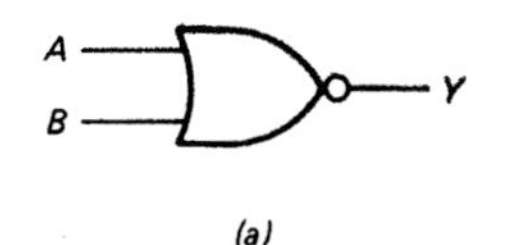

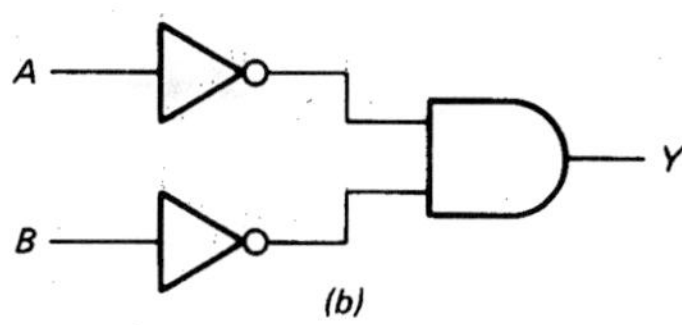

Fig. 3-3 De Morgan's first theorem: (*a*) NOR gate; (*b*) AND gate with inverted inputs.

Figure 3-3*b* has the inputs inverted before they reach the AND gate. Therefore, the boolean equation is

$$Y = \overline{A}\,\overline{B}$$

If both inputs are low in Fig. 3-3*b*, the AND gate has high inputs; therefore, the final output is high. If one or more inputs are high, one or more AND-gate inputs must be low and the final output is low. Table 3-4 summarizes these ideas.

TABLE 3-3

A	B	$\overline{A+B}$
0	0	1
0	1	0
1	0	0
1	1	0

TABLE 3-4

A	B	$\overline{A}\,\overline{B}$
0	0	1
0	1	0
1	0	0
1	1	0

Compare Tables 3-3 and 3-4. They're identical. This means that the two circuits are logically equivalent; given the same inputs, the outputs are the same. In other words, the circuits of Fig. 3-3 are interchangeable.

De Morgan discovered the foregoing equivalence long before logic circuits were invented. His first theorem says

$$\overline{A + B} = \overline{A}\,\overline{B} \tag{3-4}$$

The left member of this equation represents Fig. 3-3*a*; the right member, Fig. 3-3*b*. Equation 3-4 says that Fig. 3-3*a* and *b* are equivalent (interchangeable).

Bubbled AND Gate

Figure 3-4*a* shows an AND gate with inverted inputs. This circuit is so widely used that the abbreviated logic symbol of Fig. 3-4*b* has been adopted. Notice that the inverter triangles have been deleted and the bubbles moved to the

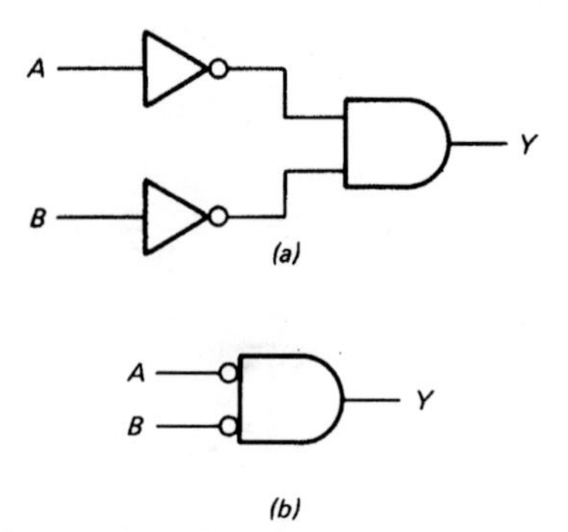

Fig. 3-4 AND gate with inverted inputs: (*a*) circuit; (*b*) abbreviated symbol.

AND-gate inputs. From now on, we will refer to Fig. 3-4*b* as a *bubbled* AND *gate;* the bubbles are a reminder of the inversion that takes place before ANDing.

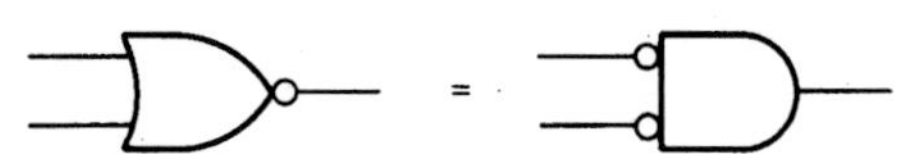

Fig. 3-5 De Morgan's first theorem.

Figure 3-5 is a graphic summary of De Morgan's first theorem. A NOR gate and a bubbled AND gate are equivalent. As shown later, because the circuits are interchangeable, you can often reduce complicated logic circuits to simpler forms.

More than Two Inputs

When 3 inputs are involved, De Morgan's first theorem is written

$$\overline{A + B + C} = \overline{A}\,\overline{B}\,\overline{C} \tag{3-5}$$

For 4 inputs

$$\overline{A + B + C + D} = \overline{A}\,\overline{B}\,\overline{C}\,\overline{D} \tag{3-6}$$

In both cases, the theorem says that the complement of a sum equals the product of the complements.

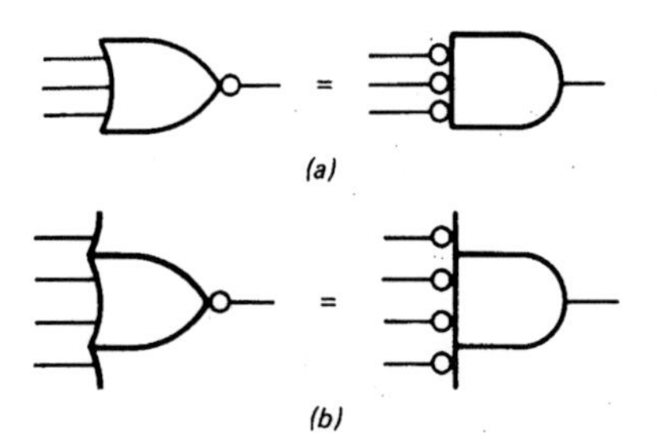

Fig. 3-6 De Morgan's first theorem: (*a*) 3-input circuits; (*b*) 4-input circuits.

Here's what really counts. Equation 3-5 says that a 3-input NOR gate and a 3-input bubbled AND gate are equivalent (see Fig. 3-6*a*). Equation 3-6 means that a 4-input NOR gate and a 4-input bubbled AND gate are equivalent (Fig. 3-6*b*). Memorize these equivalent circuits; they are a visual statement of De Morgan's first theorem.

Notice in Fig. 3-6*b* how the input edges of the NOR gate and the bubbled AND gate have been extended. This is common drafting practice when there are many input signals. The same idea applies to any type of gate.

EXAMPLE 3-1

Prove that Fig. 3-7*a* and *c* are equivalent.

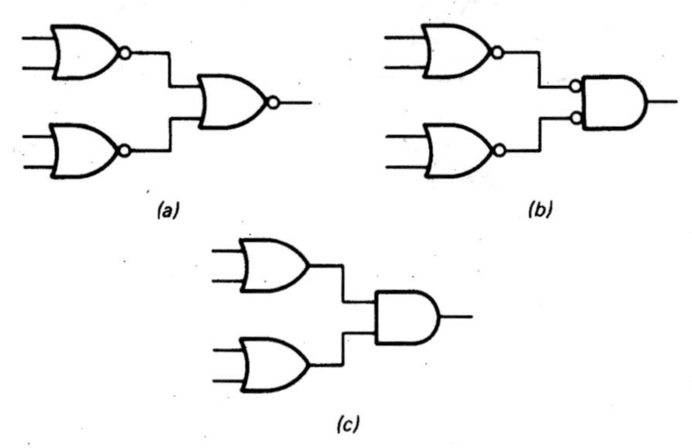

Fig. 3-7 Equivalent De Morgan circuits.

SOLUTION

The final NOR gate in Fig. 3-7*a* is equivalent to a bubbled AND gate. This allows us to redraw the circuit as shown in Fig. 3-7*b*.

Double inversion produces noninversion; therefore, each double inversion in Fig. 3-7*b* cancels out, leaving the simplified circuit of Fig. 3-7*c*. Figure 3-7*a* and *c* are therefore equivalent.

Remember the idea. Given a logic circuit, you can replace any NOR gate by a bubbled AND gate. Then any double inversion (a pair of bubbles in a series path) cancels out. Sometimes you wind up with a simpler logic circuit than you started with; sometimes not.

But the point remains. De Morgan's first theorem enables you to rearrange a logic circuit with the hope of finding a simpler equivalent circuit or perhaps getting more insight into how the original circuit works.

3-3 NAND GATES

The NAND gate has two or more input signals but only one output signal. All input signals must be high to get a low output.

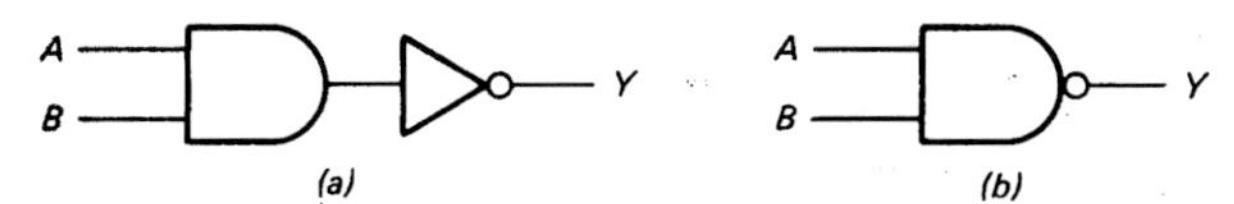

Fig. 3-8 NAND gate: (*a*) logical meaning; (*b*) standard symbol.

Two-Input Gate

Figure 3-8*a* shows the logical structure of a NAND gate, an AND gate followed by an inverter. Therefore, the final output is NOT the AND of the inputs. Originally called a NOT-AND gate, the circuit is now referred to as a NAND gate.

Figure 3-8*b* is the standard symbol for a NAND gate. The inverter triangle has been deleted and the bubble moved to the AND-gate output. If one or more inputs are low, the result of ANDing is low; therefore, the final inverted output is high. Only when all inputs are high does the ANDing produce a high signal; then the final output is low.

Table 3-5 summarizes the action of a 2-input NAND gate. As shown, the NAND gate recognizes any input word with one or more 0s. That is, one or more low inputs produce a high output. The boolean equation for a 2-input NAND gate is

$$Y = \overline{AB} \tag{3-7}$$

Read this as "Y equals NOT AB." If you use this equation, remember that the ANDing is done first then the inversion.

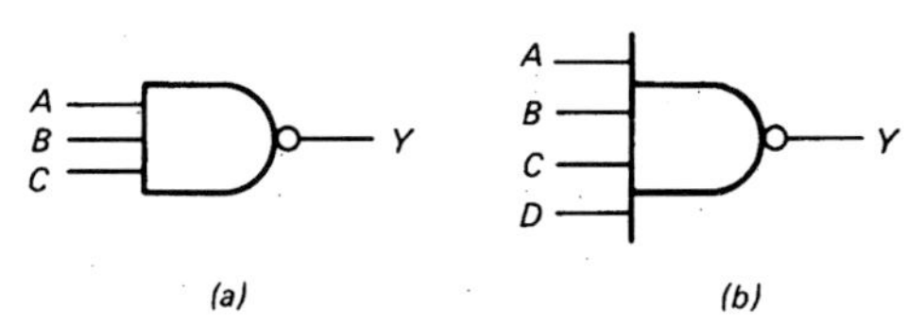

Fig. 3-9 NAND gates: (*a*) 3-input; (*b*) 4-input.

Three-Input Gate

Regardless of how many inputs a NAND gate has, it's still logically equivalent to an AND gate followed by an inverter. For example, Fig. 3-9*a* shows a 3-input NAND gate. The inputs are ANDed, and the product is inverted. Therefore, the boolean equation is

$$Y = \overline{ABC} \tag{3-8}$$

Here is the analysis of Fig. 3-9*a*. If one or more inputs are low, the result of ANDing is low; therefore, the final output is high. If all inputs are high, the ANDing gives a high signal; so the final output is low.

Table 3-6 is the truth table for a 3-input NAND gate. As indicated, the circuit recognizes words with one or more 0s. This means that one or more low inputs produce a high output.

TABLE 3-5. TWO-INPUT NAND GATE

A	B	$\overline{AB}$
0	0	1
0	1	1
1	0	1
1	1	0

TABLE 3-6. THREE-INPUT NAND GATE

A	B	C	$\overline{ABC}$
0	0	0	1
0	0	1	1
0	1	0	1
0	1	1	1
1	0	0	1
1	0	1	1
1	1	0	1
1	1	1	0

Four-Input Gate

Figure 3-9*b* is the symbol for a 4-input NAND gate. The inputs are ANDed, and the result is inverted. Therefore, the boolean equation is

$$Y = \overline{ABCD} \tag{3-9}$$

If you construct the truth table, you will have input words from 0000 to 1111. All words from 0000 through 1110 produce a 1 output; only the word 1111 gives a 0 output.

3-4 DE MORGAN'S SECOND THEOREM

The proof of De Morgan's second theorem is similar to the proof given for the first theorem. What follows is a brief explanation.

The Second Theorem

When two inputs are used, De Morgan's second theorem says that

$$\overline{AB} = \overline{A} + \overline{B} \tag{3-10}$$

In words, the complement of a product equals the sum of the complements. The left member of this equation represents a NAND gate (Fig. 3-10*a*); the right member stands

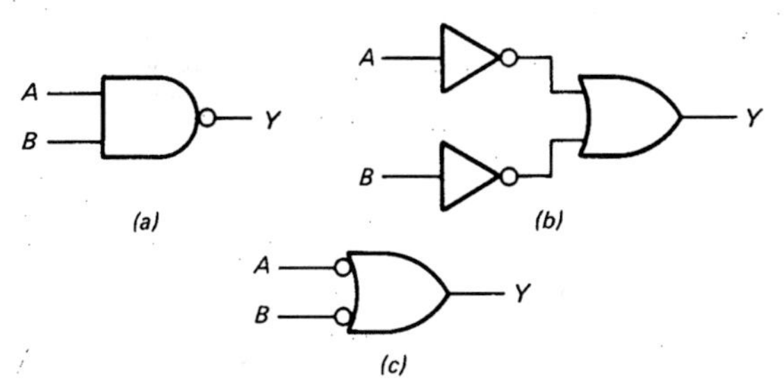

Fig. 3-10 De Morgan's second theorem: (*a*) NAND gate; (*b*) OR gate with inverted inputs; (*c*) bubbled OR gate.

for an OR gate with inverted inputs (Fig. 3-10*b*). Therefore, De Morgan's second theorem boils down to the fact that Fig. 3-10*a* and *b* are equivalent.

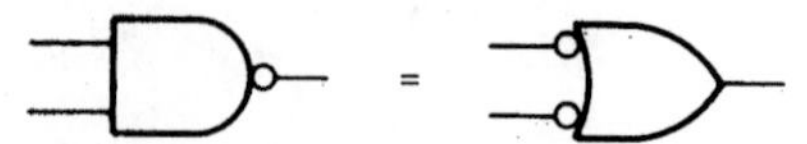

Fig. 3-11 De Morgan's second theorem.

Bubbled OR Gate

The circuit of Fig. 3-10*b* is so widely used that the abbreviated logic symbol of Fig. 3-10*c* has been adopted. From now on we will refer to Fig. 3-10*c* as a *bubbled* OR gate; the bubbles are a reminder of the inversion that takes place before ORing.

Figure 3-11 is a visual statement of De Morgan's second theorem: a NAND gate and a bubbled OR gate are equivalent. This equivalence allows you to replace one circuit by the other whenever desired. This may lead to a simpler logic circuit or give you more insight into how the original circuit works.

More than Two Inputs

When 3 inputs are involved, De Morgan's second theorem is written

$$\overline{ABC} = \overline{A} + \overline{B} + \overline{C} \tag{3-11}$$

If 4 inputs are used,

$$\overline{ABCD} = \overline{A} + \overline{B} + \overline{C} + \overline{D} \tag{3-12}$$

These equations say that the complement of a product equals the sum of the complements.

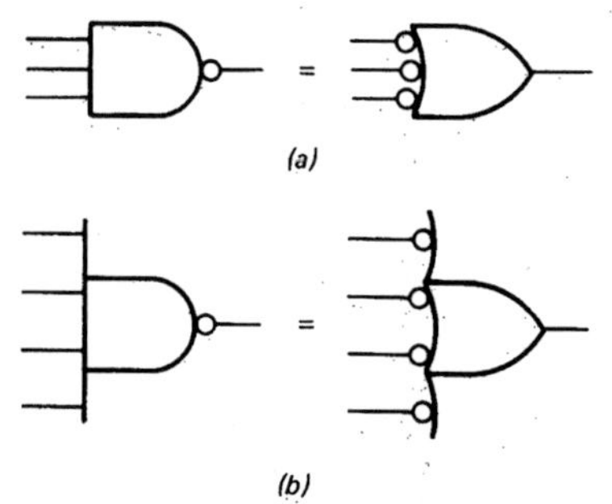

Fig. 3-12 De Morgan's second theorem: (*a*) 3-input circuits; (*b*) 4-input circuits.

Figure 3-12 is a visual summary of the second theorem. Whether 3 or 4 inputs are involved, a NAND gate and a bubbled OR gate are equivalent (interchangeable).

EXAMPLE 3-2

Prove that Fig. 3-13*a* and *c* are equivalent.

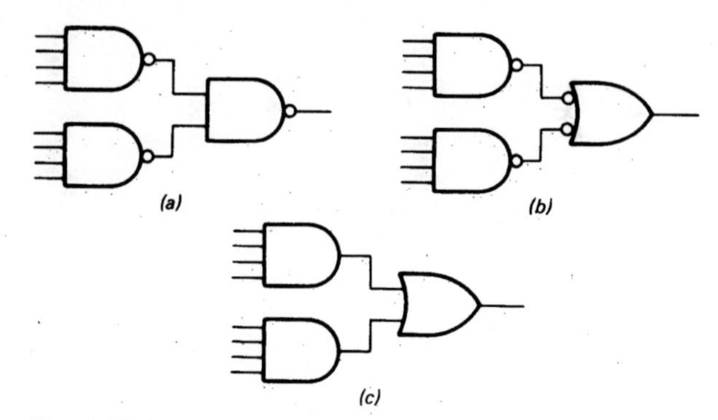

Fig. 3-13 Equivalent circuits.

SOLUTION

Replace the final NAND gate in Fig. 3-13*a* by a bubbled OR gate. This gives Fig. 3-13*b*. The double inversions cancel out, leaving the simplified circuit of Fig. 3-13*c*. Figure 3-13*a* and *c* are therefore equivalent. Driven by the same inputs, either circuit produces the same output as the other. So if you're loaded with NAND gates, build Fig. 3-13*a*. If your shelves are full of AND and OR gates, build Fig. 3-13*c*.

Incidentally, most people find Fig. 3-13*b* easier to analyze than Fig. 3-13*a*. For this reason, if you build Fig. 3-13*a*, draw the circuit like Fig. 3-13*b*. Anyone who sees Fig. 3-13*b* on a schematic diagram knows that the bubbled OR gate is the same as a NAND gate and that the built-up circuit is two NAND gates working into a NAND gate.

EXAMPLE 3-3

Figure 3-14 shows a circuit called a *control matrix*. At first, it looks complicated, but on closer inspection it is relatively simple because of the repetition of NAND gates. De Morgan's theorem tells us that NAND gates driving NAND gates are equivalent to AND gates driving OR gates.

The upper set of inputs T_1 to T_6 are called *timing signals;* only one of them is high at a time. T_1 goes high first, then T_2, then T_3, and so on. These signals control the rate and sequence of computer operations.

The lower set of inputs *LDA, ADD, SUB,* and *OUT* are computer instructions; only one of them is high at a time. The outputs C_P, E_P, L_M, . . . , to L_O control different registers in the computer.

Answer the following questions about the control matrix:

a. Which outputs are high when T_1 is high?
b. If T_4 and *LDA* are high, which outputs are high?
c. When T_6 and *SUB* are high, which outputs are high?

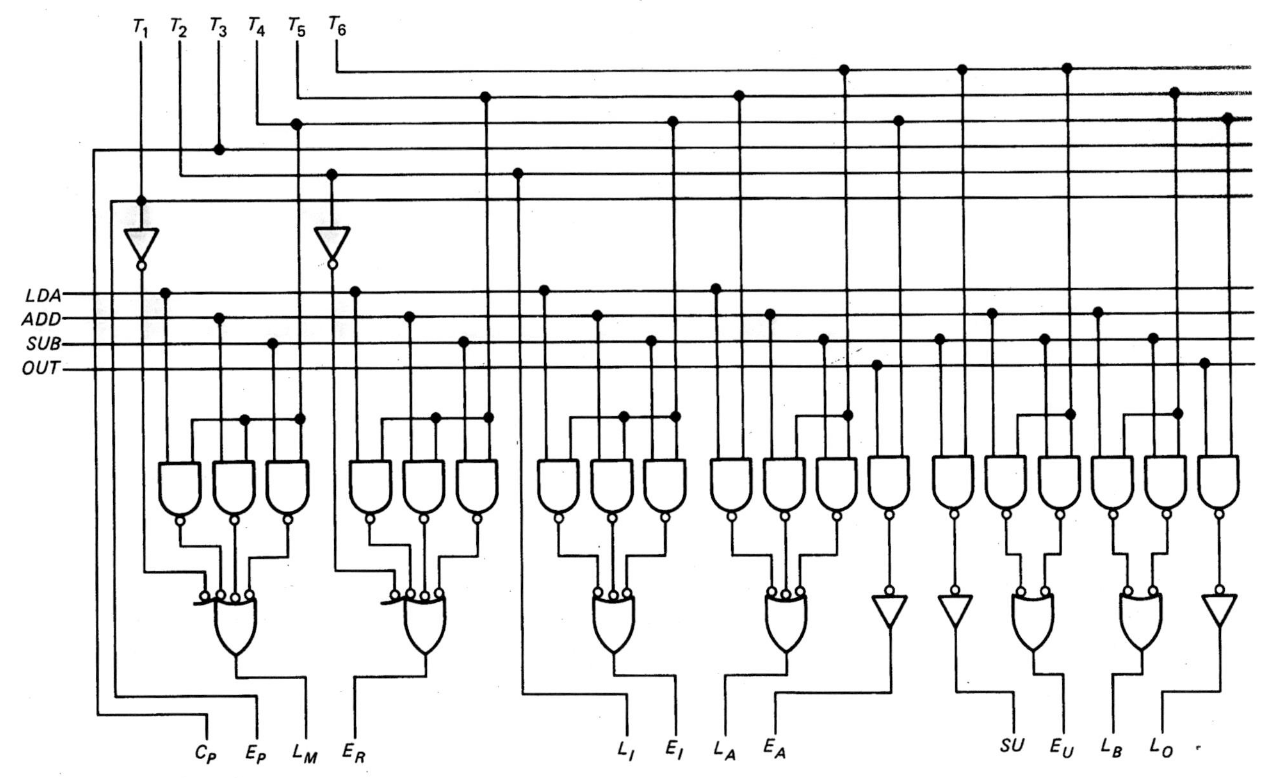

Fig. 3-14 Control matrix.

SOLUTION

a. Visualize T_1 high. You can quickly check out each gate and realize that E_P and L_M are the only high outputs.
b. This time T_4 and LDA are high. Check each gate and you can see that L_M and E_I are the only high outputs.
c. When T_6 and SUB are high, the high outputs are L_A, S_U, and E_U.

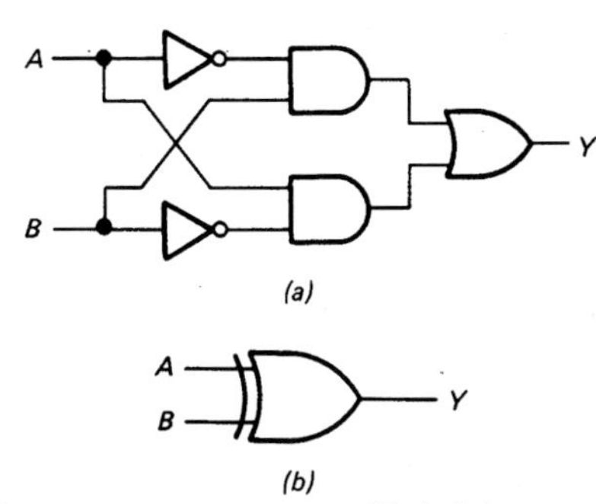

Fig. 3-15 (*a*) EXCLUSIVE-OR gate. (*b*) A 2-input EXCLUSIVE-OR gate.

3-5 EXCLUSIVE-OR GATES

An OR gate recognizes words with one or more 1s. The EXCLUSIVE-OR gate is different; it recognizes only words that have an odd number of 1s.

Two Inputs

Figure 3-15*a* shows one way to build an EXCLUSIVE-OR gate, abbreviated XOR. The upper AND gate forms the product $\overline{A}B$, and the lower AND gate gives $A\overline{B}$. Therefore, the boolean equation is

$$Y = \overline{A}B + A\overline{B} \qquad (3\text{-}13)$$

Here's what the circuit does. In Fig. 3-15*a* two low inputs mean both AND gates have low outputs; so the final output is low. If A is low and B is high, the upper AND gate has a high output; therefore, the final output is high. Likewise, a high A and low B result in a final output that is high. If both inputs are high, both AND gates have low outputs and the final output is low.

Table 3-7 shows the truth table for a 2-input EXCLUSIVE-OR gate. The output is high when A or B is high but not both; this is why the circuit is known as an EXCLUSIVE-OR gate. In other words, the output is a 1 only when the inputs are different.

TABLE 3-7. TWO-INPUT XOR GATE

A	B	$\bar{A}B + A\bar{B}$
0	0	0
0	1	1
1	0	1
1	1	0

Logic Symbol and Boolean Sign

Figure 3-15*b* is the standard symbol for a 2-input XOR gate. Whenever you see this symbol, remember the action: the inputs must be different to get a high output.

A word equation for Fig. 3-15*b* is

$$Y = A \text{ XOR } B \qquad (3\text{-}14)$$

In boolean algebra the sign $\oplus$ stands for XOR addition. This means that Eq. 3-14 can be written

$$Y = A \oplus B \qquad (3\text{-}15)$$

Read this as "Y equals A XOR B."

Given the inputs, you can substitute and solve for the output. For instance, if both inputs are low,

$$Y = 0 \oplus 0 = 0$$

because 0 XORed with 0 gives 0. If one input is low and the other high,

$$Y = 0 \oplus 1 = 1$$

because 0 XORed with 1 produces 1. And so on.

Here's a summary of the four possible XOR additions:

$$0 \oplus 0 = 0$$
$$0 \oplus 1 = 1$$
$$1 \oplus 0 = 1$$
$$1 \oplus 1 = 0$$

Remember these four results; we will be using XOR addition when we get to arithmetic circuits.

Four Inputs

In Fig. 3-16*a* the upper gate produces $A \oplus B$, while the lower gate gives $C \oplus D$. The final gate XORs both of these sums to get

$$Y = (A \oplus B) \oplus (C \oplus D) \qquad (3\text{-}16)$$

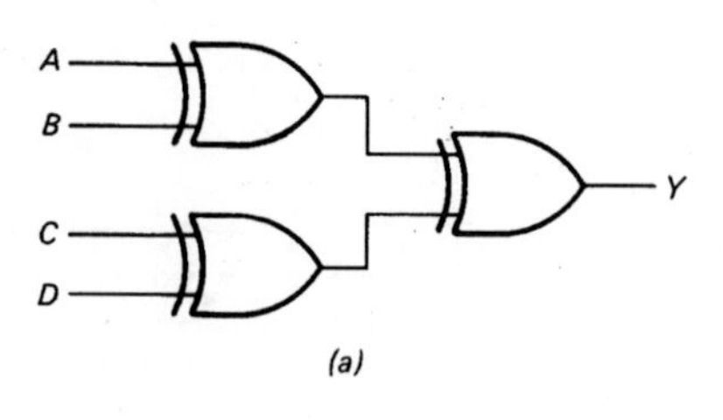

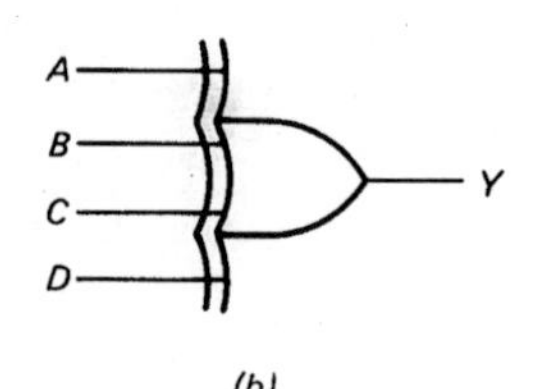

Fig. 3-16 A 4-input EXCLUSIVE-OR gate: (*a*) circuit with 2-input XOR gates; (*b*) logic symbol.

It's possible to substitute input values into the equation and solve for the output. For instance, if A through C are low and D is high,

$$\begin{aligned} Y &= (0 \oplus 0) \oplus (0 \oplus 1) \\ &= 0 \oplus 1 \\ &= 1 \end{aligned}$$

One way to get the truth table is to plow through all the input possibilities.

Alternatively, you can analyze Fig. 3-16*a* as follows. If all inputs are 0s, the first two gates have 0 outputs; so the final gate has a 0 output. If A to C are 0s and D is a 1, the upper gate has a 0 output, the lower gate has a 1 output, and the final gate has a 1 output. In this way, you can analyze the circuit action for all input words.

Table 3-8 summarizes the action. Here is an important property: each input word with an odd number of 1s produces a 1 output. For instance, the first input word to produce a 1 output is 0001; this word has an odd number of 1s. The next word with a 1 output is 0010; again an odd number of 1s. A 1 output also occurs for these words: 0100, 0111, 1000, 1011, 1101, and 1110, all of which have an odd number of 1s.

The circuit of Fig. 3-16*a* recognizes words with an odd number of 1s; it disregards words with an even number of 1s. Figure 3-16*a* is a 4-input XOR gate. In this book, we will use the abbreviated symbol of Fig. 3-16*b* to represent a 4-input XOR gate. When you see this symbol, remember the action: the circuit recognizes words with an odd number of 1s.

Any Number of Inputs

Using 2-input XOR gates as building blocks, we can make XOR gates with any number of inputs. For example, Fig.

TABLE 3-8. FOUR-INPUT XOR GATE

Comment	*A*	*B*	*C*	*D*	*Y*
Even	0	0	0	0	0
Odd	0	0	0	1	1
Odd	0	0	1	0	1
Even	0	0	1	1	0
Odd	0	1	0	0	1
Even	0	1	0	1	0
Even	0	1	1	0	0
Odd	0	1	1	1	1
Odd	1	0	0	0	1
Even	1	0	0	1	0
Even	1	0	1	0	0
Odd	1	0	1	1	1
Even	1	1	0	0	0
Odd	1	1	0	1	1
Odd	1	1	1	0	1
Even	1	1	1	1	0

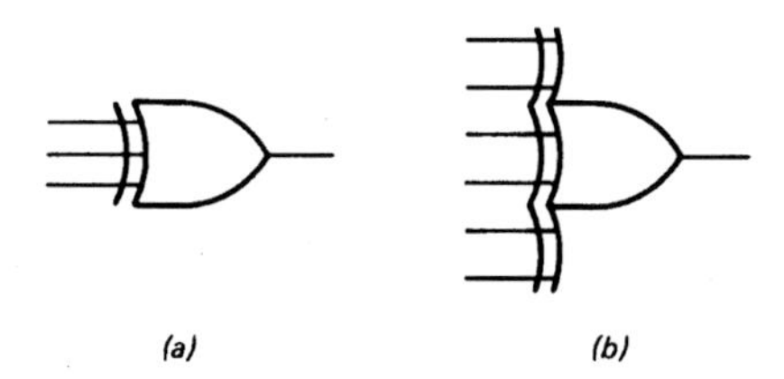

Fig. 3-17 XOR gates: (*a*) 3-input; (*b*) 6-input.

3-17*a* shows the abbreviated symbol for a 3-input XOR gate, and Fig. 3-17*b* is the symbol for a 6-input XOR gate. The final output of any XOR gate is the XOR sum of the inputs:

$$Y = A \oplus B \oplus C \cdots \qquad (3\text{-}17)$$

What you have to remember for practical work is this: an XOR gate, no matter how many inputs, recognizes only words with an odd number of 1s.

Parity

Even parity means a word has an even number of 1s. For instance, 110011 has even parity because it contains four 1s. *Odd parity* means a word has an odd number of 1s. As an example, 110001 has odd parity because it contains three 1s.

Here are two more examples:

1111 0000 1111 0011 (Even parity)
1111 0000 1111 0111 (Odd parity)

The first word has even parity because it contains ten 1s; the second word has odd parity because it contains eleven 1s.

XOR gates are ideal for testing the parity of a word. XOR gates recognize words with an odd number of 1s. Therefore, even-parity words produce a low output and odd-parity words produce a high output.

EXAMPLE 3-4

What is the output of Fig. 3-18 for each of these input words?

a. 1010 1100 1000 1100
b. 1010 1100 1000 1101

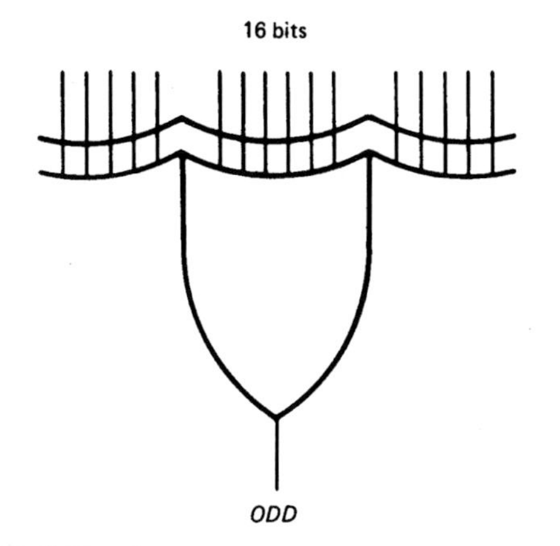

Fig. 3-18 Odd-parity tester.

SOLUTION

a. The word has seven 1s, an odd number. Therefore, the output signal is

$$ODD = 1$$

b. The word has eight 1s, an even number. Now

$$ODD = 0$$

This is an example of an odd-parity tester. An even-parity word produces a low output. An odd-parity word results in a high output.

EXAMPLE 3-5

The 7-bit register of Fig. 3-19 stores the letter A in ASCII form. What does the 8-bit output word equal?

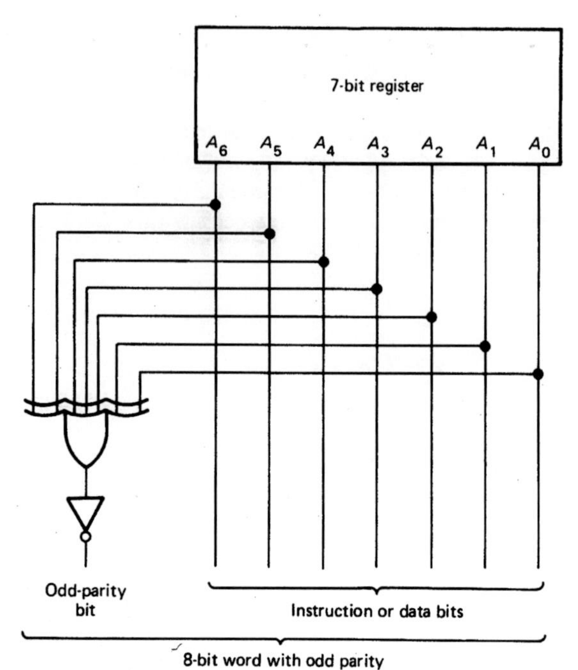

Fig. 3-19 Odd-parity generator.

SOLUTION

The ASCII code for letter A is

100 0001

(see Table 1-6 for the ASCII code). This word has an even parity, which means that the XOR gate has a 0 output. Because of the inverter, the overall output of the circuit is the 8-bit word

1100 0001

Notice that this has odd parity.

The circuit is called an *odd-parity generator* because it produces an 8-bit output word with odd parity. If the register word has even parity, 0 comes out of the XOR gate and the odd-parity bit is 1. On the other hand, if the register word has odd parity, a 1 comes out of the XOR gate and the odd-parity bit is 0. No matter what the register contents, the odd-parity bit and the register bits form a new 8-bit word that has odd parity.

What is the practical application? Because of transients, noise, and other disturbances, 1-bit errors sometimes occur in transmitted data. For instance, the letter A may be transmitted over phone lines in ASCII form:

100 0001 (A)

Somewhere along the line, one of the bits may be changed. If the X_1 bit changes, the received data will be

100 0011 (C)

Because of the 1-bit error, we receive letter C when letter A was actually sent.

One solution is to transmit an odd-parity bit along with the data word and have an XOR gate test each received word for odd parity. For instance, with a circuit like Fig. 3-19 the letter A would be transmitted as

1100 0001

An XOR gate will test this word when it is received. If no error has occurred, the XOR gate will recognize the word. On the other hand, if a 1-bit error has crept in, the XOR gate will disregard the received word and the data can be rejected.

A final point. When errors come, they are usually 1-bit errors. This is why the method described catches most of the errors in transmitted data.

EXAMPLE 3-6

What does the circuit of Fig. 3-20 do?

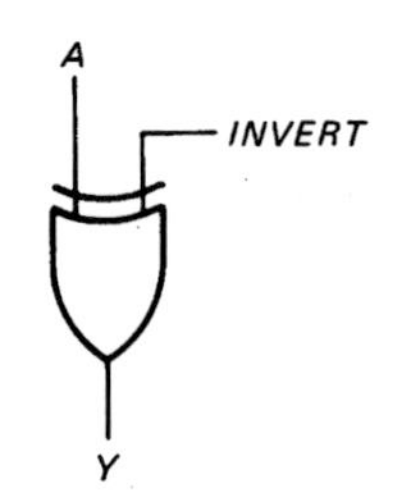

Fig. 3-20

SOLUTION

When $INVERT = 0$ and $A = 0$,

$$Y = 0 \oplus 0 = 0$$

When $INVERT = 0$ and $A = 1$,

$$Y = 0 \oplus 1 = 1$$

In either case, the output is the same as A; that is,

$$Y = A$$

for a low $INVERT$ signal.

On the other hand, when $INVERT = 1$ and $A = 0$,

$$Y = 1 \oplus 0 = 1$$

When $INVERT = 1$ and $A = 1$,

$$Y = 1 \oplus 1 = 0$$

This time, the output is the complement of A. As an equation,

$$Y = \overline{A}$$

for a high *INVERT* signal.

To summarize, the circuit of Fig. 3-20 does either of two things. It transmits A when *INVERT* is 0 and $\overline{A}$ when INVERT is 1.

3-6 THE CONTROLLED INVERTER

The preceding example suggests the idea of a *controlled inverter,* a circuit that transmits a binary word or its *1's complement.*

The 1's Complement

Complement each bit in a word and the new word you get is the 1's complement. For instance, given

1100 0111

the 1's complement is

0011 1000

Each bit in the original word is inverted to get the 1's complement.

The Circuit

The XOR gates of Fig. 3-21 form a *controlled inverter* (sometimes called a programmed inverter). This circuit can transmit the register contents or the 1's complement of the register contents. As demonstrated in Example 3-6, each XOR gate acts like this. A low *INVERT* results in

$$Y_n = A_n$$

and a high *INVERT* gives

$$Y_n = \overline{A}_n$$

So each bit is either transmitted or inverted before reaching the final output.

Visualize the register contents as a word $A_7A_6 \cdots A_0$ and the final output as a word $Y_7Y_6 \cdots Y_0$. Then a low *INVERT* means

$$Y_7Y_6 \cdots Y_0 = A_7A_6 \cdots A_0$$

On the other hand, a high *INVERT* results in

$$Y_7Y_6 \cdots Y_0 = \overline{A}_7\overline{A}_6 \cdots \overline{A}_0$$

As a concrete example, suppose the register word is

$$A_7A_6 \cdots A_0 = 1110\ 0110$$

Then, a low *INVERT* gives an output word of

$$Y_7Y_6 \cdots Y_0 = 1110\ 0110$$

and a high *INVERT* produces

$$Y_7Y_6 \cdots Y_0 = 0001\ 1001$$

The controlled inverter of Fig. 3-21 is important. Later you will see how it is used in solving arithmetic and logic problems. For now, all you need to remember is the key idea. The output word from a controlled inverter equals the

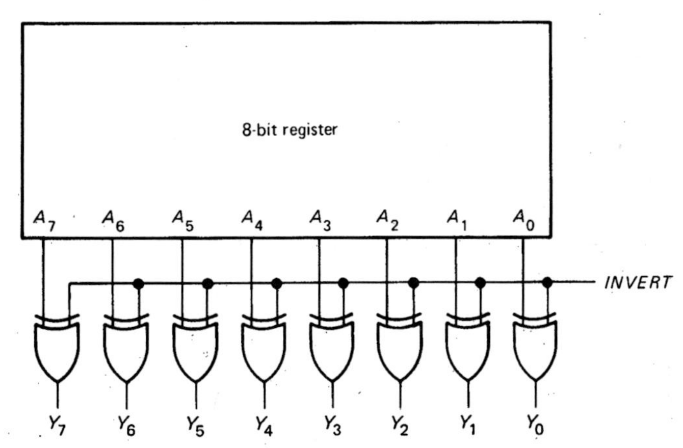

Fig. 3-21 Controlled inverter.

input word when *INVERT* is low; the output word equals the 1's complement when *INVERT* is high.

Boldface Notation

After you understand an idea, it simplifies discussions and equations if you use a symbol, letter, or other sign to represent the idea. From now on, boldface letters will stand for binary words.

For instance, instead of writing

$$A_7A_6 \cdots A_0 = 1110\ 0110$$

we can write

$$\mathbf{A} = 1110\ 0110$$

Likewise, instead of

$$Y_7Y_6 \cdots Y_0 = 0001\ 1001$$

the simpler equation

$$\mathbf{Y} = 0001\ 1001$$

can be used.

This is another example of chunking. We are replacing long strings like $A_7A_6 \cdots A_0$ and $Y_7Y_6 \cdots Y_0$ by **A** and **Y**. This chunked notation will be convenient when we get to computer analysis.

This is how to summarize the action of a controlled inverter:

$$\mathbf{Y} = \begin{cases} \mathbf{A} & \text{when } INVERT = 0 \\ \overline{\mathbf{A}} & \text{when } INVERT = 1 \end{cases}$$

(*Note:* A boldface letter with an overbar means that each bit in the word is complemented; if **A** is a word, $\overline{\mathbf{A}}$ is its 1's complement.)

3-7 EXCLUSIVE-NOR GATES

The EXCLUSIVE-NOR gate, abbreviated XNOR, is logically equivalent to an XOR gate followed by an inverter. For example, Fig. 3-22*a* shows a 2-input XNOR gate. Figure 3-22*b* is an abbreviated way to draw the same circuit.

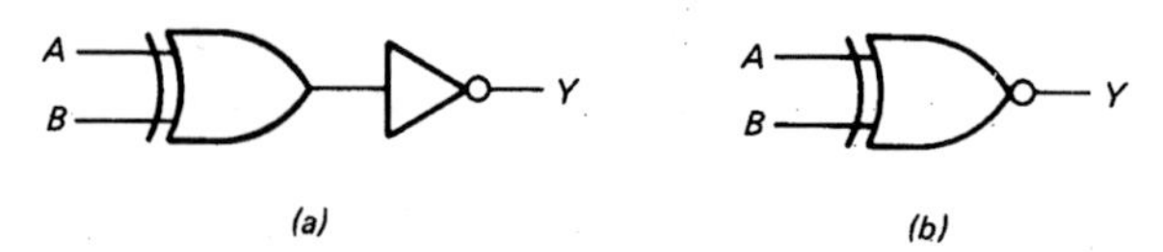

Fig. 3-22 A 2-input XNOR gate: (*a*) circuit; (*b*) abbreviated symbol.

TABLE 3-9. TWO-INPUT XNOR GATE

A	*B*	*Y*
0	0	1
0	1	0
1	0	0
1	1	1

Because of the inversion on the output side, the truth table of an XNOR gate is the complement of an XOR truth table. As shown in Table 3-9, the output is high when the inputs are the same. For this reason, the 2-input XNOR gate is ideally suited for *bit comparison*, recognizing when two input bits are identical. (Example 3-7 tells you more about bit comparison.)

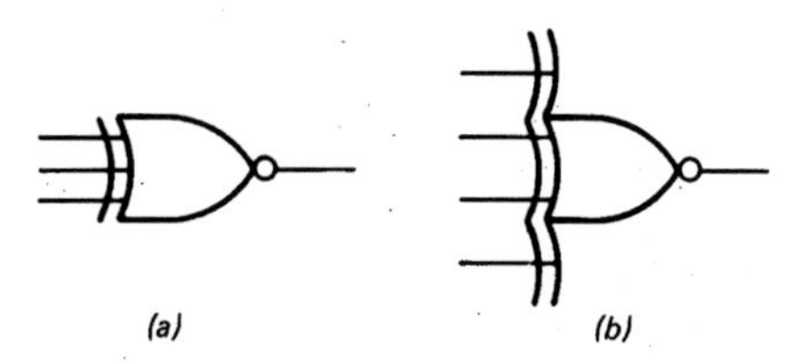

Fig. 3-23 XNOR gates: (*a*) 3-input; (*b*) 4-input.

Figure 3-23*a* is the symbol for a 3-input XNOR gate, and Fig. 3-23*b* is the 4-input XNOR gate. Because of the inversion on the output side, these XNOR gates perform the complementary function of XOR gates. Instead of recognizing odd-parity words, XNOR gates recognize even-parity words.

EXAMPLE 3-7

What does the circuit of Fig. 3-24 do?

SOLUTION

The circuit is a *word comparator;* it recognizes two identical words. Here is how it works. The leftmost XNOR gate compares A_5 and B_5; if they are the same, Y_5 is a 1. The second XNOR gate compares A_4 and B_4; if they are the same, Y_4 is a 1. In turn, the remaining XNOR gates compare the bits that are left, producing a 1 output for equal bits and a 0 output for unequal bits.

If the words **A** and **B** are identical, all XNOR gates have high outputs and the AND gate has a high *EQUAL*. If words **A** and **B** differ in one or more bit positions, the AND gate has a low *EQUAL*.

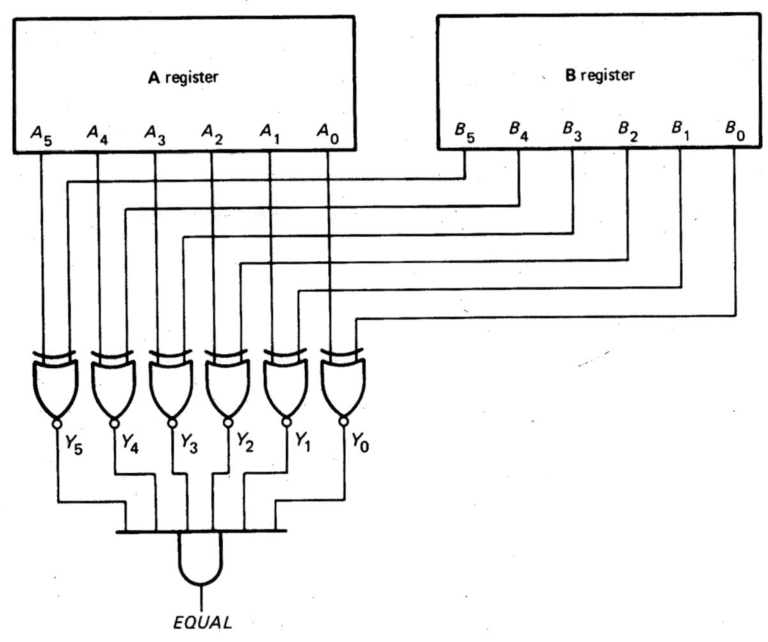

Fig. 3-24 Word comparator.

GLOSSARY

controlled inverter This circuit produces the 1's complement of the input word. One application is binary subtraction. It is sometimes called a programmed inverter.

De Morgan's theorems The first theorem says that a NOR gate is equivalent to a bubbled AND gate. The second theorem says that a NAND gate is equivalent to a bubbled OR gate.

even parity An even number of 1s in a binary word.

NAND *gate* Equivalent to an AND gate followed by an inverter. All inputs must be high to get a low output.

NOR *gate* Equivalent to an OR gate followed by an inverter. All inputs must be low to get a high output.

odd parity An odd number of 1s in a binary word.

parity generator A circuit that produces either an odd- or even-parity bit to go along with the data.

XNOR *gate* Equivalent to an EXCLUSIVE-OR gate followed by an inverter. The output is high only when the input word has even parity.

XOR *gate* An EXCLUSIVE-OR gate. It has a high output only when the input word has odd parity. For a 2-input XOR gate, the output is high only when the inputs are different.

SELF-TESTING REVIEW

Read each of the following and provide the missing words. Answers appear at the beginning of the next question.

1. A NOR gate has two or more input signals. All inputs must be ________ to get a high output. A NOR gate recognizes only the input word whose bits are ________. The NOR gate is logically equivalent to an OR gate followed by an ________.
2. (*low, 0s, inverter*) De Morgan's first theorem says that a NOR gate is equivalent to a bubbled ________ gate.
3. (AND) A NAND gate is equivalent to an AND gate followed by an inverter. All inputs must be ________ to get a low output. De Morgan's second theorem says that a NAND gate is equivalent to a bubbled ________ gate.
4. (*high,* OR) An XOR gate recognizes only words with an ________ number of 1s. The 2-input XOR gate has a high output only when the input bits are ________. XOR gates are ideal for testing parity because even-parity words produce a ________ output and odd-parity words produce a ________ output.
5. (*odd, different, low, high*) An odd-parity generator produces an odd-parity bit to go along with the data.

The parity of the transmitted data is ________. An XOR gate can test each received word for parity, rejecting words with ________ parity.

6. (*odd, even*) A controlled inverter is a logic circuit that transmits a binary word or its ________ complement.

7. (*1's*) The EXCLUSIVE-NOR gate is equivalent to an XOR gate followed by an inverter. Because of this, even-parity words produce a high output.

PROBLEMS

3-1. In Fig. 3-25*a* the two inputs are connected together. If *A* is low, what is *Y*? If *A* is high, what is *Y*? Does the circuit act like a noninverter or an inverter?

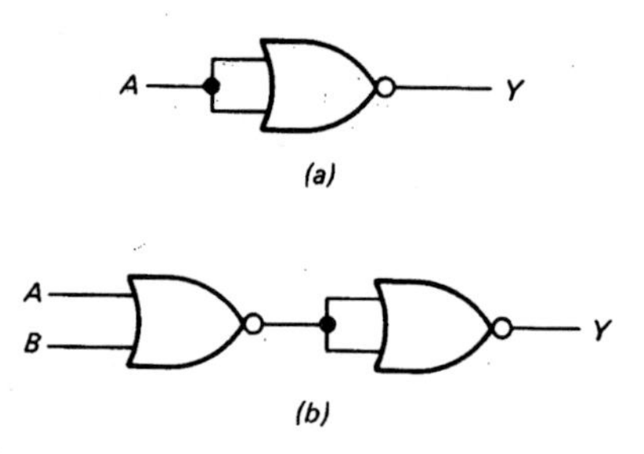

Fig. 3-25

3-2. What is the output in Fig. 3-25*b* if both inputs are low? If one is low and the other high? If both are high? Does the circuit act like an OR gate or an AND gate?

3-3. Figure 3-26 shows a NOR-gate *crossbar switch*. If all *X* and *Y* inputs are high, which of the *Z* outputs is high? If all inputs are high except X_1 and Y_2, which *Z* output is high? If X_2 and Y_0 are low and all other inputs are high, which *Z* output is high?

3-4. In Fig. 3-26, you want Z_7 to be 1 and all other *Z* outputs to be 0. What values must the *X* and *Y* inputs have?

3-5. The outputs in Fig. 3-27 are cross-coupled back to the inputs of the NOR gates. If $R = 0$ and $S = 1$, what do Q and $\overline{Q}$ equal?

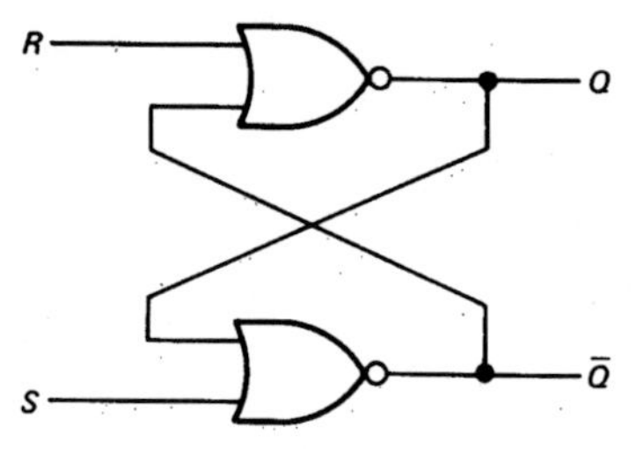

Fig. 3-27 Cross-coupled NOR gates.

3-6. If $R = 1$ and $S = 0$ in Fig. 3-27, what does Q equal? $\overline{Q}$?

3-7. Prove that Fig. 3-28*a* and *b* are equivalent.

3-8. What is the output in Fig. 3-28*a* if all inputs are 0s. If all inputs are 1s?

3-9. What is the output in Fig. 3-28*b* if all inputs are 0s. If all inputs are 1s?

3-10. A NOR has 6 inputs. How many input words are in its truth table? What is the only input word that produces a 1 output?

3-11. In Fig. 3-28*a* how many input words are there in the truth table?

3-12. What is the output in Fig. 3-29 if all inputs are low? If all inputs are high?

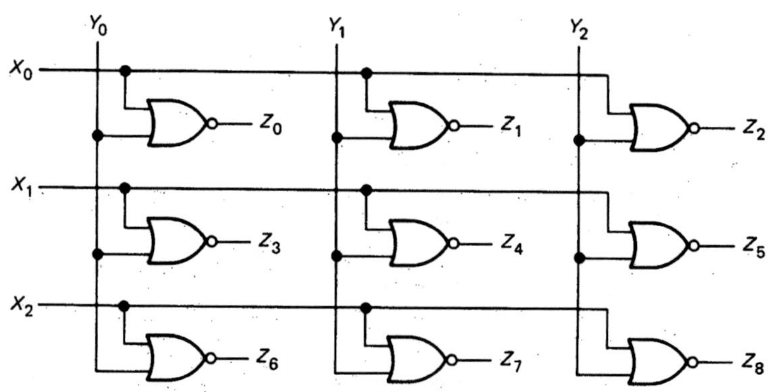

Fig. 3-26 NOR-gate crossbar switch.

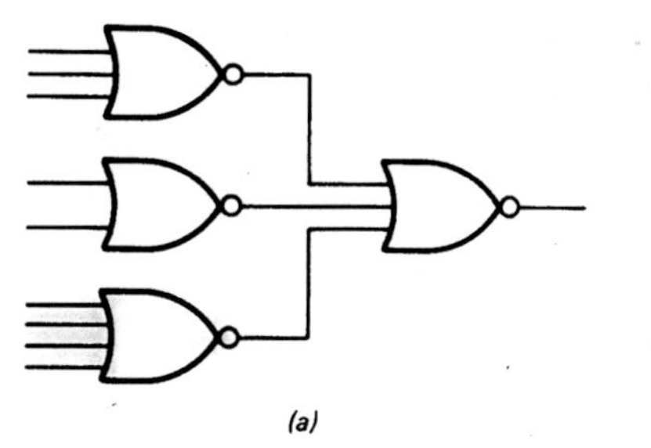
(a)

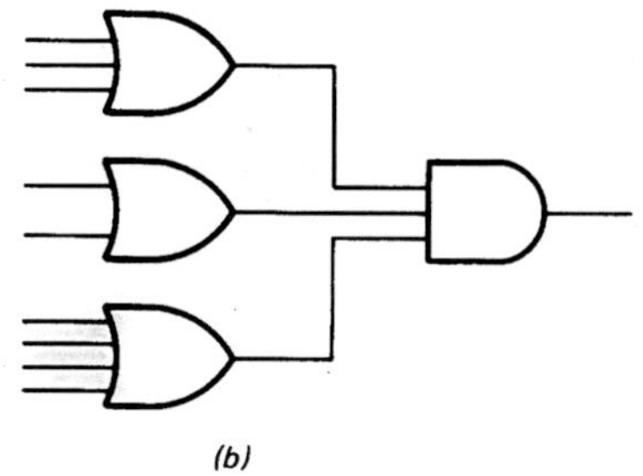
(b)

Fig. 3-28

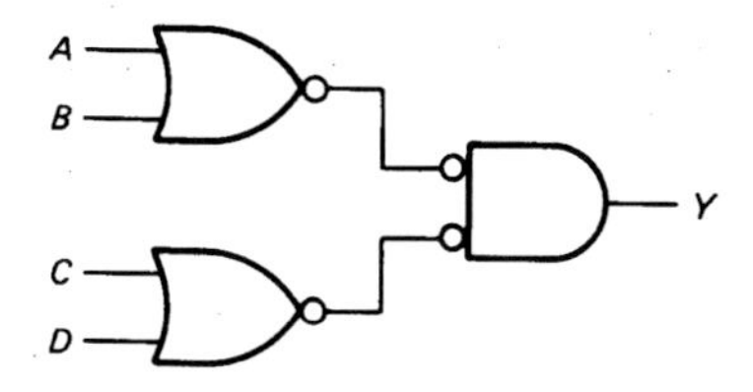

Fig. 3-29

3-13. How many words are in the truth table of Fig. 3-29. What is the value of Y for each of the following?

a. ABCD = 0011
b. ABCD = 0110
c. ABCD = 1001
d. ABCD = 1100

3-14. Which ABCD input words does the circuits of Fig. 3-29 recognize?

3-15. In Fig. 3-30a the two inputs are connected together. If $A = 0$ what does Y equal? If $A = 1$, what does Y equal? Does the circuit act like a noninverter or an inverter?

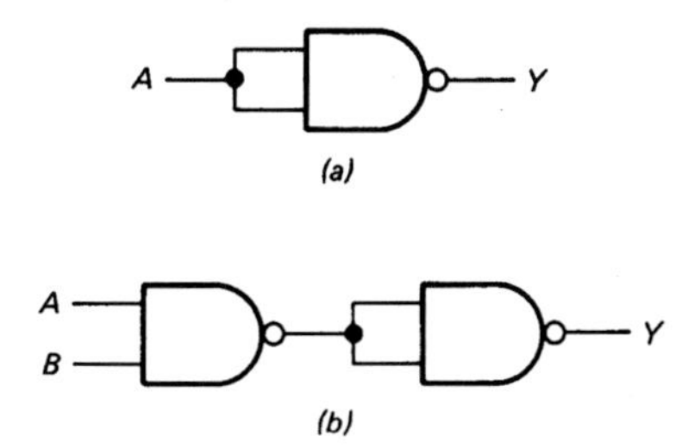

Fig. 3-30

3-16. What is the output in Fig. 3-30b if both inputs are low? If one input is low and the other high? If both are high? Does the circuit act like an OR gate or an AND gate?

3-17. Suppose the NOR gates of Fig. 3-26 are replaced by NAND gates. Then you've got a NAND-gate crossbar switch.

a. If all X and Y inputs are low, which Z output is low?
b. If all inputs are low except X_2 and Y_1, which Z output is low?
c. If all inputs are low except X_0 and Y_2, which Z output is low?
d. To get a low Z_8 output, which inputs must be high?

3-18. In Fig. 3-31, what are the outputs if $R = 0$ and $S = 1$?

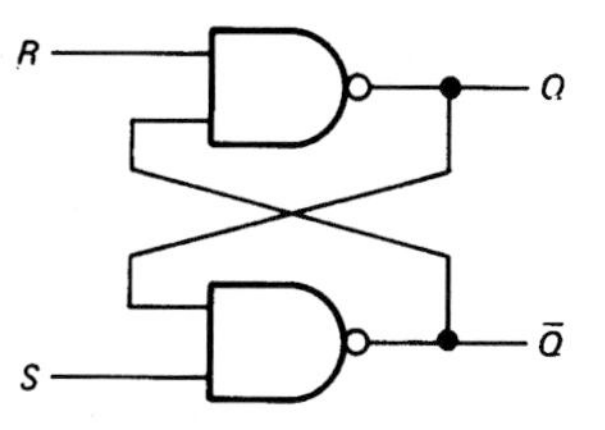

Fig. 3-31 Cross-coupled NAND gates.

3-19. If $R = 1$ and $S = 0$ in Fig. 3-31, what does Q equal? $\overline{Q}$?

3-20. What is the output in Fig. 3-32a if all inputs are 0s? If all inputs are 1s?

3-21. How many input words are there in the truth table of Fig. 3-32a?

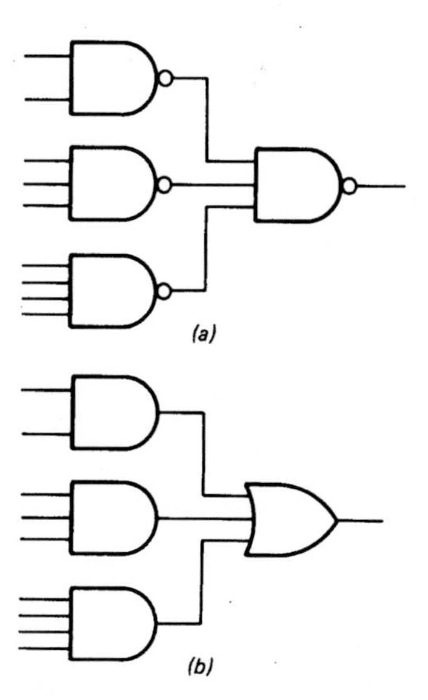

Fig. 3-32

3-22. Prove that Fig. 3-32*a* and *b* are equivalent.

3-23. What is the output in Fig. 3-33 if all inputs are low? If they are all high?

3-24. How many words are in the truth table of Fig. 3-33? What does *Y* equal for each of the following:

a. ABCDE = 00111
b. ABCDE = 10110
c. ABCDE = 11010
d. ABCDE = 10101

3-25. In Fig. 3-34 the inputs are T_4, *JMP*, *JAM*, *JAZ*, A_M, and A_Z; the output is L_P. What is the output for each of these input conditions?

a. All inputs are 0s.
b. All inputs are low except T_4 and *JMP*.
c. All inputs are low except T_4, *JAZ*, and A_Z.
d. The only high inputs are T_4, *JAM*, and A_M.

3-26. Figure 3-35 shows the control matrix discussed in Example 3-3. Only one of the timing signals T_1 to T_6 is high at a time. Also, only one of the instructions, *LDA* to *OUT*, is high at a time. Which are the high outputs for each of the following conditions?

a. T_1 high
b. T_2 high
c. T_3 high
d. T_4 and *LDA* high
e. T_5 and *LDA* high
f. T_4 and *ADD* high
g. T_5 and *ADD* high
h. T_6 and *ADD* high
i. T_4 and *SUB* high
j. T_5 and *SUB* high
k. T_6 and *SUB* high
l. T_4 and *OUT* high

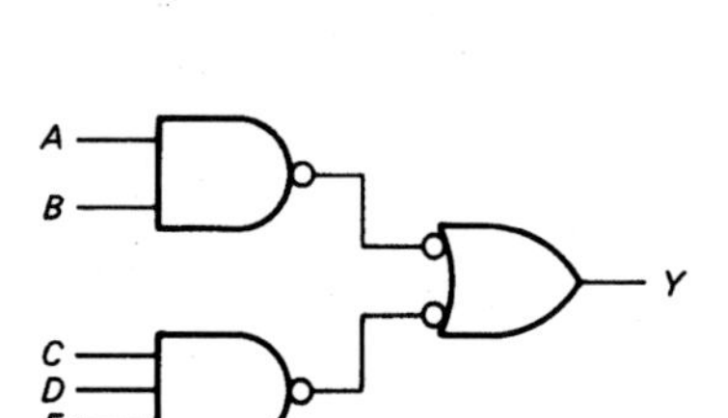

Fig. 3-33

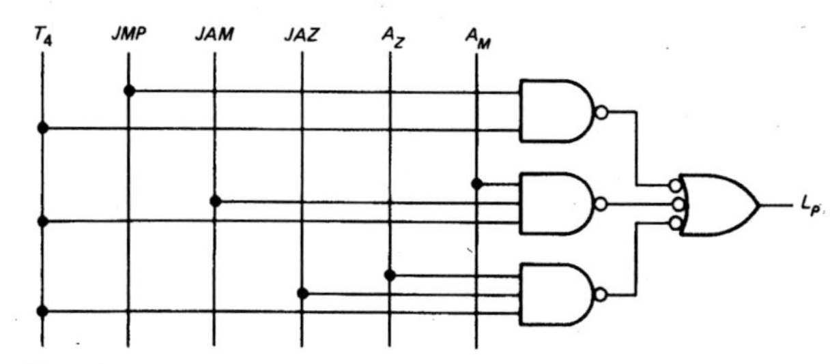

Fig. 3-34

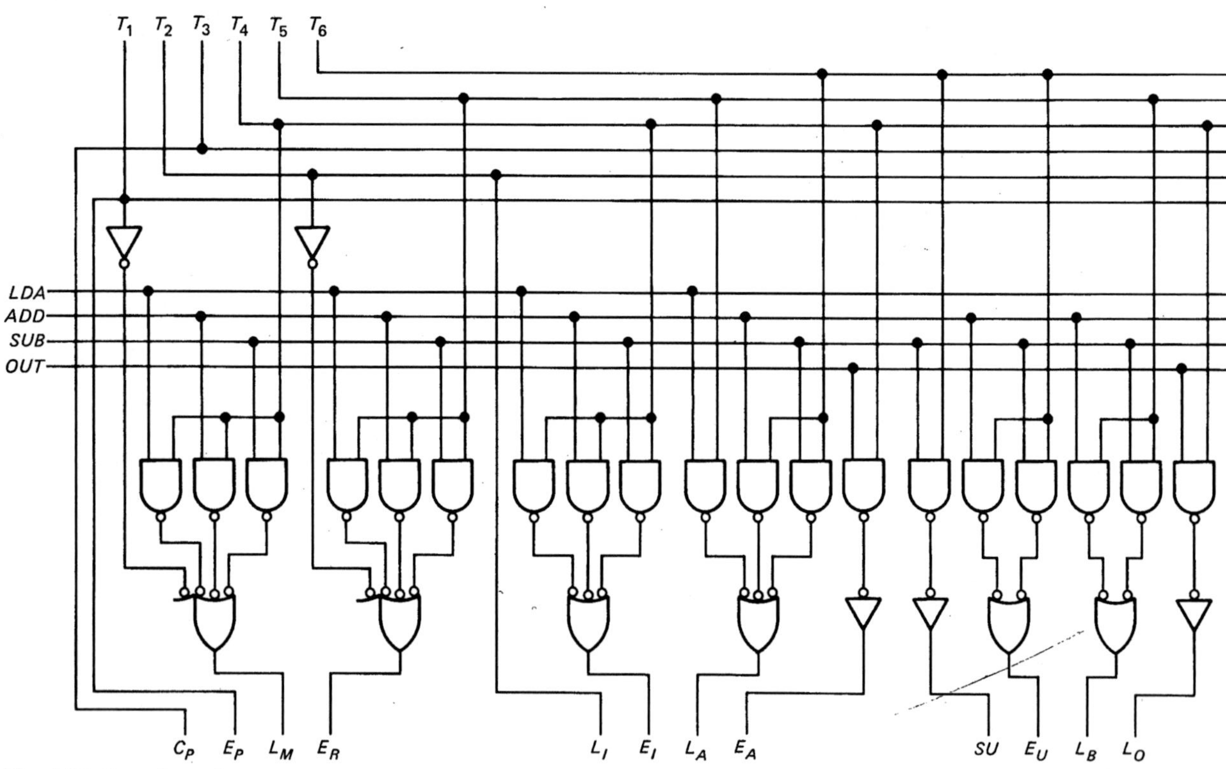

Fig. 3-35 Control matrix.

3-27. Figure 3-36 shows a binary–to–Gray-code converter. (Gray code is a special code used in analog-to-digital conversions.) The input word is $X_4X_3 \cdots X_0$, and the output word is $Y_4Y_3 \cdots Y_0$. What does the output word equal for each of these inputs?

a. $X_4X_3 \cdots X_0 = 10011$
b. $X_4X_3 \cdots X_0 = 01110$
c. $X_4X_3 \cdots X_0 = 10101$
d. $X_4X_3 \cdots X_0 = 11100$

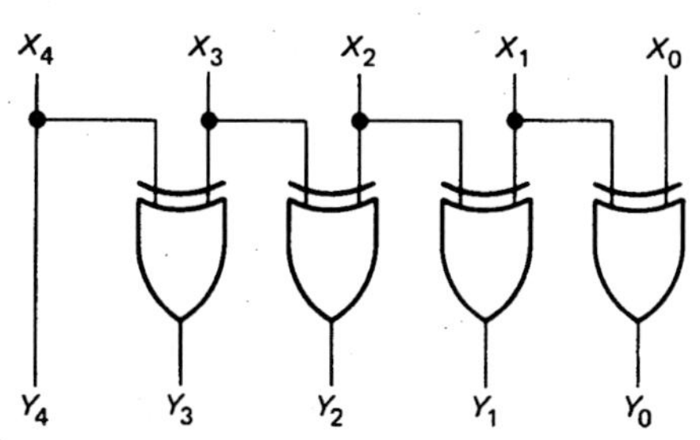

Fig. 3-36 Binary–to–Gray-code converter.

3-28. How many input words are there in the truth table of an 8-input XOR gate?

3-29. How can you modify Fig. 3-19 so that it produces an 8-bit output word with even parity?

3-30. In the controlled inverter of Fig. 3-21, what is the output word *Y* for each of these conditions?

a. **A** = 1100 1111 and *INVERT* = 0
b. **A** = 0101 0001 and *INVERT* = 1
c. **A** = 1110 1000 and *INVERT* = 1
d. **A** = 1010 0101 and *INVERT* = 0

3-31. The inputs *A* and *B* of Fig. 3-37 produce outputs of *CARRY* and *SUM*. What are the values of *CARRY* and *SUM* for each of these inputs?

a. $A = 0$ and $B = 0$
b. $A = 0$ and $B = 1$
c. $A = 1$ and $B = 0$
d. $A = 1$ and $B = 1$

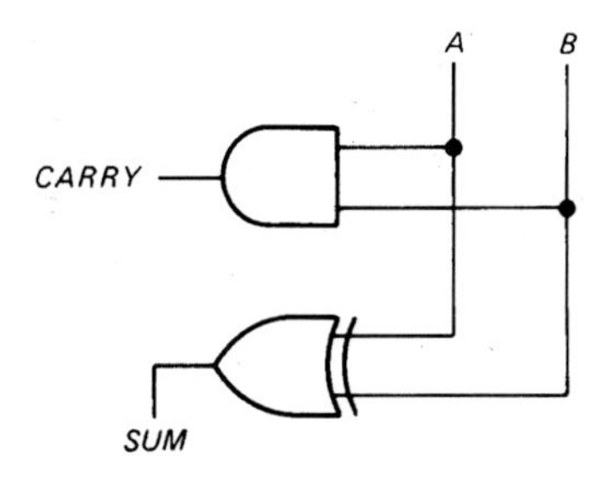

Fig. 3-37

3-32. In Fig. 3-37, what is the boolean equation for *CARRY*? For *SUM*?

3-33. What is the 1's complement for each of these numbers?

a. 1100 0011
b. 1010 1111 0011
c. 1110 0001 1010 0011
d. 0000 1111 0010 1101

3-34. What is the output of a 16-input XNOR gate for each of these input words?

a. 0000 0000 0000 1111
b. 1111 0101 1110 1100
c. 0101 1100 0001 0011
d. 1111 0000 1010 0110

3-35. The boolean equation for a certain logic circuit is $Y = AB + CD + AC$. What does Y equal for each of the following:

a. ABCD = 0000
b. ABCD = 0101
c. ABCD = 1010
d. ABCD = 1001

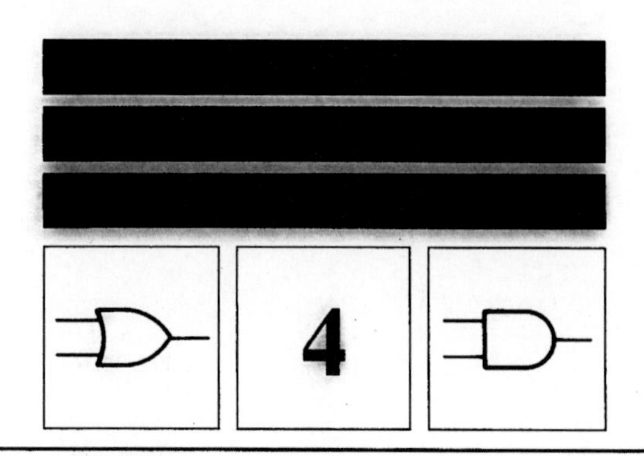

TTL CIRCUITS

In 1964 Texas Instruments introduced *transistor-transistor logic* (TTL), a widely used family of digital devices. TTL is fast, inexpensive, and easy to use. This chapter concentrates on TTL because once you are familiar with it, you can branch out to other logic families and technologies.

4-1 DIGITAL INTEGRATED CIRCUITS

Using advanced photographic techniques, a manufacturer can produce miniature circuits on the surface of a *chip* (a small piece of semiconductor material). The finished network is so small you need a microscope to see the connections. Such a circuit is called an *integrated circuit* (IC) because the components (transistors, diodes, resistors) are an integral part of the chip. This is different from a discrete circuit, in which the components are individually connected during assembly.

Levels of Integration

Small-scale integration (SSI) refers to ICs with fewer than 12 gates on the same chip. *Medium-scale integration* (MSI) means from 12 to 100 gates per chip. And *large-scale integration* (LSI) refers to more than 100 gates per chip. The typical microcomputer has its microprocessor, memory, and I/O circuits on LSI chips; a number of SSI and MSI chips are used to support the LSI chips.

Technologies and Families

The two basic technologies for manufacturing digital ICs are *bipolar* and *MOS*. The first fabricates bipolar transistors on a chip; the second, MOSFETS. Bipolar technology is preferred for SSI and MSI because it is faster. MOS technology dominates the LSI field because more MOSFETs can be packed on the same chip area.

A digital family is a group of compatible devices with the same logic levels and supply voltages ("compatible" means that you can connect the output of one device to the input of another). Compatibility permits a large number of different combinations.

Bipolar Families

In the bipolar category are these basic families:

DTL	Diode-transistor logic
TTL	Transistor-transistor logic
ECL	Emitter-coupled logic

DTL uses diodes and transistors; this design, once popular, is now obsolete. TTL uses transistors almost exclusively; it has become the most popular family of SSI and MSI chips. ECL, the fastest logic family, is used in high-speed applications.

MOS Families

In the MOS category are these families:

PMOS	p-Channel MOSFETs
NMOS	n-Channel MOSFETs
CMOS	Complementary MOSFETs

PMOS, the oldest and slowest type, is becoming obsolete. NMOS dominates the LSI field, being used for microprocessors and memories. CMOS, a push-pull arrangement of n- and p-channel MOSFETs, is extensively used where low power consumption is needed, as in pocket calculators, digital wristwatches, etc.

4-2 7400 DEVICES

The 7400 series, a line of TTL circuits introduced by Texas Instruments in 1964, has become the most widely used of all bipolar ICs. This TTL family contains a variety of SSI and MSI chips that allow you to build all kinds of digital circuits and systems.

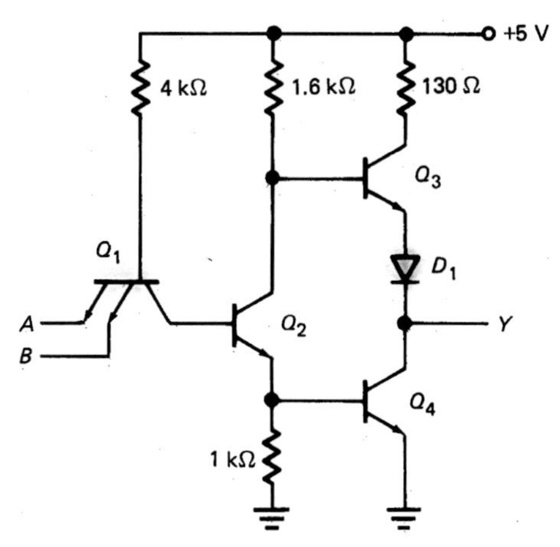

Fig. 4-1 Standard TTL NAND gate.

Standard TTL

Figure 4-1 shows a TTL NAND gate. The multiple-emitter input transistor is typical of all the gates and circuits in the 7400 series. Each emitter acts like a diode; therefore, Q_1 and the 4-kΩ resistor act like a 2-input AND gate. The rest of the circuit inverts the signal; therefore, the overall circuit acts like a 2-input NAND gate.

The output transistors (Q_3 and Q_4) form a totem-pole connection, typical of most TTL devices. Either one or the other is on. When Q_3 is on, the output is high; when Q_4 is on, the output is low. The advantage of a totem-pole connection is its low output impedance.

Ideally, the input voltages A and B are either low (grounded) or high (5 V). If A or B is low, Q_1 saturates. This reduces the base voltage of Q_2 to almost zero. Therefore, Q_2 cuts off, forcing Q_4 to cut off. Under these conditions, Q_3 acts like an emitter follower and couples a high voltage to the output.

On the other hand, when both A and B are high, the collector diode of Q_1 goes into forward conduction; this forces Q_2 and Q_4 into saturation, producing a low output. Table 4-1 summarizes all input and output conditions.

Incidentally, without diode D_1 in the circuit, Q_3 would conduct slightly when the output is low. To prevent this, the diode is inserted; its voltage drop keeps the base-emitter diode of Q_3 reverse-biased. In this way, only Q_4 conducts when the output is low.

TABLE 4-1. TWO-INPUT NAND GATE

A	B	Y
0	0	1
0	1	1
1	0	1
1	1	0

Totem-Pole Output

Why are totem-pole transistors used? Because they produce a low output impedance. Either Q_3 acts like an emitter follower (high output) or Q_4 is saturated (low output). Either way, the output impedance is very low. This is important because it reduces the switching time. In other words, when the output changes from low to high, or vice versa, the low output impedance implies a short RC time constant; this short time constant means that the output voltage can change quickly from one state to the other.

Propagation Delay Time and Power Dissipation

Two quantities needed for our later discussions are power dissipation and propagation delay time. A standard TTL gate has a power dissipation of about 10 mW. It may vary from this value because of signal levels, tolerances, etc., but on the average, it's 10 mW per gate.

The propagation delay time is the amount of time it takes for the output of a gate to change after the inputs have changed. The propagation delay time of a TTL gate is in the vicinity of 10 ns.

Device Numbers

By varying the design of Fig. 4-1 manufacturers can alter the number of inputs and the logic function. The multiple-emitter inputs and the totem-pole outputs are still used, no matter what the design. (The only exception is an open collector, discussed later.)

Table 4-2 lists some of the 7400-series TTL gates. For instance, the 7400 is a chip with four 2-input NAND gates in one package. Similarly, the 7402 has four 2-input NOR gates, the 7404 has six inverters, and so on.

TABLE 4-2. STANDARD TTL

Device number	Description
7400	Quad 2-input NAND gates
7402	Quad 2-input NOR gates
7404	Hex inverter
7408	Quad 2-input AND gates
7410	Triple 3-input NAND gates
7411	Triple 3-input AND gates
7420	Dual 4-input NAND gates
7421	Dual 4-input AND gates
7427	Triple 3-input NOR gates
7430	8-input NAND gate
7486	Quad 2-input XOR gates

5400 Series

Any device in the 7400 series works over a temperature range of 0° to 70°C and over a supply range of 4.75 to 5.25 V. This is adequate for commercial applications. The 5400 series, developed for the military applications, has the same logic functions as the 7400 series, except that it works over a temperature range of −55 to 125°C and over a supply range of 4.5 to 5.5 V. Although 5400-series devices can replace 7400-series devices, they are rarely used commercially because of their much higher cost.

High-Speed TTL

The circuit of Fig. 4-1 is called *standard TTL*. By decreasing the resistances a manufacturer can lower the internal time constants; this decreases the propagation delay time. The smaller resistances, however, increase the power dissipation. This variation is known as *high-speed TTL*. Devices of this type are numbered 74H00, 74H01, 74H02, and so on. A high-speed TTL gate has a power dissipation around 22 mW and a propagation delay time of approximately 6 ns.

Low-Power TTL

By increasing the internal resistances a manufacturer can reduce the power dissipation of TTL gates. Devices of this type are called *low-power TTL* and are numbered 74L00, 74L01, 74L02, etc. These devices are slower than standard TTL because of the larger internal time constants. A low-power TTL gate has a power dissipation of approximately 1 mW and a propagation delay time around 35 ns.

Schottky TTL

With standard TTL, high-speed TTL, and low-power TTL, the transistors go into saturation causing extra carriers to flood the base. If you try to switch this transistor from saturation to cutoff, you have to wait for the extra carriers to flow out of the base; the delay is known as the *saturation delay time*.

One way to reduce saturation delay time is with Schottky TTL. The idea is to fabricate a Schottky diode along with each bipolar transistor of a TTL circuit, as shown in Fig. 4-2. Because the Schottky diode has a forward voltage of only 0.4 V, it prevents the transistor from saturating fully.

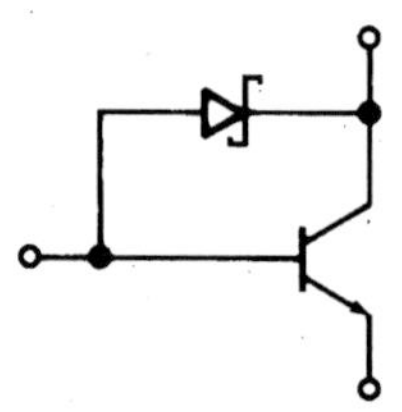

Fig. 4-2 Schottky diode prevents transistor saturation.

This virtually eliminates saturation delay time, which means better switching speed. This variation is called *Schottky TTL;* the devices are numbered 74S00, 74S01, 74S02, and so forth.

Schottky TTL devices are very fast, capable of operating reliably at 100 MHz. The 74S00 has a power dissipation around 20 mW per gate and a propagation delay time of approximately 3 ns.

Low-Power Schottky TTL

By increasing internal resistances as well as using Schottky diodes manufacturers have come up with the best compromise between low power and high speed: *low-power Schottky TTL*. Devices of this type are numbered 74LS00, 74LS01, 74LS02, etc. A low-power Schottky gate has a power dissipation of around 2 mW and a propagation delay time of approximately 10 ns, as shown in Table 4-3.

Standard TTL and low-power Schottky TTL are the mainstays of the digital designer. In other words, of the five TTL types listed in Table 4-3, standard TTL and low-power Schottky TTL have emerged as the favorites of the digital designers. You will see them used more than any other bipolar types.

4-3 TTL CHARACTERISTICS

7400-series devices are guaranteed to work reliably over a temperature range of 0 to 70°C and over a supply range of 4.75 to 5.25 V. In the discussion that follows, *worst case* means that the parameters (characteristics like maximum input current, minimum output voltage, and so on) are measured under the worst conditions of temperature and voltage—maximum temperature and minimum voltage for some parameters, minimum temperature and maximum voltage for others, or whatever combination produces the worst values.

Floating Inputs

When a TTL input is low or grounded, a current I_E (conventional direction) exists in the emitter, as shown in

TABLE 4-3. TTL POWER-DELAY VALUES

Type	Power, mW	Delay time, ns
Low-power	1	35
Low-power Schottky	2	10
Standard	10	10
High-speed	22	6
Schottky	20	3

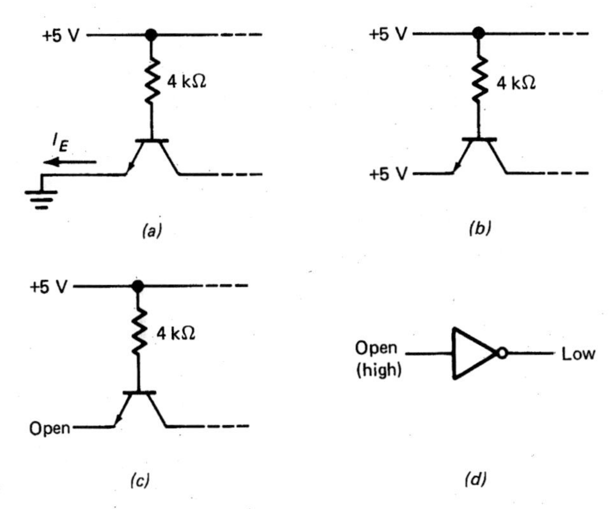

Fig. 4-3 Open or floating input is the same as a high input.

Fig. 4-3*a*. On the other hand, when a TTL input is high (Fig. 4-3*b*), the emitter diode cuts off and the emitter current is approximately zero.

When a TTL input is floating (unconnected), as shown in Fig. 4-3*c*, no emitter current is possible. Therefore, a floating TTL input is equivalent to a high input. In other words, Fig. 4-3*c* produces the same output as Fig. 4-3*b*. This is important to remember. In building circuits *any floating TTL input will act like a high input.*

Figure 4-3*d* emphasizes the point. The input is floating and is equivalent to a high input; therefore, the output of the inverter is low.

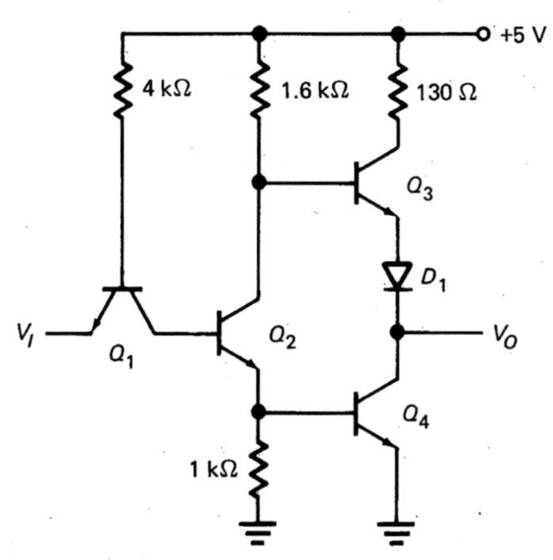

Fig. 4-4 TTL inverter.

Worst-Case Input Voltages

Figure 4-4 shows a TTL inverter with an input voltage of V_I and an output voltage of V_O. When V_I is 0 V (grounded), the output voltage is high. With TTL devices, we can raise V_I to 0.8 V and still have a high output. The maximum low-level input voltage is designated V_{IL}. Data sheets list this worst-case low input as

$$V_{IL} = 0.8 \text{ V}$$

Take the other extreme. Suppose V_I is 5 V in Fig. 4-4. This is a high input; therefore, the output of the inverter is low. V_I can decrease all the way down to 2 V, and the output will still be low. Data sheets list this worst-case high input as

$$V_{IH} = 2 \text{ V}$$

In other words, any input voltage from 2 to 5 V is a high input for TTL devices.

Worst-Case Output Voltages

Ideally, 0 V is the low output, and 5 V is the high output. We cannot attain these ideal values because of internal voltage drops. When the output is low in Fig. 4-4, Q_4 is saturated and has a small voltage drop across it. With TTL devices, any voltage from 0 to 0.4 V is a low output.

When the output is high, Q_3 acts like an emitter follower. Because of the drop across Q_3, D_1, and the 130-Ω resistor, the output is less than 5 V. With TTL devices, a high output is between 2.4 and 3.9 V, depending on the supply voltage, temperature, and load.

This means that the worst-case output values are

$$V_{OL} = 0.4 \text{ V} \qquad V_{OH} = 2.4 \text{ V}$$

Table 4-4 summarizes the worst-case values. Remember that they are valid over the temperature range (0 to 70°C) and supply range (4.75 to 5.25 V).

Compatibility

The values shown in Table 4-4 indicate that TTL devices are compatible. This means that the output of a TTL device can drive the input of another TTL device, as shown in Fig. 4-5*a*. To be specific, Fig. 4-5*b* shows a low TTL output (0 to 0.4 V). This is low enough to drive the second TTL device because any input less than 0.8 V is a low input.

TABLE 4-4. TTL STATES (WORST CASE)

	Output, V	Input, V
Low	0.4	0.8
High	2.4	2

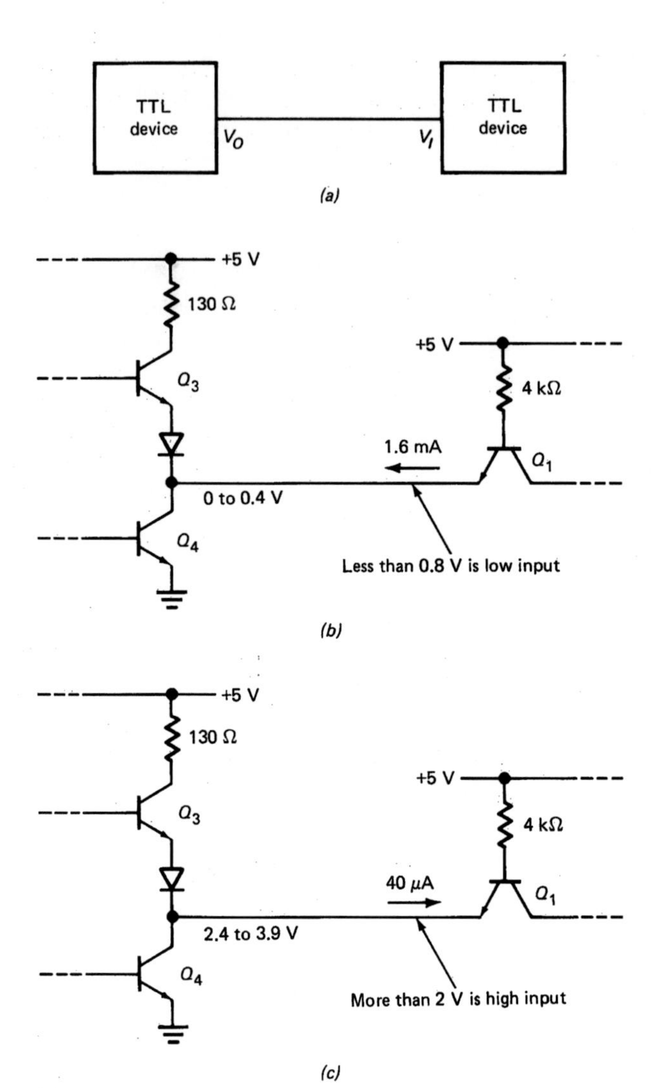

Fig. 4-5 Sourcing and sinking current.

Similarly, Fig. 4-5*c* shows a high TTL output (2.4 to 3.9 V). This is more than enough to drive the second TTL because any input greater than 2 V is a high input.

Noise Margin

In the worst case, there is a margin of 0.4 V between the driver and the load in Fig. 4-5*b* and *c*. This difference, called the *noise margin*, represents protection against noise. In other words, the connecting wire between a TTL driver and a TTL load may pick up stray noise voltages. As long as these induced voltages are less than 0.4 V, we get no false triggering of the TTL load.

Sourcing and Sinking

When a standard TTL output is low (Fig. 4-5*b*), an emitter current of approximately 1.6 mA (worst case) exists in the direction shown. The charges flow from the emitter of Q_1 to the collector of Q_4. Because it is saturated, Q_4 acts like a *current sink;* charges flow through it to ground like water flowing down a drain.

On the other hand, when a standard TTL output is high (Fig. 4-5*c*), a reverse emitter current of 40 μA (worst case) exists in the direction shown. Charges flow from Q_3 to the emitter of Q_1. In this case, Q_3 is acting like a *source*.

Data sheets lists the worst-case input currents as

$$I_{IL} = -1.6 \text{ mA} \qquad I_{IH} = 40 \ \mu\text{A}$$

The minus sign indicates that the current is out of the device; plus means the current is into the device. All data sheets use this convention.

Standard Loading

A TTL device can source current (high output) or it can sink current (low output). Data sheets of standard TTL devices indicate that any 7400-series device can sink up to 16 mA, designated as

$$I_{OL} = 16 \text{ mA}$$

and can source up to 400 μA, designated

$$I_{OH} = -400 \ \mu\text{A}$$

(Again, a minus sign means that the current is out of the device and a plus sign means that it's into the device.)

A single TTL load has a low-level input current of 1.6 mA (Fig., 4-5*b*) and a high-level input current of 40 μA (Fig. 4-5*c*). Since the maximum output currents are 10 times as large, we can connect up to 10 TTL emitters to any TTL output.

Figure 4-6*a* illustrates a low output. Here you see the TTL driver sinking 16 mA, the sum of 10 TTL load currents. In this state, the output voltage is guaranteed to be 0.4 V or less. If you try connecting more than 10 emitters, the output voltage may rise above 0.4 V.

Figure 4-6*b* shows a high output with the driver sourcing 400 μA for 10 TTL loads of 40 μA each. For this maximum loading, the output voltage is guaranteed to be 2.4 V or more under worst-case conditions.

Loading Rules

The maximum number of TTL emitters that can be reliably driven under worst-case conditions is called the *fanout*. With standard TTL, the fanout is 10, as shown in Fig. 4-6. Sometimes, we may want to use a standard TTL device to drive low-power Schottky devices. In this case, the fanout increases because low-power Schottky devices have less input current.

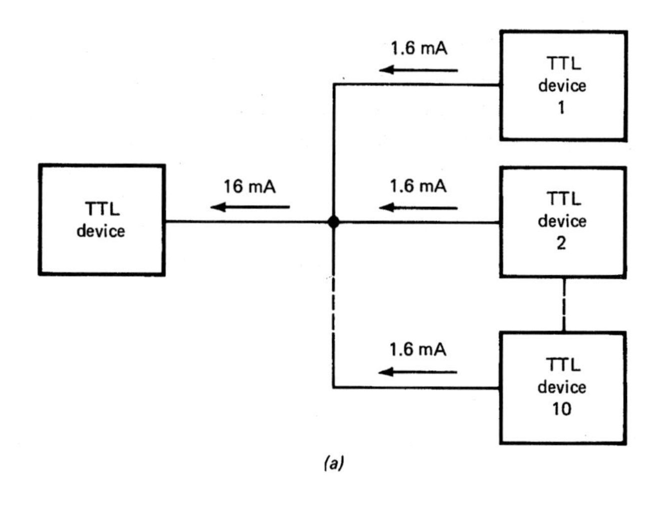

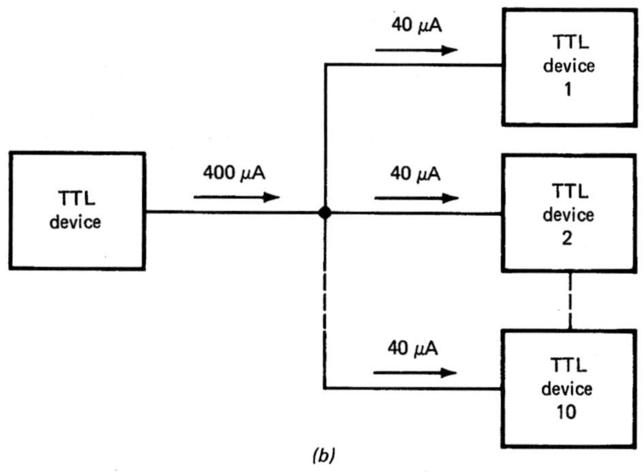

Fig. 4-6 Fanout of standard TTL devices: (*a*) low output; (*b*) high output.

By examining data sheets for the different TTL types we can calculate the fanout for all possible combinations. Table 4-5 summarizes these fanouts, which may be useful if you ever have to mix TTL types.

Read Table 4-5 as follows. The series numbers have been abbreviated; 74 stands for 7400 series, 74H for 74H00 series, and so forth. Drivers are on the left and loads on the right. Pick the driver, pick the load, and read the fanout at the intersection of the two. For instance, the fanout of a standard device (74) driving low-power Schottky devices (74LS) is 20. As another example, the fanout of a low-power device (74L) driving high-speed devices (74H) is only 1.

TABLE 4-5. FANOUTS

TTL driver	TTL load				
	74	74H	74L	74S	74LS
74	10	8	40	8	20
74H	12	10	50	10	25
74L	2	1	20	1	10
74S	12	10	100	10	50
74LS	5	4	40	4	20

4-4 TTL OVERVIEW

Let's take a look at the logic functions available in the 7400 series. This overview will give you an idea of the variety of gates and circuits found in the TTL family. As guide, Appendix 3 lists some of the 7400-series devices. You will find it useful when looking for a device number or logic function.

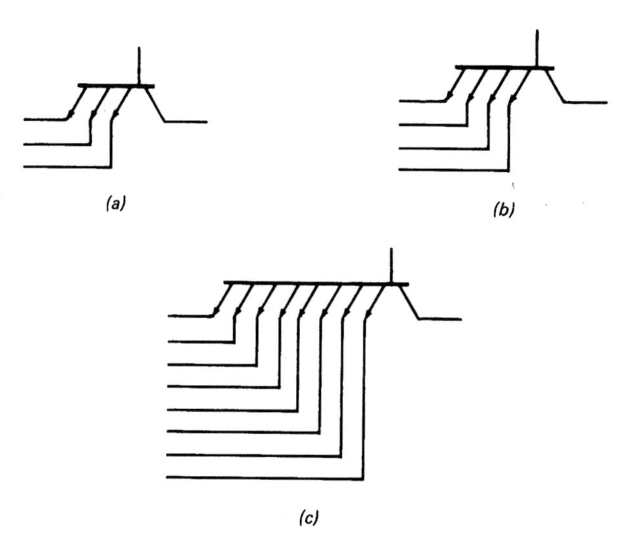

Fig. 4-7 Three, four, and eight inputs.

NAND Gates

To begin with, the NAND gate is the backbone of the entire series. All devices in the 7400 series are derived from the 2-input NAND gate shown in Fig. 4-1. To produce 3-, 4-, and 8-input NAND gates the manufacturer uses 3-, 4-, and 8-emitter transistors, as shown in Fig. 4-7. Because they are so basic, NAND gates are the least expensive devices in the 7400 series.

NOR Gates

To get other logic functions the manufacturer modifies the basic NAND-gate design. For instance, Fig. 4-8 shows a 2-input NOR gate. Q_1, Q_2, Q_3, and Q_4 are the same as in the basic design. Q_5 and Q_6 have been added to produce ORing. Notice that Q_2 and Q_6 are in parallel, the key to the ORing followed by inversion to get NORing.

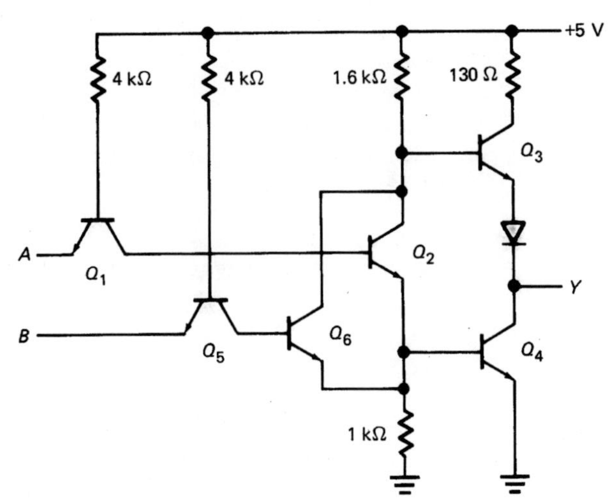

Fig. 4-8 TTL NOR gate.

When A and B are both low, Q_1 and Q_5 are saturated; this cuts off Q_2 and Q_6. Then Q_3 acts like an emitter follower and we get a high output.

If A or B or both are high, Q_1 or Q_5 or both are cut off, forcing Q_2 or Q_6 or both to turn on. When this happens, Q_4 saturates and pulls the output down to a low voltage.

With more transistors, manufacturers can produce 3- and 4-input NOR gates. (A TTL 8-input NOR gate is not available.)

AND and OR Gates

To produce the AND function, another common-emitter stage is inserted before the totem-pole output of the basic NAND gate design. The extra inversion converts the NAND gate to an AND gate. Similarly, another CE stage can be inserted before the totem-pole output of Fig. 4-8; this converts the NOR gate to an OR gate.

Buffer-Drivers

A *buffer* is a device that isolates two other devices. Typically, a buffer has a high input impedance and a low output impedance. In terms of digital ICs, this means a low input current and a high output current.

Since the output current of a standard TTL gate can be 10 times the input current, a basic gate does a certain amount of buffering (isolating). But it's only when the manufacturer optimizes the design for high output currents that we call a device a buffer or driver.

As an example, the 7437 is a quad 2-input NAND buffer, meaning four 2-input NAND gates optimized to get high output currents. Each gate has the following worst-case values of input and output currents:

$$I_{IL} = -1.6 \text{ mA} \qquad I_{IH} = 40 \ \mu\text{A}$$
$$I_{OL} = 48 \text{ mA} \qquad I_{OH} = -1.2 \text{ mA}$$

The input currents are the same as those of a standard NAND gate, but the output currents are 3 times as high, which means that the 7437 can drive heavier loads.

Appendix 3 includes several other buffer-drivers.

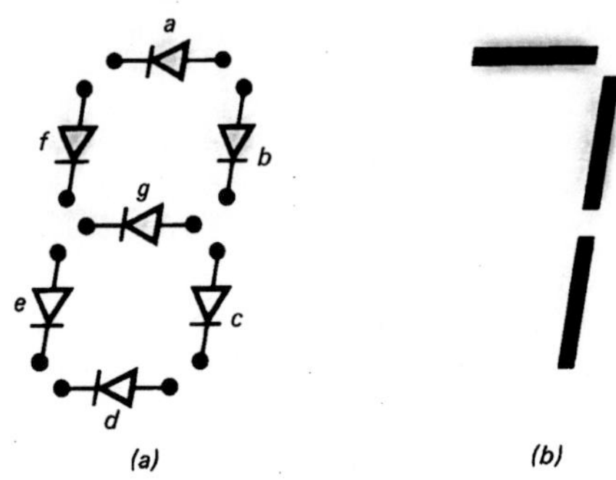

Fig. 4-9 Seven-segment display.

Encoders and Decoders

A number of TTL chips are available for encoding and decoding data. For instance, the 74147 is a decimal-to-BCD encoder. It has 10 input lines (decimal) and 4 output lines (BCD). As another example, the 74154 is a 1-of-16 decoder. It has 4 input lines (binary) and 16 output lines (hexadecimal).

Seven-segment decoders (7446, 7447, etc.) are useful for decimal displays. They convert a BCD nibble into an output that can drive a seven-segment display. Figure 4-9*a* illustrates the idea behind a seven-segment LED display. It has seven separate LEDs that allow you to display any digit between 0 and 9. To display a 7, the decoder will turn on LEDs *a*, *b*, and *c* (Fig. 4-9*b*).

Seven-segment displays are not limited to decimal numbers. For instance, in some microprocessor trainers, seven-segment displays are used to indicate hexadecimal digits. Digits A, C, E, and F are displayed in uppercase form; digit B is shown as a lowercase b (LEDs *c*, *d*, *e*, *f*, *g*); and digit D as a lowercase d (LEDs *b*, *c*, *d*, *e*, *g*).

Schmitt Triggers

When a computer is running, the outputs of gates are rapidly switching from one state to another. If you look at these signals with an oscilloscope, you see signals that ideally resemble rectangular waves like Fig. 4-10*a*.

When digital signals are transmitted and later received, they are often corrupted by noise, attenuation, or other factors and may wind up looking like the ragged waveform shown in Fig. 4-10*b*. If you try to use these nonrectangular signals to drive a gate or other digital device, you get unreliable operation.

This is where the *Schmitt trigger* comes in. It designed to clean up ragged looking pulses, producing almost vertical

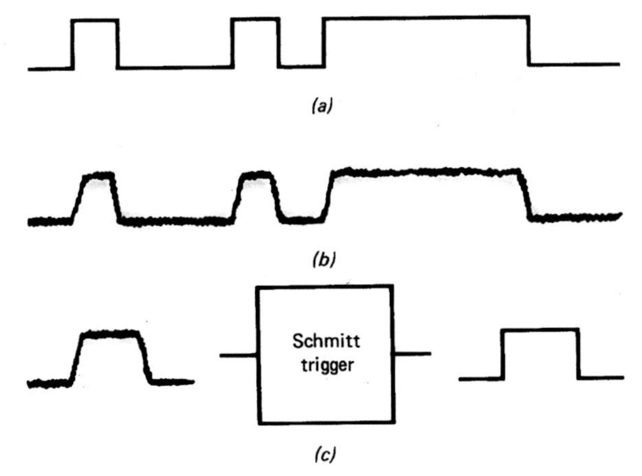

Fig. 4-10 Schmitt trigger produces rectangular output.

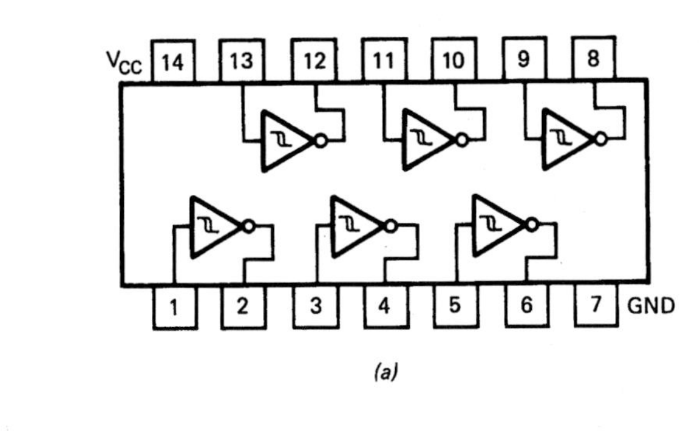

Fig. 4-11 (*a*) Hex Schmitt-trigger inverters; (*b*) 4-input NAND Schmitt trigger; (*c*) 2-input NAND Schmitt trigger.

transitions between the low and high state, and vice versa (Fig. 4-10*c*). In other words, the Schmitt trigger produces a rectangular output, regardless of the input waveform.

The 7414 is a hex Schmitt-trigger inverter, meaning six Schmitt-trigger inverters in one package like Fig. 4-11*a*. Notice the hysteresis symbol inside each inverter; it designates the Schmitt-trigger function.

Two other TTL Schmitt triggers are available. The 7413 is a dual 4-input NAND Schmitt trigger, two Schmitt-trigger gates like Fig. 4-11*b*. The 74132 is a quad 2-input NAND Schmitt trigger, four Schmitt-trigger gates like Fig. 4-11*c*.

Other Devices

The 7400 series also includes a number of other devices that you will find useful, such as AND-OR-INVERT gates (discussed in the next section), latches and flip-flops (Chap. 7), registers and counters (Chap. 8), and memories (Chap. 9).

4-5 AND-OR-INVERT GATES

Figure 4-12*a* shows an AND-OR circuit. Figure 4-12*b* shows the De Morgan equivalent circuit, a NAND-NAND network. In either case, the boolean equation is

$$Y = AB + CD \quad (4\text{-}1)$$

Since NAND gates are the preferred TTL gates, we would build the circuit of Fig. 4-12*b*. NAND-NAND circuits like this are important because with them you can build any desired logic circuit (discussed in Chap. 5).

TTL Devices

Is there any TTL device with the output given by Eq. 4-1? Yes, there are some AND-OR gates but they are not easily derived from the basic NAND-gate design. The gate that is easy to derive and comes close to having an expression like Eq. 4-1 is the AND-OR-INVERT gate shown in Fig. 4-12*c*. In other words, a variety of circuits like this are available on chips. Because of the inversion, the output has an equation of

$$Y = \overline{AB + CD} \quad (4\text{-}2)$$

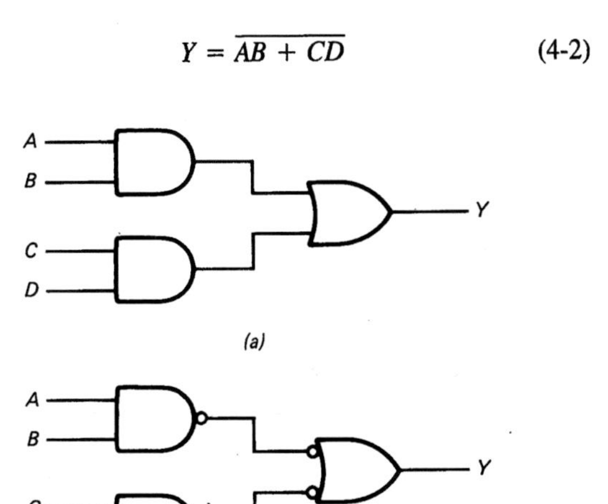

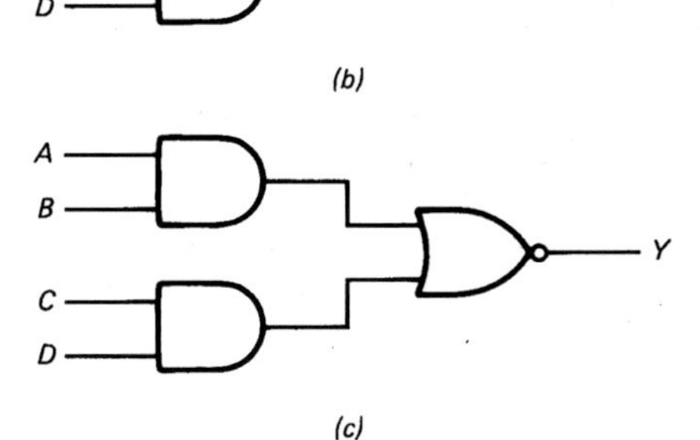

Fig. 4-12 (*a*) AND-OR circuit; (*b*) NAND-NAND circuit; (*c*) AND-OR-INVERT circuit.

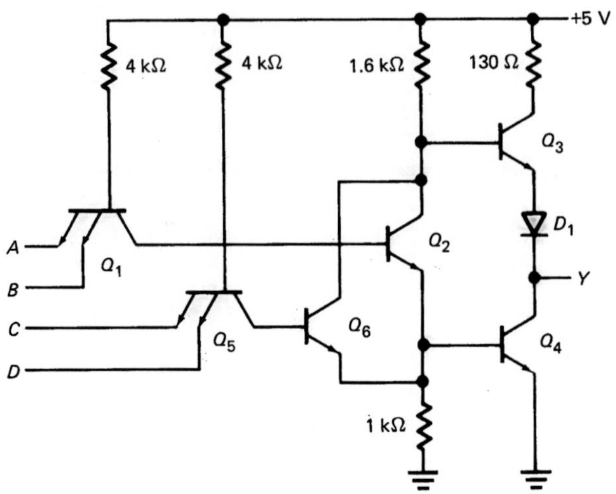

Fig. 4-13 AND-OR-INVERT schematic diagram.

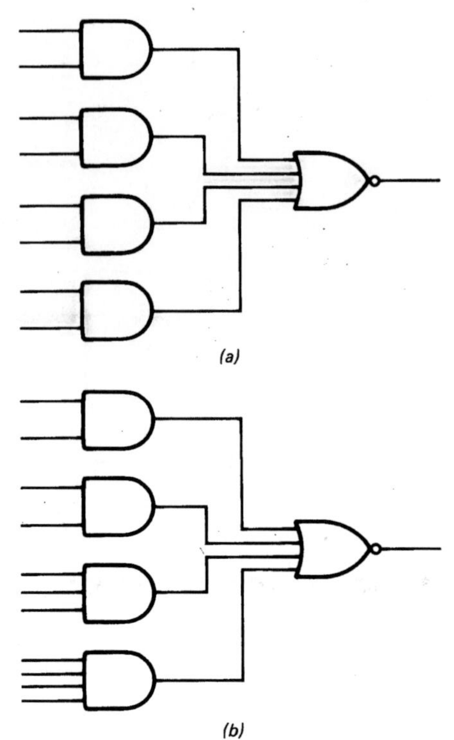

Fig. 4-14 Examples of AND-OR-INVERT circuits.

Figure 4-13 shows the schematic diagram of a TTL AND-OR-INVERT gate. Q_1, Q_2, Q_3, and Q_4 form the basic 2-input NAND gate of the 7400 series. By adding Q_5 and Q_6 we convert the basic NAND gate to an AND-OR-INVERT gate.

Q_1 and Q_5 act like 2-input AND gates; Q_2 and Q_6 produce ORing and inversion. Because of this, the circuit is logically equivalent to Fig. 4-12*c*.

In Table 4-6, listing the AND-OR-INVERT gates available in the 7400 series, 2-wide means two AND gates across, 4-wide means four AND gates across, and so on. For instance, the 7454 is a 2-input 4-wide AND-OR-INVERT gate like Fig. 4-14*a*; each AND gate has two inputs (2-input) and there are four AND gates (4-wide). Figure 4-14*b* shows the 7464; it is a 2-2-3-4-input 4-wide AND-OR-INVERT gate.

When we want the output given by Eq. 4-1, we can connect the output of a 2-input 2-wide AND-OR-INVERT gate to another inverter. This cancels out the internal inversion, giving us the equivalent of an AND-OR circuit (Fig. 4-12*a*) or a NAND-NAND network (Fig. 4-12*b*).

Expandable AND-OR-INVERT Gates

The widest AND-OR-INVERT gate available in the 7400 series is 4-wide. What do we do when we need a 6- or 8-wide circuit? One solution is to use an *expandable* AND-OR-INVERT gate.

Figure 4-15*a* shows the schematic diagram of an expandable AND-OR-INVERT gate. The only difference between this and the preceding AND-OR-INVERT gate (Fig. 4-13) is collector and emitter tie points brought outside the package. Since Q_2 and Q_6 are the key to the ORing operation, we are being given access to the internal ORing function. By connecting other gates to these new inputs we can expand the width of the AND-OR-INVERT gate.

Figure 4-15*b* shows the logic symbol for an expandable AND-OR-INVERT gate. The arrow input represents the emitter, and the bubble stands for the collector. Table 4-7 lists the expandable AND-OR-INVERT gates in the 7400 series.

Expanders

What do we connect to the collector and emitter inputs of an expandable gate? The output of an *expander* like Fig. 4-16*a*. The input transistor acts like a 4-input AND gate. The output transistor is a phase splitter; it produces two

TABLE 4-6. AND-OR-INVERT GATES

Device	Description
7451	Dual 2-input 2-wide
7454	2-input 4-wide
7459	Dual 2-3 input 2-wide
7464	2-2-3-4 input 4-wide

TABLE 4-7. EXPANDABLE AND-OR-INVERT GATES

Device	Description
7450	Dual 2-input 2-wide
7453	2-input 4-wide
7455	4-input 2-wide

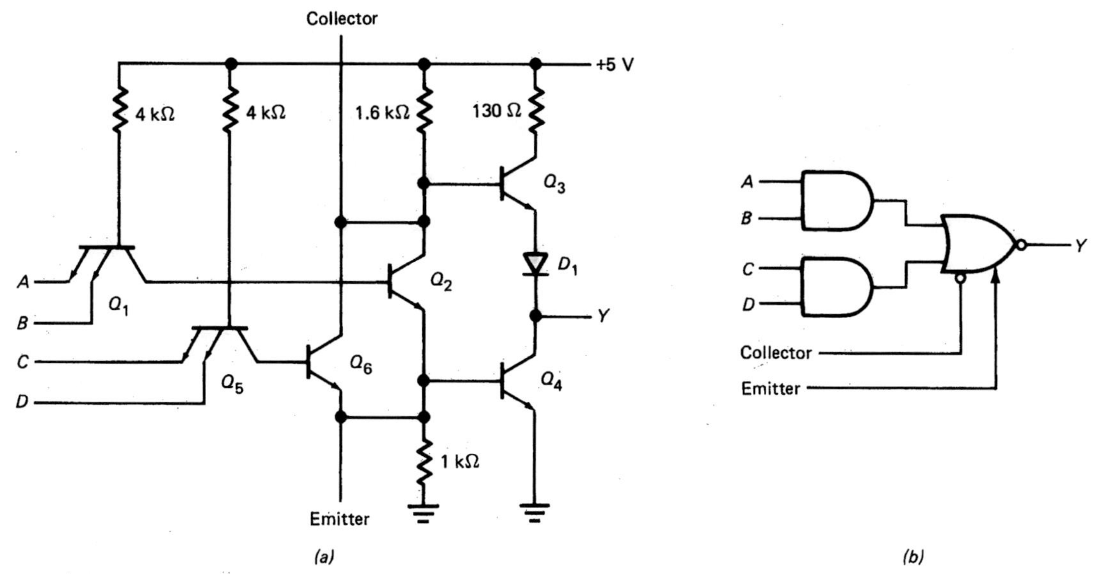

Fig. 4-15 (*a*) Expandable AND-OR-INVERT gate; (*b*) logic symbol.

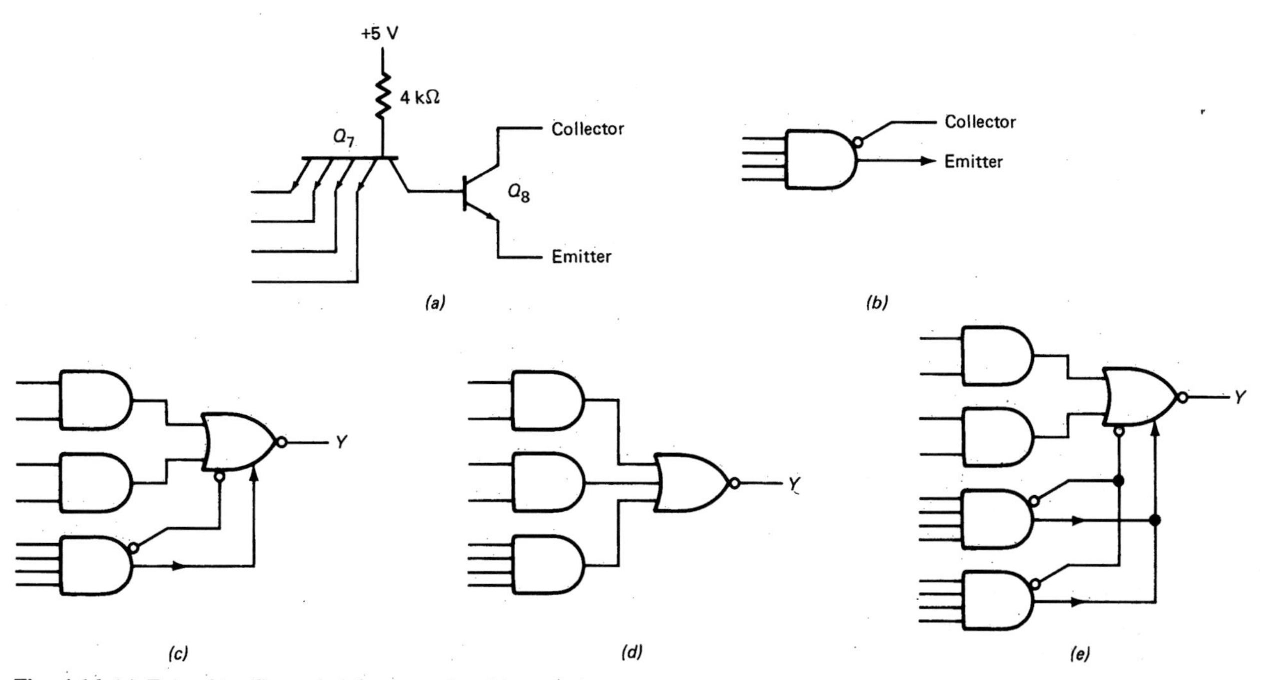

Fig. 4-16 (*a*) Expander; (*b*) symbol for expander; (*c*) expander driving expandable AND-OR-INVERT gate; (*d*) AND-OR-INVERT circuit; (*e*) expandable AND-OR-INVERT with two expanders.

output signals, one in phase (emitter) and the other inverted (collector). Figure 4-16*b* shows the symbol of a 4-input expander.

Visualize the outputs of Fig. 4-16*a* connected to the collector and emitter inputs of Fig. 4-15*a*. Then Q_8 is in parallel with Q_2 and Q_6. Figure 4-16*c* shows the logic circuit. This means that the expander outputs are being ORed with the signals of the AND-OR-INVERT gate. In other words, Fig. 4-16*c* is equivalent to the AND-OR-INVERT circuit of Fig. 4-16*d*.

We can connect more expanders. Figure 4-16*e* shows two expanders driving the expandable gate. Now we have a 2-2-4-4-input 4-wide AND-OR-INVERT circuit.

The 7460 is a dual 4-input expander. The 7450, a dual expandable AND-OR-INVERT gate, is designed for use with up to four 7460 expanders. This means that we can add two more expanders in Fig. 4-16*e* to get a 2-2-4-4-4-4-input 6-wide AND-OR-INVERT circuit.

4-6 OPEN-COLLECTOR GATES

Instead of a totem-pole output, some TTL devices have an *open-collector* output. This means they use only the lower transistor of a totem-pole pair. Figure 4-17*a* shows a 2-input NAND gate with an open-collector output. Because the collector of Q_4 is open, a gate like this won't work properly until you connect an external *pull-up* resistor, shown in Fig. 4-17*b*.

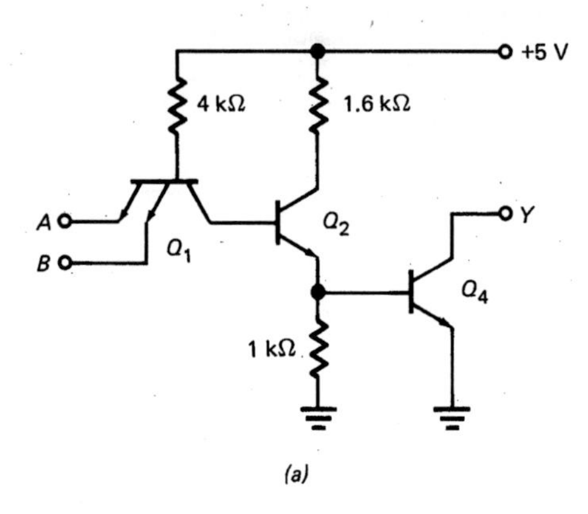

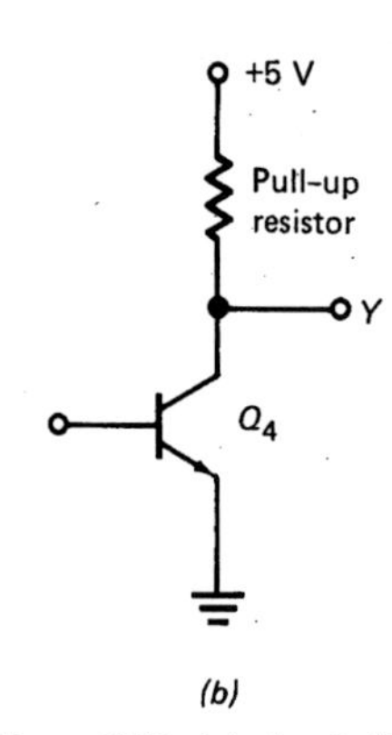

Fig. 4-17 Open-collector TTL: (*a*) circuit; (*b*) with pull-up resistor.

The outputs of open-collector gates can be wired together and connected to a common pull-up resistor. This is known as WIRE-OR. The big disadvantage of open-collector gates is their slow switching speed.

Open-collector gates are virtually obsolete because a new device called the *three-state switch* appeared in the early 1970s. Section 8-8 discusses three-state switches in detail.

4-7 MULTIPLEXERS

Multiplex means "many into one." A *multiplexer* is a circuit with many inputs but only one output. By applying control signals we can steer any input to the output.

Data Selection

Figure 4-18 shows a 16-to-1 multiplexer, also called a *data selector*. The input data bits are D_0 to D_{15}. Only one of these is transmitted to the output. Control word ABCD determines which data bit is passed to the output. For instance, when

$$ABCD = 0000$$

the upper AND gate is enabled but all other AND gates are disabled. Therefore, data bit D_0 is transmitted to the output, giving

$$Y = D_0$$

If the control word is changed to

$$ABCD = 1111$$

the bottom gate is enabled and all other gates are disabled. In this case,

$$Y = D_{15}$$

Boolean Function Generator

Digital design often starts with a truth table. The problem then is to come up with an equivalent logic circuit. Multiplexers give us a simple way to transform a truth table into an equivalent logic circuit. The idea is to use input data bits that are equal to the desired output bits of the truth table.

For example, look at the truth table of Table 4-8. When the input word ABCD is 0000, the output is 0; when ABCD

TABLE 4-8

A	*B*	*C*	*D*	*Y*
0	0	0	0	0
0	0	0	1	1
0	0	1	0	0
0	0	1	1	0
0	1	0	0	0
0	1	0	1	0
0	1	1	0	1
0	1	1	1	1
1	0	0	0	0
1	0	0	1	0
1	0	1	0	0
1	0	1	1	0
1	1	0	0	0
1	1	0	1	0
1	1	1	0	1
1	1	1	1	0

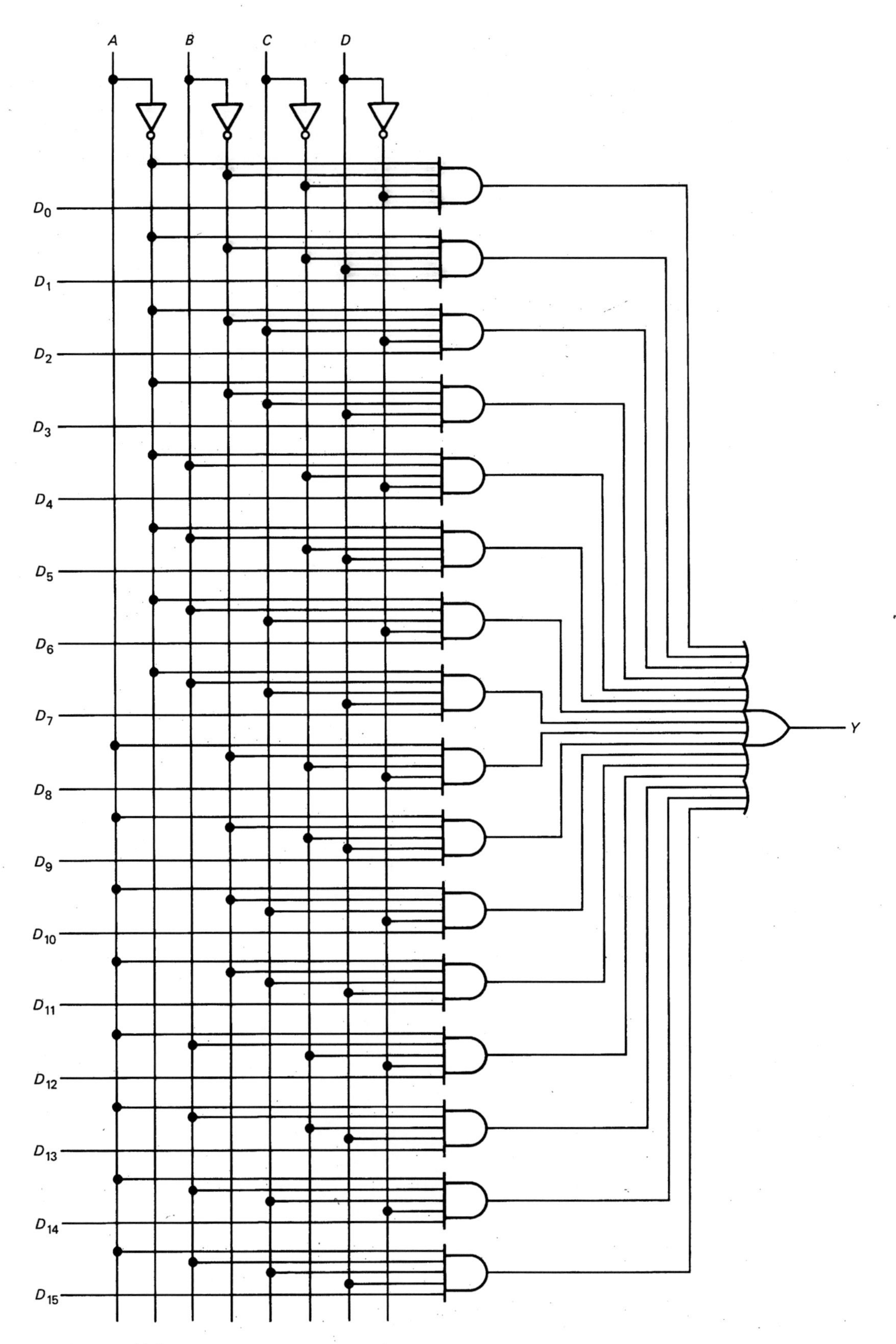

Fig. 4-18 A 16-to-1 multiplexer.

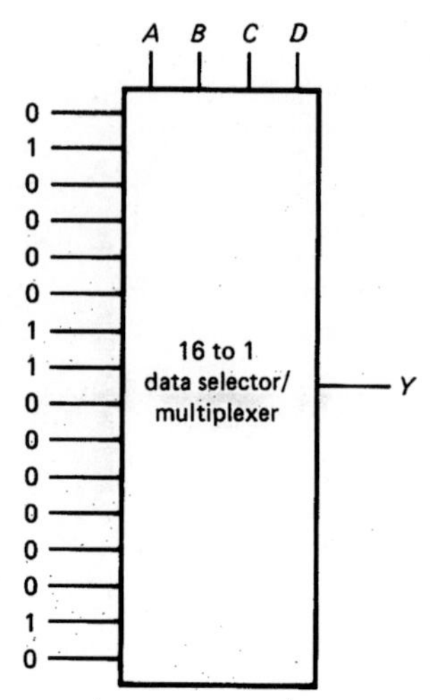

Fig. 4-19 Generating a boolean function.

= 0001, the output is 1; when ABCD = 0010, the output is 0; and so on. Figure 4-19 shows how to set up a multiplexer with the foregoing truth table. When ABCD = 0000, data bit 0 is steered to the output; when ABCD = 0001, data bit 1 is steered to the output; when ABCD = 0010, data bit 0 is steered to the output; and so forth. As a result, the truth table of this circuit is the same as Table 4-8.

Universal Logic Circuit

The 74150 is a 16-to-1 multiplexer. This TTL device is a universal logic circuit because you can use it to get the hardware equivalent of any four-variable truth table. In other words, by changing the input data bits the same IC can be made to generate thousands of different truth tables.

Multiplexing Words

Figure 4-20 illustrates a *word multiplexer* that has two input words and one output word. The input word on the left is $L_3L_2L_1L_0$ and the one on the right is $R_3R_2R_1R_0$. The control signal labeled *RIGHT* selects the input word that will be transmitted to the output. When *RIGHT* is low, the four NAND gates on the left are activated; therefore,

$$\textbf{OUT} = L_3L_2L_1L_0$$

When *RIGHT* is high,

$$\textbf{OUT} = R_3R_2R_1R_0$$

The 74157 is TTL multiplexer with an equivalent circuit like Fig. 4-20. Appendix 3 lists other multiplexers available in the 7400 series.

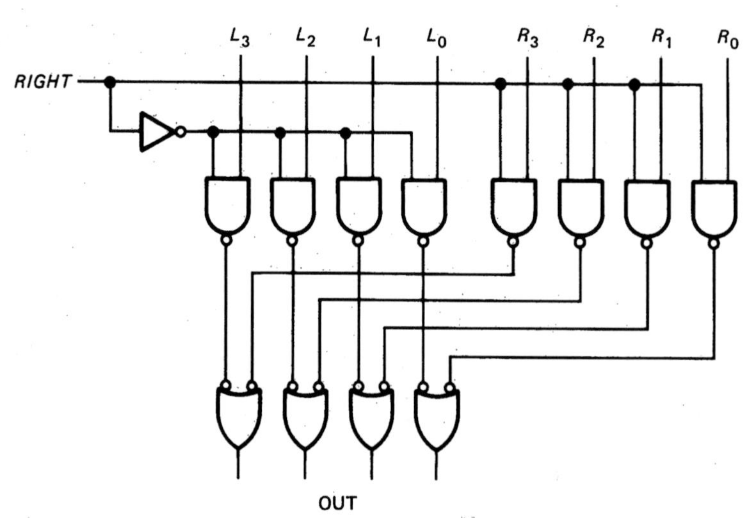

Fig. 4-20 Nibble multiplexer.

GLOSSARY

bipolar Having two types of charge carriers: free electrons and holes.

chip A small piece of semiconductor material. Sometimes, chip refers an IC device including its pins.

fanout The maximum number of TTL loads that a TTL device can drive reliably over the specified temperature range.

low-power Schottky TTL A modification of standard TTL

in which larger resistances and Schottky diodes are used. The increased resistances decrease the power dissipation, and the Schottky diodes increase the speed.

multiplexer A circuit with many inputs but only one output. Control signals select which input reaches the output.

noise margin The amount of noise voltage that causes unreliable operation. With TTL it is 0.4 V. As long as noise voltages induced on connecting lines are less than 0.4 V, the TTL devices will work reliably.

saturation delay time The time delay encountered when a transistor tries to come out of the saturation region. When the base drive switches from high to low, a transistor cannot instantaneously come out of saturation; extra carriers that flooded the base region must first flow out of the base.

Schmitt trigger A digital circuit that produces a rectangular output from any input large enough to drive the Schmitt trigger. The input waveform may be sinusoidal, triangular, distorted, and so on. The output is always rectangular.

sink A place where something is absorbed. When saturated, the lower transistor in a totem-pole output acts like a current sink because conventional charges flow through the transistor to ground.

source A place where something originates. The upper transistor of a totem-pole output acts like a source because charges flow out of its emitter into the load.

standard TTL The initial TTL design with resistance values that produce a power dissipation of 10 mW per gate and a propagation delay time of 10 ns.

SELF-TESTING REVIEW

Read each of the following and provide the missing words. Answers appear at the beginning of the next question.

1. Small-scale integration, abbreviated ________, refers to fewer than 12 gates on the same chip. Medium-scale integration (MSI) means 12 to 100 gates per chip. And large-scale integration (LSI) refers to more than ________ gates per chip.
2. (*SSI, 100*) The two basic technologies for digital ICs are bipolar and MOS. Bipolar technology is preferred for ________ and ________, whereas MOS technology is better suited to LSI. The reason MOS dominates the LSI field is that more ________ can be fabricated on the same chip area.
3. (*SSI, MSI, MOSFETs*) Some of the bipolar families include DTL, TTL, and ECL. ________ has become the most widely used bipolar family. ________ is the fastest logic family; it's used in high-speed applications.
4. (*TTL, ECL*) Some of the MOS families are PMOS, NMOS, and CMOS. ________ dominates the LSI field, and ________ is used extensively where lowest power consumption is necessary.
5. (*NMOS, CMOS*) The 7400 series, also called standard TTL, contains a variety of SSI and ________ chips that allow us to build all kinds of digital circuits and systems. Standard TTL has a multiple-emitter input transistor and a ________ output. The totem-pole output produces a low output impedance in either state.
6. (*MSI, totem-pole*) Besides standard TTL, there is high-speed TTL, low-power TTL, Schottky TTL, and low-power ________ TTL. Standard TTL and low-power ________ TTL have become the favorites of digital designers, used more than any other bipolar families.
7. (*Schottky, Schottky*) 7400-series devices are guaranteed to work reliably over a ________ range of 0 to 70°C and over a voltage range of 4.75 to 5.25 V. A floating TTL input has the same effect as a ________ input.
8. (*temperature, high*) A ________ TTL device can sink up to 16 mA and can source up to 400 μA. The maximum number of TTL loads a TTL device can drive is called the ________. With standard TTL, the fanout equals ________.
9. (*standard, fanout, 10*) A buffer is a device that isolates other devices. Typically, a buffer has a high input impedance and a ________ output impedance. In terms of digital ICs, this means a ________ input current and a high output current capability.
10. (*low, low*) A Schmitt trigger is a digital circuit that produces a ________ output regardless of the input waveform. It is used to clean up ragged looking pulses that have been distorted during transmission from one place to another.
11. (*rectangular*) A multiplexer is a circuit with many inputs but only one output. It is also called a data selector because data can be steered from one of the inputs to the output. A 74150 is a 16-to-1 multiplexer. With this TTL device you can implement the logic circuit for any four-variable truth table.

PROBLEMS

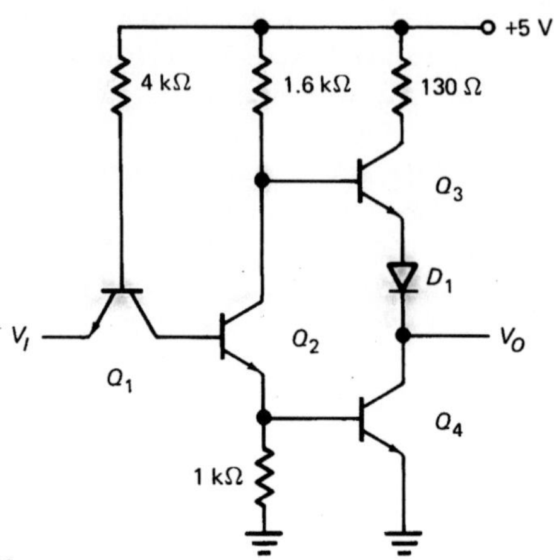

Fig. 4-21

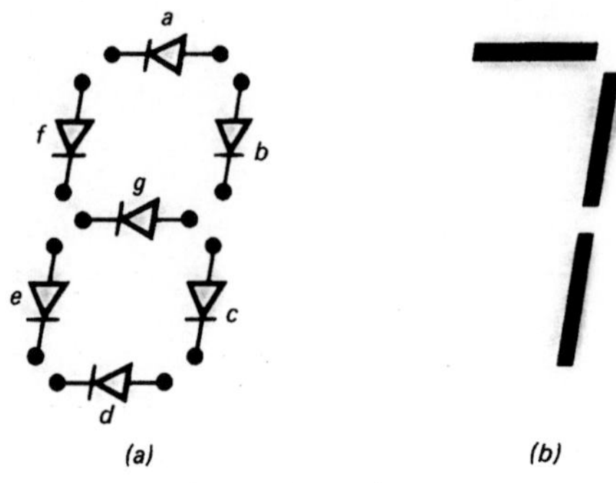

Fig. 4-22

4-1. In Fig. 4-21 a grounded input means that almost the entire supply voltage appears across the 4-kΩ resistor. Allowing 0.7 V for the emitter-base voltage of Q_1, how much input emitter current is there with a grounded input? The supply voltage can be as high as 5.25 V and the 4-kΩ resistance can be a low as 3.28 kΩ. What is the input emitter current in this case?

4-2. What is the fanout of a 74S00 device when it drives low-power TTL loads?

4-3. What is the fanout of a low-power Schottky device driving standard TTL devices?

4-4. Section 4-4 gave the input and output currents for a 7437 buffer. What is the fanout of a 7437 when it drives standard TTL loads?

4-5. A seven-segment decoder is driving a LED display like Fig. 4-22*a*. Which LEDs are on when digit 8 appears? Which LEDs are on when digit 4 appears?

4-6. Section 4-7 described the 74150, a 16-to-1 multiplexer. Refer to Fig. 4-23 and indicate the values the D_0 to D_{15} inputs of a 74150 should have to reproduce the following truth table: The output is high when ABCD = 0000, 0100, 0111, 1100, and 1111; the output is low for all other inputs.

4-7. What is propagation delay?

4-8. Why are 5400 series devices not normally used in commercial applications?

4-9. What do Schottky devices virtually eliminate which makes their high switching speeds possible?

4-10. What is the noise margin of TTL devices?

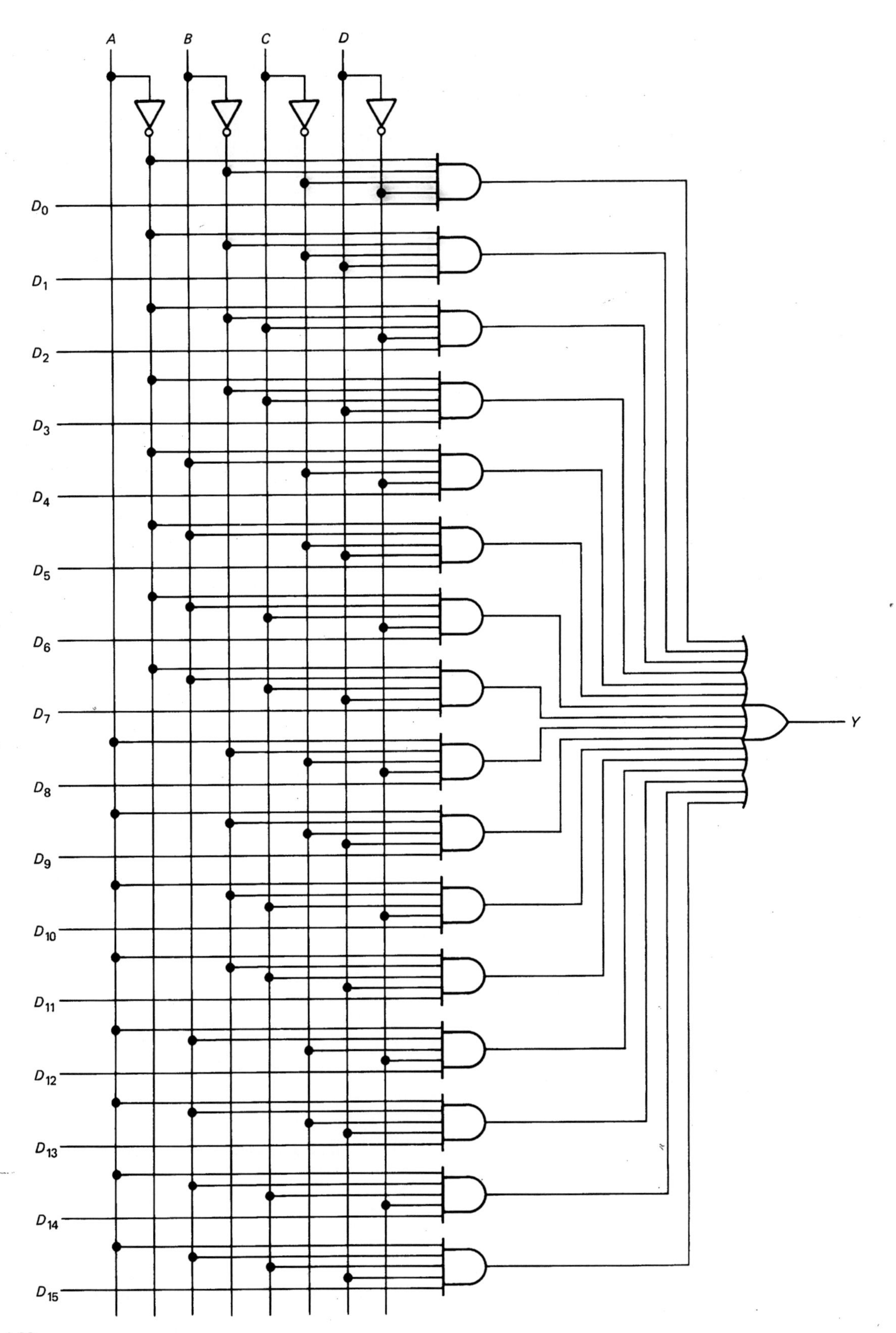

Fig. 4-23

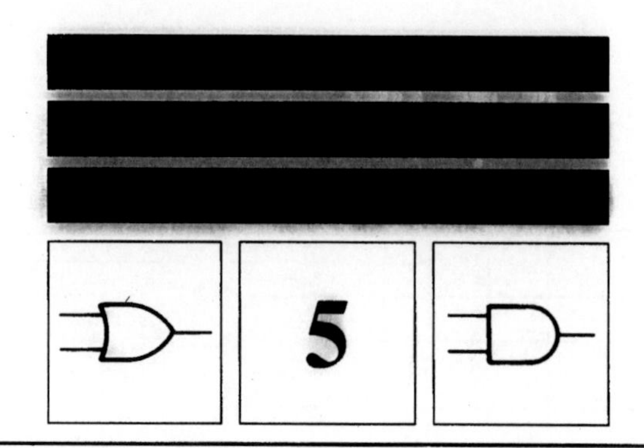

Boolean Algebra and Karnaugh Maps

This chapter discusses boolean algebra and *Karnaugh maps,* topics needed by the digital designer. Digital design usually begins by specifying a desired output with a truth table. The question then is how to come up with a logic circuit that has the same truth table. Boolean algebra and Karnaugh maps are the tools used to transform a truth table into a practical logic circuit.

5-1 BOOLEAN RELATIONS

What follows is a discussion of basic relations in boolean algebra. Many of these relations are the same as in ordinary algebra, which makes remembering them easy.

Commutative, Associative, and Distributive Laws

Given a 2-input OR gate, you can transpose the input signals without changing the output (see Fig. 5-1*a*). In boolean terms

$$A + B = B + A \qquad (5\text{-}1)$$

Similarly, you can transpose the input signals to a 2-input AND gate without affecting the output (Fig. 5-1*b*). The boolean equivalent of this is

$$AB = BA \qquad (5\text{-}2)$$

The foregoing relations are called *commutative laws.*

The next group of rules are called the *associative laws.* The associative law for ORing is

$$A + (B + C) = (A + B) + C \qquad (5\text{-}3)$$

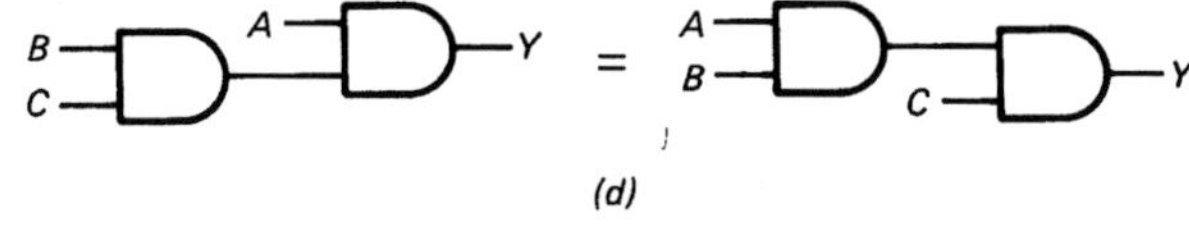

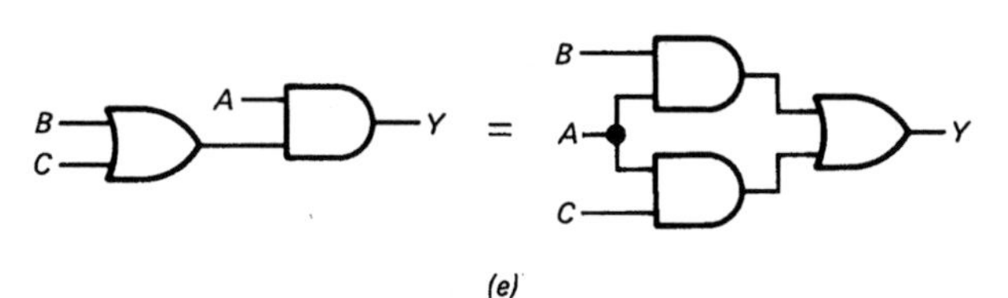

Fig. 5-1 Commutative, associative, and distributive laws.

Figure 5-1*c* illustrates this rule. The idea is that how you group variables in an ORing operation has no effect on the output. For either gate in Fig. 5-1*c* the output is

$$Y = A + B + C$$

Similarly, the associative law for ANDing is

$$A(BC) = (AB)C \tag{5-4}$$

Figure 5-1*d* illustrates this rule. How you group variables in ANDing operations has no effect on the output. For either gate of Fig. 5-1*d* the output is

$$Y = ABC$$

The *distributive law* states that

$$A(B + C) = AB + AC \tag{5-5}$$

This is easy to remember because it's identical to ordinary algebra. Figure 5-1*e* shows the meaning in terms of gates.

OR Operations

The next four boolean relations are about OR operations. Here is the first:

$$A + 0 = A \tag{5-6}$$

This says that a variable ORed with 0 equals the variable. For better grasp of this idea, look at Fig. 5-2*a*. (The solid arrow stands for "implies.") The two cases on the left imply the case on the right. In other words, if the variable is 0, the output is 0 (left gate); if the variable is 1, the output is 1 (middle gate); therefore, a variable ORed with 0 equals the variable (right gate).

Another boolean relation is

$$A + A = A \tag{5-7}$$

which is illustrated in Fig. 5-2*b*. You can see what happens. If *A* is 0, the output is 0; if *A* is 1, the output is 1; therefore, a variable ORed with itself equals the variable.

Figure 5-2*c* shows the next boolean rule:

$$A + 1 = 1 \tag{5-8}$$

In a nutshell, if one input to an OR gate is 1, the output is 1 regardless of the other input.

Finally, we have

$$A + \bar{A} = 1 \tag{5-9}$$

shown in Fig. 5-2*d*. In this case, a variable ORed with its complement equals 1.

AND Operations

The first AND relation to know about is

$$A \cdot 1 = A \tag{5-10}$$

illustrated in Fig. 5-3*a*. If *A* is 0, the output is 0; if *A* is 1, the output is 1; therefore, a variable ANDed with 1 equals the variable.

Another relation is

$$A \cdot A = A \tag{5-11}$$

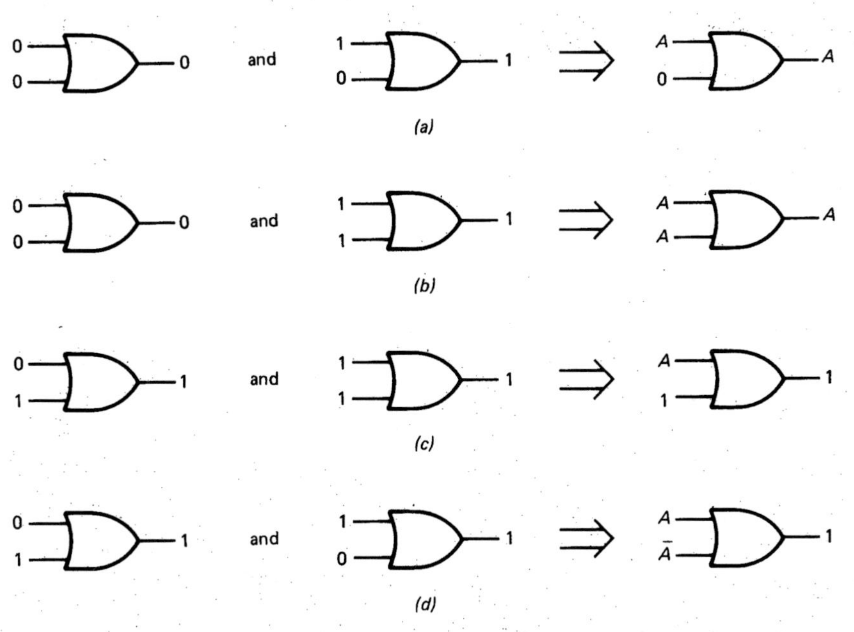

Fig. 5-2 OR relations.

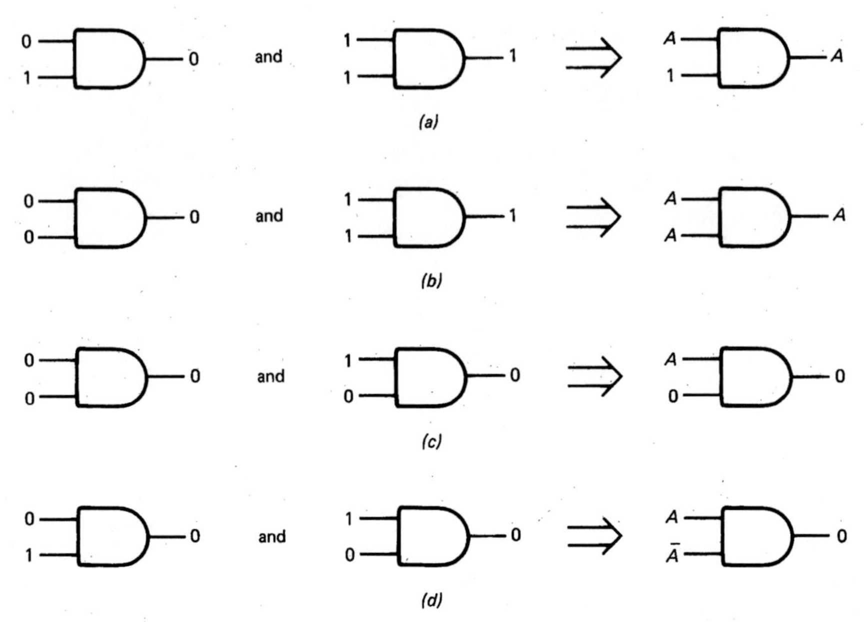

Fig. 5-3 AND relations.

shown in Fig. 5-3*b*. In this case, a variable ANDed with itself equals the variable.

Figure 5-3*c* illustrates this relation

$$A \cdot 0 = 0 \qquad (5\text{-}12)$$

The rule is clear. If one input to an AND gate is 0, the output is 0 regardless of the other input.

The last AND rule is

$$A \cdot \bar{A} = 0 \qquad (5\text{-}13)$$

As shown in Fig. 5-3*d*, a variable ANDed with its complement produces a 0 output.

Double Inversion and De Morgan's Theorems

The double-inversion rule is

$$\bar{\bar{A}} = A \qquad (5\text{-}14)$$

which says that the double complement of a variable equals the variable. Finally, there are the De Morgan theorems discussed in Chap. 3:

$$\overline{A + B} = \bar{A}\bar{B} \qquad (5\text{-}15)$$
$$\overline{AB} = \bar{A} + \bar{B} \qquad (5\text{-}16)$$

You should memorize Eqs. 5-1 to 5-16 because they are used frequently in design work.

Duality Theorem

We state the *duality theorem* without proof. Starting with a boolean relation, you can derive another boolean relation by

1. Changing each OR sign to an AND sign
2. Changing each AND sign to an OR sign
3. Complementing each 0 and 1

For instance, Eq. 5-6 says that

$$A + 0 = A$$

The dual relation is

$$A \cdot 1 = A$$

This is obtained by changing the OR sign to an AND sign, and by complementing the 0 to get a 1.

The duality theorem is useful because it sometimes produces a new boolean relation. For example, Eq. 5-5 states that

$$A(B + C) = AB + AC$$

By changing each OR and AND operation we get the dual relation

$$A + BC = (A + B)(A + C)$$

This is a new boolean relation, not previously discussed. (If you want to prove it, construct the truth table for the left and right members of the equation. The two truth tables will be identical.)

Summary

For future reference, here are some boolean relations and their duals:

$A + B = B + A$	$AB = BA$
$A + (B + C) = (A + B) + C$	$A(BC) = (AB)C$
$A(B + C) = AB + AC$	$A + BC = (A + B)(A + C)$
$A + 0 = A$	$A \cdot 1 = A$
$A + 1 = 1$	$A \cdot 0 = 0$
$A + A = A$	$AA = A$
$A + \overline{A} = 1$	$A\overline{A} = 0$
$\overline{\overline{A}} = A$	$\overline{\overline{A}} = A$
$\overline{A + B} = \overline{A}\,\overline{B}$	$\overline{AB} = \overline{A} + \overline{B}$
$A + AB = A$	$A(A + B) = A$
$A + \overline{A}B = A + B$	$A(\overline{A} + B) = AB$

5-2 SUM-OF-PRODUCTS METHOD

Digital design often starts by constructing a truth table with a desired output (0 or 1) for each input condition. Once you have this truth table, you transform it into an equivalent logic circuit. This section discusses the *sum-of-products method,* a way of deriving a logic circuit from a truth table.

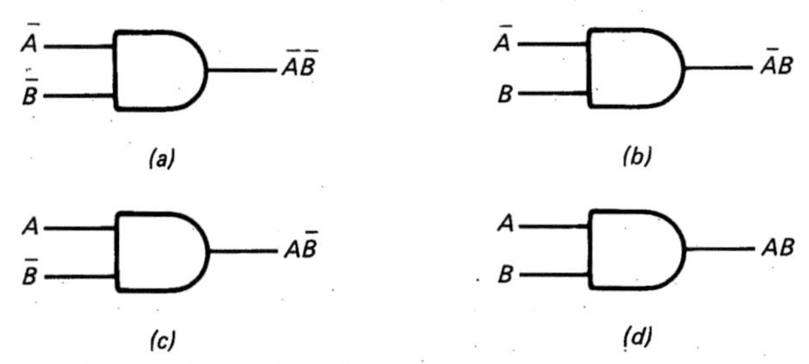

Fig. 5-4 Fundamental products.

Fundamental Products

Figure 5-4 shows the four possible ways to AND two input signals and their complements. In Fig. 5-4*a* the inputs are $\overline{A}$ and $\overline{B}$. Therefore, the output is

$$Y = \overline{A}\,\overline{B}$$

The output is high only when $A = 0$ and $B = 0$.

Figure 5-4*b* shows another possibility. Here the inputs are $\overline{A}$ and B; so the output is

$$Y = \overline{A}B$$

TABLE 5-1. TWO VARIABLES

A	B	Fundamental product
0	0	$\overline{A}\,\overline{B}$
0	1	$\overline{A}B$
1	0	$A\overline{B}$
1	1	AB

In this case, the output is 1 only when $A = 0$ and $B = 1$. In Fig. 5-4*c* the inputs are A and $\overline{B}$. The output

$$Y = A\overline{B}$$

is high only when $A = 1$ and $B = 0$. Finally, in Fig. 5-4*d* the inputs are A and B. The output

$$Y = AB$$

is 1 only when $A = 1$ and $B = 1$.

Table 5-1 summarizes the four possible ways to AND two signals in complemented or uncomplemented form. The logical products $\overline{A}\,\overline{B}$, $\overline{A}B$, $A\overline{B}$, and AB are called *fundamental products* because each produces a high output for its corresponding input. For instance, $\overline{A}\,\overline{B}$ is a 1 when A is 0 and B is 0, $\overline{A}B$ is a 1 when A is 0 and B is 1, and so forth.

Three Variables

A similar idea applies to three signals in complemented and uncomplemented form. Given A, B, C, and their complements, there are eight fundamental products: $\overline{A}\,\overline{B}\,\overline{C}$, $\overline{A}\,\overline{B}C$, $\overline{A}B\overline{C}$, $\overline{A}BC$, $A\overline{B}\,\overline{C}$, $A\overline{B}C$, $AB\overline{C}$, and ABC. Table 5-2 lists each input possibility and its fundamental product. Again notice this property: each fundamental product is high for the corresponding input. This means that $\overline{A}\,\overline{B}\,\overline{C}$ is a 1 when A is 0, B is 0, and C is 0; $\overline{A}\,\overline{B}C$ is a 1 when A is 0, B is 0, and C is 1; and so on.

TABLE 5-2. THREE VARIABLES

A	B	C	Fundamental product
0	0	0	$\overline{A}\,\overline{B}\,\overline{C}$
0	0	1	$\overline{A}\,\overline{B}C$
0	1	0	$\overline{A}B\overline{C}$
0	1	1	$\overline{A}BC$
1	0	0	$A\overline{B}\,\overline{C}$
1	0	1	$A\overline{B}C$
1	1	0	$AB\overline{C}$
1	1	1	ABC

Four Variables

When there are 4 input variables, there are 16 possible input conditions, 0000 to 1111. The corresponding fundamental products are from $\overline{A}\,\overline{B}\,\overline{C}\,\overline{D}$ through $ABCD$. Here is a quick way to find the fundamental product for any input condition. Whenever the input variable is 0, the same variable is complemented in the fundamental product. For instance, if the input condition is 0110, the fundamental product is $\overline{A}BC\overline{D}$. Similarly, if the input is 0100, the fundamental product is $\overline{A}B\overline{C}\,\overline{D}$.

Deriving a Logic Circuit

To get from a truth table to an equivalent logic circuit OR the fundamental products for each input condition that produces a high output. For example, suppose you have a truth table like Table 5-3. The fundamental products are listed for each high output. By ORing these products you get the boolean equation

$$Y = \overline{A}B\overline{C} + A\overline{B}C + AB\overline{C} + ABC \qquad (5\text{-}17)$$

This equation implies four AND gates driving an OR gate. The first AND gate has inputs of $\overline{A}$, B, and $\overline{C}$; the second AND gate has inputs of A, $\overline{B}$, and C; the third AND gate has inputs of A, B, and $\overline{C}$; the fourth AND gate has inputs of A, B, and C. Figure 5-5 shows the corresponding logic circuit. This AND-OR circuit has the same truth table as Table 5-3.

As another example of the sum-of-products method, look at Table 5-4. Find each output 1 and write its fundamental product. The resulting products are $\overline{A}\,\overline{B}CD$, $\overline{A}BCD$, and $A\overline{B}\,\overline{C}D$. This means that the boolean equation is

$$Y = \overline{A}\,\overline{B}CD + \overline{A}BCD + A\overline{B}\,\overline{C}D \qquad (5\text{-}18)$$

This equation implies that three AND gates are driving an OR gate. The first AND gate has inputs of $\overline{A}$, $\overline{B}$, C, and D; the second has inputs of $\overline{A}$, B, C, and D; the third has inputs of A, $\overline{B}$, $\overline{C}$, and D. Figure 5-6 is the equivalent logic circuit.

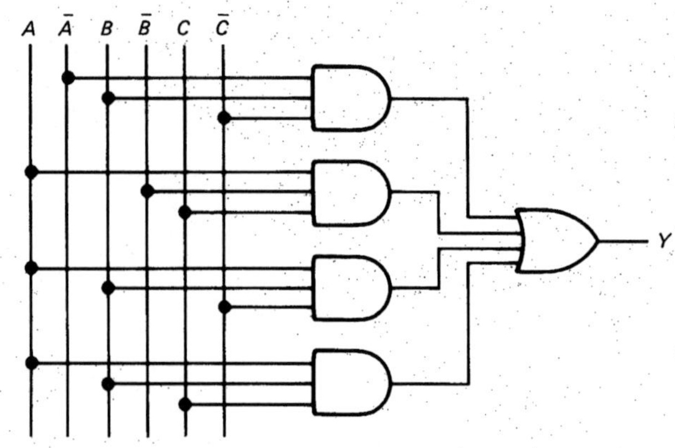

Fig. 5-5 Sum-of-products circuit.

TABLE 5-3

A	B	C	Y
0	0	0	0
0	0	1	0
0	1	0	$1 \rightarrow \overline{A}B\overline{C}$
0	1	1	0
1	0	0	0
1	0	1	$1 \rightarrow A\overline{B}C$
1	1	0	$1 \rightarrow AB\overline{C}$
1	1	1	$1 \rightarrow ABC$

TABLE 5-4

A	B	C	D	Y
0	0	0	0	0
0	0	0	1	0
0	0	1	0	0
0	0	1	1	1
0	1	0	0	0
0	1	0	1	0
0	1	1	0	0
0	1	1	1	1
1	0	0	0	0
1	0	0	1	1
1	0	1	0	0
1	0	1	1	0
1	1	0	0	0
1	1	0	1	0
1	1	1	0	0
1	1	1	1	0

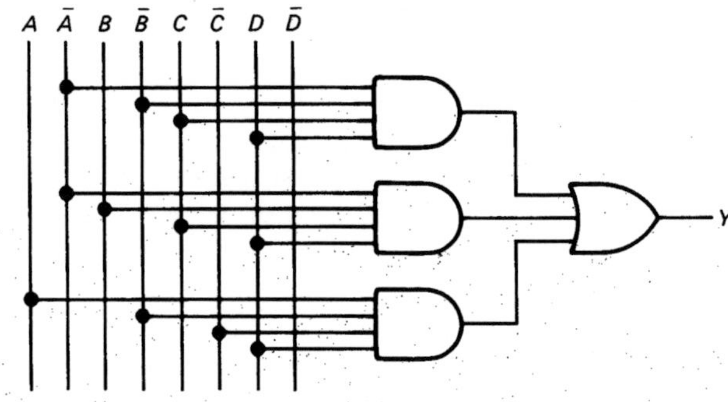

Fig. 5-6

The sum-of-products method always works. You OR the fundamental products of each high output in the truth table. This gives an equation which you can transform into an AND-OR network that is the circuit equivalent of the truth table.

5-3 ALGEBRAIC SIMPLIFICATION

After obtaining a sum-of-products equation as described in the preceding section, the thing to do is to simplify the circuit if possible. One way to do this is with boolean algebra. Here is the approach. Starting with the boolean equation for the sum-of-products circuit, you try to rearrange and simplify the equation as much as possible using the boolean rules of Sec. 5-1. The simplified boolean equation means a simpler logic circuit. This section will give you examples.

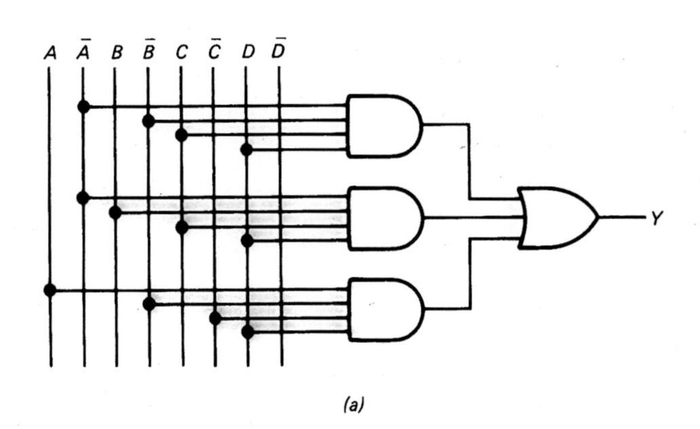

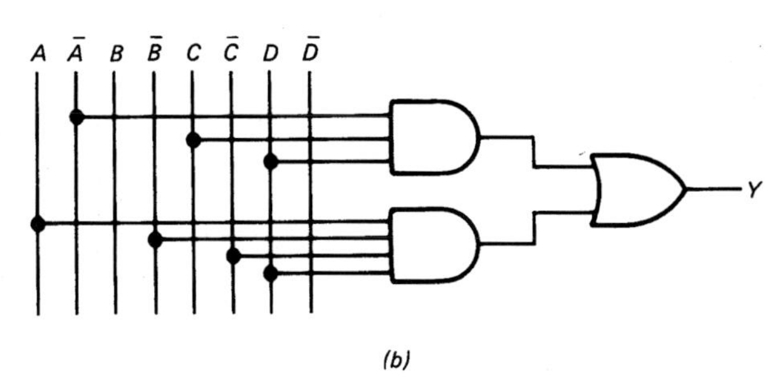

Fig. 5-7

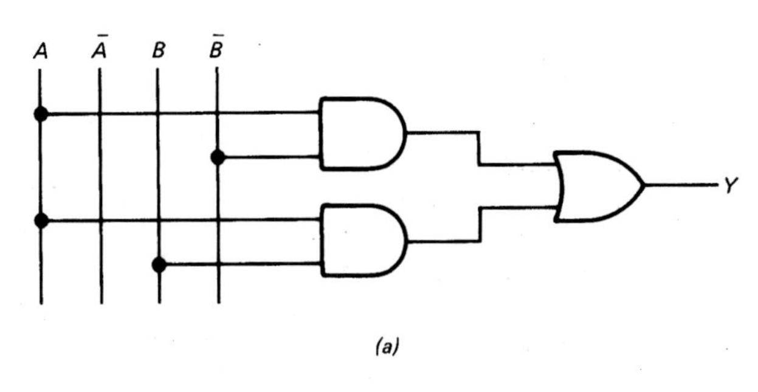

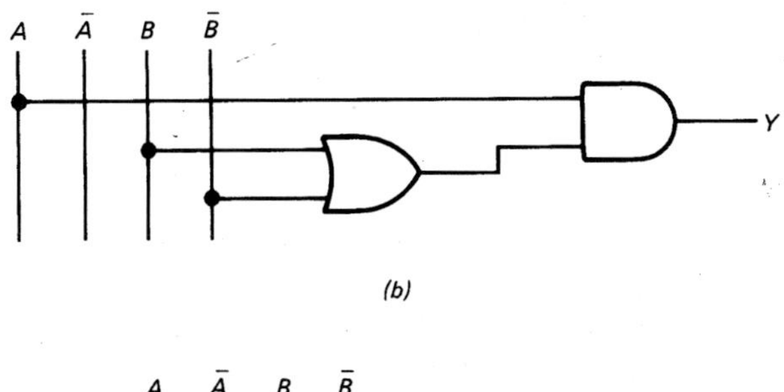

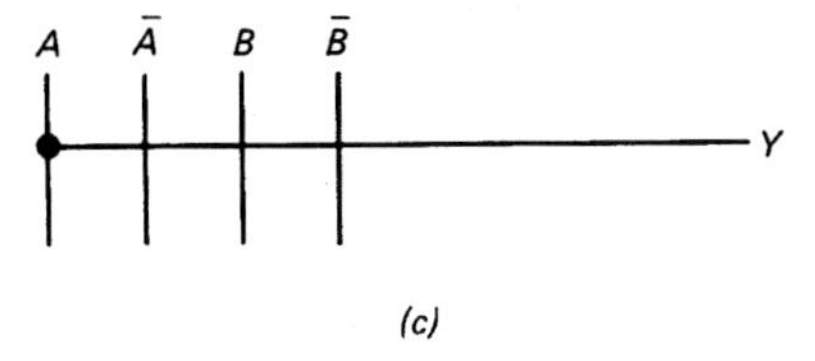

Fig. 5-8

Gate Leads

A preliminary guide for comparing the simplicity of one logic circuit with another is to count the number of *input gate leads;* the circuit with fewer input gate leads is usually easier to build. For instance, the AND-OR circuit of Fig. 5-7*a* has a total of 15 input gate leads (4 on each AND gate and 3 on the OR gate). The AND-OR circuit of Fig. 5-7*b*, on the other hand, has a total of 9 input gate leads. The AND-OR circuit of Fig. 5-7*b* is simpler than the AND-OR circuit of Fig. 5-7*a* because it has fewer input gate leads.

A *bus* is a group of wires carrying digital signals. The 8-bit bus of Fig. 5-7*a* transmits variables A, B, C, D and their complements $\bar{A}$, $\bar{B}$, $\bar{C}$, and $\bar{D}$. In the typical microcomputer, the microprocessor, memory, and I/O units exchange data by means of buses.

Factoring to Simplify

One way to reduce the number of input gate leads is to factor the boolean equation if possible. For instance, the boolean equation

$$Y = A\bar{B} + AB \tag{5-19}$$

has the equivalent logic circuit shown in Fig. 5-8*a*. This circuit has six input gate leads. By factoring Eq. 5-19 we get

$$Y = A(\bar{B} + B)$$

The equivalent logic circuit for this is shown in Fig. 5-8*b*; it has only four input gate leads.

Recall that a variable ORed with its complement always equals 1; therefore,

$$Y = A(\bar{B} + B) = A \cdot 1 = A$$

To get this output, all we need is a connecting wire from the input to the output, as shown in Fig. 5-8*c*. In other words, we don't need any gates at all.

Another Example

Here is another example of how factoring can simplify a boolean equation and its corresponding logic circuit. Suppose we are given

$$Y = AB + AC + BD + CD \tag{5-20}$$

In this equation, two variables at a time are being ANDed. The logical products are then ORed to get the final output. Figure 5-9*a* shows the corresponding logic circuit. It has 12 input gate leads.

We can factor and rearrange Eq. 5-20 as

$$Y = A(B + C) + D(B + C)$$

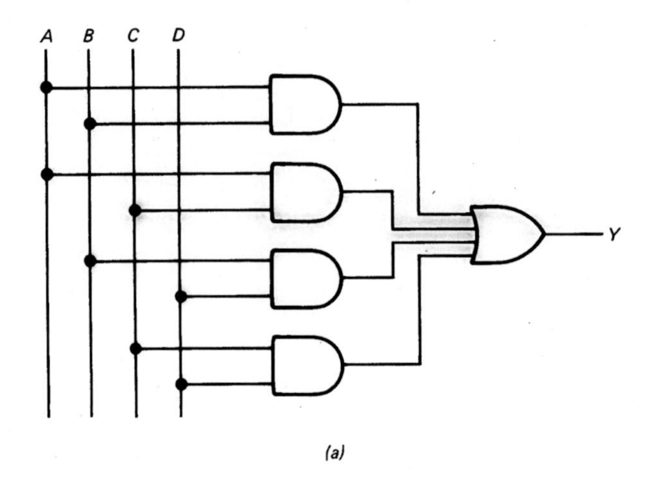

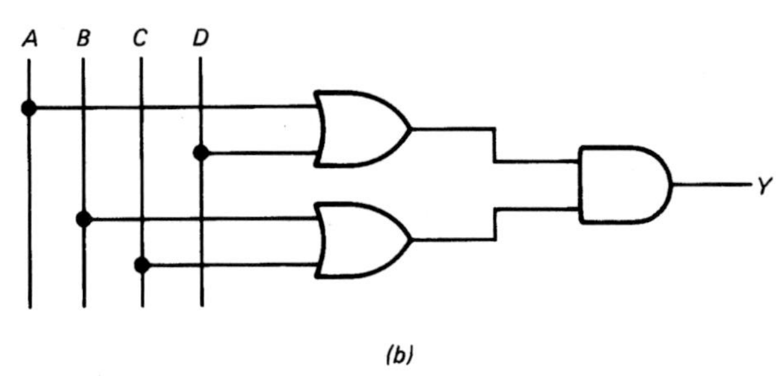

Fig. 5-9

or as

$$Y = (A + D)(B + C) \qquad (5\text{-}21)$$

In this case, the variables are first ORed, then the logical sums are ANDed. Figure 5-9*b* illustrates the logic circuit. Notice it has only six input gate leads and is simpler than the circuit of Fig. 5-9*a*.

Final Example

In Sec. 5-2 we derived this sum-of-products equation from a truth table:

$$Y = \bar{A}\bar{B}CD + \bar{A}BCD + A\bar{B}\bar{C}\bar{D} \qquad (5\text{-}22)$$

Figure 5-7*a* shows the sum-of-products circuit. It has 15 input gate leads. We can factor the equation as

$$Y = \bar{A}CD(\bar{B} + B) + A\bar{B}\bar{C}\bar{D}$$

or as

$$Y = \bar{A}CD + A\bar{B}\bar{C}\bar{D} \qquad (5\text{-}23)$$

Figure 5-7*b* shows the equivalent logic circuit; it has only nine input gate leads.

In general, one approach in digital design is to transform a truth table into a sum-of-products equation, which you then simplify as much as possible to get a practical logic circuit.

5-4 KARNAUGH MAPS

Many engineers and technicians don't simplify equations with boolean algebra. Instead, they use a method based on *Karnaugh maps*. This section tells you how to construct a Karnaugh map.

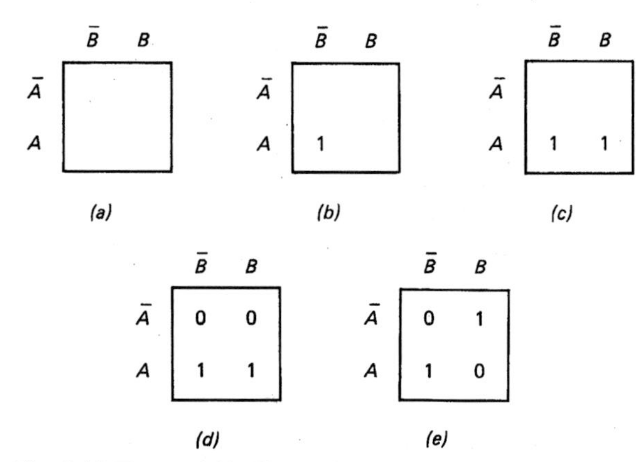

Fig. 5-10 Two-variable Karnaugh map.

Two-Variable Map

Suppose you have a truth table like Table 5-5. Here's how to construct the Karnaugh map. Begin by drawing Fig. 5-10*a*. Note the order of the variables and their complements; the vertical column has $\bar{A}$ followed by A, and the horizontal row has $\bar{B}$ followed by B.

Next, look for output 1s in Table 5-5. The first 1 output to appear is for the input of $A = 1$ and $B = 0$. The fundamental product for this is $A\bar{B}$. Now, enter a 1 on the Karnaugh map as shown in Fig. 5-10*b*. This 1 represents the product $A\bar{B}$ because the 1 is in the A row and the $\bar{B}$ column.

Similarly, Table 5-5 has an output 1 appearing for an input of $A = 1$ and $B = 1$. The fundamental product for this is AB. When you enter a 1 on the Karnaugh map to represent AB, you get the map of Fig. 5-10*c*.

The final step in the construction of the Karnaugh map is to enter 0s in the remaining spaces. Figure 5-10*d* shows how the Karnaugh map looks in its final form.

Here's another example of a two-variable map. In the truth table of Table 5-6, the fundamental products are $\bar{A}B$ and $A\bar{B}$. When 1s are entered on the Karnaugh map for these products and 0s for the remaining spaces, the completed map looks like Fig. 5-10*e*.

TABLE 5-5

A	B	Y
0	0	0
0	1	0
1	0	1
1	1	1

TABLE 5-6

A	B	Y
0	0	0
0	1	1
1	0	1
1	1	0

(a)

	$\overline{C}$	C
$\overline{A}\,\overline{B}$		
$\overline{A}B$		
AB		
$A\overline{B}$		

(b)

	$\overline{C}$	C
$\overline{A}\,\overline{B}$		
$\overline{A}B$	1	
AB	1	1
$A\overline{B}$		

(c)

	$\overline{C}$	C
$\overline{A}\,\overline{B}$	0	0
$\overline{A}B$	1	0
AB	1	1
$A\overline{B}$	0	0

Fig. 5-11 Three-variable Karnaugh map.

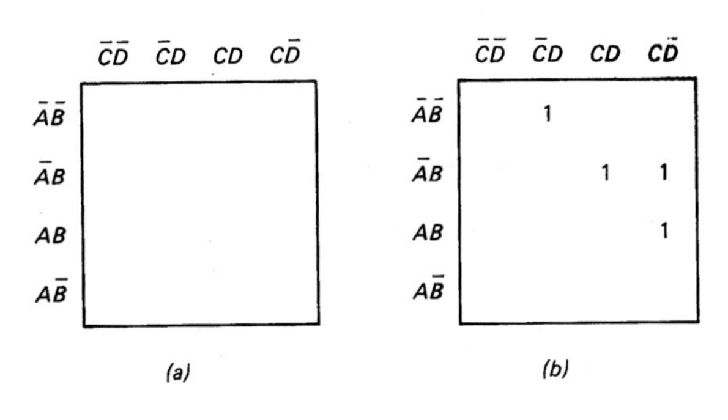

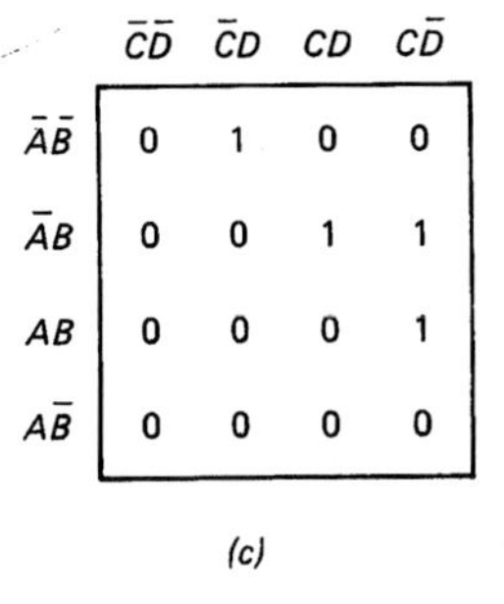

Fig. 5-12 Four-variable Karnaugh map.

Three-Variable Map

Suppose you have a truth table like Table 5-7. Begin by drawing Fig. 5-11*a*. It is especially important to notice the order of the variables and their complements. The vertical column is labeled $\overline{A}\,\overline{B}$, $\overline{A}B$, AB, and $A\overline{B}$. This order is not a binary progression; instead it follows the order of 00, 01, 11, and 10. The reason for this is explained in the derivation of the Karnaugh method; briefly, it's done so that only one variable changes from complemented to uncomplemented form (or vice versa).

Next, look for output 1s in Table 5-7. The fundamental products for these 1 outputs are $\overline{A}B\overline{C}$, $AB\overline{C}$, and ABC. Enter these 1s on the Karnaugh map (Fig. 5-11*b*). The final step is to enter 0s in the remaining spaces (Fig. 5-11*c*). This Karnaugh map is useful because it shows the fundamental products needed for the sum-of-products circuit.

TABLE 5-7

A	B	C	Y
0	0	0	0
0	0	1	0
0	1	0	1
0	1	1	0
1	0	0	0
1	0	1	0
1	1	0	1
1	1	1	1

Four-Variable Map

Many MSI circuits process binary words of 4 bits each (nibbles). For this reason, logic circuits are often designed to handle four variables (or their complements). This is why the four-variable map is the most important.

Here's an example of constructing a four-variable map. Suppose you have the truth table of Table 5-8. The first step is to draw the blank map of Fig. 5-12*a*. Again, notice the progression. The vertical column is labeled $\overline{A}\,\overline{B}$, $\overline{A}B$,

TABLE 5-8

A	B	C	D	Y
0	0	0	0	0
0	0	0	1	1
0	0	1	0	0
0	0	1	1	0
0	1	0	0	0
0	1	0	1	0
0	1	1	0	1
0	1	1	1	1
1	0	0	0	0
1	0	0	1	0
1	0	1	0	0
1	0	1	1	0
1	1	0	0	0
1	1	0	1	0
1	1	1	0	1
1	1	1	1	0

AB, and $A\bar{B}$. The horizontal row is labeled $\bar{C}\bar{D}$, $\bar{C}D$, CD, and $C\bar{D}$.

In Table 5-8 the output 1s have these fundamental products: $\bar{A}\bar{B}\bar{C}D$, $\bar{A}BC\bar{D}$, $\bar{A}BCD$, and $ABC\bar{D}$. After entering 1s on the Karnaugh map, you will have Fig. 5-12*b*. The final step of filling in 0s results in the completed map of Fig. 5-12*c*.

5-5 PAIRS, QUADS, AND OCTETS

There is a way of using the Karnaugh map to get simplified logic circuits. But before you can understand how this is done, you will have to learn the meaning of *pairs, quads,* and *octets*.

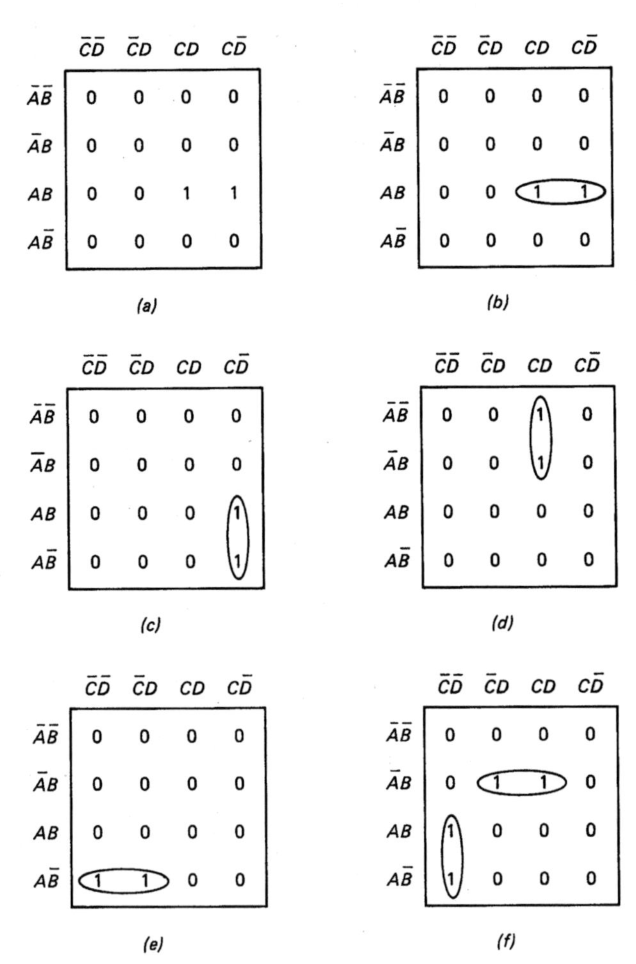

Fig. 5-13 Pairs on a Karnaugh map.

Pairs

The map of Fig. 5-13*a* contains a *pair* of 1s that are horizontally adjacent. The first 1 represents the product $ABCD$; the second 1 stands for the product $ABC\bar{D}$. As we move from the first 1 to the second 1, only one variable goes from uncomplemented to complemented form (D to $\bar{D}$). The other variables don't change form (A, B, and C remain uncomplemented). Whenever this happens, you can eliminate the variable that changes form.

Algebraic Proof

The sum-of-products equation corresponding to Fig. 5-13*a* is

$$Y = ABCD + ABC\bar{D}$$

which factors into

$$Y = ABC(D + \bar{D})$$

Since D is ORed with $\bar{D}$, the equation reduces to

$$Y = ABC$$

A pair of adjacent 1s is like those of Fig. 5-13*a* always means that the sum-of-products equation will have a variable and a complement that drop out.

For easy identification, it is customary to encircle a pair of adjacent 1s, as shown in Fig. 5-13*b*. Then when you look at the map, you can tell at a glance that one variable and its complement will drop out of the boolean equation. In other words, an encircled pair of 1s like those of Fig. 5-13*b* no longer stands for the ORing of two separate products, $ABCD$ and $ABC\bar{D}$. The encircled pair should be visualized instead as representing a single reduced product ABC.

Here's another example. Figure 5-13*c* shows a pair of 1s that are vertically adjacent. These 1s correspond to the product $ABC\bar{D}$ and $A\bar{B}C\bar{D}$. Notice that only one variable changes from uncomplemented to complemented form (B to $\bar{B}$); all other variables retain their original form. Therefore, B and $\bar{B}$ drop out. This means that the encircled pair of Fig. 5-13*c* represents $AC\bar{D}$.

From now on, whenever you see a pair of adjacent 1s, eliminate the variable that goes from complemented to uncomplemented form. A glance at Fig. 5-13*d* indicates that B changes form; therefore, the pair of 1s represents $\bar{A}CD$. Likewise, D changes form in Fig. 5-13*e*; so the pair of 1s stands for $A\bar{B}\bar{C}$.

If more than one pair exists on a Karnaugh map, you can OR the simplified products to get the boolean equation. For instance, the lower pair of Fig. 5-13*f* represents $A\bar{C}\bar{D}$. The upper pair stands for $\bar{A}BD$. The corresponding boolean equation for this map is

$$Y = A\bar{C}\bar{D} + \bar{A}BD$$

The Quad

A *quad* is a group of four 1s that are end to end, as shown in Fig. 5-14*a*, or in the form of a square, as shown in Fig.

5-14*b*. When you see a quad, always encircle it because it leads to a simpler product. In fact, a quad means that two variables and their complements drop out of the boolean equation.

Here's why a quad eliminates two variables. Visualize the four 1s of Fig. 5-14*a* as two pairs (Fig. 5-14*c*). The first pair represents $AB\overline{C}$; the second pair stands for ABC. The boolean equation for these two pairs is

$$Y = AB\overline{C} + ABC$$

This factors into

$$Y = AB(\overline{C} + C)$$

which reduces to

$$Y = AB$$

So the quad of Fig. 5-14*a* represents a product where two variables and their complements drop out.

A similar proof applies to all quads. There's no need to go through the algebra again. Merely determine which variables go from complemented to uncomplemented form; these are the variables that drop out.

For instance, look at the quad of Fig. 5-14*b*. Pick any 1 as a starting point. When you move horizontally, D is the variable that changes form. When you move vertically, B changes form. Therefore, the simplified equation is

$$Y = AC$$

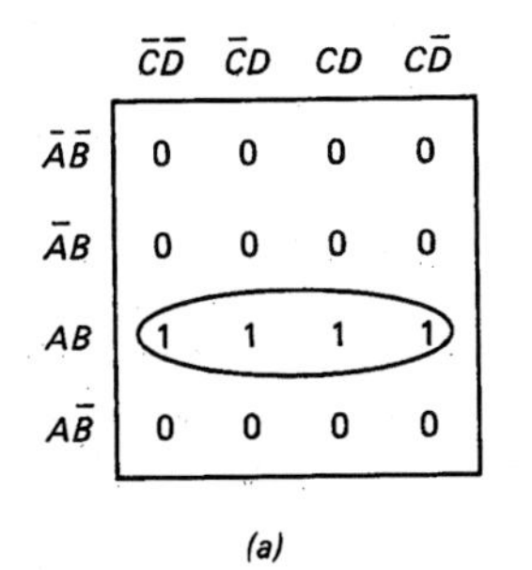

(a)

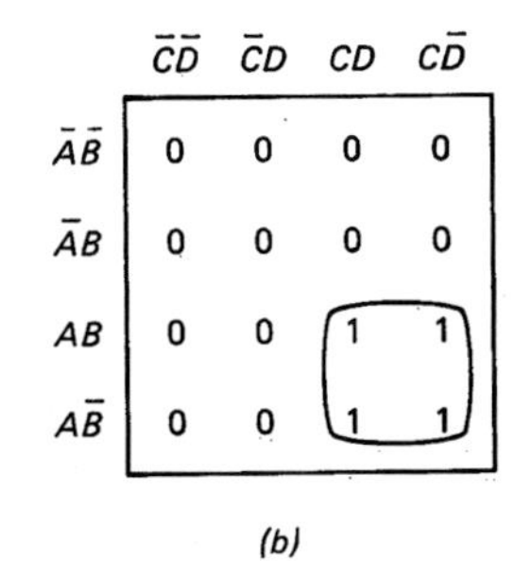

(b)

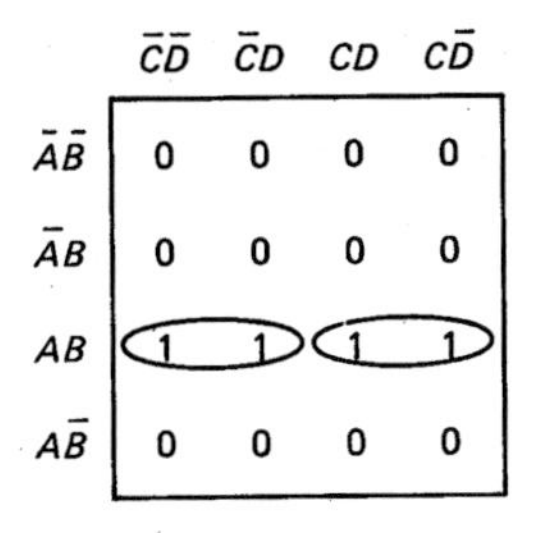

(c)

Fig. 5-14 Quads on a Karnaugh map.

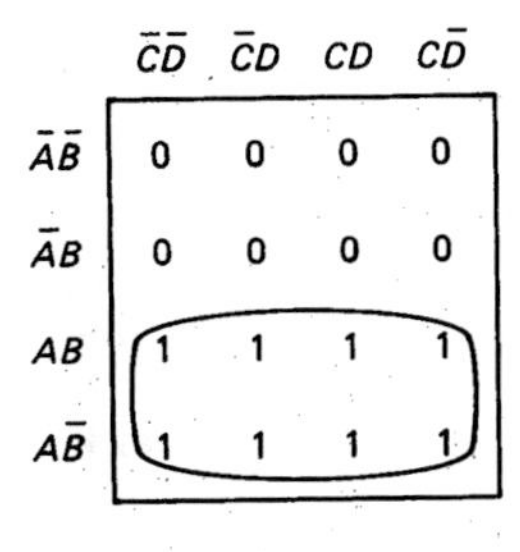

(a)

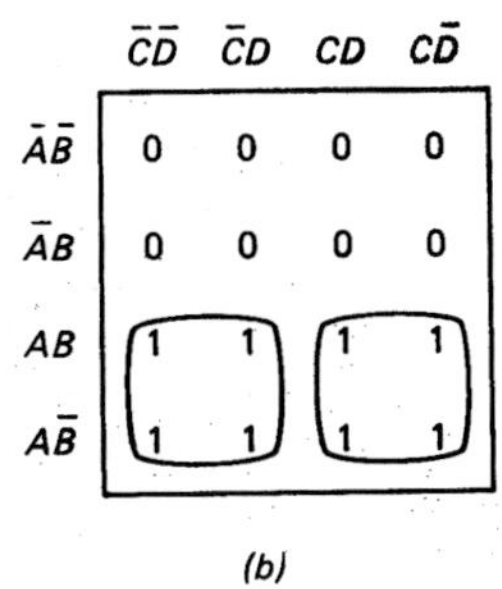

(b)

Fig. 5-15 Octets on a Karnaugh map.

The Octet

An *octet* is a group of eight adjacent 1s like those of Fig. 5-15*a*. An octet always eliminates three variables and their complements. Here's why. Visualize the octet as two quads (Fig. 5-15*b*). The equation for these two quads is

$$Y = A\overline{C} + AC$$

Factoring gives

$$Y = A(\overline{C} + C)$$

But this reduces to

$$Y = A$$

So the octet of Fig. 5-15*a* means that three variables and their complements drop out of the corresponding product.

A similar proof applies to any octet. From now on, don't bother with the algebra. Just step through the 1s of the octet and determine which three variables change form. These are the variables that drop out.

5-6 KARNAUGH SIMPLIFICATIONS

You have seen how a pair eliminates one variable, a quad eliminates two variables, and an octet eliminates three variables. Because of this, you should encircle the octets first, the quads second, and the pairs last. In this way, the greatest simplification takes place.

An Example

Suppose you've translated a truth table into the Karnaugh map shown in Fig. 5-16*a*. Look for octets first. There are none. Next, look for quads. There are two. Finally, look for pairs. There is one. If you do it correctly, you arrive at Fig. 5-16*b*.

The pair represents the simplified product $\overline{A}\overline{B}D$, the lower quad stands for $A\overline{C}$, and the quad on the right

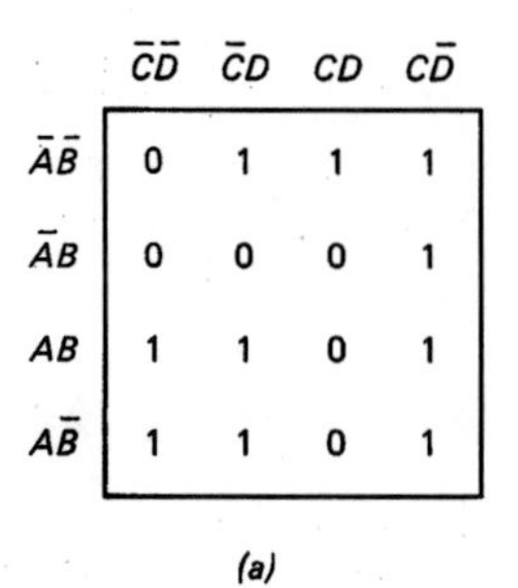

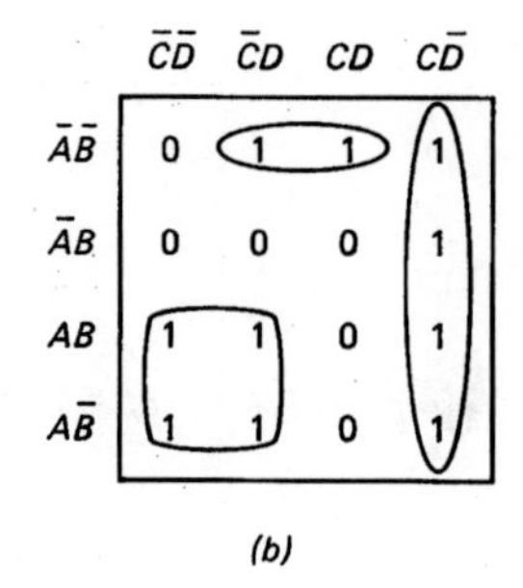

Fig. 5-16

represents $C\bar{D}$. By ORing these simplified products, you get the boolean equation for the map

$$Y = \bar{A}\bar{B}D + A\bar{C} + C\bar{D} \tag{5-24}$$

Overlapping Groups

When you encircle groups, you are allowed to use the same 1 more than once. Figure 5-17*a* illustrates the idea. The simplified equation for the overlapping groups is

$$Y = A + B\bar{C}D \tag{5-25}$$

It is valid to encircle the 1s as shown in Fig. 5-17*b*, but then the isolated 1 results in a more complicated equation:

$$Y = A + \bar{A}B\bar{C}D$$

This requires a more complicated logic circuit than Eq. 5-25. So always overlap groups if possible; that is, use the 1s more than once to get the largest groups you can.

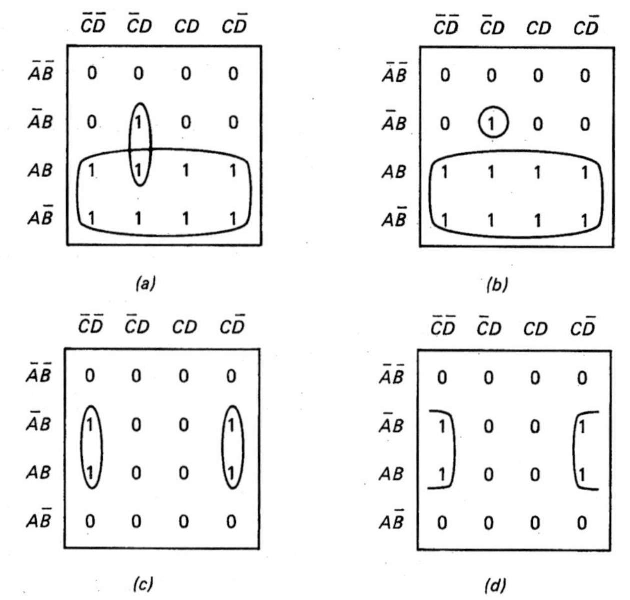

Fig. 5-17 Overlapping and rolling.

Rolling the Map

Another thing to know about is rolling. In Fig. 5-17*c*, the pairs result in the equation

$$Y = B\bar{C}\bar{D} + BC\bar{D} \tag{5-26}$$

Visualize picking up the Karnaugh map and rolling it so that the left side touches the right side. If you're visualizing correctly, you will realize the two pairs actually form a quad. To indicate this, draw half circles around each pair, as shown in Fig. 5-17*d*. From this viewpoint, the quad of Fig. 5-17*d* has the equation

$$Y = B\bar{D} \tag{5-27}$$

Why is rolling valid? Because Eq. 5-26 can be simplified to Eq. 5-27. Here's the proof. Start with Eq. 5-26:

$$Y = B\bar{C}\bar{D} + BC\bar{D}$$

This factors into

$$Y = B\bar{D}(\bar{C} + C)$$

which reduces to

$$Y = B\bar{D}$$

This final equation represents a rolled quad like Fig. 5-17*d*. Therefore, 1s on the edges of a Karnaugh map can be grouped with 1s on opposite edges.

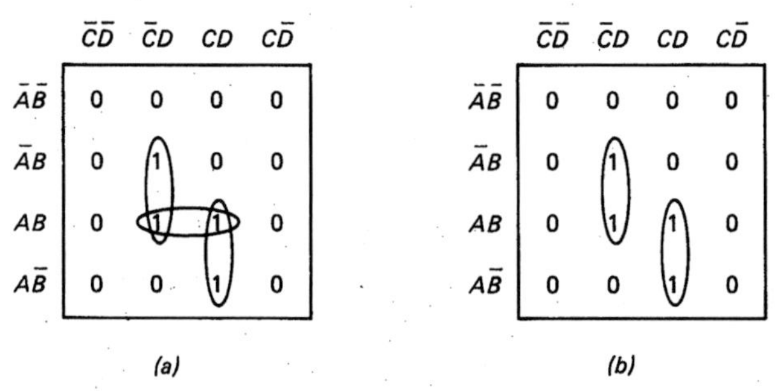

Fig. 5-18 Redundant group.

Redundant Groups

After you finish encircling groups, there is one more thing to do before writing the simplified boolean equation: eliminate any group whose 1s are completely overlapped by other groups. (A group whose 1s are all overlapped by other groups is called a *redundant group*.)

Here is an example. Suppose you have encircled the three pairs shown in Fig. 5-18*a*. The boolean equation then is

$$Y = B\bar{C}D + ABD + ACD$$

At this point, you should check to see if there are any redundant groups. Notice that the 1s in the inner pair are completely overlapped by the outside pairs. Because of this, the inner pair is a redundant pair and can be eliminated to get the simpler map of Fig. 5-18*b*. The equation for this map is

$$Y = B\overline{C}D + ACD$$

Since this is a simpler equation, it means a simpler logic circuit. This is why you should eliminate redundant groups if they exist.

Summary

Here's a summary of how to use the Karnaugh map to simplify logic circuits:

1. Enter a 1 on the Karnaugh map for each fundamental product that corresponds to 1 output in the truth table. Enter 0s elsewhere.
2. Encircle the octets, quads, and pairs. Remember to roll and overlap to get the largest groups possible.
3. If any isolated 1s remain, encircle them.
4. Eliminate redundant groups if they exist.
5. Write the boolean equation by ORing the products corresponding to the encircled groups.
6. Draw the equivalent logic circuit.

EXAMPLE 5-1

What is the simplified boolean equation for the Karnaugh map of Fig. 5-19*a*?

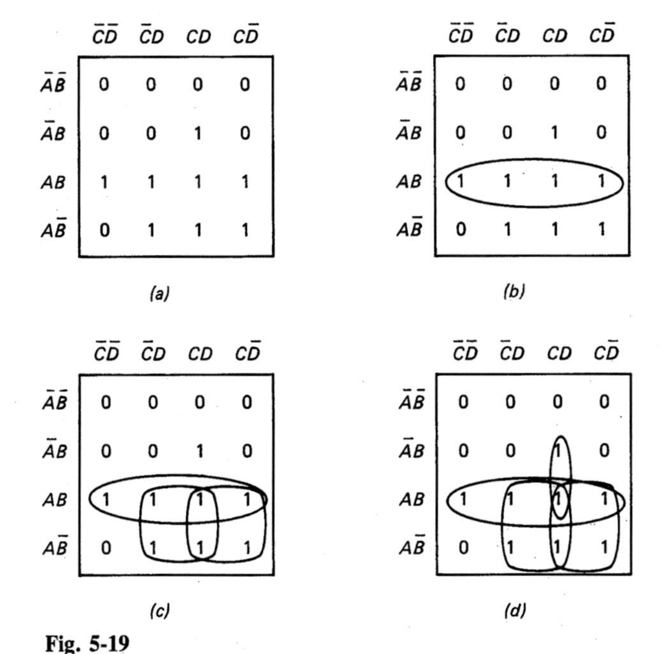

Fig. 5-19

SOLUTION

There are no octets, but there is a quad, as shown in Fig. 5-19*b*. By overlapping we can find two more quads (Fig. 5-19*c*). Finally, overlapping gives us the pair of Fig. 5-19*d*.

The horizontal quad of Fig. 5-19*d* corresponds to a simplified product of *AB*. The square quad on the right corresponds to *AC*, while the one on the left stands for *AD*. The pair represents *BCD*. By ORing these products we get the simplified equation

$$Y = AB + AC + AD + BCD \qquad (5\text{-}28)$$

Figure 5-20 shows the equivalent logic circuit.

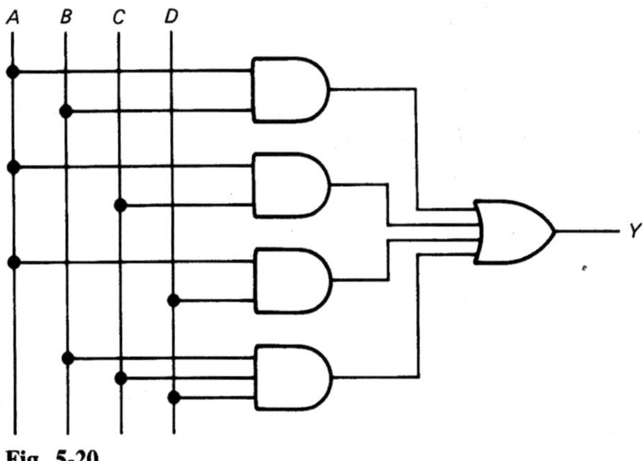

Fig. 5-20

EXAMPLE 5-2

As you know from Chap. 4, the NAND gate is the least expensive gate in the 7400 series. Because of this, AND-OR circuits are usually built as equivalent NAND-NAND circuits.

Convert the AND-OR circuit of Fig. 5-20 to a NAND-NAND circuit using 7400-series devices.

SOLUTION

Replace each AND gate of Fig. 5-20 by a NAND gate and replace the final OR gate by a NAND gate. Figure 5-21 is the De Morgan equivalent of Fig. 5-20. As shown, we can build the circuit with a 7400, a 7410, and a 7420.

5-7 DON'T-CARE CONDITIONS

Sometimes, it doesn't matter what the output is for a given input word. To indicate this, we use an X in the truth table instead of a 0 or a 1. For instance, look at Table 5-9. The

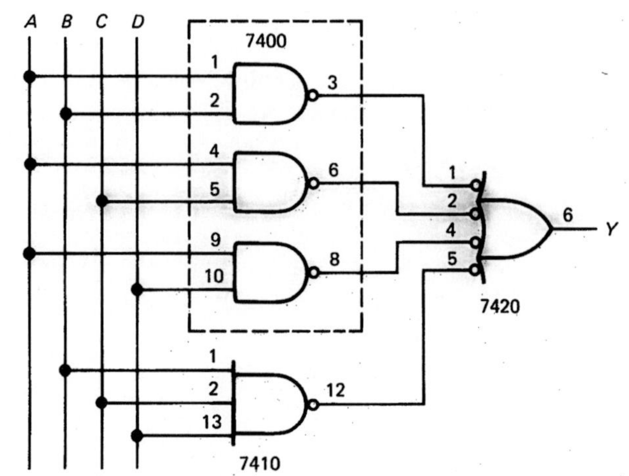

Fig. 5-21 NAND-NAND circuit using TTL gates.

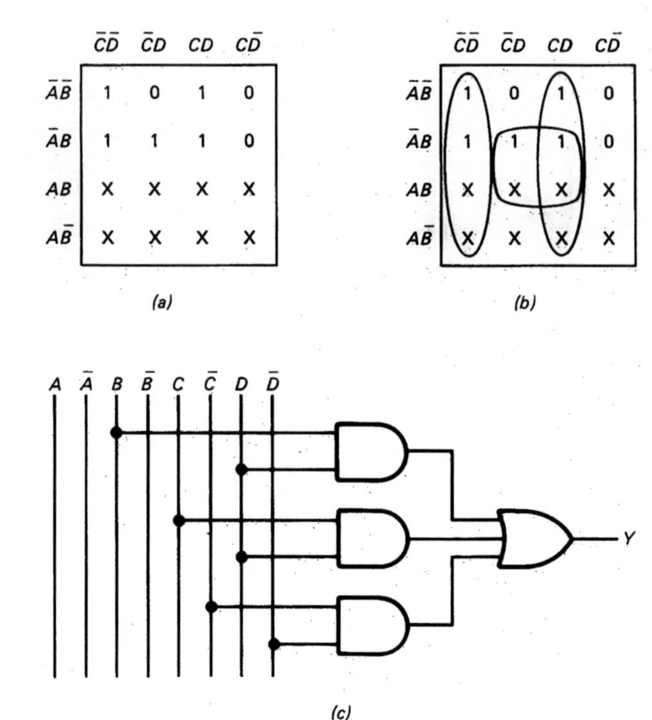

Fig. 5-22 Don't cares.

output is an X for any input word from 1000 through 1111. The X's are called *don't cares* because they can be treated either as 0s or 1s, whichever leads to a simpler circuit.

Figure 5-22*a* shows the Karnaugh map for Table 5-9. X's are used for $A\overline{B}\overline{C}\overline{D}$, $A\overline{B}\overline{C}D$, $A\overline{B}C\overline{D}$, $A\overline{B}CD$, $AB\overline{C}\overline{D}$, $AB\overline{C}D$, $ABC\overline{D}$, and $ABCD$ because these are don't cares in the truth table. Figure 5-22*b* shows the most efficient way to encircle the groups. Notice two crucial ideas. First, we visualize all X's as 1s and try to form the largest groups that include the real 1s. This gives us three quads. Second, we visualize all remaining X's as 0s. In this way, the X's are used to the best advantage. We are free to do this because the don't cares can be either 0s or 1s, whichever we prefer.

TABLE 5-9

A	*B*	*C*	*D*	*Y*
0	0	0	0	1
0	0	0	1	0
0	0	1	0	0
0	0	1	1	1
0	1	0	0	1
0	1	0	1	1
0	1	1	0	0
0	1	1	1	1
1	0	0	0	X
1	0	0	1	X
1	0	1	0	X
1	0	1	1	X
1	1	0	0	X
1	1	0	1	X
1	1	1	0	X
1	1	1	1	X

Figure 5-22*b* implies the simplified boolean equation

$$Y = BD + \overline{C}\overline{D} + CD$$

Figure 5-22*c* is the simplified logic circuit. This AND-OR network has nine input gate leads.

EXAMPLE 5-3

Recall that BCD numbers express each decimal digit as a nibble: 0 to 9 are encoded as 0000 to 1001. Especially important, nibbles 1010 to 1111 are never used in a BCD system.

Table 5-10 shows a truth table for use in a BCD system. As you see, don't cares appear for 1010 through 1111. Construct the Karnaugh map and show the simplified logic circuit.

SOLUTION

Figure 5-23*a* illustrates the Karnaugh map. The largest group we can form is the pair shown in Fig. 5-23*b*. The boolean equation is

$$Y = BCD$$

Figure 5-23*c* is the simplified logic circuit.

TABLE 5-10

A	B	C	D	Y
0	0	0	0	0
0	0	0	1	0
0	0	1	0	0
0	0	1	1	0
0	1	0	0	0
0	1	0	1	0
0	1	1	0	0
0	1	1	1	1
1	0	0	0	0
1	0	0	1	0
1	0	1	0	X
1	0	1	1	X
1	1	0	0	X
1	1	0	1	X
1	1	1	0	X
1	1	1	1	X

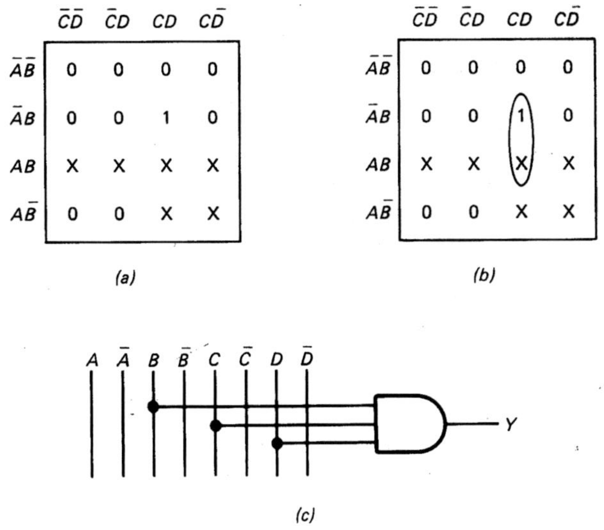

	$\bar{C}\bar{D}$	$\bar{C}D$	CD	$C\bar{D}$
$\bar{A}\bar{B}$	0	0	0	0
$\bar{A}B$	0	0	1	0
AB	X	X	X	X
$A\bar{B}$	0	0	X	X

(a)

	$\bar{C}\bar{D}$	$\bar{C}D$	CD	$C\bar{D}$
$\bar{A}\bar{B}$	0	0	0	0
$\bar{A}B$	0	0	1	0
AB	X	X	X	X
$A\bar{B}$	0	0	X	X

(b)

A $\bar{A}$ B $\bar{B}$ C $\bar{C}$ D $\bar{D}$ Y

(c)

Fig. 5-23 Don't cares in a BCD system.

GLOSSARY

bus A group of wires carrying digital signals.
don't care An output that may be either low or high without affecting the operation of the system.
fundamental product The logical product of variables and complements that produces a high output for a given input condition.
Karnaugh map A graphical display of the fundamental products in a truth table.
octet A group of eight adjacent 1s on a Karnaugh map.
pair A group of two adjacent 1s on a Karnaugh map. These 1s may be horizontally or vertically aligned.
quad A group of four adjacent 1s on a Karnaugh map.
redundant group A group of 1s on a Karnaugh map all of which are overlapped by other groups.
sum-of-products circuit An AND-OR circuit obtained by ORing the fundamental products that produce output 1s in a truth table.

SELF-TESTING REVIEW

Read each of the following and provide the missing words. Answers appear at the beginning of the next question.

1. Digital design often starts by constructing a ________ table. By ORing the ________ products, you get a sum-of-products equation.
2. (*truth, fundamental*) A preliminary guide for comparing the simplicity of logic circuits is to count the number of input ________ leads.
3. (*gate*) A bus is a group of ________ carrying digital signals. In the typical microcomputer, the microprocessor, memory, and I/O units communicate via buses.
4. (*wires*) One way to simplify the sum-of-products equation is to use boolean algebra. Another way is the ________ map.
5. (*Karnaugh*) A pair eliminates one variable, a ________ eliminates two variables, and an octet eliminates ________ variables. Because of this, you should encircle the ________ first, the quads next, and the pairs last.
6. (*quad, three, octets*) NAND-NAND circuits are equivalent to AND-OR circuits. This is important because ________ gates are the least expensive gates in the 7400 series.
7. (*NAND*) When a truth table has don't cares, we enter X's on the Karnaugh map. These can be treated as 0s or 1s, whichever leads to a simpler logic circuit.

PROBLEMS

5-1. What are the fundamental products for each of the inputs words ABCD = 0010, ABCD = 1101, ABCD = 1110?

5-2. A truth table has output 1s for each of these inputs:
a. ABCD = 0011
b. ABCD = 0101
c. ABCD = 1000
d. ABCD = 1101
What are the fundamental products?

5-3. Draw the logic circuit for this boolean equation:

$$Y = \overline{A}\,\overline{B}\,\overline{C}D + A\overline{B}\,\overline{C}D + AB\overline{C}D + ABC\overline{D}$$

5-4. Output 1s appear in the truth table for these input conditions: ABCD = 0001, ABCD = 0110, and ABCD = 1110. What is the sum-of-products equation?

5-5. Draw the AND-OR circuit for

$$Y = A\overline{B}\,\overline{C}\,\overline{D} + AB\overline{C}\,\overline{D} + ABCD$$

How many input gate leads does this circuit have?

5-6. A truth table has output 1s for these inputs: ABCD = 0011, ABCD = 0110, ABCD = 1001, and ABCD = 1110. Draw the Karnaugh map showing the fundamental products.

5-7. A truth table has four input variables. The first eight outputs are 0s, and the last eight outputs are 1s. Draw the Karnaugh map.

5-8. Draw the Karnaugh map for the Y_3 output of Table 5-11. Simplify as much as possible; then draw the logic circuit.

5-9. Use the Karnaugh map to work out the simplified logic circuit for the Y_2 output of Table 5-11.

5-10. Repeat Prob. 5-9 for the Y_1 output.

5-11. Repeat Prob. 5-9 for the Y_0 output.

5-12. Use the Karnaugh map to work out the simplified logic circuit for the Y_3 output of Table 5-12.

5-13. Repeat Prob. 5-12 for the Y_2 output.

5-14. Repeat Prob. 5-12 for the Y_1 output.

5-15. Repeat Prob. 5-12 for Y_0 output.

5-16. A + 0 = ?

5-17. A · 1 = ?

5-18. A + 1 = ?

5-19. A · 0 = ?

5-20. Use the duality theorem to derive another boolean relation from:

$$A + \overline{A}B = A + B$$

5.21. Use the commutative law to complete the following equations.
a. A + B =
b. AB =

5.22 Use the associative law to complete the following equations.
a. A + (B + C) =
b. A(BC) =

5.23 Use the distributive law to complete the equation A(B + C) =

TABLE 5-11

A	B	C	D	Y_3	Y_2	Y_1	Y_0
0	0	0	0	1	0	1	0
0	0	0	1	0	1	0	1
0	0	1	0	0	1	1	1
0	0	1	1	1	0	0	1
0	1	0	0	0	0	1	1
0	1	0	1	1	0	0	0
0	1	1	0	1	1	1	0
0	1	1	1	1	1	1	1
1	0	0	0	0	0	0	0
1	0	0	1	0	0	0	1
1	0	1	0	1	0	1	1
1	0	1	1	0	1	0	0
1	1	0	0	0	1	1	0
1	1	0	1	1	0	1	0
1	1	1	0	1	1	0	0
1	1	1	1	1	1	0	1

TABLE 5-12

A	B	C	D	Y_3	Y_2	Y_1	Y_0
0	0	0	0	1	0	1	0
0	0	0	1	0	1	0	1
0	0	1	0	0	1	1	1
0	0	1	1	1	0	0	1
0	1	0	0	0	0	1	1
0	1	0	1	1	0	0	0
0	1	1	0	1	1	1	0
0	1	1	1	1	1	1	1
1	0	0	0	0	0	0	0
1	0	0	1	0	0	0	1
1	0	1	0	X	X	X	X
1	0	1	1	X	X	X	X
1	1	0	0	X	X	X	X
1	1	0	1	X	X	X	X
1	1	1	0	X	X	X	X
1	1	1	1	X	X	X	X

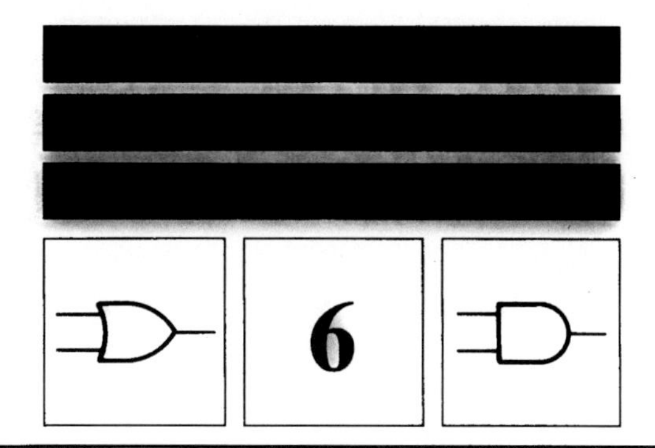

ARITHMETIC-LOGIC UNITS

The arithmetic-logic unit (ALU) is the number-crunching part of a computer. This means not only arithmetic operations but logic as well (OR, AND, NOT, and so forth). In this chapter you will learn how the ALU adds and subtracts binary numbers. Later chapters will discuss the logic operations.

6-1 BINARY ADDITION

ALUs don't process decimal numbers; they process binary numbers. Before you can understand the circuits inside an ALU, you must learn how to add binary numbers. There are five basic cases that must be understood before going on.

Case 1

When no pebbles are added to no pebbles, the total is no pebbles. As a word equation,

None + none = none

With binary numbers, this equation is written as

$$0 + 0 = 0$$

Case 2

If no pebbles are added to one pebble, the total is one pebble:

None + ● = ●

In terms of binary numbers,

$$0 + 1 = 1$$

Case 3

Addition is commutative. This means you can transpose the numbers of the preceding case to get

● + none = ●

or

$$1 + 0 = 1$$

Case 4

Next, one pebble added to one pebble gives two pebbles:

As a binary equation,

$$1 + 1 = 10$$

To avoid confusion with decimal numbers, read this as "one plus one equals one-zero." An alternative way of reading the equation is "one plus one equals zero, carry one."

Case 5

One pebble plus one pebble plus one pebble gives a total of three pebbles:

● + ● + ● = ●●●

The binary equation is

$$1 + 1 + 1 = 11$$

Read this as "one plus one plus one equals one-one." Alternatively, "one plus one plus one equals one, carry one."

Rules to Remember

The foregoing cases are all you need for more complicated binary addition. Therefore, memorize these five rules:

$$0 + 0 = 0 \qquad (6\text{-}1)$$
$$0 + 1 = 1 \qquad (6\text{-}2)$$
$$1 + 0 = 1 \qquad (6\text{-}3)$$
$$1 + 1 = 10 \qquad (6\text{-}4)$$
$$1 + 1 + 1 = 11 \qquad (6\text{-}5)$$

Larger Binary Numbers

Column-by-column addition applies to binary numbers as well as decimal. For example, suppose you have this problem in binary addition:

$$\begin{array}{r} 11100 \\ +\ 11010 \\ \hline ? \end{array}$$

Start with the least significant column to get

$$\begin{array}{r} 11100 \\ +\ 11010 \\ \hline 0 \end{array}$$

Here, 0 + 0 gives 0.

Next, add the bits of the second column as follows:

$$\begin{array}{r} 11100 \\ +\ 11010 \\ \hline 10 \end{array}$$

This time, 0 + 1 results in 1.

The third column gives

$$\begin{array}{r} 11100 \\ +\ 11010 \\ \hline 110 \end{array}$$

In this case, 1 + 0 produces 1.

The fourth column results in

$$\begin{array}{r} 11100 \\ +\ 11010 \\ \hline 0110 \end{array} \qquad \text{(carry 1)}$$

As you see, 1 + 1 equals 0 with a carry of 1.

Finally, the last column gives

$$\begin{array}{r} 11100 \\ +\ 11010 \\ \hline 110110 \end{array}$$

Here, 1 + 1 + 1 (carry) produces 11, recorded as 1 with a carry to the next higher column.

EXAMPLE 6-1

Add the binary numbers 01010111 and 00110101.

SOLUTION

This is the problem:

$$\begin{array}{r} 01010111 \\ +\ 00110101 \\ \hline ? \end{array}$$

If you add the bits column by column as previously demonstrated, you will get

$$\begin{array}{r} 01010111 \\ +\ 00110101 \\ \hline 10001100 \end{array}$$

Expressed in hexadecimal numbers, the foregoing addition is

$$\begin{array}{r} 57 \\ +\ 35 \\ \hline 8C \end{array}$$

For clarity, we can use subscripts:

$$\begin{array}{r} 57_{16} \\ +\ 35_{16} \\ \hline 8C_{16} \end{array}$$

In microprocessor work, it is more convenient to use the letter H to signify hexadecimal numbers. In other words, the usual way to express the foregoing addition is

$$\begin{array}{r} 57H \\ +\ 35H \\ \hline 8CH \end{array}$$

6-2 BINARY SUBTRACTION

To subtract binary numbers, we need to discuss four cases.

Case 1: $0 - 0 = 0$
Case 2: $1 - 0 = 1$
Case 3: $1 - 1 = 0$
Case 4: $10 - 1 = 1$

The last result represents

●● − ● = ●

which makes sense.

To subtract larger binary numbers, subtract column by column, borrowing from the next higher column when necessary. For instance, in subtracting 101 from 111, proceed like this:

$$\begin{array}{r} 7 \\ -\ 5 \\ \hline 2 \end{array} \qquad \begin{array}{r} 111 \\ -\ 101 \\ \hline 010 \end{array}$$

Starting on the right, 1 − 1 gives 0; then, 1 − 0 is 1; finally, 1 − 1 is 0.

Here is another example: subtract 1010 from 1101.

$$\begin{array}{r} 13 \\ -\ 10 \\ \hline 3 \end{array} \qquad \begin{array}{r} 1101 \\ -\ 1010 \\ \hline 0011 \end{array}$$

In the least significant column, 1 − 0 is 1. In the second column, we have to borrow from the next higher column; then, 10 − 1 is 1. In the third column, 0 (after borrow) − 0 is 0. In the fourth column, 1 − 1 = 0.

Direct subtraction like the foregoing has been used in computers; however, it is possible to subtract in a different way. Later sections of this chapter will show you how.

6-3 HALF-ADDERS

Figure 6-1 is a *half-adder*, a logic circuit that adds 2 bits. Notice the outputs: *SUM* and *CARRY*. The boolean equations for these outputs are

$$SUM = A \oplus B \qquad (6\text{-}6)$$
$$CARRY = AB \qquad (6\text{-}7)$$

The *SUM* output is *A* XOR *B*; the *CARRY* output is *A* AND *B*. Therefore, *SUM* is a 1 when *A* and *B* are different; *CARRY* is a 1 when *A* and *B* are 1s.

Table 6-1 summarizes the operation. When *A* and *B* are 0s, the *SUM* is 0 with a *CARRY* of 0. When *A* is 0 and *B* is 1, the *SUM* is 1 with a *CARRY* of 0. When *A* is 1 and *B* is 0, the *SUM* equals 1 with a *CARRY* of 0. Finally, when *A* is 1 and *B* is 1, the *SUM* is 0 with a *CARRY* of 1.

The logic circuit of Fig. 6-1 does electronically what we do mentally when we add 2 bits. Applications for the half-adder are limited. What we need is a circuit that can add 3 bits at a time.

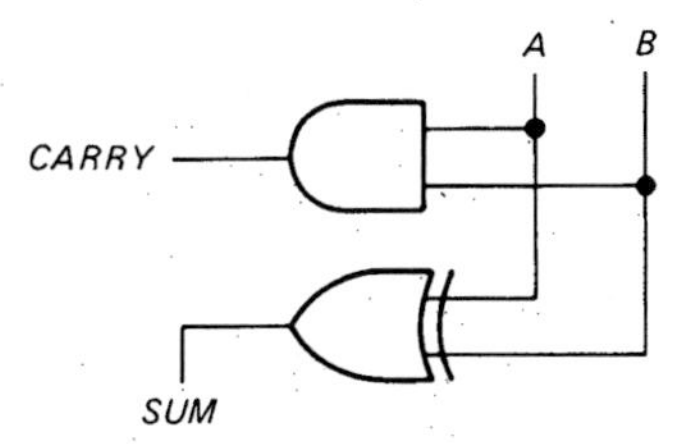

Fig. 6-1 Half-adder.

TABLE 6-1. HALF-ADDER

A	*B*	*CARRY*	*SUM*
0	0	0	0
0	1	0	1
1	0	0	1
1	1	1	0

6-4 FULL ADDERS

Figure 6-2 shows a *full adder*, a logic circuit that can add 3 bits. Again there are two outputs, *SUM* and *CARRY*. The boolean equations are

$$SUM = A \oplus B \oplus C \qquad (6\text{-}8)$$
$$CARRY = AB + AC + BC \qquad (6\text{-}9)$$

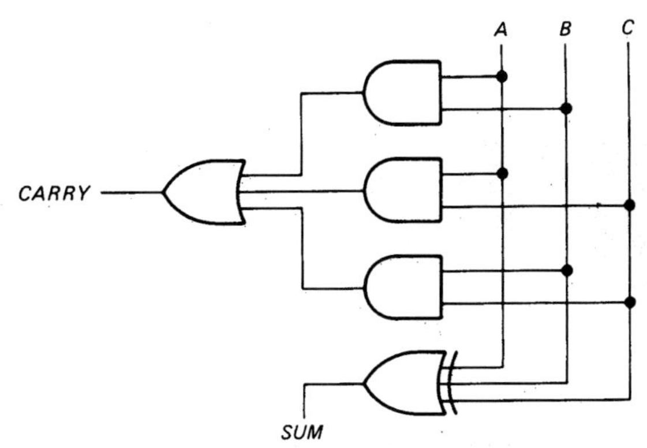

Fig. 6-2 Full adder.

In this case, *SUM* equals *A* XOR *B* XOR *C*; *CARRY* equals *AB* OR *AC* OR *BC*. Therefore, *SUM* is 1 when the number of input 1s is odd; *CARRY* is a 1 when two or more inputs are 1s.

Table 6-2 summarizes the circuit action. *A*, *B*, and *C* are the bits being added. If you check each entry, you will see that the circuit adds 3 bits at a time and comes up with the correct answer.

TABLE 6-2. FULL ADDER

A	*B*	*C*	*CARRY*	*SUM*
0	0	0	0	0
0	0	1	0	1
0	1	0	0	1
0	1	1	1	0
1	0	0	0	1
1	0	1	1	0
1	1	0	1	0
1	1	1	1	1

Here's the point. The circuit of Fig. 6-2 does electronically what we do mentally when we add 3 bits. The full adder can be cascaded to add large binary numbers. The next section tells you how.

6-5 BINARY ADDERS

Figure 6-3 shows a *binary adder,* a logic circuit that can add two binary numbers. The block on the right (labeled HA) represents a half-adder. The inputs are A_0 and B_0; the outputs are S_0 (*SUM*) and C_1 (*CARRY*). All other blocks are full adders (abbreviated FA). Each of these full adders has three inputs (A_n, B_n, and C_n) and two outputs.

The circuit adds two binary numbers. In other words, it carries out the following addition:

$$\begin{array}{r} A_3A_2A_1A_0 \\ +\ B_3B_2B_1B_0 \\ \hline C_4S_3S_2S_1S_0 \end{array}$$

Here's an example. Suppose **A** = 1100 and **B** = 1001. Then the problem is

$$\begin{array}{r} 1100 \\ +\ 1001 \\ \hline ? \end{array}$$

Figure 6-4 shows the binary adder with the same inputs, 1100 and 1001. The half-adder produces a sum of 1 and carry of 0, the first full adder produces a sum of 0 and a carry of 0, the second full adder produces a sum of 1 and a carry of 0, and the third full adder produces a sum of 0 and a carry of 1. The overall output is 10101, the same answer we would get with pencil and paper.

By using more full adders, we can build binary adders of any length. For example, to add 16-bit numbers, we need 1 half-adder and 15 full adders. From now on, we will use the abbreviated symbol of Fig. 6-5 to represent a binary adder of any length. Notice the solid arrows, the standard way to indicate words in motion. In Fig. 6-5, words **A** and **B** are added to get a sum of **S** plus a final *CARRY*.

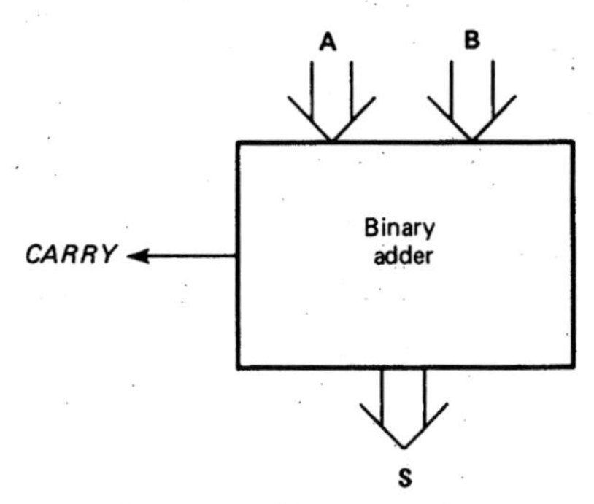

Fig. 6-5 Symbol for binary adder.

EXAMPLE 6-2

Find the output in Fig. 6-5 if the two input words are

A = 0000 0001 0000 1100
B = 0000 0000 0100 1001

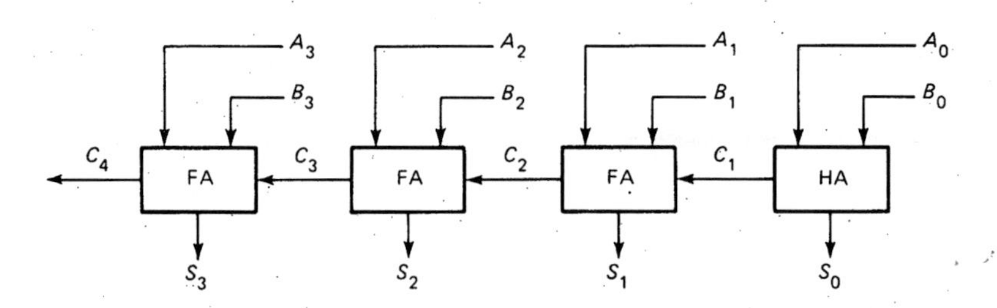

Fig. 6-3 Binary adder.

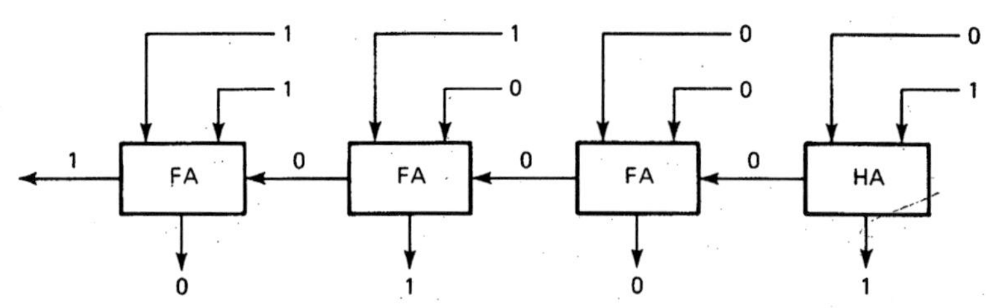

Fig. 6-4 Adding 12 and 9 to get 21.

SOLUTION

The binary adder adds the two inputs to get

$$\begin{array}{r} 0000\ 0001\ 0000\ 1100 \\ +\ \underline{0000\ 0000\ 0100\ 1001} \\ 0000\ 0001\ 0101\ 0101 \end{array}$$

In hexadecimal form, the foregoing addition is

$$\begin{array}{r} 010CH \\ +\ \underline{0049H} \\ 0155H \end{array}$$

6-6 SIGNED BINARY NUMBERS

The negative decimal numbers are -1, -2, -3, and so on. One way to code these as binary numbers is to convert the *magnitude* (1, 2, 3, . . .) to its binary equivalent and prefix the sign. With this approach, -1, -2, and -3 becomes -001, -010, and -011. It's customary to use 0 for the + sign and 1 for the − sign. Therefore, -001, -010, and -011 are coded as 1001, 1010, and 1011.

The foregoing numbers have the *sign bit* followed by the magnitude bits. Numbers in this form are called *signed binary numbers* or *sign-magnitude numbers*. For larger decimal numbers you need more than 4 bits. But the idea is still the same: the leading bit represents the sign and the remaining bits stand for the magnitude.

EXAMPLE 6-3

Express each of the following as 16-bit signed binary numbers.

a. +7
b. −7
c. +25
d. −25

SOLUTION

a. $+7 = 0000\ 0000\ 0000\ 0111$
b. $-7 = 1000\ 0000\ 0000\ 0111$
c. $+25 = 0000\ 0000\ 0001\ 1001$
d. $-25 = 1000\ 0000\ 0001\ 1001$

No subscripts are used in these equations because it's clear from the context that decimal numbers are being expressed in binary form. Nevertheless, you can use subscripts if you prefer. The first equation can be written as

$$+7_{10} = 0000\ 0000\ 0000\ 0111_2$$

the next equation as

$$-7_{10} = 1000\ 0000\ 0000\ 0111_2$$

and so forth.

EXAMPLE 6-4

Convert the following signed binary numbers to decimal numbers:

a. 0000 0000 0000 1001
b. 1000 0000 0000 1111
c. 1000 0000 0011 0000
d. 0000 0000 1010 0101

SOLUTION

As usual, the leading bit gives the sign and the remaining bits give the magnitude.

a. $0000\ 0000\ 0000\ 1001 = +9$
b. $1000\ 0000\ 0000\ 1111 = -15$
c. $1000\ 0000\ 0011\ 0000 = -48$
d. $0000\ 0000\ 1010\ 0101 = +165$

6-7 2's COMPLEMENT

Sign-magnitude numbers are easy to understand, but they require too much hardware for addition and subtraction. This has led to the widespread use of complements for binary arithmetic.

Definition

Recall that a high invert signal to a controlled inverter produces the 1's complement. For instance, if

$$\mathbf{A} = 0111 \qquad (6\text{-}10a)$$

the 1's complement is

$$\overline{\mathbf{A}} = 1000 \qquad (6\text{-}10b)$$

The 2's complement is defined as the new word obtained by adding 1 to 1's complement. As an equation,

$$\mathbf{A}' = \overline{\mathbf{A}} + 1 \qquad (6\text{-}11)$$

where $\mathbf{A}'$ = 2's complement
$\overline{\mathbf{A}}$ = 1's complement

Here are some examples. If

$$\mathbf{A} = 0111$$

the 1's complement is

$$\overline{A} = 1000$$

and the 2's complement is

$$A' = 1001$$

In terms of a binary odometer, the 2's complement is the next reading after the 1's complement.

Another example. If

$$A = 0000\ 1000$$

then

$$\overline{A} = 1111\ 0111$$

and

$$A' = 1111\ 1000$$

Double Complement

If you take the 2's complement twice, you get the original word back. For instance, if

$$A = 0111$$

the 2's complement is

$$A' = 1001$$

If you take the 2's complement of this, you get

$$A'' = 0111$$

which is the original word.

In general, this means that

$$A'' = A \qquad (6\text{-}12)$$

Read this as "the double complement of **A** equals **A**." Because of this property, the 2's complement of a binary number is equivalent to the negative of a decimal number. This idea is explained in the following discussion.

Back to the Odometer

Chapter 1 used an odometer to introduce binary numbers. The discussion was about positive numbers only. But odometer readings can also indicate negative numbers. Here's how.

If a car has a binary odometer, all bits eventually reset to 0s. A few readings before and after a complete reset look like this:

1101
1110
1111
0000 (RESET)
0001
0010
0011

1101 is the reading 3 miles before reset, 1110 occurs 2 miles before reset, and 1111 indicates 1 mile before reset. Then, 0001 is the reading 1 mile after reset, 0010 occurs 2 miles after reset, and 0011 indicates 3 miles after reset.

"Before" and "after" are synonymous with "negative" and "positive." Figure 6-6 illustrates this idea with the number line learned in basic algebra: 0 marks the origin, positive decimal numbers are on the right, and negative decimal numbers are on the left. The odometer readings are the binary equivalent of positive and negative decimal numbers: 1101 is the binary equivalent of −3, 1110 stands for −2, 1111 for −1; 0000 for 0; 0001 for +1; 0010 for +2, and 0011 for +3.

The odometer readings of Fig. 6-6 demonstrate how positive and negative numbers are stored in a typical microcomputer. Positive decimal numbers are expressed in sign-magnitude form, but negative decimal numbers are represented as 2's complements. As before, positive numbers have a leading sign bit of 0, and negative numbers have a leading sign bit of 1.

2's Complement Same as Decimal Sign Change

Taking the 2's complement of a binary number is the same as changing the sign of the equivalent decimal number. For example, if

$$A = 0001 \qquad (+1 \text{ in Fig. 6-6})$$

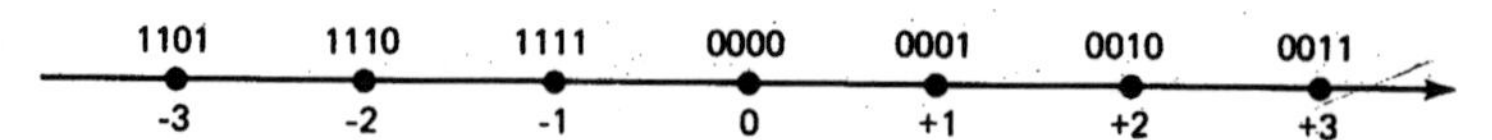

Fig. 6-6 Decimal numbers and odometer readings.

taking the 2's complement gives

$$A' = 1111 \quad (-1 \text{ in Fig. 6-6})$$

Similarly, if

$$A = 0010 \quad (+2 \text{ in Fig. 6-6})$$

then the 2's complement is

$$A' = 1110 \quad (-2 \text{ in Fig. 6-6})$$

Again, if

$$A = 0011 \quad (+3 \text{ in Fig. 6-6})$$

the 2's complement is

$$A' = 1101 \quad (-3 \text{ in Fig. 6-6})$$

The same principle applies to binary numbers of any length: taking the 2's complement of any binary number is the same as changing the sign of the equivalent decimal number. As will be shown later, this property allows us to use a binary adder for both addition and subtraction.

Summary

Here are the main things to remember about 2's complement representation:

1. The leading bit is the sign bit; 0 for plus, 1 for minus.
2. Positive decimal numbers are in sign-magnitude form.
3. Negative decimal numbers are in 2's-complement form.

EXAMPLE 6-5

What is the 2's complement of this word?

$$A = 0011\ 0101\ 1001\ 1100$$

SOLUTION

The 2's complement is

$$A' = 1100\ 1010\ 0110\ 0100$$

EXAMPLE 6-6

What is the binary form of +5 and −5 in 2's-complement representation? Express the answers as 8-bit numbers.

SOLUTION

Decimal +5 is expressed in sign-magnitude form:

$$+5 = 0000\ 0101$$

On the other hand, −5 appears as the 2's complement:

$$-5 = 1111\ 1011$$

EXAMPLE 6-7

What is the 2's-complement representation of −24 in a 16-bit microcomputer?

SOLUTION

Start with the positive form:

$$+24 = 0000\ 0000\ 0001\ 1000$$

Then take the 2's complement to get the negative form:

$$-24 = 1111\ 1111\ 1110\ 1000$$

EXAMPLE 6-8

What decimal number does this represent in 2's-complement representation?

$$1111\ 0001$$

SOLUTION

Start by taking the 2's complement to get

$$0000\ 1111$$

This represents +15. Therefore, the original number is

$$1111\ 0001 = -15$$

6-8 2's-COMPLEMENT ADDER-SUBTRACTER

Early computers used signed binary for both positive and negative numbers. This led to complicated arithmetic circuits. Then, engineers discovered that the 2's-complement representation could greatly simplify arithmetic hardware.

This is why 2's-complement adder-subtracters are now the most widely used arithmetic circuits.

Addition

Figure 6-7 shows a 2's-complement adder-subtracter, a logic circuit that can add or subtract binary numbers. Here's how it works. When *SUB* is low, the *B* bits pass through the controlled inverter without inversion. Therefore, the full adders produce the sum

$$\mathbf{S} = \mathbf{A} + \mathbf{B} \tag{6-13}$$

Incidentally, as indicated in Fig. 6-7, the final *CARRY* is not used. This is because S_3 is the sign bit and S_2 to S_0 are the numerical bits. The final *CARRY* therefore has no significance at this time.

Subtraction

When *SUB* is high, the controlled inverter produces the 1's complement. Furthermore, the high *SUB* adds a 1 to the first full adder. This addition of 1 to the 1's complement forms the 2's complement of **B**. In other words, the controlled inverter produces $\overline{\mathbf{B}}$, and adding 1 results in $\mathbf{B}'$.

The output of the full adders is

$$\mathbf{S} = \mathbf{A} + \mathbf{B}' \tag{6-14}$$

which is equivalent to

$$\mathbf{S} = \mathbf{A} - \mathbf{B} \tag{6-15}$$

because the 2's complement is equivalent to a sign change.

EXAMPLE 6-9

A 7483 is a TTL circuit with four full adders. This means that it can add nibbles (4-bit numbers).

Figure 6-8 shows a TTL adder-subtracter. The *CARRY* out (pin 14) of the least significant nibble is used as the *CARRY* in (pin 13) for the most significant nibble. This allows the two 7483s to add 8-bit numbers. Two 7486s form the controlled inverter needed for subtraction.

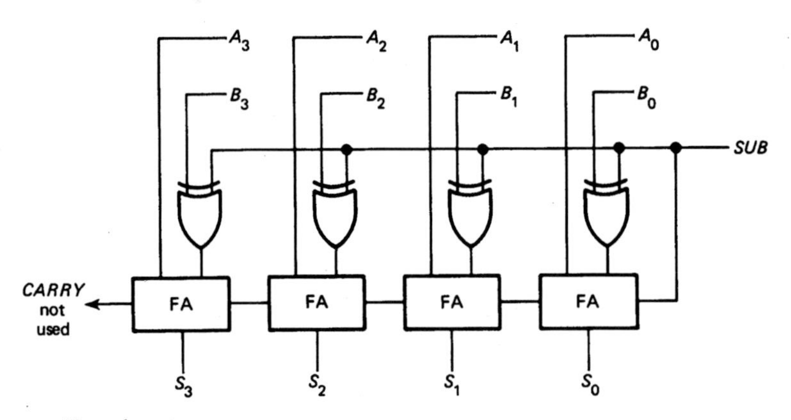

Fig. 6-7 A 2's-complement adder-subtracter.

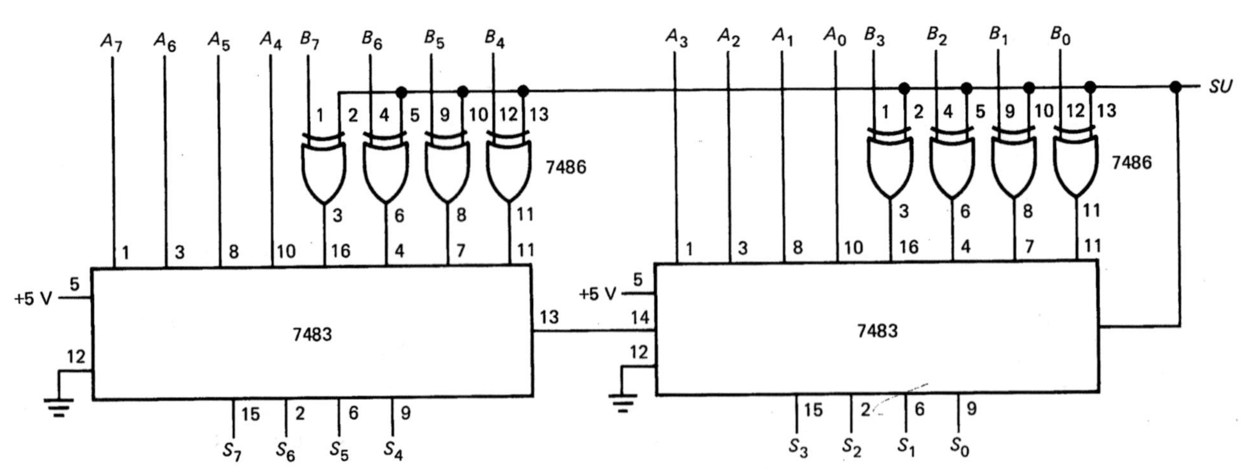

Fig. 6-8 TTL adder-subtracter.

Suppose the circuit has these inputs:

$$\mathbf{A} = 0001\ 1000$$
$$\mathbf{B} = 0001\ 0000$$

If *SUB* = 0, what is the output of the adder-subtracter?

SOLUTION

When *SUB* is 0, the adder-subtracter adds the two inputs as follows:

$$\begin{array}{r} 0001\ 1000 \\ +\ 0001\ 0000 \\ \hline 0010\ 1000 \end{array}$$

Therefore, the output is 0010 1000. Notice that the decimal equivalent of the foregoing addition is

$$\begin{array}{r} 24 \\ +\ 16 \\ \hline 40 \end{array}$$

EXAMPLE 6-10

Repeat the preceding example for *SUB* = 1.

SOLUTION

When *SUB* is 1, the adder-subtracter subtracts the inputs by adding the 2's complement as follows:

$$\begin{array}{r} 0001\ 1000 \\ +\ 1111\ 0000 \\ \hline 0000\ 1000 \end{array}$$

The decimal equivalent is

$$\begin{array}{r} 24 \\ +\ -16 \\ \hline 8 \end{array}$$

EXAMPLE 6-11

In Fig. 6-8, what are the largest positive and negative sums we can get?

SOLUTION

The largest positive output is

0111 1111

which represents decimal +127. The largest negative output is

1000 0000

which represents −128. With 8 bits, therefore, all answers must lie between −128 and +127. If you try to, add numbers with a sum outside this range, you get an *overflow* into the sign-bit position, causing an error.

Chapter 12 discusses the overflow problem in more detail. All you have to remember for now is that an overflow or error will occur if the true sum lies outside the range of −128 to +127.

GLOSSARY

ALU Arithmetic-logic unit. The ALU carries out arithmetic and logic operations.
binary adder A logic circuit that can add two binary numbers.
full adder A logic circiut that can add 3 bits.
half-adder A logic circuit that adds 2 bits.
overflow In 2's-complement representation, a carry into the sign-bit position, which results in an error. For an 8-bit adder-substracter, the true sum must lie between −128 and +127 to avoid overflow.
signed binary A system in which the leading bit represents the sign and the remaining bits the magnitude of the number. Also called sign magnitude.
2's complement The new number you get when you take the 1's complement and then add 1.

SELF-TESTING REVIEW

Read each of the following and provide the missing words. Answers appear at the beginning of the next question.

1. The ALU carries out arithmetic and ________ operations (OR, AND, NOT, etc.). It processes ________ numbers rather than decimal numbers.
2. (*logic, binary*) A half-adder adds ________ bits. A full adder adds ________ bits, producing a SUM and a ________.
3. (*two, three,* CARRY) A binary adder is a logic cicuit that can add ________ binary numbers at a time. The 7483 is a TTL binary adder. It can add two 4-bit binary numbers.
4. (*two*) With signed binary numbers, also known as sign-magnitude numbers, the leading bit stands for the ________ and the remaining bits for the ________.
5. (*sign, magnitude*) Signed binary numbers require too much hardware. This has led to the use of ________ complements to represent negative numbers. To get the 2's complement of a binary number, you first take the ________ complement, then add ________.
6. (*2's, 1's, 1*) If you take the 2's complement twice, you get the original binary number back. Because of this property, taking the ________ complement of a binary number is equivalent to changing the sign of a decimal number.
7. (*2's*) In a microcomputer positive numbers are represented in ________ form and negative numbers in 2's-complement form. The leading bit still represents the ________.
8. (*sign-magnitude, sign*) A 2's-complement adder-subtracter can add or subtract binary numbers. Sign-magnitude numbers represent ________ decimal numbers, and 2's complements stand for ________ decimal numbers. You can tell one from the other by the leading bit, which represents the ________.
9. (*positive, negative, sign*) With 2's-complement representation and an 8-bit adder-subtracter no overflow is possible if the true sum is between −128 and +127.

PROBLEMS

6-1. Add these 8-bit numbers:
  a. 0001 0000 and 0000 1000
  b. 0001 1000 and 0000 1100
  c. 0001 1100 and 0000 1110
  d. 0010 1000 and 0011 1011
  After you have each binary sum, convert it to hexadecimal form.

6-2. Add these 16-bit numbers:

  1000 0001 1100 1001
  + 0011 0011 0001 0111

  Express the answer in hexadecimal form.

6-3. In each of the following, convert to binary to do the addition, then convert the answer back to hexadecimal:
  a. 2CH + 4FH = ?
  b. 5EH + 1AH = ?
  c. 3BH + 6DH = ?
  d. A5H + 2CH = ?

6-4. Convert each of the following decimal numbers to an 8-bit sign-magnitude number:
  a. +27
  b. −27
  c. +80
  d. −80
  After you have the sign-magnitude numbers, convert them to hexadecimal form.

6-5. Convert each of these sign-magnitude numbers to its decimal equivalent:
  a, 0001 1110
  b. 1000 0111
  c. 1001 1100
  d. 0011 0001

6-6. The following hexadecimal numbers represent sign-magnitude numbers. Convert each to its decimal equivalent.
  a. 8FH
  b. 3AH
  c. 7FH
  d. FFH

6-7. Find the 2's complements:
  a. 0000 0111
  b. 1111 1111
  c. 1111 1101
  d. 1110 0001
  Express your answers in hexadecimal form.

6-8. Convert each of the following to binary. Then take the 2's complement:
  a. 4CH
  b. 8DH
  c. CBH
  d. FFH

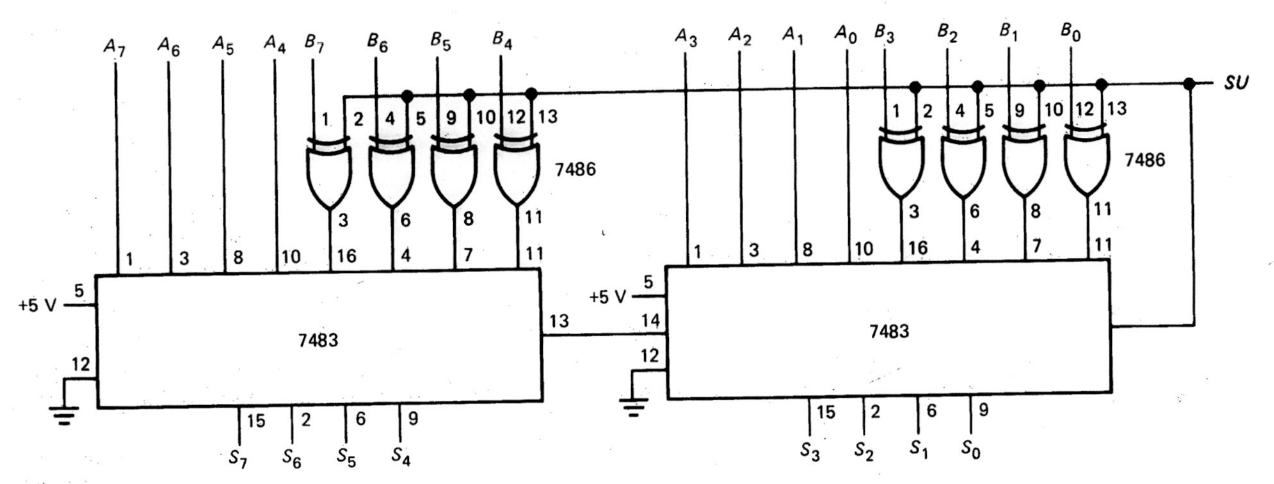

Fig. 6-9

After you have the 2's complements, convert them to hexadecimal form.

6-9. An 8-bit microprocessor uses 2's-complement representation. How do the following decimal numbers appear:

a. −19
b. −48
c. +37
d. −33

Express your answers in binary and hexadecimal form.

6-10. The output of an ALU is EEH. What decimal number does this represent in 2's-complement representation?

6-11. Suppose the inputs to Fig. 6-9 are **A** = 3CH and **B** = 5FH. What is the output for a low *SUB?* A high *SUB?* Express your final answers in hexadecimal form.

6-12. In Fig. 6-9 which of the following inputs cause an overflow when *SUB* is low?

a. 2DH and 4BH
b. 8FH and C3H
c. 5EH and B8H
d. 23H and 14H

6-13. Why are applications for the half-adder limited, what does the full adder do which makes it more useful than the half-adder, and what can be done with a full adder as a result of this feature?

6-14. Since sign-magnitude numbers are fairly easy to understand, why has the 2's-complement system become so widespread?

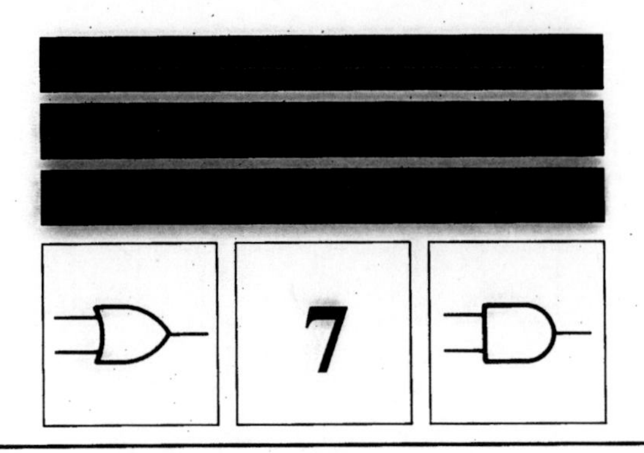

FLIP-FLOPS

Gates are decision-making elements. As shown in the preceding chapter, they can perform binary addition and subtraction. But decision-making elements are not enough. A computer also needs *memory elements,* devices that can store a binary digit. This chapter is about memory elements called *flip-flops*.

7-1 *RS* LATCHES

A flip-flop is a device with two stable states; it remains in one of these states until triggered into the other. The *RS* latch, discussed in this section, is one of the simplest flip-flops.

Transistor Latch

In Fig. 7-1*a* each collector drives the opposite base through a 100-kΩ resisitor. In a circuit like this, one of the transistors is saturated and the other is cut off.

For instance, if the right transistor is saturated, its collector voltage is approximately 0 V. This means that there is no base drive for the left transistor, so it cuts off and its collector voltage approaches +5 V. This high voltage produces enough base current in the right transistor to sustain its saturation. The overall circuit is *latched* with the left transistor cut off (dark shading) and the right transistor saturated. Q is approximately 0 V.

By a similar argument, if the left transistor is saturated, the right transistor is cut off. Figure 7-1*b* illustrates this other state. Q is approximately 5 V for this condition.

Output Q can be low or high, binary 0 or 1. If latched as shown in Fig. 7-1*a*, the circuit is storing a binary 0 because

$$Q = 0$$

On the other hand, when latched as shown in Fig. 7-1*b*, the circuit stores a binary 1 because

$$Q = 1$$

Control Inputs

To control the bit stored in the latch, we can add the inputs shown in Fig. 7-1*c*. These control inputs will be either low (0 V) or high (+5 V). A high *set* input S forces the left transistor to saturate. As soon as the left transistor saturates, the overall circuit latches and

$$Q = 1$$

Once set, the output will remain a 1 even though the S input goes back to 0 V.

A high *reset* input R drives the right transistor into saturation. Once this happens, the circuit latches and

$$Q = 0$$

The output stays latched in the 0 state, even though the R input returns to a low.

In Fig. 7-1*c*, Q represents the stored bit. A complementary output $\overline{Q}$ is available from the collector of the left transistor. This may or may not be used, depending on the application.

Truth Table

Table 7-1 summarizes the operation of the transistor latch. With both control inputs low, no change can occur in the output and the circuit remains latched in its last state. This condition is called the *inactive state* because nothing changes.

TABLE 7-1. TRANSISTOR LATCH

R	S	Q	Comments
0	0	NC	No change
0	1	1	Set
1	0	0	Reset
1	1	*	Race

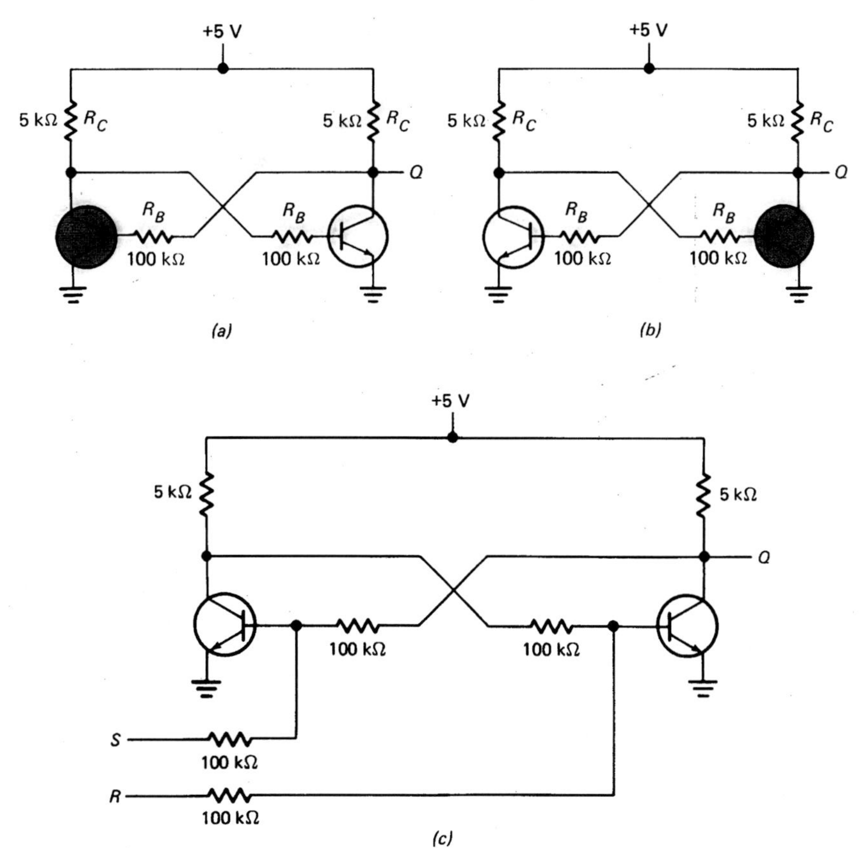

Fig. 7-1 (*a*) Latched state; (*b*) alternative state; (*c*) trigger inputs.

When *R* is low and *S* is high, the circuit sets the *Q* output to a high. On the other hand, if *R* is high and *S* is low, the *Q* output resets to a low.

Race Condition

Look at the last entry in Table 7-1. *R* and *S* are high simultaneously. This is called a *race condition;* it is never used because it leads to unpredictable operation.

Here's why. If both control inputs are high, both transistors saturate. When the *R* and *S* inputs return to low, both transistors try to come out of saturation. It is a race between the transistors to see which one desaturates first. The faster transistor (the one with the shorter saturation delay time) will win the race and latch the circuit. If the faster transistor is on the left side of Fig. 7-1*c*, the *Q* output will be low. If the faster transistor is on the right side, the *Q* output will go high. In mass production, either transistor can be faster; therefore, the *Q* output is unpredictable. This is why the race condition must be avoided.

Here's how to recognize a race condition. If simultaneously changing both inputs to a memory element leads to an unpredictable output, you've got a race condition. With the transistor latch, $R = 1$ and $S = 1$ is a race condition because simultaneously returning *R* and *S* to 0 forces *Q* into a random state.

From now on, an asterisk in a truth table (see Table 7-1) indicates a race condition, sometimes called a forbidden or invalid state.

NOR Latches

A discrete circuit like Fig. 7-1*c* is rarely used because we are in the age of integrated circuits. Nowadays, you build *RS* latches with NOR gates or NAND gates.

Figure 7-2*a* shows how it's done with NOR gates. Figure 7-2*b* is the De Morgan equivalent. As shown in Table 7-2, a low *R* and a low *S* give us the inactive state; the circuit stores or remembers. A low *R* and a high *S* represent the set state, while a high *R* and a low *S* give the reset state. Finally, a high *R* and a high *S* produce a race condition; therefore, we must avoid $R = 1$ and $S = 1$ when using a NOR latch.

Figure 7-2*c* is a *timing diagram;* it shows how the input signals interact to produce the output signal. As you see, the *Q* output goes high when *S* goes high. *Q* remains high after *S* goes low. *Q* returns to low when *R* goes high, and stays low after *R* returns to low.

TABLE 7-2. NOR LATCH

R	*S*	*Q*	Comment
0	0	NC	No change
0	1	1	Set
1	0	0	Reset
1	1	*	Race

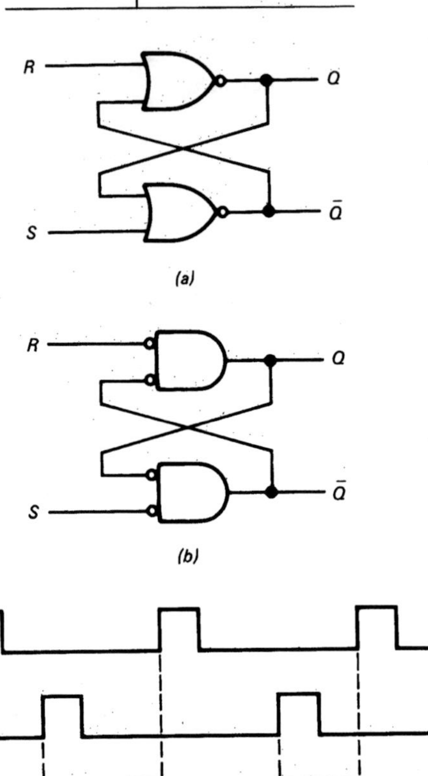

Fig. 7-2 (*a*) NOR latch; (*b*) De Morgan equivalent; (*c*) timing diagram.

NAND Latches

If you prefer using NAND gates, you can build an *RS* latch as shown in Fig. 7-3*a*. Sometimes it is convenient to draw the De Morgan equivalent shown in Fig. 7-3*b*. In either case, a low *R* and a high *S* set *Q* to high; a high *R* and a low *S* reset *Q* to low.

Because of the NAND-gate inversion, the inactive and race conditions are reversed. In other words, $R = 1$ and $S = 1$ becomes the inactive state; $R = 0$ and $S = 0$ becomes the race condition (see Table 7-3). Therefore, whenever you use a NAND latch, you must avoid having both inputs low at the same time. (To remember the race condition for a NAND latch, glance at Fig. 7-3*b*. If $R = 0$ and $S = 0$, then $Q = 1$ and $\bar{Q} = 1$; both outputs are the same, indicating an invalid condition.)

TABLE 7-3. NAND LATCH

R	*S*	*Q*	Comment
0	0	*	Race
0	1	1	Set
1	0	0	Reset
1	1	NC	No change

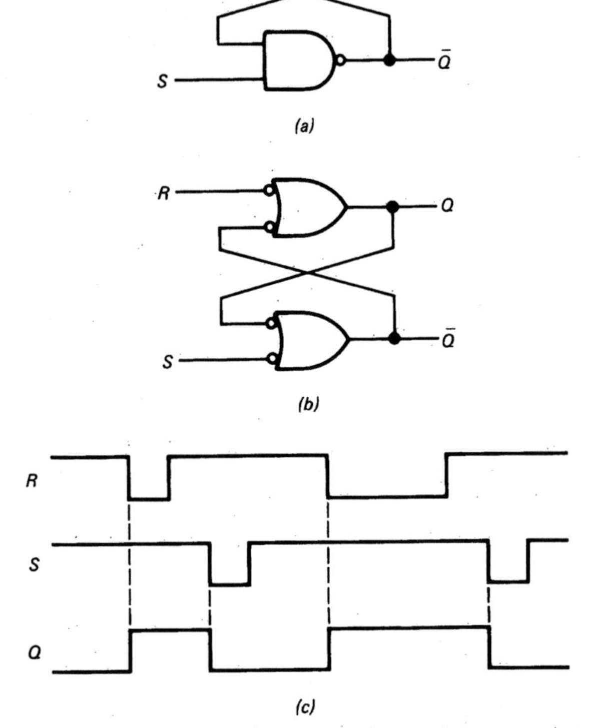

Fig. 7-3 (*a*) NAND latch; (*b*) De Morgan equivalent; (*c*) timing diagram.

Figure 7-3*c* shows the timing diagram for a NAND latch. *R* and *S* are normally high to avoid the race condition. Only one of them goes low at any time. As you see, the *Q* output goes high whenever *R* goes low; the *Q* output goes low whenever *S* goes low.

Switch Debouncers

RS latches are often used as *switch debouncers*. Whenever you throw a switch from the open to the closed position, the contacts bounce and the switch alternately makes and breaks for a few milliseconds before finally settling in the closed position. One way to eliminate the effects of contact bounce is to use an *RS* latch in conjunction with the switch. The following example explains the idea.

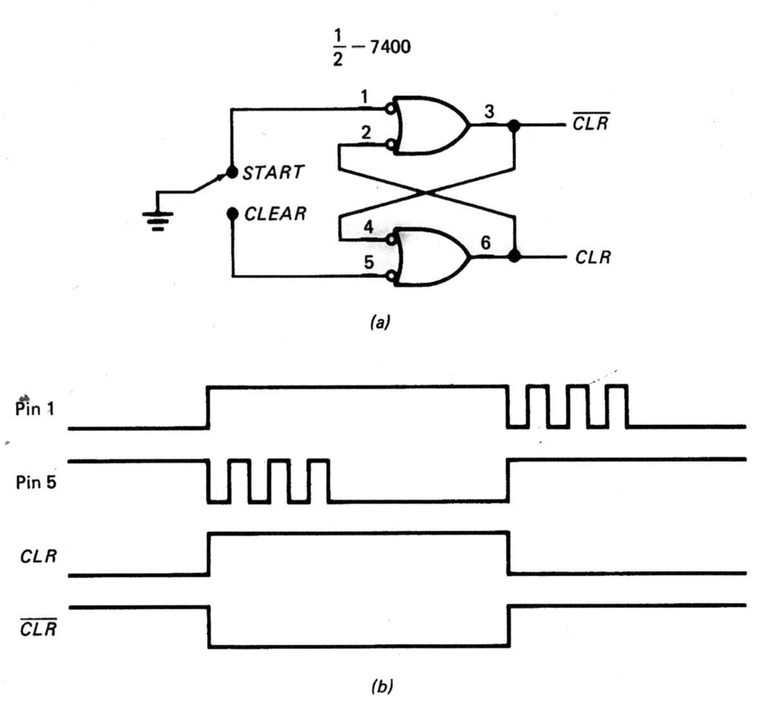

Fig. 7-4 Switch debouncer.

EXAMPLE 7-1

Figure 7-4*a* shows a switch debouncer. What does it do?

SOLUTION

As discussed in Chap. 4, floating TTL inputs are equivalent to high inputs. With the switch in the START position, pin 1 is low and pin 5 is high; therefore, $\overline{CLR}$ is high and *CLR* is low. When the switch is thrown to the CLEAR position, pin 1 goes high, as shown in Fig. 7-4*b*. Because of contact bounce, pin 5 goes alternately low and high for a few milliseconds before settling in the low state, symbolized by the ideal pulses of Fig. 7-4*b*. The first time pin 5 goes low, the latch sets, *CLR* going high and $\overline{CLR}$ going low. Subsequent bounces have no effect on *CLR* and $\overline{CLR}$ because the latch stays set.

Similarly, when the switch is thrown back to START, pin 1 bounces low and high for a while. The first time pin 1 goes low, *CLR* goes back to low and $\overline{CLR}$ to high. Later bounces have no effect on *CLR* and $\overline{CLR}$.

Registers need clean signals like *CLR* and $\overline{CLR}$ of Fig. 7-4*b* to operate properly. If the bouncing signals on pins 1 and 5 drove the registers, the operation would be erratic. This is why you often see *RS* latches used as switch debouncers.

7-2 LEVEL CLOCKING

Computers use thousands of flip-flops. To coordinate the overall action, a square-wave signal called the *clock* is sent to each flip-flop. This signal prevents the flip-flops from changing states until the right time.

Clocked Latch

In Fig. 7-5*a* a pair of NAND gates drive a NAND latch. *S* and *R* signals drive the input gates. To avoid confusion, the inner control signals are labeled R' and S'. The NAND latch works as previously described; a low R' and a high S' set Q to 1, whereas a high R' and a low S' reset Q to 0. Furthermore, a low R' and S' represent the race condition; therefore, R' and S' are normally high when the latch is inactive. Because of the inversion through the input NAND gates, the *S* input has to drive the upper NAND input and the *R* input must drive the lower NAND input.

Double Inversions Cancel

When analyzing the operation of this and similar circuits, remember that a double inversion (two bubbles in a series path) cancels out; this makes it appear as though two AND gates drove OR gates, as shown in Fig. 7-5*b*. In this way, you can see at a glance that a high *S* and high *CLK* force

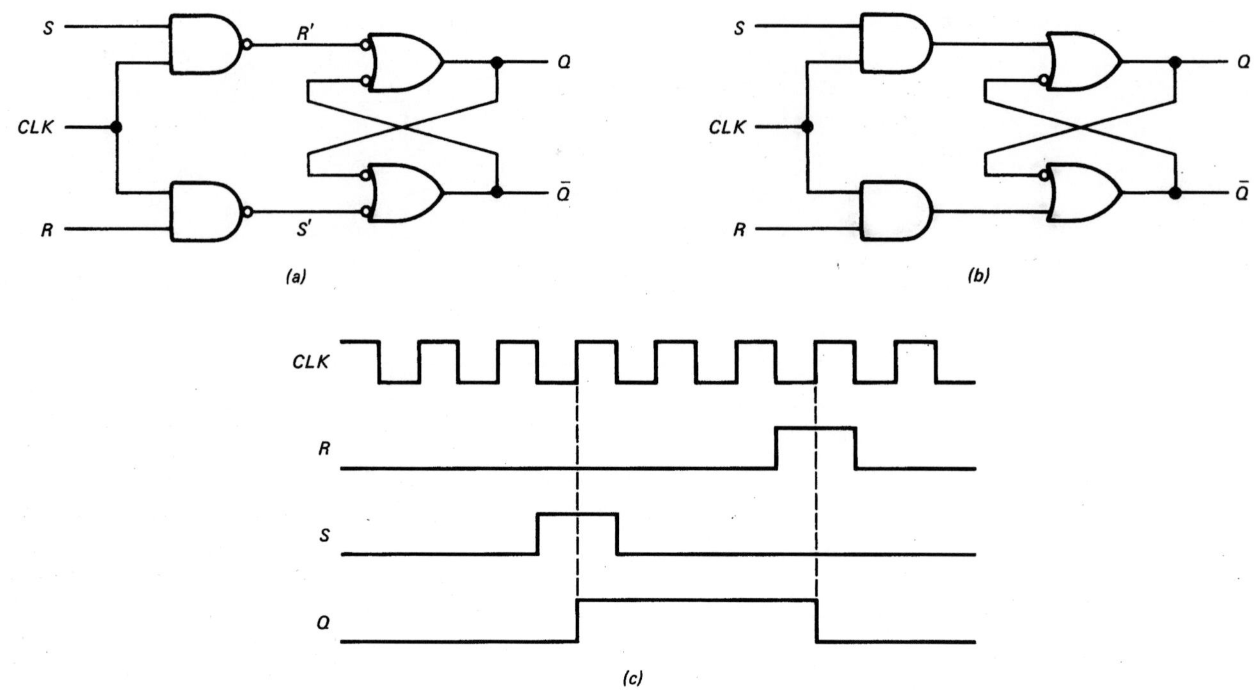

Fig. 7-5 (*a*) Clocked latch; (*b*) equivalent circuit; (*c*) timing diagram.

Q to go high. In other words, even though you are looking at Fig. 7-5*a*, in your mind you should see Fig. 7-5*b*.

Positive Clocking

In Fig. 7-5*a* the clock is a square-wave signal. Because the clock (abbreviated CLK) drives both NAND gates, a low *CLK* prevents *S* and *R* from controlling the latch. If a high *S* and a low *R* drive the gate inputs, the latch must wait until the clock goes high before *Q* can be set to 1. Similarly, given a low *S* and a high *R*, the latch must wait for a high *CLK* before *Q* can reset to 0. This is an example of *positive clocking,* making a latch wait until the clock signal is high before the output can change.

Negative clocking is similar. Visualize an inverter between CLK and the input gates of Fig. 7-5*a*. In this case, the latch must wait until *CLK* is low before the output can change.

Positive and negative clocking are often called *level clocking* because the flip-flop responds to the level (high or low) of the clock signal. Level clocking is the simplest way to control flip-flops with a clock. Later, we will discuss more advanced methods called edge triggering and master-slave clocking.

Race Condition

What about the race condition? When the clock is low in Fig. 7-5*a*, *R'* and *S'* are high, which is a stable condition. The only way to get a race condition is to have a high *CLK,* high *R,* and high *S*. Therefore, normal operation of this circuit requires that *R* and *S* never both be high when the clock goes high.

Timing Diagram and Truth Table

Figure 7-5*c* shows the timing diagram. *Q* goes high when *S* is high and *CLK* goes high. *Q* returns to the low state when *R* is high and *CLK* goes high. Using a common *CLK* signal to drive many flip-flops allows us to synchronize the operation of the different sections of a computer.

Table 7-4 summarizes the operation of the clocked NAND latch. When the clock is low, the output is latched in its last state. When the clock goes high, the circuit will set if *S* is high or reset if *R* is high. *CLK, R,* and *S* all high is a race condition, which is never used deliberately.

TABLE 7-4. CLOCKED NAND LATCH

CLK	*R*	*S*	*Q*
0	0	0	NC
0	0	1	NC
0	1	0	NC
0	1	1	NC
1	0	0	NC
1	0	1	1
1	1	0	0
1	1	1	*

7-3 *D* LATCHES

Since the *RS* flip-flop is susceptible to a race condition, we will modify the design to eliminate the possibility of a race condition. The result is a new kind of flip-flop known as a *D latch.*

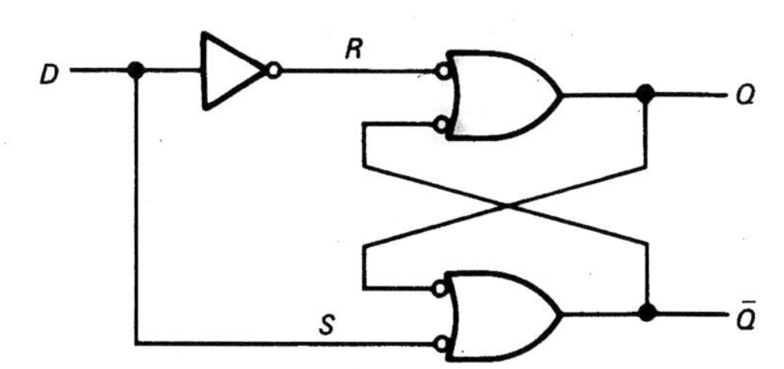

Fig. 7-6 *D* latch.

Unclocked

Figure 7-6 shows one way to build a *D* latch. Because of the inverter, data bit *D* drives the *S* input of a NAND latch and the complement $\overline{D}$ drives the *R* input. Therefore, a high *D* sets the latch, and a low *D* resets it. Table 7-5 summarizes the operation of the *D* latch. Especially important, there is no race condition in this truth table. The inverter guarantees that *S* and *R* will always be in opposite states; therefore, it's impossible to set up a race condition in the *D* latch.

The *D* latch of Fig. 7-6 is unclocked; it will set or reset as soon as *D* goes high or low. An unclocked flip-flop like this is almost never used.

TABLE 7-5. UNCLOCKED *D* LATCH

D	*Q*
0	0
1	1

TABLE 7-6. CLOCKED *D* LATCH

CLK	*D*	*Q*
0	X	NC
1	0	0
1	1	1

Clocked

Figure 7-7*a* is level-clocked. A low *CLK* disables the input gates and prevents the latch from changing states. In other words, while *CLK* is low, the latch is in the inactive state and the circuit stores or remembers. When *CLK* is high, *D* controls the output. A high *D* sets the latch, while a low *D* resets it.

Table 7-6 summarizes the operation. X represents a don't-care condition; it stands for either 0 or 1. While *CLK* is low, the output cannot change, no matter what *D* is. When *CLK* is high, however, the output equals the input

$$Q = D$$

Figure 7-7*b* shows a timing diagram. If the clock is low, the circuit is latched and the *Q* output cannot be changed. While the clock is high, however, *Q* equals *D*; when *D* goes high, *Q* goes high; when *D* goes low, *Q* goes low. The latch is *transparent*, meaning that the output follows the value of *D* while the clock is high.

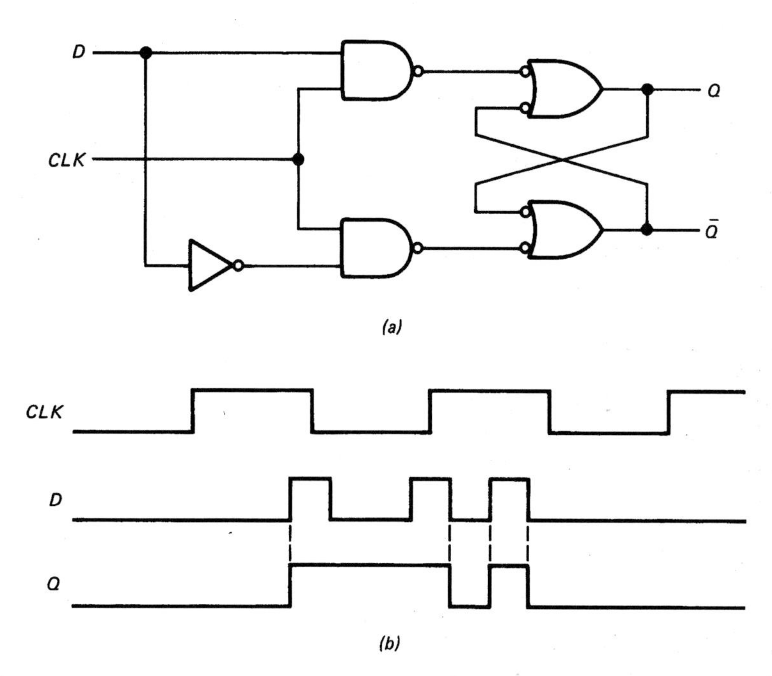

Fig. 7-7 Clocked *D* latch.

Disadvantage

Because the D latch is level-clocked, it has a serious disadvantage. While the clock is high, the output follows the value of D. Transparent latches may be all right in some applications but not in the computer circuits we will be discussing. To be truly useful, the circuit of Fig. 7-7*a* needs a slight modification.

7-4 EDGE-TRIGGERED *D* FLIP-FLOPS

Now we're ready to talk about the most common type of D flip-flop. What a practical computer needs is a D flip-flop that samples the data bit at a unique instant.

Edge Triggering

Figure 7-8*a* shows an *RC* circuit at the input of a D flip-flop. By deliberate design, the *RC* time constant is much smaller than the clock's pulse width. Because of this, the capacitor can charge fully when *CLK* goes high; this exponential charging produces a narrow positive voltage spike across the resistor. Later, the trailing edge of the clock pulse results in a narrow negative spike.

The narrow positive spike enables the input gates for an instant; the narrow negative spike does nothing. The effect is to activate the input gates during the positive spike, equivalent to sampling the value of D for an instant. At this unique time, D and its complement hit the flip-flop inputs, forcing Q to set or reset.

TABLE 7-7. EDGE-TRIGGERED *D* FLIP-FLOP

CLK	*D*	*Q*
0	X	NC
1	X	NC
↓	X	NC
↑	0	0
↑	1	1

This kind of operation is called *edge triggering* because the flip-flop responds only when the clock is changing states. The triggering in Fig. 7-8*a* occurs on the positive-going edge of the clock; this is why it's referred to as *positive-edge triggering*.

Figure 7-8*b* illustrates the action. The crucial idea is that the output changes only on the rising edge of the clock. In other words, data is stored only on the positive-going edge.

Table 7-7 summarizes the operation of the positive-edge-triggered D flip-flop. The up and down arrows represent the rising and falling edges of the clock. The first three entries indicate that there's no output change when the clock is low, high, or on its negative edge. The last two entries indicate an output change on the positive edge of the clock. In other words, input data D is stored only on the positive-going edge of the clock.

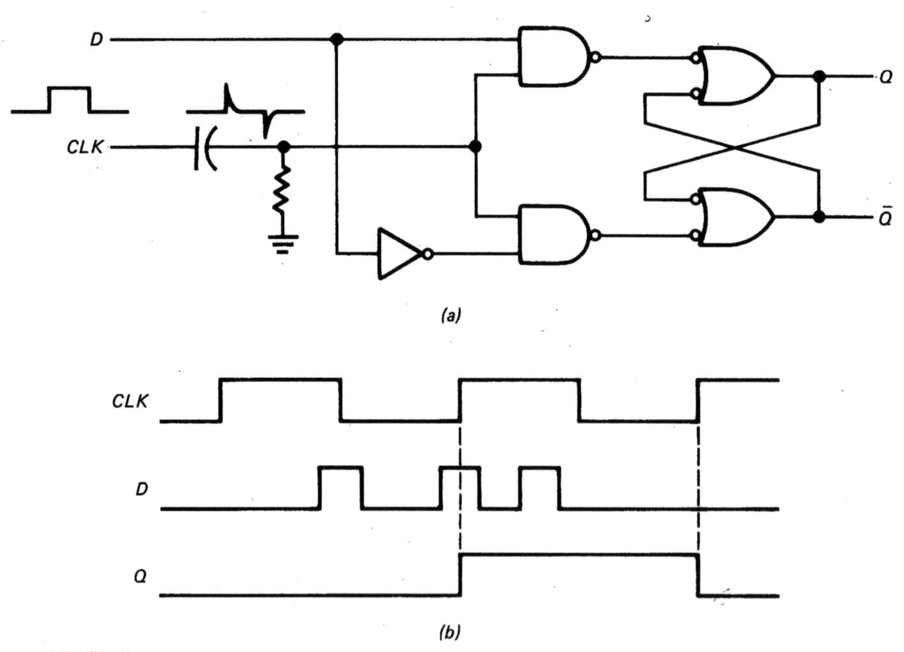

Fig. 7-8 Edge-triggered D flip-flop.

Edge Triggering versus Level Clocking

When a circuit is edge-triggered, the output can change only on the rising (or falling) edge of the clock. But when the circuit is level-clocked, the output can change while the clock is high (or low). With edge triggering, the output can change only at one instant during the clock cycle; with level clocking, the output can change during an entire half cycle of the clock.

Preset and Clear

When power is first applied, flip-flops come up in random states. To get some computers started, an operator has to push a master reset button. This sends a *clear* (reset) signal to all flip-flops. Also, it is necessary in some computers to *preset* (synonymous with "set") certain flip-flops before a computer run.

Figure 7-9 shows how to include both functions in a *D* flip-flop. The edge triggering is the same as previously described. In addition, the AND gates allow us to slip in a low *PRESET* or low *CLEAR* when desired. A low *PRESET* forces *Q* to equal 1; a low *CLEAR* resets *Q* to 0.

Table 7-8 summarizes the circuit action. When *PRESET* and *CLEAR* are both low, we get a race condition; therefore, *PRESET* and *CLEAR* should be kept high when inactive. Take *PRESET* low by itself and you set the flip-flop; take *CLEAR* low by itself and you reset the flip-flop. As shown in the remaining entries, the output changes only on the positive-going edge of the clock.

Preset is sometimes called *direct set*, and clear is sometimes called *direct reset*. The word "direct" means unclocked. For instance, the clear signal may come from a push button; regardless of what the clock is doing, the output will reset when the operator pushes the clear button.

The preset and clear inputs override the other inputs; they have first priority. For example, when *PRESET* goes low, the *Q* output goes high and stays there no matter what the *D* and *CLK* inputs are doing. The output will remain high as long as *PRESET* is low. Therefore, the normal procedure in presetting is to take the *PRESET* low temporarily, then return it to high. Similarly, for the clear function: take *CLEAR* low briefly to reset the flip-flop, then take it back to high to allow the circuit to operate.

TABLE 7-8. *D* FLIP-FLOP WITH PRESET AND CLEAR

PRESET	*CLEAR*	*CLK*	*D*	*Q*
0	0	X	X	*
0	1	X	X	1
1	0	X	X	0
1	1	0	X	NC
1	1	1	X	NC
1	1	↓	X	NC
1	1	↑	0	0
1	1	↑	1	1

Direct-Coupled Edge-Triggered *D* Flip-Flop

Integrated *D* flip-flops do not use *RC* circuits to get narrow spikes because capacitors are difficult to fabricate on a chip. Instead, a variety of direct-coupled designs is used. As an example, Fig. 7-10 shows a positive-edge-triggered *D* flip-flop. This direct-coupled circuit has no capacitors, only NAND gates. The analysis is too long and complicated to go into here, but the idea is the same as previously discussed. The circuit responds only during the brief instant the clock switches from low to high. That is, data bit *D* is stored only on the positive-going edge of the clock.

Logic Symbol

Figure 7-11 is the symbol of a positive-edge-triggered *D* flip-flop. The *CLK* input has a small triangle, a reminder of the edge triggering. When you see this schematic symbol, remember what it means: the *D* input is stored on the rising edge of the clock.

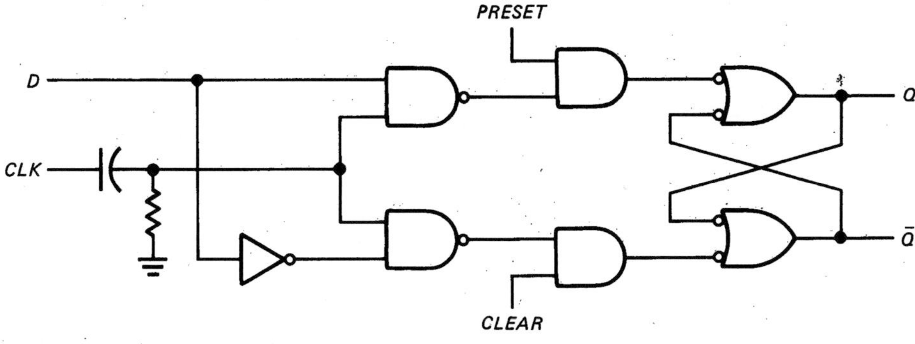

Fig. 7-9 Edge-triggered *D* flip-flop with preset and clear.

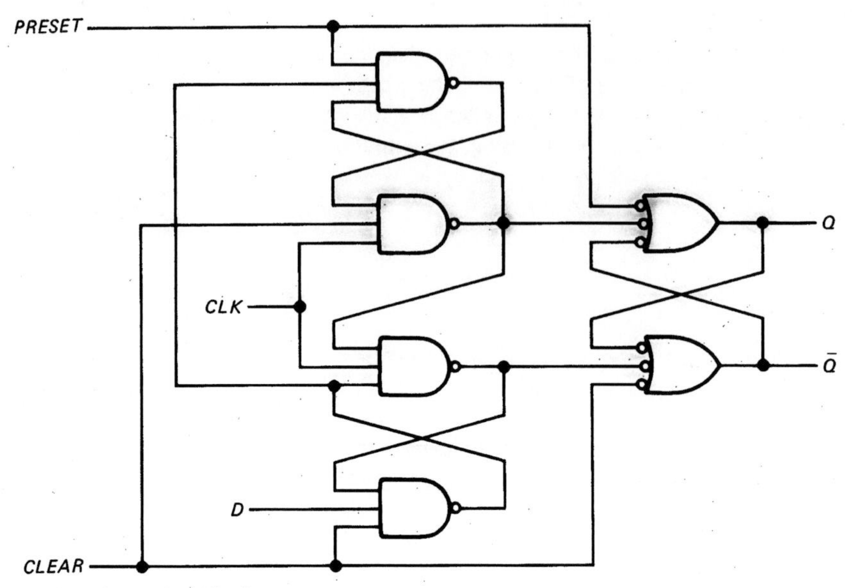

Fig. 7-10 Direct-coupled edge-triggered D flip-flop.

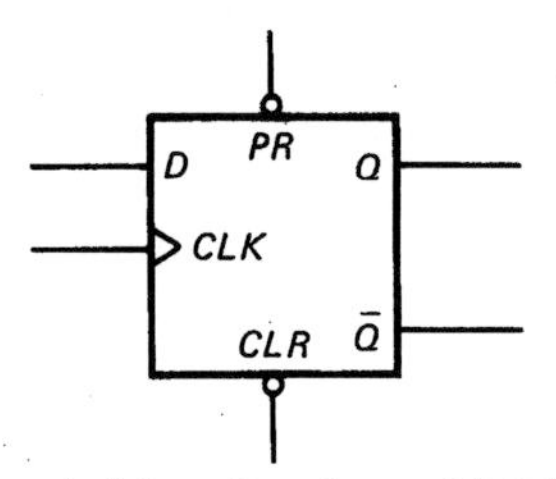

Fig. 7-11 Logic symbol for edge-triggered D flip-flop.

Figure 7-11 also includes preset (*PR*) and clear (*CLR*) inputs. The bubbles indicate an *active low state*. In other words, the preset and clear inputs are high when inactive. To preset the flip-flop, the preset input must go low temporarily and then be returned to high. Similarly, to reset the flip-flop, the clear input must go low, then back to high.

The same idea applies to circuits discussed later. A bubble at an input means an active low state: the input has to go low to produce an effect. When no bubble is present, the input has to go high to have an effect.

Propagation Delay Time

Diodes and transistors cannot switch states instantaneously. It always takes a small amount of time to turn a diode on or off. Likewise, it takes a time for a transistor to switch from saturation to cutoff or vice versa. For bipolar diodes and transistors, switching time is in the nanosecond region.

Switching time is the main cause of *propagation delay time* t_p. This represents the amount of time it takes for the output of a gate or flip-flop to change states. For instance, if the data sheet of a D flip-flop indicates a t_p of 10 ns, it takes approximately 10 ns for Q to change states after D has been sampled by the clock edge.

Propagation delay time is so small that it's negligible in many applications, but in high-speed circuits you have to take it into account. If a flip-flop has a t_p of 10 ns, this means that you have to wait 10 ns before the output can trigger another circuit.

Setup Time

Stray capacitance at the D input (plus other factors) makes it necessary for data bit D to be at the input before the *CLK* edge arrives. The *setup time* t_{setup} is the minimum length of time the data bit must be present before the *CLK* edge hits.

For instance, if the data sheet of a D flip-flop indicates a t_{setup} of 15 ns, the data bit to be stored must be at the D input at least 15 ns before the *CLK* edge arrives; otherwise, the IC manufacturer does not guarantee correct sampling and storing.

Hold Time

Furthermore, data bit D has to be held long enough for the internal transistors to switch states. Only after the transition is assured can we allow data bit D to change. *Hold time* t_{hold} is the minimum length of time the data bit must be present after the *CLK* edge has struck.

For example, if t_{setup} is 15 ns and t_{hold} is 5 ns, the data bit has to be at the D input at least 15 ns before the *CLK* edge arrives and held at least 5 ns after the *CLK* edge hits.

7-5 EDGE-TRIGGERED JK FLIP-FLOPS

The next chapter shows you how to build a counter, the electronic equivalent of a binary odometer. When it comes to circuits that count, the *JK flip-flop* is the ideal memory element to use.

Circuit

Figure 7-12*a* shows one way to build a *JK* flip-flop. As before, an *RC* circuit with a short time constant converts the rectangular *CLK* pulse to narrow spikes. Because of the double inversion through the NAND gates, the circuit is positive-edge-triggered. In other words, the input gates are enabled only on the rising edge of the clock.

Inactive

The *J* and *K* inputs are control inputs; they determine what the circuit will do on the positive clock edge. When *J* and *K* are low, both input gates are disabled and the circuit is inactive at all times including the rising edge of the clock.

Reset

When *J* is low and *K* is high, the upper gate is disabled; so there's no way to set the flip-flop. The only possibility is reset. When *Q* is high, the lower gate passes a reset trigger as soon as the positive clock edge arrives. This forces *Q* to become low. Therefore, $J = 0$ and $K = 1$ means that a rising clock edge resets the flip-flop.

Set

When *J* is high and *K* is low, the lower gate is disabled; so it's impossible to reset the flip-flop. But you can set the flip-flop as follows. When *Q* is low, $\overline{Q}$ is high; therefore, the upper gate passes a set trigger on the positive clock edge. This drives *Q* into the high state. That is, $J = 1$ and $K = 0$ means that the next positive clock edge sets the flip-flop.

Toggle

When *J* and *K* are both high, it is possible to set or reset the flip-flop, depending on the current state of the output. If *Q* is high, the lower gate passes a reset trigger on the

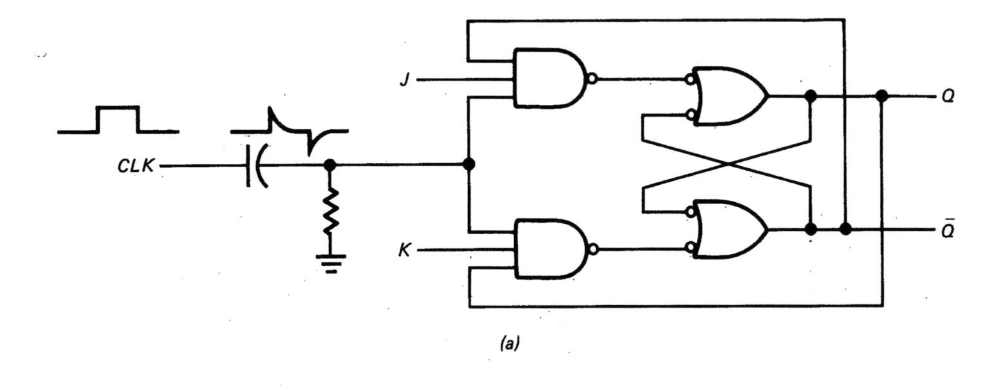

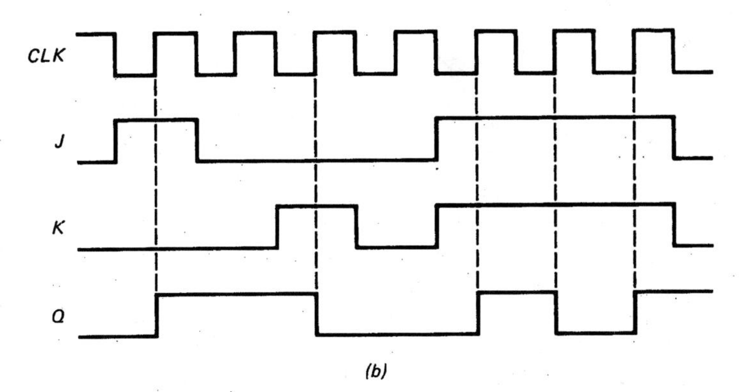

Fig. 7-12 (*a*) Edge-triggered *JK* flip-flop; (*b*) timing diagram.

TABLE 7-9. POSITIVE-EDGE-TRIGGERED *JK* FLIP-FLOP

CLK	*J*	*K*	*Q*
0	X	X	NC
1	X	X	NC
↓	X	X	NC
X	0	0	NC
↑	0	1	0
↑	1	0	1
↑	1	1	Toggle

next positive clock edge. On the other hand, when Q is low, the upper gate passes a set trigger on the next positive clock edge. Either way, Q changes to the complement of the last state. Therefore, $J = 1$ and $K = 1$ means that the flip-flop will *toggle* on the next positive clock edge. ("Toggle" means switch to opposite state.)

Timing Diagram

The timing diagram of Fig. 7-12*b* is a visual summary of the action. When J is high and K is low, the rising clock edge sets Q to high. On the other hand, when J is low and K is high, the rising clock edge resets Q to low. When J and K are high simultaneously, the output toggles on each rising clock edge.

Truth Table

Table 7-9 summarizes the operation. The circuit is inactive when the clock is low, high, or on its negative edge. Likewise, the circuit is inactive when J and K are both low. Output changes occur only on the rising edge of the clock, as indicated by the last three entries of the table. The output either resets, sets, or toggles.

Racing

The *JK* flip-flop shown in Fig. 7-12*a* has to be edge-triggered to avoid oscillations. Why? Assume that the circuit is level-clocked. In other words, assume that we remove the *RC* circuit and run the clock straight into the gates. With a high J, high K, and high *CLK*, the output will toggle. New outputs are then fed back to the input gates. After two propagation times (input and output gates), the output toggles again. And once more, new outputs return to the input gates. In this way, the output can toggle repeatedly as long as the clock is high. That is, we get oscillations during the positive half cycle of the clock. Toggling more than once during a clock cycle is called *racing*.

Now assume that we put the *RC* circuit back in and return to edge triggering. Propagation delay time prevents the *JK* flip-flop from racing. Here's why. In Fig. 7-12*a* the outputs change after the positive clock edge has struck. By the time the new Q and $\overline{Q}$ signals return to the input gates, the positive spikes have decayed to zero. This is why we get only one toggle during each clock cycle.

For instance, if the total propagation delay time from input to output is 20 ns, the outputs change approximately 20 ns after the rising edge of the clock. If the spikes are narrower than 20 ns, the returning Q and $\overline{Q}$ arrive too late to cause false triggering.

Symbols

As previously mentioned, capacitors are too difficult to fabricate on a chip. This is why manufacturers prefer direct-coupled designs for edge-triggered *JK* flip-flops. Such designs are too complicated to reproduce here, but you can find them in manufacturers' IC data books.

Figure 7-13*a* is the standard symbol for a positive-edge-triggered *JK* flip-flop of any design.

Figure 7-13*b* is the symbol for a *JK* flip-flop with the preset and clear functions. As usual, *PR* and *CLR* have active low states. This means that they are normally high and taken low temporarily to preset or clear the circuit.

Figure 7-13*c* is another commercially available *JK* flip-flop. The bubble on the clock input is the standard way to indicate negative-edge triggering. As shown in Table 7-10, the output can change only on the *falling* edge of the clock. The timing diagram of Fig. 7-13*d* emphasizes this negative-edge triggering.

7-6 JK MASTER-SLAVE FLIP-FLOP

Figure 7-14 shows a *JK master-slave* flip-flop, another way to avoid racing. A master-slave flip-flop is a combination of two clocked latches; the first is called the *master*, and the second is the *slave*. Notice that the master is positively

TABLE 7-10. NEGATIVE-EDGE-TRIGGERED *JK* FLIP-FLOP

CLK	*J*	*K*	*Q*
0	X	X	NC
1	X	X	NC
↑	X	X	NC
X	0	0	NC
↓	0	1	0
↓	1	0	1
↓	1	1	Toggle

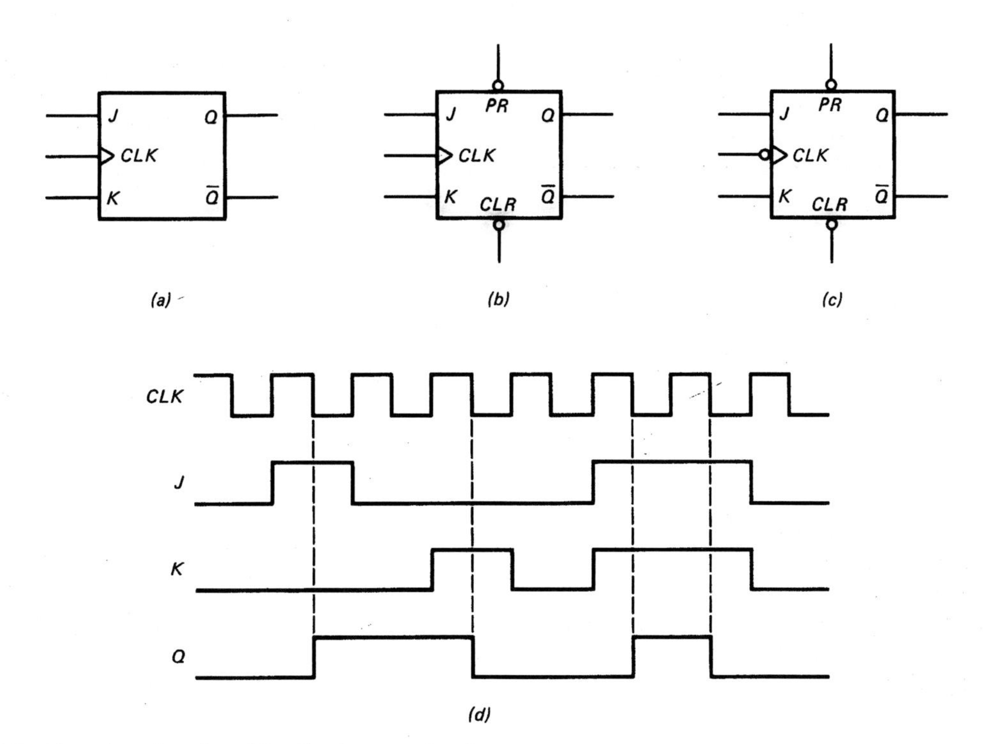

Fig. 7-13 (*a*) Positive-edge triggering; (*b*) active low preset and clear; (*c*) negative-edge triggering; (*d*) timing diagram.

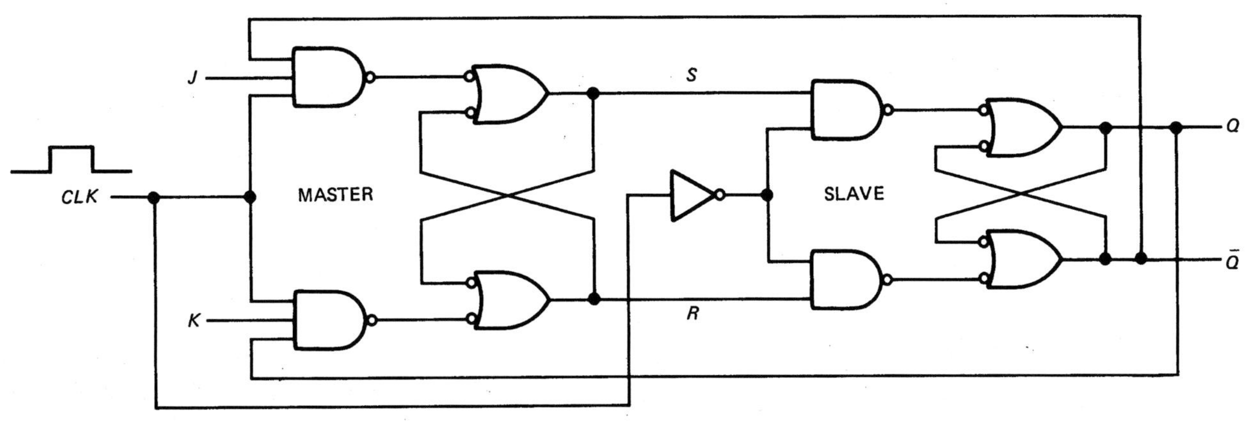

Fig. 7-14 Master-slave *JK* flip-flop.

clocked but the slave is negatively clocked. This implies the following:

1. While the clock is high, the master is active and the slave is inactive.
2. While the clock is low, the master is inactive and the slave is active.

Set

To start the analysis, let's assume low Q and high $\overline{Q}$. For an input condition of high J, low K, and high CLK, the master goes into the set state, producing high S and low R. Nothing happens to the Q and $\overline{Q}$ outputs because the slave is inactive while the clock is high. When the clock goes low, however, the high S and low R force the slave into the set state, producing a high Q and a low $\overline{Q}$.

There are two distinct steps in setting the final Q output. First, the master is set while the clock is high. Second, the slave is set while the clock is low. This action is sometimes called *cocking* and *triggering*. You cock the master during the positive half cycle of the clock, and you trigger the slave during the negative half cycle of the clock.

Reset

When the slave is set, Q is high and $\overline{Q}$ is low. For the input condition of low J, high K, and high CLK, the master will reset, forcing S to go low and R to go high. Again, no changes can occur in Q and $\overline{Q}$ because the slave is inactive while the clock is high. When the clock returns to the low state, the low S and high R force the slave to reset; this forces Q to go low and $\overline{Q}$ to go high.

Again, notice the cocking and triggering. This is the key idea behind the master-slave flip-flop. Every action of the master with a high CLK is copied by the slave when the clock goes low.

Toggle

If the J and K inputs are both high, the master toggles once while the clock is high; the slave then toggles once when the clock goes low. No matter what the master does, the slave copies it. If the master toggles into the set state, the slave toggles into the set state. If the master toggles into the reset state, the slave toggles into the reset state.

Level Clocking

The master-slave flip-flop is level-clocked in Fig. 7-14. While the clock is high, therefore, any changes in J and K can affect the S and R outputs. For this reason, you normally keep J and K constant during the positive half cycle of the clock. After the clock goes low, the master becomes inactive and you can allow J and K to change.

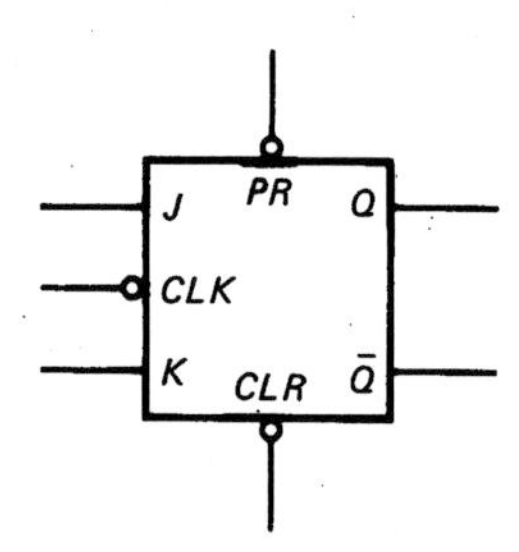

Fig. 7-15 Symbol for master-slave *JK* flip-flop.

Symbol

Figure 7-15 shows the symbol for a *JK* master-slave flip-flop with preset and clear functions. The bubble on the *CLK* input reminds us that the output changes when the clock goes low.

Truth Table

Table 7-11 summarizes the operation of a *JK* master-slave flip-flop. A low *PR* and low *CLR* produces a race condition; therefore, *PR* and *CLR* are normally kept at a high voltage when inactive. To clear, you take *CLR* low; to preset, you take *PR* low. In either case, you return them to high when ready to run.

TABLE 7-11. MASTER-SLAVE FLIP-FLOP

PR	*CLR*	*CLK*	*J*	*K*	*Q*
0	0	X	X	X	*
0	1	X	X	X	1
1	0	X	X	X	0
1	1	X	0	0	NC
1	1	⎍	0	1	0
1	1	⎍	1	0	1
1	1	⎍	1	1	Toggle

As before, low J and low K produce an inactive state, regardless of the what the clock is doing. If K goes high by itself, the next clock pulse resets the flip-flop. If J goes high by itself, the next clock pulse sets the flip-flop. When J and K are both high, each clock pulse produces one toggle.

EXAMPLE 7-2

Figure 7-16*a* shows a *clock generator*. What does it do when $\overline{HLT}$ is high?

SOLUTION

To begin with, the 555 is an IC that can generate a rectangular output when connected as shown in Fig. 7-16*a*. The frequency of the output is

$$f = \frac{1.44}{(R_A + 2R_B)C}$$

The duty cycle (ratio of high state to period) is

$$D = \frac{R_A + R_B}{R_A + 2R_B}$$

With the values shown in Fig. 7-16*a* the frequency of the output is

$$f = \frac{1.44}{(36\text{ k}\Omega + 36\text{ k}\Omega)(0.01\ \mu\text{F})} = 2\text{ kHz}$$

and the duty cycle is

$$D = \frac{36\text{ k}\Omega + 18\text{ k}\Omega}{36\text{ k}\Omega + 36\text{ k}\Omega} = 0.75$$

which is equivalent to 75 percent.

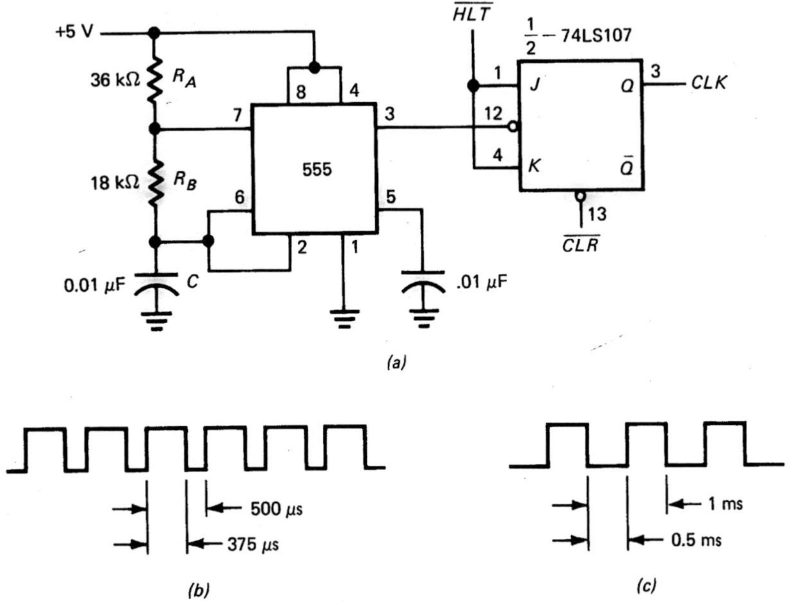

Fig. 7-16 Clock generator: (*a*) circuit; (*b*) 555 output; (*c*) *JK* flip-flop output.

Figure 7-16*b* illustrates how the output (pin 3) of the 555 looks. Note how the signal is high for 75 percent of the cycle. This unsymmetrical output drives the clock input of a *JK* master-slave flip-flop.

The *JK* master-slave flip-flop toggles once per input cycle; therefore, its output has a frequency of 1 kHz and a duty cycle of 50 percent. One of the reasons for using the flip-flop is to get the symmetrical output shown in Fig. 7-16*c*.

Another reason for using the flip-flop is to control the starting phase of the clock. A computer run starts with $\overline{CLR}$ going momentarily low, then back to high. This resets the flip-flop, forcing *CLK* to go low. Therefore, the starting phase of the *CLK* signal is always low. You will see the clock generator of Fig. 7-16*a* again in Chap. 10; remember that the *CLK* signal has a frequency of 1 kHz, a duty cycle of 50 percent, and starting phase of low.

GLOSSARY

contact bounce The making and breaking of contacts for a few milliseconds after a switch closes.

edge triggering Changing the output state of a flip-flop on the rising or falling edge of a clock pulse.

flip-flop A two-state circuit that can remain in either state indefinitely. Also called a bistable multivibrator. An external trigger can change the output state.

hold time The minimum amount of time the input signals must be held constant after the clock edge has struck. After a clock edge strikes a flip-flop, the internal transistors need time to change from one state to another. The input control signals (*D*, or *J* and *K*) must be held constant while these internal transistors are switching over.

latch The simplest type of flip-flop, consisting of two cross-coupled NAND or NOR latches.

level clocking A type of triggering in which the output of a flip-flop responds to the level (high or low) of the clock signal. With positive level clocking, for example, the output can change at any time during the positive half cycle.

master-slave triggering A type of triggering using two cascaded latches called the master and the slave. The master is cocked during the positive half cycle of the clock, and the slave is triggered during the negative half cycle.

propagation delay time The time it takes for the output of a gate or flip-flop to change after the inputs have changed.

race condition An undesirable condition which may exist in a system when two or more inputs change simultaneously. If the final output depends on which input changes first, a race condition exists.

setup time The minimum amount of time the inputs to a flip-flop must be present before the clock edge arrives.

toggle Change of the output to the opposite state in a *JK* flip-flop.

SELF-TESTING REVIEW

Read each of the following and provide the missing words. Answers appear at the beginning of the next question.

1. A flip-flop is a _________ element that stores a binary digit as a low or high voltage. With an *RS* latch a high *S* and a low *R* sets the output to _________; a low *S* and a high *R* _________ the output to low.
2. (*memory, high, reset*) With a NAND latch a low *R* and a low *S* produce a _________ condition. This is why *R* and *S* are kept high when inactive. One use for latches is switch debouncers; they eliminate the effects of _________ bounce.
3. (*race, contact*) Computers use thousands of flip-flops. To coordinate the overall action, a common signal called the _________ is sent to each flip-flop. With positive clocking the clock signal must be _________ for the flip-flop to respond. Positive and negative clocking are also called level clocking because the flip-flop responds to the _________ of the clock, either high or low.
4. (*clock, high, level*) In a *D* latch, data bit *D* drives the *S* input of a latch, and the complement $\overline{D}$ drives the *R* input; therefore, a high *D* _________ the latch and a low *D* resets it. Since *R* and *S* are always in opposite states in a *D* latch, the _________ condition is impossible.
5. (*sets, race*) With a positive-edge-triggered *D* flip-flop, the data bit is sampled and stored on the _________ edge of the clock pulse. Preset and clear inputs are often called _________ set and _________ reset. These inputs override the other inputs; they have first priority. When preset goes low, the *Q* output goes _________ and stays there no matter what the *D* and *CLK* inputs are doing.
6. (*rising, direct, direct, high*) In a flip-flop, propagation delay time is the amount of time it takes for the _________ to change after the clock edge has struck. Setup time is the amount of time an input signal must be present _________ the clock edge strikes. Hold time is the amount of time an input signal must be present _________ the clock edge strikes.
7. (*output, before, after*) In a positive-edge-triggered *JK* flip-flop, a low *J* and a low *K* produce the _________ state. A high *J* and a high *K* mean that the output will _________ on the rising edge of the clock.
8. (*inactive, toggle*) With a *JK* master-slave flip-flop the master is cocked when the clock is _________, and the slave is triggered when the clock is _________. This type of flip-flop is usually level-clocked instead of edge-triggered. For this reason, *J* and *K* are normally kept _________ while the clock is high.
9. (*high, low, constant*) Since capacitors are too difficult to fabricate on an IC chip, manufacturers rely on various direct-coupled designs for *D* flip-flops and *JK* flip-flops.

PROBLEMS

7-1. The waveforms of Fig. 7-17 drive a clocked *RS* latch (Fig. 7-5*a*). If *Q* is low before time *A*,

a. At what point does *Q* become a 1?

b. When does *Q* reset to 0?

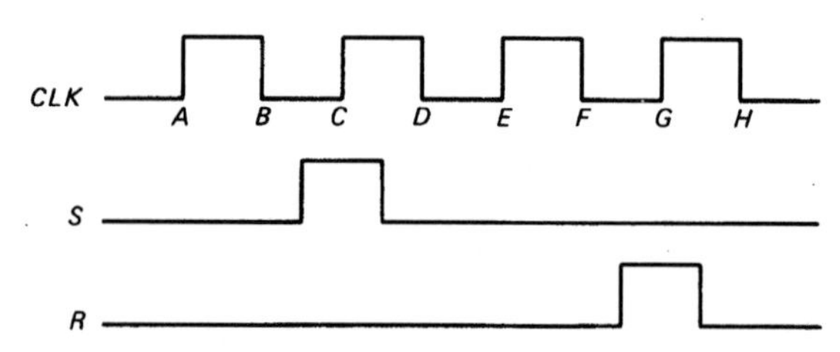

Fig. 7-17

7-2. A *D* flip-flop has these specifications:

$$t_{setup} = 10 \text{ ns}$$
$$t_{hold} = 5 \text{ ns}$$
$$t_p = 30 \text{ ns}$$

a. How far ahead of the rising clock edge must the data bit be applied to the *D* input to ensure correct storage?

b. After the rising clock edge, how long must you wait before letting the data bit change?

c. How long after the rising clock edge will *Q* change?

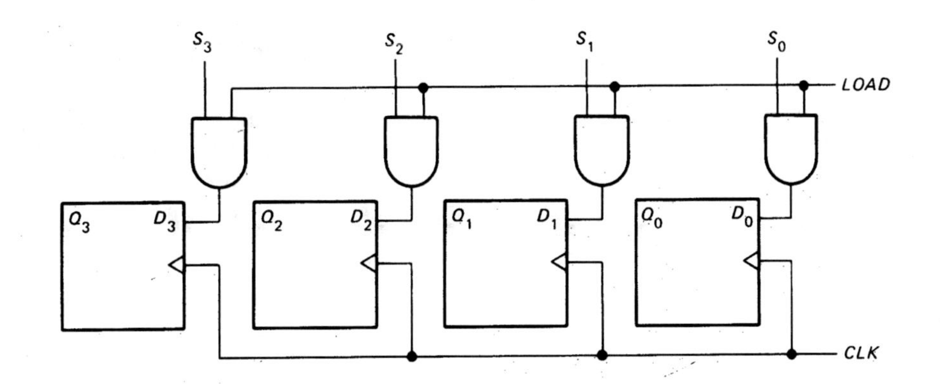

Fig. 7-18

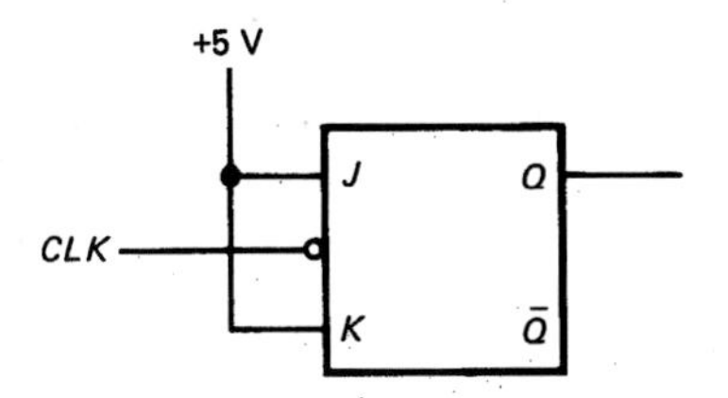

Fig. 7-19

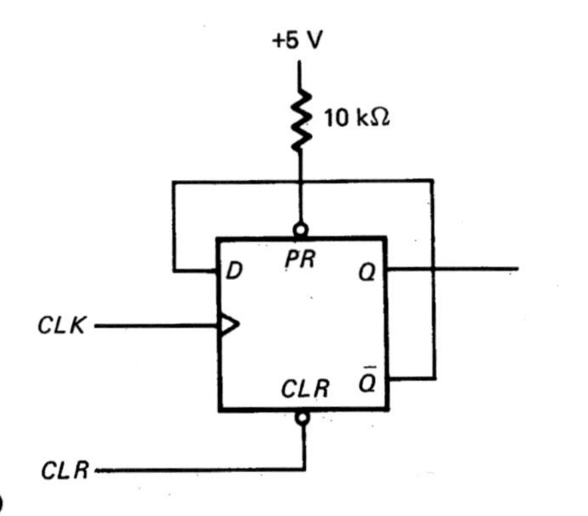

Fig. 7-20

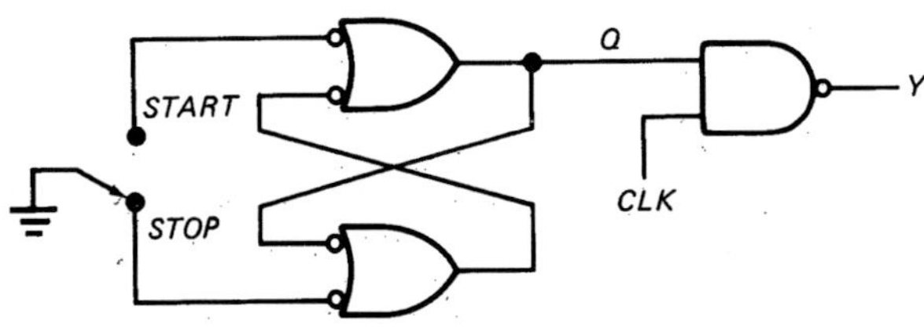

Fig. 7-21

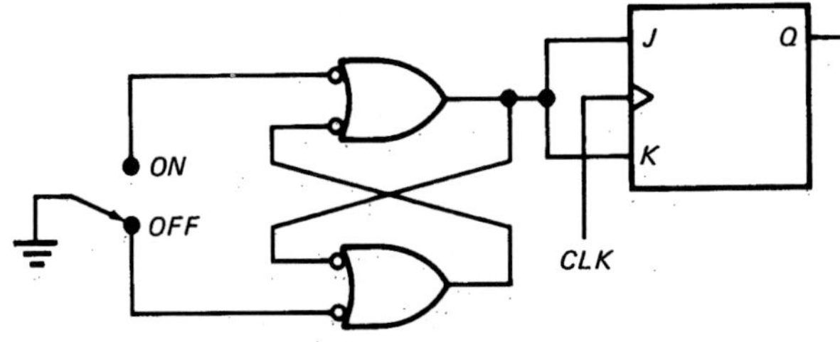

Fig. 7-22

7-3. In Fig. 7-18, the data word to be stored is

$$S = 1001$$

a. If *LOAD* is low, what does *Q* equal after the positive clock edge?
b. If *LOAD* is high, what does *Q* equal after the positive clock edge.

7-4. The clock of Fig. 7-19 has a frequency of 1 MHz, and the flip-flop has a propagation delay time of 25 ns.

a. What is the period of the clock?
b. The frequency of the *Q* output? Its period?
c. How long after the negative clock edge does the *Q* output change?

7-5. The clock has a frequency of 6 MHz in Fig. 7-19. What is the frequency of the *Q* output? This circuit is sometimes called a divide-by-2 circuit. Explain why.

7-6. In Fig. 7-20, *CLR* is taken low temporarily, then high. Draw the timing diagram. If the clock has a frequency of 1 MHz, what is the frequency of the *Q* output? Is this a divide-by-2 circuit?

7-7. Figure 7-21 shows a NAND latch used as a switch debouncer. With the switch in the STOP position, what do *Q* and *Y* equal? If the switch is thrown to the START position, what do *Q* and *Y* equal?

7-8. The clock has a frequency of 1 MHz in Fig. 7-22. With the switch in the OFF position, what is the frequency of the *Q* output? If the switch is thrown to the ON position, what is the frequency of the *Q* output?

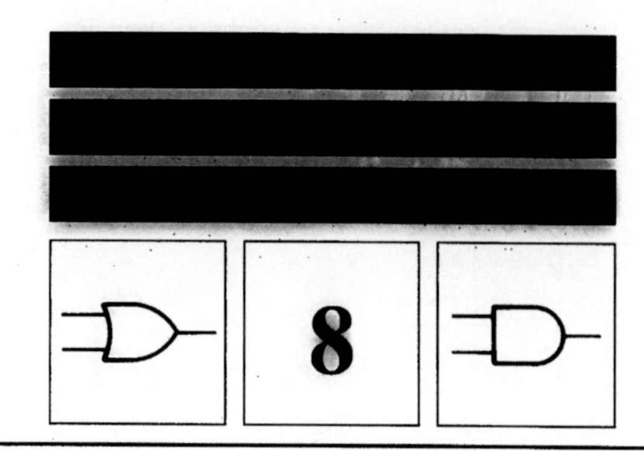

REGISTERS AND COUNTERS

A *register* is a group of memory elements that work together as a unit. The simplest registers do nothing more than store a binary word; others modify the stored word by shifting its bits left or right or by performing other operations to be discussed in this chapter. A *counter* is a special kind of register, designed to count the number of clock pulses arriving at its input. This chapter discusses some basic registers and counters used in microcomputers.

8-1 BUFFER REGISTERS

A *buffer register* is the simplest kind of register; all it does is store a digital word.

Basic Idea

Figure 8-1 shows a buffer register built with positive-edge-triggered *D* flip-flops. The *X* bits set up the flip-flops for loading. Therefore, when the first positive clock edge arrives, the stored word becomes $Q_3Q_2Q_1Q_0 = X_3X_2X_1X_0$. In chunked notation,

$$\mathbf{Q} = \mathbf{X}$$

The circuit is too primitive to be of any use. What it needs is some control over the *X* bits, some way of holding them off until we're ready to store them.

Controlled

Figure 8-2 is more like it. This is a controlled buffer register with an active-high *CLR*. Therefore, when *CLR* goes high, all flip-flops reset and the stored word becomes

$$\mathbf{Q} = 0000$$

When *CLR* returns low, the register is ready for action.

LOAD is a control input; it determines what the circuit does. When *LOAD* is low, the *X* bits cannot reach the flip-flops. At the same time, the inverted signal $\overline{LOAD}$ is high; this forces each flip-flop output to feed back to its data input. When each rising clock edge arrives, data is circulated or retained. In other words, the register contents are unchanged when *LOAD* is low.

When *LOAD* goes high, the *X* bits are transmitted to the data inputs. After a short setup time, the flip-flops are ready for loading. With the arrival of the positive clock edge, the *X* bits are loaded and the stored word becomes

$$Q_3Q_2Q_1Q_0 = X_3X_2X_1X_0$$

If *LOAD* returns to low, the foregoing word is stored indefinitely; this means that the *X* bits can change without affecting the stored word.

EXAMPLE 8-1

Chapter 10 discusses the *SAP* (simple-as-possible) computer. This educational computer has three generations, SAP-1, SAP-2, and SAP-3. Figure 8-3 shows the output register of the SAP-1 computer. The 74LS173 chips are controlled buffer registers, similar to Fig. 8-2. What does the circuit do?

SOLUTION

To begin with, it is an 8-bit buffer register built with TTL chips. Each chip handles 4 bits of input word **X**. The upper nibble $X_7X_6X_5X_4$ goes to pins 14, 13, 12, and 11 of C22; the lower nibble $X_3X_2X_1X_0$ goes to pins 14, 13, 12, and 11 of the C23.

Output word **Q** drives an 8-bit LED display. The upper nibble $Q_7Q_6Q_5Q_4$ comes out of pins 3, 4, 5, and 6 of C22; the lower nibble $Q_3Q_2Q_1Q_0$ comes out of pins 3, 4, 5, and 6 of C23. The typical high-state output of a 74LS173 is 3.5 V, and the typical LED drop is 1.5 V. Since each current-limiting resistance is 1 kΩ, the high-state current is approximately 2 mA for each output pin.

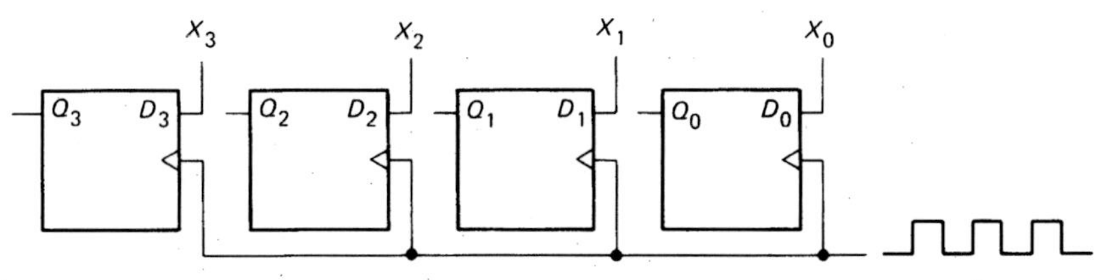

Fig. 8-1 Buffer register.

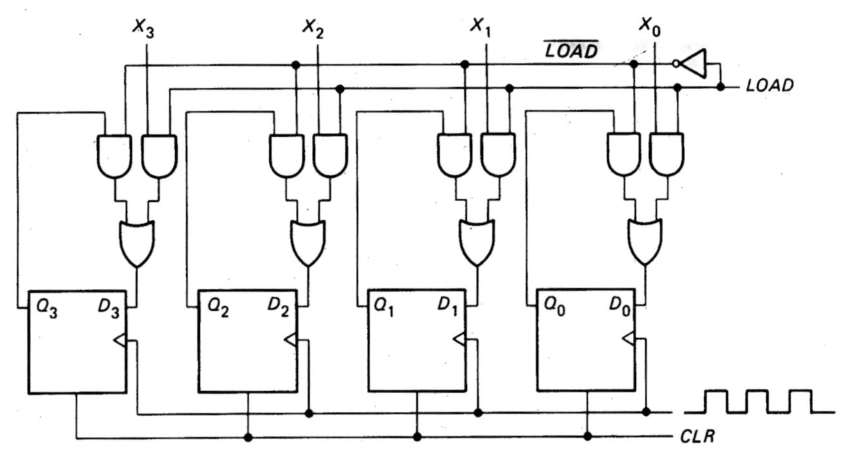

Fig. 8-2 Controlled buffer register.

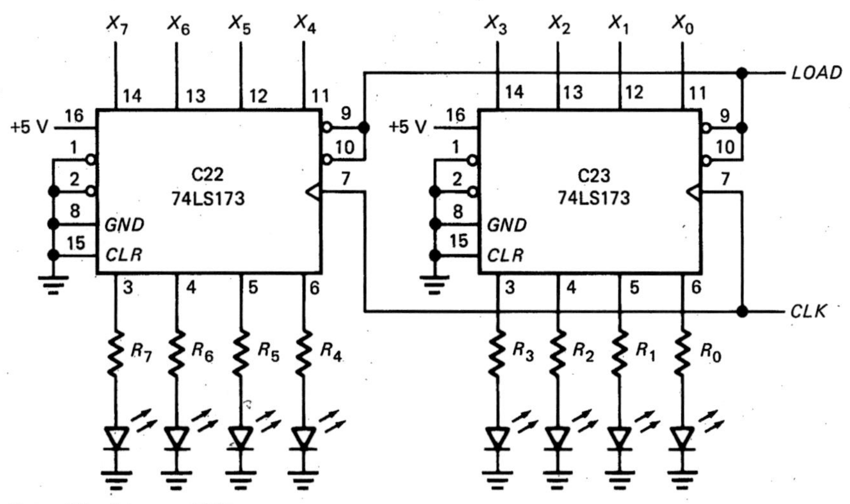

Fig. 8-3 SAP-1 output register.

The 74LS173 requires a 5-V supply for pin 16 and a ground return on pin 8. The SAP-1 output register never needs clearing; this is why the CLR input (pin 15) is made inactive by tying it to ground. In a 74LS173, pins 9 and 10 are separate *LOAD* controls. Because SAP-1 needs only a single *LOAD* control, pins 9 and 10 are tied together. The bubbles on pins 9 and 10 indicate an active low state; this means that *LOAD* must be low for the positive clock edge to store the input word. See Appendix 4 for a more detailed description of the 74LS173.

The action of the circuit is straightforward. While *LOAD* is high, the register contents are unchanged even though the clock is running. To change the stored word, *LOAD* must go low. Then the next rising clock edge loads the *X* bits into the register. As soon as this happens, the LED display shows the new contents.

8-2 SHIFT REGISTERS

A *shift register* moves the stored bits left or right. This bit shifting is essential for certain arithmetic and logic operations used in microcomputers.

Shift Left

Figure 8-4 is a *shift-left* register. As shown, D_{in} sets up the right flip-flop, Q_0 sets up the second flip-flop, Q_1 the third, and so on. When the next positive clock edge strikes, therefore, the stored bits move one position to the left.

As an example, here's what happens with $D_{in} = 1$ and

$$Q = 0000$$

All data inputs except the one on the right are 0s. The arrival of the first rising clock edge sets the right flip-flop, and the stored word becomes

$$Q = 0001$$

This new word means D_1 now equals 1, as well as D_0. When the next positive clock edge hits, the Q_1 flip-flop sets and the register contents become

$$Q = 0011$$

The third positive clock edge results in

$$Q = 0111$$

and the fourth rising clock edge gives

$$Q = 1111$$

Hereafter, the stored word is unchanged as long as $D_{in} = 1$.

Suppose D_{in} is now changed to 0. Then, successive clock pulses produce these register contents:

$$Q = 1110$$
$$Q = 1100$$
$$Q = 1000$$
$$Q = 0000$$

As long as $D_{in} = 0$, subsequent clock pulses have no further effect.

The timing diagram of Fig. 8-5 summarizes the foregoing discussion.

Shift Right

Figure 8-6 is a *shift-right register*. As shown, each Q output sets up the D input of the preceding flip-flop. When the rising clock edge arrives, the stored bits move one position to the right.

Here's an example with $D_{in} = 1$ and

$$Q = 0000$$

All data inputs except the one on the left are 0s. The first positive clock edge sets the left flip-flop and the stored word becomes

$$Q = 1000$$

With the appearance of this word, D_3 and D_2 are 1s. The second rising clock edge gives

$$Q = 1100$$

The third clock pulse gives

$$Q = 1110$$

and the fourth clock pulse gives

$$Q = 1111$$

8-3 CONTROLLED SHIFT REGISTERS

A *controlled shift register* has control inputs that determine what it does on the next clock pulse.

SHL Control

Figure 8-7 shows how the shift-left operation can be controlled. *SHL* is the control signal. When *SHL* is low, the inverted signal $\overline{SHL}$ is high. This forces each flip-flop output to feed back to its data input. Therefore, the data is retained in each flip-flop as the clock pulses arrive. In this way, a digital word can be stored indefinitely.

When *SHL* goes high, D_{in} sets up the right flip-flop, Q_0 sets up the second flip-flop, Q_1 the third flip-flop, and so on. In this mode, the circuit acts like a shift-left register. Each positive clock edge shifts the stored bits one position to the left.

Serial Loading

Serial loading means storing a word in the shift register by entering 1 bit per clock pulse. To store a 4-bit word, we need four clock pulses. For instance, here's how to serially store the word

$$X = 1010$$

With *SHL* high in Fig. 8-7, make $D_{in} = 1$ for the first clock pulse, $D_{in} = 0$ for the second clock pulse, $D_{in} = 1$

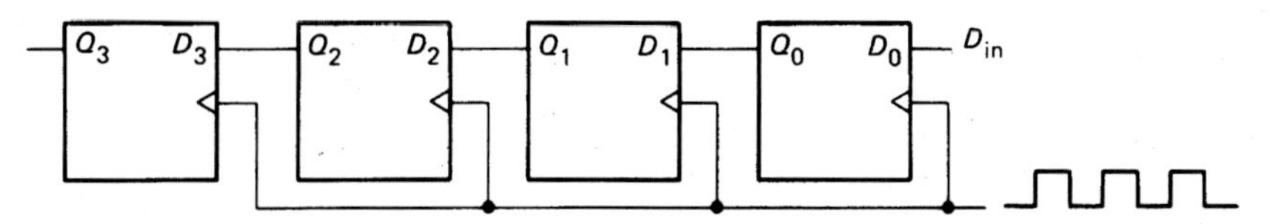

Fig. 8-4 Shift-left register.

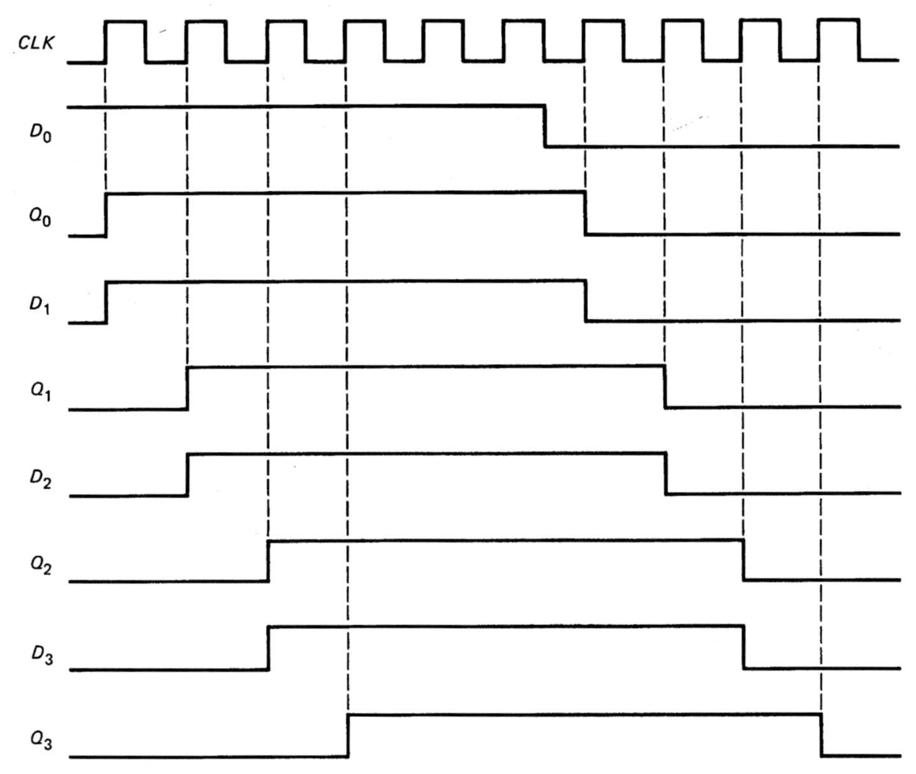

Fig. 8-5 Shift-left timing diagram.

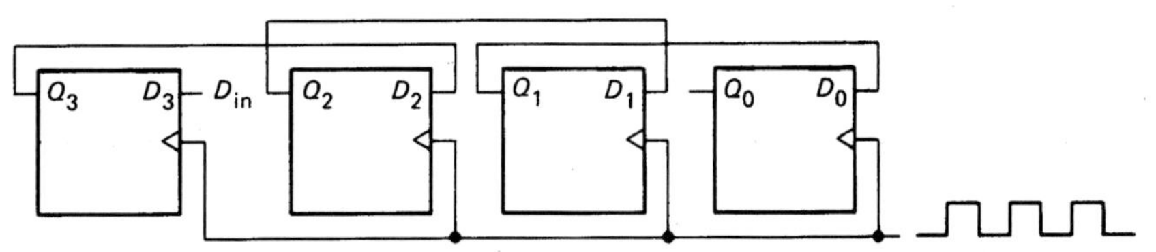

Fig. 8-6 Shift-right register.

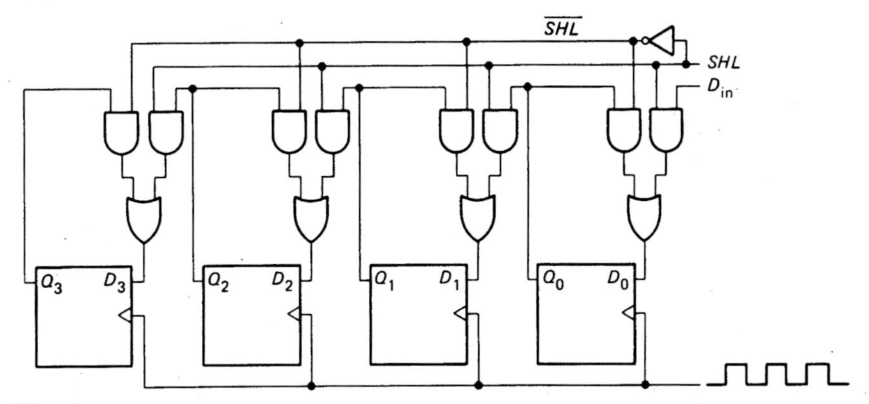

Fig. 8-7 Controlled shift register.

for the third clock pulse, and D_{in} = 0 for the fourth clock pulse. If the register is clear before the first clock pulse, the successive register contents look like this:

$$
\begin{aligned}
\mathbf{Q} &= 0001 \quad (D_{in} = 1\text{: first clock pulse}) \\
\mathbf{Q} &= 0010 \quad (D_{in} = 0\text{: second clock pulse}) \\
\mathbf{Q} &= 0101 \quad (D_{in} = 1\text{: third clock pulse}) \\
\mathbf{Q} &= 1010 \quad (D_{in} = 0\text{: fourth clock pulse})
\end{aligned}
$$

In this way, data is entered serially into the right end of the register and shifted left until all 4 bits have been stored. After the last bit is entered, *SHL* is taken low to freeze the register contents.

Parallel Loading

Figure 8-8 is another step in the evolution of shift registers. The circuit can load *X* bits directly into the flip-flops, the same as a buffer register. This kind of entry is called *parallel* or *broadside loading;* it takes only one clock pulse to store a digital word.

If *LOAD* and *SHL* are low, the output of the NOR gate is high and flip-flop outputs return to their data inputs. This forces the data to be retained in each flip-flop as the positive clock edges arrive. In other words, the register is inactive when *LOAD* and *SHL* are low, and the contents are stored indefinitely.

When *LOAD* is low and *SHL* is high, the circuit acts like a shift-left register, as previously described. On the other hand, when *LOAD* is high and *SHL* is low, the circuit acts like a buffer register because the *X* bits set up the flip-flops for broadside loading. (Having *LOAD* and *SHL* simultaneously high is forbidden because it's impossible to do both operations on a single clock edge.)

By adding more flip-flops we can build a controlled shift register of any length. And with more gates, the shift-right operation can be included. As an example, the 74198 is a TTL 8-bit bidirectional shift register. It can broadside load, shift left, or shift right.

8-4 RIPPLE COUNTERS

A *counter* is a register capable of counting the number of clock pulses that have arrived at its clock input. In its simplest form it is the electronic equivalent of a binary odometer.

The Circuit

Figure 8-9*a* shows a counter built with *JK* flip-flops. Since the *J* and *K* inputs are returned to a high voltage, each flip-flop will toggle when its clock input receives a negative edge.

Here's how the counter works. Visualize the *Q* outputs as a binary word

$$\mathbf{Q} = Q_3Q_2Q_1Q_0$$

Q_3 is the most significant bit (MSB), and Q_0 is the least significant bit (LSB). When *CLR* goes low; all flip-flops reset. This results in a digital word of

$$\mathbf{Q} = 0000$$

When *CLR* returns to high, the counter is ready to go. Since the LSB flip-flop receives each clock pulse, Q_0 toggles once per negative clock edge, as shown in the timing diagram of Fig. 8-9*b*. The remaining flip-flops toggle less often because they receive their negative edges from the preceding flip-flops.

For instance, when Q_0 goes from 1 back to 0, the Q_1 flip-flop receives a negative edge and toggles. Likewise,

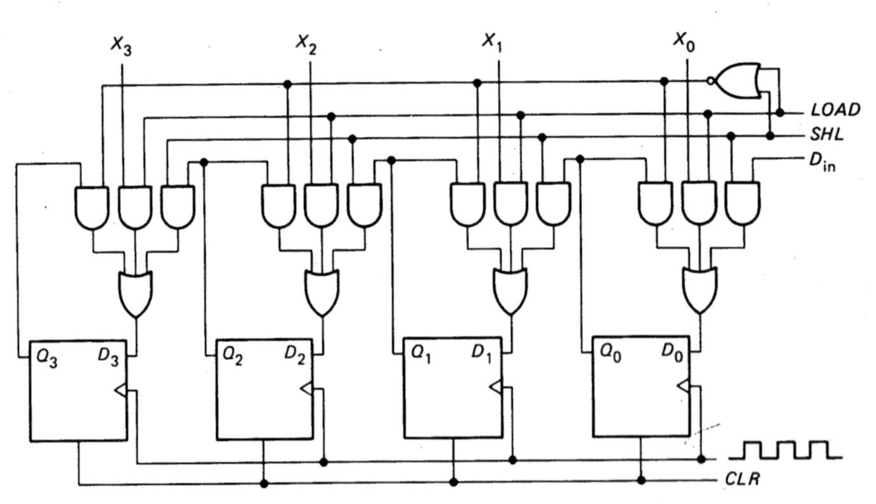

Fig. 8-8 Shift register with broadside load.

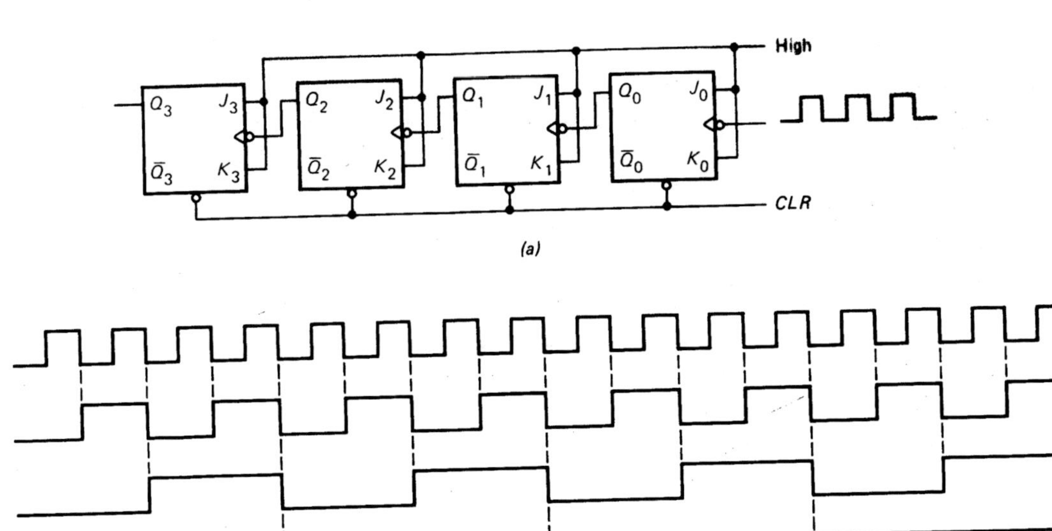

Fig. 8-9 (*a*) Ripple counter; (*b*) timing diagram.

when Q_1 changes from 1 back to 0, the Q_2 flip-flop gets a negative edge and toggles. And when Q_2 goes from 1 to 0, the Q_3 flip-flop toggles. In other words, whenever a flip-flop resets to 0, the next higher flip-flop toggles (see Fig. 8-9*b*).

What does this remind you of? Reset and carry! Each flip-flop acts like a wheel in a binary odometer; whenever it resets to 0, it sends a carry to the next higher flip-flop. Therefore, the counter of Fig. 8-9*a* is the electronic equivalent of a binary odometer.

Counting

If *CLR* goes low then high, the register contents of Fig. 8-9*a* become

$$\mathbf{Q} = 0000$$

When the first clock pulse hits the LSB flip-flop, Q_0 becomes a 1. So the first output word is

$$\mathbf{Q} = 0001$$

When the second clock pulse arrives, Q_0 resets and carries; therefore, the next output word is

$$\mathbf{Q} = 0010$$

The third clock pulse advances Q_0 to 1; this gives

$$\mathbf{Q} = 0011$$

The fourth clock pulse forces the Q_0 flip-flop to reset and carry. In turn, the Q_1 flip-flop resets and carries. The resulting output word is

$$\mathbf{Q} = 0100$$

The fifth clock pulse gives

$$\mathbf{Q} = 0101$$

The sixth gives

$$\mathbf{Q} = 0110$$

and the seventh gives

$$\mathbf{Q} = 0111$$

On the eighth clock pulse, Q_0 resets and carries, Q_1 resets and carries, Q_2 resets and carries, and Q_3 advances to 1. So the output word becomes

$$\mathbf{Q} = 1000$$

The ninth clock pulse gives

$$\mathbf{Q} = 1001$$

The tenth gives

$$\mathbf{Q} = 1010$$

and so on.

TABLE 8-1. RIPPLE COUNTER

Count	$Q_3Q_2Q_1Q_0$
0	0 0 0 0
1	0 0 0 1
2	0 0 1 0
3	0 0 1 1
4	0 1 0 0
5	0 1 0 1
6	0 1 1 0
7	0 1 1 1
8	1 0 0 0
9	1 0 0 1
10	1 0 1 0
11	1 0 1 1
12	1 1 0 0
13	1 1 0 1
14	1 1 1 0
15	1 1 1 1

The last word is

$$Q = 1111$$

corresponding to the fifteenth clock pulse. The next clock pulse resets all flip-flops. Therefore, the counter resets to

$$Q = 0000$$

and the cycle repeats.

Table 8-1 summarizes the operation of the counter. *Count* represents the number of clock pulses that have arrived. As you see, the counter output is the binary equivalent of the decimal count.

Frequency Division

Each flip-flop in Fig. 8-9*a* divides the clock frequency by a factor of 2. This is why a flip-flop is sometimes called a *divide-by-2 circuit.* Since each flip-flop divides the clock frequency by 2, *n* flip-flops divide the clock frequency by 2^n.

The timing diagram of Fig. 8-9*b* illustrates the divide-by-2 action. Q_0 is one-half the clock frequency, Q_1 is one-fourth the clock frequency, Q_2 is one-eighth the clock frequency, and Q_3 is one-sixteenth of the clock frequency. In other words,

1 flip-flop divides by 2
2 flip-flops divide by 4
3 flip-flops divide by 8
4 flip-flops divide by 16

and

n flip-flops divide by 2^n

Ripple Counter

The counter of Fig. 8-9*a* is known as a *ripple counter* because the carry moves through the flip-flops like a ripple on water. In other words, the Q_0 flip-flop must toggle before the Q_1 flip-flop, which in turn must toggle before the Q_2 flip-flop, which in turn must toggle before the Q_3 flip-flop. The worst case occurs when the stored word changes from 0111 to 1000, or from 1111 to 0000. In either case, the carry has to move all the way to the MSB flip-flop. Given a t_p of 10 ns per flip-flop, it takes 40 ns for the MSB to change.

By adding more flip-flops to the left end of Fig. 8-9*a* we can build a ripple counter of any length. Eight flip-flops give an 8-bit ripple counter, twelve flip-flops result in a 12-bit ripple counter, and so on.

Controlled Counter

A controlled counter counts clock pulses only when commanded to do so. Figure 8-10 shows how it's done. The *COUNT* signal can be low or high. Since it conditions the *J* and *K* inputs, *COUNT* controls the action of the counter, forcing it to either do nothing or to count clock pulses.

When *COUNT* is low, the *J* and *K* inputs are low; therefore, all flip-flops remain latched in spite of the clock pulses driving the counter.

On the other hand, when *COUNT* is high, the *J* and *K* inputs are high. In this case, the counter works as previously described; each negative clock edge increments the stored count by 1.

EXAMPLE 8-2

As mentioned earlier, the program and data are stored in the memory before a computer run. The program is a list of instructions telling the computer how to process the data.

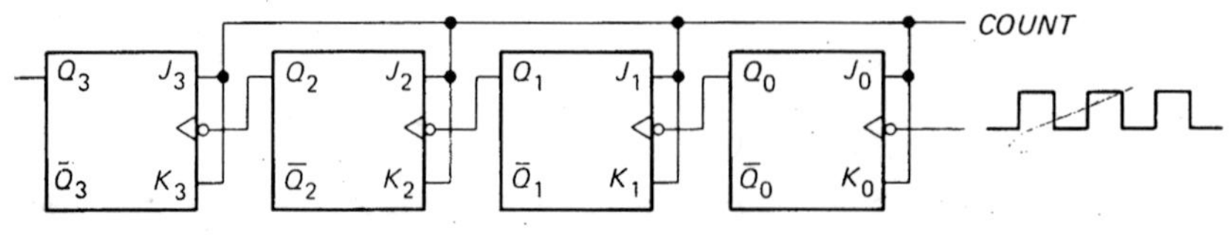

Fig. 8-10 Controlled ripple counter.

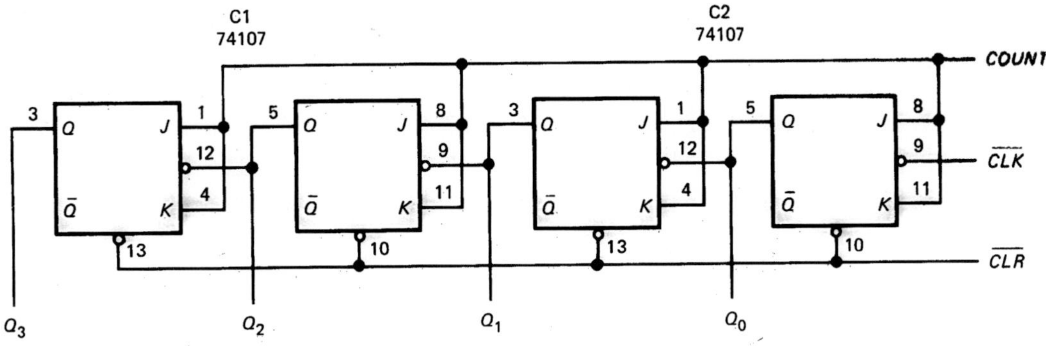

Fig. 8-11 SAP-1 program counter.

Every microcomputer has a *program counter* to keep track of the instruction being executed.

Figure 8-11 shows part of the program counter used in SAP-1. What does it do?

SOLUTION

To begin with, let's find out why the $\overline{CLR}$ and $\overline{CLK}$ signals are shown as complements. Signals are often available in complemented and uncomplemented form. The switch debouncer of Fig. 7-4*a* has two outputs, *CLR* and $\overline{CLR}$. In SAP-1 the *CLR* signal goes to any circuit that uses an active high clear and the $\overline{CLR}$ signal to any circuit with an active low clear. This is why $\overline{CLR}$ goes to the counter of Fig. 8-11; it has an active low clear. A similar idea applies to the clock signal.

The 74107 is a dual *JK* master-slave flip-flop. The SAP-1 program counter uses two 74107s. Although not shown, pin 14 ties to the 5-V supply, and pin 7 is the chip ground. Because master-slave flip-flops are used, a high $\overline{CLK}$ cocks the master and a low $\overline{CLK}$ triggers the slave.

Before a computer run, the operator pushes a clear button that sends a low $\overline{CLR}$ to the program counter. This resets its count to

$$Q = 0000$$

When the operator releases the button, $\overline{CLR}$ goes high and the computer run begins.

After the first instruction has been fetched from the memory, *COUNT* goes high for one clock pulse and the count becomes

$$Q = 0001$$

This count indicates that the first instruction has been fetched from the memory. (Later you will see how the computer executes the first instruction.)

After the first instruction has been executed, the computer fetches the second instruction in the memory. Once again, *COUNT* goes high for one clock pulse, producing a new count of

$$Q = 0010$$

The program counter now indicates that the second instruction has been fetched from the memory.

Each time a new instruction is fetched from the memory, the program counter is incremented to produce the next higher count. In this way, the computer can keep track of which instruction it's working on.

8-5 SYNCHRONOUS COUNTERS

When the carry has to propagate through a chain of n flip-flops, the overall propagation delay time is nt_p. For this reason ripple counters are too slow for some applications. To get around the ripple-delay problem, we can use a *synchronous counter*.

The Circuit

Figure 8-12 shows one way to build a synchronous counter with positive-edge-triggered flip-flops. This time, clock pulses drive all flip-flops in parallel. Because of the simultaneous clocking, the correct binary word appears after one propagation delay time rather than four.

The least significant flip-flop has its J and K inputs tied to a high voltage; therefore, it responds to each positive clock edge. But the remaining flip-flops can respond to the positive clock edge only under certain conditions. As shown in Fig. 8-12, the Q_1 flip-flop toggles on the positive clock edge only when Q_0 is a 1. The Q_2 flip-flop toggles only when Q_1 and Q_0 are 1s. And the Q_3 flip-flop toggles only when Q_2, Q_1, and Q_0 are 1s. In other words, a flip-flop toggles on the next positive clock edge if all lower bits are 1s.

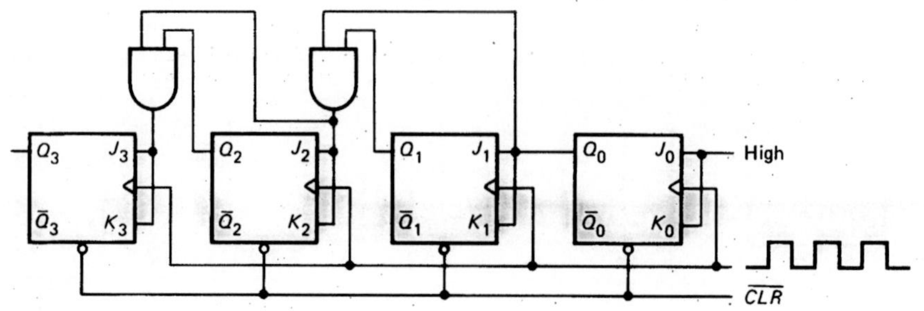

Fig. 8-12 Synchronous counter.

Here's the counting action. A low $\overline{CLR}$ resets the counter to

$$\mathbf{Q} = 0000$$

When the $\overline{CLR}$ line goes high, the counter is ready to go. The first positive clock edge sets Q_0 to get

$$\mathbf{Q} = 0001$$

Since Q_0 is now 1, the Q_1 flip-flop is conditioned to toggle on the next positive clock edge.

When the second positive clock edge arrives, Q_1 and Q_0 simultaneously toggle and the output word becomes

$$\mathbf{Q} = 0010$$

The third positive clock edge advances the count by 1:

$$\mathbf{Q} = 0011$$

Because Q_1 and Q_0 are now 1s, the Q_2, Q_1, and Q_0 flip-flops are conditioned to toggle on the next positive clock edge. When the fourth positive clock edge arrives, Q_2, Q_1, and Q_0 toggle simultaneously, and after one propagation delay time the output word becomes

$$\mathbf{Q} = 0100$$

The successive **Q** words are 0101, 0110, 0111, and so on up to 1111 (equivalent to decimal 15). The next positive clock edge resets the counter, and the cycle repeats.

By adding more flip-flops and gates we can build synchronous counters of any length. The advantage of a synchronous counter is its speed; it takes only one propagation delay time for the correct binary count to appear after the clock edge hits.

Controlled Counter

Figure 8-13 shows how to build a *controlled synchronous counter*. A low *COUNT* disables all flip-flops. When *COUNT* is high, the circuit becomes a synchronous counter; each positive clock edge advances the count by 1.

8-6 RING COUNTERS

Instead of counting with binary numbers, a *ring counter* uses words that have only a single high bit.

Circuit

Figure 8-14 is a *ring counter* built with *D* flip-flops. The Q_0 output sets up the D_1 input, the Q_1 output sets up the D_2 input, and so on. Therefore, a ring counter resembles a

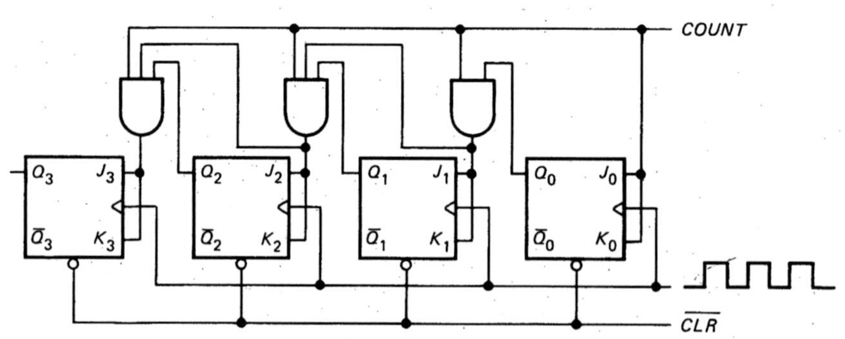

Fig. 8-13 Controlled synchronous counter.

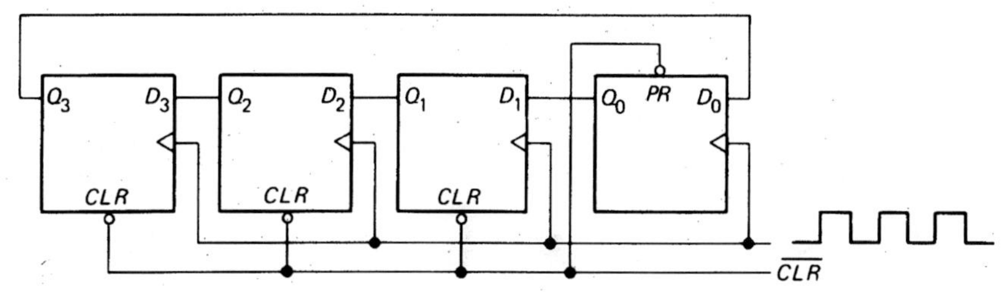

Fig. 8-14 Ring counter.

shift-left register because the bits are shifted left one position per positive clock edge. But the circuit differs because the final output is fed back to the D_0 input. This kind of action is called *rotate left;* bits are shifted left and fed back to the input.

When $\overline{CLR}$ goes low then back to high, the initial output word is

$$\mathbf{Q} = 0001$$

The first positive clock edge shifts the MSB into the LSB position; the other bits shift left one position. Therefore, the output word becomes

$$\mathbf{Q} = 0010$$

The second positive clock edge causes another rotate left and the output word changes to

$$\mathbf{Q} = 0100$$

After the third positive clock edge, the output word is

$$\mathbf{Q} = 1000$$

The fourth positive clock edge starts the cycle over because the rotate left produces

$$\mathbf{Q} = 0001$$

The stored 1 bit follows a circular path, moving left through the flip-flops until the final flip-flop sends it back to the first flip-flop. This is why the circuit is called a ring counter.

More Bits

Add more flip-flops and you can build a ring counter of any length. With six flip-flops we get a 6-bit ring counter. Again, the $\overline{CLR}$ signal resets all flip-flops except the LSB flip-flop. Therefore, the successive ring words are

$$\mathbf{Q} = 000001 \quad (0)$$
$$\mathbf{Q} = 000010 \quad (1)$$
$$\mathbf{Q} = 000100 \quad (2)$$
$$\mathbf{Q} = 001000 \quad (3)$$
$$\mathbf{Q} = 010000 \quad (4)$$
$$\mathbf{Q} = 100000 \quad (5)$$

Each of the foregoing words has only 1 high bit. The initial word stands for decimal 0 and the final word for decimal 5. If a ring counter has n flip-flops, therefore, the final ring word represents decimal $n - 1$.

Applications

Ring counters cannot compete with ripple and synchronous counters when it comes to ordinary counting, but they are invaluable when it's necessary to control a sequence of operations. Because each ring word has only 1 high bit, you can activate one of several devices.

For instance, suppose the six small boxes (A to F) of Fig. 8-15 are digital circuits that can be turned on by a high Q bit. When $\overline{CLR}$ goes low, Q_0 goes high and activates device A. After $\overline{CLR}$ returns to high, successive clock pulses turn on each device for a short time. In other words, as the stored 1 bit shifts left, it turns on B to F in sequence, and then the cycle starts over.

Many digital circuits participate during a computer run. To fetch and execute instructions, a computer has to activate

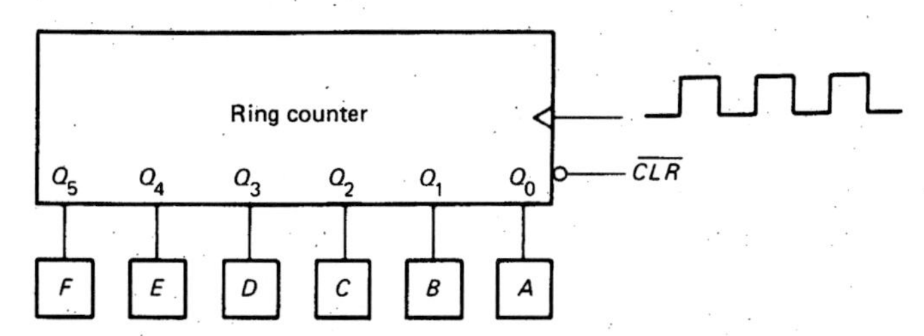

Fig. 8-15 Controlling a sequence of operations

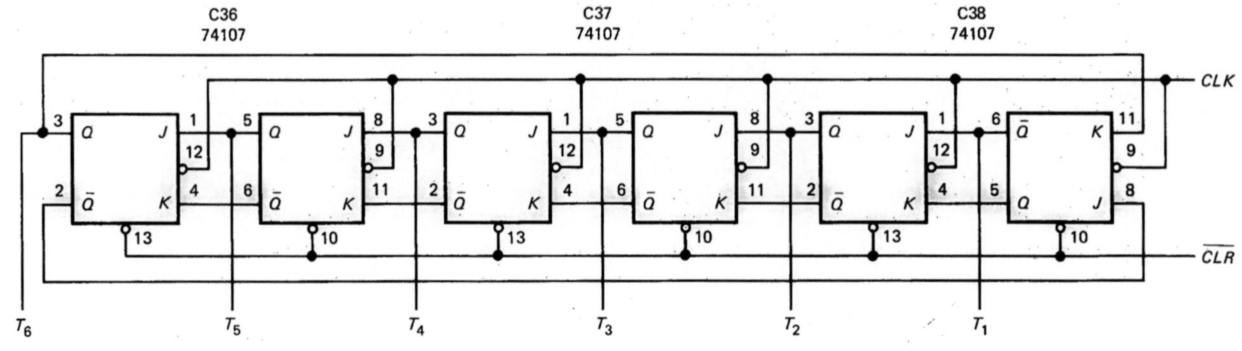

Note: Pin 14 is connected to +5 V, and pin 7 is grounded.

Fig. 8-16 SAP-1 ring counter.

these circuits at precisely the right time and in the right sequence. This is where ring counters shine; they produce the ring words for timing different operations during a computer run.

EXAMPLE 8-3

Figure 8-16 shows the ring counter used in the SAP-1 computer. T_6 to T_1 are called *timing* signals because they control a sequence of digital operations. What does this ring counter do?

SOLUTION

The 74107 is a dual *JK* master-slave flip-flop, previously used in the SAP-1 program counter (Example 8-2). The flip-flops are connected in a rotate-left mode. Since the 74107 does not have a preset input, the Q_0 flip-flop is inverted so that its $\overline{Q}$ output drives the *J* input of the Q_1 flip-flop. In this way, a low $\overline{CLR}$ produces the initial timing word

$$\mathrm{T_6T_5T_4T_3T_2T_1} = 000001$$

In chunked form

$$\mathbf{T} = 000001$$

Because of the master-slave action, a complete clock pulse is needed to produce the next ring word. After $\overline{CLR}$ returns high, the successive clock pulses produce the timing words

$$\mathbf{T} = 000010$$
$$\mathbf{T} = 000100$$
$$\mathbf{T} = 001000$$
$$\mathbf{T} = 010000$$
$$\mathbf{T} = 100000$$

Then the cycle repeats.

EXAMPLE 8-4

The clock frequency in Fig. 8-16 is 1 kHz. $\overline{CLR}$ goes low then high. Show the timing diagram.

SOLUTION

Figure 8-17 is the timing diagram. Since the clock has a frequency of 1 kHz, it has a period of 1 ms. This is the amount of time between successive negative clock edges. Each negative clock edge produces the next ring word. When its turn comes, each timing signal goes high for 1 ms.

Notice that the *CLK* signal of Fig. 8-17 is the input to the ring counter of Fig. 8-16, whereas the complement $\overline{CLK}$ is the input to the program counter of Fig. 8-11. This half-cycle difference is deliberate. The reason is given in Chap. 10, which explains how the timing signals of Fig. 8-17 control circuits that fetch and execute each program instruction.

8-7 OTHER COUNTERS

The *modulus* of a counter is the number of output states it has. A 4-bit ripple counter has a modulus of 16 because it has 16 distinct states numbered from 0000 to 1111. By changing the design we can produce a counter with any desired modulus.

Mod-10 Counter

Figure 8-18*a* shows a way to build a modulus-10 (or mod-10) counter. The circuit counts from 0000 to 1001, as before. However, on the tenth clock pulse, the counter

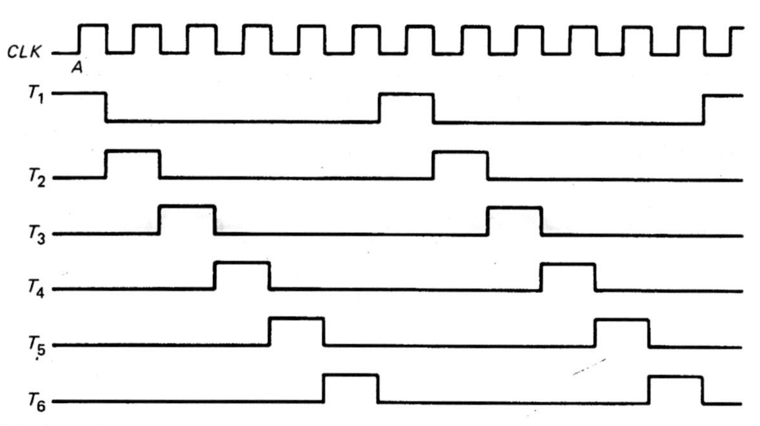

Fig. 8-17 SAP-1 clock and timing pulses.

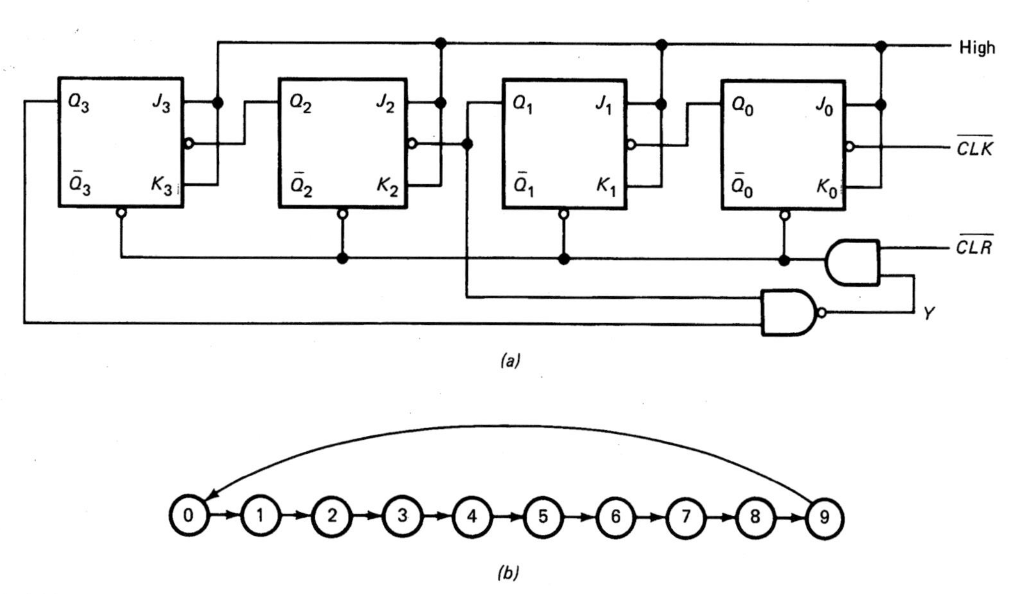

Fig. 8-18 Mod-10 counter.

generates its own clear signal and the count jumps back to 0000. In other words, the count sequence is

$$
\begin{aligned}
\mathbf{Q} &= 0000 \quad (0)\\
\mathbf{Q} &= 0001 \quad (1)\\
\mathbf{Q} &= 0010 \quad (2)\\
\mathbf{Q} &= 0011 \quad (3)\\
\mathbf{Q} &= 0100 \quad (4)\\
\mathbf{Q} &= 0101 \quad (5)\\
\mathbf{Q} &= 0110 \quad (6)\\
\mathbf{Q} &= 0111 \quad (7)\\
\mathbf{Q} &= 1000 \quad (8)\\
\mathbf{Q} &= 1001 \quad (9)\\
\mathbf{Q} &= 0000 \quad (0)
\end{aligned}
$$

As you see, the circuit skips states 10 to 15 (1010 through 1111). The counting sequence is summarized by the *state diagram* of Fig. 8-18*b*.

Why does the counter skip the states from 10 to 15? Because of the AND gate, the counter can be reset by a low $\overline{CLR}$ or a low Y. Initially, $\overline{CLR}$ goes low to produce

$$\mathbf{Q} = 0000$$

When $\overline{CLR}$ returns to high, the counter is ready for action.

The output of the NAND gate is

$$Y = \overline{Q_3 Q_1}$$

This output is high for the first nine states (0000 to 1001). Nothing unusual happens when the circuit is counting from 0 to 9. On the tenth clock pulse, however, the **Q** word becomes

$$\mathbf{Q} = 1010$$

which means that Q_3 and Q_1 are high. Almost immediately, Y goes low, forcing the counter to reset to

$$\mathbf{Q} = 0000$$

Y then goes high, and the counter is ready to start over.

Since it takes 10 clock pulses to reset the counter, the output frequency of the Q_3 flip-flop is one-tenth of the clock frequency. This is why a mod-10 counter is also known as a *divide-by-10 circuit*.

A mod-10 counter like Fig. 8-18*a* is often called a *decade counter*. Because it counts from 0 to 9, it is a natural choice in BCD applications like frequency counters, digital voltmeters, and electronic wristwatches.

To get any other modulus, we can use the same basic idea. For instance, to get a mod-12 counter, we can drive the NAND gate of Fig. 8-18*a* with Q_3 and Q_2. Then the circuit counts from 0 to 11 (0000 to 1011). On the next clock pulse, Q_3 and Q_2 are high, which clears the counter. (What is the modulus if Q_3 and Q_0 drive the NAND gate?)

Down Counter

All the counters discussed so far have counted upward, toward higher numbers. Figure 8-19 shows a *down counter;* it counts from 1111 to 0000. Each flip-flop toggles when its clock input goes from 1 to 0. This is equivalent to an uncomplemented output going from 0 to 1. For instance, the Q_1 flip-flop toggles when $\overline{Q}_0$ goes from 1 to 0; this is equivalent to Q_0 going from 0 to 1.

A preset signal generated elsewhere is available in either uncomplemented or complemented form; *PRE* goes to all circuits with an active-high preset; $\overline{PRE}$ goes to all circuits with an active-low preset. Initially, the preset signal $\overline{PRE}$ goes low in Fig. 8-19, producing an output word of

$$\mathbf{Q} = 1111 \qquad (15)$$

When $\overline{PRE}$ goes high, the action starts. Notice that Q_0 toggles once per clock pulse. In the following discussion, a *positive toggle* means a change from 0 to 1, a *negative toggle* means a change from 1 to 0.

The first clock pulse produces a negative toggle in Q_0; nothing else happens:

$$\mathbf{Q} = 1110 \qquad (14)$$

The second clock pulse produces a positive toggle in Q_0, which produces a negative toggle in Q_1:

$$\mathbf{Q} = 1101 \qquad (13)$$

On the third clock pulse, Q_0 toggles negatively, and

$$\mathbf{Q} = 1100 \qquad (12)$$

On the fourth clock pulse, Q_0 toggles positively, Q_1 toggles positively, and Q_2 toggles negatively:

$$\mathbf{Q} = 1011 \qquad (11)$$

You should have the idea by now. The circuit is counting down, from 15 to 0. When it reaches 0,

$$\mathbf{Q} = 0000$$

On the next clock pulse, all flip-flops toggle positively to get

$$\mathbf{Q} = 1111$$

and the cycle repeats.

Up-Down Counter

Figure 8-20 shows how to build an *up-down counter*. The flip-flop outputs are connected to *steering* networks. An *UP* control signal produces either down counting or up counting. If the *UP* signal is low, $\overline{Q}_2$, $\overline{Q}_1$, and $\overline{Q}_0$ are transmitted to the clock inputs; this results in a down counter. On the other hand, when *UP* is high, Q_2, Q_1, and Q_0 drive the clock inputs and the circuit becomes an up counter.

Presettable Counter

In a *presettable counter,* the count starts at a number greater than zero. Figure 8-21*a* shows a presettable counter; the count begins with $P_3P_2P_1P_0$, a number between 0000 and 1111.

To start the analysis, look at the *LOAD* control line. When it is low, all NAND gates have high outputs; therefore,

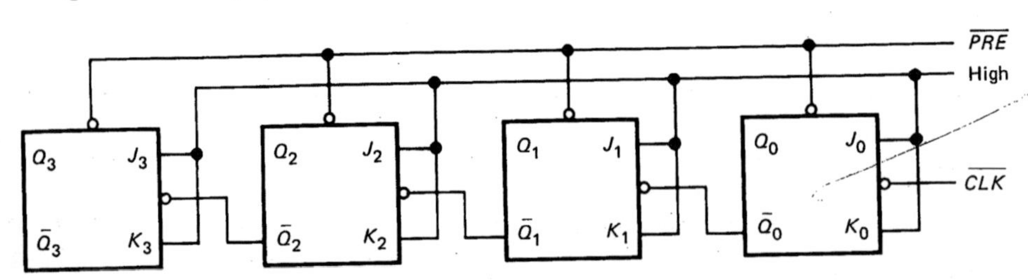

Fig. 8-19 Down counter.

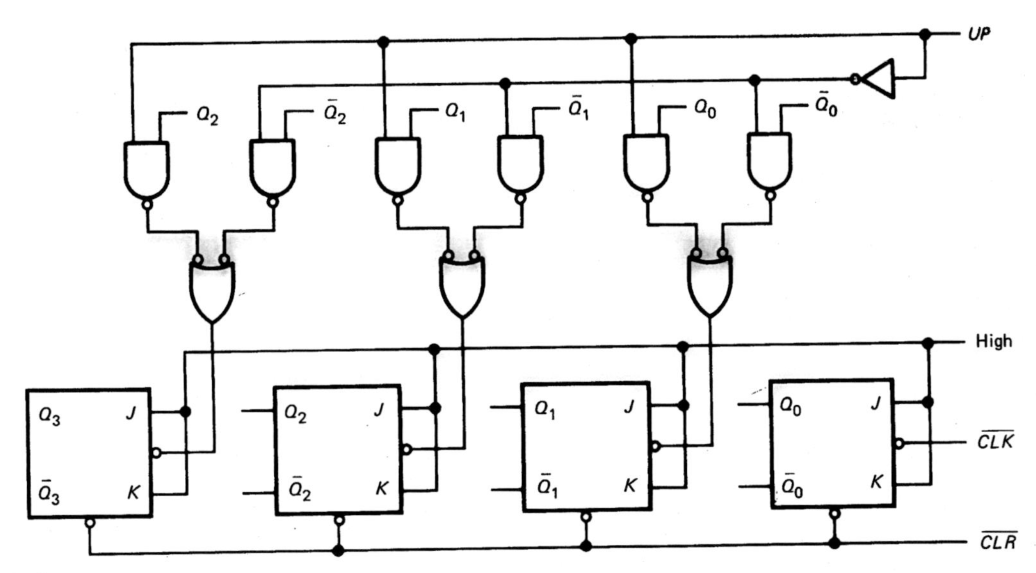

Fig. 8-20 Up-down counter.

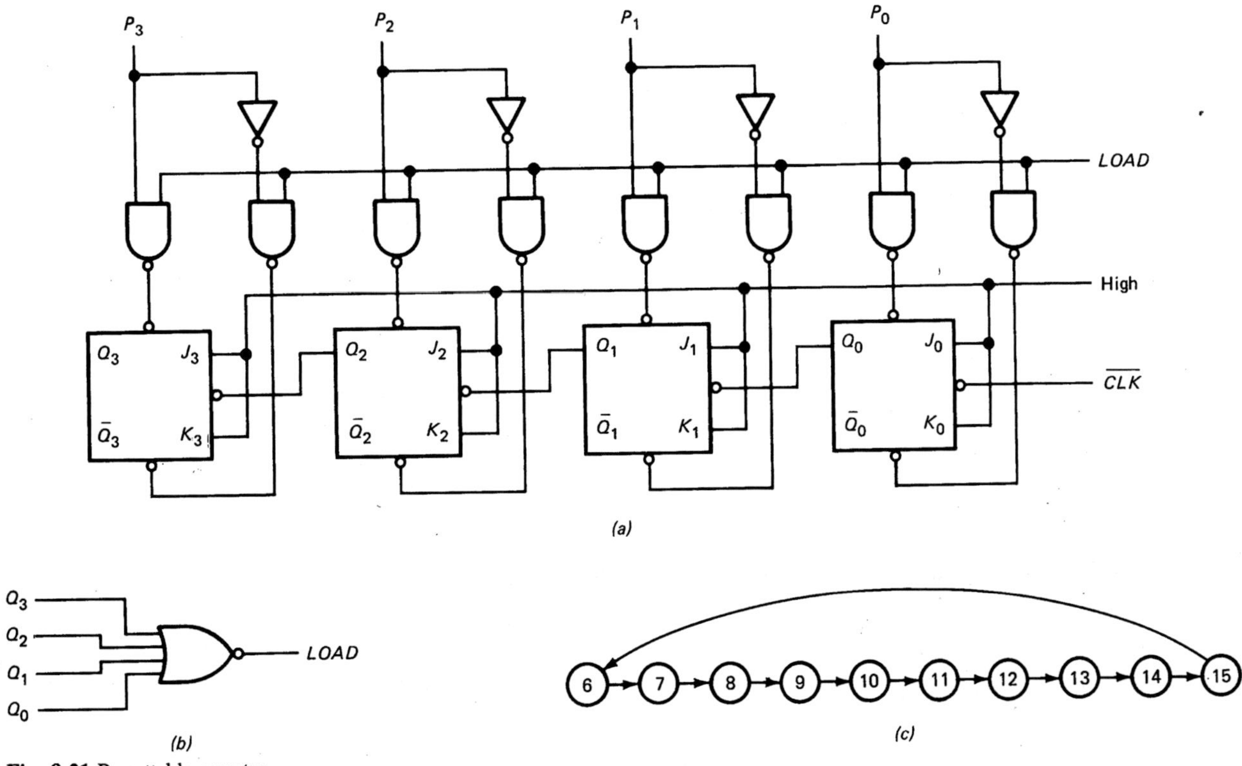

Fig. 8-21 Presettable counter.

the preset and clear inputs of all flip-flops are inactive. In this case, the circuit counts upward, as previously described. The data inputs P_3 to P_0 have no effect because the NAND gates are disabled.

When the *LOAD* line is high, the data inputs and their complements pass through the NAND gates and preset the counter to $P_3P_2P_1P_0$. As an example, suppose the preset input is

$$P_3P_2P_1P_0 = 0110$$

Because of the two left NAND gates, the low P_3 produces a high preset and a low clear for the Q_3 flip-flop; this clears

Q_3 to a 0. By a similar argument, the high P_2 sets Q_2, the high P_1 sets Q_1, and the low P_0 clears Q_0. Therefore, the counter is preset to

$$\mathbf{Q} = 0110$$

When *LOAD* returns to low, the circuit reverts to a counter. Successive clock pulses produce

$$\mathbf{Q} = 0111$$
$$\mathbf{Q} = 1000$$
$$\mathbf{Q} = 1001$$

up to a maximum count of

$$\mathbf{Q} = 1111$$

The next clock pulse resets the counter to

$$\mathbf{Q} = 0000$$

In summary,

1. When *LOAD* is low, the circuit counts.
2. When *LOAD* is high, the counter presets to $P_3P_2P_1P_0$.

Programmable Modulus

The most important use of a presettable counter is *programming a modulus*. Here's the idea. Let's add the NOR gate of Fig. 8-21*b* to the presettable counter of Fig. 8-21*a*. Then the *Q* outputs drive the NOR gate, and the NOR gate controls the *LOAD* line of the presettable counter. Because a NOR gate recognizes a word with all 0s and disregards all others, *LOAD* is high for $\mathbf{Q} = 0000$ and low for all other words. This means that the circuit presets when $\mathbf{Q} = 0000$ and counts when $\mathbf{Q}$ is 0001 to 1111.

If the preset input is 0110, successive clock pulses produce 0111, 1000, 1001, . . . , reaching a maximum value of

$$\mathbf{Q} = 1111$$

The next clock pulse resets the count to

$$\mathbf{Q} = 0000$$

Almost immediately, however, the NOR-gate outputs goes high, and the data inputs preset the counter to

$$\mathbf{Q} = 0110$$

In other words, the counter effectively skips states 0 to 5, illustrated by the state diagram of Fig. 8-21*c*.

Figure 8-21*c* shows 10 distinct states; by presetting 0110, we have programmed the counter to become a mod-10 counter. If we change the preset input, we get a different modulus. In general,

$$M = N - P \tag{8-1}$$

where M = modulus of preset counter
N = natural modulus
P = preset count

The natural modulus equals 2^n where n is the number of flip-flops in the counter. So four flip-flops give a natural modulus of 16, eight give a natural modulus of 256, and so on.

As an example, if you preset 82 into a preset counter with eight flip-flops, the modulus is

$$M = 256 - 82 = 174$$

In other words, this preset counter is equivalent to a divide-by-174 circuit.

TTL Counters

Table 8-2 lists some TTL counters. The 7490 is an industry standard, a widely used decade counter. This ripple counter has two sections, a divide-by-2 and a divide-by-5. This allows you to divide by 2, to divide by 5, or to cascade both sections to divide by 10.

The 7492 is a mod-12 ripple counter, organized in two sections by divide-by-2 and divide-by-6. This allows you to divide by 2, divide by 6, or cascade to divide by 12. The 7493 is a mod-16 ripple counter, with two sections of divide-by-2 and divide-by-8.

The 74160 and 74161 are presettable synchronous counters, the first being a decade counter and the second a divide-by-16 counter. Finally, the 74190 and 74191 are up-down presettable counters.

This is a sample of basic TTL counters; others are listed in Appendix 3.

TABLE 8-2. TTL COUNTERS

Number	Type
7490	Decade
7492	Divide-by-12
7493	Divide-by-16
74160	Presettable decade
74161	Presettable divide-by-16
74190	Up-down presettable decade
74191	Up-down presettable divide-by-16

8-8 THREE-STATE REGISTERS

The *three-state switch,* a development of the early 1970s, has greatly simplified computer wiring and design because it's ideal for *bus-organized computers* (the common type nowadays).

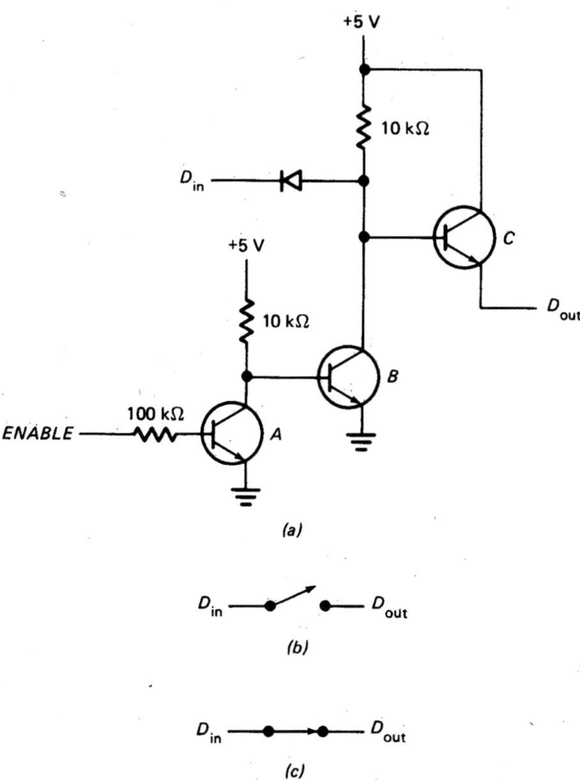

Fig. 8-22 (*a*) Three-state switch; (*b*) floating or high-impedance state; (*c*) output equals input.

Three-State Switch

Figure 8-22*a* is an example of a three-state switch. The *ENABLE* input can be low or high. When it's low, transistor A cuts off and transistor B saturates. This pulls the base of transistor C down to ground, opening its base-emitter diode. As a result, D_{out} floats. This floating state is equivalent to an *open* switch (Fig. 8-22*b*).

On the other hand, when *ENABLE* is high, transistor A saturates and transistor B cuts off. Now, the transistor C acts like an emitter follower, and the overall circuit is equivalent to a *closed* switch (Fig. 8-22*c*). In this case,

$$D_{out} = D_{in}$$

This means that D_{out} is low or high, the same as D_{in}.

Table 8-3 summarizes the action. When *ENABLE* is low, D_{in} is a don't care and D_{out} is open or floating. When *ENABLE* is high, the circuit acts like a noninverting buffer because D_{out} equals D_{in}.

TABLE 8-3. NORMALLY OPEN

ENABLE	D_{in}	D_{out}
0	X	Open
1	0	0
1	1	1

Commercial three-state switches are much more complicated than Fig. 8-22*a* (a totem-pole output and other enhancements are added). But simple as it is, Fig. 8-22*a* captures the key idea of a three-state switch; the output can be in any of three states: low, high, or floating (sometimes called the *high-impedance state* because the Thevenin impedance is high).

Three-state switches are also known as *Tri-state switches.* (Tri-state is a trademark name used by National Semiconductor, the originator of three-state TTL logic.)

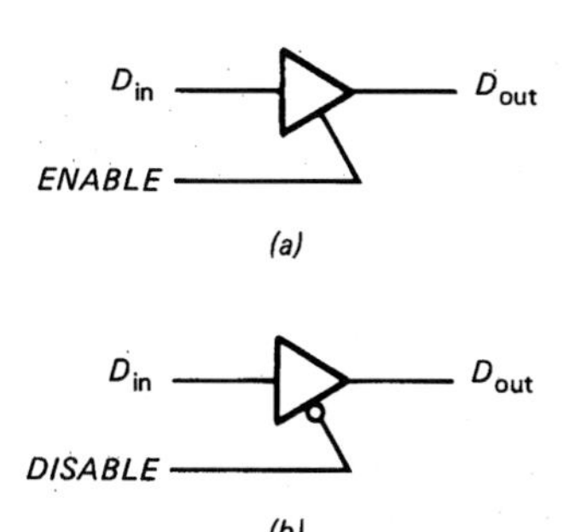

Fig. 8-23 (*a*) Normally open switch; (*b*) normally closed switch.

Normally Open Switch

Figure 8-23*a* is the symbol for a three-state noninverting buffer. When you see this symbol, remember the action: a low *ENABLE* means that the output is floating; a high *ENABLE* means that the output is 0 or 1, the same as the input. Think of this switch as *normally open;* to close it, you have to apply a high *ENABLE.*

In the 7400 series, the 74126 is a quad three-state normally open switch. This means four switches like Fig. 8-23*a* in one package. The SAP-1 computer uses five 74126s.

Normally Closed Switch

Figure 8-23*b* is different. This is the symbol for a *normally closed switch* because the control input *DISABLE* is active low. In other words, the switch is closed when *DISABLE* is low, and open when *DISABLE* is high. Table 8-4 summarizes the operation.

The 74125 is a quad three-state normally closed switch (four switches like Fig. 8-23*b* in one package).

TABLE 8-4. NORMALLY CLOSED

DISABLE	D_{in}	D_{out}
0	0	0
0	1	1
1	X	Open

Three-State Buffer Register

The main application of three-state switches is to convert the two-state output of a register to a three-state output. For instance, Fig. 8-24 shows a three-state buffer register, so called because of the three-state switches on the output lines. When *ENABLE* is low, the *Y* outputs float. But when *ENABLE* is high, the *Y* outputs equal the *Q* outputs; therefore,

$$\mathbf{Y} = \mathbf{Q}$$

You already know how the rest of the circuit works; it's the controlled buffer register discussed earlier. When *LOAD* is low, the contents of the register are unchanged. When *LOAD* is high, the next positive clock edge loads $X_3X_2X_1X_0$ into the register.

8-9 BUS-ORGANIZED COMPUTERS

A *bus* is a group of wires that transmit a binary word. In Fig. 8-25, vertical wires W_3, W_2, W_1, and W_0 are a bus; these wires are a common transmission path between the three-state registers. The input data bits for register A come from the **W** bus; at the same time, the three-state output of register A connects back to the **W** bus. Similarly, the other registers have their inputs and outputs connected to the **W** bus.

In Fig. 8-25 all control signals are in uncomplemented form; this means that the registers have active high inputs. In other words, a load input (L_A to L_D) must be high to set up for loading, and an enable signal (E_A to E_D) must be high to connect an output to the bus.

Register Transfers

The beauty of bus organization is the ease of transferring a word from one register to another. To begin with, the same clock signal drives all registers, but nothing happens until you apply high control inputs. In other words, as long as all *LOAD* and *ENABLE* inputs are low, the registers are isolated from the bus.

To transfer a word from one register to another, make the appropriate control inputs high. For instance, here's how to transfer the contents of register A to the register D. Make E_A and L_D high; then the contents of register A appear on the bus and register D is set up for loading. When the next positive clock edge arrives, word **A** is stored in register D.

Here is another example. Suppose the following words are stored in the registers:

$$\mathbf{A} = 0011$$
$$\mathbf{B} = 0110$$
$$\mathbf{C} = 1001$$
$$\mathbf{D} = 1100$$

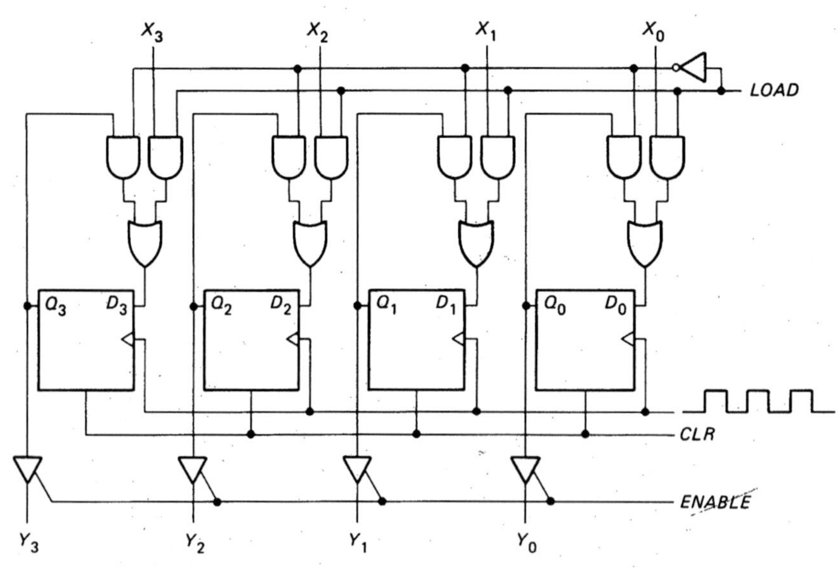

Fig. 8-24 Three-state buffer register.

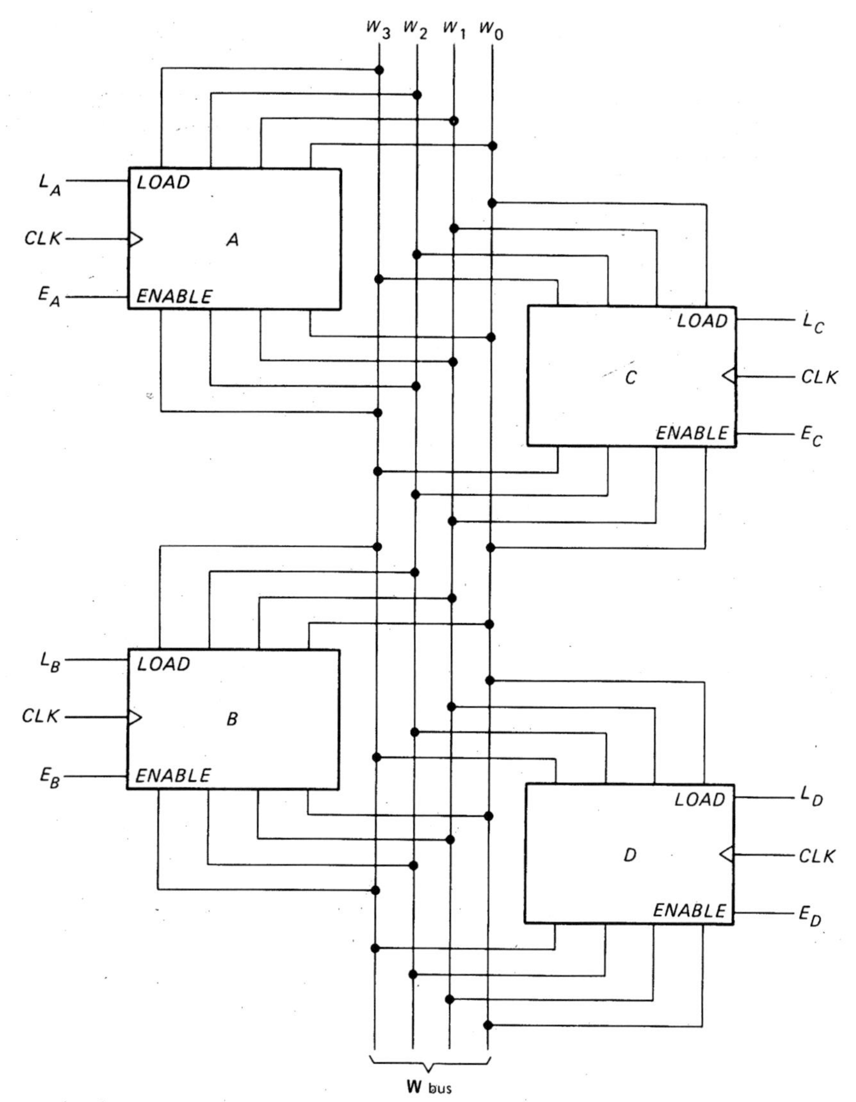

Fig. 8-25 Registers connected to bus.

To transfer word **C** into register B, make E_C and L_B high. The high E_C closes the three-state switches of register C, placing word **C** on the bus. The high L_B sets up register B for loading. When the next positive clock edge arrives, word **C** is stored in register B, and the new words are

$$\mathbf{A} = 0011$$
$$\mathbf{B} = 1001$$
$$\mathbf{C} = 1001$$
$$\mathbf{D} = 1100$$

The whole point of bus organization (connecting the registers to a common word path) is to simplify the wiring and operation of computers. As you will see in Chap. 10, SAP-1 is a bus-organized computer of incredible simplicity made possible by the three-state switch.

Simplified Drawings

Figure 8-25 shows a 4-bit bus. The same idea applies to any number of bits. For example, a 16-bit bus has 16 wires, each carrying 1 bit of a word. By connecting the inputs and outputs of 16-bit registers to this bus, we can transfer 16-bit words from one register to another.

Drawings get very messy unless we simplify the appearance of the bus. Figure 8-26 shows an abbreviated form of Fig. 8-25. The solid arrows represents words going into and out of registers. The solid bar represents the W bus.

EXAMPLE 8-5

Figure 8-27 shows part of the SAP-1 computer. Describe the circuitry.

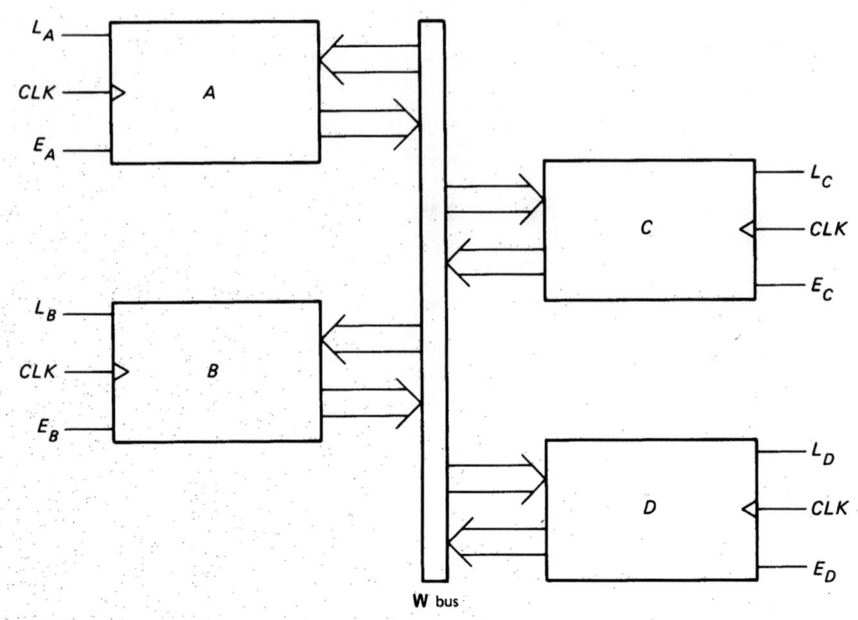

Fig. 8-26 Simplified bus diagram.

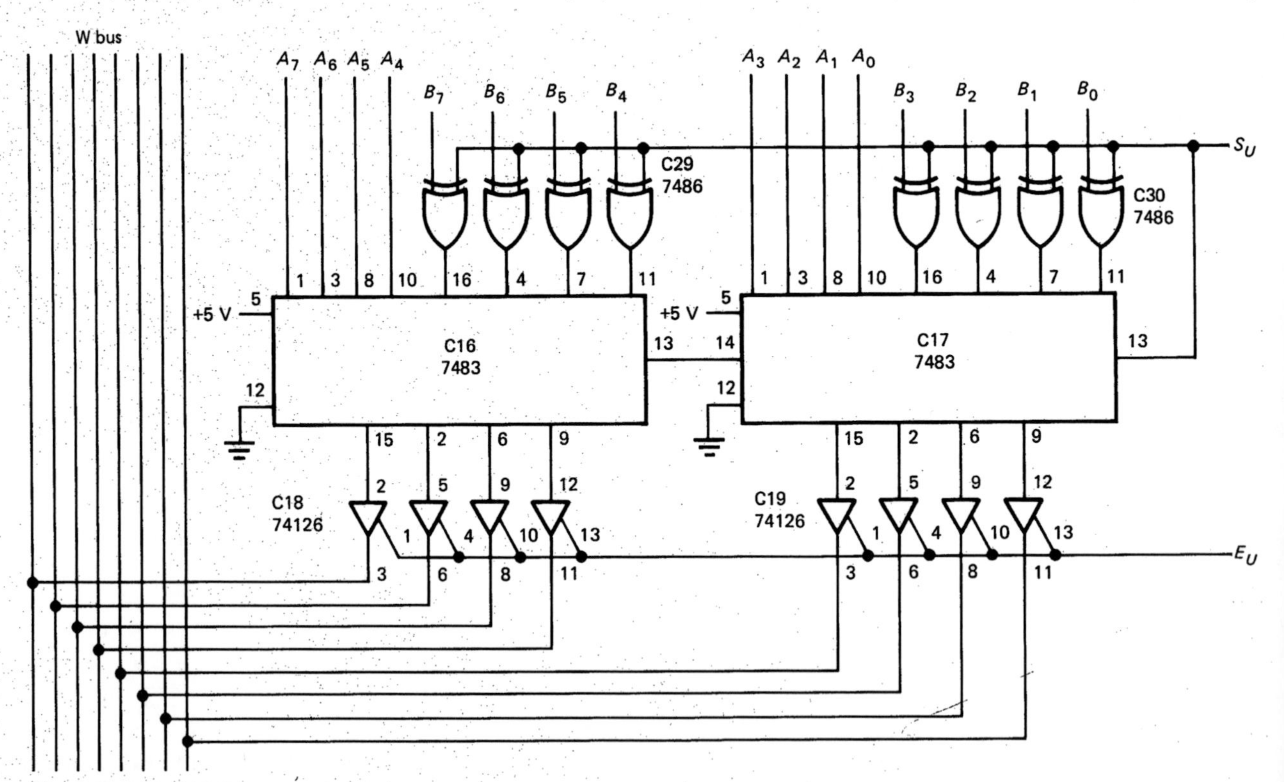

Fig. 8-27 SAP-1 ALU connected to bus.

SOLUTION

As discussed in Sec. 6-8, the 7483 is a 4-bit adder. The two 7483s of Fig. 8-27 are the ALU of the SAP-1 computer. The inputs to this ALU are the words

$$\mathbf{A} = A_7A_6A_5A_4A_3A_2A_1A_0$$
$$\mathbf{B} = B_7B_6B_5B_4B_3B_2B_1B_0$$

A pair of 7486s allow us to complement the **B** input for subtraction.

The sum (S_U low) or difference (S_U high) appears at the output (pins 15, 2, 6, 9 of C16 and pins 15, 2, 6, 9 of C17). Three-state switches (C18 and C19) connect the ALU output to the W bus when E_U is high. If E_U is low, the 74126s are open and the ALU output is isolated from the bus.

EXAMPLE 8-6

Figure 8-28 shows the *instruction register* (C8 and C9) of the SAP-1 computer. What does this 8-bit register do?

SOLUTION

Example 8-1 introduced the 74LS173. As you may recall, pins 9 and 10 are tied together and control the *LOAD* function. Because of the bubble, a low $\overline{L}_I$ is needed to set up the registers for loading. When $\overline{L}_I$ is low, the next positive clock edge loads the data on the bus into the instruction register.

The output of the instruction register is split; the upper nibble $I_7I_6I_5I_4$ goes to the *instruction decoder,* a circuit that will be discussed in Chap. 10. The lower nibble out of the instruction register goes back to the W bus.

The 74LS173 is a 4-bit three-state buffer register; it has internal three-state switches controlled by pins 1 and 2. The bubbles on pins 1 and 2 indicate active-low inputs; therefore, the output of C9 is connected to the bus when $\overline{E}_I$ is low and disconnected when $\overline{E}_I$ is high.

Notice that pins 1 and 2 of C8 are grounded; this means that the upper nibble is always a two-state output. In other words, the 74LS173 can be used as an ordinary two-state register by grounding pins 1 and 2. (This was done in Example 8-1, where we used two 74LS173s for the output register to drive an 8-bit LED display.)

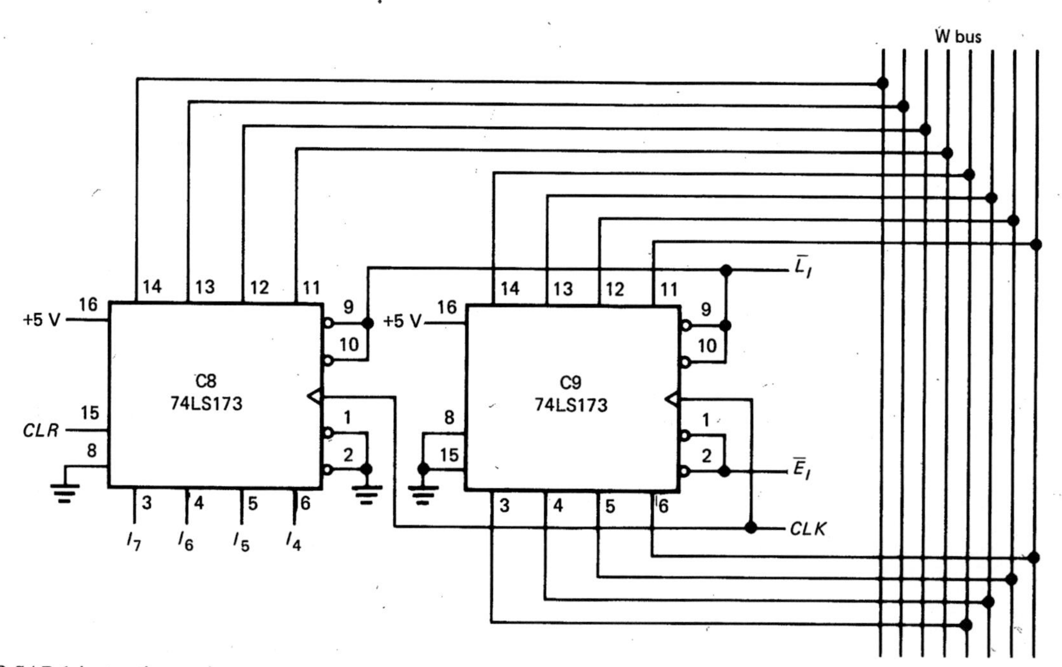

Fig. 8-28 SAP-1 instruction register.

GLOSSARY

buffer register A register that temporarily stores a word during data processing.

bus A group of wires used as a common word path by several registers.

modulus The number of stable states a counter has.

parallel entry Loading all bits of a word in parallel during one clock pulse. Also called broadside loading.

presettable counter A counter that allows you to preset a

number from which the count begins. Sometimes called a programmable counter.
register A group of memory elements that store a word.
ring counter A counter producing words with 1 high bit, which shifts one position per clock pulse.
ripple counter A counter with cascaded flip-flops. This means that the carry has to propagate in series through the flip-flops.
serial entry Loading a word into a shift register 1 bit per clock pulse
shift register A register that can shift the stored bits one position to the left or right.
synchronous counter A counter in which the clock drives each flip-flop to eliminate the ripple delay.
three-state switch A noninverting buffer that can be closed or opened by a control signal. Also called a Tri-state switch.

SELF-TESTING REVIEW

Read each of the following and provide the missing words. Answers appear at the beginning of the next question.

1. When the *LOAD* input of a buffer register is active, the input word is stored on the next positive ________ edge. If *LOAD* then becomes inactive, the input word can change without effecting the ________ word.
2. (*clock, stored*) A shift register moves the ________ left or right. Serial loading means storing a word in a shift register by entering ________ bit per clock pulse. With parallel or broadside loading, it takes only one ________ pulse to load the input word.
3. (*bits, 1, clock*) One flip-flop divides the clock frequency by a factor of ________. Two flip-flops divide by 4, three flip-flops by 8, and four flip-flops by ________. In general, *n* flip-flops divide by 2^n.
4. (*2, 16*) In a ripple counter, the carry has to propagate through all the flip-flops to reach the MSB flip-flop. The overall propagation delay time is ________. A controlled counter counts ________ pulses only when the *COUNT* signal is active. The clock signal drives each flip-flop of a ________ counter.
5. (nt_p, *clock, synchronous*) Instead of counting with binary numbers, a ring counter uses words that have a single high ________. A ring counter is ideal for timing a sequence of digital operations.
6. (*bit*) The modulus of a counter is the number of stable output ________ it has. A mod-10 counter can divide the clock frequency by a factor of ________.
7. (*states, 10*) An up-down counter can count up or down. A presettable counter starts the count from a ________ number. This allows us to program the ________. If the modulus is *M*, a presettable counter is equivalent to a divide-by-*M* circuit.
8. (*preset, modulus*) A three-state switch has an output that is either low, high, or ________. Two types are available; normally open and normally closed. The main use of three-state switches is to convert the ________ output of a register to a three-state output.
9. (*floating, two-state*) A bus is a group of wires used by three-state registers as a common word path. Bus-organized computers, the common type nowadays, have several registers connected to one or more buses. Instructions and data travel along these buses as they move from one register to another.

PROBLEMS

8-1. Figure 8-29 shows an output register. Before time *A* the data word to be loaded is

$$\mathbf{X} = 1000\ 1101$$

and the LED display is

$$\mathbf{Q} = 0001\ 0111$$

a. What is the LED display at time *D*?
b. What is the LED display at time *F*?

8-2. The data sheet of a 74173 gives these values:

$t_{setup} = 17$ ns (*L_O* input)
$t_{setup} = 10$ ns (Data)
$t_{hold} = 2$ ns (*L_O* input)
$t_{hold} = 10$ ns (Data)

a. In Fig. 8-29, how far ahead of point *E* must the *X* bits be applied to ensure accurate loading?
b. Suppose the clock has a frequency of 1 MHz

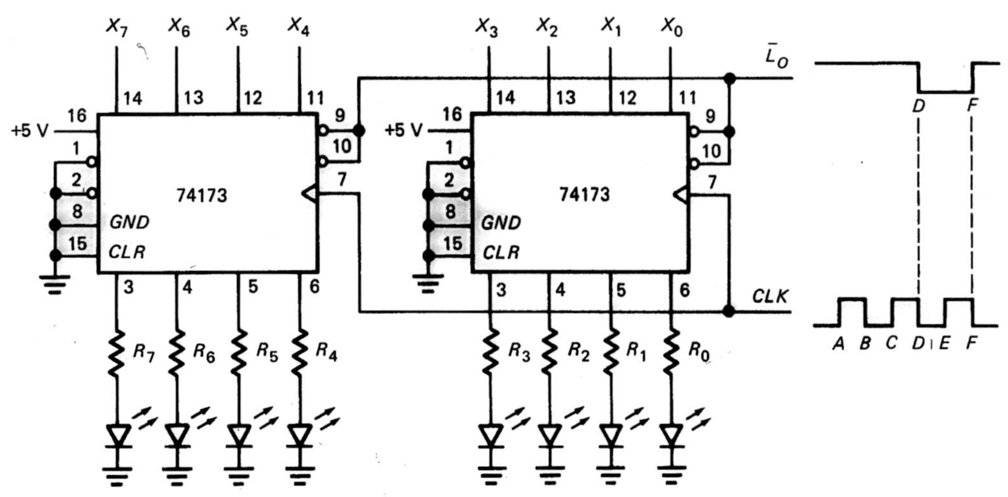

Fig. 8-29

and the X bits are applied at the point D. Is the setup time sufficient for the data inputs?

c. How long must you wait after point E before removing the X bits or letting them change?

8-3. Each output pin of a 74173 can source up to 5.2 mA. In Fig. 8-29 suppose the high output voltage is 3.5 V and the LED drop is 1.5 V. To get more light out of the LEDs, we want to reduce the current-limiting resistors. What is the minimum allowable resistance?

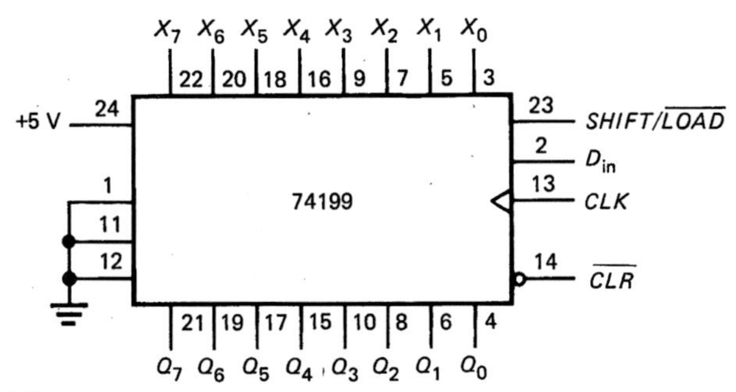

Fig. 8-30

8-4. A 74199 is an 8-bit shift-left register with a single control signal, as shown in Fig. 8-30. When $SHIFT/\overline{LOAD}$ is low, the circuit loads the **X** word on the next positive clock edge. When $SHIFT/\overline{LOAD}$ is high, the register shifts the bits to the left.

a. To clear the register, should $\overline{CLR}$ be low or high? When you are ready to run, what should $\overline{CLR}$ be?

b. Is the **X** word loaded on the positive or negative edge of the clock?

c. If **X** = 0100 1011, D_{in} = 0, and $SHIFT/\overline{LOAD}$ = 0, what does the **Q** output word equal after two positive clock edges?

d. If **X** = 0100 1011, D_{in} = 0, and $SHIFT/\overline{LOAD}$ = 1, what does the **Q** output word equal after two positive clock edges?

8-5. The clock frequency is 2 MHz. How long will it take to serially load the shift register of Fig. 8-30?

8-6. In Fig. 8-30, **Q** = 0001 0110. If $SHIFT/\overline{LOAD}$ is high and D_{in} is high, what does **Q** equal after three clock pulses?

8-7. Data from a satellite is received in serial form (1 bit after another). If this data is coming at a 5-MHz rate and if the clock frequency is 5 MHz, how long will it take to serially load a word in a 32-bit shift register?

8-8. A ripple counter has 16 flip-flops, each with a propagation delay time of 25 ns. If the count is

$$\mathbf{Q} = 0111\ 1111\ 1111\ 1111$$

how long after the next active clock edge before

$$\mathbf{Q} = 1000\ 0000\ 0000\ 0000$$

8-9. What is the maximum decimal count for the counter of the preceding problem?

8-10. When pins 1 and 12 of a 7490 are tied together as shown in Fig. 8-31, the divide-by-2 and divide-by-5 sections are cascaded to get a mod-10 counter. Pin 14 is the input and pin 11 is the output of each 7490. As a result, each 7490 acts like a divide-by-10 circuit and the overall circuit divides by 1,000.

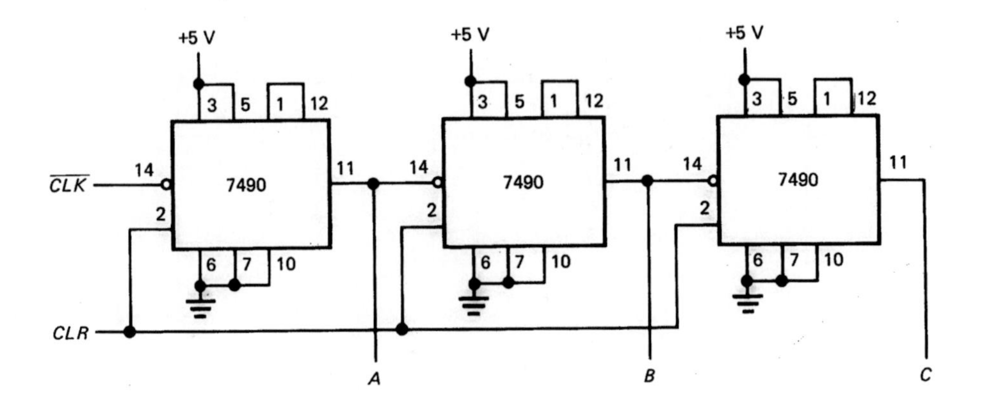

Fig. 8-31

If the clock has a frequency of 5 MHz, what is the frequency of A? Of B? Of C?

8-11. The clock signal driving a 6-bit ring counter has a frequency of 1 MHz. How long is each timing bit high? How long does it take to cycle through all the ring words?

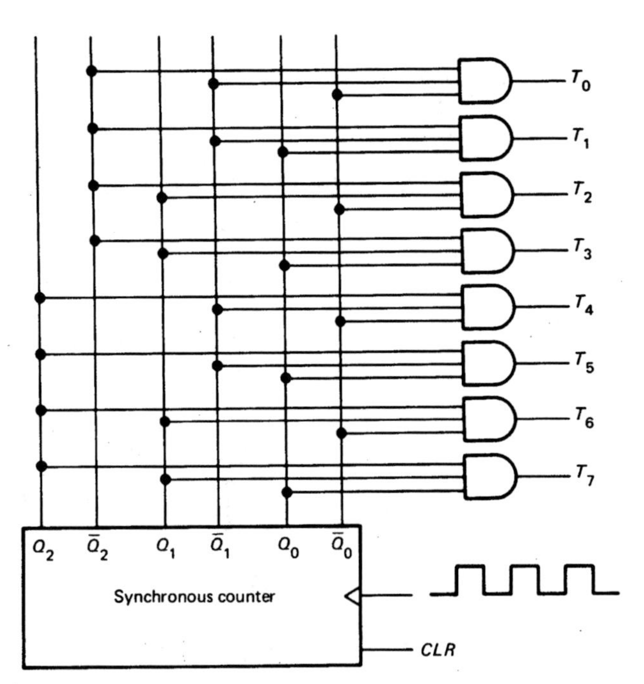

Fig. 8-32

8-12. Figure 8-32 shows another way to produce ring words. After the circuit is cleared,

$$\mathbf{Q} = Q_2Q_1Q_0 = 000$$

Since the AND gates are a 1-of-8 decoder, the first timing word is

$$\mathbf{T} = 0000\ 0001$$

What does **T** equal for each of the following:

a. **Q** = 001
b. **Q** = 010
c. **Q** = 101
d. **Q** = 111

8-13. If the clock frequency is 5 MHz in Fig. 8-32, how long does it take to produce all the ring words? How long is each timing bit high?

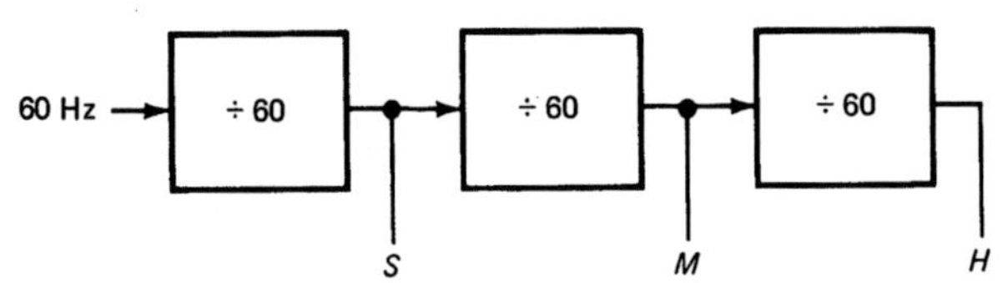

Fig. 8-33

8-14. In a digital clock, the 60-Hz line frequency is divided down to lower frequencies, as shown in Fig. 8-33. What are the frequency and period of the S output? Of the M output? Of the H output?

8-15. You have an unlimited number of the following ICs to work with: 7490, 7492, and 7493. Which of these would you use to build the divide-by-60 circuits of Fig. 8-33?

8-16. A presettable counter has eight flip-flops. If the preset number is 125, what is the modulus?

8-17. Given a presettable 8-bit counter, what number would you preset to get a divide-by-120 circuit?

8-18. In Fig. 8-34, we want to transfer the contents of register D to register C. Which are the *ENABLE* and *LOAD* inputs you should make high?

8-19. Look at Fig. 8-35 and answer each of these questions.

a. To add the inputs and put the answer on the bus, what should S_U and E_U be?
b. To subtract the inputs and put the answer on the bus, what should S_U and E_U be?
c. To isolate the ALU from the bus, what should E_U be?

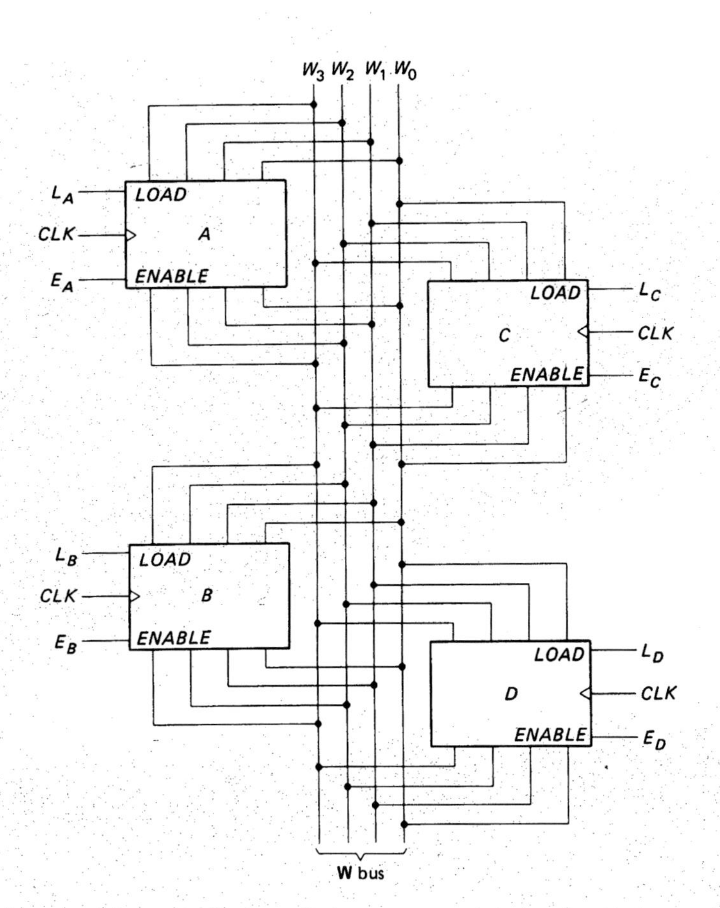

Fig. 8-34

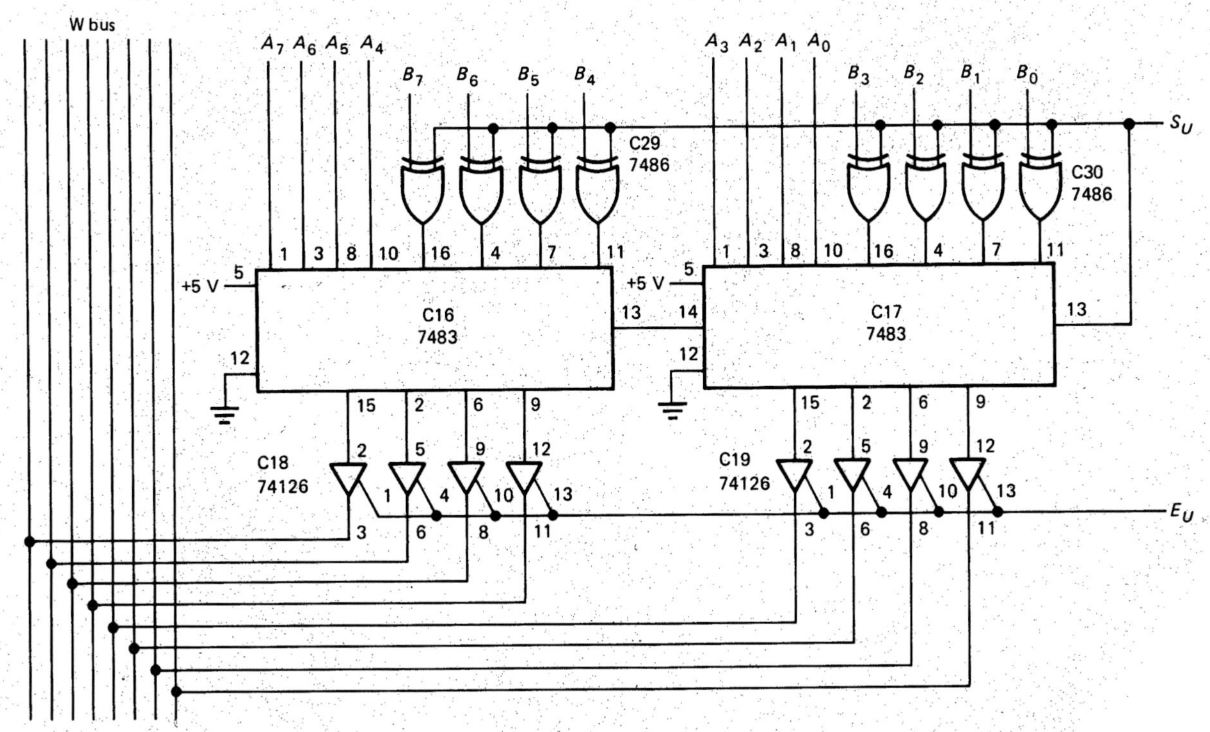

Fig. 8-35

MEMORIES

The *memory* of a computer is where the program and data are stored before the calculations begin. During a computer run, the control section may store partial answers in the memory, similar to the way we use paper to record our work. The memory is therefore one of the most active parts of a computer, storing not only the program and data but processed data as well.

The memory is equivalent to thousands of registers, each storing a binary word. The latest generation of computers relies on semiconductor memories because they are less expensive and easier to work with than core memories. A typical microcomputer has a semiconductor memory with up to 655,360 memory locations, each capable of storing 1 byte of information.

9-1 ROMS

A *read-only memory* (ROM) is the simplest kind of memory. It is equivalent to a group of registers, each permanently storing a word. By applying control signals, we can *read* the word in any memory location. (''Read'' means to make the contents of the memory location appear at the output terminals of the ROM.)

Diode ROM

Figure 9-1 shows one way to build a ROM. Each horizontal row is a register or memory location. The R_0 register

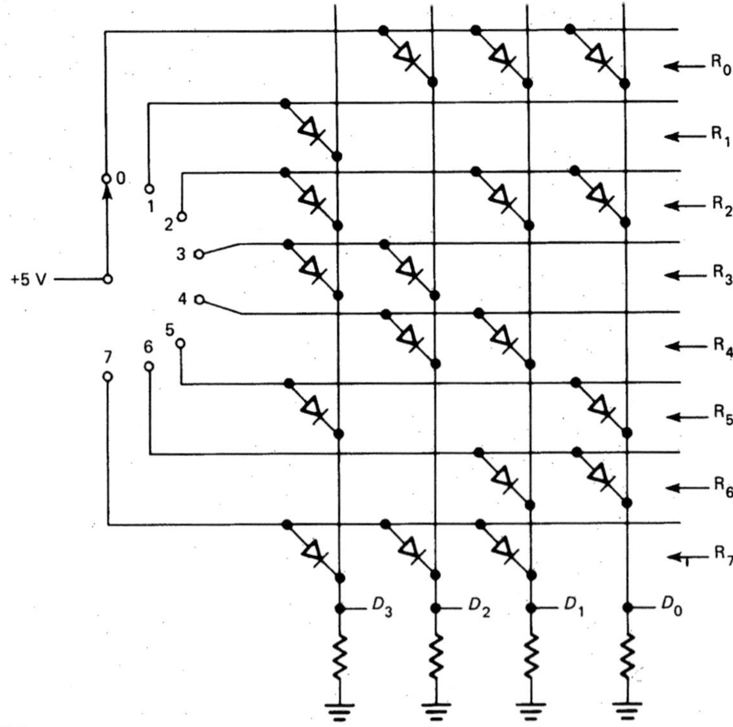

Fig. 9-1 Simple diode ROM.

TABLE 9-1. DIODE ROM

Register	Address	Word
R_0	0	0111
R_1	1	1000
R_2	2	1011
R_3	3	1100
R_4	4	0110
R_5	5	1001
R_6	6	0011
R_7	7	1110

contains three diodes, the R_1 register has one diode, and so on. The output of the ROM is the word

$$\mathbf{D} = D_3D_2D_1D_0$$

In switch position 0, a high voltage turns on the diodes in the R_0 register; all other diodes are off. This means that a high output appears at D_2, D_1, and D_0. Therefore, the word stored at memory location 0 is

$$\mathbf{D} = 0111$$

What happens if the switch is moved to position 1? The diode in the R_1 register conducts, forcing D_3 to go high. Because all other diodes are off, the output from the ROM becomes

$$\mathbf{D} = 1000$$

So the contents of memory location 1 are 1000.

As you move the switch to other positions, you will read the contents of the other memory locations. Table 9-1 shows these contents, which you can check by analyzing Fig. 9-1.

With discrete circuits we can change the contents of a memory location by adding or removing diodes. With integrated circuits, the manufacturer stores the words at the time of fabrication. In either case, the words are permanently stored once the diodes are wired in place.

Addresses

The *address* and *contents* of a memory location are two different things. As shown in Table 9-1, the address of a memory location is the same as the subscript of the register storing the word. This is why register 0 has an address of 0 and contents of 0111; register 1 has an address of 1 and contents of 1000; register 2 has an address of 2 and contents of 1011; and so on.

The idea of addresses applies to ROMs of any size. For example, a ROM with 256 memory locations has decimal addresses running from 0 to 255. A ROM with 1,024 memory locations has decimal addresses from 0 to 1,023.

On-Chip Decoding

Rather than switch-select the memory location, as shown in Fig. 9-1, IC manufacturers use *on-chip decoding*. Figure 9-2 gives you the idea. The three input pins (A_2, A_1, and A_0) supply the binary address of the stored word. Then a 1-of-8 decoder produces a high output to one of the registers.

For instance, if

$$\mathbf{ADDRESS} = A_2A_1A_0 = 100$$

the 1-of-8 decoder applies a high voltage to the R_4 register, and the ROM output is

$$\mathbf{D} = 0110$$

If you change the address word to

$$\mathbf{ADDRESS} = 110$$

you will read the contents of memory location 6, which is

$$\mathbf{D} = 0011$$

The circuit of Fig. 9-2 is a 32-bit ROM organized as 8 words of 4 bits each. It has three address (input) lines and four data (output) lines. This is a very small ROM compared with commercially available ROMs.

Number of Address Lines

With on-chip decoding, n address lines can select 2^n memory locations. For instance, we need 3 address lines in Fig. 9-2 to access 8 memory locations. Similarly, 4 address lines can access 16 memory locations, 8 address lines can access 256 memory locations, and so on.

9-2 PROMS AND EPROMS

With a ROM, you have to send a list of data to be stored in the different memory locations to the manufacturer, who then produces a *mask* (a photographic template of the circuit) used in mass production of your ROMs. In fabricating ROMs the manufacturer may use bipolar transistors or MOSFETs. But the idea is still basically the same; the transistors or MOSFETs act like the diodes of Fig. 9-2.

Programmable

A *programmable* ROM (PROM) is different. It allows the user to store the data. An instrument called a *PROM programmer* does the storing by "burning in." (Fusible links at the bit locations can be burned open by high currents.) With a PROM programmer, the user can burn in the program and data. Once this has been done, the programming is permanent. In other words, the stored contents cannot be erased.

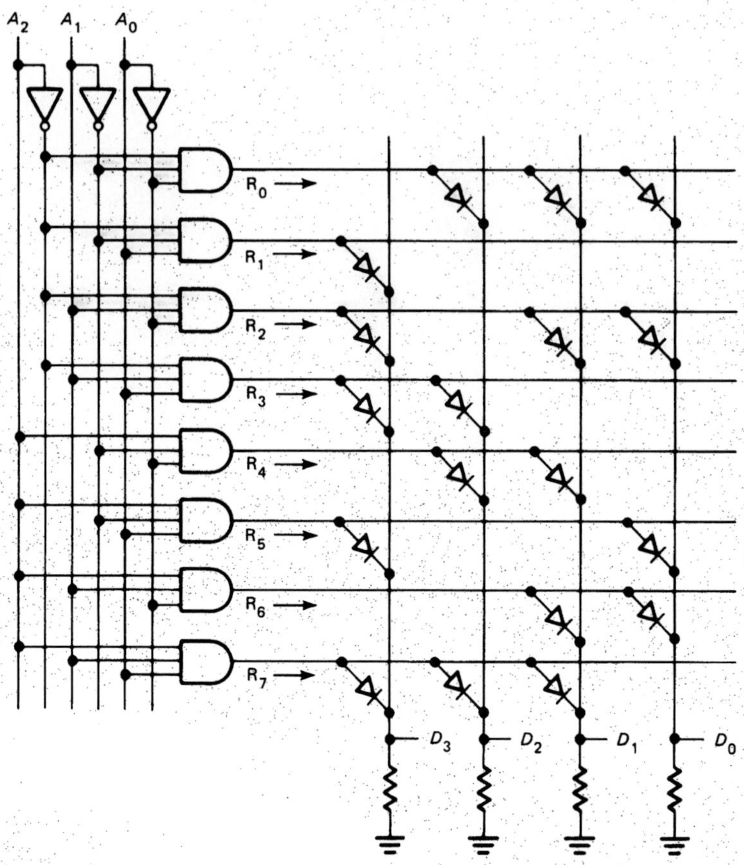

Fig. 9-2 ROM with on-chip decoding.

Erasable

The *erasable PROM* (EPROM) uses MOSFETs. Data is stored with a PROM programmer. Later, data can be erased with ultraviolet light. The light passes through a window in the IC package to the chip, where it releases stored charges. The effect is to wipe out the stored contents. In other words, the EPROM is ultraviolet-light-erasable and electrically reprogrammable.

The EPROM is helpful in design and development. The user can erase and store until the program and data are perfected. Then the program and data can be sent to an IC manufacturer who makes a ROM mask for mass production.

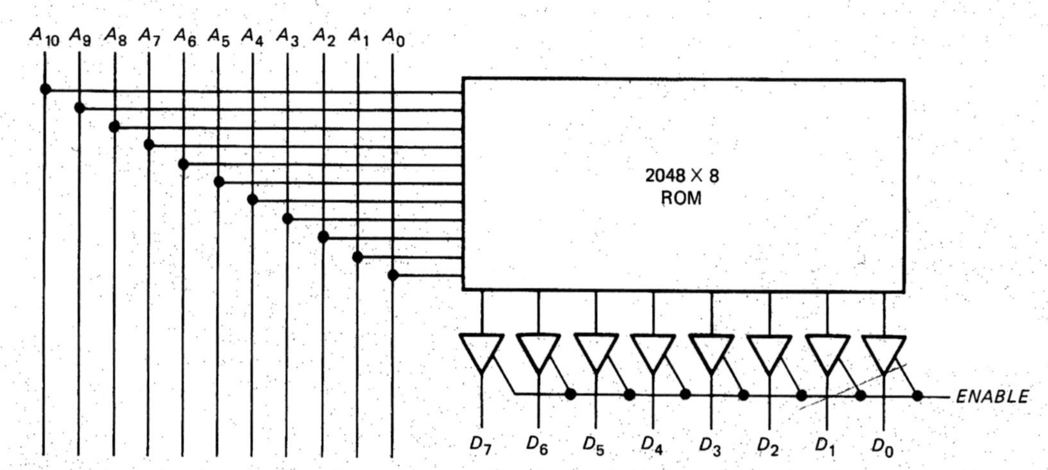

Fig. 9-3. Three-state ROM.

EEPROM

Another type of reprogrammable ROM device is the EEPROM (Electrically Erasable Programmable Read Only Memory), which is nonvolatile like EPROM but does not require ultraviolet light to be erased. It can be completely erased or have certain bytes changed, using electrical pulses. Individual bytes (or any number of bytes) can be changed using a programmer designed for use with EEPROMs. Individual bytes can also be changed by the host circuit after the EEPROM has been installed.

EEPROM is useful when data being gathered by the circuit must be stored by the system. Writing to EEPROM is slower than writing to RAM, so it cannot be used in high-speed circuits.

Unlimited READ cycles are possible; however, EEPROM will eventually wear out from repeated ERASE cycles. Since the life of typical EEPROMS allows thousands of erase cycles, this is usually not a problem.

There are matching EEPROM replacements for most EPROMs. The EEPROM uses an 8 digit in the part number whereas EPROM uses a 7 digit. For example, the 2816 EEPROM can replace the 2716 EPROM.

Manufactured Devices

With large-scale integration, manufacturers can fabricate ROMs, PROMs, and EPROMs that store thousands of words. For instance, the 8355 is a 16,384-bit ROM organized as 2,048 words of 8 bits each. It has 11 address lines and 8 data lines.

As another example, the 2764 is 65,536-bit EPROM organized as 8,192 words of 8 bits each. It has 13 address lines and 8 data lines.

Access Time

The *access time* of a memory is the time it takes to read a stored word after applying address bits. Since bipolar transistors are faster than MOSFETs, bipolar memories have faster access times than MOS memories. For instance, the 3636 is a bipolar PROM with an access time of 80 ns; the 2716 is a MOS EPROM with an access time of 450 ns. You have to pay for the speed; a bipolar memory is more expensive than a MOS memory, so it's up to the designer to decide which type to use in a specific application.

Three-State Memories

By adding three-state switches to the data lines of a memory we can get a three-state output. As an example, Fig. 9-3 shows a 16,384-bit ROM organized as 2,048 words of 8 bits each. It has 11 address lines and 8 data lines. A low *ENABLE* opens all switches and floats the output lines. On the other hand, a high *ENABLE* allows the addressed word to reach the final output.

Most of the commercially available ROMs, PROMs, and EPROMs have three-state outputs. In other words, they have built-in three-state switches that allow you to connect or disconnect the output lines from a data bus. More will be said about this later.

Nonvolatile Memory

ROMs, PROMs, and EPROMs are *nonvolatile memories.* This means that they retain the stored data even when the power to the device is shut off. Not all memories are like this, as will be explained in Sec. 9-3.

EXAMPLE 9-1

A 16 × 8 ROM stores these words in its first four locations:

$$\mathbf{R_0} = 1110\ 0010 \qquad \mathbf{R_2} = 0011\ 1100$$
$$\mathbf{R_1} = 0101\ 0111 \qquad \mathbf{R_3} = 1011\ 1111$$

Express the stored contents in hexadecimal notation.

SOLUTION

In hexadecimal shorthand, the stored contents are

$$\mathbf{R_0} = \text{E2H} \qquad \mathbf{R_2} = \text{3CH}$$
$$\mathbf{R_1} = \text{57H} \qquad \mathbf{R_3} = \text{BFH}$$

9-3 RAMS

A *random-access memory* (RAM), or a *read-write* memory, is equal to a group of addressable registers. After supplying an address, you can read the stored contents of the memory location or write new contents into the memory location.

Core RAMs

The core RAM was the workhorse of earlier computers. It has the advantage of being nonvolatile; even though you shut off the power, a core RAM continues to store data. The disadvantage of core RAMs is that they are expensive and harder to work with than semiconductor memories.

Semiconductor RAMs

Semiconductor RAMs may be *static* or *dynamic*. The static RAM uses bipolar or MOS flip-flops; data is retained indefinitely as long as power is applied to the flip-flops. On the other hand, a dynamic RAM uses MOSFETs and capacitors that store data. Because the capacitor charge leaks off, the stored data must be *refreshed* (recharged) every few milliseconds. In either case, the RAMs are volatile; turn off the power and you lose the stored data.

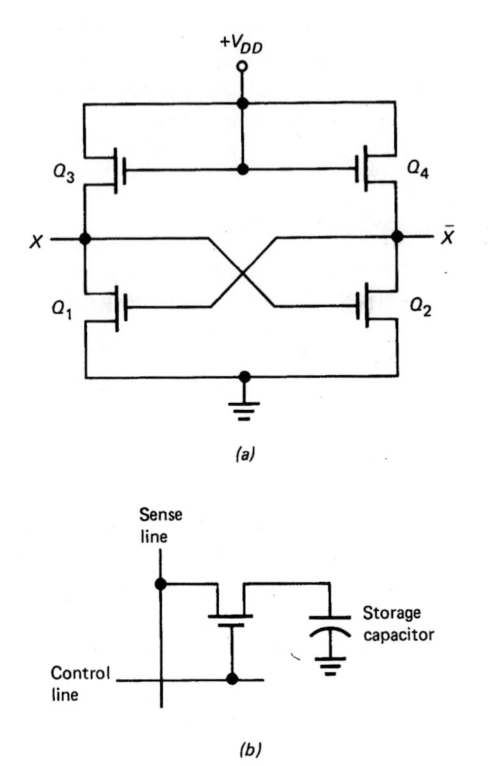

Fig. 9-4 (*a*) Static cell; (*b*) dynamic cell.

Static RAM

Figure 9-4*a* shows one of the flip-flops used in a static MOS RAM. Q_1 and Q_2 act like switches. Q_3 and Q_4 are active loads, meaning that they behave like resistors. The circuit action is similar to the transistor latch discussed in Sec. 7-1. Either Q_1 conducts and Q_2 is cut off or vice versa. A static RAM will contain thousands of flip-flops like this, one for each stored bit. As long as power is applied, the flip-flop remains latched and can store the bit indefinitely.

Dynamic RAM

Figure 9-4*b* shows one of the memory elements (called *cells*) in a dynamic RAM. When the *sense* and *control* lines go high, the MOSFET conducts and charges the capacitor. When the sense and control lines go low, the MOSFET opens and the capacitor retains its charge. In this way, it can store 1 bit. A dynamic RAM may contain thousands of memory cells like Fig. 9-4*b*. Since only a single MOSFET and capacitor are needed, the dynamic RAM contains more memory cells than a comparable static RAM. In other words, a dynamic RAM has more memory locations than a static RAM of the same physical size.

The disadvantage of the dynamic RAM is the need to refresh the capacitor charge every few milliseconds. This complicates the design problem because more circuitry is needed. In short, it's much simpler to work with static RAMs than dynamic RAMs. The remainder of this book emphasizes static RAMs.

Three-State RAMs

Many of the commercially available RAMs, either static or dynamic, have three-state outputs. In other words, the manufacturer includes three-state switches on the chip so that you can connect or disconnect the output lines of the RAM from a data bus.

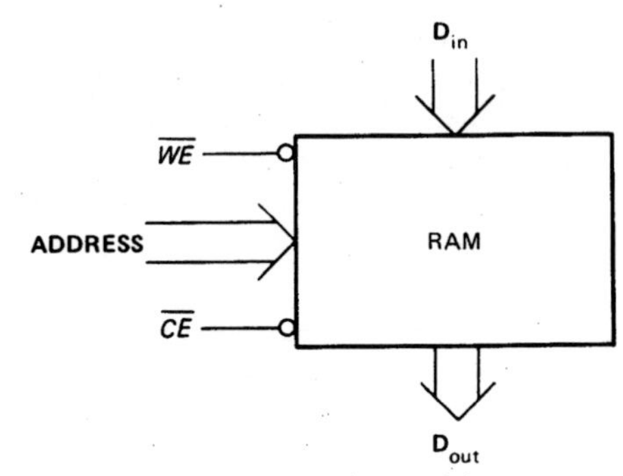

Fig. 9-5 Static RAM with inverted control inputs.

Figure 9-5 shows a static RAM and typical input signals. The **ADDRESS** bits select the memory location; control signals $\overline{WE}$ and $\overline{CE}$ select a write, read, or do nothing operation. $\overline{WE}$ is known as the *write-enable signal*, and $\overline{CE}$ is called the *chip-enable signal*. Notice that the control inputs are active low.

Table 9-2 summarizes the operation of the static RAM. Here's what happens. A low $\overline{CE}$ and low $\overline{WE}$ produce a write operation. This means that the input data $\mathbf{D}_{in}$ is stored in the addressed memory location. The three-state output data lines are floating during this write operation.

When $\overline{CE}$ is low and $\overline{WE}$ is high, we get a read operation. The contents of the addressed memory location appear on the data output lines because the internal three-state switches are closed at this time.

The final possibility is $\overline{CE}$ high. This is a holding pattern where nothing happens. Internal data at all memory locations is frozen or unchanged. Notice that the output data lines are floating.

TABLE 9-2. STATIC RAM

$\overline{CE}$	$\overline{WE}$	Operation	Output
0	0	Write	Floating
0	1	Read	Connected
1	X	Hold	Floating

Bubble Memories

A *bubble memory* sandwiches a thin film of magnetic material between two permanent bias magnets. Logical 1s and 0s are represented by magnetic bubbles in this thin film. The details of how a bubble memory works are too complicated to go into here. What is worth knowing is that bubble memories are nonvolatile and capable of storing huge amounts of data. For instance, the INTEL 7110 is a bubble memory that can store approximately 1 million bits. One disadvantage is they have slow access times.

EXAMPLE 9-2

Figure 9-6 shows the pin configuration of a 74189, a Schottky TTL static RAM with three-state outputs. This 64-bit RAM is organized as 16 words of 4 bits each. It has an access time of 35 ns. What are the different pin functions?

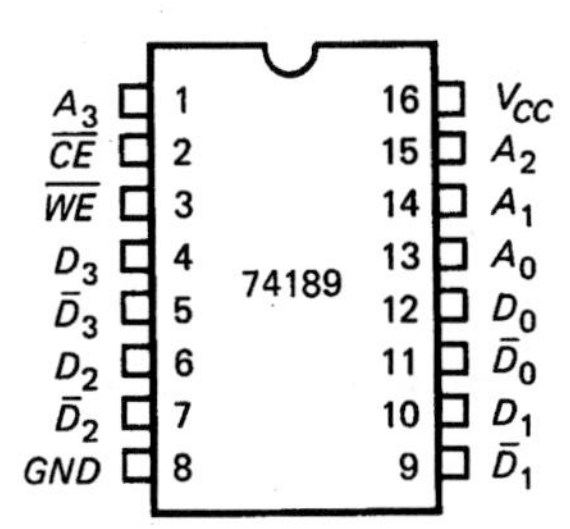

Fig. 9-6 Pinout for 74189.

SOLUTION

To begin with, 4 address bits can access $2^4 = 16$ words. This is why the 74189 needs 4 address bits to select the desired memory location.

The **ADDRESS** bits go to pin 1 (A_3), pin 15 (A_2), pin 14 (A_1), and pin 13 (A_0). The data inputs are pin 4 (D_3), pin 6 (D_2), pin 10 (D_1), and pin 12 (D_0). Because of the TTL design, the data is stored as the complement of the input bits. This is why the data outputs are pin 5 ($\bar{D}_3$), pin 7 ($\bar{D}_2$), pin 9 ($\bar{D}_1$), and pin 11 ($\bar{D}_0$).

The chip enable is pin 2, and the write enable is pin 3. These control signals work as previously described. $\overline{CE}$ and $\overline{WE}$ must be low for a write operation; $\overline{CE}$ must be low and $\overline{WE}$ high for a read, and $\overline{CE}$ must be high to do nothing.

Pin 16 gets the supply voltage, which is +5 V, and pin 8 is grounded.

9-4 A SMALL TTL MEMORY

Figure 9-7 shows a modified version of the SAP-1 memory. Two 74189s (see Appendix 4) are used to get a 16 × 8 memory. This means that we can store 16 words of 8 bits each. The bubbles on the output data pins (pins 5, 7, 9, 11) remind us that the stored data bits are the complements of the input data bits.

Addressing the Memory

The address bits come from an address-switch register (A_3, A_2, A_1, A_0). By setting the switches we can input any address from 0000 to 1111. As noted at the bottom of Fig. 9-7, an up address switch is equal to a 1. Therefore, the address with all switches up is 1111.

Setting Up Data

The data inputs come from the two other switch registers. The upper input nibble is D_7, D_6, D_5, and D_4. The lower input nibble is D_3, D_2, D_1, and D_0. By setting the data switches we can input any data word from 0000 0000 to 1111 1111, equivalent to 00H to FFH. The note at the bottom of Fig. 9-7 indicates that an up data switch produces an input 0 or an output 1. In other words, a data switch must be up to store a 1.

Programming the Memory

To *program the memory* (this means to store instruction and data words), the RUN-PROG switch must be in the PROG position. This grounds pin 2 ($\overline{CE}$) of each 74189. When the READ-WRITE switch is thrown to WRITE, pin 3 ($\overline{WE}$) is grounded and the complement of the input data word is written into the addressed memory location.

For instance, suppose we want to store the following words:

Address	Data
0000	0000 1111
0001	0010 1110
0010	0001 1101
0011	1110 1000

Begin by placing the RUN-PROG switch in the PROG position. To store the first data word at address 0000, set the switches as follows:

Address	Data
DDDD	DDDD UUUU

where D stands for down and U for up. When the READ-WRITE switch is thrown to WRITE, 0000 1111 is written into memory location 0000. The READ-WRITE switch is then returned to READ in preparation for the next WRITE operation.

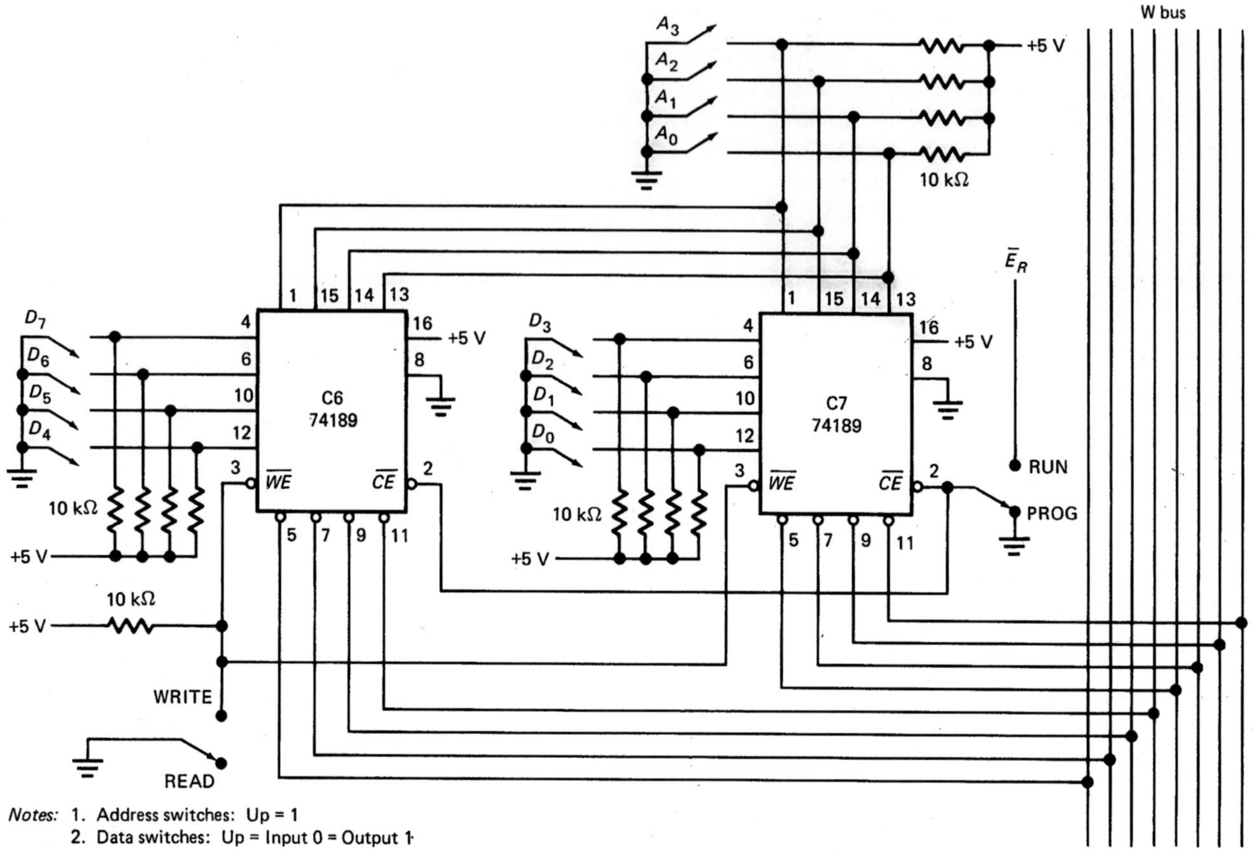

Notes: 1. Address switches: Up = 1
2. Data switches: Up = Input 0 = Output 1

Fig. 9-7 Modified SAP-1 read-write memory.

To load the second word at address 0001, set the address and data switches as follows:

Address	**Data**
DDDU	DDUD UUUD

When the READ-WRITE switch is thrown to WRITE, the data word 0010 1110 is stored at memory location 0001.

Continuing like this, we can program the memory with the remaining words.

The SAP-1 memory is slightly different from Fig. 9-7 and will be discussed in Chap. 10. What we have discussed here, however, gives you an example of how a program and data can be entered into a memory before a computer run.

9-5 HEXADECIMAL ADDRESSES

During a computer run, the CPU sends binary addresses to the memory, where read or write operations occur. These address words may contain 16 or more bits. There's no need for us to get bogged down with long strings of binary numbers. We can chunk those 0s and 1s into neat strings of hexadecimal numbers. Using hexadecimal shorthand is standard in microprocessor work.

Typical microcomputers have an address bus with 16 address lines. The words on this bus have the binary format of

ADDRESS = XXXX XXXX XXXX XXXX

For convenience, we can chunk this into its equivalent hexadecimal form. For instance, instead of writing

ADDRESS = 0101 1110 0111 1100

we can write

ADDRESS = 5E7CH

The 16 address lines can access 2^{16} memory locations, equivalent to 65,536 words. The hexadecimal addresses are from 0000H to FFFFH. In microcomputers using 8-bit microprocessors, 1 byte is stored in each memory location. Figure 9-8 illustrates how to visualize such a memory. The first memory location has an address of 0000H, the second memory location an address of 0001H, the third an address of 0002H, and so on. Moving toward higher memory, we eventually reach FFFDH, FFFEH, and FFFFH.

Notice that 1 byte is stored in each memory location. This is common in products using an 8-bit microprocessor like the Z80 and 6808. In other words, it is common for 8-bit microprocessor–based products to have a maximum memory of 64K (1K = 1,024 bytes).

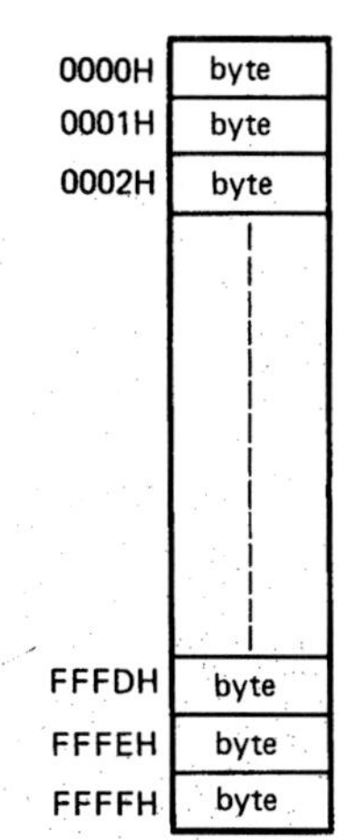

Fig. 9-8 Memory layout.

GLOSSARY

access time The time it takes to read the contents of a memory location after it has been addressed.

address A way of specifying the location of data in memory, similar to a house address.

dynamic memory A memory that relies on a MOSFET switch to charge a capacitor. This memory is highly volatile because not only must the power be kept on, but the capacitor charge must also be refreshed every few milliseconds.

EPROM Erasable programmable read-only memory, a device that is ultraviolet-erasable and electrically reprogrammable.

nonvolatile A type of memory in which the stored data is not lost when the power is turned off.

PROM Programmable read-only memory. With a PROM programmer, you can burn in your own programs and data.

RAM Random-access memory. It is also called a read-write memory because you can read the contents of a memory location or write new contents into it.

ROM Read-only memory. (ROM rhymes with Mom.) This device provides nonvolatile storage of programs and data. You can access any memory location by supplying its address.

static RAM A volatile memory using bipolar or MOSFET flip-flops. It is easy to work with. Refreshing data is unnecessary. You simply supply address and control bits for a read or write operation.

volatile A type of memory in which data stored in the memory is lost when the power is turned off.

SELF-TESTING REVIEW

Read each of the following and provide the missing words. Answers appear at the beginning of the next question.

1. The memory of a computer is where the ________ and ________ are stored before the calculations begin. During a computer run, partial answers may also be stored in the ________.
2. (*program, data, memory*) A read-only memory or ________ is equivalent to a group of memory locations, each permanently storing a word. The ________ is the only one who can store programs and data in a ROM.
3. (*ROM, manufacturer*) The ________ and contents of a memory location are two different things. Because the address is in binary form, the manufacturer uses on-chip decoding to access the memory location. With on-chip decoding, *n* address lines can access ________ memory locations.
4. (*address*, 2^n) The PROM allows users to store their own programs and data. An instrument called a PROM ________ does the storing or burning in. Once this is done, the programming is permanent.
5. (*programmer*) The ________ is ultraviolet-light-erasable and electrically programmable. This allows the user to erase and store until programs and data are perfected.
6. (*EPROM*) The ________ time of a memory is the

time it takes to read the contents of a memory location. Bipolar memories are faster than ________ memories but more expensive.

7. (*access, MOS*) ROMs, PROMs, and EPROMs are ________ memories. This means that they retain stored data even though the power is turned off. Core RAMs are also ________, but they are becoming obsolete.
8. (*nonvolatile, nonvolatile*) Semiconductor RAM memories may be static or ________. Both are volatile. The first type uses bipolar or MOS flip-flops, which means that data is stored as long as power is applied. The second type uses MOSFETs and capacitors to store data, which must be ________ every few milliseconds.
9. (*dynamic, refreshed*) The memory cell of a dynamic RAM is simpler and smaller than the memory cell of a ________ RAM. Because of this, the dynamic RAM can contains more memory cells than a ________ RAM of the same chip size.
10. (*static, static*) The ________ bits of a static RAM select the memory location. The write enable ($\overline{WE}$) and chip enable ($\overline{CE}$) select a write, read, or do-nothing. When $\overline{WE}$ and $\overline{CE}$ are both low, you get a ________ operation. When $\overline{WE}$ is high and $\overline{CE}$ is low, you get a ________ operation. $\overline{CE}$ high is the inactive state.
11. (*address, write, read*) During a computer run, the CPU sends binary addresses to the ________, where read or write operations occur. Typical microcomputers have an address bus with ________ bits.
12. (*memory, 16*) An address bus with 16 bits can access a maximum of 65,536 memory locations. The hexadecimal addresses of these memory locations are from 0000H to FFFFH. First-generation microcomputers store 1 byte in each memory location, which implies a maximum memory of 64K.

PROBLEMS

9-1. How many memory locations can 14 address bits access?

9-2. The 2708 is an 8,192-bit EPROM organized as a 1,024 × 8 memory. How many address pins does it have?

9-3. The 2732 is a 4,096 × 8 EPROM. How many address lines does it have?

9-4. An 8156 is a 2,048-bit static RAM with 256 words of 8 bits each. How many address lines does this RAM have?

9-5. Use U (up) and D (down) to program the TTL memory of Fig. 9-9 with the following data:

Address	Data
0000	1000 1001
0001	0111 1100
0010	0011 0110
0011	0010 0011
0100	0001 0111
0101	0101 1111
0110	1110 1101
0111	1111 1000

Show your answer by converting each 0 to a D and each 1 to a U.

9-6. The following data is to be programmed into the TTL memory of Fig. 9-9:

Address	Data
0H	EEH
1H	5CH
2H	26H
3H	6AH
4H	FDH
5H	15H
6H	94H
7H	C3H

Convert these hexadecimal addresses and contents to ups (U) and downs (D) as described in Sec. 9-4.

9-7. Address 2000H contains the byte 3FH. What is the decimal equivalent of 3FH?

9-8. In a 32K memory, the hexadecimal addresses are from 0000H to 7FFFH. What is the decimal equivalent of the highest address?

9-9. What is the highest address in a 48K memory? Express the answer in hexadecimal and decimal form.

9-10. A byte is stored at hexadecimal location 6F9EH. What is the decimal address? (Use Appendix 2.)

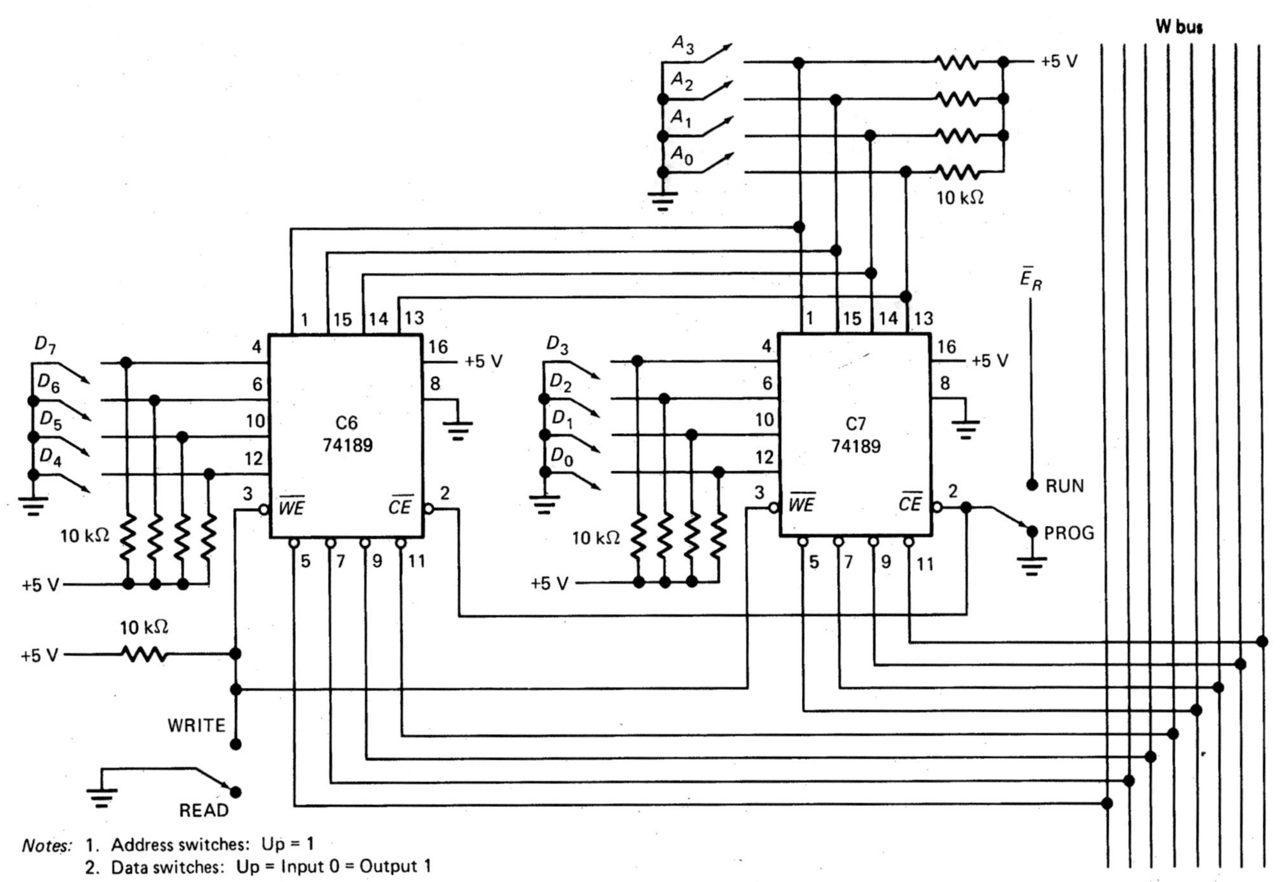

Fig. 9-9

9-11. Here is some data stored in a memory:

Address	Data
8E00H	2FH
8E01H	D4H
8E02H	CFH
8E03H	6EH
8E04H	53H
8E05H	7AH

a. What is the decimal equivalent of each stored byte? (Use Appendix 2.)
b. What is the decimal equivalent of the highest address?

9-12. Suppose there are four different memories with the following capacities:

Memory A = 16K
Memory B = 32K
Memory C = 48K
Memory D = 64K

All memories start with hexadecimal address 0000H.

a. How many bytes can memory C store? Express the answer in decimal.
b. What is the highest decimal address in memory A?
c. We want to store a byte at address C300H. Which memory must we use?
d. What is the highest hexadecimal address for each memory?

9-13. What kind of memory can be programmed and then erased with ultraviolet light, so that it can be reprogrammed?

9-14. What kind of memory can be programmed and then erased with electrical pulses, so that it can be reprogrammed?

9-15. What kind of nonvolatile memory can have individual bytes reprogrammed without erasing the entire chip?

PART 2
SAP
(SIMPLE-AS-POSSIBLE) COMPUTERS

10

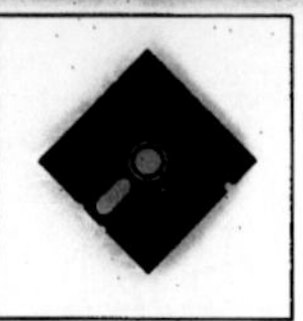

SAP-1

The SAP (Simple-As-Possible) computer has been designed for you, the beginner. The main purpose of SAP is to introduce all the crucial ideas behind computer operation without burying you in unnecessary detail. But even a simple computer like SAP covers many advanced concepts. To avoid bombarding you with too much all at once, we will examine three different generations of the SAP computer.

SAP-1 is the first stage in the evolution toward modern computers. Although primitive, SAP-1 is a big step for a beginner. So, dig into this chapter; master SAP-1, its architecture, its programming, and its circuits. Then you will be ready for SAP-2.

10-1 ARCHITECTURE

Figure 10-1 shows the *architecture* (structure) of SAP-1, a bus-organized computer. All register outputs to the W bus are three-state; this allows orderly transfer of data. All other register outputs are two-state; these outputs continuously drive the boxes they are connected to.

The layout of Fig. 10-1 emphasizes the registers used in SAP-1. For this reason, no attempt has been made to keep all control circuits in one block called the control unit, all input-output circuits in another block called the I/O unit, etc.

Many of the registers of Fig. 10-1 are already familiar from earlier examples and discussions. What follows is a brief description of each box; detailed explanations come later.

Program Counter

The program is stored at the beginning of the memory with the first instruction at binary address 0000, the second instruction at address 0001, the third at address 0010, and so on. The *program counter,* which is part of the control unit, counts from 0000 to 1111. Its job is to send to the memory the address of the next instruction to be fetched and executed. It does this as follows.

The program counter is reset to 0000 before each computer run. When the computer run begins, the program counter sends address 0000 to the memory. The program counter is then incremented to get 0001. After the first instruction is fetched and executed, the program counter sends address 0001 to the memory. Again the program counter is incremented. After the second instruction is fetched and executed, the program counter sends address 0010 to the memory. In this way, the program counter is keeping track of the next instruction to be fetched and executed.

The program counter is like someone pointing a finger at a list of instructions, saying do this first, do this second, do this third, etc. This is why the program counter is sometimes called a *pointer;* it points to an address in memory where something important is being stored.

Input and MAR

Below the program counter is the *input* and *MAR* block. It includes the address and data switch registers discussed in Sec. 9-4. These switch registers, which are part of the input unit, allow you to send 4 address bits and 8 data bits to the RAM. As you recall, instruction and data words are written into the RAM before a computer run.

The *memory address register* (MAR) is part of the SAP-1 memory. During a computer run, the address in the program counter is latched into the MAR. A bit later, the MAR applies this 4-bit address to the RAM, where a read operation is performed.

The RAM

The *RAM* is a 16 × 8 static TTL RAM. As discussed in Sec. 9-4, you can program the RAM by means of the address and data switch registers. This allows you to store a program and data in the memory before a computer run.

During a computer run, the RAM receives 4-bit addresses from the MAR and a read operation is performed. In this way, the instruction or data word stored in the RAM is placed on the W bus for use in some other part of the computer.

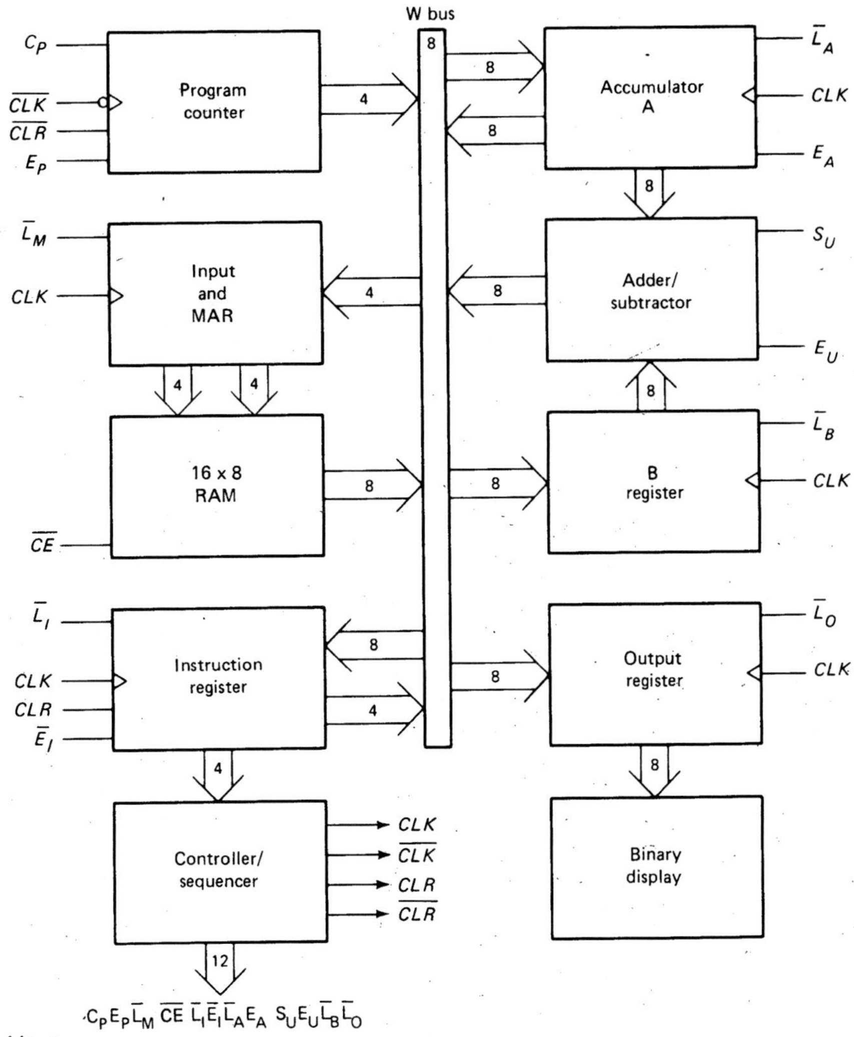

Fig. 10-1 SAP-1 architecture.

Instruction Register

The *instruction register* is part of the control unit. To fetch an instruction from the memory the computer does a memory read operation. This places the contents of the addressed memory location on the W bus. At the same time, the instruction register is set up for loading on the next positive clock edge.

The contents of the instruction register are split into two nibbles. The upper nibble is a two-state output that goes directly to the block labeled "Controller-sequencer." The lower nibble is a three-state output that is read onto the W bus when needed.

Controller-Sequencer

The lower left block contains the *controller-sequencer*. Before each computer run, a $\overline{CLR}$ signal is sent to the program counter and a *CLR* signal to the instruction register. This resets the program counter to 0000 and wipes out the last instruction in the instruction register.

A clock signal *CLK* is sent to all buffer registers; this synchronizes the operation of the computer, ensuring that things happen when they are supposed to happen. In other words, all register transfers occur on the positive edge of a common *CLK* signal. Notice that a $\overline{CLK}$ signal also goes to the program counter.

The 12 bits that come out of the controller-sequencer form a word controlling the rest of the computer (like a supervisor telling others what to do.) The 12 wires carrying the control word are called the *control bus*.

The control word has the format of

$$\text{CON} = C_PE_P\overline{L}_M\overline{CE} \quad \overline{L}_I\overline{E}_I\overline{L}_AE_A \quad S_UE_U\overline{L}_B\overline{L}_O$$

This word determines how the registers will react to the next positive *CLK* edge. For instance, a high E_P and a low

$\overline{L}_M$ mean that the contents of the program counter are latched into the MAR on the next positive clock edge. As another example, a low $\overline{CE}$ and a low $\overline{L}_A$ mean that the addressed RAM word will be transferred to the accumulator on the next positive clock edge. Later, we will examine the timing diagrams to see exactly when and how these data transfers take place.

Accumulator

The *accumulator* (A) is a buffer register that stores intermediate answers during a computer run. In Fig. 10-1 the accumulator has two outputs. The two-state output goes directly to the adder-subtracter. The three-state output goes to the W bus. Therefore, the 8-bit accumulator word continuously drives the adder-subtracter; the same word appears on the W bus when E_A is high.

The Adder-Subtracter

SAP-1 uses a 2's-complement *adder-subtracter*. When S_U is low in Fig. 10-1, the sum out of the adder-subtracter is

$$\mathbf{S} = \mathbf{A} + \mathbf{B}$$

When S_U is high, the difference appears:

$$\mathbf{A} = \mathbf{A} + \mathbf{B}'$$

(Recall that the 2's complement is equivalent to a decimal sign change.)

The adder-subtracter is *asynchronous* (unclocked); this means that its contents can change as soon as the input words change. When E_U is high, these contents appear on the W bus.

B Register

The *B register* is another buffer register. It is used in arithmetic operations. A low $\overline{L}_B$ and positive clock edge load the word on the W bus into the B register. The two-state output of the B register drives the adder-subtracter, supplying the number to be added or subtracted from the contents of the accumulator.

Output Register

Example 8-1 discussed the output register. At the end of a computer run, the accumulator contains the answer to the problem being solved. At this point, we need to transfer the answer to the outside world. This is where the *output register* is used. When E_A is high and $\overline{L}_O$ is low, the next positive clock edge loads the accumulator word into the output register.

The output register is often called an *output port* because processed data can leave the computer through this register. In microcomputers the output ports are connected to *interface circuits* that drive peripheral devices like printers, cathode-ray tubes, teletypewriters, and so forth. (An interface circuit prepares the data to drive each device.)

Binary Display

The *binary display* is a row of eight light-emitting diodes (LEDs). Because each LED connects to one flip-flop of the output port, the binary display shows us the contents of the output port. Therefore, after we've transferred an answer from the accumulator to the output port, we can see the answer in binary form.

Summary

The SAP-1 control unit consists of the program counter, the instruction register, and the controller-sequencer that produces the control word, the clear signals, and the clock signals. The SAP-1 ALU consists of an accumulator, an adder-subtracter, and a B register. The SAP-1 memory has the MAR and a 16 × 8 RAM. The I/O unit includes the input programming switches, the output port, and the binary display.

10-2 INSTRUCTION SET

A computer is a useless pile of hardware until someone programs it. This means loading step-by-step instructions into the memory before the start of a computer run. Before you can program a computer, however, you must learn its *instruction set*, the basic operations it can perform. The SAP-1 instruction set follows.

LDA

As described in Chap. 9, the words in the memory can be symbolized by $\mathbf{R}_0$, $\mathbf{R}_1$, $\mathbf{R}_2$, etc. This means that $\mathbf{R}_0$ is stored at address 0H, $\mathbf{R}_1$ at address 1H, $\mathbf{R}_2$ at address 2H, and so on.

LDA stands for "load the accumulator." A complete LDA instruction includes the hexadecimal address of the data to be loaded. LDA 8H, for example, means "load the accumulator with the contents of memory location 8H." Therefore, given

$$\mathbf{R}_8 = 1111\ 0000$$

the execution of LDA 8H results in

$$\mathbf{A} = 1111\ 0000$$

Similarly, LDA AH means "load the accumulator with the contents of memory location AH," LDA FH means "load the accumulator with the contents of memory location FH," and so on.

ADD

ADD is another SAP-1 instruction. A complete ADD instruction includes the address of the word to be added. For instance, ADD 9H means "add the contents of memory location 9H to the accumulator contents"; the sum replaces the original contents of the accumulator.

Here's an example. Suppose decimal 2 is in the accumulator and decimal 3 is in memory location 9H. Then

$$\mathbf{A} = 0000\ 0010$$
$$\mathbf{R_9} = 0000\ 0011$$

During the execution of ADD 9H, the following things happen. First, $\mathbf{R_9}$ is loaded into the B register to get

$$\mathbf{B} = 0000\ 0011$$

and almost instantly the adder-subtracter forms the sum of **A** and **B**

$$\mathbf{SUM} = 0000\ 0101$$

Second, this sum is loaded into the accumulator to get

$$\mathbf{A} = 0000\ 0101$$

The foregoing routine is used for all ADD instructions; the addressed RAM word goes to the B register and the adder-subtracter output to the accumulator. This is why the execution of ADD 9H adds $\mathbf{R_9}$ to the accumulator contents, the execution of ADD FH adds $\mathbf{R_F}$ to the accumulator contents, and so on.

SUB

SUB is another SAP-1 instruction. A complete SUB instruction includes the address of the word to be subtracted. For example, SUB CH means "subtract the contents of memory location CH from the contents of the accumulator"; the difference out of the adder-subtracter then replaces the original contents of the accumulator.

For a concrete example, assume that decimal 7 is in the accumulator and decimal 3 is in memory location CH. Then

$$\mathbf{A} = 0000\ 0111$$
$$\mathbf{R_C} = 0000\ 0011$$

The execution of SUB CH takes place as follows. First, $\mathbf{R_C}$ is loaded into the B register to get

$$\mathbf{B} = 0000\ 0011$$

and almost instantly the adder-subtracter forms the difference of **A** and **B**:

$$\mathbf{DIFF} = 0000\ 0100$$

Second, this difference is loaded into the accumulator and

$$\mathbf{A} = 0000\ 0100$$

The foregoing routine applies to all SUB instructions; the addressed RAM word goes to the B register and the adder-subtracter output to the accumulator. This is why the execution of SUB CH subtracts $\mathbf{R_C}$ from the contents of the accumulator, the execution of SUB EH subtracts $\mathbf{R_E}$ from the accumulator, and so on.

OUT

The instruction OUT tells the SAP-1 computer to transfer the accumulator contents to the output port. After OUT has been executed, you can see the answer to the problem being solved.

OUT is complete by itself; that is, you do not have to include an address when using OUT because the instruction does not involve data in the memory.

HLT

HLT stands for halt. This instruction tells the computer to stop processing data. HLT marks the end of a program, similar to the way a period marks the end of a sentence. You must use a HLT instruction at the end of every SAP-1 program; otherwise, you get computer trash (meaningless answers caused by runaway processing).

HLT is complete by itself; you do not have to include a RAM word when using HLT because this instruction does not involve the memory.

Memory-Reference Instructions

LDA, ADD, and SUB are called *memory-reference instructions* because they use data stored in the memory. OUT and HLT, on the other hand, are not memory-reference instructions because they do not involve data stored in the memory.

Mnemonics

LDA, ADD, SUB, OUT, and HLT are the instruction set for SAP-1. Abbreviated instructions like these are called *mnemonics* (memory aids). Mnemonics are popular in computer work because they remind you of the operation that will take place when the instruction is executed. Table 10-1 summarizes the SAP-1 instruction set.

The 8080 and 8085

The 8080 was the first widely used microprocessor. It has 72 instructions. The 8085 is an enhanced version of the 8080 with essentially the same instruction set. To make SAP practical, the SAP instructions will be upward com-

TABLE 10-1. SAP-1 INSTRUCTION SET

Mnemonic	Operation
LDA	Load RAM data into accumulator
ADD	Add RAM data to accumulator
SUB	Subtract RAM data from accumulator
OUT	Load accumulator data into output register
HLT	Stop processing

patible with the 8080/8085 instruction set. In other words, the SAP-1 instructions LDA, ADD, SUB, OUT, and HLT are 8080/8085 instructions. Likewise, the SAP-2 and SAP-3 instructions will be part of the 8080/8085 instruction set. Learning SAP instructions is getting you ready for the 8080 and 8085, two widely used microprocessors.

EXAMPLE 10-1

Here's a SAP-1 program in mnemonic form:

Address	Mnemonics
0H	LDA 9H
1H	ADD AH
2H	ADD BH
3H	SUB CH
4H	OUT
5H	HLT

The data in higher memory is

Address	Data
6H	FFH
7H	FFH
8H	FFH
9H	01H
AH	02H
BH	03H
CH	04H
DH	FFH
EH	FFH
FH	FFH

What does each instruction do?

SOLUTION

The program is in the low memory, located at addresses 0H to 5H. The first instruction loads the accumulator with the contents of memory location 9H, and so the accumulator contents become

$$A = 01H$$

The second instruction adds the contents of memory location AH to the accumulator contents to get a new accumulator total of

$$A = 01H + 02H = 03H$$

Similarly, the third instruction add the contents of memory location BH

$$A = 03H + 03H = 06H$$

The SUB instruction subtracts the contents of memory location CH to get

$$A = 06H - 04H = 02H$$

The OUT instruction loads the accumulator contents into the output port: therefore, the binary display shows

0000 0010

The HLT instruction stops the data processing.

10-3 PROGRAMMING SAP-1

To load instruction and data words into the SAP-1 memory we have to use some kind of code that the computer can interpret. Table 10-2 shows the code used in SAP-1. The number 0000 stands for LDA, 0001 for ADD, 0010 for SUB, 1110 for OUT, and 1111 for HLT. Because this code tells the computer which operation to perform, it is called an *operation code* (op code).

As discussed earlier, the address and data switches of Fig. 9-7 allow you to program the SAP-1 memory. By design, these switches produce a 1 in the up position (U)

TABLE 10-2. SAP-1 OP CODE

Mnemonic	Op code
LDA	0000
ADD	0001
SUB	0010
OUT	1110
HLT	1111

and a 0 in the down position (D). When programming the data switches with an instruction, the op code goes into the upper nibble, and the *operand* (the rest of the instruction) into the lower nibble.

For instance, suppose we want to store the following instructions:

Address	Instruction
0H	LDA FH
1H	ADD EH
2H	HLT

First, convert each instruction to binary as follows:

LDA FH = 0000 1111
ADD EH = 0001 1110
HLT = 1111 XXXX

In the first instruction, 0000 is the op code for LDA, and 1111 is the binary equivalent of FH. In the second instruction, 0001 is the op code for ADD, and 1110 is the binary equivalent of EH. In the third instruction, 1111 is the op code for HLT, and XXXX are don't cares because the HLT is not a memory-reference instruction.

Next, set up the address and data switches as follows:

Address	Data
DDDD	DDDD UUUU
DDDU	DDDU UUUD
DDUD	UUUU XXXX

After each address and data word is set, you press the write button. Since D stores a binary 0 and U stores a binary 1, the first three memory locations now have these contents:

Address	Contents
0000	0000 1111
0001	0001 1110
0010	1111 XXXX

A final point. *Assembly language* involves working with mnemonics when writing a program. *Machine language* involves working with strings of 0s and 1s. The following examples bring out the distinction between the two languages.

EXAMPLE 10-2

Translate the program of Example 10-1 into SAP-1 machine language.

SOLUTION

Here is the program of Example 10-1:

Address	Instruction
0H	LDA 9H
1H	ADD AH
2H	ADD BH
3H	SUB CH
4H	OUT
5H	HLT

This program is in assembly language as it now stands. To get it into machine language, we translate it to 0s and 1s as follows:

Address	Instruction
0000	0000 1001
0001	0001 1010
0010	0001 1011
0011	0010 1100
0100	1110 XXXX
0101	1111 XXXX

Now the program is in machine language.

Any program like the foregoing that's written in machine language is called an *object program*. The original program with mnemonics is called a *source program*. In SAP-1 the operator translates the source program into an object program when programming the address and data switches.

A final point. The four MSBs of a SAP-1 machine-language instruction specify the operation, and the four LSBs give the address. Sometimes we refer to the MSBs as the *instruction field* and to the LSBs as the *address field*. Symbolically,

Instruction = XXXX XXXX

Instruction field (first XXXX)
Address field (second XXXX)

EXAMPLE 10-3

How would you program SAP-1 to solve this arithmetic problem?

$$16 + 20 + 24 - 32$$

The numbers are in decimal form.

SOLUTION

One way is to use the program of the preceding example, storing the data (16, 20, 24, 32) in memory locations 9H to CH. With Appendix 2, you can convert the decimal data into hexadecimal data to get this assembly-language version:

Address	Contents
0H	LDA 9H
1H	ADD AH
2H	ADD BH
3H	SUB CH
4H	OUT
5H	HLT
6H	XX
7H	XX
8H	XX
9H	10H
AH	14H
BH	18H
CH	20H

The machine-language version is

Address	Contents
0000	0000 1001
0001	0001 1010
0010	0001 1011
0011	0010 1100
0100	1110 XXXX
0101	1111 XXXX
0110	XXXX XXXX
0111	XXXX XXXX
1000	XXXX XXXX
1001	0001 0000
1010	0001 0100
1011	0001 1000
1100	0010 0000

Notice that the program is stored ahead of the data. In other words, the program is in low memory and the data in high memory. This is essential in SAP-1 because the program counter points to address 0000 for the first instruction, 0001 for the second instruction, and so forth.

EXAMPLE 10-4

Chunk the program and data of the preceding example by converting to hexadecimal shorthand.

SOLUTION

Address	Contents
0H	09H
1H	1AH
2H	1BH
3H	2CH
4H	EXH
5H	FXH
6H	XXH
7H	XXH
8H	XXH
9H	10H
AH	14H
BH	18H
CH	20H

This version of the program and data is still considered machine language.

Incidentally, negative data is loaded in 2's-complement form. For example, −03H is entered as FDH.

10-4 FETCH CYCLE

The *control unit* is the key to a computer's automatic operation. The control unit generates the control words that fetch and execute each instruction. While each instruction is fetched and executed, the computer passes through different *timing states* (*T* states), periods during which register contents change. Let's find out more about these *T* states.

Ring Counter

Earlier, we discussed the SAP-1 ring counter (see Fig. 8-16 for the schematic diagram). Figure 10-2*a* symbolizes the ring counter, which has an output of

$$\mathbf{T} = T_6T_5T_4T_3T_2T_1$$

At the beginning of a computer run, the ring word is

$$\mathbf{T} = 000001$$

Successive clock pulses produce ring words of

$$\mathbf{T} = 000010$$
$$\mathbf{T} = 000100$$
$$\mathbf{T} = 001000$$
$$\mathbf{T} = 010000$$
$$\mathbf{T} = 100000$$

Then, the ring counter resets to 000001, and the cycle repeats. Each ring word represents one *T* state.

Figure 10-2*b* shows the timing pulses out of the ring counter. The initial state T_1 starts with a negative clock edge and ends with the next negative clock edge. During this *T* state, the T_1 bit out of the ring counter is high.

During the next state, T_2 is high; the following state has a high T_3; then a high T_4; and so on. As you can see, the

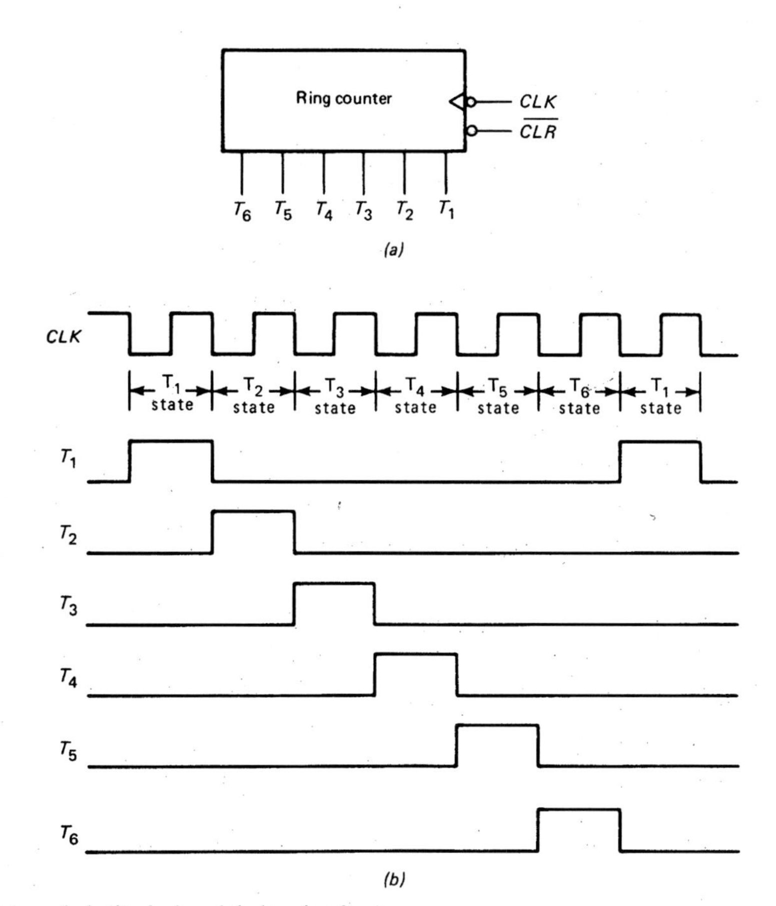

Fig. 10-2 Ring counter: (*a*) symbol; (*b*) clock and timing signals.

ring counter produces six T states. Each instruction is fetched and executed during these six T states.

Notice that a positive *CLK* edge occurs midway through each T state. The importance of this will be brought out later.

Address State

The T_1 state is called the *address state* because the address in the program counter (PC) is transferred to the memory address register (MAR) during this state. Figure 10-3*a* shows the computer sections that are active during this state (active parts are light; inactive parts are dark).

During the address state, E_P and $\overline{L}_M$ are active; all other control bits are inactive. This means that the controller-sequencer is sending out a control word of

$$\mathbf{CON} = C_P E_P \overline{L}_M \overline{CE} \quad \overline{L}_I \overline{E}_I \overline{L}_A E_A \quad S_U E_U \overline{L}_B \overline{L}_O$$
$$= 0\,1\,0\,1 \quad 1\,1\,1\,0 \quad 0\,0\,1\,1$$

during this state.

Increment State

Figure 10-3*b* shows the active parts of SAP-1 during the T_2 state. This state is called the *increment state* because the program counter is incremented. During the increment state, the controller-sequencer is producing a control word of

$$\mathbf{CON} = C_P E_P \overline{L}_M \overline{CE} \quad \overline{L}_I \overline{E}_I \overline{L}_A E_A \quad S_U E_U \overline{L}_B \overline{L}_O$$
$$= 1\,0\,1\,1 \quad 1\,1\,1\,0 \quad 0\,0\,1\,1$$

As you see, the C_P bit is active.

Memory State

The T_3 state is called the *memory state* because the addressed RAM instruction is transferred from the memory to the instruction register. Figure 10-3*c* shows the active parts of SAP-1 during the memory state. The only active control bits during this state are $\overline{CE}$ and $\overline{L}_I$, and the word out of the controller-sequencer is

$$\mathbf{CON} = C_P E_P \overline{L}_M \overline{CE} \quad \overline{L}_I \overline{E}_I \overline{L}_A E_A \quad S_U E_U \overline{L}_B \overline{L}_O$$
$$= 0\,0\,1\,0 \quad 0\,1\,1\,0 \quad 0\,0\,1\,1$$

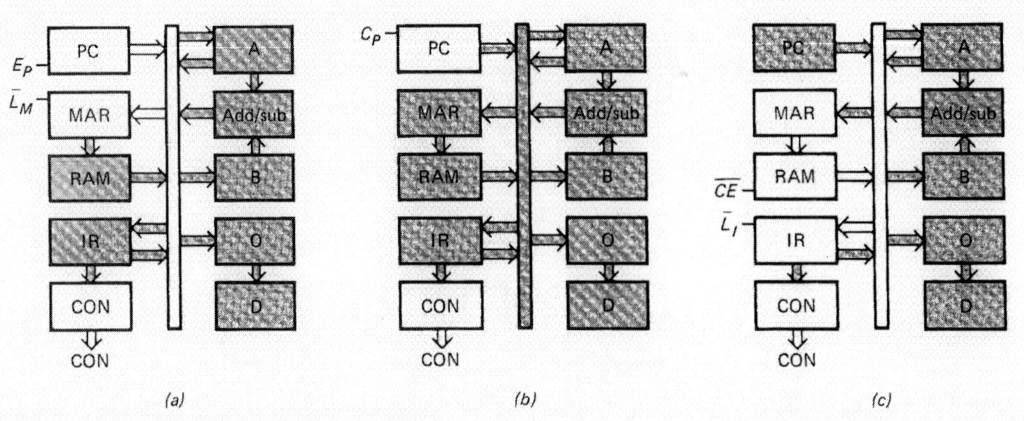

Fig. 10-3 Fetch cycle: (*a*) T_1 state; (*b*) T_2 state; (*c*) T_3 state.

Fetch Cycle

The address, increment, and memory states are called the *fetch cycle* of SAP-1. During the address state, E_P and $\overline{L}_M$ are active; this means that the program counter sets up the MAR via the W bus. As shown earlier in Fig. 10-2*b*, a positive clock edge occurs midway through the address state; this loads the MAR with the contents of the PC.

C_P is the only active control bit during the increment state. This sets up the program counter to count positive clock edges. Halfway through the increment state, a positive clock edge hits the program counter and advances the count by 1.

During the memory state, $\overline{CE}$ and $\overline{L}_I$ are active. Therefore, the addressed RAM word sets up the instruction register via the W bus. Midway through the memory state, a positive clock edge loads the instruction register with the addressed RAM word.

10-5 EXECUTION CYCLE

The next three states (T_4, T_5, and T_6) are the execution cycle of SAP-1. The register transfers during the execution cycle depend on the particular instruction being executed. For instance, LDA 9H requires different register transfers than ADD BH. What follows are the *control routines* for different SAP-1 instructions.

LDA Routine

For a concrete discussion, let's assume that the instruction register has been loaded with LDA 9H:

$$\textbf{IR} = 0000\ 1001$$

During the T_4 state, the instruction field 0000 goes to the controller-sequencer, where it is decoded; the address field 1001 is loaded into the MAR. Figure 10-4*a* shows the active parts of SAP-1 during the T_4 state. Note that $\overline{E}_I$ and $\overline{L}_M$ are active; all other control bits are inactive.

During the T_5 state, $\overline{CE}$ and $\overline{L}_A$ go low. This means that the addressed data word in the RAM will be loaded into the accumulator on the next positive clock edge (see Fig. 10-4*b*).

T_6 is a no-operation state. During this third execution state, all registers are inactive (Fig. 10-4*c*). This means that the controller-sequencer is sending out a word whose bits are all inactive. *Nop* (pronounced *no op*) stands for "no operation." The T_6 state of the LDA routine is a nop.

Figure 10-5 shows the timing diagram for the fetch and LDA routines. During the T_1 state, E_P and $\overline{L}_M$ are active; the positive clock edge midway through this state will transfer the address in the program counter to the MAR. During the T_2 state, C_P is active and the program counter is incremented on the positive clock edge. During the T_3 state, $\overline{CE}$ and $\overline{L}_I$ are active; when the positive clock edge occurs, the addressed RAM word is transferred to the instruction register. The LDA execution starts with the T_4 state, where $\overline{L}_M$ and $\overline{E}_I$ are active; on the positive clock edge the address field in the instruction register is transferred to the MAR. During the T_5 state, $\overline{CE}$ and $\overline{L}_A$ are active; this means that the addressed RAM data word is transferred to the accumulator on the positive clock edge. As you know, the T_6 state of the LDA routine is a nop.

ADD Routine

Suppose at the end of the fetch cycle the instruction register contains ADD BH:

$$\textbf{IR} = 0001\ 1011$$

During the T_4 state the instruction field goes to the controller-sequencer and the address field to the MAR (see Fig. 10-6*a*). During this state $\overline{E}_I$ and $\overline{L}_M$ are active.

Control bits $\overline{CE}$ and $\overline{L}_B$ are active during the T_5 state. This allows the addressed RAM word to set up the B

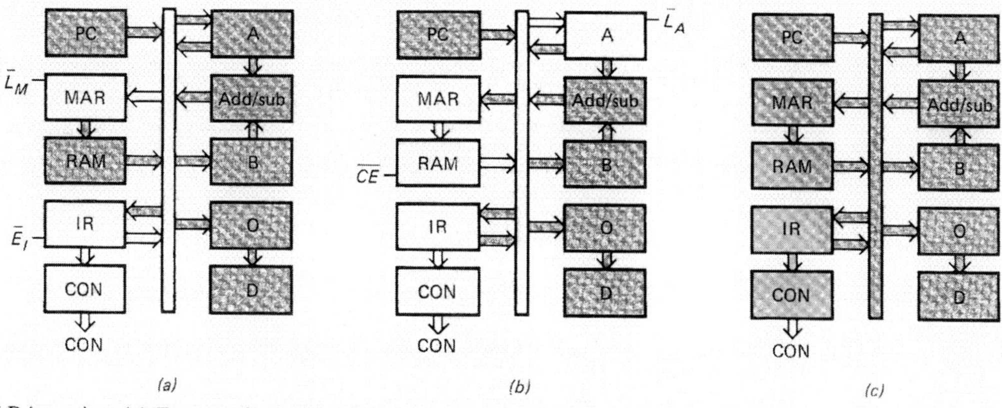

Fig. 10-4 LDA routine: (*a*) T_4 state; (*b*) T_5 state; (*c*) T_6 state.

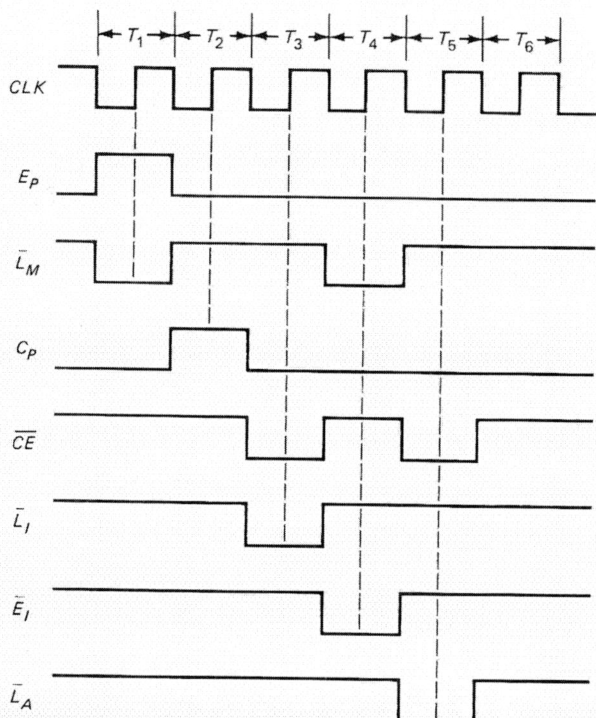

Fig. 10-5 Fetch and LDA timing diagram.

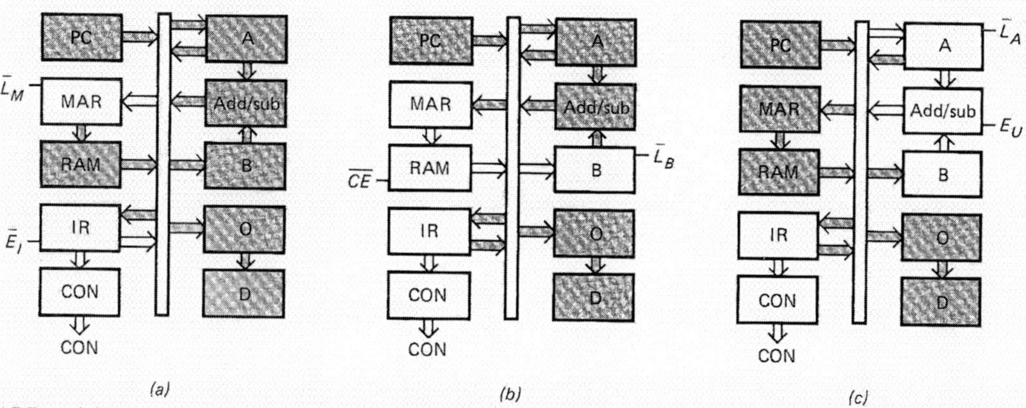

Fig. 10-6 ADD and SUB routines: (*a*) T_4 state; (*b*) T_5 state; (*c*) T_6 state.

register (Fig. 10-6*b*). As usual, loading takes place midway through the state when the positive clock edge hits the *CLK* input of the B register.

During the T_6 state, E_U and $\overline{L}_A$ are active; therefore, the adder-subtracter sets up the accumulator (Fig. 10-6*c*). Halfway through this state, the positive clock edge loads the sum into the accumulator.

Incidentally, setup time and propagation delay time prevent racing of the accumulator during this final execution state. When the positive clock edge hits in Fig. 10-6*c*, the accumulator contents change, forcing the adder-subtracter contents to change. The new contents return to the accumulator input, but the new contents don't get there until two propagation delays after the positive clock edge (one for the accumulator and one for the adder-subtracter). By then it's too late to set up the accumulator. This prevents accumulator racing (loading more than once on the same clock edge).

Figure 10-7 shows the timing diagram for the fetch and ADD routines. The fetch routine is the same as before: the T_1 state loads the PC address into the MAR; the T_2 state increments the program counter; the T_3 state sends the addressed instruction to the instruction register.

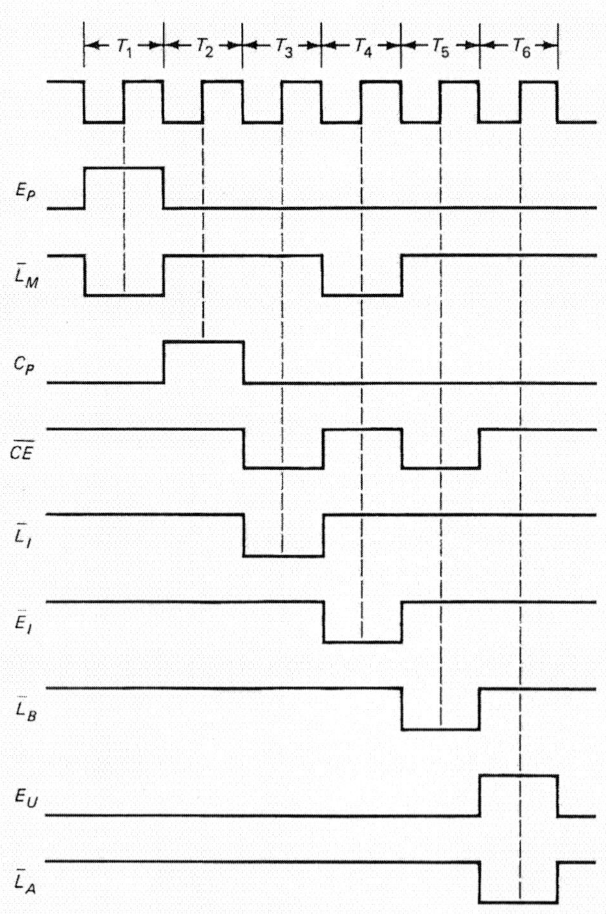

Fig. 10-7 Fetch and ADD timing diagram.

During the T_4 state, $\overline{E}_I$ and $\overline{L}_M$ are active; on the next positive clock edge, the address field in the instruction register goes to the MAR. During the T_5 state, $\overline{CE}$ and $\overline{L}_B$ are active; therefore, the addressed RAM word is loaded into the B register midway through the state. During the T_6 state, E_U and $\overline{L}_A$ are active; when the positive clock edge hits, the sum out of the adder-subtracter is stored in the accumulator.

SUB Routine

The SUB routine is similar to the ADD routine. Figure 10-6*a* and *b* show the active parts of SAP-1 during the T_4 and T_5 states. During the T_6 state, a high S_U is sent to the adder-subtracter of Fig. 10-6*c*. The timing diagram is almost identical to Fig. 10-7. Visualize S_U low during the T_1 to T_5 states and S_U high during the T_6 state.

OUT Routine

Suppose the instruction register contains the OUT instruction at the end of a fetch cycle. Then

$$\text{IR} = 1110\ \text{XXXX}$$

The instruction field goes to the controller-sequencer for decoding. Then the controller-sequencer sends out the control word needed to load the accumulator contents into the output register.

Figure 10-8 shows the active sections of SAP-1 during the execution of an OUT instruction. Since E_A and $\overline{L}_O$ are active, the next positive clock edge loads the accumulator contents into the output register during the T_4 state. The T_5 and T_6 states are nops.

Figure 10-9 is the timing diagram for the fetch and OUT routines. Again, the fetch cycle is same: address state, increment state, and memory state. During the T_4 state, E_A and $\overline{L}_O$ are active; this transfers the accumulator word to the output register when the positive clock edge occurs.

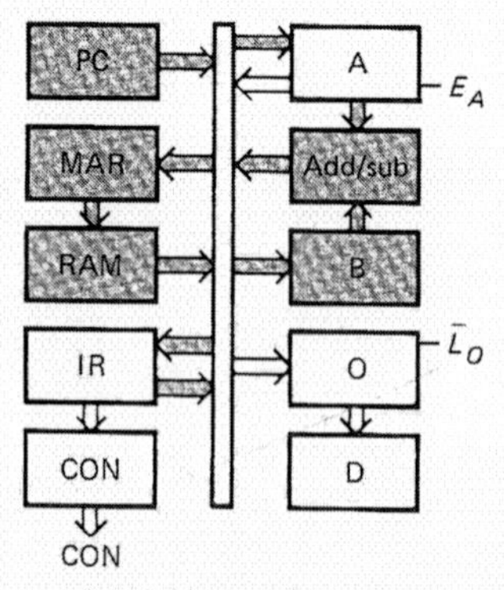

Fig. 10-8 T_4 state of OUT instruction.

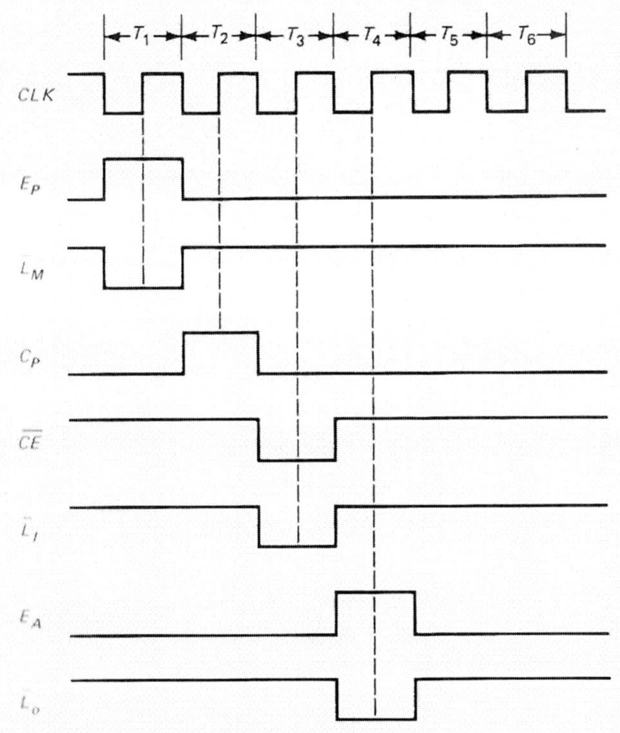

Fig. 10-9 Fetch and OUT timing diagram.

HLT

HLT does not require a control routine because no registers are involved in the execution of an HLT instruction. When the IR contains

$$\text{IR} = 1111\ \text{XXXX}$$

the instruction field 1111 signals the controller-sequencer to stop processing data. The controller-sequencer stops the computer by turning off the clock (circuitry discussed later).

Machine Cycle and Instruction Cycle

SAP-1 has six T states (three fetch and three execute). These six states are called a *machine cycle* (see Fig. 10-10*a*). It takes one machine cycle to fetch and execute each instruction. The SAP-1 clock has a frequency of 1 kHz, equivalent to a period of 1 ms. Therefore, it takes 6 ms for a SAP-1 machine cycle.

SAP-2 is slightly different because some of its instructions take more than one machine cycle to fetch and execute. Figure 10-10*b* shows the timing for an instruction that requires two machine cycles. The first three T states are the fetch cycle; however, the execution cycle requires the next nine T states. This is because a two-machine-cycle instruction is more complicated and needs those extra T states to complete the execution.

The number of T states needed to fetch and execute an instruction is called the *instruction cycle*. In SAP-1 the instruction cycle equals the machine cycle. In SAP-2 and other microcomputers the instruction cycle may equal two or more machine cycles, as shown in Fig. 10-10*b*.

The instruction cycles for the 8080 and 8085 take from one to five machine cycles (more on this later).

EXAMPLE 10-5

The 8080/8085 programming manual says that it takes thirteen T states to fetch and execute the LDA instruction.

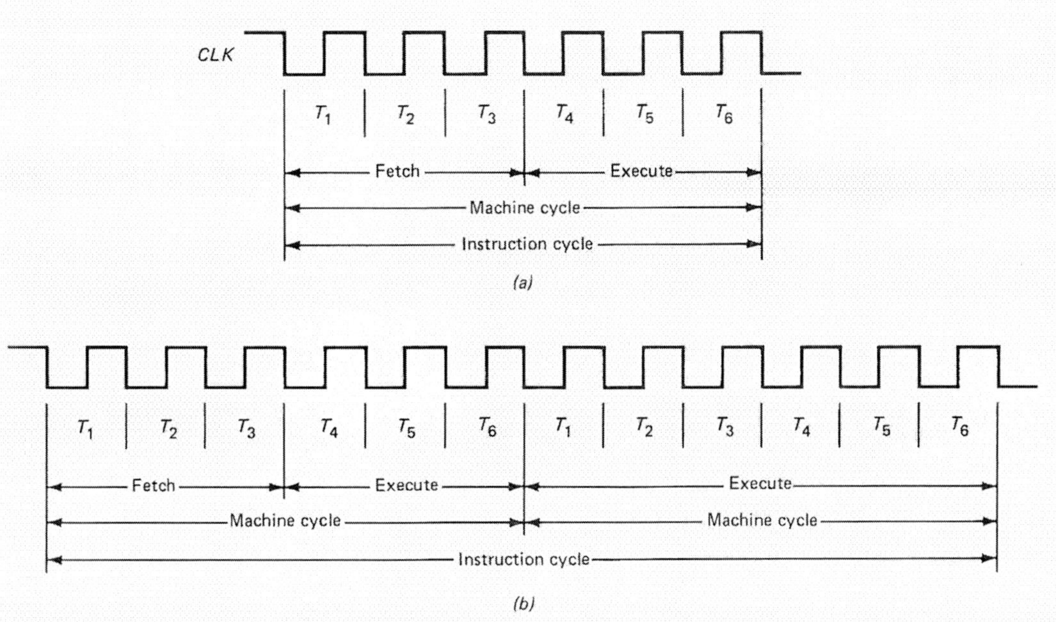

Fig. 10-10 (*a*) SAP-1 instruction cycle; (*b*) instruction cycle with two machine cycles.

If the system clock has a frequency of 2.5 MHz, how long is an instruction cycle?

SOLUTION

The period of the clock is

$$T = \frac{1}{f} = \frac{1}{2.5 \text{ MHz}} = 400 \text{ ns}$$

Therefore, each T state lasts 400 ns. Since it takes thirteen T states to fetch and execute the LDA instruction, the instruction cycle lasts for

$$13 \times 400 \text{ ns} = 5{,}200 \text{ ns} = 5.2 \ \mu\text{s}$$

EXAMPLE 10-6

Figure 10-11 shows the six T states of SAP-1. The positive clock edge occurs halfway through each state. Why is this important?

SOLUTION

SAP-1 is a *bus-organized computer* (the common type nowadays). This allows its registers to communicate via the W bus. But reliable loading of a register takes place only when the setup and hold times are satisfied. Waiting half a cycle before loading the register satisfies the setup time; waiting half a cycle after loading satisfies the hold time. This is why the positive clock edge is designed to strike the registers halfway through each T state (Fig. 10-11).

There's another reason for waiting half a cycle before loading a register. When the *ENABLE* input of the sending register goes active, the contents of this register are suddenly dumped on the W bus. Stray capacitance and lead inductance prevent the bus lines from reaching their correct voltage levels immediately. In other words, we get transients on the W bus and have to wait for them to die out to ensure valid data at the time of loading. The half-cycle delay before clocking allows the data to settle before loading.

10-6 THE SAP-1 MICROPROGRAM

We will soon be analyzing the schematic diagram of the SAP-1 computer, but first we need to summarize the execution of SAP-1 instructions in a neat table called a *microprogram*.

Microinstructions

The controller-sequencer sends out control words, one during each T state or clock cycle. These words are like directions telling the rest of the computer what to do. Because it produces a small step in the data processing, each control word is called a *microinstruction*. When looking at the SAP-1 block diagram (Fig. 10-1), we can visualize a steady stream of microinstructions flowing out of the controller-sequencer to the other SAP-1 circuits.

Macroinstructions

The instructions we have been programming with (LDA, ADD, SUB, . . .) are sometimes called *macroinstructions* to distinguish them from microinstructions. Each SAP-1 macroinstruction is made up of three microinstructions. For example, the LDA macroinstruction consists of the microinstructions in Table 10-3. To simplify the appearance of these microinstructions, we can use hexadecimal chunking as shown in Table 10-4.

Table 10-5 shows the SAP-1 microprogram, a listing of each macroinstruction and the microinstructions needed to carry it out. This table summarizes the execute routines for the SAP-1 instructions. A similar table can be used with more advanced instruction sets.

10-7 THE SAP-1 SCHEMATIC DIAGRAM

In this section we examine the complete schematic diagram for SAP-1. Figures 10-12 to 10-15 show all the chips, wires, and signals. You should refer to these figures throughout the following discussion. Appendix 4 gives additional details for some of the more complicated chips.

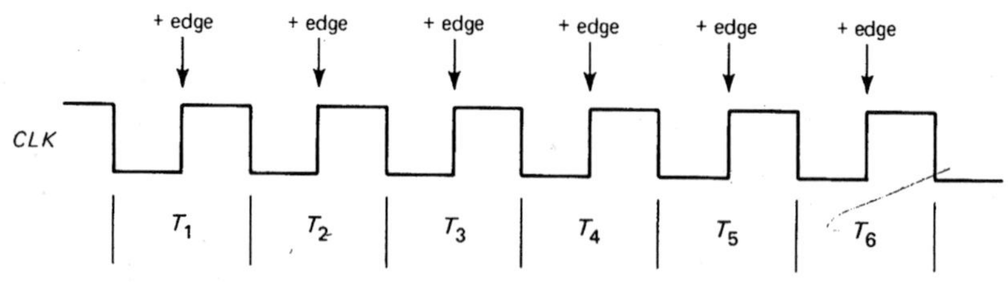

Fig. 10-11 Positive clock edges occur midway through T states.

TABLE 10-3

Macro	State	$C_P E_P \overline{L}_M \overline{CE}$	$\overline{L}_I \overline{E}_I \overline{L}_A E_A$	$S_U E_U \overline{L}_B \overline{L}_O$	Active
LDA	T_4	0 0 0 1	1 0 1 0	0 0 1 1	$\overline{L}_M$, $\overline{E}_I$
	T_5	0 0 1 0	1 1 0 0	0 0 1 1	$\overline{CE}$, $\overline{L}_A$
	T_6	0 0 1 1	1 1 1 0	0 0 1 1	None

TABLE 10-4

Macro	State	CON	Active
LDA	T_4	1A3H	$\overline{L}_M$, $\overline{E}_I$
	T_5	2C3H	$\overline{CE}$, $\overline{L}_A$
	T_6	3E3H	None

TABLE 10-5. SAP-1 MICROPROGRAM†

Macro	State	CON	Active
LDA	T_4	1A3H	$\overline{L}_M$, $\overline{E}_I$
	T_5	2C3H	$\overline{CE}$, $\overline{L}_A$
	T_6	3E3H	None
ADD	T_4	1A3H	$\overline{L}_M$, $\overline{E}_I$
	T_5	2E1H	$\overline{CE}$, $\overline{L}_B$
	T_6	3C7H	$\overline{L}_A$, E_U
SUB	T_4	1A3H	$\overline{L}_M$, $\overline{E}_I$
	T_5	2E1H	$\overline{CE}$, $\overline{L}_B$
	T_6	3CFH	$\overline{L}_A$, S_U, E_U
OUT	T_4	3F2H	E_A, $\overline{L}_O$
	T_5	3E3H	None
	T_6	3E3H	None

† CON = $C_P E_P \overline{L}_M \overline{CE}$ $\overline{L}_I \overline{E}_I \overline{L}_A E_A$ $S_U E_U \overline{L}_B \overline{L}_O$.

Program Counter

Chips C1, C2, and C3 of Fig. 10-12 are the *program counter*. Chip C1, a 74LS107, is a dual *JK* master-slave flip-flop, that produces the upper 2 address bits. Chip C2, another 74LS107, produces the lower 2 address bits. Chip C3 is a 74LS126, a quad three-state normally open switch; it gives the program counter a three-state output.

At the start of a computer run, a low $\overline{CLR}$ resets the program counter to 0000. During the T_1 state, a high E_P places the address on the W bus. During the T_2 state, a high C_P is applied to the program counter; midway through this state, the negative $\overline{CLK}$ edge (equivalent to positive *CLK* edge) increments the program counter.

The program counter is inactive during the T_3 to T_6 states.

MAR

Chip C4, a 74LS173, is a 4-bit buffer register; it serves as the MAR. Notice that pins 1 and 2 are grounded; this converts the three-state output to a two-state output. In other words, the output of the MAR is not connected to the W bus, and so there's no need to use the three-state output.

2-to-1 Multiplexer

Chip C5 is a 74LS157, a 2-to-1 nibble *multiplexer*. The left nibble (pins 14, 11, 5, 2) comes from the address switch register (S_1). The right nibble (pins 13, 10, 6, 3) comes from the MAR. The RUN-PROG switch (S_2) selects the nibble to reach to the output of C5. When S_2 is in the PROG position, the nibble out of the address switch register is selected. On the other hand, when S_2 is the RUN position, the output of the MAR is selected.

16 × 8 RAM

Chips C6 and C7 are 74189s. Each chip is a 16 × 4 static RAM. Together, they give us a 16 × 8 *read-write memory*. S_3 is the data switch register (8 bits), and S_4 is the read-write switch (a push-button switch). To program the memory, S_2 is put in the PROG position; this takes the $\overline{CE}$ input low (pin 2). The address and data switches are then set to the correct address and data words. A momentary push of the read-write switch takes $\overline{WE}$ low (pin 3) and loads the memory.

After the program and data are in memory, the RUN-PROG switch (S_2) is put in the RUN position in preparation for the computer run.

Instruction Register

Chips C8 and C9 are 74LS173s. Each chip is a 4-bit three-state buffer register. The two chips are the *instruction register*. Grounding pins 1 and 2 of C8 converts the three-state output to a two-state output, $I_7I_6I_5I_4$. This nibble goes to the instruction decoder in the controller-sequencer. Signal $\overline{E}_I$ controls the output of C9, the lower nibble in the instruction register. When $\overline{E}_I$ is low, this nibble is placed on the W bus.

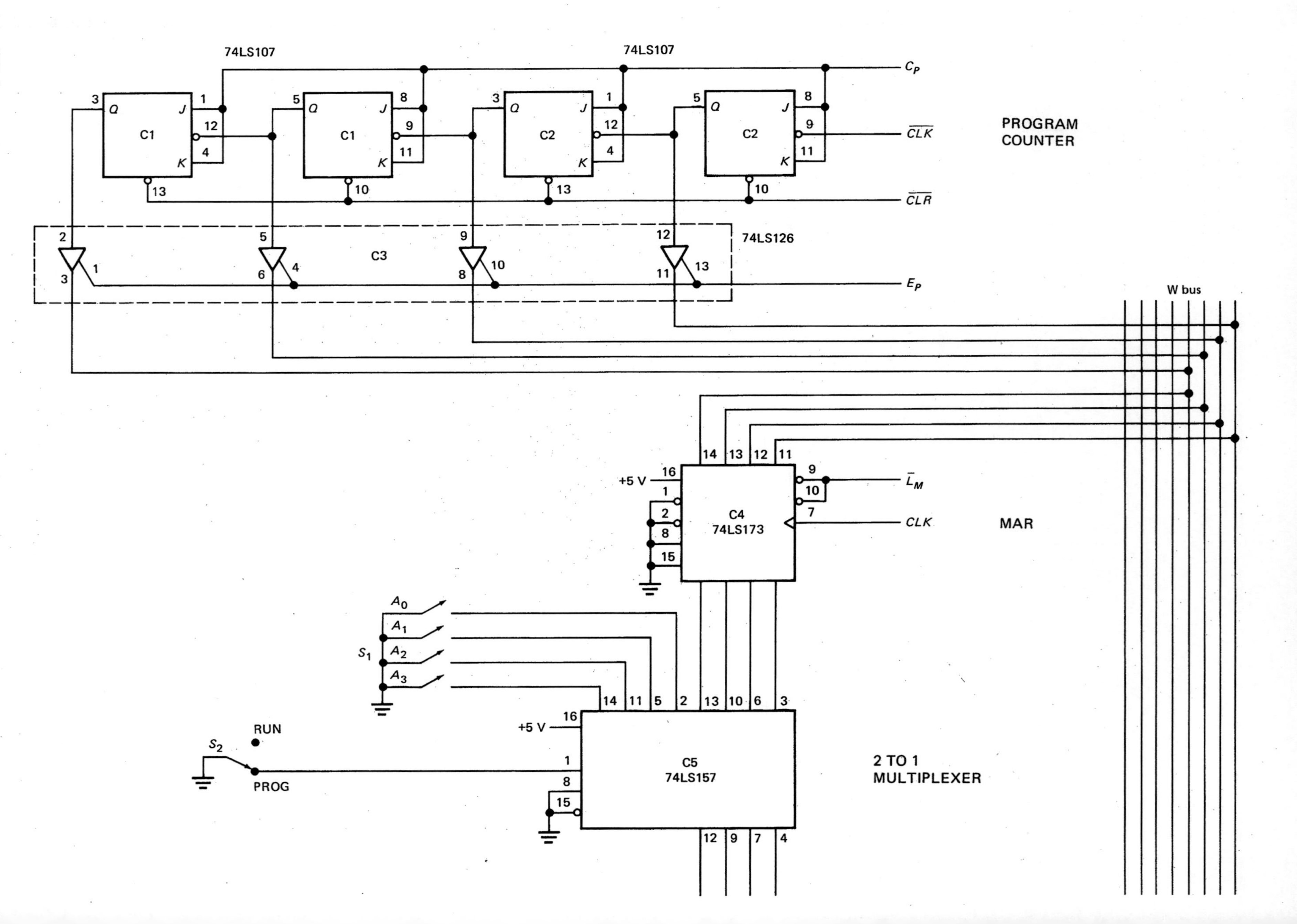

74LS107
74LS107
C1
C1
C2
C2
C_P
$\overline{CLK}$
$\overline{CLR}$
PROGRAM COUNTER
74LS126
C3
E_P
W bus
+5 V
C4 74LS173
$\bar{L}_M$
CLK
MAR
A_0
A_1
A_2
A_3
S_1
RUN
S_2
PROG
+5 V
C5 74LS157
2 TO 1 MULTIPLEXER

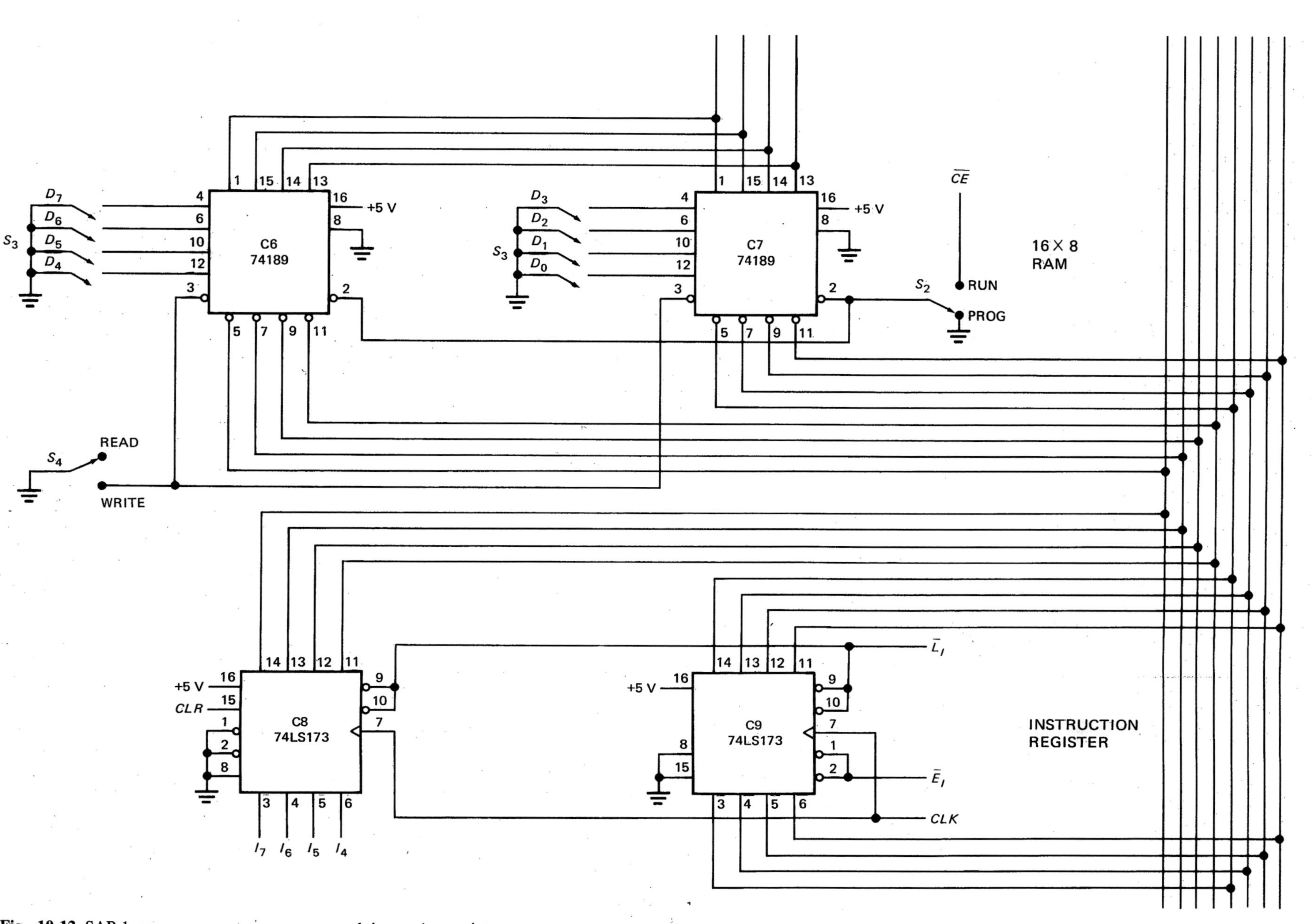

Fig. 10-12 SAP-1 program counter, memory, and instruction register.

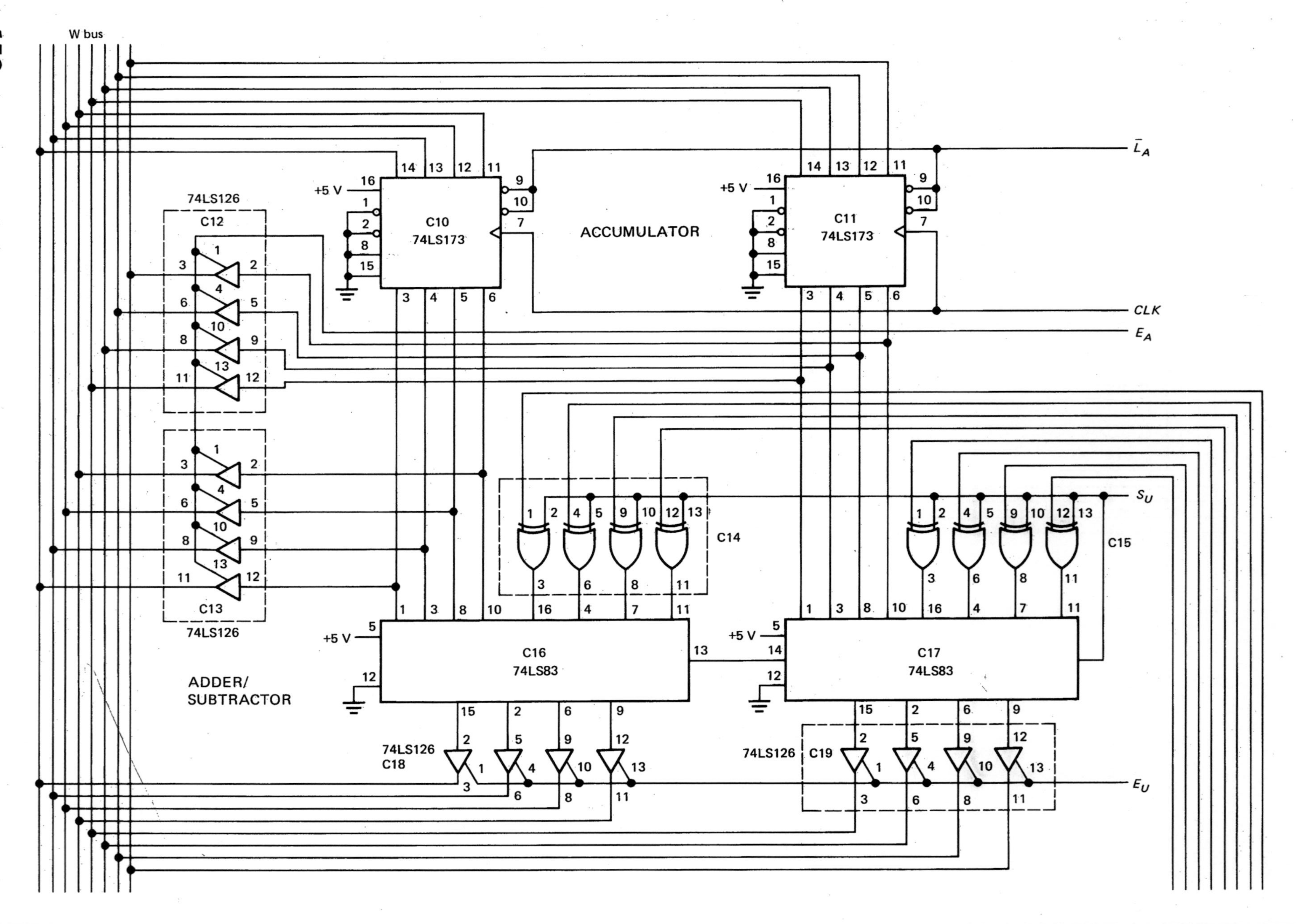

W bus
ACCUMULATOR
C10
74LS173
C11
74LS173
+5 V
$\bar{L}_A$
CLK
E_A
74LS126
C12
C13
74LS126
C14
C15
S_U
C16
74LS83
C17
74LS83
ADDER/
SUBTRACTOR
74LS126
C18
74LS126
C19
E_U

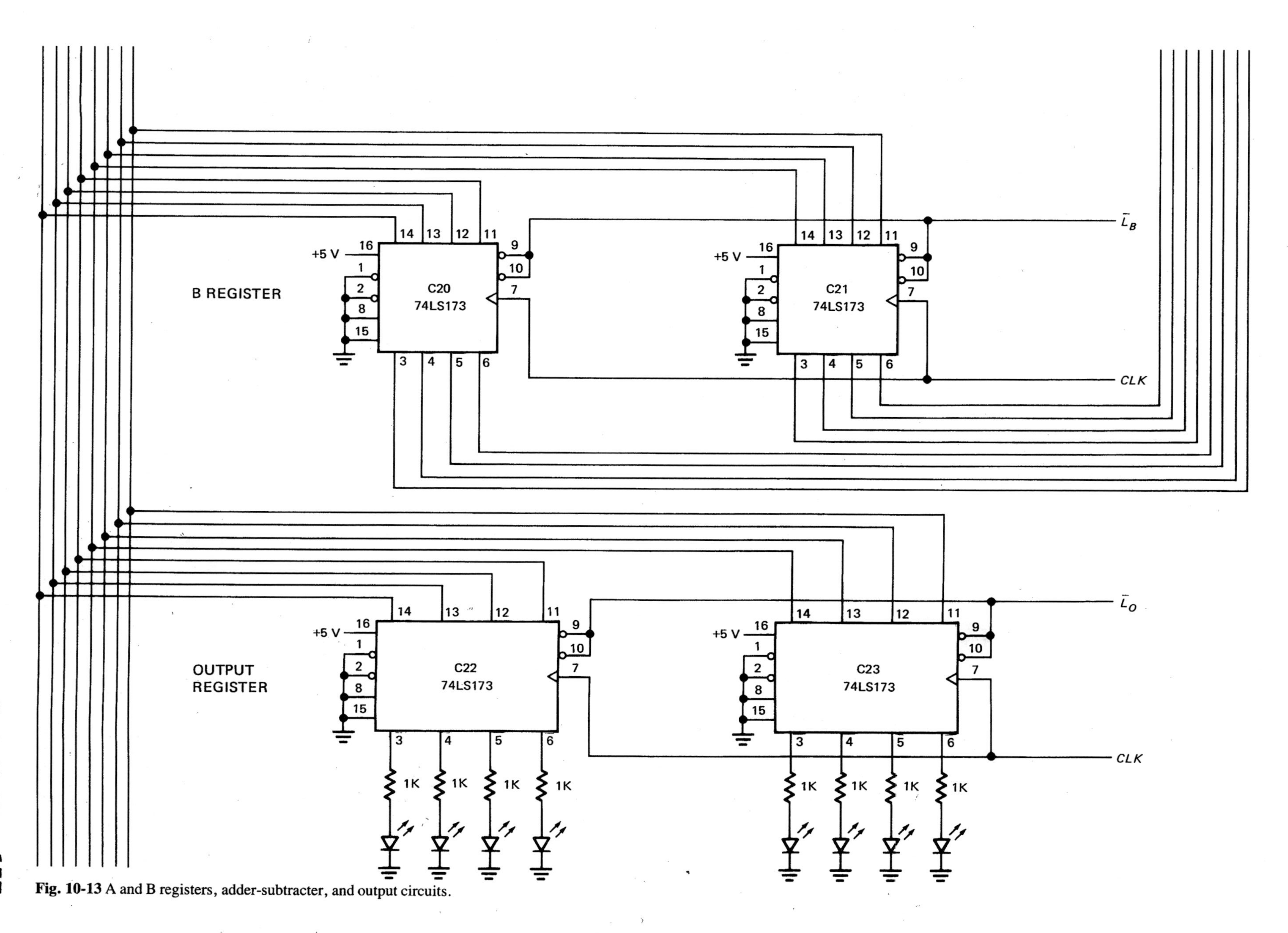

Fig. 10-13 A and B registers, adder-subtracter, and output circuits.

Accumulator

Chips C10 and C11, 74LS173s, are the *accumulator* (see Fig. 10-13). Pins 1 and 2 are grounded on both chips to produce a two-state output for the adder-subtracter. Chips C12 and C13 are 74LS126s; these three-state switches place the accumulator contents on the W bus when E_A is high.

Adder-subtracter

Chips C14 and C15 are 74LS86s. These EXCLUSIVE-OR gates are a controlled inverter. When S_U is low, the contents of the B register are transmitted. When S_U is high, the 1's complement is transmitted and a 1 is added to the LSB to form the 2's complement.

Chips C16 and C17 are 74LS83s. These 4-bit full adders combine to produce an 8-bit sum or difference. Chips C18 and C19, which are 74LS126s, convert this 8-bit answer into a three-state output for driving the W bus.

B Register and Output Register

Chips C20 and C21, which are 74LS173s, form the *B register*. It contains the data to be added or subtracted from the accumulator. Grounding pins 1 and 2 of both chips produces a two-state output for the adder-subtracter.

Chips C22 and C23 are 74LS173s and form the output register. It drives the binary display and lets us see the processed data.

Clear-Start Debouncer

In Fig. 10-14, the *clear-start debouncer* produces two outputs: CLR for the instruction register and $\overline{CLR}$ for the program counter and ring counter. $\overline{CLR}$ also goes to C29, the clock-start flip-flop. S_5 is a push-button switch. When depressed, it goes to the CLEAR position, generating a high CLR and a low $\overline{CLR}$. When S_5 is released, it returns to the START position, producing a low CLR and a high $\overline{CLR}$.

Notice that half of C24 is used for the clear-start debouncer and the other half for the single-step debouncer. Chip C24 is a 7400, a quad 2-input NAND gate.

Single-Step Debouncer

SAP-1 can run in either of two modes, manual or automatic. In the manual mode, you press and release S_6 to generate one clock pulse. When S_6 is depressed, CLK is high; when released, CLK is low. In other words, the *single-step debouncer* of Fig. 10-14 generates the T states one at a time as you press and release the button. This allows you to step through the different T states while troubleshooting or debugging. (Debugging means looking for errors in your program. You troubleshoot hardware and debug software.)

Manual-Auto Debouncer

Switch S_7 is a single-pole double-throw (SPDT) switch that can remain in either the MANUAL position or the AUTO position. When in MANUAL, the single-step button is active. When in AUTO, the computer runs automatically. Two of the NAND gates in C26 are used to debounce the MANUAL-AUTO switch. The other two NAND C26 gates are part of a NAND-NAND network that steers the single-step clock or the automatic clock to the final CLK and $\overline{CLK}$ outputs.

Clock Buffers

The output of pin 11, C26, drives the *clock buffers*. As you see in Fig. 10-14, two inverters are used to produce the final CLK output and one inverter to produce the $\overline{CLK}$ output. Unlike most of the other chips, C27 is standard TTL rather than a low-power Schottky (see SAP-1 Parts List, Appendix 5). Standard TTL is used because it can drive 20 low-power Schottky TTL loads, as indicated in Table 4-5.

If you check the data sheets of the 74LS107 and 74LS173 for input currents, you will be able to count the following low-power Schottky (LS) TTL loads on the clock and clear signals:

$$CLK = 19 \text{ LS loads}$$
$$\overline{CLK} = 2 \text{ LS loads}$$
$$CLR = 1 \text{ LS load}$$
$$\overline{CLR} = 20 \text{ LS loads}$$

This means that the CLK and $\overline{CLK}$ signals out of C27 (standard TTL) are adequate to drive the low-power Schottky TTL loads. Also, the CLR and $\overline{CLR}$ signals out of C24 (standard TTL) can drive their loads.

Clock Circuits and Power Supply

Chip C28 is a 555 timer. This IC produces a rectangular 2-kHz output with a 75 percent duty cycle. As previously discussed, a *start-the-clock flip-flop* (C29) divides the signal down to 1 kHz and at the same time produces a 50 percent duty cycle.

The *power supply* consists of a full-wave bridge rectifier working into a capacitor-input filter. The dc voltage across the 1,000-μF capacitor is approximately 20 V. Chip C30, an LM340T-5, is a voltage regulator that produces a stable output of +5 V.

Instruction Decoder

Chip C31, a hex inverter, produces complements of the op-code bits, $I_7I_6I_5I_4$ (see Fig. 10-15). Then chips C32, C33, and C34 decode the op code to produce five output signals: LDA, ADD, SUB, OUT, and $\overline{HLT}$. Remember:

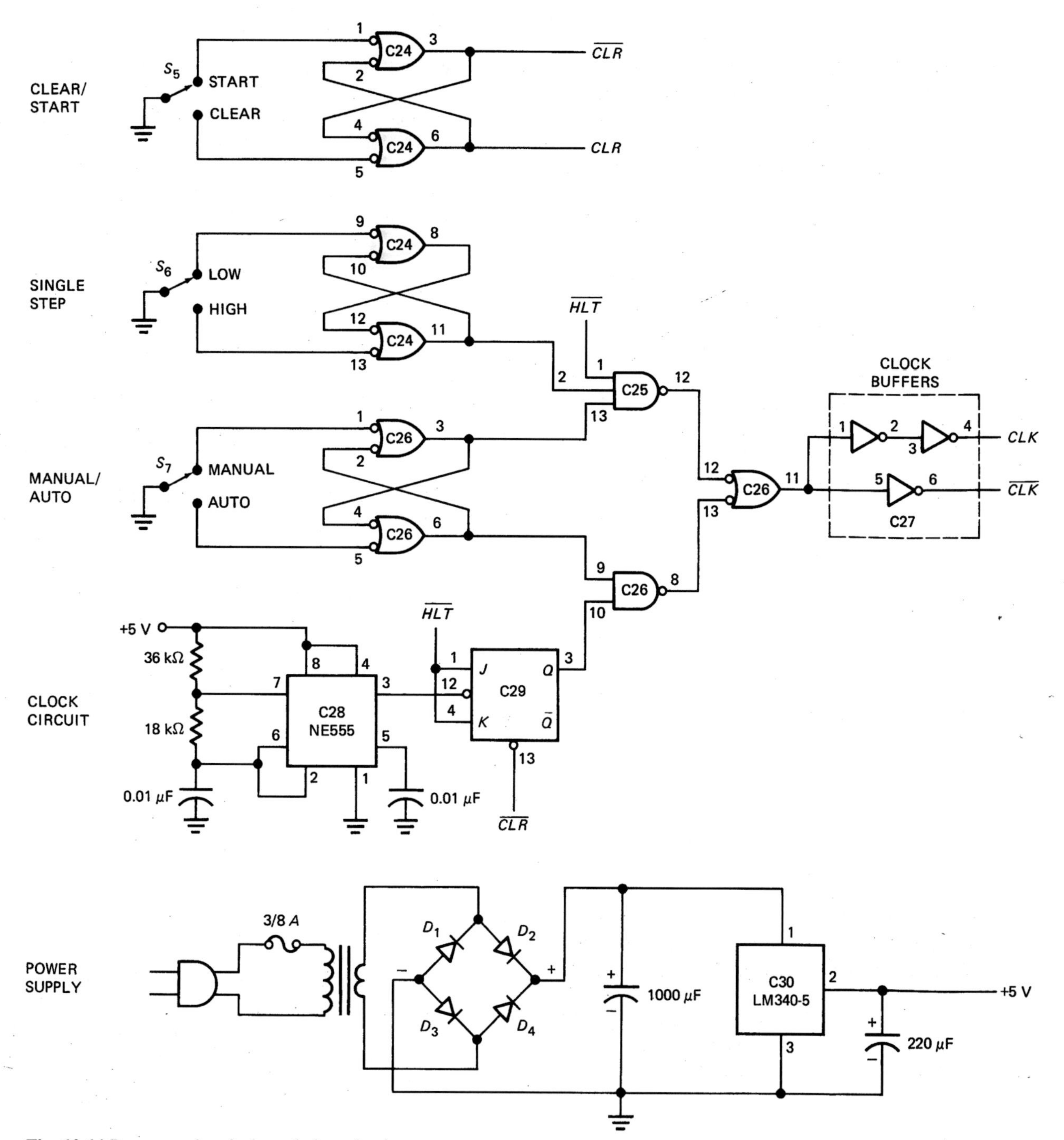

Fig. 10-14 Power supply, clock, and clear circuits.

only one of these is active at a time. ($\overline{HLT}$ is active low; all the others are active high.)

When the HLT instruction is in the *instruction register*, bits $I_7I_6I_5I_4$ are 1111 and $\overline{HLT}$ is low. This signal returns to C25 (single-step clock) and C29 (automatic clock). In either MANUAL or AUTO mode, the clock stops and the computer run ends.

Ring Counter

The ring counter, sometimes called a *state counter*, consists of three chips, C36, C37, and C38. Each of these chips is a 74LS107, a dual *JK* master-slave flip-flop. This counter is reset when the clear-start button (S_5) is pressed. The Q_0 flip-flop is inverted so that its $\overline{Q}$ output (pin 6, C38) drives

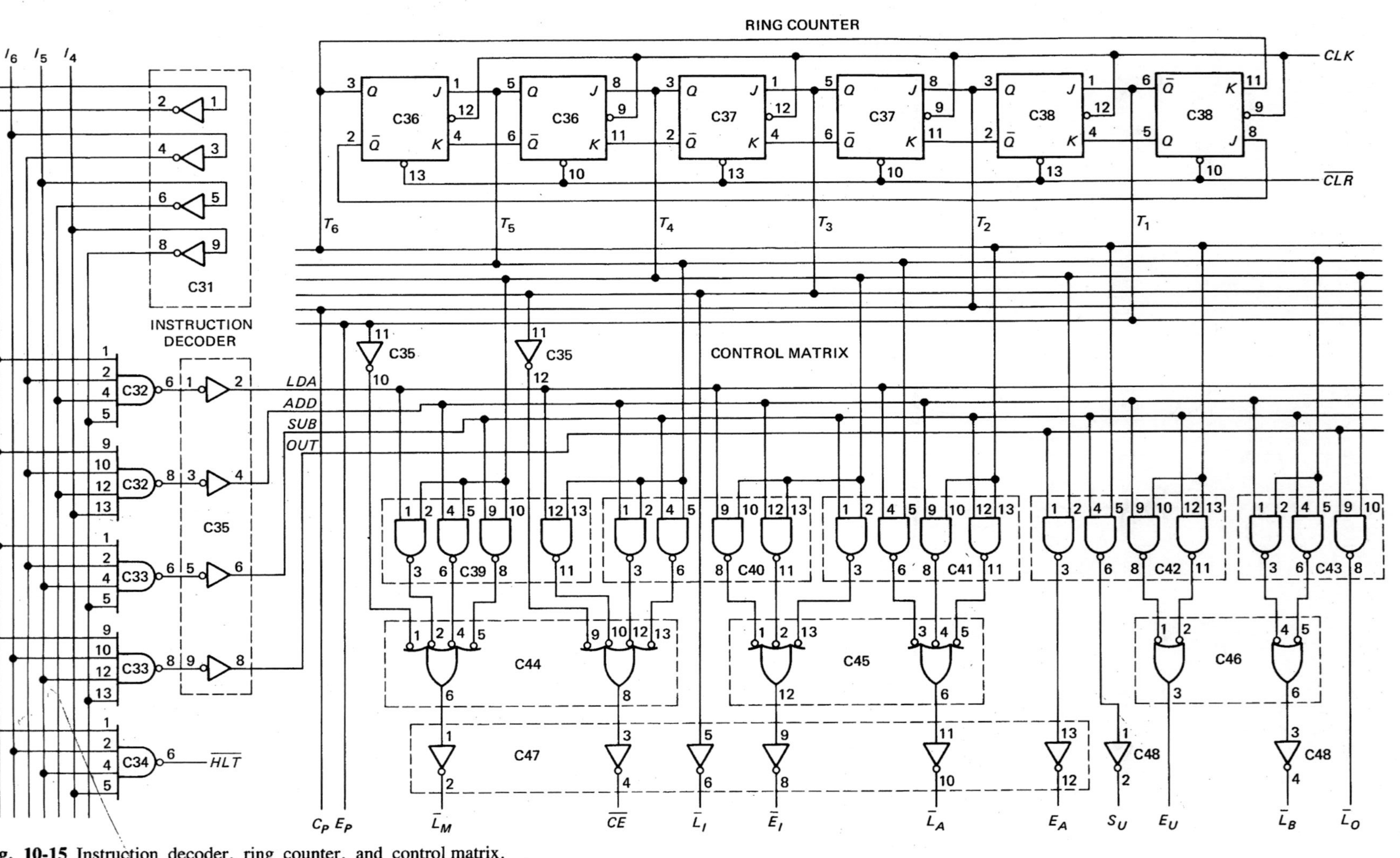

Fig. 10-15 Instruction decoder, ring counter, and control matrix.

the J input of the Q_1 flip-flop (pin 1, C38). Because of this, the T_1 output is initially high.

The CLK signal drives an active low input. This means that the negative edge of the CLK signal initiates each T state. Half a cycle later, the positive edge of the CLK signal produces register loading, as previously described.

Control Matrix

The *LDA*, *ADD*, *SUB*, and *OUT* signals from the instruction decoder drive the *control matrix*, C39 to C48. At the same time, the ring-counter signals, T_1 to T_6, are driving the matrix (a circuit receiving two groups of bits from different sources). The matrix produces **CON**, a 12-bit microinstruction that tells the rest of the computer what to do.

In Fig. 10-15, T_1 goes high, then T_2, then T_3, and so on. Analyze the control matrix and here is what you will find. A high T_1 produces a high E_P and a low $\overline{L}_M$ (address state); a high T_2 results in a high C_P (increment state); and a high T_3 produces a low $\overline{CE}$ and a low $\overline{L}_I$ (memory state). The first three T states, therefore, are always the fetch cycle in SAP-1. In chunked notation, the **CON** words for the fetch cycle are

State	CON	Active Bits
T_1	5E3H	E_P, $\overline{L}_M$
T_2	BE3H	C_P
T_3	263H	$\overline{CE}$, $\overline{L}_I$

During the execution states, T_4 through T_6 go high in succession. At the same time, only one of the decoded signals (*LDA* through *OUT*) is high. Because of this, the matrix automatically steers active bits to the correct output control lines.

For instance, when *LDA* is high, the only enabled 2-input NAND gates are the first, fourth, seventh, and tenth. When T_4 is high, it activates the first and seventh NAND gates, resulting in low $\overline{L}_M$ and low $\overline{E}_I$ (load MAR with address field). When T_5 is high, it activates the fourth and tenth NAND gates, producing a low $\overline{CE}$ and a low $\overline{L}_A$ (load RAM data into accumulator). When T_6 goes high, none of the control bits are active (nop).

You should analyze the action of the control matrix during the execution states of the remaining possibilities: high *ADD*, high *SUB*, and high *OUT*. Then you will agree the control matrix can generate the ADD, SUB, and OUT microinstructions shown in Table 10-5 (SAP-1 microprogram).

Operation

Before each computer run, the operator enters the program and data into the SAP-1 memory. With the program in low memory and the data in high memory, the operator presses and releases the clear button. The CLK and $\overline{CLK}$ signals drive the registers and counters. The microinstruction out of the controller-sequencer determines what happens on each positive CLK edge.

Each SAP-1 machine cycle begins with a fetch cycle. T_1 is the address state, T_2 is the increment state, and T_3 is the memory state. At the end of the fetch cycle, the instruction is stored in the instruction register. After the instruction field has been decoded, the control matrix automatically generates the correct execution routine. Upon completion of the execution cycle, the ring counter resets and the next machine cycle begins.

The data processing ends when a HLT instruction is loaded into the instruction register.

10-8 MICROPROGRAMMING

The control matrix of Fig. 10-15 is one way to generate the microinstructions needed for each execution cycle. With larger instruction sets, the control matrix becomes very complicated and requires hundreds or even thousands of gates. This is why *hardwired control* (matrix gates soldered together) forced designers to look for an alternative way to produce the control words that run a computer.

Microprogramming is the alternative. The basic idea is to store microinstructions in a ROM rather than produce them with a control matrix. This approach simplifies the problem of building a controller-sequencer.

Storing the Microprogram

By assigning addresses and including the fetch routine, we can come up with the SAP-1 microinstructions shown in Table 10-6. These microinstructions can be stored in a *control ROM* with the fetch routine at addresses 0H to 2H, the LDA routine at addresses 3H to 5H, the ADD routine at 6H to 8H, the SUB routine at 9H to BH, and the OUT routine at CH to EH.

To access any routine, we need to supply the correct addresses. For instance, to get the ADD routine, we need to supply addresses 6H, 7H, and 8H. To get the OUT routine, we supply addresses CH, DH, and EH. Therefore, accessing any routine requires three steps:

1. Knowing the starting address of the routine
2. Stepping through the routine addresses
3. Applying the addresses to the control ROM.

Address ROM

Figure 10-16 shows how to microprogram the SAP-1 computer. It has an *address* ROM, a *presettable* counter, and a *control* ROM. The address ROM contains the starting addresses of each routine in Table 10-6. In other words,

TABLE 10-6. SAP-1 CONTROL ROM

Address	Contents†	Routine	Active
0H	5E3H	Fetch	E_P, $\overline{L}_M$
1H	BE3H		C_P
2H	263H		$\overline{CE}$, $\overline{L}_I$
3H	1A3H	LDA	$\overline{L}_M$, $\overline{E}_I$
4H	2C3H		$\overline{CE}$, $\overline{L}_A$
5H	3E3H		None
6H	1A3H	ADD	$\overline{L}_M$, $\overline{E}_I$
7H	2E1H		$\overline{CE}$, $\overline{L}_B$
8H	3C7H		$\overline{L}_A$, E_U
9H	1A3H	SUB	$\overline{L}_M$, $\overline{E}_I$
AH	2E1H		$\overline{CE}$, $\overline{L}_B$
BH	3CFH		$\overline{L}_A$, S_U, E_U
CH	3F2H	OUT	E_A, $\overline{L}_O$
DH	3E3H		None
EH	3E3H		None
FH	X	X	Not used

† CON = $C_P E_P \overline{L}_M \overline{CE}$ $\overline{L}_I \overline{E}_I \overline{L}_A E_A$ $S_U E_U \overline{L}_B \overline{L}_O$.

TABLE 10-7. ADDRESS ROM

Address	Contents	Routine
0000	0011	LDA
0001	0110	ADD
0010	1001	SUB
0011	XXXX	None
0100	XXXX	None
0101	XXXX	None
0110	XXXX	None
0111	XXXX	None
1000	XXXX	None
1001	XXXX	None
1010	XXXX	None
1011	XXXX	None
1100	XXXX	None
1101	XXXX	None
1110	1100	OUT
1111	XXXX	None

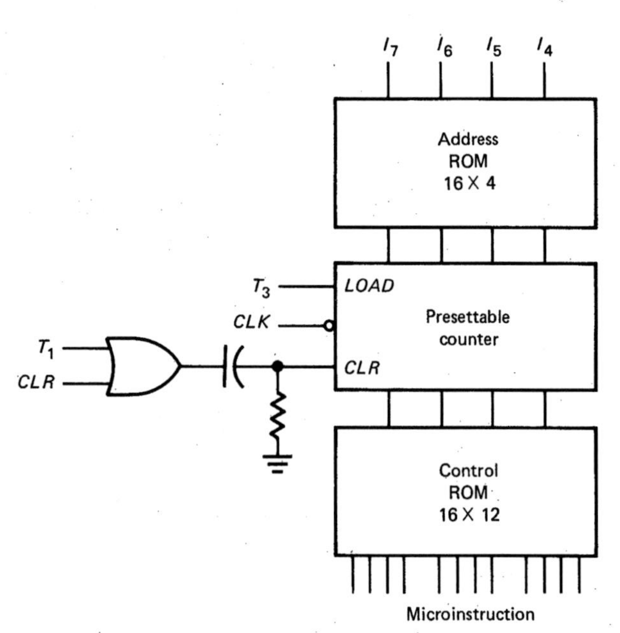

Fig. 10-16 Microprogrammed control of SAP-1.

the address ROM contains the data listed in Table 10-7. As shown, the starting address of the LDA routine is 0011, the starting address of the ADD routine is 0110, and so on.

When the op-code bits $I_7I_6I_5I_4$ drive the address ROM, the starting address is generated. For instance, if the ADD instruction is being executed, $I_7I_6I_5I_4$ is 0001. This is the input to the address ROM; the output of this ROM is 0110.

Presettable Counter

When T_3 is high, the load input of the *presettable counter* is high and the counter loads the starting address from the address ROM. During the other T states, the counter counts.

Initially, a high *CLR* signal from the clear-start debouncer is differentiated to get a narrow positive spike. This resets the counter. When the computer run begins, the counter output is 0000 during the T_1 state, 0001 during the T_2 state, and 0010 during the T_3 state. Every fetch cycle is the same because 0000, 0001, and 0010 come out of the counter during states T_1, T_2, and T_3.

The op code in the instruction register controls the execution cycle. If an ADD instruction has been fetched, the $I_7I_6I_5I_4$ bits are 0001. These op-code bits drive the address ROM, producing an output of 0110 (Table 10-7). This starting address is the input to the presettable counter. When T_3 is high, the next negative clock edge loads 0110 into the presettable counter. The counter is now preset, and counting can resume at the starting address of the ADD routine. The counter output is 0110 during the T_4 state, 0111 during the T_5 state, and 1000 during the T_6 state.

When the T_1 state begins, the leading edge of the T_1 signal is differentiated to produce a narrow positive spike which resets the counter to 0000, the starting address of the fetch routine. A new machine cycle then begins.

Control ROM

The control ROM stores the SAP-1 microinstructions. During the fetch cycle, it receives addresses 0000, 0001, and 0010. Therefore, its outputs are

5E3H
BE3H
263H

These microinstructions, listed in Table 10-6, produce the address state, increment state, and memory state.

If an ADD instruction is being executed, the control ROM receives addresses 0110, 0111, and 1000 during the execution cycle. Its outputs are

1A3H
2E1H
3C7H

These microinstructions carry out the addition as previously discussed.

For another example, suppose the OUT instruction is being executed. Then the op code is 1110 and the starting address is 1100 (Table 10-7). During the execution cycle, the counter output is 1100, 1101, and 1110. The output of the control ROM is 3F2H, 3E3H, and 3E3H (Table 10-6). This routine transfers the accumulator contents to the output port.

Variable Machine Cycle

The microinstruction 3E3H in Table 10-6 is a nop. It occurs once in the LDA routine and twice in the OUT routine. These nops are used in SAP-1 to get a *fixed machine cycle* for all instructions. In other words, each machine cycle takes exactly six T states, no matter what the instruction. In some computers a fixed machine cycle is an advantage. But when speed is important, the nops are a waste of time and can be eliminated.

One way to speed up the operation of SAP-1 is to skip any T state with a nop. By redesigning the circuit of Fig. 10-16 we can eliminate the nop states. This will shorten the machine cycle of the LDA instruction to five states (T_1, T_2, T_3, T_4, and T_5). It also shortens the machine cycle of the OUT instruction to four T states (T_1, T_2, T_3, and T_4).

Figure 10-17 shows one way to get a *variable machine cycle*. With an LDA instruction, the action is the same as before during the T_1 to T_5 states. When the T_6 state begins, the control ROM produces an output of 3E3H (the nop microinstruction). The NAND gate detects this nop instantly and produces a low output signal $\overline{NOP}$. $\overline{NOP}$ is fed back to the ring counter through an AND gate, as shown in Fig. 10-18. This resets the ring counter to the T_1 state, and a new machine cycle begins. This reduces the machine cycle of the LDA instruction from six states to five.

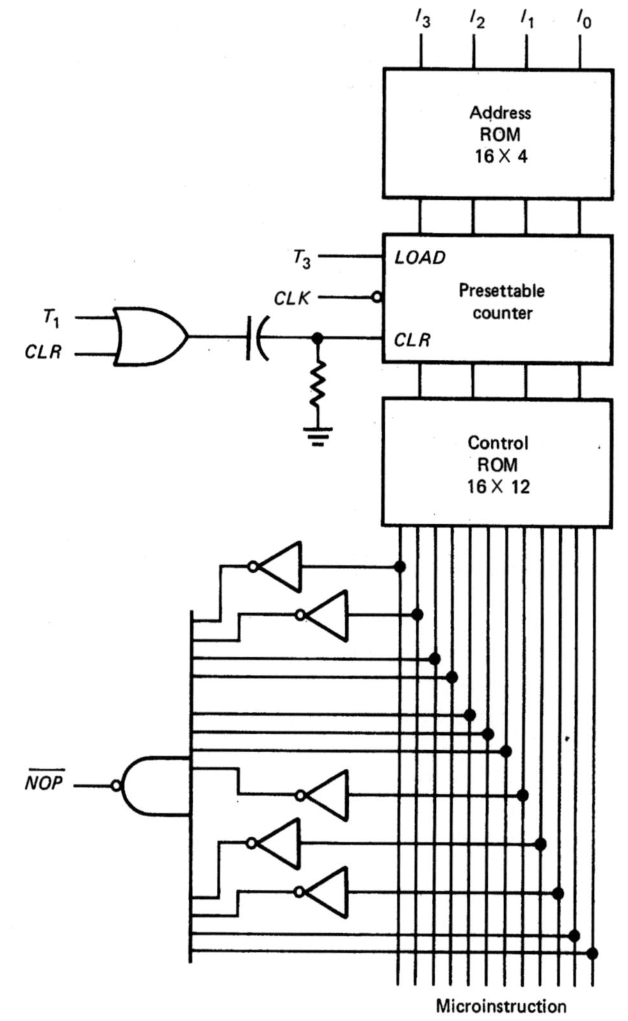

Fig. 10-17 Variable machine cycle.

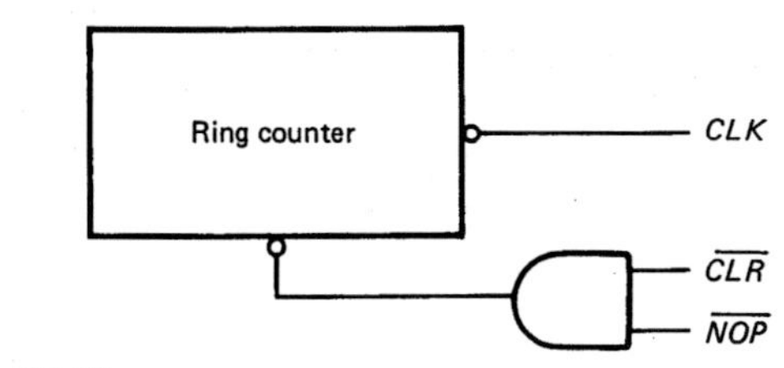

Fig. 10-18

With the OUT instruction, the first nop occurs in the T_5 state. In this case, just after the T_5 state begins, the control ROM produces an output of 3E3H, which is detected by the NAND gate. The low $\overline{NOP}$ signal then resets the ring counter to the T_1 state. In this way, we have reduced the machine cycle of the OUT instruction from six states to four.

Variable machine cycles are commonly used with microprocessors. In the 8085, for example, the machine cycles take from two to six *T* states because all unwanted nop states are ignored.

Advantages

One advantage of microprogramming is the elimination of the instruction decoder and control matrix; both of these become very complicated for larger instruction sets. In other words, it's a lot easier to store microinstructions in a ROM than it is to wire an instruction decoder and control matrix.

Furthermore, once you wire an instruction decoder and control matrix, the only way you can change the instruction set is by disconnecting and rewiring. This is not necessary with microprogrammed control; all you have to do is change the control ROM and the starting-address ROM. This is a big advantage if you are trying to upgrade equipment sold earlier.

Summary

In conclusion, most modern microprocessors use microprogrammed control instead of hardwired control. The microprogramming tables and circuits are more complicated than those for SAP-1, but the idea is the same. Microinstructions are stored in a control ROM and accessed by applying the address of the desired microinstruction.

GLOSSARY

address state The T_1 state. During this state, the address in the program counter is transferred to the MAR.
accumulator The place where answers to arithmetic and logic operations are accumulated. Sometimes called the A register.
assembly language The mnemonics used in writing a program.
B register An auxiliary register that stores the data to be added or subtracted from the accumulator.
fetch cycle The first part of the instruction cycle. During the fetch cycle, the address is sent to the memory, the program counter is incremented, and the instruction is transferred from the memory to the instruction register.
increment state The T_2 state. During this state, the program counter is incremented.
instruction cycle All the states needed to fetch and execute an instruction.
instruction register The register that receives the instruction from the memory.
instruction set The instructions a computer responds to.
LDA Mnemonic for load the accumulator.
machine cycle All the states generated by the ring counter.
machine language The strings of 0s and 1s used in a program.
macroinstruction One of the instructions in the instruction set.
MAR Memory address register. This register receives the address of the data to be accessed in memory. The MAR supplies this address to the memory.
memory-reference instruction An instruction that calls for a second memory operation to access data.
memory state The T_3 state. During this state, the instruction in the memory is transferred to the instruction register.
microinstruction A control word out of the controller-sequencer. The smallest step in the data processing.
nop No operation. A state during which nothing happens.
output register The register that receives processed data from the accumulator and drives the output display of SAP-1. Also called an output port.
object program A program written in machine language.
op code Operation code. That part of the instruction which tells the computer what operation to perform.
program counter A register that counts in binary. Its contents are the address of the next instruction to be fetched from the memory.
RAM Random-access memory. A better name is read-write memory. The RAM stores the program and data needed for a computer run.
source program A program written in mnemonics.

SELF-TESTING REVIEW

Read each of the following and provide the missing words. Answers appear at the beginning of the next question.

1. The ________ counter, which is part of the control unit, counts from 0000 to 1111. It sends to the memory the ________ of the next instruction.
2. (*program, address*) The MAR, or ________ register, latches the address from the program counter. A bit later, the MAR applies this address to the ________, where a read operation is performed.
3. (*memory-address, RAM*) The instruction register is

part of the control unit. The contents of the ________ register are split into two nibbles. The upper nibble goes to the ________.

4. (*instruction, controller-sequencer*) The controller-sequencer produces a 12-bit word that controls the rest of the computer. The 12 wires carrying this ________ word are called the control ________.
5. (*control, bus*) The ________ is a buffer register that stores sums or differences. Its two-state output goes to the adder-subtracter. The ________ produces the sum when S_U is low and the difference when S_U is high. The output register is sometimes called an output ________.
6. (*accumulator, adder-subtracter, port*) The SAP-1 ________ set is LDA, ADD, SUB, OUT, and HLT. LDA, ADD, and SUB are called ________ instructions because they use data stored in the memory.
7. (*instruction, memory-reference*) The 8080 was the first widely used microprocessor. The ________ is an enhanced version of the 8080 with essentially the same instruction set.
8. (*8085*) LDA, ADD, SUB, OUT, and HLT are coded as 4-bit strings of 0s and 1s. This code is called the ________ code. ________ language uses mnemonics when writing a program. ________ language uses strings of 0s and 1s.
9. (*op, Assembly, Machine*) SAP-1 has ________ T states, periods during which register contents change. The ring counter, or ________ counter, produces these T states. These six T states represent one machine cycle. In SAP-1 the instruction cycle has only one machine cycle. In microprocessors like the 8080 and the 8085, the ________ cycle may have from one to five machine cycles.
10. (*six, state, instruction*) The controller-sequencer sends out control words, one during each T state or clock cycle. Each control word is called a ________. Instructions like LDA, ADD, SUB, etc. are called ________. Each SAP-1 macroinstruction is made up of three ________.
11. (*microinstruction, macroinstructions, microinstructions*) With larger instruction sets, the control matrix becomes very complicated. This is why hardwired control is being replaced by ________. The basic idea is to store the ________ in a control ROM.
12. (*microprogramming, microinstructions*) SAP-1 uses a fixed machine cycle for all instructions. In other words, each machine cycle takes exactly six T states. Microprocessors like the 8085 have variable machine cycles because all unwanted nop states are eliminated.

PROBLEMS

10-1. Write a SAP-1 program using mnemonics (similar to Example 10-1) that will display the result of

$$5 + 4 - 6$$

Use addresses DH, EH, and FH for the data.

10-2. Convert the assembly language of Prob. 10-1 into SAP-1 machine language. Show the answer in binary form and in hexadecimal form.

10-3. Write an assembly-language program that performs this operation:

$$8 + 4 - 3 + 5 - 2$$

Use addresses BH to FH for the data.

10-4. Convert the program and data of Prob. 10-3 into machine language. Express the result in both binary and hexadecimal form.

10-5. Figure 10-19 shows the timing diagram for the ADD instruction. Draw the timing diagram for the SUB instruction.

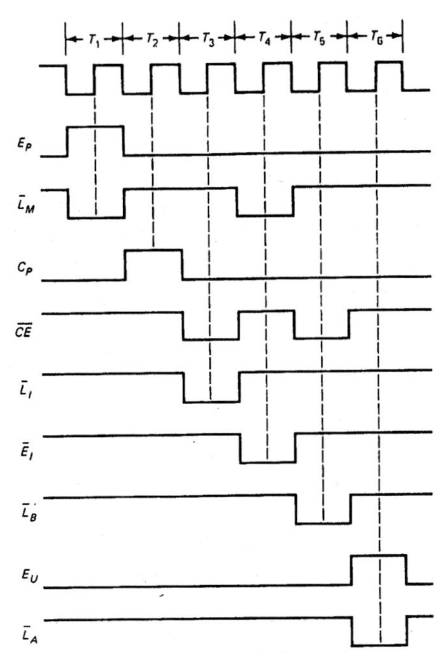

Fig. 10-19

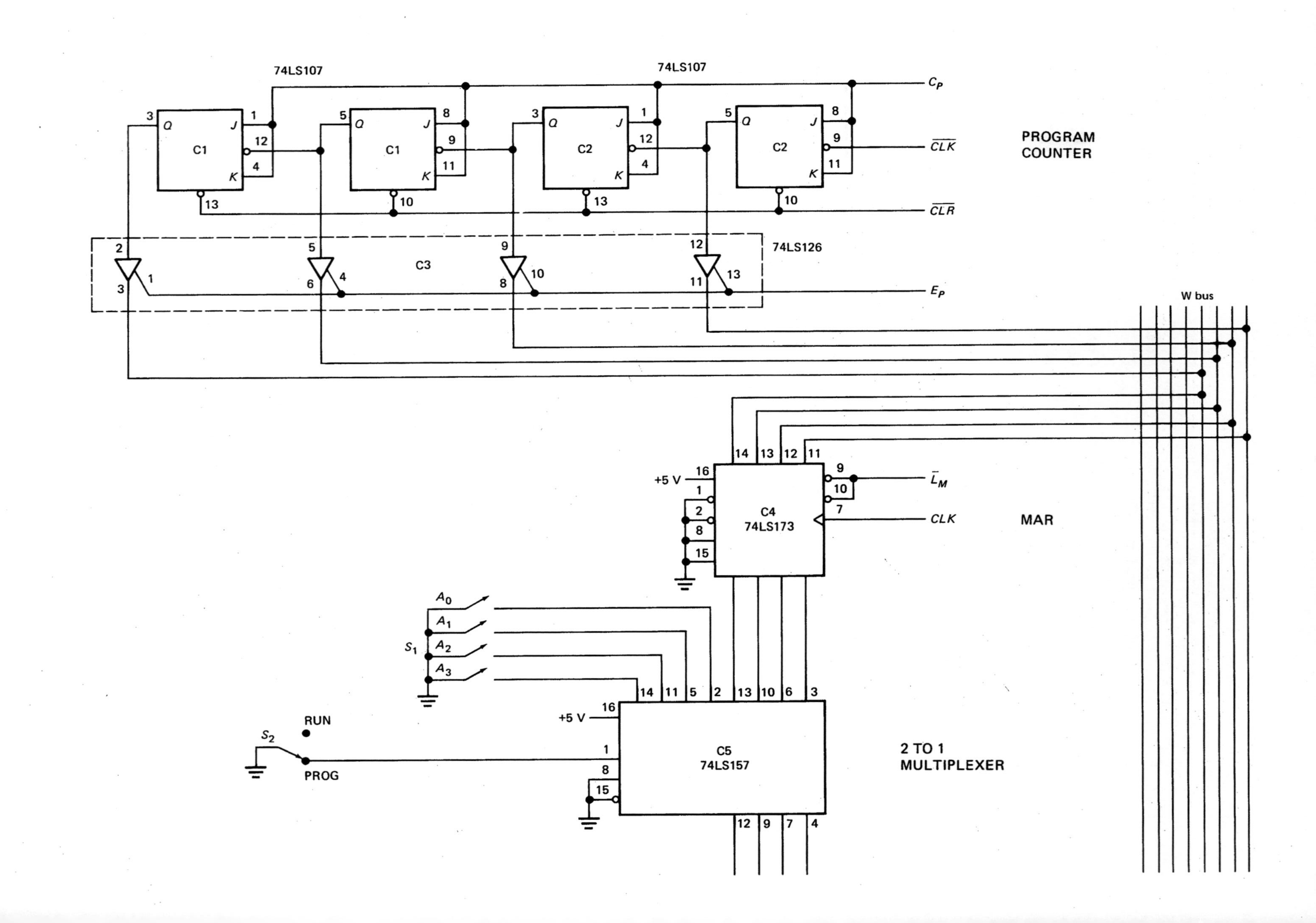

74LS107
74LS107
C1
C1
C2
C2
C_P
$\overline{CLK}$
$\overline{CLR}$
PROGRAM COUNTER
C3
74LS126
E_P
W bus
C4
74LS173
+5 V
$\bar{L}_M$
CLK
MAR
A_0
A_1
A_2
A_3
S_1
RUN
S_2
PROG
+5 V
C5
74LS157
2 TO 1 MULTIPLEXER

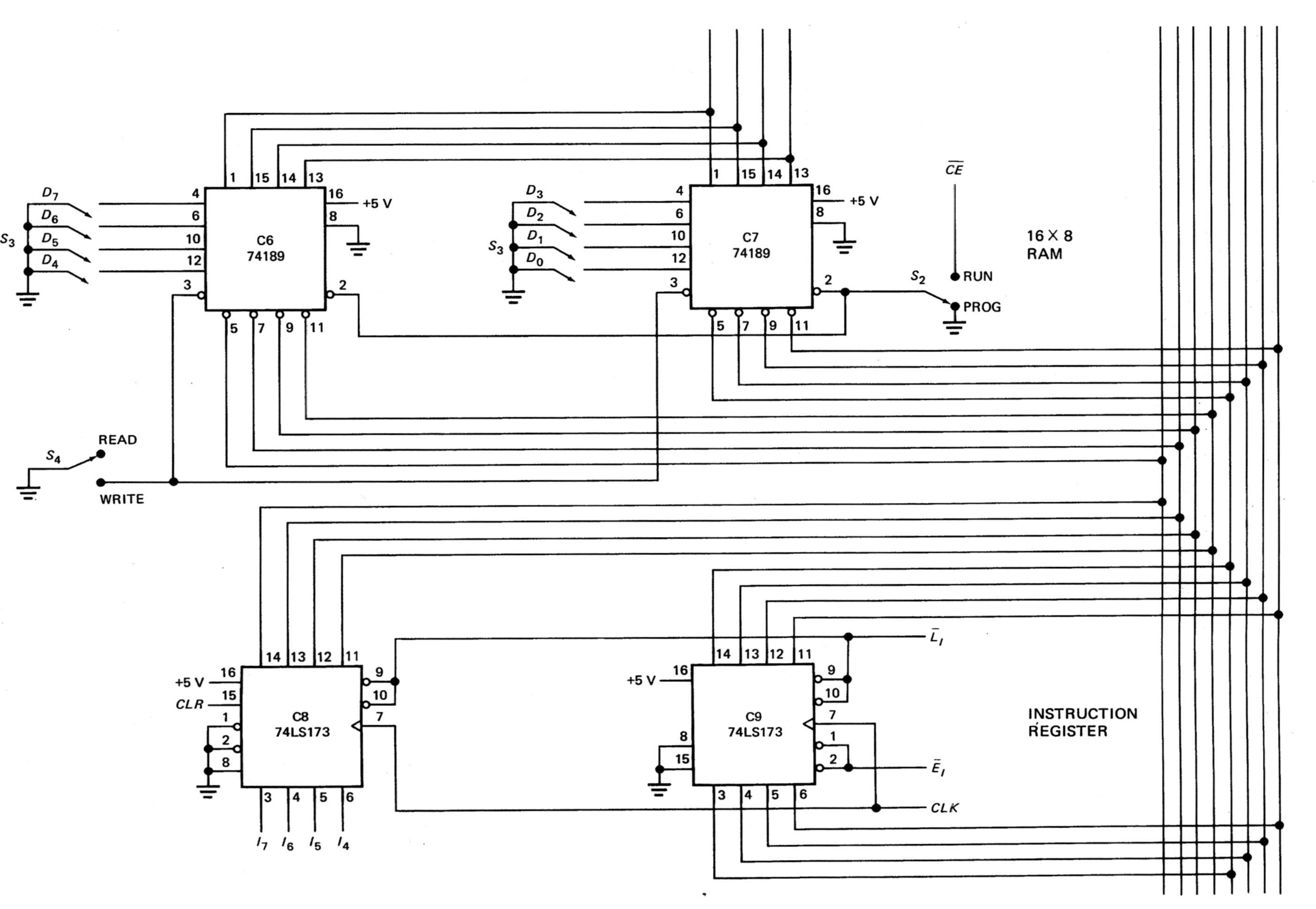

16 X 8
RAM
C6
74189
C7
74189
C8
74LS173
C9
74LS173
INSTRUCTION
REGISTER
D_7
D_6
D_5
D_4
D_3
D_2
D_1
D_0
S_3
S_2
S_4
READ
WRITE
$\overline{CE}$
RUN
PROG
+5 V
CLR
$\bar{L}_I$
$\bar{E}_I$
CLK
I_7
I_6
I_5
I_4

Fig. 10-20

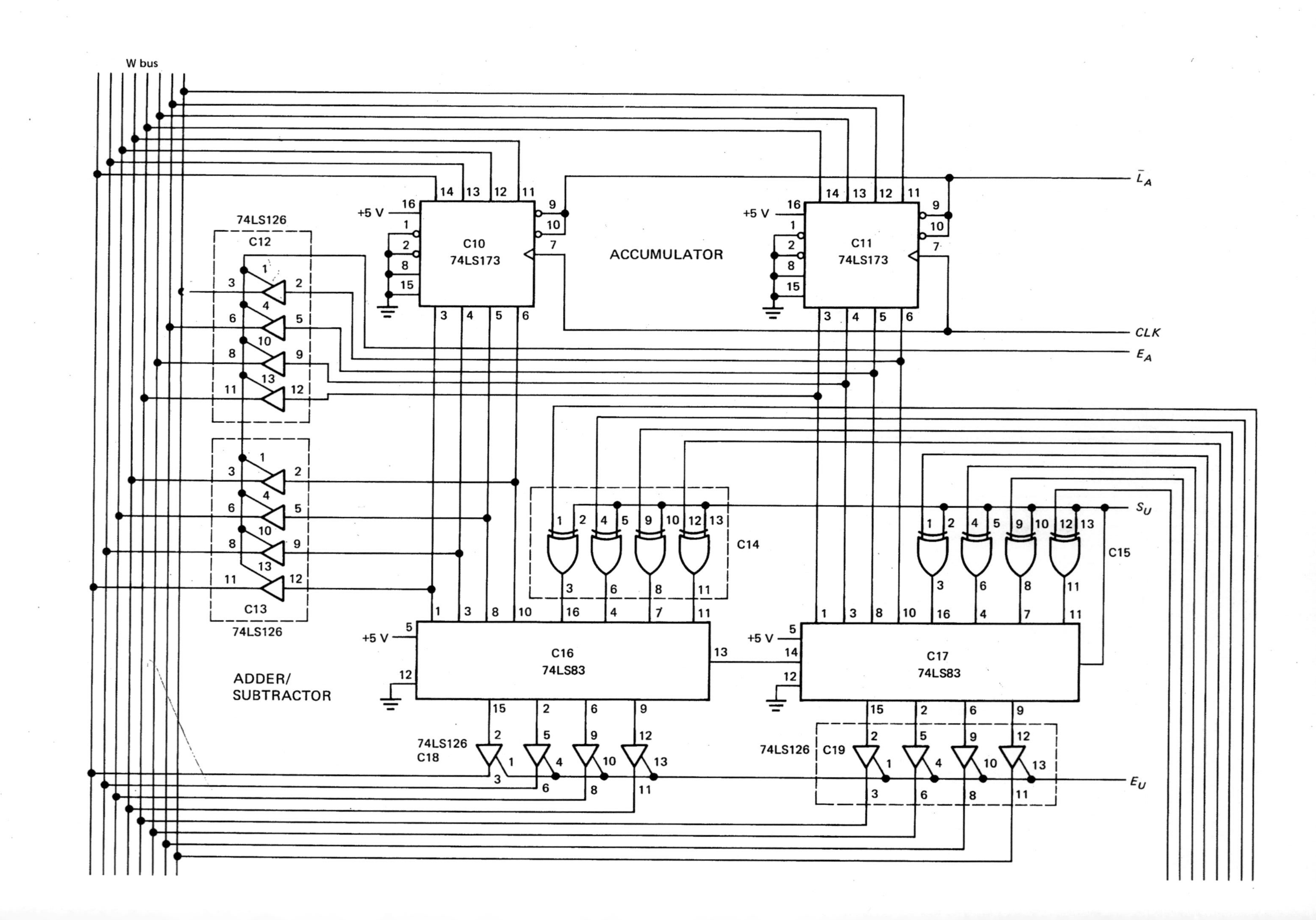

W bus
74LS126
C12
C10
74LS173
ACCUMULATOR
C11
74LS173
$\bar{L}_A$
CLK
E_A
C13
74LS126
C14
C15
S_U
ADDER/
SUBTRACTOR
C16
74LS83
C17
74LS83
74LS126
C18
74LS126
C19
E_U
+5 V

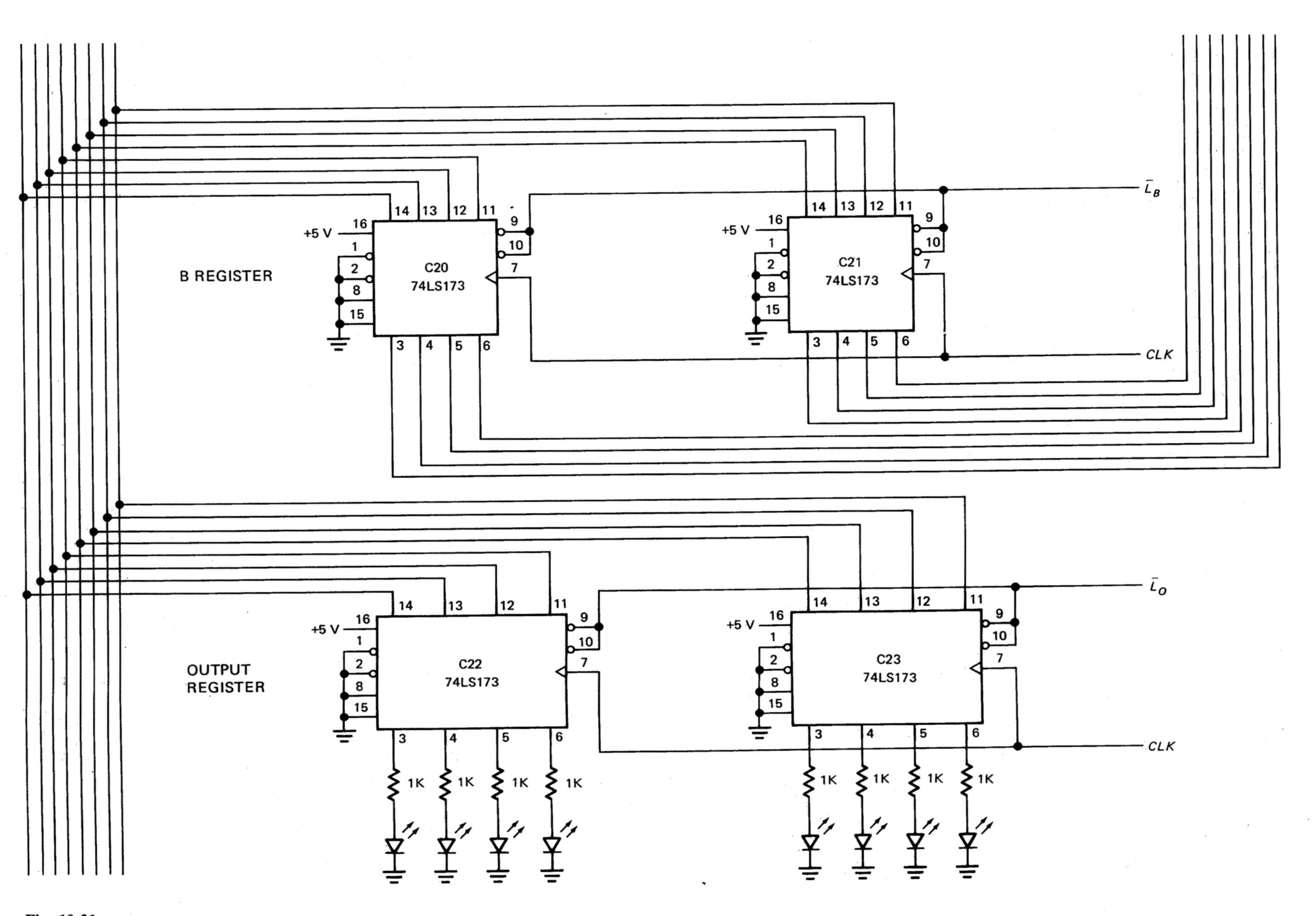
B REGISTER
C20
74LS173
C21
74LS173
+5 V
$\bar{L}_B$
CLK
OUTPUT
REGISTER
C22
74LS173
C23
74LS173
$\bar{L}_O$
1K

Fig. 10-21

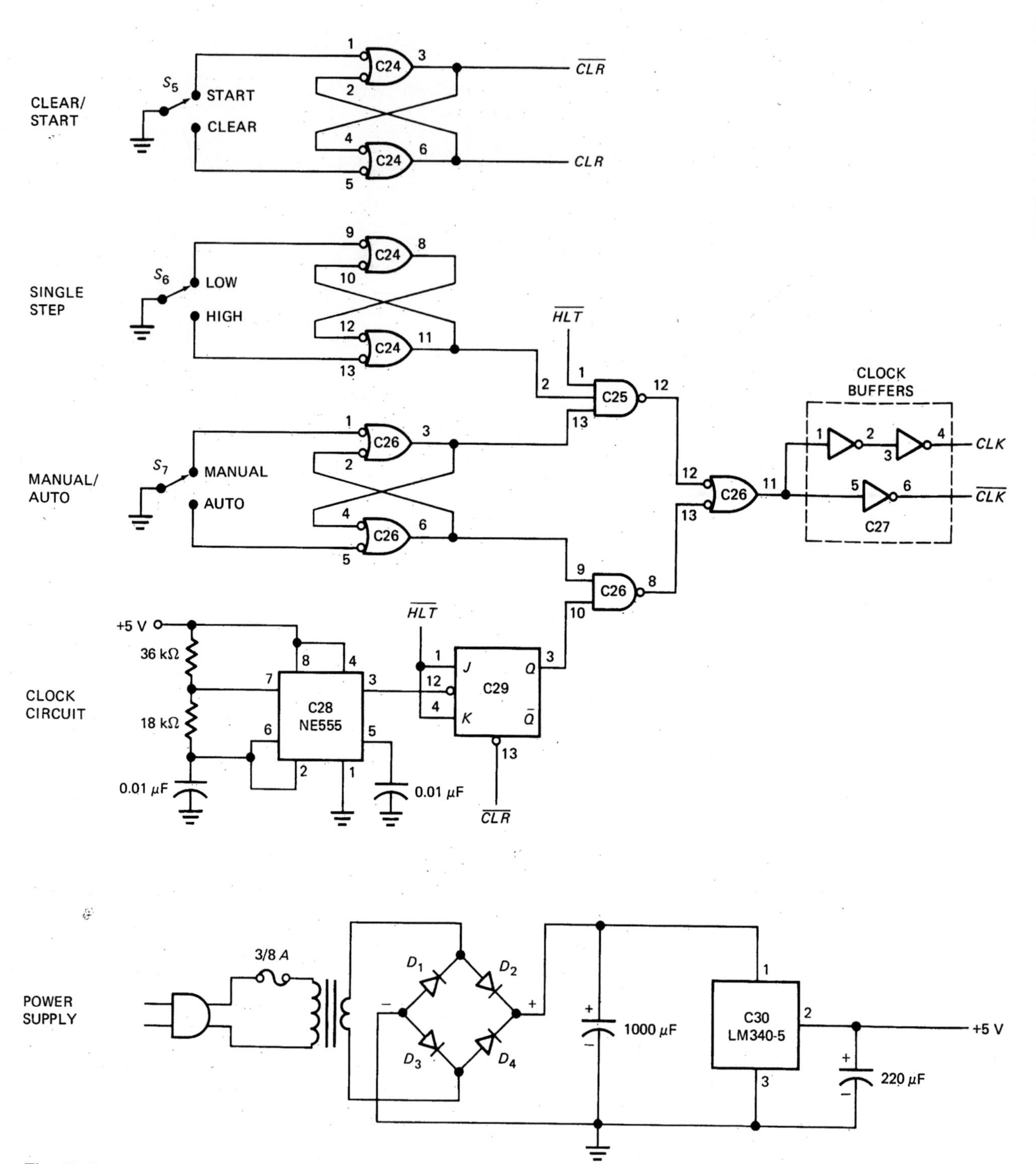

CLEAR/
START
S_5
START
CLEAR
C24
$\overline{CLR}$
CLR
SINGLE
STEP
S_6
LOW
HIGH
$\overline{HLT}$
C25
CLOCK
BUFFERS
MANUAL/
AUTO
S_7
MANUAL
AUTO
C26
CLK
$\overline{CLK}$
C27
CLOCK
CIRCUIT
+5 V
36 kΩ
18 kΩ
0.01 μF
C28
NE555
C29
J
K
Q
$\bar{Q}$
$\overline{CLR}$
POWER
SUPPLY
3/8 A
D_1
D_2
D_3
D_4
1000 μF
C30
LM340-5
220 μF
+5 V

Fig. 10-22

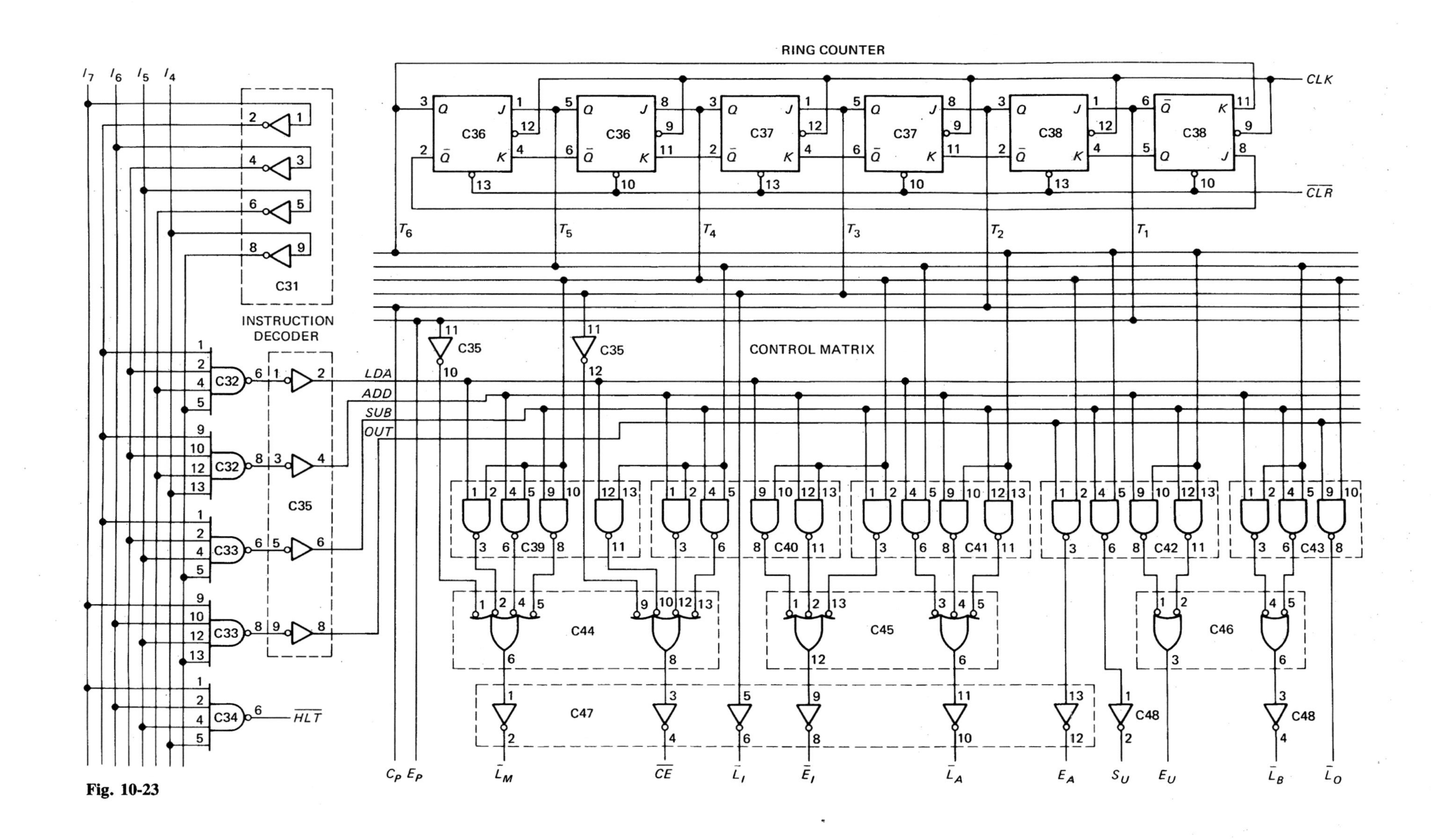

RING COUNTER
CLK
CLR
C36
C37
C38
T1
T2
T3
T4
T5
T6
I7 I6 I5 I4
C31
INSTRUCTION DECODER
C32
C33
C34
C35
HLT
LDA
ADD
SUB
OUT
CONTROL MATRIX
C39
C40
C41
C42
C43
C44
C45
C46
C47
C48
CP EP
LM
CE
LI
EI
LA
EA
SU
EU
LB
LO

Fig. 10-23

10-6. Suppose an 8085 uses a clock frequency of 3 MHz. The ADD instruction of an 8085 takes four *T* states to fetch and execute. How long is this?

10-7. What are the SAP-1 microinstructions for the LDA routine? For the SUB routine? Express the answers in binary and hexadecimal form.

10-8. Suppose we want to transfer the contents of the accumulator to the B register. This requires a new microinstruction. What is this microinstruction? Express your answer in hexadecimal and binary form.

10-9. Look at Fig. 10-20 and answer the following questions:

a. Are the contents of the program counter changed on the positive or negative edge of the $\overline{CLK}$ signal? At this instant, is the CLK signal on its rising or falling edge?

b. To increment the program counter, does C_P have to be low or high?

c. To clear the program counter, does $\overline{CLR}$ have to be low or high?

d. To place the contents of the program counter on the W bus, should E_P be low or high?

10-10. Refer to Fig. 10-21:

a. If $\overline{L}_A$ is high, what happens to the accumulator contents on the next positive clock edge?

b. If **A** = 0010 1100 and **B** = 1100 1110, what is on the W bus if E_A is high?

c. If **A** = 0000 1111, **B** = 0000 0001, and S_U = 1, what is on the W bus when E_U is high?

10-11. Answer the following questions for Fig. 10-22:

a. With S_5 in the CLEAR position, is the $\overline{CLR}$ output low or high?

b. With S_6 in the LOW position, is the output low or high for pin 11, C24?

c. To have a clock signal at pin 3 of C29, should $\overline{HLT}$ be low or high?

10-12. Refer to Fig. 10-23 to answer the following:

a. If $I_7I_6I_5I_4$ = 1110, only one of the output pins in C35 is high. Which pin is this? (Disregard pins 10 and 12.)

b. $\overline{CLR}$ goes low. Which is the timing signal (T_1 to T_6) that goes high?

c. LDA and T_5 are high. Is the voltage low or high at pin 6, C45?

d. ADD and T_4 are high. Is the signal low or high at pin 12, C45?

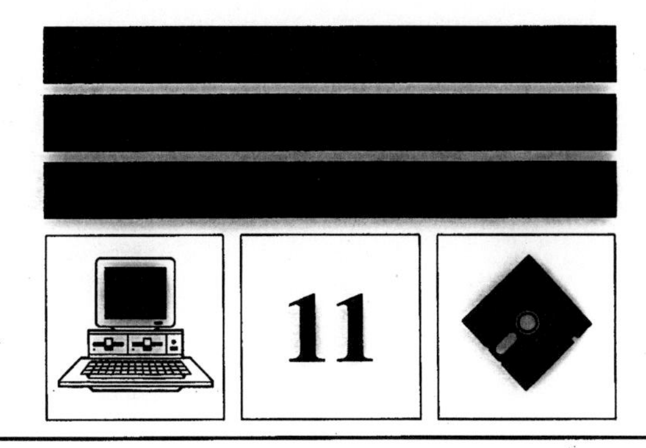

SAP-2

SAP-1 is a computer because it stores a program and data before calculations begin; then it automatically carries out the program instructions without human intervention. And yet, SAP-1 is a primitive computing machine. It compares to a modern computer the way a Neanderthal human would compare to a modern person. Something is missing, something found in every modern computer.

SAP-2 is the next step in the evolution toward modern computers because it includes *jump* instructions. These new instructions force the computer to repeat or skip part of a program. As you will discover, jump instructions open up a whole new world of computing power.

11-1 BIDIRECTIONAL REGISTERS

To reduce the wiring capacitance of SAP-2, we will run only one set of wires between each register and the bus. Figure 11-1*a* shows the idea. The input and output pins are shorted; only one group of wires is connected to the bus.

Does this shorting the input and output pins ever cause trouble? No. During a computer run, either *LOAD* or *ENABLE* may be active, but not both at the same time. An active *LOAD* means that a binary word flows from the bus to the register input; during a load operation, the output lines are floating. On the other hand, an active *ENABLE* means that a binary word flows from the register to the bus; in this case, the input lines float.

The IC manufacturer can internally connect the input and output pins of a three-state register. This not only reduces the wiring capacitance; it also reduces the number of I/O pins. For instance, Fig. 11-1*b* has four I/O pins instead of eight.

Figure 11-1*c* is the symbol for a three-state register with internally connected input and output pins. The double-headed arrow reminds us that the path is *bidirectional*; data can move either way.

11-2 ARCHITECTURE

Figure 11-2 shows the architecture of SAP-2. All register outputs to the W bus are three-state; those not connected to the bus are two-state. As before, the controller-sequencer sends control signals (not shown) to each register. These control signals load, enable, or otherwise prepare the register for the next positive clock edge. A brief description of each box is given now.

Input Ports

SAP-2 has two input ports, numbered 1 and 2. A hexadecimal keyboard encoder is connected to port 1. It allows us to enter hexadecimal instructions and data through port 1. Notice that the hexadecimal keyboard encoder sends a *READY* signal to bit 0 of port 2. This signal indicates when the data in port 1 is valid.

Also notice the *SERIAL IN* signal going to pin 7 of port 2. A later example will show you how to convert serial input data to parallel data.

Program Counter

This time, the program counter has 16 bits; therefore, it can count from

$$\mathbf{PC} = 0000\ 0000\ 0000\ 0000$$

to

$$\mathbf{PC} = 1111\ 1111\ 1111\ 1111$$

This is equivalent to 0000H to FFFFH, or decimal 0 to 65,535.

A low $\overline{CLR}$ signal resets the PC before each computer run; so the data processing starts with the instruction stored in memory location 0000H.

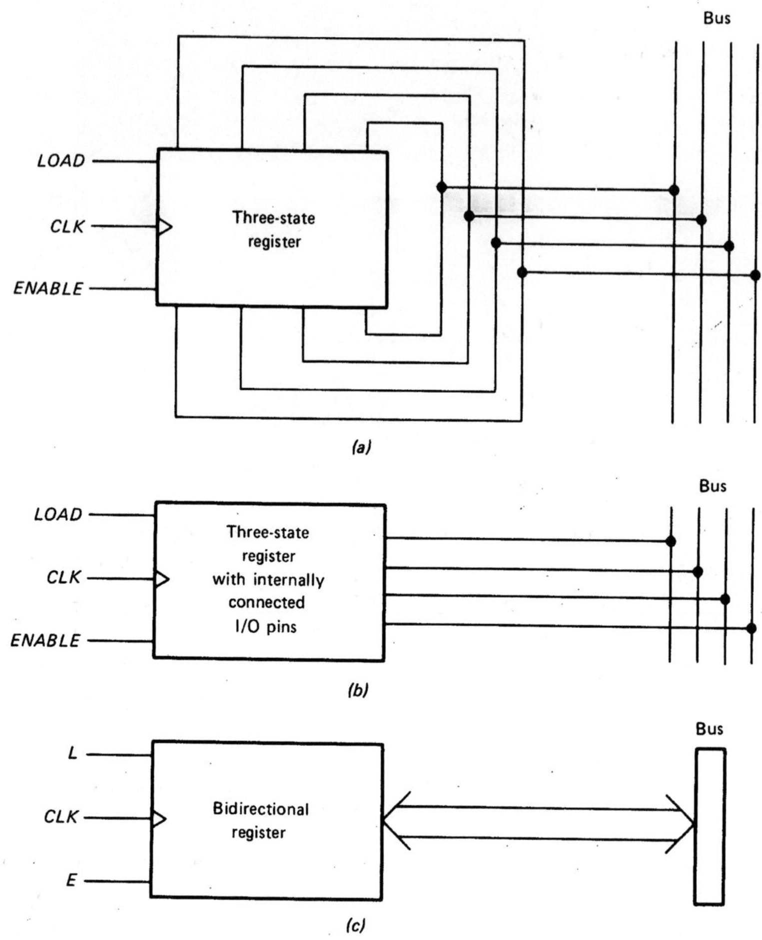

Fig. 11-1 Bidirectional register.

MAR and Memory

During the fetch cycle, the MAR receives 16-bit addresses from the program counter. The two-state MAR output then addresses the desired memory location. The memory has a 2K ROM with addresses of 0000H to 07FFH. This ROM contains a program called a *monitor* that initializes the computer on power-up, interprets the keyboard inputs, and so forth. The rest of the memory is a 62K RAM with addresses from 0800H to FFFFH.

Memory Data Register

The memory data register (MDR) is an 8-bit buffer register. Its output sets up the RAM. The memory data register receives data from the bus before a write operation, and it sends data to the bus after a read operation.

Instruction Register

Because SAP-2 has more instructions than SAP-1, we will use 8 bits for the op code rather than 4. An 8-bit op code can accommodate 256 instructions. SAP-2 has only 42 instructions, so there will be no problem coding them with 8 bits. Using an 8-bit op code also allows upward compatibility with the 8080/8085 instruction set because it is based on an 8-bit op code. As mentioned earlier, all SAP instructions are identical with 8080/8085 instructions.

Controller-Sequencer

The controller-sequencer produces the control words or microinstructions that coordinate and direct the rest of the computer. Because SAP-2 has a bigger instruction set, the controller-sequencer has more hardware. Although the CON word is bigger, the idea is the same: the control word or microinstruction determines how the registers react to the next positive clock edge.

Accumulator

The two-state output of the accumulator goes to the ALU; the three-state output to the W bus. Therefore, the 8-bit word in the accumulator continuously drives the ALU, but this same word appears on the bus only when E_A is active.

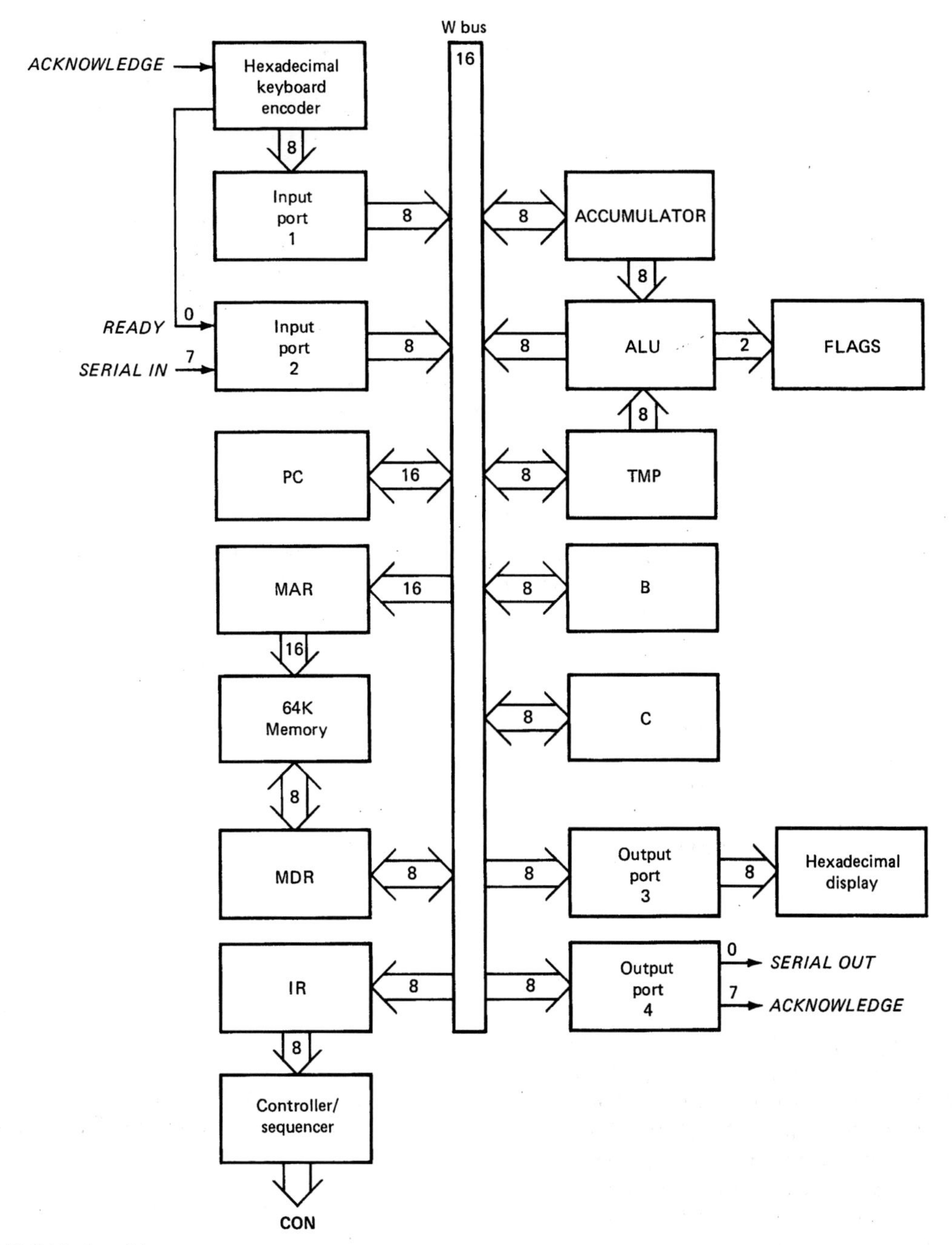

Fig. 11-2 SAP-2 block architecture.

ALU and Flags

Standard ALUs are commercially available as integrated circuits. These ALUs have 4 or more control bits that determine the arithmetic or logic operation performed on words **A** and **B.** The ALU used in SAP-2 includes arithmetic and logic operations.

In this book a *flag* is a flip-flop that keeps track of a changing condition during a computer run. The SAP-2 computer has two flags. The *sign flag* is set when the accumulator contents become negative during the execution of some instructions. The *zero flag* is set when the accumulator contents become zero.

TMP, B, and C Registers

Instead of using the B register to hold the data being added or subtracted from the accumulator, a *temporary* (TMP) register is used. This allows us more freedom in using the B register. Besides the TMP and B registers, SAP-2 includes a C register. This gives us more flexibility in moving data during a computer run.

Output Ports

SAP-2 has two output ports, numbered 3 and 4. The contents of the accumulator can be loaded into port 3, which drives a hexadecimal display. This allows us to see the processed data.

The contents of the accumulator can also be sent to port 4. Notice that pin 7 of port 4 sends an *ACKNOWLEDGE* signal to the hexadecimal encoder. This *ACKNOWLEDGE* signal and the *READY* signal are part of a concept called *handshaking,* to be discussed later.

Also notice the *SERIAL OUT* signal from pin 0 of port 4; one of the examples will show you how to convert parallel data in the accumulator into serial output data.

11-3 MEMORY-REFERENCE INSTRUCTIONS

The SAP-2 fetch cycle is the same as before. T_1 is the address state, T_2 is the increment state, and T_3 is the memory state. All SAP-2 instructions therefore use the memory during the fetch cycle because a program instruction is transferred from the memory to the instruction register.

During the execution cycle, however, the memory may or may not be used; it depends on the type of instruction that has been fetched. A memory-reference instruction (MRI) is one that uses the memory during the execution cycle.

The SAP-2 computer has an instruction set with 42 instructions. What follows is a description of the memory-reference instructions.

LDA and STA

LDA has the same meaning as before: *load the accumulator* with the addressed memory data. The only difference is that more memory locations can be accessed in SAP-2 because the addresses are from 0000H to FFFFH. For example, LDA 2000H means to load the accumulator with the contents of memory location 2000H.

To distinguish the different parts of an instruction, the mnemonic is sometimes called the *op code* and the rest of the instruction is known as the *operand.* With LDA 2000H, LDA is the op code and 2000H is the operand. Therefore, "op code" has a double meaning in microprocessor work; it may stand for the mnemonic or for the binary code used to represent the mnemonic. The intended meaning is clear from the context.

STA is a mnemonic for *store the accumulator.* Every STA instruction needs an address. STA 7FFFH means to store the accumulator contents at memory location 7FFFH. If

$$A = 8AH$$

the execution of STA 7FFFH stores 8AH at address 7FFFH.

MVI

MVI is the mnemonic for *move immediate.* It tells the computer to load a designated register with the byte that immediately follows the op code. For instance,

MVI A,37H

tells the computer to load the accumulator with 37H. After this instruction has been executed, the binary contents of the accumulator are

$$A = 0011\ 0111$$

You can use MVI with the A, B, and C registers. The formats for these instructions are

MVI A,byte
MVI B,byte
MVI C,byte

Op Codes

Table 11-1 shows the op codes for the SAP-2 instruction set. These are the 8080/8085 op codes. As you can see, 3A is the op code for LDA, 32 is the op code for STA, etc. Refer to this table in the remainder of this chapter.

EXAMPLE 11-1

Show the mnemonics for a program that loads the accumulator with 49H, the B register with 4AH, and the C register with 4BH; then have the program store the accumulator data at memory location 6285H.

SOLUTION

Here's one program that will work:

Mnemonics

MVI A,49H
MVI B,4AH
MVI C,4BH
STA 6285H
HLT

The first three instructions load 49H, 4AH, and 4BH into the A, B, and C registers. STA 6285H stores the accumulator contents at 6285H.

Note the use of HLT in this program. It has the same meaning as before: halt the data processing.

TABLE 11-1. SAP-2 OP CODES

Instruction	Op Code	Instruction	Op Code
ADD B	80	MOV B,A	47
ADD C	81	MOV B,C	41
ANA B	A0	MOV C,A	4F
ANA C	A1	MOV C,B	48
ANI byte	E6	MVI A,byte	3E
CALL address	CD	MVI B,byte	06
CMA	2F	MVI C,byte	0E
DCR A	3D	NOP	00
DCR B	05	ORA B	B0
DCR C	0D	ORA C	B1
HLT	76	ORI byte	F6
IN byte	DB	OUT byte	D3
INR A	3C	RAL	17
INR B	04	RAR	1F
INR C	0C	RET	C9
JM address	FA	STA address	32
JMP address	C3	SUB B	90
JNZ address	C2	SUB C	91
JZ address	CA	XRA B	A8
LDA address	3A	XRA C	A9
MOV A,B	78	XRI byte	EE
MOV A,C	79		

EXAMPLE 11-2

Translate the foregoing program into 8080/8085 machine language using the op codes of Table 11-1. Start with address 2000H.

SOLUTION

Address	Contents	Symbolic
2000H	3EH	MVI A,49H
2001H	49H	
2002H	06H	MVI B,4AH
2003H	4AH	
2004H	0EH	MVI C,4BH
2005H	4BH	
2006H	32H	STA 6285H
2007H	85H	
2008H	62H	
2009H	76H	HLT

There are a couple of new ideas in this machine-language program. With the

MVI A,49H

instruction, notice that the op code goes into the first address and the byte into the second address. This is true of all 2-byte instructions: op code into the first available memory location and byte into the next.

The instruction

STA 6285H

is a 3-byte instruction (1 byte for the op code and 2 for the address). The op code for STA is 32H. This byte goes into the first available memory location, which is 2006H. The address 6285H has 2 bytes. The lower byte 85H goes into the next memory location, and the upper byte 62H into the next location.

Why does the address get programmed with the lower byte first and the upper byte second? This is a peculiarity of the original 8080 design. To keep upward compatibility, the 8085 and some other microprocessors use the same scheme: lower byte into lower memory, upper byte into upper memory.

The last instruction HLT has an op code of 76H, stored in memory location 2009H.

In summary, the MVI instructions are 2-byte instructions, the STA is a 3-byte instruction, and the HLT is a 1-byte instruction.

11-4 REGISTER INSTRUCTIONS

Memory-reference instructions are relatively slow because they require more than one memory access during the instruction cycle. Furthermore, we often want to move data directly from one register to another without having to go through the memory. What follows are some of the SAP-2 register instructions, designed to move data from one register to another in the shortest possible time.

MOV

MOV is the mnemonic for *move*. It tells the computer to move data from one register to another. For instance,

MOV A,B

tells the computer to move the data in the B register to the accumulator. The operation is nondestructive, meaning that the data in B is copied but not erased. For example, if

A = 34H and **B** = 9DH

then the execution of MOV A,B results in

A = 9DH
B = 9DH

You can move data between the A, B, and C registers. The formats for all MOV instructions are

MOV A,B
MOV A,C
MOV B,A
MOV B,C
MOV C,A
MOV C,B

These instructions are the fastest in the SAP-2 instruction set, requiring only one machine cycle.

ADD and SUB

ADD stands for add the data in the designated register to the accumulator. For instance,

ADD B

means to add the contents of the B register to the accumulator. If

A = 04H and **B** = 02H

then the execution of ADD B results in

A = 06H

Similarly, SUB means subtract the data in the designated register from the accumulator. SUB C will subtract the contents of the C register from the accumulator.

The formats for the ADD and SUB instructions are

ADD B
ADD C
SUB B
SUB C

INR and DCR

Many times we want to increment or decrement the contents of one of the registers. INR is the mnemonic for *increment;* it tells the computer to increment the designated register. DCR is the mnemonic for *decrement,* and it instructs the computer to decrement the designated register. The formats for these instructions are

INR A
INR B
INR C
DCR A
DCR B
DCR C

As an example, if

B = 56H and **C** = 8AH

then the execution of INR B results in

B = 57H

and the execution of a DCR C produces

C = 89H

EXAMPLE 11-3

Show the mnemonics for adding decimal 23 and 45. The answer is to be stored at memory location 5600H. Also, the answer incremented by 1 is to be stored in the C register.

SOLUTION

As shown in Appendix 2, decimal 23 and 45 are equivalent to 17H and 2DH. Here is a program that will do the job:

Mnemonics

MVI A,17H
MVI B,2DH
ADD B
STA 5600H
INR A
MOV C,A
HLT

EXAMPLE 11-4

To *hand-assemble* a program means to translate a source program into a machine-language program by hand rather than machine. Hand-assemble the program of the preceding example starting at address 2000H.

SOLUTION

Address	Contents	Symbolic
2000H	3EH	MVI A,17H
2001H	17H	
2002H	06H	MVI B,2DH
2003H	2DH	
2004H	80H	ADD B
2005H	32H	STA 5600H
2006H	00H	
2007H	56H	
2008H	3CH	INR A
2009H	4FH	MOV C,A
200AH	76H	HLT

Notice that the ADD, INR, MOV, and HLT instructions are 1-byte instructions; the MVI instructions are 2-byte instructions, and the STA is a 3-byte instruction.

11-5 JUMP AND CALL INSTRUCTIONS

SAP-2 has four *jump* instructions; these can change the program sequence. In other words, instead of fetching the next instruction in the usual way, the computer may jump or *branch* to another part of the program.

JMP

To begin with, JMP is the mnemonic for jump; it tells the computer to get the next instruction from the designated memory location. Every JMP instruction includes an address that is loaded into the program counter. For instance,

JMP 3000H

tells the computer to get the next instruction from memory location 3000H.

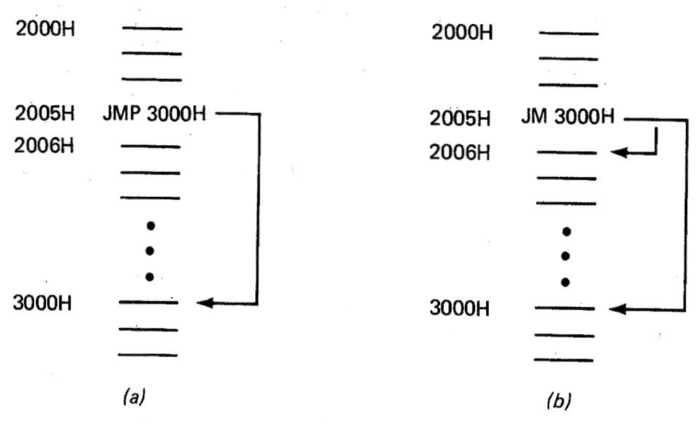

Fig. 11-3 (*a*) Unconditional jump; (*b*) conditional jump.

Here is what happens. Suppose JMP 3000H is stored at 2005H, as shown in Fig. 11-3*a*. At the end of the fetch cycle, the program counter contains

$$\mathbf{PC} = 2006\text{H}$$

During the execution cycle, the JMP 3000H loads the program counter with the designated address:

$$\mathbf{PC} = 3000\text{H}$$

When the next fetch cycle begins, the next instruction comes from 3000H rather than 2006H (see Fig. 11-3*a*).

JM

SAP-2 has two flags called the *sign flag* and the *zero flag*. During the execution of some instructions, these flags will be set or reset, depending on what happens to the accumulator contents. If the accumulator contents become negative, the sign flag will be set; otherwise, the sign flag is cleared. Symbolically,

$$S = \begin{cases} 0 & \text{if } A \geqslant 0 \\ 1 & \text{if } A < 0 \end{cases}$$

where S stands for sign flag. The sign flag will remain set or clear until another operation that affects the flag.

JM is a mnemonic for *jump if minus;* the computer will jump to a designated address if and only if the sign flag is set. As an example, suppose a JM 3000H is stored at 2005H. After this instruction has been fetched,

$$\mathbf{PC} = 2006\text{H}$$

If $S = 1$, the execution of JM 3000H loads the program counter with

$$\mathbf{PC} = 3000\text{H}$$

Since the program counter now points to 3000H, the next instruction will come from 3000H.

If the jump condition is not met ($S = 0$), the program counter is unchanged during the execution cycle. Therefore, when the next fetch cycle begins, the instruction is fetched from 2006H.

Figure 11-3*b* symbolizes the two possibilities for a JM instruction. If the minus condition is satisfied, the computer jumps to 3000H for the next instruction. If the minus condition is not satisfied, the program *falls through* to the next instruction.

JZ

The other flag affected by accumulator operations is the zero flag. During the execution of some instructions, the accumulator will become zero. To record this event, the zero flag is set; if the accumulator contents do not go to zero, the zero flag is reset. Symbolically,

$$Z = \begin{cases} 0 & \text{when } A \neq 0 \\ 1 & \text{when } A = 0 \end{cases}$$

JZ is the mnemonic for *jump if zero;* it tells the computer to jump to the designated address only if the zero flag is set. Suppose a JZ 3000H is stored at 2005H. If $Z = 1$ during the exection of JZ 3000H, the next instruction is fetched from 3000H. On the other hand, if $Z = 0$, the next instruction will come from 2006H.

JNZ

JNZ stands for *jump if not zero*. In this case, we get a jump when the zero flag is clear and no jump when it is set. Suppose a JNZ 7800H is stored at 2100H. If $Z = 0$, the next instruction will come from 7800H; however, if $Z = 1$, the program falls through to the instruction at 2101H.

JM, JZ, and JNZ are called *conditional jumps* because the program jump occurs only if certain conditions are satisfied. On the other hand, JMP is *unconditional;* once this instruction is fetched, the execution cycle always jumps the program to the specified address.

CALL and RET

A *subroutine* is a program stored in the memory for possible use in another program. Many microcomputers have subroutines for finding sines, cosines, tangents, logarithms, square roots, etc. These subroutines are part of the software supplied with the computer.

CALL is the mnemonic for *call the subroutine*. Every CALL instruction must include the starting address of the desired subroutine. For instance, if a square-root subroutine starts at address 5000H and a logarithm subroutine at 6000H, the execution of

CALL 5000H

will jump to the square-root subroutine. On the other hand, a

CALL 6000H

produces a jump to the logarithm subroutine.

RET stands for *return*. It is used at the end of every subroutine to tell the computer to go back to the original program. A RET instruction is to a subroutine as a HLT is to a program. Both tell the computer that something is finished. If you forget to use a RET at the end of a subroutine, the computer cannot get back to the original program and you will get computer trash.

When a CALL is executed in the SAP-2 computer, the contents of the program counter are automatically saved in memory locations FFFEH and FFFFH (the last two memory locations). The CALL address is then loaded into the program counter, so that execution begins with the first instruction in the subroutine. After the subroutine is finished, the RET instruction causes the address in memory locations FFFEH and FFFFH to be loaded back into the program counter. This returns control to the original program.

Figure 11-4 shows the program flow during a subroutine. The CALL 5000H sends the computer to the subroutine located at 5000H. After this subroutine has been completed, the RET sends the computer back to the instruction following the CALL.

CALL is unconditional, like JMP. Once a CALL has been fetched into the instruction register, the computer will jump to the starting address of the subroutine.

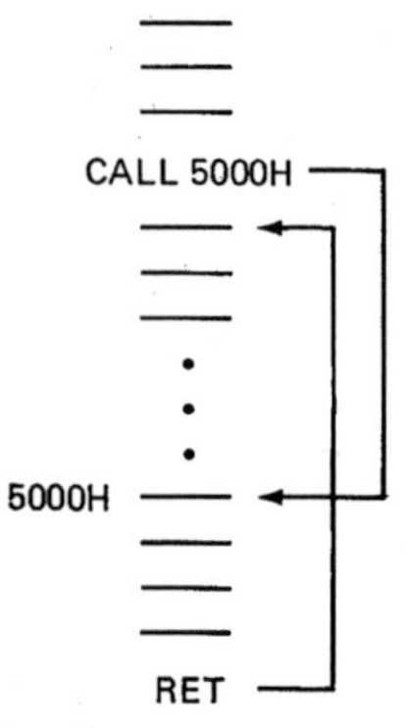

Fig. 11-4 CALL instruction.

More on Flags

The sign or zero flag may be set or reset during certain instructions. Table 11-2 lists the SAP-2 instructions that can affect the flags. All these instructions use the accumulator during the execution cycle. If the accumulator goes negative or zero while one of these instructions is being executed, the sign or zero flag will be set.

For instance, suppose the instruction is ADD C. The contents of the C register are added to the accumulator contents. If the accumulator contents become negative or zero in the process, the sign or zero flag will be set.

A word about the INR and DCR instructions. Since these instructions use the accumulator to add or subtract 1 from the designated register, they also affect the flags. For instance, to execute a DCR C, the contents of the C register are decremented by sending these contents to the accumulator, subtracting 1, and sending the result back to the C register. If the accumulator goes negative while the DCR C is executed, the sign flag is set; if the accumulator goes to zero, the zero flag is set.

TABLE 11-2. INSTRUCTIONS AFFECTING FLAGS

Instruction	Flags Affected
ADD	S, Z
SUB	S, Z
INR	S, Z
DCR	S, Z
ANA	S, Z
ORA	S, Z
XRA	S, Z
ANI	S, Z
ORI	S, Z
XRI	S, Z

EXAMPLE 11-5

Hand-assemble the following program starting at address 2000H:

```
MVI C,03H
DCR C
JZ 0009H
JMP 0002H
HLT
```

SOLUTION

Address	Contents	Symbolic
2000H	0EH	MVI C,03H
2001H	03H	
2002H	0DH	DCR C
2003H	CAH	JZ 2009H
2004H	09H	
2005H	20H	
2006H	C3H	JMP 2002H
2007H	02H	
2008H	20H	
2009H	76H	HLT

EXAMPLE 11-6

In the foregoing program, how many times is the DCR instruction executed?

SOLUTION

Figure 11-5 illustrates the program flow. Here is what happens. The MVI C,03H instruction loads the C register with 03H. DCR C reduces the contents to 02H. The contents are greater than zero; therefore, the zero flag is reset, and the JZ 2009H is ignored. The JMP 2002H returns the computer to the DCR C instruction.

The second time the DCR C is executed, the contents drop to 01H; the zero flag is still reset. JZ 2009H is again ignored, and the JMP 2002H returns the computer to DCR C.

The third DCR C reduces the contents to zero. This time the zero flag is set, and the JZ 2009H jumps the program to HLT instruction.

A *loop* is part of a program that is repeated. In this example, we have passed through the loop (DCR C and JZ 2009H) 3 times, as shown in Fig. 11-5. Note that the number of passes through the loop equals the number initially loaded into the C register. If we change the first instruction to

MVI C,07H

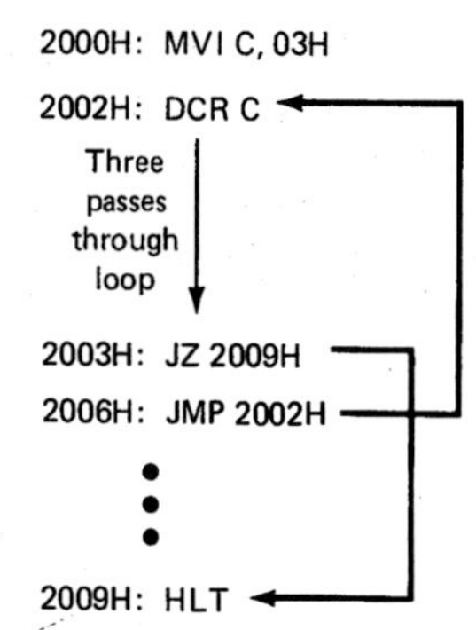

Fig. 11-5 Looping.

the computer will loop 7 times. Similarly, if we wanted to pass through the loop 200 times (equivalent to C8H), the first instruction would be

MVI C,C8H

The C register acts like a presettable down counter. This is why it is sometimes referred to as a *counter*.

The point to remember is this. We can set up a loop by using an MVI, DCR, JZ, and JMP in a program. The number loaded into the designated register (the counter) determines the number of passes through the loop. If we put new instructions inside the loop, these added instructions will be executed *X* times, the number preset into the counter.

EXAMPLE 11-7

When you buy a microcomputer, you often purchase software to do different jobs. One of the programs you can buy is an *assembler*. The assembler allows you to write programs in mnemonic form. Then the assembler converts these mnemonics into machine language. In other words, if you have an assembler, you no longer have to hand-assemble your programs; the computer does the work for you.

Show the assembly-language version of the program in Example 11-5. Include *labels* and *comments*.

SOLUTION

Label	Instruction	Comment
	MVI C,03H	;Load counter with decimal.3
REPEAT:	DCR C	;Decrement counter
	JZ END	;Test for zero
	JMP REPEAT	;Do it again
END:	HLT	

When you write a program, it helps to include your own comments about what the instruction is supposed to do. These comments jog your memory if you have to read the program months later. The first comment reminds us that we are presetting the down counter with decimal 3, the second comment reminds us that we are decrementing the counter, the third comment tells us that we are testing for zero before jumping, and the fourth comment tells us that the program will loop back.

When the assembler converts your source program into an object program, it ignores everything after the semicolon. Why? Because that's the way the assembler program is written. The semicolon is a coded way to tell the computer that your personal comments follow. (Remember the ASCII code. 3BH is the ASCII for a semicolon. When the assembler encounters 3BH in your source programs, it knows comments follow.)

Labels are another programming aid used with jumps and calls. When we write an assembly-language program, we often have no idea what address to use in a jump or call instruction. By using a label instead of a numerical address we can write programs that make sense to us. The assembler will keep track of our labels and automatically assign the correct addresses to them. This is a great laborsaving feature of an assembler.

For instance, when the assembler converts the foregoing program to machine language, it will replace JZ by CA (op code of Table 11-1) and END by the address of the HLT instruction. Likewise, it will replace JMP by C3 (op code) and REPEAT by the address of the DCR C instruction. The assembler determines the addresses of the HLT and JMP by counting the number of bytes needed by all instructions and figuring out where the HLT and DCR C instructions will be in the final assembled program.

All you have to remember is that you can make up any label you want for jump and call instructions. The same label followed by a colon is placed in front of the instruction you are trying to jump to. When the assembler converts your program into machine language, the colon tells it a label is involved.

One more point about labels. With SAP-2, the labels can be from one to six characters, the first of which must be a letter. Labels are usually words or abbreviations, but numbers can be included. The following are examples of acceptable labels:

REPEAT
DELAY
RDKBD
A34
B12C3

The first two are words; the third is an abbreviation for read the keyboard. The last two are labels that include numbers. The restrictions on length (no more than six characters) and starting character (must be letter) are typical of commercially available assemblers.

EXAMPLE 11-8

Show a program that multiplies decimal 12 and 8.

SOLUTION

The hexadecimal equivalents of 12 and 8 are 0CH and 08H. Let us set up a loop that adds 12 to the accumulator during each pass. If the computer loops 8 times, the accumulator contents will equal 96 (decimal) at the end of the looping.

Here's one assembly-language program that will do the job:

Label	Mnemonic	Comment
	MVI A,00H	;Clear accumulator
	MVI B,0CH	;Load decimal 12 into B
	MVI C,08H	;Preset counter with 8
REPEAT:	ADD B	;Add decimal 12
	DCR C	;Decrement the counter
	JZ DONE	;Test for zero
	JMP REPEAT	;Do it again
DONE:	HLT	;Stop it

The comments tell most of the story. First, we clear the accumulator. Next, we load decimal 12 into the B register. Then the counter is preset to decimal 8. These first three instructions are part of the initialization before entering a loop.

The ADD B begins the loop by adding decimal 12 to accumulator. The DCR C reduces the count to 7. Since the zero flag is clear, JZ DONE is ignored the first time through and the program flow returns to the ADD B instruction.

You should be able to see what will happen. ADD B is inside the loop and will be executed 8 times. After eight passes through the loop, the zero flag is set; then the JZ DONE will take the program out of the loop to the HLT instruction.

Since 12 is added 8 times,

$$12 + 12 + 12 + 12 + 12 + 12 + 12 + 12 = 96$$

(Because decimal 96 is equivalent to hexadecimal 60, the accumulator contains 0110 0000.) Repeated addition like this is equivalent to multiplication. In other words, adding 12 eight times is identical to 12×8. Most microprocessors do not have multiplication hardware; they only have an adder-subtracter like the SAP computer. Therefore, with the typical microprocessor, you have to use some form of programmed multiplication such as repeated addition.

EXAMPLE 11-9

Modify the foregoing multiply program by using a JNZ instead of a JZ.

SOLUTION

Look at this:

Label	Mnemonic	Comment
	MVI A,00H	;Clear accumulator
	MVI B,0CH	;Load decimal 12 into B
	MVI C,08H	;Preset counter with 8
REPEAT:	ADD B	;Add decimal 12
	DCR C	;Decrement the counter
	JNZ REPEAT	;Test for zero
	HLT	;Stop it

This is simpler. It eliminates one JMP instruction and one label. As long as the counter is greater than zero, the JNZ will force the computer to loop back to REPEAT. When the counter drops to zero, the program will fall through the JNZ to the HLT.

EXAMPLE 11-10

Hand-assemble the foregoing program starting at address 2000H.

SOLUTION

Address	Contents	Symbolic
2000H	3EH	MVI A,00H
2001H	00H	
2002H	06H	MVI B,0CH
2003H	0CH	
2004H	0EH	MVI, C,08H
2005H	08H	
2006H	80H	ADD B
2007H	0DH	DCR C
2008H	C2H	JNZ 2006H
2009H	06H	
200AH	20H	
200BH	76H	HLT

The first three instructions initialize the registers before the multiplication begins. If we change the initial values, we can multiply other numbers.

EXAMPLE 11-11

Change the multiplication part of the foregoing program into a subroutine located at starting address F006H.

SOLUTION

Address	Contents	Symbolic
F006H	80H	ADD B
F007H	0DH	DCR C
F008H	C2H	JNZ F006H
F009H	06H	
F00AH	F0H	
F00BH	C9H	RET

Here's what happened. The initializing instructions depend on the numbers we are multiplying, so they don't belong in the subroutine. The subroutine should contain only the multiplication part of the program.

In relocating the program we *mapped* (converted) addresses 2006H–200BH to F006H–F00BH. Also, the HLT was changed to a RET to get us back to the original program.

EXAMPLE 11-12

The multiply subroutine of the preceding example is used in the following program. What does the program do?

```
MVI A,00H
MVI B,10H
MVI C,0EH
CALL F006H
HLT
```

SOLUTION

Hexadecimal 10H is equivalent to decimal 16, and hexadecimal 0EH is equivalent to decimal 14. The first three instructions clear the accumulator, load the B register with decimal 16, and preset the counter to decimal 14. The CALL sends the computer to the multiply subroutine of the preceding example. When the RET is executed, the accumulator contents are E0H, which is equivalent to 224.

Incidentally, a *parameter* is a piece of data that the subroutine needs to work properly. The multiply subroutine located at F006H needs three parameters to work properly (*A*, *B*, and *C*). We pass these parameters to the multiply subroutine by clearing the accumulator, loading the B register with the multiplicand, and presetting the C register with the multiplier. In other words, we set A = 00H, B = 10H, and C = 0EH. Passing data to a subroutine in this way is called *register parameter passing*.

11-6 LOGIC INSTRUCTIONS

A microprocessor can do logic as well as arithmetic. What follows are the SAP-2 logic instructions. Again, they are a subset of the 8080/8085 instructions.

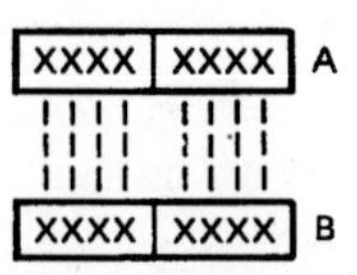

Fig. 11-6 Logic instructions are bitwise.

CMA

CMA stands for "complement the accumulator." The execution of a CMA inverts each bit in the accumulator, producing the 1's complement.

ANA

ANA means to AND *the accumulator* contents with the designated register. The result is stored in the accumulator. For instance,

ANA B

means to AND the contents of the accumulator with the contents of the B register. The ANDing is done on a bit-by-bit basis. For example, suppose the two registers contain

$$A = 1100\ 1100 \qquad (11\text{-}1)$$

and

$$B = 1111\ 0001 \qquad (11\text{-}2)$$

The execution of an ANA B results in

$$A = 1100\ 0000$$

Notice that the ANDing is bitwise, as illustrated in Fig. 11-6. The ANDing is done on pairs of bits; A_7 is ANDed with B_7, A_6 with B_6, A_5 with B_5, and so on, with the result stored in the accumulator.

Two ANA instructions are available in SAP-2: ANA B and ANA C. Table 11-1 shows the op codes.

ORA

ORA is the mnemonic for OR *the accumulator* with the designated register. The two ORA instructions in SAP-2 are ORA B and ORA C. As an example, if the accumulator and B register contents are given by Eqs. 11-1 and 11-2, then executing ORA B gives

$$A = 1111\ 1101$$

XRA

XRA means XOR *the accumulator* with the designated register. The SAP-2 instruction set contains XRA B and XRA C. If the accumulator and B contents are given by Eqs. 11-1 and 11-2, the execution of XRA B produces

$$A = 0011\ 1101$$

ANI

SAP-2 also has immediate logic instructions. ANI means AND *immediate*. It tells the computer to AND the accumulator contents with the byte that immediately follows the op code. For instance, if

$$A = 0101\ 1110$$

the execution of ANI C7H will AND

0101 1110 with 1100 0111

to produce new accumulator contents of

$$A = 0100\ 0110$$

ORI

ORI is the mnemonic for OR *immediate*. The accumulator contents are ORed with the byte that follows the op code. If

$$A = 0011\ 1000$$

the execution of ORI 5AH will OR

0011 1000 with 0101 1010

to produce new accumulator contents of

0111 1010

XRI

XRI means XOR *immediate*. If

$$A = 0001\ 1100$$

the execution of XRI D4H will XOR

0001 1100 with 1101 0100

to produce

$$A = 1100\ 1000$$

11-7 OTHER INSTRUCTIONS

This section looks at the last of the SAP-2 instructions. Since these instructions don't fit any particular category, they are being collected here in a miscellaneous group.

NOP

NOP stands for *no operation*. During the execution of a NOP, all *T* states are do nothings. Therefore, no register changes occur during a NOP.

The NOP instruction is used to waste time. It takes four *T* states to fetch and execute the NOP instruction. By repeating a NOP a number of times, we can delay the data processing, which is useful in timing operations. For instance, if we put a NOP inside a loop and execute it 100 times, we create a time delay of 400 *T* states.

HLT

We have already used this. HLT stands for *halt*. It ends the data processing.

IN

IN is the mnemonic for *input*. It tells the computer to transfer data from the designated port to the accumulator. Since there are two input ports, you have to designate which one is being used. The format for an input operation is

IN byte

For instance,

IN 02H

means to transfer the data in port 2 to the accumulator.

OUT

OUT stands for *output*. When this instruction is executed, the accumulator word is loaded into the designated output port. The format for this instruction is

OUT byte

Since the output ports are numbered 3 and 4 (Fig. 11-2), you have to specify which port is to be used. For instance,

OUT 03H

will transfer the contents of the accumulator to port 3.

RAL

RAL is the mnemonic for *rotate the accumulator left*. This instruction will shift all bits to the left and move the MSB into the LSB position, as illustrated in Fig. 11-7*a*. As an example, suppose the contents of the accumulator are

A = 1011 0100

Executing the RAL will produce

A = 0110 1001

As you see, all bits moved left, and the MSB went to the LSB position.

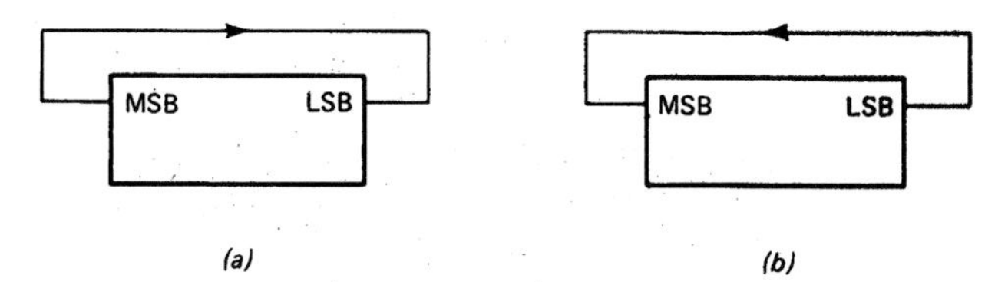

Fig. 11-7 Rotate instructions: (*a*) RAL; (*b*) RAR.

RAR

RAR stands for *rotate the accumulator right*. This time, the bits shift to the right, the LSB going to the MSB position, as shown in Fig. 11-7*b*. If

A = 1011 0100

the execution of a RAR will result in

A = 0101 1010

EXAMPLE 11-13

The bits in a byte are numbered 7 to 0 (MSB to LSB). Show a program that can input a byte from port 2 and determine if bit 0 is a 1 or a 0. If the bit is a 1, the program is to load the accumulator with an ASCII Y (yes). If the bit is a 0, the program should load the accumulator with an ASCII N (no). The yes or no answer is to be sent to output port 3.

SOLUTION

Label	Mnemonic	Comment
	IN 02H	;Get byte from port 2
	ANI 01H	;Isolate bit 0
	JNZ YES	;Jump if bit 0 is a 1
	MVI A,4EH	;Load N into accumulator
	JMP DONE	;Skip next instruction
YES:	MVI A,59H	;Load Y into accumulator
DONE:	OUT 03H	;Send answer to port 3
	HLT	

The IN 02H transfers the contents of input port 2 to the accumulator to get

$$A = A_7A_6A_5A_4A_3A_2A_1A_0$$

The immediate byte in ANI 01H is

0000 0001

This byte is called a *mask* because its 0s will mask or blank out the corresponding high bits in the accumulator. In other words, after the execution of ANI 01H the accumulator contents are

$$A = 0000\ 000A_0$$

If A_0 is 1, the JNZ YES will produce a jump to the MVI A,59H; this loads a 59H (the ASCII for Y) into the accumulator. If A_0 is 0, the program falls through to the MVI A,4EH. This loads the accumulator with the ASCII for N.

The OUT 03H loads the answer, either ASCII Y or N, into port 3. The hexadecimal display therefore shows either 59H or 4EH.

EXAMPLE 11-14

Instead of a parallel output at port 3, we want a serial output at port 4. Modify the foregoing program so that it converts the answer (59H or 4EH) into a serial output at bit 0, port 4.

SOLUTION

Label	Mnemonic	Comment
	IN 02H	
	ANI 01H	
	JNZ YES	
	MVI A,4EH	
	JMP DONE	
YES:	MVI A,59H	
DONE:	MVI C,08H	;Load counter with 8
AGAIN:	OUT 04H	;Send LSB to port 4
	RAR	;Position next bit
	DCR C	;Decrement count
	JNZ AGAIN	;Test count
	HLT	

In converting from parallel to serial data, the A_0 bit is sent first, then the A_1 bit, then the A_2 bit, and so on.

EXAMPLE 11-15

Handshaking is an interaction between a CPU and a peripheral device that takes place during an I/O data transfer.

In SAP-2 the handshaking takes place as follows. After you enter two digits (1 byte) into the hexadecimal encoder of Fig. 11-2, the data is loaded into port 1; at the same time, a *high READY* bit is sent to port 2.

Before accepting input data, the CPU checks the *READY* bit in port 2. If the *READY* bit is low, the CPU waits. If the *READY* bit is high, the CPU loads the data in port 1. After the data transfer is finished, the CPU sends a high *ACKNOWLEDGE* signal to the hexadecimal keyboard encoder; this resets the *READY* bit to 0. The *ACKNOWLEDGE* bit then is reset to low.

After you key in a new byte, the cycle starts over with new data going to the port 1 and a high *READY* bit to port 2.

The sequence of SAP-2 handshaking is

1. *READY* bit (bit 0, port 2) goes high.
2. Input the data in port 1 to the CPU.
3. *ACKNOWLEDGE* bit (bit 7, port 4) goes high to reset *READY* bit.
4. Reset the *ACKNOWLEDGE* bit.

Write a program that inputs a byte of data from port 1 using handshaking. Store the byte in the B register.

SOLUTION

Label	Mnemonic	Comment
STATUS:	IN 02H	;Input byte from port 2
	ANI 01H	;Isolate *READY* bit
	JZ STATUS	;Jump back if not ready
	IN 01H	;Transfer data in port 1
	MOV B,A	;Transfer from A to B
	MVI A,80H	;Set *ACKNOWLEDGE* bit
	OUT 04H	;Output high *ACKNOWLEDGE*
	MVI A,00H	;Reset *ACKNOWLEDGE* bit
	OUT 04H	;Output low *ACKNOWLEDGE*
	HLT	

If the *READY* bit is low, the ANI 01H will force the accumulator contents to go to zero. The JZ STATUS therefore will loop back to IN 02H. This looping will continue until the *READY* bit is high, indicating valid data in port 1.

When the *READY* bit is high, the program falls through the JZ STATUS to the IN 01H. This transfers a byte from port 1 to the accumulator. The MOV sends the byte to the B register. The MVI A,80H sets the *ACKNOWLEDGE* bit

(bit 7). The OUT 04H sends this high *ACKNOWLEDGE* to the hexadecimal encoder where the internal hardware resets the *READY* bit. Then the *ACKNOWLEDGE* bit is reset in preparation for the next input cycle.

11-8 SAP-2 SUMMARY

This section summarizes the SAP-2 *T* states, flags, and addressing modes.

T States

The SAP-2 controller-sequencer is microprogrammed with a variable machine cycle. This means that some instructions take longer than others to execute. As you recall, the idea behind microprogramming is to store the control routines in a ROM and access them as needed.

Table 11-3 shows each instruction and the number of *T* states needed to execute it. For instance, it takes four *T* states to execute the ADD B instruction, seven to execute the ANI byte, eighteen to execute the CALL, and so on. Knowing the number of *T* states is important in timing applications.

Notice that the JM instruction has *T* states of 10/7. This means it takes 10 *T* states when a jump occurs but only 7 without the jump. The same idea applies to the other conditional jumps; 10 *T* states for a jump, 7 with no jump.

Flags

As you know, the accumulator goes negative or zero during the execution of some instructions. This affects the sign and zero flags. Figure 11-8 shows the circuits used in SAP-2 to set the flags.

When the accumulator contents are negative, the leading bit A_7 is a 1. This sign bit drives the lower AND gate. When the accumulator contents are zero, all bits are zero and the output of the NOR gate is a 1. This NOR output drives the upper AND gate. If gating signal L_F is high, the flags will be updated to reflect the sign and zero condition of the accumulator. This means the Z_{FLAG} will be high when the accumulator contents are zero; the S_{FLAG} will be high when the accumulator contents are negative.

Not all instructions affect the flags. As shown in Table 11-3, the instructions that update the flags are ADD, ANA, ANI, DCR, INR, ORA, ORI, SUB, XRA, and XRI. Why only these instructions? Because the L_F signal of Fig. 11-8 is high only when these instructions are executed. This is accomplished by microprogramming an L_F bit for each instruction. In other words, in the control ROM we store a high L_F bit for the foregoing instructions, and a low L_F bit for all others.

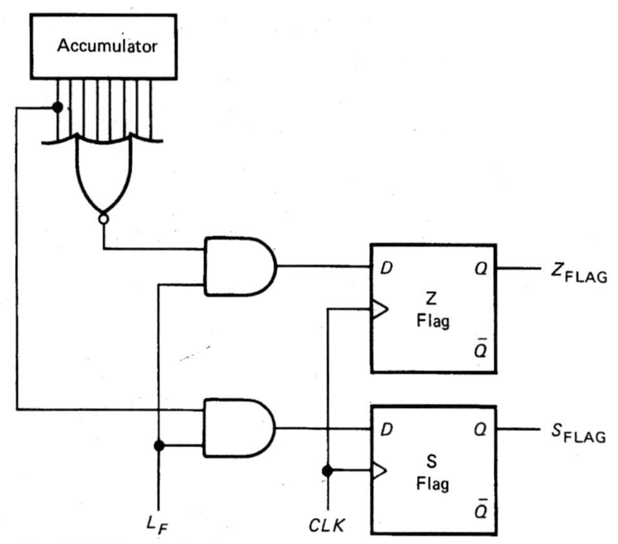

Fig. 11-8 Setting the flags.

Conditional Jumps

As mentioned earlier, the conditional jumps take ten *T* states when the jump occurs but only seven *T* states when no jump take place. Briefly, this is accomplished as follows. During the execution cycle the address ROM sends the computer to the starting address of a conditional-jump microroutine. The initial microinstruction looks at the flags and judges whether or not to jump. If a jump is indicated, the microroutine continues; otherwise, it is aborted and the computer begins a new fetch cycle.

Addressing Modes

The SAP-2 instructions access data in different ways. It is the operand that tells us how the data is to be accessed. For instance, the first instructions discussed were

LDA address
STA address

These are examples of *direct* addressing because we specify the address where the data is to be found.

Immediate addressing is different. Instead of giving an address for the data, we give the data itself. For instance,

MVI A,byte

accesses the data to be loaded into the accumulator by using the byte in memory that immediately follows the op code. Table 11-3 shows the other immediate instructions.

An instruction like

MOV A,B

TABLE 11-3. SAP-2 INSTRUCTION SET

Instruction	Op Code	*T* States	Flags	Addressing	Bytes
ADD B	80	4	S, Z	Register	1
ADD C	81	4	S, Z	Register	1
ANA B	A0	4	S, Z	Register	1
ANA C	A1	4	S, Z	Register	1
ANI byte	E6	7	S, Z	Immediate	2
CALL address	CD	18	None	Immediate	3
CMA	2F	4	None	Implied	1
DCR A	3D	4	S, Z	Register	1
DCR B	05	4	S, Z	Register	1
DCR C	0D	4	S, Z	Register	1
HLT	76	5	None	—	1
IN byte	DB	10	None	Direct	2
INR A	3C	4	S, Z	Register	1
INR B	04	4	S, Z	Register	1
INR C	0C	4	S, Z	Register	1
JM address	FA	10/7	None	Immediate	3
JMP address	C3	10	None	Immediate	3
JNZ address	C2	10/7	None	Immediate	3
JZ address	CA	10/7	None	Immediate	3
LDA address	3A	13	None	Direct	3
MOV A,B	78	4	None	Register	1
MOV A,C	79	4	None	Register	1
MOV B,A	47	4	None	Register	1
MOV B,C	41	4	None	Register	1
MOV C,A	4F	4	None	Register	1
MOV C,B	48	4	None	Register	1
MVI A,byte	3E	7	None	Immediate	2
MVI B,byte	06	7	None	Immediate	2
MVI C,byte	0E	7	None	Immediate	2
NOP	00	4	None	—	1
ORA B	B0	4	S, Z	Register	1
ORA C	B1	4	S, Z	Register	1
ORI byte	F6	7	S, Z	Immediate	2
OUT byte	D3	10	None	Direct	2
RAL	17	4	None	Implied	1
RAR	1F	4	None	Implied	1
RET	C9	10	None	Implied	1
STA address	32	13	None	Direct	3
SUB B	90	4	S, Z	Register	1
SUB C	91	4	S, Z	Register	1
XRA B	A8	4	S, Z	Register	1
XRA C	A9	4	S, Z	Register	1
XRI byte	EE	7	S, Z	Immediate	2

is an example of *register addressing*. The data to be loaded is stored in a CPU register rather than in the memory. Register addressing has the advantage of speed because fewer *T* states are needed for this type of instruction.

Implied addressing means that the location of the data is contained within the op code itself. For instance,

RAL

tells us to rotate the accumulator bits left. The data is in the accumulator; this is why no operand is needed with implied addressing.

Bytes

Each instruction occupies a number of bytes in the memory. SAP-2 instructions are either 1, 2, or 3 bytes long. Table 11-3 shows the length of each instruction. As you see, ADD instructions are 1-byte instructions, ANI instructions are 2-byte instructions, CALLs are 3-byte instructions, and so forth.

EXAMPLE 11-16

SAP-2 has a clock frequency of 1 MHz. This means that each T state has a duration of 1 μs. How long does it take to execute the following SAP-2 subroutine?

Label	Mnemonic	Comment
	MVI C,46H	;Preset count to decimal 70
AGAIN:	DCR C	;Count down
	JNZ AGAIN	;Test count
	NOP	;Delay
	RET	

SOLUTION

The MVI is executed once to initialize the count. The DCR is executed 70 times. The JNZ jumps back 69 times and falls through once. With the number of T states given in Table 11-3, we can calculate the total execution time of the subroutine as follows:

MVI:	1 × 7 × 1 μs =	7 μs	
DCR:	70 × 4 × 1 μs =	280	
JNZ:	69 × 10 × 1 μs =	690	(jump)
JNZ:	1 × 7 × 1 μs =	7	(no jump)
NOP:	1 × 4 × 1 μs =	4	
RET:	1 × 10 × 1 μs =	10	
		998 μs ≈ 1 ms	

As you see, the total time needed to execute the subroutine is approximately 1 ms.

A subroutine like this can produce a time delay of 1 ms whenever it is called. There are many applications where you need a delay.

According to Table 11-3, the instructions in the foregoing subroutine have the following byte lengths:

Instruction	MVI	DCR	JNZ	NOP	RET
Bytes	2	1	3	1	1

The total byte length of the subroutine is 8. As part of the SAP-2 software, the foregoing subroutine can be assembled and relocated at addresses F010H to F017H. Hereafter, the execution of a CALL F010H will produce a time delay of 1 ms.

EXAMPLE 11-17

How much time delay does this SAP-2 subroutine produce?

Label	Mnemonic	Comment
	MVI B,0AH	;Preset B counter with decimal 10
LOOP1:	MVI C,47H	;Preset C counter with decimal 71
LOOP2:	DCR C	;Count down on C
	JNZ LOOP2	;Test for C count of zero
	DCR B	;Count down on B
	JNZ LOOP1	;Test for B count of zero
	RET	

SOLUTION

This subroutine has two loops, one inside the other. The inner loop consists of DCR C and JNZ LOOP2. This inner loop produces a time delay of

DCR C:	71 × 4 × 1 μs =	284 μs	
JNZ LOOP2:	70 × 10 × 1 μs =	700	(jump)
JNZ LOOP2:	1 × 7 × 1 μs =	7	(no jump)
		991 μs	

When the C count drops to zero, the program falls through the JNZ LOOP2. The B counter is decremented, and the JNZ LOOP1 sends the program back to the MVI C,47H. Then we enter LOOP2 for a second time. Because LOOP2 is inside LOOP1, LOOP2 will be executed 10 times and the overall time delay will be approximately 10 ms.

Here are the calculations for the overall subroutine:

MVI B,0AH:	1 × 7 × 1 μs =	7 μs	
MVI C,47H:	10 × 7 × 1 μs =	70	
LOOP2:	10 × 991 μs =	9,910	
DCR B:	10 × 4 × 1 μs =	40	
JNZ LOOP1:	9 × 10 × 1 μs =	90	(jump)
JNZ LOOP1:	1 × 7 × 1 μs =	7	(no jump)
RET:	1 × 10 × 1 μs =	10	
		10,134 μs ≈ 10 ms	

This SAP-2 subroutine has a byte length of

$$2 + 2 + 1 + 3 + 1 + 3 + 1 = 13$$

It can be assembled and located at addresses F020H to F02CH. From now on, a CALL F020H will produce a time delay of approximately 10 ms.

By changing the first instruction to

MVI B,64H

the B counter is preset with decimal 100. In this case, the inner loop is executed 100 times and the overall time delay is approximately 100 ms. This 100-ms subroutine can be located at addresses F030H to F03CH.

EXAMPLE 11-18

Here is a subroutine with three loops *nested* one inside the other. How much time delay does it produce?

Label	Mnemonic	Comment
	MVI A,0AH	;Preset A counter with decimal 10
LOOP1:	MVI B,64H	;Preset B counter with decimal 100
LOOP2:	MVI C,47H	;Preset C counter with decimal 71
LOOP3:	DCR C	;Count down C
	JNZ LOOP3	;Test C for zero
	DCR B	;Count down B
	JNZ LOOP2	;Test B for zero
	DCR A	;Count down A
	JNZ LOOP1	;Test A for zero
	RET	

SOLUTION

LOOP3 still takes approximately 1 ms to get through. LOOP2 makes 100 passes through LOOP3, so it takes about 100 ms to complete LOOP2. LOOP1 makes 10 passes through LOOP2; therefore, it takes around 1 s to complete the overall subroutine.

What do we have? A 1-s subroutine. It will be located in F040H to F052H. To produce a 1-s time delay, we would use a CALL F040H.

By changing the initial instruction to

MVI A,64H

LOOP1 will make 100 passes through LOOP2, which makes 100 passes through LOOP3. The resulting subroutine can be located at F060H to F072H and will produce a time delay of 10 s.

Table 11-4 summarizes the SAP-2 time delays. With these subroutines, we can produce delays from 1 ms to 10 s.

TABLE 11-4. SAP-2 SUBROUTINES

Label	Starting Address	Delay	Registers Used
D1MS	F010H	1 ms	C
D10MS	F020H	10 ms	B, C
D100MS	F030H	100 ms	B, C
D1SEC	F040H	1 s	A, B, C
D10SEC	F060H	10 s	A, B, C

EXAMPLE 11-19

The traffic lights on a main road show green for 50 s, yellow for 6 s, and red for 30 s. Bits 1, 2, and 3 of port 4 are the control inputs to peripheral equipment that runs these traffic lights. Write a program that produces time delays of 50, 6, and 30 s for the traffic lights.

SOLUTION

Label	Mnemonic	Comment
AGAIN:	MVI A,32H	;Preset counter with decimal 50
	STA SAVE	;Save accumulator contents
	MVI A,02H	;Set bit 1
	OUT 04H	;Turn on green light
LOOPGR:	CALL D1SEC	;Call 1-s subroutine
	LDA SAVE	;Load current A count
	DCR A	;Decrement A count
	STA SAVE	;Save reduced A count
	JNZ LOOPGR	;Test for zero
	MVI A,06H	;Preset counter with decimal 6
	STA SAVE	
	MVI A,04H	;Set bit 2
	OUT 04H	;Turn on yellow light
LOOPYE:	CALL D1SEC	
	LDA SAVE	
	DCR A	
	STA SAVE	
	JNZ LOOPYE	
	MVI A,1EH	;Preset counter with decimal 30
	STA SAVE	
	MVI A,08H	;Set bit 3
	OUT 04H	;Turn on red light
LOOPRE:	CALL D1SEC	
	LDA SAVE	
	DCR A	
	STA SAVE	
	JNZ LOOPRE	
	JMP AGAIN	
SAVE:	Data	

Let's go through the green part of the program; the yellow and red are similar. The green starts with MVI A,32H, which loads decimal 50 into the accumulator. The STA SAVE will store this initial value in a memory location called SAVE. The MVI A,02H sets bit 1 in the accumulator; then the OUT 04H transfers this high bit to port 4. Since this port controls the traffic lights, the green light comes on.

The CALL D1SEC produces a time delay of 1 s. The LDA SAVE loads the accumulator with decimal 50. The DCR A decrements the count to decimal 49. The STA SAVE stores this decimal 49. Then the JNZ LOOPGR takes the program back to the CALL D1SEC for another 1-s delay.

The CALL D1SEC is executed 50 times; therefore, the green light is on for 50 s. Then the program falls through the JNZ LOOPGR to the MVI A,06H. The yellow part of the program then begins and results in the yellow light being on for 6 s. Finally, the red part of the program is executed and the red light is on for 30 s. The JMP AGAIN repeats the whole process. In this way, the program is controlling the timing of the green, yellow, and red lights.

EXAMPLE 11-20

Middle C on a piano has a frequency of 261.63 Hz. Bit 5 of port 4 is connected to an amplifier which drives a loudspeaker. Write a program that sends middle C to the loudspeaker.

SOLUTION

To begin with, the period of middle C is

$$T = \frac{1}{f} = \frac{1}{261.63 \text{ Hz}} = 3{,}822 \ \mu\text{s}$$

What we are going to do is send to port 4 a signal like Fig. 11-9. This square wave is high for 1,911 μs and low for 1,911 μs. The overall period is 3,822 μs, and the frequency is 261.63 Hz. Because the signal is square rather than sinusoidal, it will sound distorted but it will be recognizable as middle C.

Here is a program that sends middle C to the loudspeaker.

Label	Mnemonic	Comment
LOOP1:	OUT 04H	;Send bit to speaker
	MVI C,86H	;Preset counter with decimal 134
LOOP2:	DCR C	;Count down
	JNZ LOOP2	;Test count
	CMA	;Reset bit 5
	NOP	;Fine tuning
	NOP	;Fine tuning
	JMP LOOP1	;Go back for next half cycle

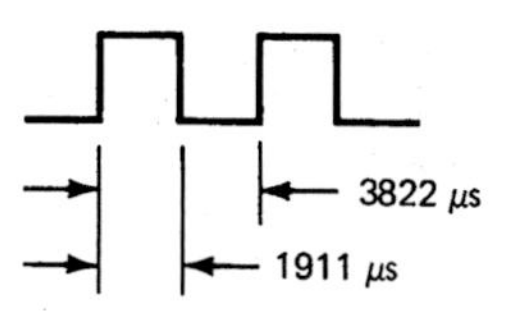

Fig. 11-9 Generating middle C note.

The OUT 04H sends a bit (either low or high) to the loudspeaker. The MVI presets the counter to decimal 134. Then comes LOOP2, the DCR and JNZ, which produces a time delay of 1,866 μs. The program then falls through to the CMA, which complements all bits in the accumulator. The two NOPs add a time delay of 8 μs. The JMP LOOP1 then takes the program back. When the OUT 04H is executed, bit 5 (complemented) goes to the loudspeaker. In this way the loudspeaker is driven into the opposite state. The execution time for both half cycles is 3,824 μs, close enough to middle C.

Here are the calculations for the time delay:

OUT 04H:	1 × 10 × 1 μs =	10 μs
MVI C,86H:	1 × 7 × 1 μs =	7
DCR C:	134 × 4 × 1 μs =	536
JNZ LOOP2:	133 × 10 × 1 μs =	1,330
JNZ LOOP2:	1 × 7 × 1 μs =	7
CMA:	1 × 4 × 1 μs =	4
2 NOPs:	2 × 4 × 1 μs =	8
JMP LOOP1:	1 × 10 × 1 μs =	10
		1,912 μs

This is the half-cycle time. The period is 3,824 μs.

EXAMPLE 11-21

Serial data is sometimes called a serial data stream because bits flow one after another. In Fig. 11-10 a serial data stream drives bit 7 of port 2 at a rate of approximately 600 bits per second. Write a program that inputs an 8-bit character in a serial data stream and stores it in memory location 2100H.

SOLUTION

Since approximately 600 bits are received each second, the period of each bit is

$$\frac{1}{600 \text{ Hz}} = 1{,}667 \ \mu\text{s}$$

The idea will be to input a bit from port 2, rotate the accumulator right, wait approximately 1,600 μs, then input another bit, rotate the accumulator right, and so on, until all bits have been received.

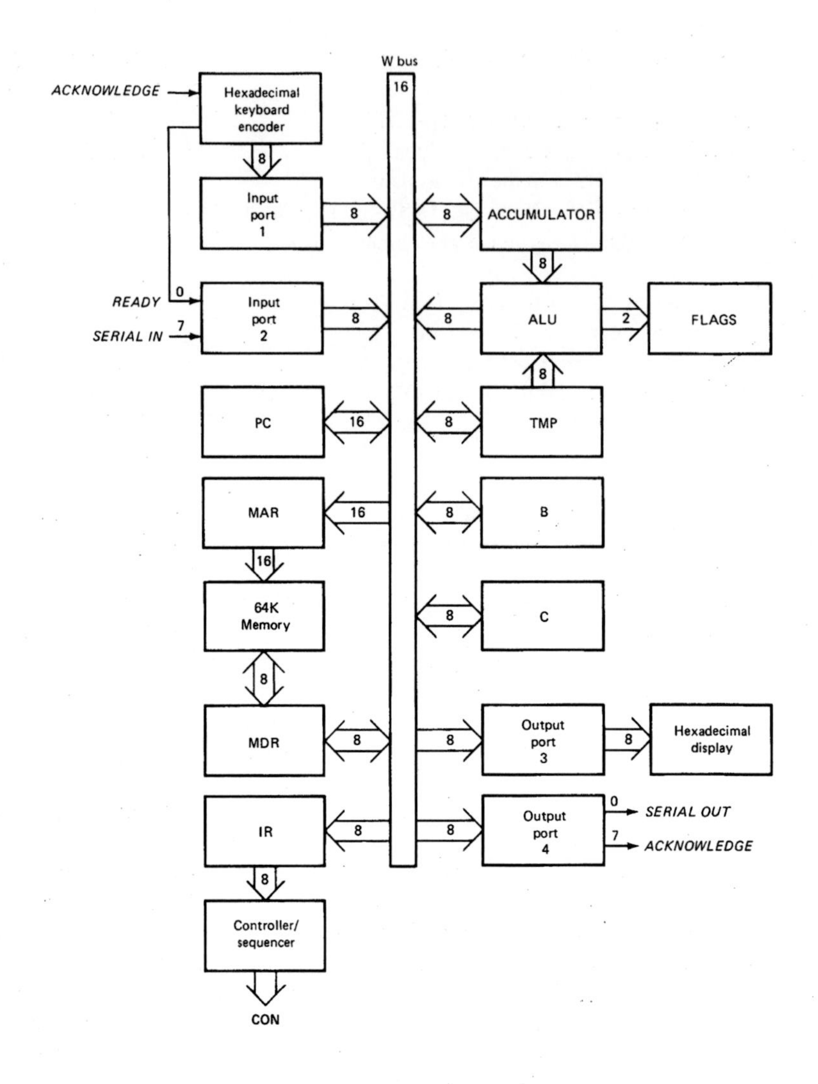

Fig. 11-10

Here is a program that will work:

Label	Mnemonic	Comment
	MVI B,00H	;Load zero into B register
	MVI C,07H	;Preset counter with decimal 7
BIT:	IN 02H	;Input data
	ANI 80H	;Isolate bit 7
	ORA B	;Update character
	RAR	;Move bits right
	MOV B,A	;Save bits in B
	MVI A,73H	;Begin a delay of 1,600 μs
DELAY:	DCR A	;Count down A
	JNZ DELAY	;Test A count for zero
	DCR C	;Count down C
	JNZ BIT	;Test C count for zero
	IN 02H	;Input last bit
	ANI 80H	;Isolate bit 7
	ORA B	
	STA 2100H	;Save character

The first instruction clears the B register. The second instruction loads decimal 7 into the C counter. The IN 02H brings in the data from port 2. The ANI mask isolates bit 7 because this is the *SERIAL IN* bit from port 2. The ORA B does nothing the first time through because B is full of 0s. The RAR moves the accumulator bits to the right. The MOV B,A stores the accumulator contents in the B register.

MVI A,73H presets the accumulator with decimal 115. Then comes a delay loop, DCR A and JNZ DELAY, that takes approximately 1,600 μs to complete.

The DCR C reduces the C count by 1, and the JNZ BIT tests the C count for zero. The program jumps back to the IN 02H to get the next bit from the serial data stream. The ANI mask isolates bit 7, which is then ORed with the contents of the B register; this combines the previous bit with the newly received bit. After another RAR, the two received bits are stored in the B register. Then comes another delay of approximately 1,600 μs.

The program continues to loop and each time a new bit is input from the serial data stream. After 7 bits have been received, the program will fall through the JNZ BIT instruction.

The last four instructions do the following. The IN 02H brings in the eighth bit. The ANI isolates bit 7. The ORA B combines this new bit with the other seven bits in the B register. At this point, all received bits are in the accumulator. The STA 2100H then stores the byte in the accumulator at 2100H.

A concrete example will help. Suppose the 8 bits being received are 57H, the ASCII code for W. The LSB is received first, the MSB last. Here is how the contents of the B register appear after the execution of the ORA B:

A = 1000 0000 (First pass through loop)
A = 1100 0000 (Second pass)
A = 1110 0000 (Third pass)
A = 0111 0000 (Fourth pass)
A = 1011 1000 (Fifth pass)
A = 0101 1100 (Sixth pass)
A = 1010 1110 (Seventh pass)
A = 0101 0111 (Final contents)

Incidentally, the ASCII code only requires 7 bits; for this reason, the eighth bit (A_7) may be set to zero or used as a parity bit.

GLOSSARY

assembler A program that converts a source program into a machine-language program.

comment Personal notes in an assembly-language program that are not assembled. They refresh the programmer's memory at a later date.

conditional jump A jump that occurs only if certain conditions are satisfied.

direct addressing Addressing in which the instruction contains the address of the data to be operated on.

flag A flip-flop that keeps track of a changing condition during a computer run.

hand assembling Translating a source program into a machine-language program by hand rather than computer.

handshaking Interaction between a CPU and a peripheral device that takes place during an I/O operation. In SAP-2 it involves *READY* and *ACKNOWLEDGE* signals.

immediate addressing Addressing in which the data to be operated on is the byte immediately following the op code of the instruction.

implied addressing Addressing in which the location of the data is contained within the mnemonic.

label A name given to an instruction in an assembly-language program. To jump to this instruction, you can use the label rather than the address. The assembler will work out the correct address of the label and will use this address in the machine-language program.

mask A byte used with an ANI instruction to blank out certain bits.

register addressing Addressing in which the data is stored in a CPU register.

relocate To move a program or subroutine to another part of the memory. In doing this, the addresses of jump instructions must be converted to new addresses.

subroutine A program stored in higher memory that can be used repeatedly as part of a main program.

SELF-TESTING REVIEW

Read each of the following and provide the missing words. Answers appear at the beginning of the next question.

1. The controller-sequencer produces ________ words or microinstructions.
2. (*control*) A flag is a ________ that keeps track of a changing condition during a computer run. The sign flag is set when the accumulator contents go negative. The ________ flag is set when the accumulator contents go to zero.
3. (*flip-flop, zero*) In coding the LDA address and STA address instructions, the ________ byte of the address is stored in lower memory, the ________ byte in upper memory.
4. (*lower, upper*) The JMP instruction changes the program sequence by jumping to another part of the program. With the JM instruction, the jump is executed only if the sign flag is ________. With the JNZ instruction, the jump is executed only if the zero flag is ________.
5. (*set, clear*) Every subroutine must terminate with a ________ instruction. This returns the program to the instruction following the CALL. The CALL instruction is unconditional; it sends the computer to the starting address of a ________.
6. (*RET, subroutine*) An assembler allows you to write programs in mnemonic form. Then the assembler

converts these mnemonics into _______ language. The assembler ignores the _______ following a semicolon and assigns addresses to the labels. Labels can be up to six characters, the first of which must be a _______.

7. (*machine, comments, letter*) Repeated addition is one way to do _______. Programmed multiplication is used in most microprocessors because their ALUs can only add and subtract.
8. (*multiplication*) A parameter is a piece of data passed to a _______. When you call a subroutine, you often need to pass _______ for the subroutine to work properly.
9. (*subroutine, parameters*) A _______ is used to isolate a bit; it does this because the ANI sets all other bits to zero.
10. (*mask*) Handshaking is an interaction between a _______ and a peripheral device. In SAP-2 the _______ bit tells the CPU whether the input data is valid or not. After the data has been transferred into the computer, the CPU sends an _______ bit to the peripheral device.
11. (CPU, *READY, ACKNOWLEDGE*) The SAP-2 computer is microprogrammed with a _______ machine cycle. This means that some instructions take longer than others to execute.
12. (*variable*) The types of addressing covered up to now are direct, immediate, register, and implied.

PROBLEMS

11-1. Write a source program that loads the accumulator with decimal 100, the B register with decimal 150, and the C register with decimal 200.

11-2. Hand-assemble the source program of the preceding problem starting at address 2000H.

11-3. Write a source program that stores decimal 50 at memory location 4000H, decimal 51 at 4001H, and decimal 52 at 4002H.

11-4. Hand-assemble the source program in the preceding problem starting at address 2000H.

11-5. Write a source program that adds decimal 68 and 34, with the answer stored at memory location 5000H.

11-6. Hand-assemble the preceding program starting at address 2000H.

11-7. Here is a program:

Label	Mnemonic
LOOP:	MVI C,78H
	DCR C
	JNZ LOOP
	HLT

a. How many times (decimal) is the DCR C executed?
b. How many times does the program jump to LOOP?
c. How can you change the program to loop 210 times?

11-8. Which of the following are valid labels?
a. G100
b. UPDATE
c. 5TIMES
d. 678RED
e. T
f. REPEAT

11-9. Write a program that multiplies decimal 25 and 7 and stores the answer at 2000H. (Use the multiply subroutine located at F006H.)

11-10. Write a program that inputs a byte from port 1 and determines if the decimal equivalent is even or odd. If the byte is even, the program is to send an ASCII E to port 3; if odd, an ASCII O.

11-11. Modify the foregoing program so that it sends the answer in serial form to bit 0 of port 4.

11-12. Write a program that inputs a byte from port 1 using handshaking. Store the byte at address 4000H.

11-13. Hand assemble the foregoing program starting at address 2000H.

11-14. Write a subroutine that produces a time delay of approximately 500 μs.

11-15. Hand-assemble the preceding program starting at address 2000H.

11-16. Write a subroutine that produces a time delay of approximately 35 ms using a SAP-2 subroutine. Hand-assemble this subroutine and locate it at starting address E000H.

11-17. Write a subroutine that produces a time delay of 50 ms. (Use a SAP-2 subroutine.) Hand-assemble the program at starting address E100H.

11-18. Write a subroutine that produces a delay of 1 min. (Use CALL F060H.)

11-19. Hand-assemble the preceding subroutine at starting addresses F080H.

11-20. The C note one octave above middle C has a frequency of 523.25 Hz. Write a program that sends this note to bit 4 of port 4.

11-21. Hand-assemble the foregoing program starting at address 2000H.

SAP-3

The SAP-3 computer is an 8-bit microcomputer that is upward-compatible with the 8085 microprocessor. In this chapter, the emphasis is on the SAP-3 instruction set. This instruction set includes all the SAP-2 instructions of the preceding chapter plus new instructions to be discussed.

Appendix 6 shows the op codes, *T* states, flags, and so forth, for the SAP-3 instructions. In the remainder of this chapter, refer to Appendix 6 as needed.

12-1 PROGRAMMING MODEL

All you need to know about SAP-3 hardware is the programming model of Fig. 12-1. This is a diagram showing the CPU registers needed by a programmer.

Some of the CPU registers are familiar from SAP-2. For instance, the program counter (PC) is a 16-bit register that can count from 0000H to FFFFH or decimal 0 to 65,535. As you know, the program counter sends out the address of the next instruction to be fetched. This address is latched into the MAR.

CPU registers A, B, and C are the same as in SAP-2. These 8-bit registers are used in arithmetic and logic operations. Since the accumulator is only 8 bits wide, the range of unsigned numbers is 0 to 255; the range of signed 2's-complement numbers is −128 to +127.

SAP-3 has additional CPU registers (D, E, H, and L) for more efficient data processing. These 8-bit registers can be loaded with MOV and MVI instructions, the same as the A, B, and C registers. Also notice the F register, which stores flag bits S, Z, and others.

Finally, there is the *stack pointer* (SP), a 16-bit register. This new register controls a portion of memory known as the *stack*. The stack and the stack pointer are discussed later in this chapter.

Figure 12-1 shows all the CPU registers needed to understand the SAP-3 instruction set. With this programming model we can discuss the SAP-3 instruction set, which is upward-compatible with the 8080 and 8085. At the end of this chapter, you will know almost all of the 8080/8085 instruction set.

12-2 MOV AND MVI

The MOV and MVI instructions work the same as in SAP-2. The only difference is more registers to choose from. The format of any move instruction is

MOV reg1, reg2

where reg1 = A, B, C, D, E, H, or L
reg2 = A, B, C, D, E, H, or L

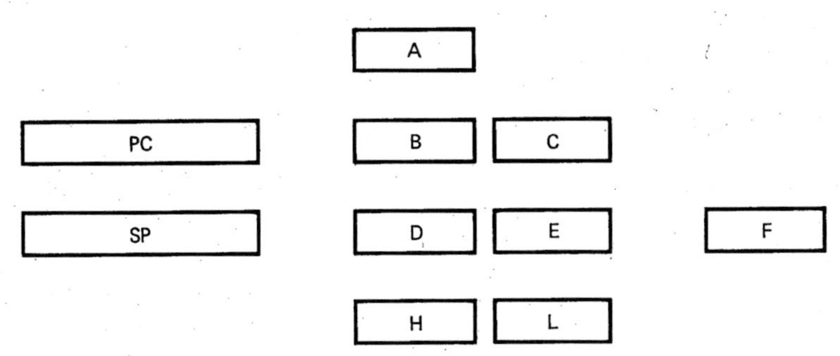

Fig. 12-1 SAP-3 programming model.

The MOV instructions send the data in reg2 to reg1. Symbolically,

$$\text{reg1} \leftarrow \text{reg2}$$

where the arrow indicates that the data in register 2 is copied nondestructively into register 1. At the end of the execution

$$\text{reg1} = \text{reg2}$$

For instance,

MOV L,A

copies A into L, so that

L = A

Similarly,

MOV E,H

gives

E = H

The immediate moves have the format of

MVI reg,byte

where reg = A, B, C, D, E, H, or L. Therefore, the execution of

MVI D,0EH

will result in

D = 0EH

Likewise,

MVI L,FFH

produces

L = FFH

What is the advantage of more CPU registers? As you may recall, MOV and MVI instructions use fewer *T* states than memory-reference instructions (MRIs). The extra CPU registers mean that we can use more MOV and MVI instructions and fewer MRIs. Because of this, SAP-3 programs can run faster than SAP-2 programs; furthermore, having more CPU registers for temporary storage simplifies program writing.

12-3 ARITHMETIC INSTRUCTIONS

Since the accumulator is only 8 bits wide, its contents can represent unsigned numbers from 0 to 255 or signed 2's complement numbers from −128 to +127. Whether signed or unsigned binary numbers are used, the programmer needs to detect *overflows*, sums or differences that lie outside the normal range of the accumulator. This is where the *carry* flag comes in.

Carry Flag

As shown in Fig. 6-7, a 4-bit adder-subtracter produces a sum $S_3S_2S_1S_0$ and a carry. In SAP-1, two 74LS83s (equivalent to eight full adders) produce an 8-bit sum and a carry. In this simple computer, the carry is disregarded. SAP-3, however, takes the carry into account.

Figure 12-2*a* shows the logic circuit used for the SAP-3 adder-subtracter. When *SUB* is low, the circuit adds the **A** and **B** inputs. If a final carry is generated, *CARRY* will be high and *CY* will be high. If there is no final carry, *CY* is low.

On the other hand, when *SUB* is high, the circuit forms the 2's complement of **B**, which is then added to **A.** Because of the final XOR gate, a high *CARRY* out of the last full-adder produces a low *CY*. If no carry occurs, *CY* is high.

In summary,

$$CY = \begin{cases} CARRY & \text{for ADD instructions} \\ \overline{CARRY} & \text{for SUB instructions} \end{cases}$$

During an add operation, *CY* is called a carry. During a subtract operation, *CY* is referred to as a *borrow*.

The 8-bit sum $S_7S_6S_5S_4S_3S_2S_1S_0$ is stored in the accumulator of Fig. 12-2*b*. The carry (or borrow) is stored in a special flip-flop called the *carry flag*, designated CY in Fig. 12-2*b*. This flag acts like the next higher bit of the accumulator. That is,

$$CY = A_8$$

Carry-Flag Instructions

There are two instructions we can use to control the carry flag. The STC instruction will set the CY flag if it is not already set. (STC stands for *set carry*.) So, if

$$CY = 0$$

the execution of a STC instruction produces

$$CY = 1$$

The other carry-flag instruction is the CMC, which stands for *complement the carry*. When executed, a CMC com-

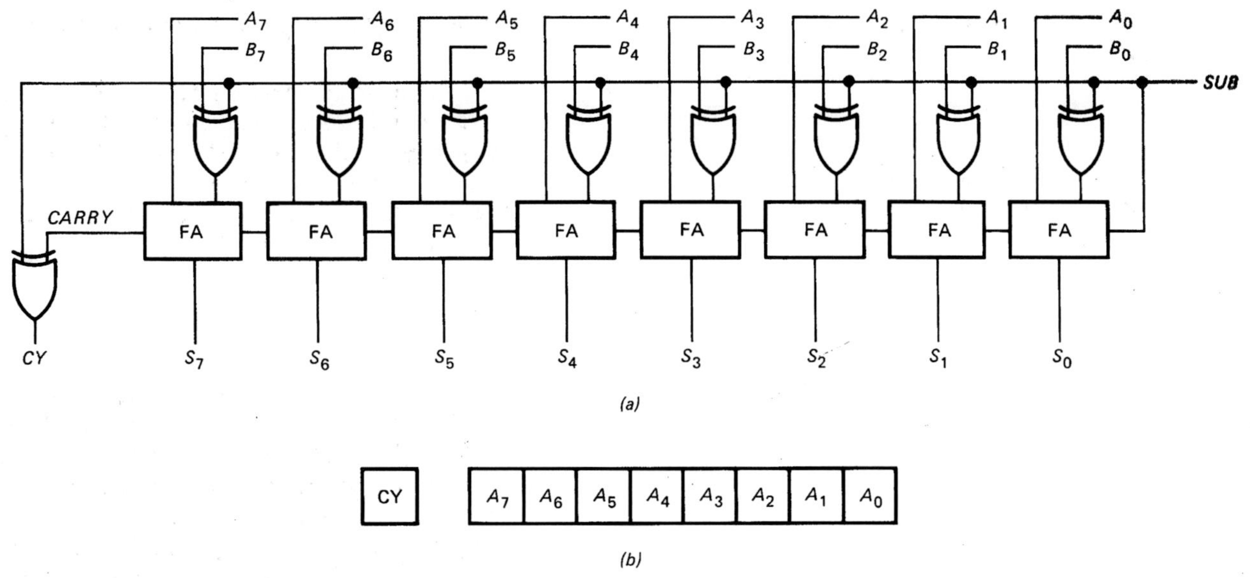

Fig. 12-2 (*a*) SAP-3 adder-subtractor (*b*) carry flag and accumulator.

plements the value of CY. If $CY = 1$, CMC produces a CY of 0. On the other hand, if $CY = 0$, CMC results in a CY of 1.

If you want to reset the carry flag and its current status is unknown, you have to set it, then complement it. That is, execution of

STC
CMC

guarantees that the final value of CY will be 0 if the initial value of CY is unknown.

ADD Instructions

The format of the ADD instruction is

ADD reg

where reg = A, B, C, D, E, H, or L. This instruction adds the contents of the specified register to the accumulator contents. The sum is stored in the accumulator and the carry flag is set or reset, depending on whether there is a final carry or not.

For instance, suppose

A = 1111 0001 and **E** = 0000 1000

The instruction

ADD E

produces the binary addition

```
  1111 0001
+ 0000 1000
-----------
  1111 1001
```

There is no final carry; therefore, at the end of the instruction cycle,

$CY = 0$ and **A** = 1111 1001

As another example, suppose

A = 1111 1111 and **L** = 0000 0001

Then executing an ADD L produces

```
  1111 1111
+ 0000 0001
-----------
1 0000 0000
```

At the end of the instruction cycle

$CY = 1$ and **A** = 0000 0000

ADC Instructions

The ADC instruction (add with carry) is formatted like this:

ADC reg

where reg = A, B, C, D, E, H, or L. This instruction adds the contents of the specified register plus the carry flag to the contents of the accumulator. Because it includes the CY flag, the ADC instruction allows us to add numbers outside the unsigned 0 to 255 range or the signed −128 to +127 range.

As an example, suppose

A = 1000 0011
E = 0001 0010

and $CY = 1$

The execution of

ADC E

produces the following addition:

```
  1000 0011
  0001 0010
+         1
  ---------
  1001 0110
```

Therefore, the new accumulator and carry flag contents are

$CY = 0$ A = 1001 0110

SUB Instructions

The SUB instruction is formatted as

SUB reg

where reg = A, B, C, D, E, H, or L. This instruction will subtract the contents of the specified register from the accumulator contents; the result is stored in the accumulator. If a final borrow occurs, the CY flag is set. If there is no borrow, the CY flag is reset. In other words, during subtraction the CY flag functions as a borrow flag.

For example, if

A = 0000 1111 and C = 0000 0001

then

SUB C

results in

```
  0000 1111
− 0000 0001
  ---------
  0000 1110
```

Notice that there is no final borrow. In terms of 2's-complement addition, the foregoing subtraction appears like this:

```
   0000 1111
+  1111 1111
  ----------
 1 0000 1110
```

The final *CARRY* is 1, but this is complemented during subtraction to get a *CY* of 0 (Fig. 12-2*a*). This is why the execution of SUB C produces

$CY = 0$ A = 0000 1110

Here is another example. If

A = 0000 1100 and C = 0001 0010

then a SUB C produces

```
   0000 1100
−  0001 0010
  ----------
 1 1111 1010
```

Notice the final borrow. This borrow occurs because the contents of the C register (decimal 18) are greater than the contents of the accumulator (decimal 12). In terms of 2's-complement arithmetic, the foregoing looks like

```
   0000 1100
+  1110 1110
  ----------
 0 1111 1010
```

In this case, *CARRY* is 0 and *CY* is 1. The final register and flag contents are

$CY = 1$ and A = 1111 1010

SBB Instructions

SBB stands for *subtract with borrow*. This instruction goes one step further than the SUB. It subtracts the contents of a specified register and the CY flag from the accumulator contents. If

A = 1111 1111
E = 0000 0010

and $CY = 1$

the instruction SBB E starts by combining **E** and *CY* to get 0000 0011 and then subtracts this from the accumulator as follows:

```
  1111 1111
− 0000 0011
  ---------
  1111 1100
```

The final contents are

$$CY = 0 \quad \text{and} \quad A = 1111\ 1100$$

EXAMPLE 12-1

In unsigned binary, 8 bits can represent 0 to 255, whereas 16 bits can represent 0 to 65,535. Show a SAP-3 program that adds 700 and 900, with the final answer stored in the H and L registers.

SOLUTION

Double bytes can represent decimal 700 and 900 as follows:

$$700_{10} = \text{02BCH} = 0000\ 0010\ 1011\ 1100_2$$
$$900_{10} = \text{0384H} = 0000\ 0011\ 1000\ 0100_2$$

Here is how to add 700 and 900:

Label	Instruction	Comment
	MVI A,00H	;Clear the accumulator
	MVI B,02H	;Store upper byte (UB) of 700
	MVI C,BCH	;Store lower byte (LB) of 700
	MVI D,03H	;Store UB of 900
	MVI E,84H	;Store LB of 900
	ADD C	;Add LB of 700
	ADD E	;Add LB of 900
	MOV L,A	;Store partial sum
	MVI A,00H	;Clear the accumulator
	ADC B	;Add UB of 700 with carry
	ADD D	;Add UB of 900
	MOV H,A	;Store partial sum
	HLT	;Stop

The first five instructions initialize registers A through E. The ADD C and ADD E add the lower bytes BCH and 84H; this addition sets the carry flag because

$$\begin{array}{r} \text{BCH} = \quad 1011\ 1100_2 \\ +\ \text{84H} = \quad 1000\ 0100_2 \\ \hline 1\ \text{40H} = 1\ 0100\ 0000_2 \end{array}$$

The sum is stored in the L register and the final carry in the CY flag.

Next, the accumulator is cleared. The ADC B adds the upper byte plus the carry flag to get

$$\begin{array}{r} \text{00H} = 0000\ 0000_2 \\ +\ \text{02H} = 0000\ 0010_2 \\ +\ \text{1H} = \qquad\quad 1_2 \\ \hline \text{03H} = 0000\ 0011_2 \end{array}$$

Then the ADD D produces

$$\begin{array}{r} \text{03H} = 0000\ 0011_2 \\ +\ \text{03H} = 0000\ 0011_2 \\ \hline \text{06H} = 0000\ 0110_2 \end{array}$$

The MOV H,A stores this upper sum in the H register.

So the program ends with the answer stored in the H and L registers as follows:

$$H = \text{06H} = 0000\ 0110_2$$

and

$$L = \text{40H} = 0100\ 0000_2$$

The complete answer is 0640H, which is equivalent to decimal 1,600.

12-4 INCREMENTS, DECREMENTS, AND ROTATES

This section is about increment, decrement, and rotate instructions. The increment and decrement are similar to those of SAP-2, but the rotates are different because of the carry flag.

Increment

The increment instruction appears as

INR reg

where reg = A, B, C, D, E, H, or L. It works as previously described. Therefore, given

$$L = 0000\ 1111$$

the execution of INR L produces

$$L = 0001\ 0000$$

The INR instruction has no effect on the carry flag, but, as before, it does affect the sign and zero flags. For instance, if

$$B = 1111\ 1111$$

and the initial flags are

$$S = 1 \quad Z = 0 \quad CY = 0$$

then INR B produces

$$B = 0000\ 0000$$
$$S = 0 \quad Z = 1 \quad CY = 0$$

As you see, the carry flag is unaffected even though the B register overflowed. At the same time, the zero flag has been set and the sign flag reset.

Decrement

The decrement is similar. It looks like

DCR reg

where reg = A, B, C, D, E, H, or L. If

E = 0111 0110

then a DCR E produces

E = 0111 0101

The DCR affects the sign and zero flags but not the carry flag. This is why the initial values may be

E = 0000 0000

$S = 0 \quad Z = 1 \quad CY = 0$

Executing a DCR E results in

E = 1111 1111

$S = 1 \quad Z = 0 \quad CY = 0$

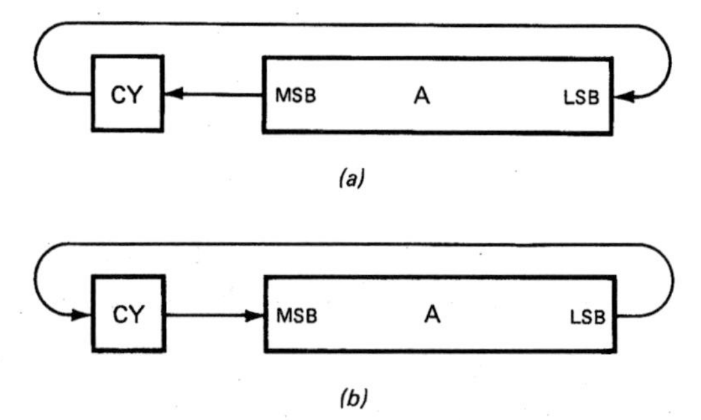

Fig. 12-3 (*a*) RAL; (*b*) RAR.

Rotate All Left

Figure 12-3*a* illustrates the RAL instruction used in SAP-3. The CY flag is included in the rotation of bits. RAL stands for rotate all left, which is a reminder that all bits including the CY flag are rotated to the left.

If the initial values are

$CY = 1$ **A** = 0111 0100

then executing a RAL instruction produces

$CY = 0$ **A** = 1110 1001

As you see, the original CY goes to the LSB position, and the original MSB goes to the CY flag.

Rotate All Right

The rotate-all-right instruction (RAR) rotates all bits including the CY flag to the right, as shown in Fig. 12-3*b*. If

$CY = 1$ **A** = 0111 0100

an RAR will result in

$CY = 0$ **A** = 1011 1010

This time, the original *CY* goes to the MSB position, and the original LSB goes into the CY flag.

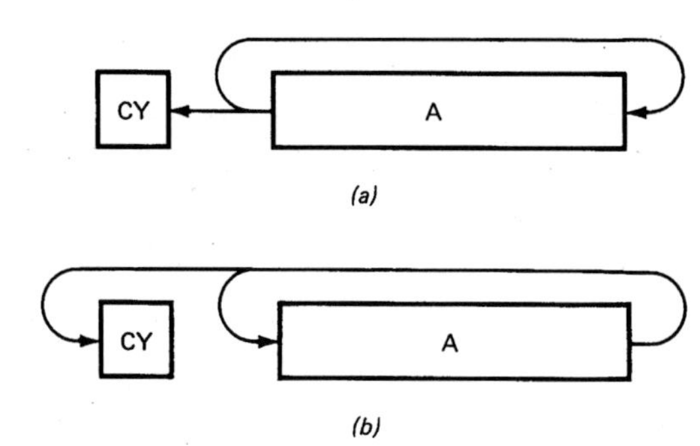

Fig. 12-4 (*a*) RLC; (*b*) RRC.

Rotate Left with Carry

Sometimes you don't want to treat the CY flag as an extension of the accumulator. In other words, you may not want to rotate all bits. Figure 12-4*a* illustrates the RLC instruction. The accumulator bits are rotated left, and the MSB is saved in the CY flag. For instance, given

$CY = 1$ **A** = 0111 0100

executing an RLC produces

$CY = 0$ **A** = 1110 1000

Rotate Right with Carry

Figure 12-4*b* shows how the RRC instruction rotates the bits. In this case, the accumulator bits are rotated right and the LSB is saved in the CY flag. So, given

$CY = 1$ **A** = 0111 0100

an RRC will result in

$CY = 0$ **A** = 0011 1010

Multiply and Divide by 2

Example 11-14 showed a program where the RAR instruction was used in converting from parallel to serial data. Parallel-to-serial conversion, and vice versa, is one of the main uses of rotate instructions.

There is another use for rotate instructions. Rotating has the effect of multiplying or dividing the accumulator contents by a factor of 2. Specifically, with the carry flag reset, an RAL has the effect of multiplying by 2, while the RAR divides by 2. This can be proved algebraically, but it's much easier to examine a few specific examples to see how it works.

Suppose

$$CY = 0 \qquad \mathbf{A} = 0000\,0111$$

Then an RAL produces

$$CY = 0 \qquad \mathbf{A} = 0000\,1110$$

The accumulator contents have changed from decimal 7 to decimal 14. The RAL has multiplied by 2.

Likewise, if

$$CY = 0 \qquad \mathbf{A} = 0010\,0001$$

then an RAL results in

$$CY = 0 \qquad \mathbf{A} = 0100\,0010$$

In this case, **A** has changed from decimal 33 to 66.

RAR instructions have the opposite effect; they divide by 2. If

$$CY = 0 \qquad \mathbf{A} = 0001\,1000$$

an RAR gives

$$CY = 0 \qquad \mathbf{A} = 0000\,1100$$

The decimal contents of the accumulator have changed from decimal 24 to 12.

Remember the basic idea. RAL instructions have the effect of multiplying by 2; RAR instructions divide by 2.

12-5 LOGIC INSTRUCTIONS

The SAP-3 logic instructions are almost the same as in SAP-2. For instance, three of the logic instructions are

ANA reg
ORA reg
XRA reg

where reg = A, B, C, D, E, H, or L. These instructions will AND, OR, or XOR the contents of the specified register with the contents of the accumulator on a bit-by-bit basis.

The only new logic instruction is the CMP, formatted as

CMP reg

where reg = A, B, C, D, E, H, or L. CMP compares the contents of the specified register with the contents of the accumulator. The zero flag indicates the outcome of this comparison as follows:

$$Z = \begin{cases} 1 & \text{if A = reg} \\ 0 & \text{if A} \neq \text{reg} \end{cases}$$

SAP-3 carries out a CMP as follows. The contents of the accumulator are copied in a temporary register. Then the contents of the specified register are subtracted from the contents of the temporary register. Since the ALU does the subtraction, the zero flag is affected. If the 2 bytes being compared are equal, the zero flag is set. If the bytes are unequal, the zero flag is reset. Because the temporary register is used, the accumulator contents are not changed by a CMP instruction.

For example, if

$$\mathbf{A} = \text{F8H}$$
$$\mathbf{D} = \text{F8H}$$

and $Z = 0$

executing a CMP D results in

$$\mathbf{A} = \text{F8H}$$
$$\mathbf{D} = \text{F8H}$$

and $Z = 1$

CMP has no effect on **A** and **D**; only the flag changes to indicate that **A** and **D** are equal. (If they were not equal, Z would be 0.)

CMP is a powerful instruction because it allows us to compare the accumulator contents with the data in a specified register. By following a CMP with a conditional zero jump, we can control loops in a new way. Later programs will show how this is done.

12-6 ARITHMETIC AND LOGIC IMMEDIATES

So far, we have introduced these arithmetic and logic instructions: ADD, ADC, SUB, SBB, ANA, ORA, XRA, and CMP. Each of these has the accumulator as an implied register; the data comes from a specified register (A, B, C, D, E, H, or L).

The immediate instructions from SAP-2 that carry over to SAP-3 are ANI, ORI, and XRI. As you know, each of these has the format of

ANI byte
ORI byte
XRI byte

where the immediate byte is ANDed, ORed, or XORed with the accumulator byte.

Besides the foregoing, SAP-3 has these immediate instructions:

ADI byte
ACI byte
SUI byte
SBI byte
CPI byte

The ADI adds the immediate byte to the accumulator byte. The ACI adds the immediate byte plus the CY flag to the accumulator byte. The SUI subtracts the immediate byte from the accumulator byte. The SBI subtracts immediate byte and the CY flag from the accumulator byte. The CPI compares the immediate byte with the accumulator byte; if the bytes are equal, the zero flag is set; if not, it is reset.

EXAMPLE 12-2

Show a program that subtracts 700 from 900 and stores the answer in the H and L registers.

SOLUTION

We need double bytes to represent 900 and 700 as follows:

$$900_{10} = 0384\text{H} = 0000\ 0011\ 1000\ 0100_2$$
$$700_{10} = 02\text{BCH} = 0000\ 0010\ 1011\ 1100_2$$

Here's the program for subtracting 700 from 900:

Label	Instruction	Comment
	MVI A, 84H	;Load LB of 900
	SUI BCH	;Subtract LB of 700
	MOV L,A	;Save lower half answer
	MVI A, 03H	;Load UB of 900
	SBI 02H	;Subtract UB of 700 with borrow
	MOV H,A	;Save upper half answer

The first two instructions subtract the lower bytes as follows:

```
  1000 0100
- 1011 1100
-----------
1 1100 1000
```

At this point,

$$CY = 1 \qquad \text{A} = \text{C8H}$$

The high CY flag indicates a borrow.

After saving C8H in the L register, the program loads the upper byte of 900 into the accumulator. The SBI is used instead of a SUI because of the borrow that occurred when subtracting the bytes. The execution of the SBI gives

```
  0000 0011
- 0000 0010
-         1
-----------
  0000 0000
```

This part of the answer is stored in the H register, so that the final contents are

$$\text{H} = 00\text{H} = 0000\ 0000_2$$
$$\text{L} = \text{C8H} = 1100\ 1000_2$$

12-7 JUMP INSTRUCTIONS

Here are the SAP-2 jump instructions that become part of the SAP-3 instruction set:

JMP address	(Unconditional jump)
JM address	(Jump if minus)
JZ address	(Jump if zero)
JNZ address	(Jump if not zero)

Here are some more SAP-3 jump instructions.

JP

JM stands for *jump if minus*. When the program encounters a JM address, it will jump to the specified address if the sign flag is set.

The JP instruction has the opposite effect. JP stands for *jump if positive* (including zero). This means that

JP address

produces a jump to the specified address if the sign flag is reset.

JC and JNC

The instruction

JC address

means to jump to the specified address if the carry flag is set. In short, JC stands for jump if carry. Similarly,

JNC address

means to jump to the specified address if the carry flag is not set. That is, jump if no carry.

Here is a program segment to illustrate JC and JNC:

Label	Instruction	Comment
	MVI A,FEH	
REPEAT:	ADI 01H	
	JNC REPEAT	
	MVI A,C4H	
	JC ESCAPE	
	.	
	.	
	.	
ESCAPE:	MOV L,A	

The MVI loads the accumulator with FEH. The ADI adds 1 to get FFH. Since no carry takes place, the JNC takes the program back to the REPEAT point, where a second ADI is executed. This time the accumulator overflows to get contents of 00H with a carry. Since the CY flag is set, the program falls through the JNC. The accumulator is loaded with C4H. Then the JC produces a jump to the ESCAPE point, where the C4H is loaded into the L register.

JPE and JPO

Besides the sign, zero, and carry flag, SAP-3 has a *parity flag* designated P. During the execution of certain instructions (like ADD, INR, etc.), the ALU result is checked for parity. If the result has an even number of 1s, the parity flag is set; if an odd number of 1s, the flag is reset.

The instruction

JPE address

produces a jump to the specified address when the parity flag is set (even parity). On the other hand,

JPO address

results in a jump when the parity flag is reset (odd parity). For instance, given these flags,

$$S = 1 \quad Z = 0 \quad CY = 0 \quad P = 1$$

the program would jump if it encountered a JPE instruction; but it would fall through a JPO instruction.

Incidentally, we now have discussed all the flags in the SAP-3 computer. For upward compatibility with the 8085 microprocessor, these flags are stored in the F register, as shown in Fig. 12-5. For instance, if the contents of the F register are

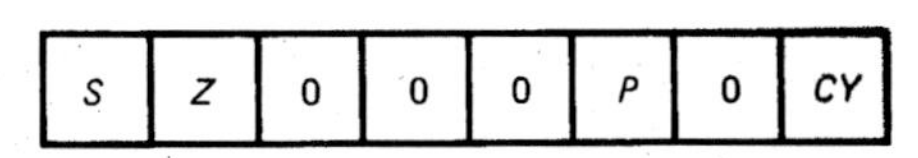

Fig. 12-5 F register stores flags.

$$\mathbf{F} = 0100\ 0101$$

then we know that the flags are

$$S = 0 \quad Z = 1 \quad P = 1 \quad CY = 1$$

EXAMPLE 12-3

What does the following program segment do?

SOLUTION

Label	Instruction	Comment
	MVI E,00H	;Initialize counter
LOOP:	INR E	;Increment counter
	MOV A,E	;Load A with E
	CPI FFH	;Compare to 255
	JNZ LOOP	;Go back if not 255

The E register is being used as a counter. It starts at 0. The first time the INR and MOV are executed

$$\mathbf{A} = 01H$$

After executing the CPI, the zero flag is 0 because 01H and FFH are unequal. The JNZ then forces the program to return to the LOOP point.

The looping will continue until the INR and MOV have been executed 255 times to get

$$\mathbf{A} = FFH$$

On this pass through the loop, the CPI sets the zero flag because the accumulator byte and the immediate byte are equal. With the zero flag set for the first time, the program falls through the JNZ instruction.

Do you see the point? The computer will loop 255 times before it falls through the JNZ. One use of this program segment is to set up a time delay. Another use is to insert additional instructions inside the loop as follows:

Label	Instruction	Comment
	MVI E,00H	
LOOP:	.	
	.	
	.	
	INR E	
	MOV A,E	
	CPI FFH	
	JNZ LOOP	

The instructions at the beginning of the loop (symbolized by dots) will be executed 255 times. If you want to change the number of passes through the loop, modify the CPI instruction as required.

12-8 EXTENDED-REGISTER INSTRUCTIONS

Some SAP-3 instructions use pairs of CPU registers to process 16-bit data. In other words, during the execution of certain instructions, the CPU registers are cascaded, as shown in Fig. 12-6. The pairing is always as shown: B with C, D with E, and H with L. What follows are the SAP-3 instructions that use *register pairs*. Throughout these instructions, you will notice the letter X, which stands for extended register, a reminder that register pairs are involved.

B	C
D	E
H	L

Fig. 12-6 Register pairs.

Load Extended Immediate

Since there are three register pairs (BC, DE, and HL), the LXI instruction can appear in any of these forms:

LXI B,dble
LXI D,dble
LXI H,dble

where B stands for BC
D stands for DE
H stands for HL
dble stands for double byte

The LXI instruction says to load the specified register pair with the double byte. For instance, if we execute

LXI B,90FFH

the B and C registers are loaded with the upper and lower bytes to get

B = 90H
C = FFH

Visualizing **B** and **C** paired off as shown in Fig. 12-6, we can write

BC = 90FFH

DAD Instructions

DAD stands for double-add. This instruction has three forms:

DAD B
DAD D
DAD H

where B stands for BC
D stands for DE
H stands for HL

The DAD instruction adds the contents of the specified register pair to the contents of the HL register pair; the result is then stored in the HL register pair. For instance, given

BC = F521H
HL = 0003H

the execution of a DAD B produces

HL = F524H

As you see, F521H and 0003H are added to get F524H. The result is stored in the HL register pair.

The DAD instruction affects the CY flag. If there is a carry out of the HL register pair, the CY flag is set; otherwise it is reset. As an example, if

DE = 0001H
HL = FFFFH

a DAD D will result in

HL = 0000H
CY = 1

Incidentally, a DAD H has the effect of adding the data in the HL register pair to itself. In other words, a DAD H doubles the value of HL. If

HL = 1234H

a DAD H results in

$$\textbf{HL} = 2468\text{H}$$

INX and DCX

INX stands for *increment the extended register,* and DCX means *decrement the extended register.* The extended increment instructions are

INX B
INX D
INX H

where B stands for BC
D stands for DE
H stands for HL

The DCX instructions have a similar format: DCX B, DCX D, and DCX H.

The INX and DCX instructions have no effect on the flags. For instance, if

$$\textbf{BC} = \text{FFFFH}$$
$$S = 1$$
$$Z = 0$$
$$P = 1$$
$$CY = 0$$

executing an INX B results in

$$\textbf{BC} = 0000\text{H}$$
$$S = 1$$
$$Z = 0$$
$$P = 1$$
$$CY = 0$$

Notice that all flags are unaffected.

In summary, the extended register instructions are LXI, DAD, INX, and DCX. Of the three register pairs, the HL combination is special. The next section tells you why.

12-9 INDIRECT INSTRUCTIONS

As discussed in Chap. 10, the program counter is an *instruction pointer;* it points to the memory location where the next instruction is stored.

The HL register pair is different; it points to memory locations where data is stored. In other words, SAP-3 has several instructions where the HL register pair acts like a *data pointer*. The following discussion clarifies the idea.

Visualizing the HL Pointer

Figure 12-7*a* shows a 64K memory; it has 65,636 memory registers or memory locations where data is stored. The first memory location is M_{0000H}, the next is M_{0001H}, and so on. The memory location with address HL is M_{HL}.

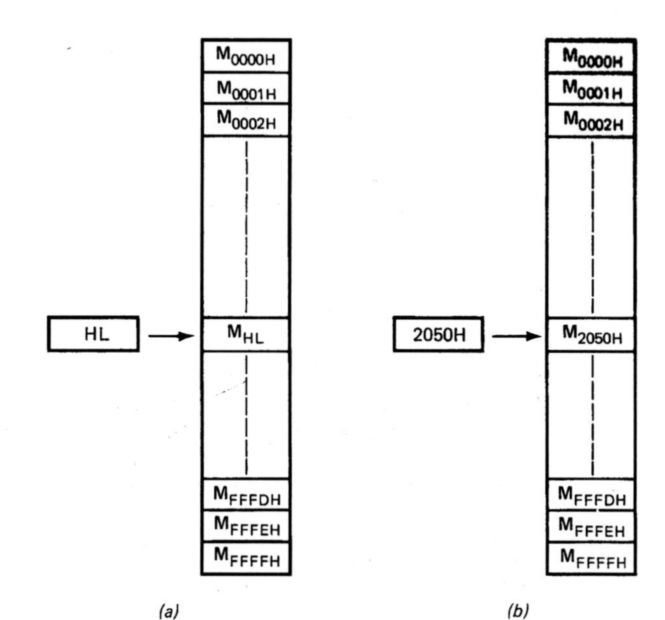

Fig. 12-7 (*a*) HL pointer; (*b*) pointing to 2050H.

With some SAP-3 instructions, the contents of the HL register pair are used as the address for data in memory. That is, the contents of the HL register pair are sent to the MAR, and then a memory read or write is performed. It's as though the HL register pair were pointing to the desired memory location, as shown in Fig. 12-7*a*.

For instance, suppose

$$\textbf{HL} = 2050\text{H}$$

If HL is acting as a pointer, its contents (2050H) are sent to the MAR during one T state. During the next T state, the memory location whose address is 2050H undergoes a read or write operation. As shown in Fig. 12-7*b* the HL register pair points to the desired memory location.

Indirect Addressing

With direct addressing like LDA 5000H and STA 6000H, the programmer knows the address of the memory location because the instruction itself directly gives the address. With instructions that use the HL pointer, however, programmers do not know the address; all they know is that the address is stored in the HL register pair. Whenever an instruction uses the HL pointer, the addressing is called *indirect addressing*.

Indirect Read

One of the indirect instructions is

MOV reg,M

where reg = A, B, C, D, E, H, or L
M = M_{HL}

This instruction says to load the specified register with the data addressed by HL. After execution of this instruction, the designated register contains M_{HL}.

For instance, if

HL = 3000H and $\mathbf{M}_{3000H}$ = 87H

executing a

MOV C,M

produces

C = 87H

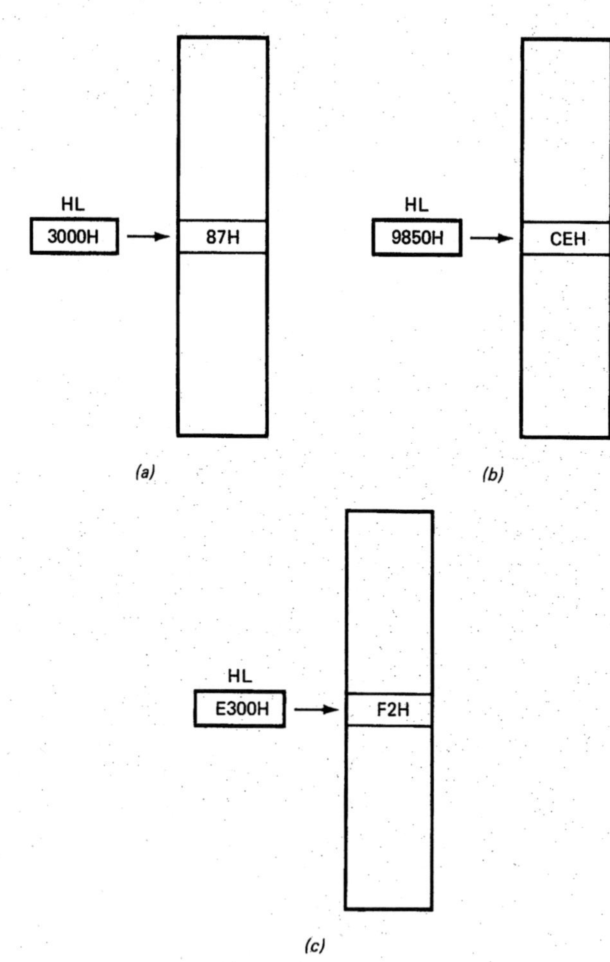

Fig. 12-8 Examples of indirect addressing.

Figure 12-8*a* shows how to visualize the MOV C,M. The HL pointer points to 87H, which is the data to be read into register C.

As another example, if

HL = 9850H and $\mathbf{M}_{9850H}$ = CEH

a MOV A,M results in

A = CEH

Figure 12-8*b* illustrates the MOV A,M. The HL pointer points to CEH, which is the data to be loaded into the A register.

Indirect Write

Here is another indirect MOV instruction:

MOV M,reg

where M = M_{HL}
reg = A, B, C, D, E, H, or L

This says to load the memory location addressed by HL with the contents of the specified register. After execution of this instruction,

$\mathbf{M}_{HL}$ = reg

As an example, if

HL = E300H
B = F2H

the execution of a MOV M,B produces

$\mathbf{M}_{E300H}$ = F2H

Figure 12-8*c* illustrates the idea.

Indirect-Immediate Instructions

Sometimes we want to write immediate data into the memory location addressed by the HL pointer. The instruction to use in this case is

MVI M,byte

Here is an example. If **HL** = 3000H, executing a

MVI M,87H

produces

$\mathbf{M}_{3000H}$ = 87H

Other Pointer Instructions

Here are more instructions using the HL pointer:

```
ADD M
ADC M
SUB M
SBB M
INR M
DCR M
ANA M
ORA M
XRA M
CMP M
```

In each of these, M is the memory location addressed by HL. Think of M as another register where data is stored. Each of the foregoing instructions operates on this data as previously described.

EXAMPLE 12-4

Suppose 256 bytes of data are stored in memory between addresses 2000H and 20FFH. Show a program that will copy these 256 bytes at addresses 3000H to 30FFH.

SOLUTION

Label	Instruction	Comment
	LXI H,1FFFH	;Initialize pointer
LOOP:	INX H	;Advance pointer
	MOV B,M	;Read byte
	MOV A,H	;Load 20H into accumulator
	ADI 10H	;Add offset to get 30H
	MOV H,A	;Offset pointer
	MOV M,B	;Write byte in new location
	SUI 10H	;Subtract offset
	MOV H,A	;Restore H for next read
	MOV A,L	;Prepare for compare
	CPI FFH	;Check for 255
	JNZ LOOP	;If not done, get next byte
	HLT	;Stop

This looping program transfers each successive byte in the 2000H–20FFH area of memory into the 3000H–30FFH area of memory. Here are the details.

The LXI initializes the pointer with address 1FFFH. The first time into the loop, the INX will advance the HL pointer to 2000H. The MOV B,M then reads the first byte into the B register. The next three instructions

```
MOV A,H
ADI 10H
MOV H,A
```

offset the HL pointer to 3000H. Then the MOV M,B writes the first byte into location 3000H. The next two instructions, SUI and MOV, restore the HL pointer to 2000H. The MOV A,L puts 00H into the accumulator. Because the CPI FFH resets the zero flag, the JNZ forces the program to return to the LOOP entry point.

On the second pass through the loop, the computer will read the byte at 2001H and it will store this byte at 3001H. The looping will continue with successive bytes being moved from the 2000H–20FFH section of memory to the 3000H–30FFH area. Since the first byte is read from 2000H, the 256th byte is read from 20FFH. After this byte is stored at 30FFH, the pointer is restored to 20FFH. The MOV A,L then loads the accumulator to get

$$A = FFH$$

This time, the CPI FFH will set the zero flag. Therefore, the program will fall through the JNZ to the HLT.

12-10 STACK INSTRUCTIONS

SAP-2 has a CALL instruction that sends the program to a subroutine. As you recall, before the jump takes place, the program counter is incremented and the address is saved at addresses FFFEH and FFFFH. The addresses FFFEH and FFFFH are set aside for the purpose of saving the return address. At the completion of a subroutine, the RET instruction loads the program counter with the return address, which allows the computer to get back to the main program.

The Stack

A *stack* is a portion of memory set aside primarily for saving return addresses. SAP-2 has a stack because addresses FFFEH and FFFFH are used exclusively for saving the return address of a subroutine call. Figure 12-9*a* shows how to visualize the SAP-2 stack.

SAP-3 is different. To begin with, the programmer decides where to locate the stack and how large to make it. As an example, Fig. 12-9*b* shows a stack between addresses 20E0H and 20FFH. This stack contains 32 memory locations for saving return addresses. Programmers can locate the stack anywhere they want in memory, but once they have set up the stack, they no longer use that portion of memory for program and data. Instead, the stack becomes a special space in memory, used for storing the return addresses of subroutine calls.

Stack Pointer

The instructions that read and write into the stack are called *stack instructions;* these include PUSH, POP, CALL, and

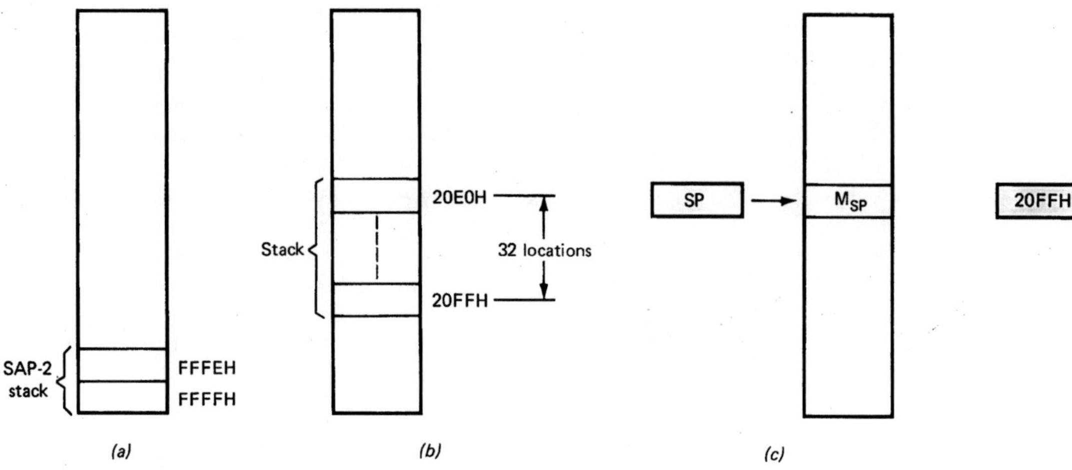

Fig. 12-9 (*a*) SAP-2 stack; (*b*) example of a stack; (*c*) stack pointer addresses the stack; (*d*) SP points to 20FFH.

others to be discussed. Stack instructions use indirect addressing because a 16-bit register called the *stack pointer* (SP) holds the address of the desired memory location. As shown in Fig. 12-9*c*, the stack pointer is similar to the HL pointer because the contents of the stack pointer indicate which memory location is to be accessed. For instance, if

SP = 20FFH

the stack pointer points to memory location M_{20FFH} (see Fig. 12-9*d*). Depending on the stack instruction, a byte is then read from, or written into, this memory location.

To initialize the stack pointer, we can use the immediate load instruction

LXI SP,dble

For instance, if we execute

LXI SP,20FFH

the stack pointer is loaded with 20FFH.

PUSH Instructions

The contents of the accumulator and the flag register are known as the *program status word* (PSW). The format for this word is

PSW = **AF**

where **A** = contents of accumulator
F = contents of flag register

The accumulator contents are the high byte, and the flag contents the low byte. When calling subroutines, we usually have to save the program status word, so that the main program can resume after the subroutine is executed. We may also have to save the contents of the other registers.

PUSH instructions allow us to save data in a stack. Here are the four PUSH instructions:

PUSH B
PUSH D
PUSH H
PUSH PSW

where B stands for BC
D stands for DE
H stands for HL
PSW stands for program status word

When a PUSH instruction is executed, the following things happen:

1. The stack pointer is decremented to get a new value of SP − 1.
2. The high byte in the specified register pair is stored in M_{SP-1}.
3. The stack pointer is decremented again to get SP − 2.
4. The low byte in the specified register pair is stored in M_{SP-2}.

Here is an example. Suppose

BC = 5612H
SP = 2100H

When a PUSH B is executed,

1. The stack pointer is decremented to get 20FFH.
2. The high byte 56H is stored at 20FFH (Fig. 12-10*a*).
3. The stack pointer is again decremented to get 20FEH.
4. The low byte 12H is stored at 20FEH (Fig. 12-10*b*).

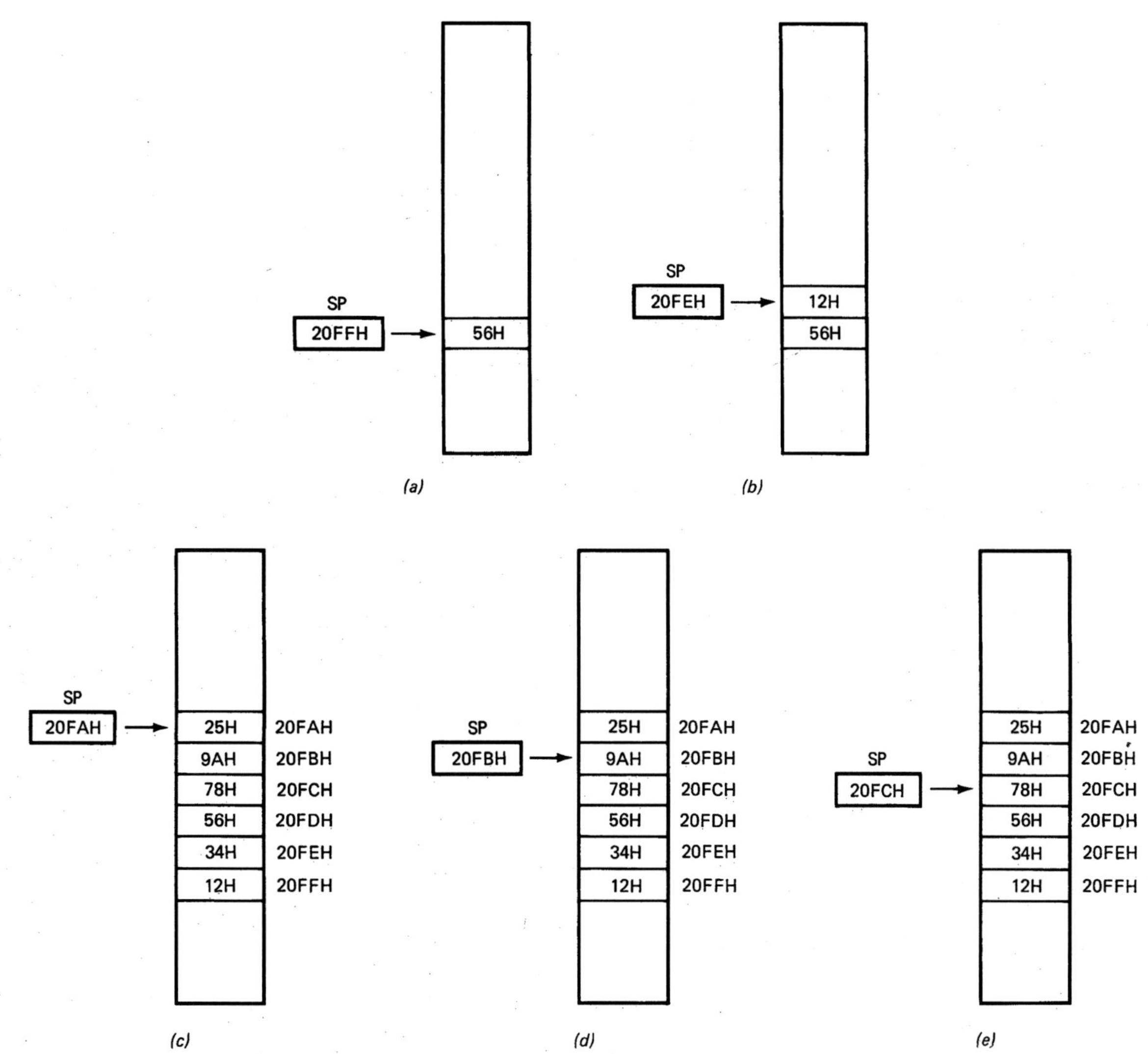

Fig. 12-10 Push operations: (*a*) high byte first; (*b*) low byte second; (*c*) 6 bytes pushed on stack; (*d*) popping a byte off the stack; (*e*) incrementing stack pointer.

Here's another example. Suppose

SP = 2100H
AF = 1234H
DE = 5678H
HL = 9A25H

then executing

```
PUSH PSW
PUSH D
PUSH H
```

loads the stack as shown in Fig. 12-10*c*. The first PUSH stores 12H at 20FFH and 34H at 20FEH. The next PUSH stores 56H at 20FDH and 78H at 20FCH. The last PUSH stores 9AH at 20FBH and 25H at 20FAH. Notice how the stack builds. Each new PUSH shoves data onto the stack.

POP Instructions

Here are four POP instructions:

```
POP B
POP D
POP H
POP PSW
```

where B stands for BC
D stands for DE
H stands for HL
PSW stands for program status word

When a POP is executed, the following happens:

1. The low byte is read from the memory location addressed by the stack pointer. This byte is stored in the lower half of the specified register pair.
2. The stack pointer is incremented.
3. The high byte is read and stored in the upper half of the specified register pair.
4. The stack pointer is incremented.

Here's an example. Suppose the stack is loaded as shown in Fig. 12-10*c* with the stack pointer at 20FAH. Then execution of POP B does the following:

1. Byte 25H is read from 20FAH (Fig. 12-10*c*) and stored in the C register.
2. The stack pointer is incremented to get 20FBH. Byte 9AH is read from 20FBH (Fig. 12-10*d*) and stored in the B register. The BC register pair now contains

BC = 9A25H

3. The stack pointer is incremented to get 20FCH (Fig. 12-10*e*).

Each time we execute a POP, 2 bytes come off the stack. If we were to execute a POP PSW and a POP H in Fig. 12-10*e*, the final register contents would be

AF = 5678H
HL = 1234H

and the stack pointer would contain

SP = 2100H

CALL and RET

The main purpose of the SAP-3 stack is to save return addresses automatically when using CALLs. When a

CALL address

is executed, the contents of the program counter are pushed onto the stack. Then the starting address of the subroutine is loaded into the program counter. In this way, the next instruction fetched is the first instruction of the subroutine. On completion of the subroutine, a RET instruction pops the return address off the stack into the program counter.

Here is an example:

Address	Instruction
2000H	LXI SP,2100H
2001H	
2002H	
2003H	CALL 8050H
2004H	
2005H	
2006H	MVI A,0EH
.	.
.	.
.	.
20FFH	HLT
.	
.	
.	
8050H	.
.	.
.	.
8059H	RET

To begin with, LXI and CALL instructions take 3 bytes each when assembled: 1 byte for the op code and 2 for the data. This is why the LXI instruction occupies 2000H to 2002H and the CALL occupies 2003H to 2005H.

The LXI loads the stack pointer with 2100H. During the execution of CALL 8050H, the address of the next instruction is saved in the stack. This address (2006H) is pushed onto the stack in the usual way; the stack pointer is decremented and the high byte 20H is stored; the stack pointer is decremented again, and the low byte 06H is stored (see Fig. 12-11*a*). The program counter is then loaded with 8050H, the starting address of the subroutine.

When the subroutine is completed, the RET instruction takes the computer back to the main program as follows. First, the low byte is popped from the stack into the lower half of the program counter; then the high byte is popped from the stack into the upper half of the program counter.

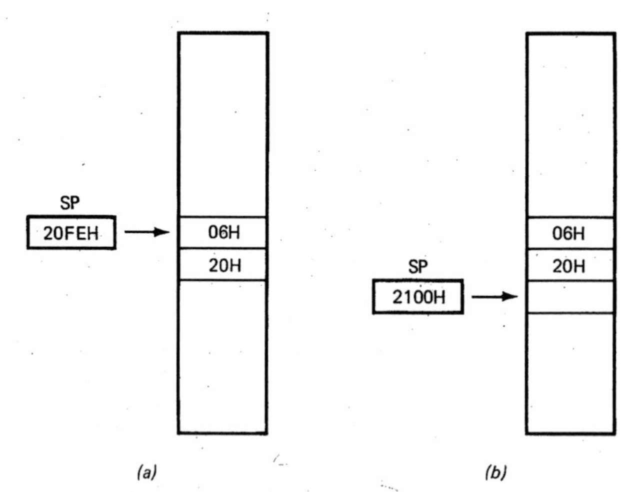

Fig. 12-11 (*a*) Saving a return address during a subroutine call; (*b*) popping the return address during a RET.

After the second increment, the stack pointer is back at 2100H, as shown in Fig. 12-11*b*.

The stack operation is automatic during CALL and RET instructions. All we have to do is initialize the setting of the stack pointer; this is purpose of the LXI SP,dble instruction. It sets the upper boundary of the stack. Then a CALL automatically pushes the return address onto the stack, and a RET automatically pops this return address off the stack.

Conditional Calls and Returns

Here is a list of the SAP-3 conditional calls:

CNZ address
CZ address
CNC address
CC address
CPO address
CPE address
CP address
CM address

They are similar to the conditional jumps discussed earlier. The CNZ branches to a subroutine only if the zero flag is reset, the CZ branches only if the zero flag is set, the CNC branches only if the carry flag is reset, and so forth.

The return from a subroutine may also be conditional. Here is a list of the conditional returns:

RNZ
RZ
RNC
RC
RPO
RPE
RP
RM

The RNZ will return only if the zero flag is reset, the RZ returns only when the zero flag is set, the RNC returns only if the carry flag is reset, and so on.

EXAMPLE 12-5

SAP-3 has a clock frequency of 1 MHz, the same as SAP-2. Write a program that provides a time delay of approximately 80 ms.

SOLUTION

Label	Mnemonic	Comment
	LXI SP,E000H	;Initialize stack pointer
	MVI E,08H	;Initialize counter
LOOP:	CALL F020H	;Delay for 10 ms
	DCR E	;Count down
	JNZ LOOP	;Test for 8 passes
	HLT	

You almost always use subroutines in complicated programs; this means that the stack will be used to save return addresses. For this reason, one of the first instructions in any program should be a LXI SP to initialize the stack pointer.

The 80-ms time delay program shown here starts with a LXI SP,E000H. This implies that the stack grows from address DFFFH toward lower memory. In other words, the stack pointer is decremented before the first push operation; this means that the stack begins at DFFFH.

The remainder of the program is straightforward. The E register is used as a counter. The program calls the 10-ms time delay 8 times. Therefore, the overall time delay is approximately 80 ms.

GLOSSARY

data pointer Another name for the HL register pair because some instructions use its contents to address the memory.
extended register A pair of CPU registers that act like a 16-bit register with certain instructions.
indirect addressing Addressing in which the address of data is contained in the HL register pair.
overflow A sum or difference that lies outside the normal range of the accumulator.
pop To read data from the stack.
push To save data in the stack.
stack A portion of memory reserved for return addresses and data.
stack pointer A 16-bit register that addresses the stack. The stack pointer must be initialized by an LXI instruction before calling subroutines.

SELF-TESTING REVIEW

Read each of the following and provide the missing words. Answers appear at the beginning of the next question.

1. An __________ is a sum or difference that lies outside the normal range of the accumulator. One way to detect an overflow is with the __________ flag.
2. (*overflow, carry*) To reset the carry flag, you may use an __________ followed by a CMC. STC stands for __________ the carry flag.
3. (*STC, set*) The ADC instruction adds the __________ flag and the contents of the specified register to the contents of the __________. SBB stands for subtract with __________.
4. (*carry, accumulator, borrow*) The RAL rotates all bits to the __________ with CY going to the LSB. RRC rotates the accumulator bits to the right with the LSB going to the carry flag.
5. (*left*) The CMP instruction compares the contents of the designated register with the contents of the accumulator. If the two are equal, the zero flag is __________. The CPI compares an immediate byte to the contents of the __________.
6. (*set, accumulator*) JM stands for jump if __________. The program will branch to a new address if the __________ flag is set. JNZ means jump if not zero. With this instruction, the program branches only if the __________ flag is reset.
7. (*minus, sign, zero*) The LXI instruction is used to load register pairs. B is paired off with C, D with E, and H with __________. The HL register pair acts like a __________ pointer with some instructions. This type of addressing is called __________.
8. (*L, data, indirect*) The stack is a portion of memory reserved primarily for return addresses. The stack pointer is a 16-bit register that addresses the stack. It is necessary to initialize the stack pointer before calling any subroutines.

PROBLEMS

12-1. Write a program that adds decimal 345 and 753. (Use immediate bytes for the data.)

12-2. Write a program that subtracts decimal 456 from 983. (Use immediate data.)

12-3. Suppose that 1,024 bytes of data are stored between addresses 5000H and 53FFH. Write a program that copies these bytes at addresses 9000H to 93FFH.

12-4. Show a program that provides a delay of approximately 35 ms. If you use the SAP subroutines of Chap. 11, start your program with LXI SP,E000H.

12-5. Write a program that sends 1, 2, 3, . . . , 255 to port 22 with a time delay of 1 ms between OUT 22 instructions. (Use a LXI SP,E000H and a CALL F010H.)

12-6. Bytes arrive a port 21H at a rate of approximately 1 per millisecond. Write a program that inputs 256 bytes and stores them at addresses 8000H to 80FFH. (Use CALL F010H.)

12-7. Suppose that 512 bytes of data are stored at addresses 6000H to 61FFH and write a program that outputs these bytes to port 22H at a rate of approximately 100 bytes per second. (Use CALL F020H.)

12-8. A peripheral device is sending serial data to bit 7 of port 21H at a rate of 1,000 bits per second. Write a program that converts any 8 bits in the serial data stream to an 8-bit parallel word, which is then sent to port 22H. (Use CALL F010H.)

12-9. Suppose that 256 bytes are stored at addresses 5000H to 50FFH and write a program that converts each of these bytes into a serial data stream at bit 0 of port 22H. Output the data at a rate of approximately 1,000 bits per second. (Use CALL F010H.)

PART 3

PROGRAMMING POPULAR MICROPROCESSORS

13

INTRODUCTION TO MICROPROCESSORS

This part of the text is designed to introduce you to some of the more popular microprocessors. The design and operation of a microprocessor are based on the digital circuits which you studied in Part 1.

You will learn the basic principles of microprocessors and how to write simple assembly language programs. In the study of computers, programming, and microprocessors, one fundamental idea emerges:

> If you do correctly a great number of very simple tasks, you will have done something complicated.

If you understand the basic principles and simple programs presented here, you will be on your way to understanding more complicated ideas.

Since the microprocessor is a "computer on a chip," it may help to take a quick look at computers before starting to study microprocessors.

13-1 COMPUTER HARDWARE

The digital circuits you studied in the first part of this text are the building blocks of a computer. In the early days of computers, digital circuits were made by using vacuum tubes and later were built with transistors. Circuits were designed which would act as the "brain" of a computer. These circuits were called the *central processing unit* (CPU). The CPU could perform basic arithmetic operations such as addition and subtraction, logic operations such as ANDing and ORing, and control operations. Thus it could process data.

A CPU cannot be used alone. There are other components which are needed to make a computer. For example, we said that a CPU can process data. Where is this data? We need memory—a place where data can be stored until the CPU needs it. And what if the CPU does a calculation and comes up with an answer? How would we know what the answer is? We need a way for the CPU to communicate with us. We need an output device. Figure 13-1 illustrates what a simple system looks like.

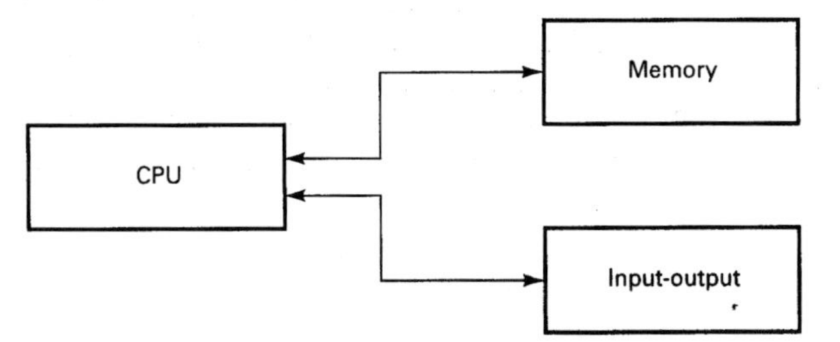

Fig. 13-1 A simplified overview of a microprocessor system.

13-2 DEFINITION OF A MICROPROCESSOR

What exactly is a microprocessor? As the name implies, it must be small (micro-) and it must be able to process data (-processor). A microprocessor is a CPU which is constructed on a single silicon chip. What, then, is a CPU? A *CPU* is an electronic circuit which can interpret and execute instructions and control input and output.

In this text, when reference is made to a microprocessor, only the microprocessor is being referred to. However, if reference is made to a computer, then we are talking about a device which contains a microprocessor and several subsystems. Figure 13-2 serves to illustrate this.

13-3 SOME COMMON USES FOR MICROPROCESSORS

Microprocessors can be found in a variety of products. Some well-known examples are computers and industrial controls. Some not-so-obvious products that use micropro-

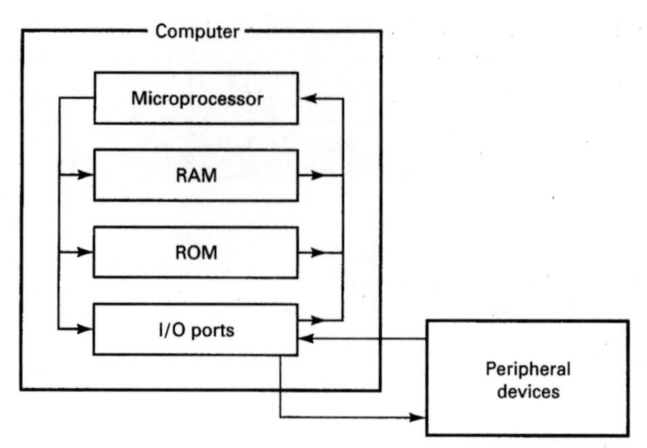

Fig. 13-2 Block diagram of a complete computer and peripherals.

cessors include answering machines, compact disk players, and automobiles.

The microprocessor supplies electronic products with a new dimension. In the past, electronic products have been able to make simple decisions because of certain kinds of circuitry and/or sensors. The microprocessor, however, has multiplied this trait many times: Some devices, most notably computers, now almost appear to think.

13-4 MICROPROCESSORS FEATURED IN THIS TEXT

It is the purpose of this book to examine the most popular 8-bit microprocessor families in addition to the 16-bit Intel 8086-8088 family.

6502 Family

The 6502 family is supported by this text. The 65C02, an advanced version of the 6502 which is used in the Apple IIc, has some additional instructions and enhanced features which can be found in the manufacturer's programming manuals.

6800 Family

The 6800/6808 is supported by this text. The 6809 is an enhanced version of the 6800. It understands all the instructions of the 6800 and includes some other advanced features.

8080/8085/Z80 Family

The 8080, 8085, and Z80 are also supported in this text. The 8080 and 8085 have exactly the same instruction set except for two additional instructions included in the 8085. The Z80 understands all the 8080/8085 instructions and has many other additional instructions.

Only those instructions common to all three microprocessors are discussed in this text. (The extended Z80 instructions are not used in the text.) This has the advantage of making it possible for students to use a mixture of 8085 and Z80 microprocessor trainers in the same class at the same time with all students on equal footing and with a minimum of confusion. Either Z80 or 8085 mnemonics can be used interchangeably for the homework problems and the object code will be the same.

8086/8088 Family

The Intel 8086/8088 is the only 16-bit microprocessor discussed in this text. This microprocessor (in addition to the 80286, 80386, and 80486) is used in the popular IBM PCs, IBM compatibles, and clones. The DOS DEBUG utility is used throughout the text. Assemblers are introduced in later chapters.

13-5 ACCESS TO MICROPROCESSORS

Developing skill in programming and interfacing microprocessors requires access to a microprocessor. Here are some ways to gain access to a microprocessor supported by this text.

Computers

The 6502 or one of its derivatives can be found in the entire line of Commodore computers including the PET, Vic-20, C-64, C-16, Plus-4, and C-128. They can also be found in the Apple II line of computers including the Apple II, II+, IIe, IIc, and IIc+. They are also included in that portion of the Laser line of computers that are Apple-compatible, including the Laser 128, Laser 128 EX, and Laser 128 EX/2. And last of all, some of the older Atari home computers contain this type of microprocessor.

The 8085 and Z80 can be found in some of the older CP/M machines. (CP/M stands for control program for microprocessors.) The Z80 was used in Radio Shack's TRS-80 line of computers and is also found in the Commodore 128 (the Commodore 128 contains two microprocessors). The Commodore 128 will also run CP/M software if that is desired.

The 8086/8088 are found in all of the IBM PCs and XTs, IBM compatibles, and clones. The 80286 is used in AT-class machines, and of course the 80386 is used in the newer 386s. These microprocessors use a superset of the 8086/8088 instructions set and can therefore also be used with this text.

Some IBM compatibles use the NEC-V20 or one of the other NEC microprocessors. These are compatible with the Intel series of microprocessors and will work equally well.

Microprocessor Trainers

Another way to gain access to a microprocessor supported by this text is through the use of a microprocessor trainer. Heathkit's ET-3400-A trainer contains a 6808 chip. E&L Instruments has the "FOX" (MT-80Z) with a Z80 microprocessor. Intel makes the SDK-85, which features the 8085 chip, and the SDK-86, which uses the 8086. Motorola makes the MEK6800D with a 6800 chip.

Software Emulation Programs

Finally, there are software emulation programs that will make a computer act as though it is using another microprocessor.

PROGRAMMING AND LANGUAGES

What is a program and why do we need one? What do we mean by *program design?* What is a *programming language?* Why do we need a language? What is a *flowchart?* How does all of this relate to electronics and digital circuits? These are some of the questions we will try to answer in this chapter.

14-1 RELATIONSHIP BETWEEN ELECTRONICS AND PROGRAMMING

A question sometimes raised by electronics students is, "Why are we learning about programming microprocessors?"

Programming is a topic which is closely related to electronics. Mathematics and physics are topics which support or undergird the subject of electronics. They form a foundation. Programming is not so much a support subject as it is a related subject. Let's take a closer look at this.

Digital Electronics and Microprocessors

What prompted the creation of digital electronics? It was the desire to make a machine without moving parts which could perform mathematical calculations. Such a machine would be much faster than any mechanical calculator. Correctly connecting enough digital logic circuits together created such a machine.

Once the calculating machine had been built, there had to be a way to tell this machine to add, or subtract, or perform some logical operation. Thus programming was born. We simply needed a way to tell the machine what to do. In the beginning, programming was done by connecting wires or patch cords. This was very slow compared to what we do today.

Over the years digital circuits became more complex, the calculating machine grew into far more than just a big calculator, and the need for ways to communicate with the machine grew. Finally, it became possible to put the entire computer "brain" on a single chip.

Until this point an electronics technician might never work on or even see a computer. However, when the "brain" could be put on a chip, and the cost was measured in dollars rather than thousands of dollars, its possibilities became endless.

Designers and engineers realized that these "brains," or microprocessors, could improve the performance of many common electronic products and could make new products economically possible. With microprocessors everywhere, the electronics technician can no longer be unaware of their operation.

The Electronic Technician and Programming

So why should a technician learn about programming? Because the technician will probably eventually work on products with microprocessors, and the microprocessor cannot be separated from its program. A microprocessor without a program would be like a resistor with no resistance or a wire with no conductivity. Without the program, a microprocessor does nothing.

Programming is now part of the overall picture that electronics is concerned with—like mathematics and physics. Some technicians will not need as much knowledge about programming as others: It depends on what your career field is. But everyone should at least be aware of the basics.

The goal of this book is to provide the digital understanding and programming experience which would be appropriate for the "typical" electronics student.

14-2 PROGRAMMING

In everyday language:

> A *program* is a very detailed list of steps which must be followed to accomplish a certain task.

A Familiar Example

We have all used this concept of programming—of following specific steps to accomplish a certain task—but have probably not thought of it in these terms. Let's look at something like taking a city bus downtown. You would be likely to

1. Wear clothes appropriate for the weather that particular day.
2. Take some money or tickets.
3. Go to a nearby bus stop.
4. Wait for the correct bus.
5. Get on.
6. Pay the driver.
7. Sit down if there were empty seats available.
8. Wait until the bus arrived in the area you wished to go to.
9. Alert the driver you wished to get off.
10. Wait for the bus to stop.
11. And finally get off.

Figure 14-1 is a flowchart (we'll talk about flowcharts in just a minute) of this process.

Unless this was your first time riding a bus, you wouldn't think about every detail because much of it is understood and is a natural part of your life. You usually dress for the weather when you go outside, and you usually take money when you go places. With a computer, though, things are different.

Very little is "natural" for a computer. The microprocessor has several temporary storage places where numbers can be kept (called *registers*). The machine can add and subtract, it can AND and OR, it can move numbers from one register to another, and it can do other simple things, but *everything must be specified!* One of the things that often surprises people learning to program microprocessors is the amount of detail which is necessary when writing a program.

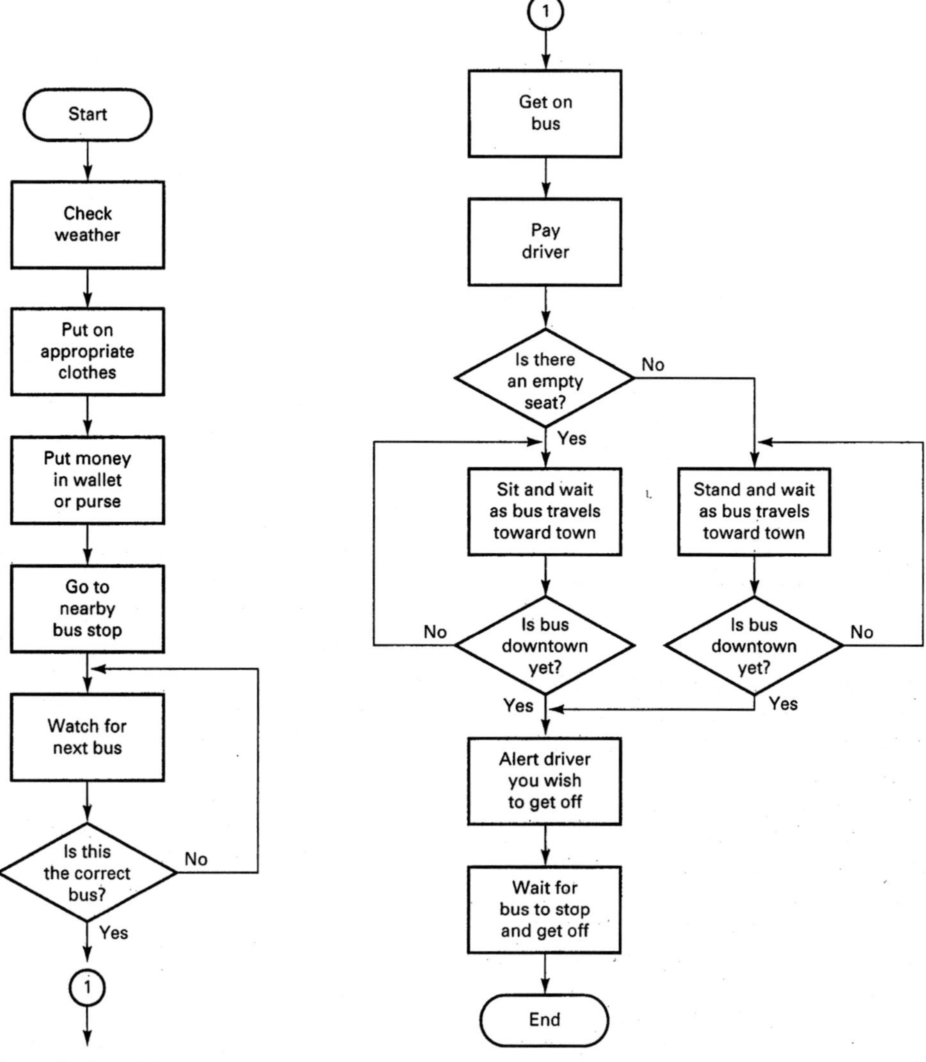

Fig. 14-1 Flowchart of a bus ride.

14-3 FUNDAMENTAL PREMISE

Before we look further at the subject of programming and flowcharts, we need to discuss a fundamental concept of programming. The concept is this:

> You cannot program the computer to do something you don't know how to do.

If you use computers only with application software (spreadsheets, word processors, and so on), this may not always be true, but if you want to program microprocessors, it is. Before you begin to think about how you will program a computer to do something, think about how you would do it yourself without a computer. After you know how you would do it, you can begin to tell the computer how it should do it.

14-4 FLOWCHARTS

When you are writing a program, it helps to have an organized way to write or express the flow of the program's logic. A flowchart is one way to do this.

Flowchart symbols

Figure 14-2 shows some common flowchart symbols. There are others, but we'll need only a few for most of the programs we'll be writing.

Straight-Line Programs

The simplest type of program is the *straight-line program.* In this type of program the steps involved follow each other, one after another, without any alternate routes or paths. Figure 14-3 is an example of a straight-line program.

This program is similar to one that might be used at the cash register of a store. It allows you to enter the price and product code of one item. The program then calculates a 5 percent sales tax, adds the tax to the original price to arrive at a total, and finally displays the total cost. The program will accept only one item, which means that it would have to be "run" again to find the total cost of a second item. Since we often buy more than one item at a time, let's look at another flowchart.

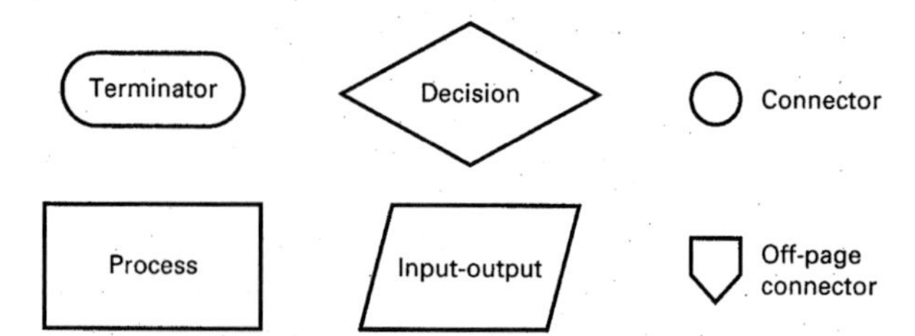

Fig. 14-2 Some flowchart symbols.

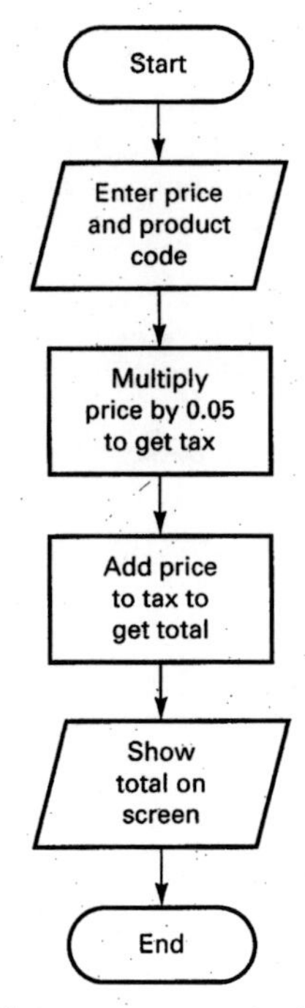

Fig. 14-3 Straight-line program to calculate sales tax and display total cost for one item.

Looping

A *loop* is a section of a program which will repeat over and over again. We can make the loop repeat indefinitely, or make it stop after a certain number of repetitions, or make it stop when some condition is met. Look at Fig. 14-4 and compare it to Fig. 14-3.

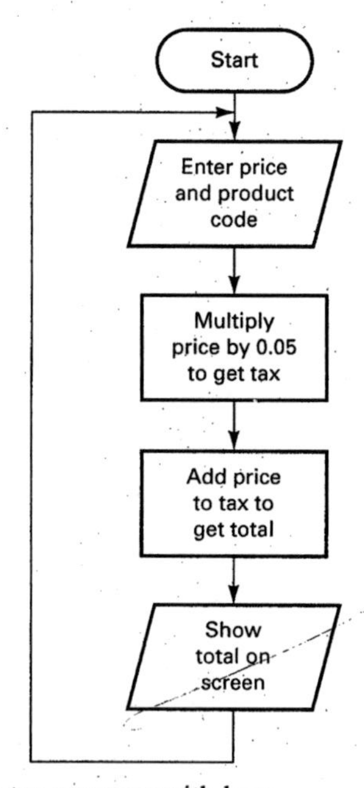

Fig. 14-4 Sales-tax program with loop.

These are almost identical, aren't they? What do you think this program will do that the one in Fig. 14-3 didn't? The answer, of course, is that this program is ready to accept a new number immediately after displaying the previous total. After you enter an item's price, the total cost is shown on the screen and the program then waits for you to enter the price of the next item.

Loops make it easier for programs to perform repetitive tasks. The program that uses loops can do the same calculations or functions over and over again.

Branching

Sometimes we want the computer program to do different things based on the situation at the time or based on the results of certain operations. We need a way to *branch* off from the main program flow. *Branching* allows us to write one program that can do different things at different times. Let's look at the sales-tax situation again. Study Fig. 14-5 at this time. This new version of the sales-tax program has a branch and a decision symbol.

Let's look at the decision symbol (diamond). If the program is to be able to take an alternate path when certain conditions exist, we must give it a chance to check for those conditions. The decision diamond represents that time. If the item is a nonfood item, it will be taxed as usual, and the program flow continues downward. If it is a food item which is not to be taxed, then we take the *branch*. The branch doesn't actually say not to tax the food item. But by making the total cost equal to the original price and bypassing the tax calculation section, we have effectively done the same thing. The total that appears will be the same as the original price, and the program will then loop back to the beginning to wait for the next item.

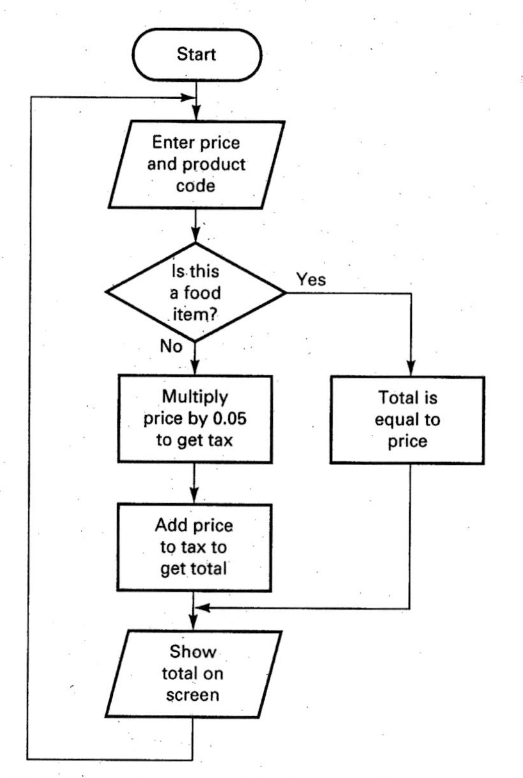

Fig. 14-5 Sales-tax program with loop and branch for non-taxable food items.

Subroutines

Sometimes we need to have the computer program take care of some intermediate task before it can continue with the main job at hand. We don't want it to branch and then end up somewhere else after the branch is finished. Rather, we want it to go to an intermediate task and then come right back to where it was before it left. This is called a *subroutine*. Looking at a subroutine will help clarify this new concept. Figure 14-6 shows our new program.

Everything is the same as in the last (Fig. 14-5) program except that we have added a subroutine which handles inventory. This subroutine is really just another small program that works along with the main one. It reduces the inventory total for this particular item by 1. If this total is less than 10, then it's time to order more. Either way, the subroutine prints a line on a printer in the administrative office with the product code and name of the product. We then "return" from the subroutine to the main program and continue where we left off.

Calling Subroutines

The act of going to a subroutine is often referred to as *calling* a subroutine, at the end of which we *return* to the main program.

The greatest advantage in having subroutines is not in calling or using them once but in using them several times in a program. You write that part of the program only once, but you can use it many times. Figure 14-7 illustrates this.

In Fig. 14-7 the boxes are not process boxes but rather representations of certain parts or modules of the whole computer program.

In this hypothetical situation there may be times when merchandise needs to be ordered other than when inventory drops below 10. For example, if a clerk finds a piece of merchandise damaged too badly to sell at a reduced price, it may simply be disposed of; however, it must be replaced to keep inventory up. The "damaged merchandise" part of the program can then *call* the "inventory-ordering subroutine" at some point.

Likewise, the store might sometimes give food or clothing to charity. This part of the program might also call the inventory-ordering subroutine to replace that merchandise.

This store's computer program uses the same subroutine in three different situations, but the programmer had to write the subroutine only once.

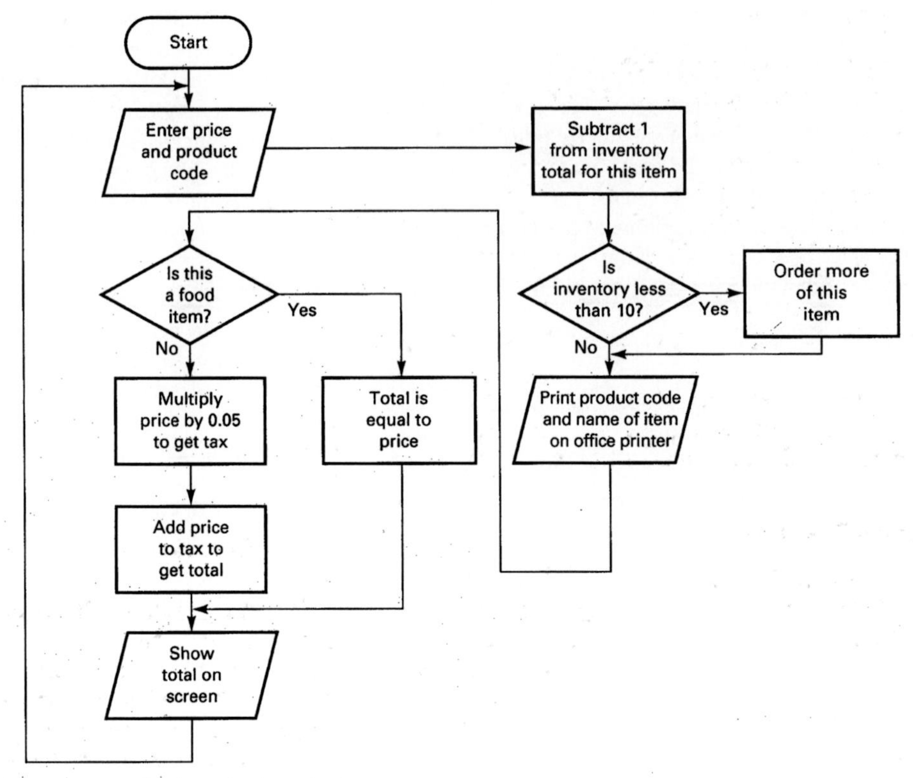

Fig. 14-6 Sales-tax program with inventory control reordering subroutine.

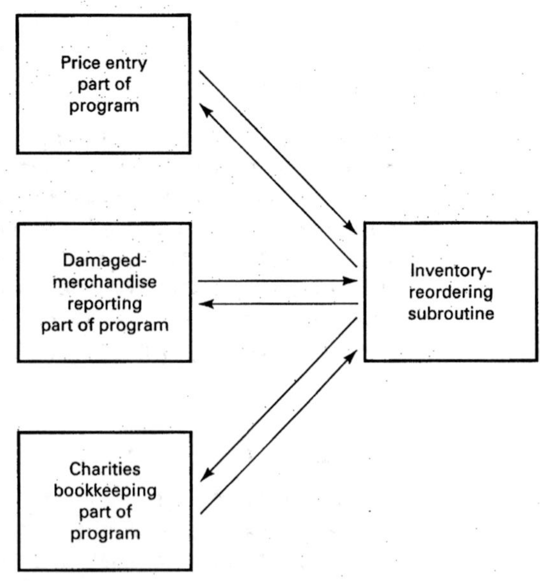

Fig. 14-7 Repetitive calling of inventory-reordering subroutine.

14-5 PROGRAMMING LANGUAGES

Now that we can define and flowchart the desired process, we need to be able to communicate this process to the computer. We need a language which the computer understands. Many languages have been developed for use with computers.

Machine Language

There is only one language the computer actually understands, and that is machine language, which consists of 1s and 0s. This binary language is fine for the computer but not for people. To have to communicate with the computer in binary, you would place in its memory a series of numbers that might look like this:

10010100
01001010
11101110
00101001

It would be nearly impossible to remember what the many different patterns of 1s and 0s meant, and the probability of making a mistake would be very high. Something better is needed.

Assembly Language

The first step toward a language that is easier for people to work with uses abbreviations to stand for different operations. For example, the instruction which tells a 6800 microprocessor to add numbers is the ADDA instruction, which stands for ADD accumulator A to a memory location.

This "language" of abbreviations is called *assembly language*. The "abbreviations" are called *mnemonics*. A mnemonic (pronounced ne-'män-ik) is something that aids the memory. Mnemonics are designed to be easy to remember and are a significant improvement over binary digits.

Machine language and assembly language are the subjects of this book. We refer to them as *low-level languages* because only very simple instructions exist.

High-Level Languages

Over the course of time, people working with computers felt it would be helpful to create languages that were more like English, so that it would not be so difficult to communicate with the computer and so that more advanced commands could be created. We call these *high-level languages*.

For example, many microprocessors do not have the ability to multiply or divide. It is obvious, however, that these are common mathematical functions that must be available to a computer programmer. In machine or assembly language one can use repeated additions to multiply or repeated subtractions to divide. This is not necessarily the best way to multiply or divide, but it is one way. In a high-level language there are "multiply" and "divide" commands. The language knows how to create the multiply and divide functions even though the microprocessor does not have these functions built in. In fact, these languages can understand English commands like *print, run, do, next,* and *end.* The microprocessor does not understand these English words, but the language changes (interprets or compiles) them into machine language before sending them to the microprocessor.

Many high-level languages have been created over the years. FORTRAN (formula translation) is a language that handles high-level mathematics very well and is designed for scientists and engineers. COBOL, which stands for common business-oriented language, is tailored to the needs of business. BASIC, which stands for beginner's all-purpose symbolic instruction code, was designed to be easy for nonprofessional programmers to learn and use. Pascal, named for the French mathematician Blaise Pascal, is designed to encourage the programmer to adhere to what are considered "correct" programming practices.

There are some languages that are somewhat "in between" the high-level and low-level languages, most notably C and FORTH. Figure 14-8 illustrates this.

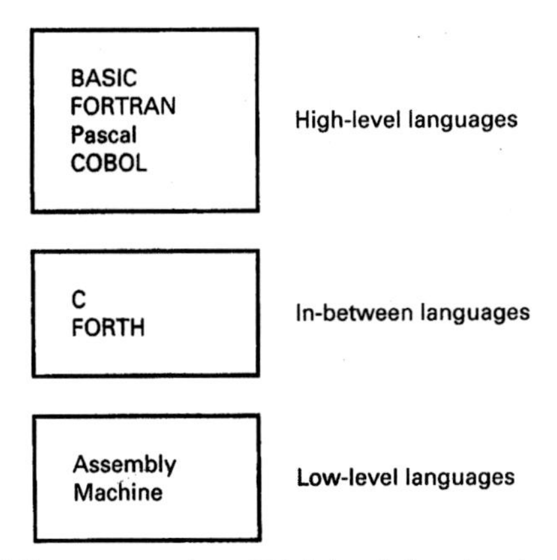

Fig. 14-8 Some examples of high-level, low-level, and in-between languages.

14-6 ASSEMBLY LANGUAGE

Let's look at the subject of assembly-language programming in a little more detail.

Machine language is the language the computer understands, but it is difficult for people to work with. Assembly language gives us the advantages of machine language without the disadvantage of doing something that seems so unnatural.

When we write in assembly language, we use abbreviations called mnemonics for certain operations or functions. The assembly language is called *source code.* It is more like English than machine language. The microprocessor, however, cannot act upon or execute mnemonics. It doesn't understand mnemonics. We need to convert the assembly language or source code into machine language or *object code.* There are a couple of ways to do this.

Manual Assembly

Let's look at manual assembly first. When using this technique, you write your program on paper using mnemonics. Then you look up each mnemonic on a chart. On the chart there will be a number which is the machine-language code for the assembly-language mnemonic. You then write down this object code so that you can later key it into the microprocessor trainer or computer. This is called *manual assembly* because you must look up the codes yourself.

Assembly with an Assembler or Monitor

The other way to create machine-language object code from assembly-language source code is through the use of a monitor or assembler. Since manual assembly involves simply looking up mnemonics on a chart, it seems reasonable that the chart could be stored in a computer and the computer could look up the mnemonics and find their corresponding object code. Though there is much more to a fairly sophisticated assembler or monitor, this is the basic idea.

A *monitor* is a program that is normally stored in ROM and gives you access to the microprocessor's various registers. It sometimes has in it a simple *assembler* to change mnemonics into machine code and a *disassembler* to change machine code back into mnemonics.

An assembler program is usually more sophisticated than a monitor and has features that are difficult to explain at this point, but suffice it to say they are for more serious programming than the monitor. A longer period of time is required to become skilled in the use of an assembler, but it is a more powerful tool.

14-7 WORKSHEETS

During the remainder of this book you will be writing assembly-language programs. In addition to the flowchart, the *worksheet* is a tool which helps you stay organized as you write programs. The worksheet is simply a form on which you can write your program. It is laid out in such a way that it's a little easier to stay neat. Figure 14-9 is a portion of such a worksheet.

Name ______________________ Date ______________

Program name ______________________ Sheet ________ of ________

Address	Obj code	Label	Mnemonic	Operand/Addr	Comment

Fig. 14-9 Example of a portion of a worksheet.

GLOSSARY

assembler A program which translates assembly language mnemonics into binary patterns (machine language).
assembly language A low-level language which uses mnemonics in place of binary patterns (machine language).
branch A section of a program which causes different actions to be taken based on conditions.
disassembler A program which translates binary patterns (machine language) into assembly language mnemonics.
loop A section of a program which will repeat over and over again.
mnemonic Something that aids the memory. Assembly language uses mnemonics, which are abbreviations for machine-language instructions.
monitor A program (usually stored in ROM) which gives the programmer access to the microprocessor's stack, accumulator, registers, and so forth. It sometimes contains a simple assembler.
straight-line program A program in which each step is followed by the next without any alternate routes or paths.
subroutine A portion of the program which is called upon to perform a specific task. When the task is finished, the main part of the program is returned to.

SELF-TESTING REVIEW

Read each of the following and provide the missing words. Answers appear at the beginning of the next question.

1. Without a __________, a microprocessor does nothing.
2. (*program*) A __________ is a very detailed list of steps which must be followed to accomplish a certain task.
3. (*program*) What is the shape of the decision symbol?

4. (*Diamond*) ________ make programs more practical for doing repetitive tasks.
5. (*Loops*) The only language a computer actually understands is ________ language.
6. (*machine*) What does COBOL stand for?
7. (*Common business-oriented language*) A program in which the steps involved occur one after the other without any alternate paths is called a ________ program.
8. (*straight-line*) A section of a program which repeats indefinitely, a certain number of times, or while or until a certain condition exists is called a ________. (*loop*)

PROBLEMS

14-1. If you want to write a program to do something, what should you think about before you try to figure out what computer instructions to use?

14-2. What is the shape of the process symbol?

14.3. What provides an alternate path for program flow based on certain conditions?

14-4. What allows program execution to go to an intermediate task and then return to the place where it was before it started the intermediate task?

14-5. What is one of the advantages of using subroutines?

14-6. What is assembly language?

14-7. What does FORTRAN stand for?

14-8. What does BASIC stand for?

14-9. What was one of the goals of the creator of the Pascal language?

14-10. What does an assembler translate source code (mnemonics) into?

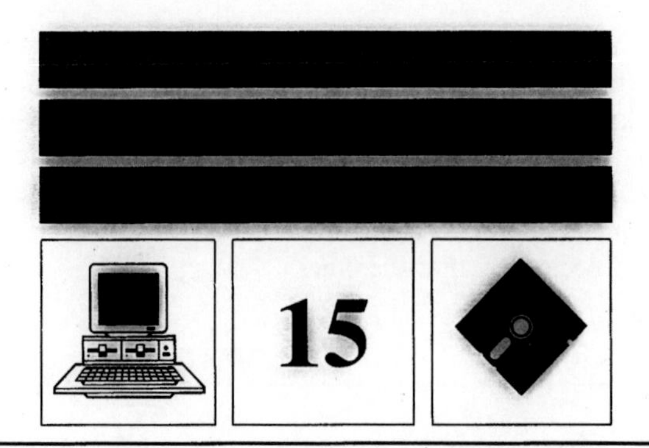

SYSTEM OVERVIEW

New Concepts

We'll begin this chapter by reviewing computer architecture. Then we'll spend the greater part of the chapter looking at microprocessor architecture in general and at the architecture of the microprocessor families supported by this text in particular.

15-1 COMPUTER ARCHITECTURE

Let's review computer architecture a little. Refer to Fig. 15-1.

Memory

We said that memory was needed so that there would be a place for data and instructions to be stored. Data and instructions which can be lost after power is removed are stored in RAM (random-access memory). Data and instructions which must never be lost, even after the power is turned off, are stored in ROM (read-only memory). Remember that ROM is a type of memory which cannot have its contents changed once the ROM chip is manufactured. PROM and EPROM are used in much the same way as ROM but can be programmed after being manufactured (PROM) or even programmed more than once (EPROM). PROM and EPROM differ from RAM in that they require special equipment to program them.

When we refer to memory in this text, we will usually be referring to RAM.

Fig. 15-1 Block diagram of a complete computer with peripheral devices. (Arrows indicate data flow.)

Addressing

Since there are many memory locations, it is necessary to have a means of referring to specific locations. This is done through addressing. Typically, memory locations are numbered from 0000 (in hexadecimal numbering) to the highest location used by that particular trainer or computer. This sequential number which is assigned to each location is its *address*. See Fig. 15-2.

A memory address is similar to the address of your home. Your house has a number or *address* assigned to it, and no other house on your street can have the same address. Inside your house are its *contents;* chairs, beds, and so on. Notice

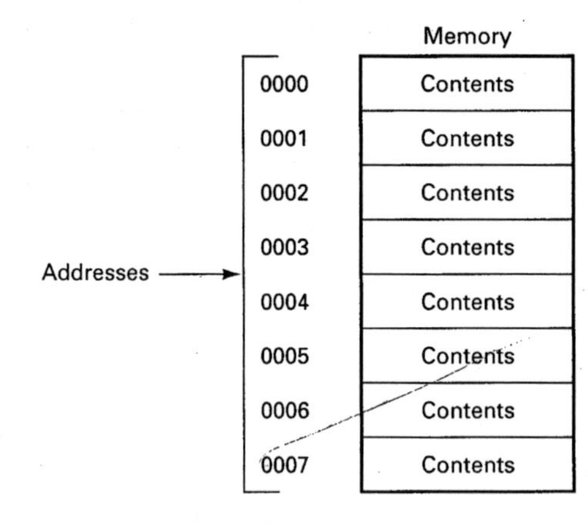

Fig. 15-2 Memory addressing.

that your home's address and your home's contents are not the same.

Each memory location has an address and contents. The address is necessary to specify which memory location to *read* information from or *write* information into. The contents is the information itself.

Address Bus

Most microprocessors can store information and instructions in a wide range of memory locations. Usually the memory locations are in a memory chip rather than in the microprocessor. The microprocessor needs a way to tell the memory chip which memory location it wants to put data into or take data from. It does this through the *address bus*. See Fig. 15-3.

The address bus is a communications link between the microprocessor and the memory chips. Physically, it is simply a group of electrical paths which are connected to RAM, ROM, and the I/O chips. Through this bus the microprocessor can specify the address of any memory location in any chip or device. Notice in Fig. 15-3 that information travels on the address bus in only one direction, from the microprocessor to memory and I/O. There are more details involved, but this is the basic idea.

Data Bus

Once the microprocessor has specified which memory location or device it wants to put data into or take data from, it then needs a set of electrical paths for this information to travel on. This set of paths is called the *data bus*.

It is this set of electrical paths that allows data to flow from one chip to the next. Notice in Fig. 15-3 that information on the data bus travels both to and from the microprocessor, memory, and I/O devices. Eight-bit microprocessors have a data bus that is 8 bits wide; 16-bit microprocessors have a data bus that is 16 bits wide. That is, the bus consists of 8 or 16 parallel connecting paths.

Addressing Range

Let's look at the normal range of addresses possible with 8-bit computers at this time.

In earlier chapters you studied the binary number system and learned that each position represents a certain power of 2. This is similar to the way each position in our decimal number system represents a certain power of 10. This is illustrated below.

Decimal	10^3	10^2	10^1	10^0
	1,000's	100's	10s	1s
Binary	2^3	2^2	2^1	2^0
	8s	4s	2s	1s

If we look at a decimal number like $9{,}999_{10}$ (the subscript 10 means that we are using a number in base 10), it not only tells us about a quantity of items, such as apples, but also tells us about possible combinations.

The number 9,999 is a *four*-digit number. Using the 10 different decimal digits from 0 through 9, and using no more than *four* digits at a time, there would be 9,999 + 1, or 10,000, possible numbers you could create. (You add the 1 because the number 0000 or simply 0 must also be included.) This can also be calculated as $10^4 = 10{,}000$.

If you were interested in giving unique addresses to 10,000 homes on the same street (quite a long street), it would be possible to do so by using only four digits. The first house would have the address 0, and then you would just continue numbering up to 9,999.

EXAMPLE 15-1

Using only three digits, how many unique addresses could you give to homes on a single street (a decimal number)?

SOLUTION

Since $10^3 = 1{,}000$, this is the number of unique addresses that are possible.

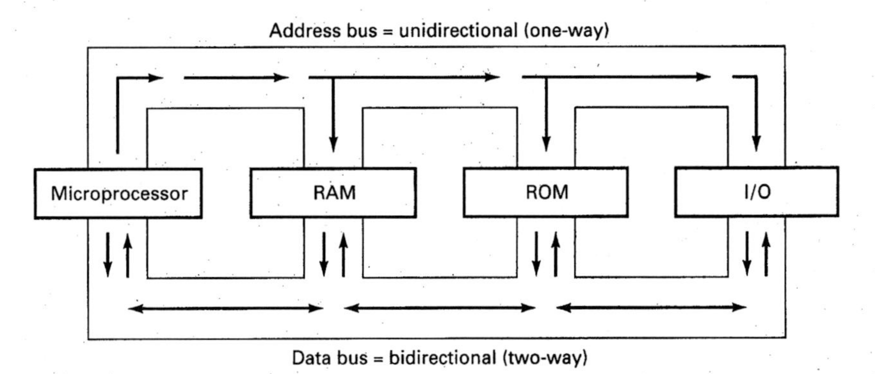

Fig. 15-3 Data bus and address bus.

Now, let's try the same problem in binary: 1111_2 is a binary number. (The subscript 2 tells us we are using base 2 or binary numbers.) The size of this number is shown below.

Binary	2^3	2^2	2^1	2^0
	8s	4s	2s	1s
	1	1	1	1

We have one 8. We have one 4. We have one 2. And we have one 1. That is, we have an 8, a 4, a 2, and a 1. If we add this up, we get

$$8 + 4 + 2 + 1 = 15$$

The number 1111_2 is the same as 15_{10} (decimal 15). This means that using only 4 binary digits or bits, there are a total of 15 + 1, or 16 unique numbers possible. This can be calculated by using $2^4 = 16$.

If you wanted to give unique binary addresses to 16 houses on the same street (not such a long street), it would be possible to do so with only 4 bits. The first house would be 0000 or simply 0, the next would be 0001, the next 0010, and so on up to 1111.

EXAMPLE 15-2

Using 12 binary digits, how many unique house addresses would be possible?

SOLUTION

$2^{12} = 4{,}096$ unique addresses

This is essentially what is necessary in the matter of addressing memory locations. The highest number that exists in binary using only 4 bits is 1111_2 (15_{10}). That means that if we had only four address lines—that is, an address bus with only four lines—we would be able to have only a maximum of 16_{10} different addresses. (0000 counts as one address.) Obviously, this is not enough. Look at Fig. 15-4. This illustrates the number of unique addresses possible with different numbers of address lines.

As can be seen in Fig. 15-4, if we decide to use only eight address lines, since we are studying 8-bit chips, we then limit ourselves to 256 memory locations. (Add the values of the first eight positions starting from the far right + 1.) This is not nearly enough. Most 8-bit chips use 2 bytes for addressing purposes, which then allows 65,536 different memory locations. (One byte is 8 bits; 2 bytes is 16 bits, which then allows 2^{16} combinations.) This is often adequate. If not, there are ways to increase this number by using a method known as bank switching.

EXAMPLE 15-3

How many memory locations could be addressed by a 10-line address bus?

SOLUTION

$2^{10} = 1{,}024$ memory locations can be addressed.

15-2 MICROPROCESSOR ARCHITECTURE

We now need to look more closely at the actual microprocessor, which is the ''brain'' of our computer. First, we will study those features which most microprocessors have in common. Then we will look at each of the microprocessor families and study their specific features.

Accumulator

One of the most often used parts of a microprocessor is the accumulator. The *accumulator* is a storage place or register which often has its contents altered in some way. For example, we can add the contents of the accumulator to the contents of a memory location. Usually the result of an operation is also placed in the accumulator. This action is illustrated in Fig. 15-5.

The microprocessor can take the contents of the accumulator and the data coming in, perform some operation on the two, and place the result back in the accumulator. There are times when no data is coming in but some operation is being performed on the contents of the accumulator only. For example, the microprocessor might find the 1's complement of the contents of the accumulator and place the result in the accumulator in place of the original number.

Some microprocessors have only one accumulator; others have more than one.

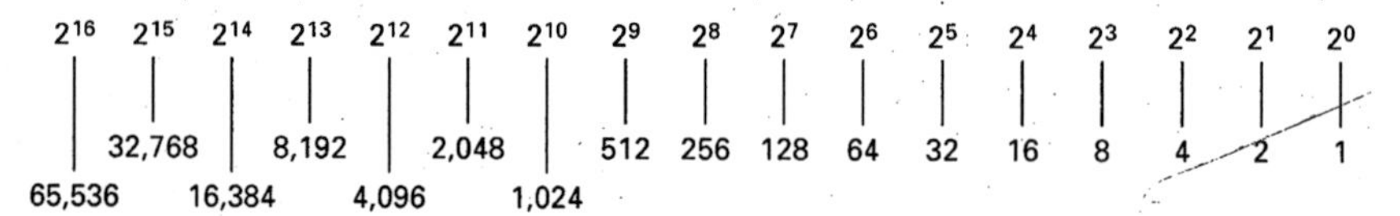

Fig. 15-4 Powers of 2. Also the number of memory addresses available with varying numbers of address lines.

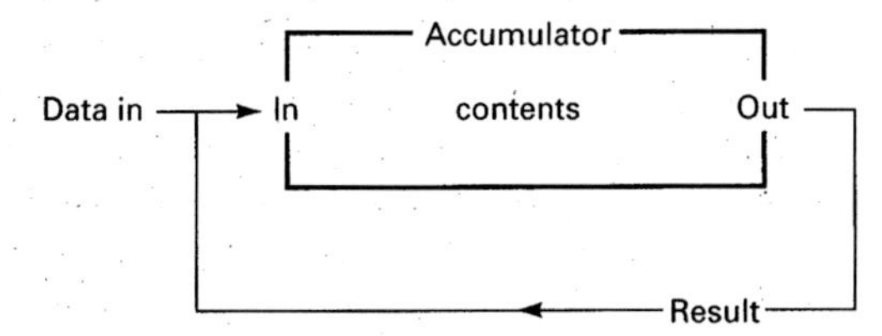

Fig. 15-5 Accumulator operation.

General-Purpose Registers

General-purpose registers are similar to the accumulator. In fact, the accumulator is a special type of register. *General-purpose registers* are temporary storage locations. They differ from the accumulator in that operations involving two pieces of data are usually not performed in them with the result going back into the register itself, as in the case of the accumulator. The microprocessor will often alter the contents of a register, however. Figure 15-6 shows the operation of a general-purpose register.

One might wonder why a microprocessor needs general-purpose registers when it has RAM to temporarily store information. The answer is speed. Data in registers can be accessed and moved much more quickly than data in RAM.

Program Counter/Instruction Pointer

We mentioned earlier that instructions are stored in memory. Considering the fact that there can be tens of thousands, hundreds of thousands, or even millions of memory locations, it's obvious that the microprocessor must keep track of the location from which it will be getting its next instruction. This is the job of the program counter.

The *program counter* is a very special register whose only job is to keep track of the location of the next instruction which the microprocessor will use. Figure 15-7 illustrates its operation.

The program counter "points" to the address of the next instruction to be retrieved and used by the microprocessor.

The act of "getting" an instruction is usually referred to as *fetching* the instruction. The period of time needed for this is often called the *fetch cycle*.

Index Registers

Another type of register is the index register. In the same way that the index of a book helps a person locate information, the index register can be used to help locate data. The *index register* is normally used as an aid in

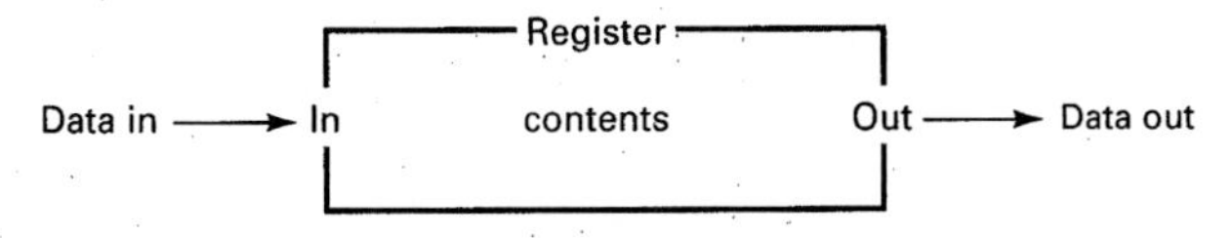

Fig. 15-6 General-purpose register operation.

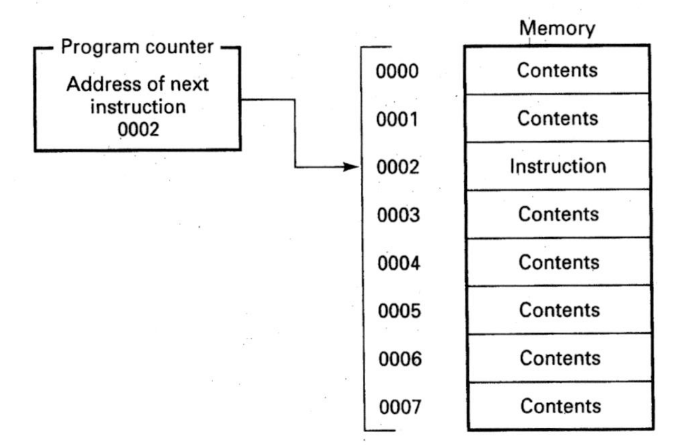

Fig. 15-7 Program counter operation.

accessing data in tables stored in memory. The index register(s) can be incremented (increased by 1) or decremented (decreased by 1) but normally does not have other arithmetic or logical capabilities.

We will look at the index register(s) more completely in later chapters.

Status Register

The *status register,* sometimes called the *condition code register,* or *flag register,* is a special register which keeps track of certain facts about the outcome of arithmetic, logical, and other operations. This register makes it possible for the microprocessor to be able to test for certain conditions and then to perform alternate functions based on those conditions. This is done through the use of *flags*.

We will now take an overall look at flags. Don't be concerned if these next few paragraphs are not completely clear at this point. They can serve as a refresher for those who may have had some experience with microprocessors in the past. And for those who are new to this subject, reading about them now will at least give you some idea of what flags are and how they are used. These concepts will be covered again in greater detail as they arise in later chapters.

The status register is divided into individual bits which have their own unique functions. Each bit is called a *flag*. Each flag keeps track of, or "flags," us concerning certain conditions. Not every operation or instruction affects every flag. Some instructions affect many flags, and some don't affect any at all. Figure 15-8 shows a model of a typical status register.

When referring to flags, the following logic is used. If some condition has come to be, or is true, the flag uses a 1 to say, "Yes, this is true or has happened." If that condition has not occurred, the flag uses a 0 to say, "No, this is not true or has not happened." Causing a flag to become 1 is called *setting* a flag. Causing a flag to become 0 is called *clearing* a flag.

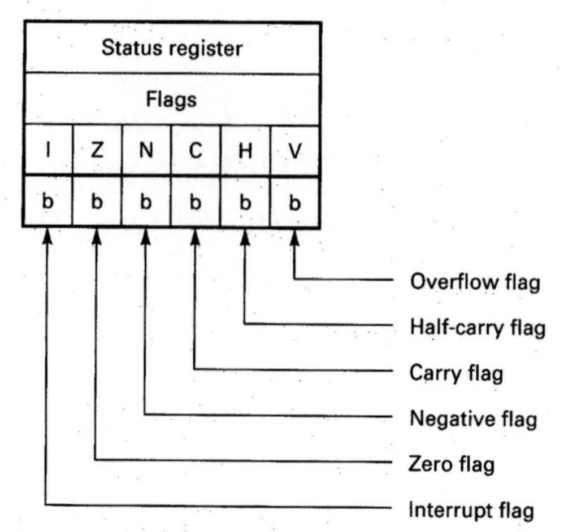

Fig. 15-8 Model of a typical status register. (b's represent bits.)

The zero flag keeps track of whether the last operation which affects this flag produced an answer of zero. This flag is set or 1 if a zero result has been produced and is cleared or 0 if a nonzero result has been produced.

The negative flag tells us if the last operation which affects this flag produced a negative number. When 8-bit signed binary numbers are used, if bit 7 (the eighth bit) of the number is 1, then the number is negative and the N flag will be set; if bit 7 of the number is 0, then the number is positive and the N flag will be cleared or 0. (This negative flag is sometimes called a sign flag and is indicated with an "S.")

The carry flag tells us if the last operation which affects this flag produced a carry from bit 7 (in 8-bit systems) of the accumulator (bit 7 is the left-most or most significant bit) into the carry bit. The carry flag also tells us if, during subtraction, a borrow into bit 7 was needed. How a borrow is indicated depends on which microprocessor is being used. See Fig. 15-9.

The half-carry flag tells us if the last operation which affects this flag was an arithmetic operation which produced a carry from bit 3 to bit 4. This feature is primarily used with BCD (binary-coded-demical) numbers.

The overflow flag tells us if the last operation which affects this flag caused a result that is outside the range of signed binary numbers for the word size being used at the time. In the case of 8-bit microprocessors, this is +127 or −128. If this range is exceeded, the overflow flag is set (1) to warn the programmer.

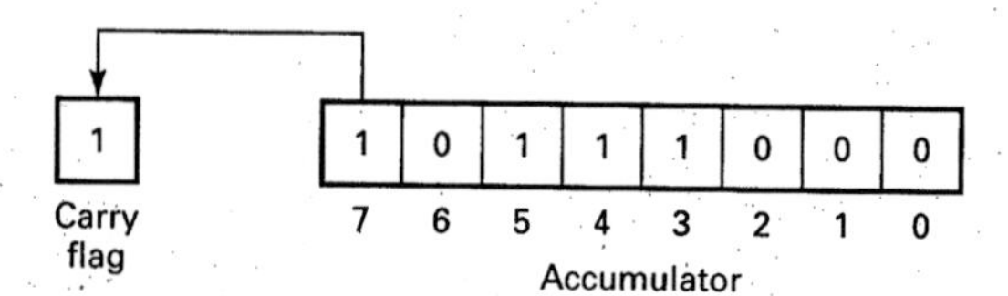

Fig. 15-9 A "carry" from bit 7 into the carry flag.

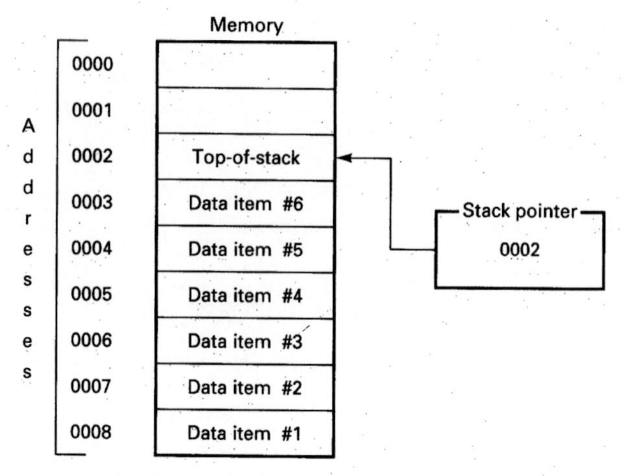

Fig. 15-10 Typical stack and stack pointer.

The interrupt (interrupt mask, interrupt flag, interrupt enable bit) prevents maskable interrupts from occurring when it is set and allows them when cleared.

Stack and Stack Pointer

The stack is a special place in memory. The *stack* is most often used to store certain critical pieces of data during subroutines and interrupts. You'll learn more about these later, but let's look at the structure of a stack at this time. Refer to Fig. 15-10.

The structure of the stack is a *first-in-last-out* (FILO) type of structure. Unlike main memory, where you can access any data item in any order, the stack is designed so that you can access only the top of the stack. If you want to place data in the stack, it must go on top; if you wish to remove data from the stack, it must be on top before it can be removed.

Let's see how the situation in Fig. 15-10 has come to be. To do that, refer to Fig. 15-11. Data item #1 is the first item we wish to place on the stack.

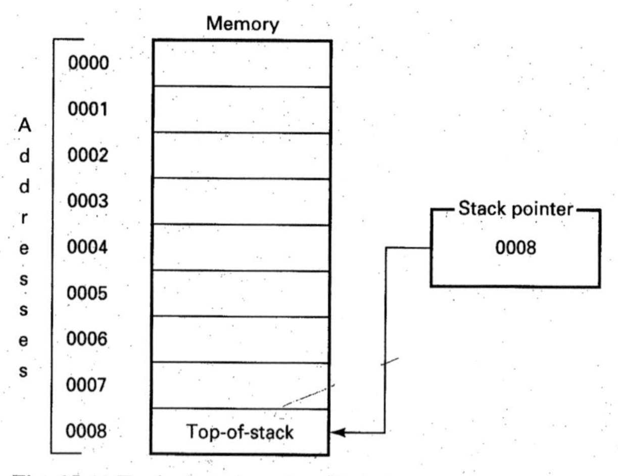

Fig. 15-11 Typical stack and stack pointer.

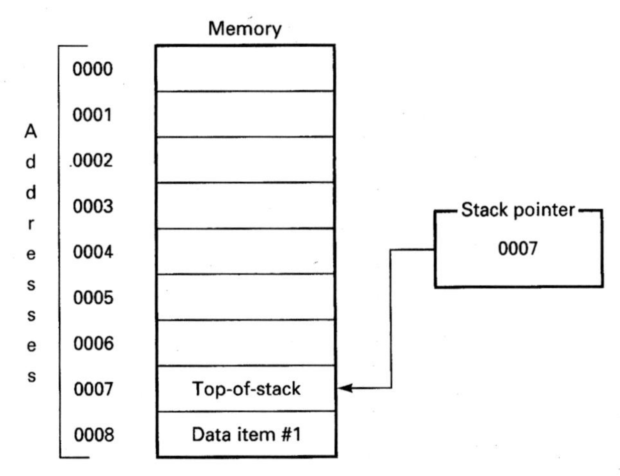

Fig. 15-12 Typical stack and stack pointer.

At this time the stack pointer is ''pointing'' to memory location 0008; therefore data item #1 will be placed in the stack at that memory location. The act of putting a piece of data in the stack is called *pushing* data onto the stack. It is as though the data is being pushed in from the top. Now look at Fig. 15-12.

We have pushed data item #1 onto the stack and the stack pointer has been decremented or decreased by one, which means that it is now pointing to memory location 0007. Location 0007 is the top-of-the-stack now. Now let's push data item #2 onto the stack. The stack will appear as it does in Fig. 15-13.

When data item #2 was *pushed* onto the stack, it went into the location the stack pointer was pointing to—which was 0007. The stack pointer was then decremented to 0006. This process will be repeated until it appears as it did in Fig. 15-10.

At some point we will need this data in the stack, so we will remove it from the top-of-the-stack. This is called *popping* or *pulling* the data from the stack. We simply

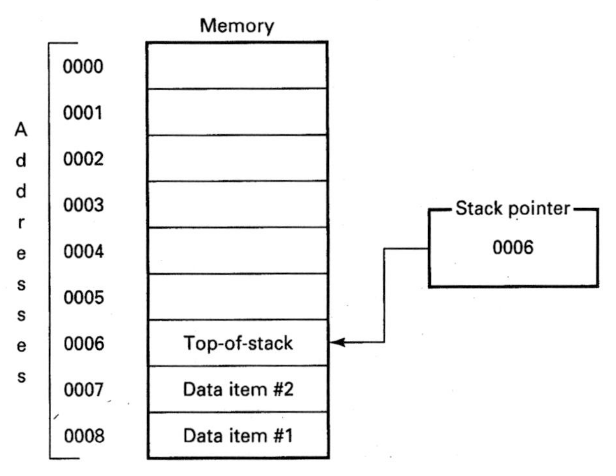

Fig. 15-13 Typical stack and stack pointer.

reverse the whole process. As each data item is removed, the stack pointer will drop, which in this case means that it will point to the next-greater memory address.

EXAMPLE 15-4

Refer to Fig. 15-13. If we *pull* data item #2 from the stack, will the stack pointer increment or decrement? What hexadecimal value will appear in the stack pointer?

SOLUTION

The stack pointer will be incremented as data item #2 is pulled from the stack. The hexadecimal value 0007 will appear in the stack pointer. In fact, the stack will appear as it did in Fig. 15-12.

Width of Registers

All registers have a maximum capacity. That is, they will only hold a certain number of bits. The width is generally 8, 16, or 32 bits.

8-Bit Registers

An 8-bit register is one that is *8 bits wide*. This means it can hold 1 byte as shown in Fig. 15-14. Most computers and trainers you will be using will not display an 8-bit register in binary. Instead, they will have a hexademical display. If you have forgotten how to convert binary to hexadecimal and hexadecimal to binary, review that section in Chap. 1.

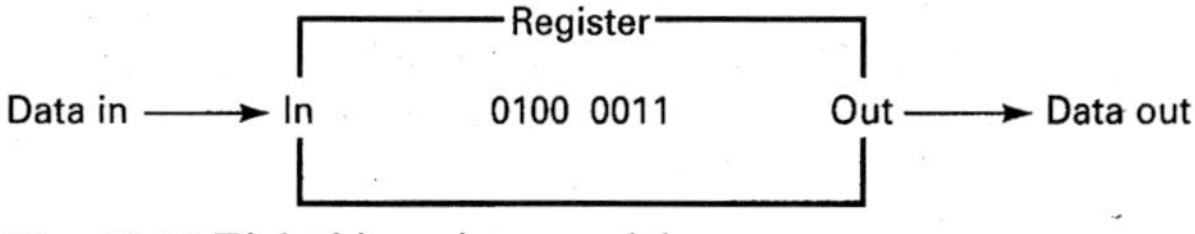

Fig. 15-14 Eight-bit register model.

It is often useful to separate the 8 bits into two groups of 4. The left group of 4 is called the upper nibble, and the right group of 4 is called the lower nibble. This is illustrated in Fig. 15-15.

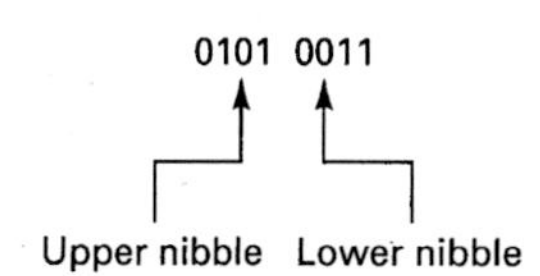

Fig. 15-15 Upper- and lower-nibble positions.

EXAMPLE 15-5

If a register contained the binary number shown in Fig. 15-16, what would appear in the hexadecimal display for that register?

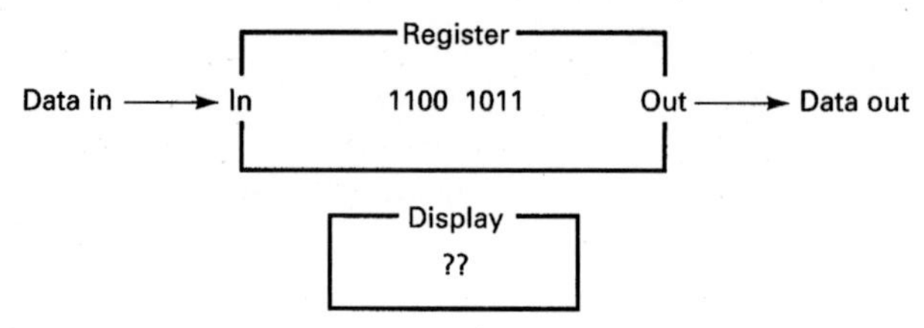

Fig. 15-16 Example A.

SOLUTION

The upper nibble, 1100, is the same as the hexadecimal digit C. The lower nibble, 1011, is the same as the hexadecimal digit B. Therefore, the hexadecimal display will show CB.

16-Bit Registers

A 16-bit register of course is *16 bits wide*. This is illustrated in Fig. 15-17. As you can see, the 16 bits are again separated into groups of 4. Each nibble, or group of 4, will be represented in the display as 1 hexadecimal digit.

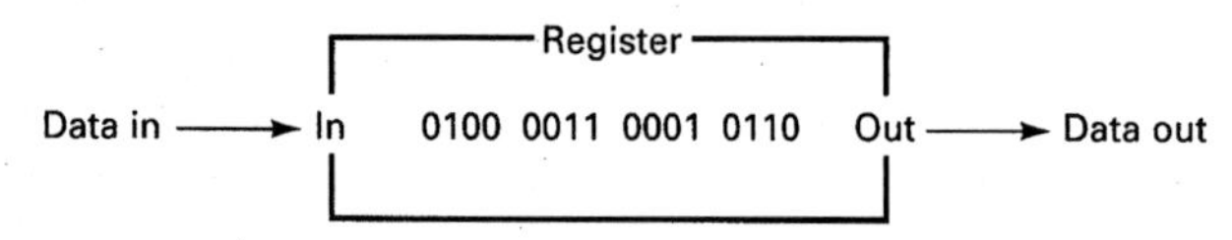

Fig. 15-17 Sixteen-bit register model.

EXAMPLE 15-6

In Fig. 15-18, what are the binary contents of the register when the display is as shown?

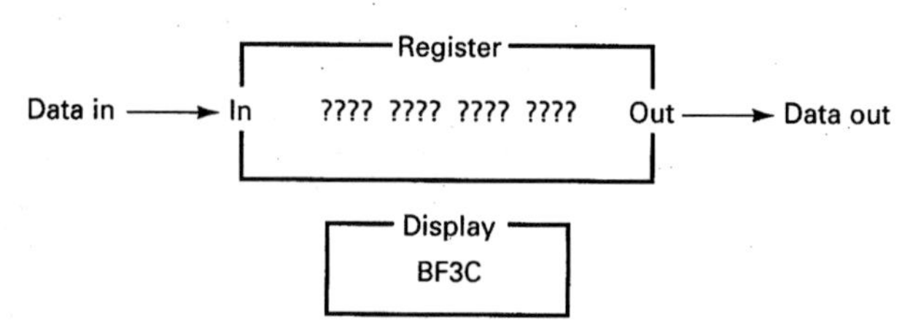

Fig. 15-18 Example B.

SOLUTION

The far left digit (also called the *most significant digit*), the B, has a binary equivalent of 1011. The F would be 1111. The 3 would be 0011. And the hexadecimal digit C would be represented by 1100 in binary. Putting the four nibbles together produces 1011 1111 0011 1100, which constitutes the binary contents of this register.

Specific Microprocessor Families

The rest of this chapter is divided into sections, each of which is devoted to one particular microprocessor family. Go to the section which discusses the microprocessor family you are using.

15-3 6502 FAMILY

Let's look at specific characteristics of the 6502 family of microprocessors.

Accumulator

The accumulator in the 6502 family of microprocessors is 8 bits wide. The 6502 has only one accumulator, unlike others which have more than one. Figure 15-19 shows what it looks like.

General-Purpose Registers

The 6502 has no general-purpose registers. The functions they perform must be accomplished in the 6502 by using the accumulator, index registers, and memory.

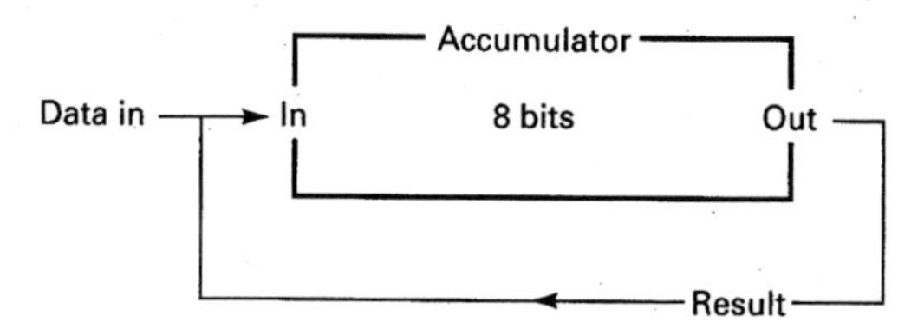

Fig. 15-19 6502 accumulator model.

Program Counter

The 6502 family program counter, as shown in Fig. 15-20, is 16 bits wide and is divided into an upper half which we have labeled PC_H (*program counter high*) and a lower half which we have labeled PC_L (*program counter low*).

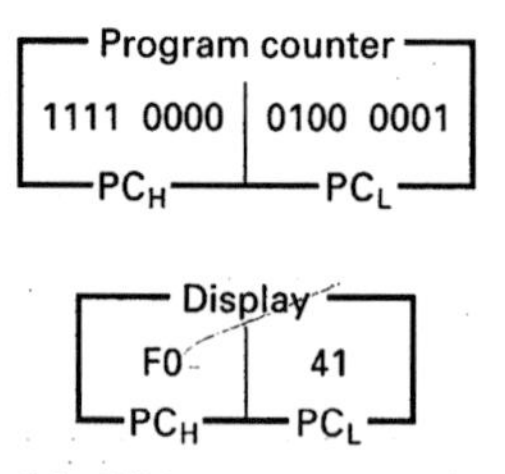

Fig. 15-20 Sixteen-bit 6502 program counter and display.

Most of the time it operates as one 16-bit counter, but there are times, particularly when subroutines are involved, when the division into 2 bytes is necessary. The display for the program counter will appear as four hexadecimal digits as shown in the figure.

Index Registers

The 6502 has two index registers. They are each 8 bits wide. One is the X index register, and the other is the Y index register.

Status Register

The 6502 status register contains 8 bits, but only 7 are actually used. The layout of this register is shown in Fig. 15-21.

The 6502 has several flags in addition to those mentioned in the New Concepts section of this chapter.

The break flag keeps track of what are called "software interrupts." When the programmer puts a BRK (**BReaK**) instruction in the program telling the microprocessor to stop, the programmer "interrupts" the program in progress. If this occurs, the break flag is set.

The decimal mode flag, when set, tells the microprocessor to assume that any numbers which it is instructed to add or subtract are BCD (binary-coded decimal) numbers instead of regular binary numbers. This will result in a BCD answer.

During addition the carry flag in the 6502 is used as described in the New Concepts section of this chapter. When a carry goes out from bit 7 of the accumulator, it goes into the carry bit. During subtraction, however, if a borrow is needed from the carry bit by bit 7, then the carry flag is cleared (0). If you think of it as though the 1 that was needed during the borrow actually came from the carry bit, it will be easier to remember. Please note that other microprocessors handle this situation with the carry flag and subtraction in just the opposite manner.

Stack and Stack Pointer

The 6502 has a stack with a maximum size of 256 bytes or memory locations. The stack pointer is 8 bits wide with a 9th bit that is always set. Figure 15-22 shows it in more detail.

The greatest memory address (lowest position) which can be designated as the top-of-the-stack is $1\ 1111\ 1111_2$, which is $01FF_{16}$. Each time another number is pushed onto the stack, the top-of-the-stack rises, which means that the stack pointer is decremented by one (since smaller-numbered memory addresses are toward the top). The smallest address which can be designated as the top-of-the-stack is $1\ 0000\ 0000_2$, which is 0100_{16}. This is not always the top; it is simply the highest position (smallest memory address) at which the top can exist.

We will look at the stack and its uses in later chapters.

Complete Model

Let's look at a complete model of the 6502 family of microprocessors. Refer to Fig. 15-23.

In our model we do not show the binary numbers that are actually in each register or location but, rather, the hexadecimal numbers which appear in the display of microprocessor trainers. The exception is the status register, in which both binary and hexadecimal are shown. The small h's and b's represent the data that would be in each register or memory location. Each "h" stands for one hexadecimal digit or nibble—which is to say, 4 bits. Each "b" stands for 1 bit. When we use this model in later chapters, we will place actual values in place of the h's and b's.

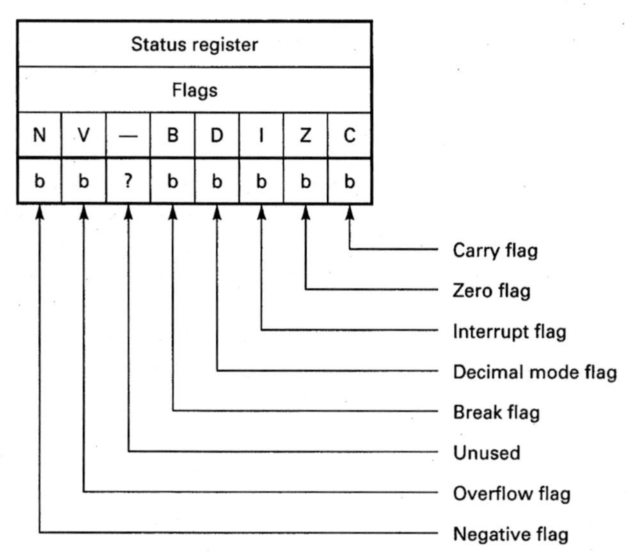

Fig. 15-21 6502 family status register. (b's represent bits.)

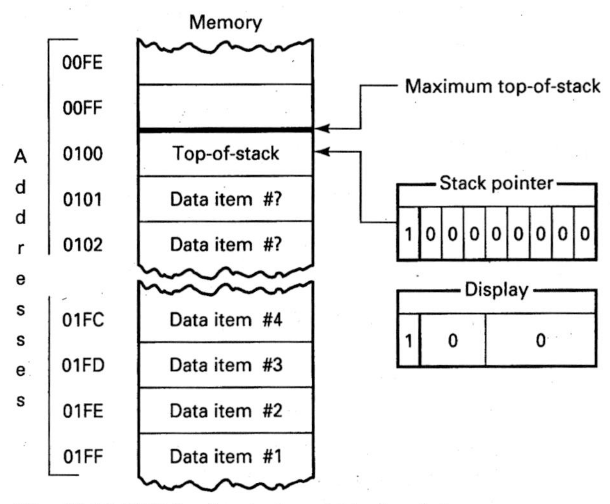

Fig. 15-22 6502 family stack and stack pointer.

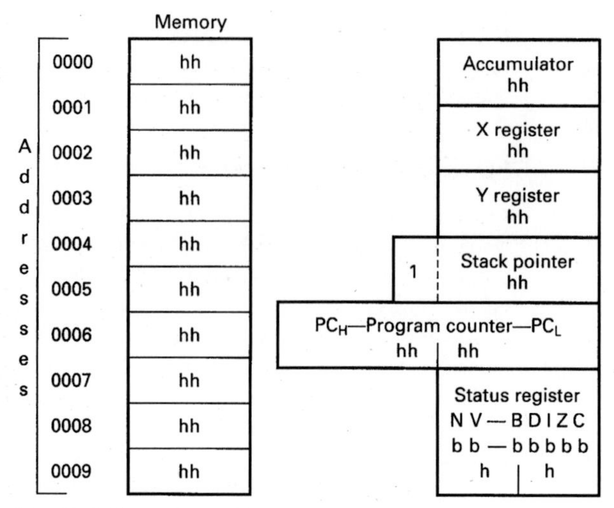

Fig. 15-23 Complete 6502 programming model.

15-4 6800/6808 FAMILY

This section covers the Motorola 6800 and 6808 microprocessors. The 6809 is an enhanced version of the 6800/6808, but most of this section can be applied to the 6809 as well. The 6809 has all of the features of the 6800 plus additional ones. The 6800 and 6808 are the primary subjects of this section, but some differences in the 6809 are mentioned.

Accumulators

The 6800/6808 microprocessors have two 8-bit accumulators. Each has the same capabilities; that is, neither is a general-purpose register. Both are true accumulators. (General-purpose registers do not have all of the features of an accumulator.) Figure 15-24 illustrates their functions.

The operation of these accumulators is the same as that described in the New Concepts section of this chapter. One note of interest concerning the 6809. It has the same 8-bit accumulators; however, it has the additional ability to treat the two as a single 16-bit accumulator known as *accumulator D* and has special instructions for such operation.

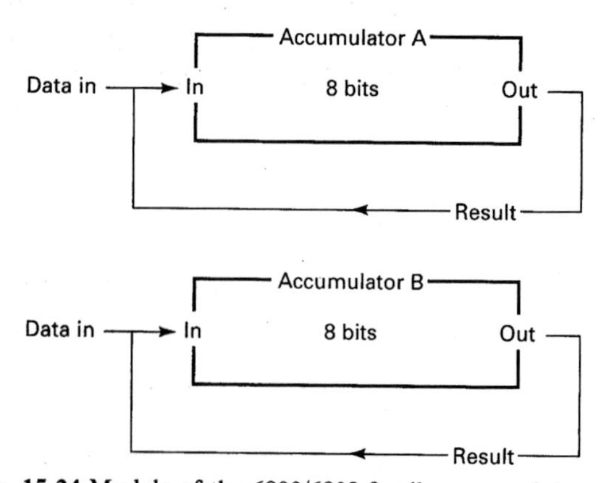

Fig. 15-24 Models of the 6800/6808 family accumulators.

General-Purpose Registers

The 6800/6808, like the 6502, has no general-purpose registers. Their functions must be performed by using the accumulators, index register, and memory.

Program Counter

The 6800, 6808, and 6809 each have 16-bit program counters. The 6800 family program counter, as shown in Fig. 15-25, is 16 bits wide but is divided into an upper half which we have labeled PC_H (for *program counter high*) and a lower half we have labeled PC_L (for *program counter low*). Most of the time it operates as one 16-bit counter, but there are times, particularly when subroutines are involved, when the division into 2 bytes is necessary. The display for the program counter will appear as four hexadecimal digits as shown in the figure.

Index Register

The 6800 and 6808 microprocessors each have one 16-bit index register called the *X index register*. The 6809 has two 16-bit registers named the *X index register* and the *Y index register*.

The 6800 family's index registers operate as described in the New Concepts section of this chapter and will be discussed in more detail in later chapters.

Condition Code Register

The 6800/6808 condition code register (called a *status register* in other microprocessors), which is shown in Fig. 15-26, is composed of 6 flags or bits in an 8-bit register. The 2 most significant bits are not used and are always set (1).

In the 6809 the 2 bits that are unused on the 6800/6808 have functions and are called the *E flag* and the *F flag*. They will not be discussed in this text.

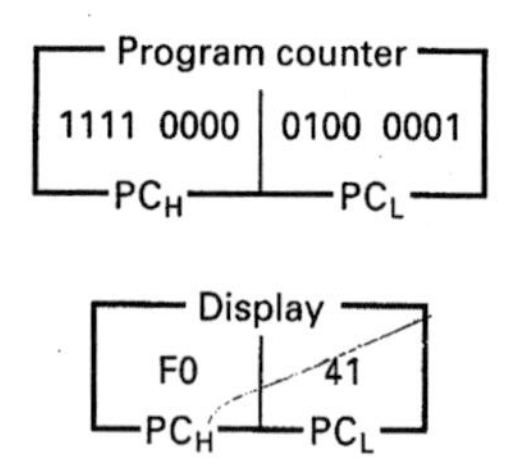

Fig. 15-25 Sixteen-bit 6800 family program counter and display.

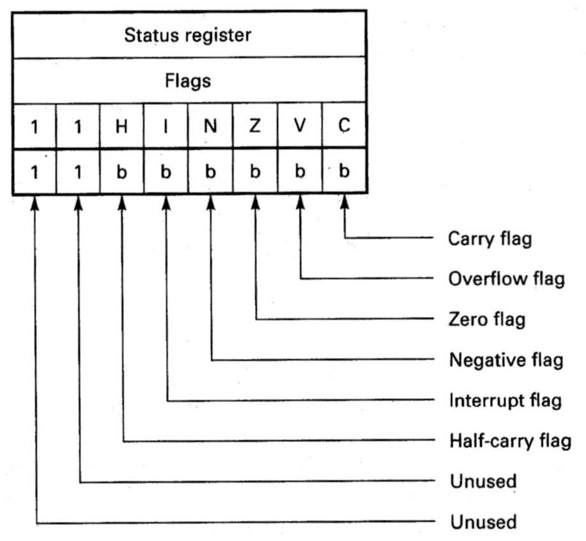

Fig. 15-26 6800/6808 status register. (b's represent bits.)

The carry flag in the 6800 family is set (1) when either a *carry or borrow* from bit 7 occurs. (The 6502 by contrast sets the flag for a carry but clears it for a borrow.)

All flags used in the 6800/6808 operate as described in the New Concepts section of this chapter.

Stack and Stack Pointer

The 6800/6808 has a 16-bit stack pointer which uses RAM for the stack itself. It operates as described in the New Concepts section of this chapter.

The 6809 has a second stack called the *user stack* which operates in a fashion similar to the first stack, which is called the *hardware stack*. The user stack is not used for interrupts and subroutines but is left free for the programmer to use.

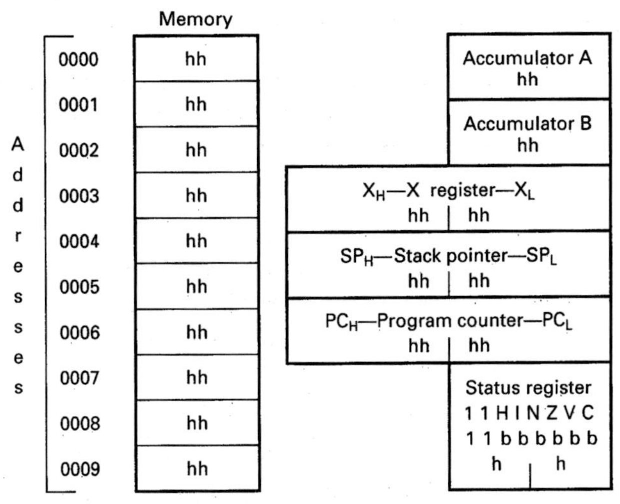

Fig. 15-27 Complete 6800/6808 programming model.

Complete Model

Let's look at a complete model of the 6800 family of microprocessors. Refer to Fig. 15-27.

In our model we do not show the binary numbers that are actually in each register or location but, rather, the hexadecimal numbers which appear in the display of microprocessor trainers. The exception is the status register in which both binary and hexadecimal are shown. The small h's and b's represent the data that would be in each register or memory location. Each "h" stands for one hexadecimal digit or nibble—which is to say, 4 bits. Each "b" stands for 1 bit. When we use this model in later chapters, we will place actual values in place of the h's and b's.

15-5 8080/8085/Z80 FAMILY

This section deals with the 8080 and 8085 microprocessors from Intel and the Z80 microprocessor manufactured by the Zilog Corp.

The 8080 and 8085 are nearly identical, the 8085 being a slightly improved version of the 8080. Except for two instructions, the instruction sets for the two chips are identical.

The Z80 is a considerably enhanced version of the 8080. It understands all the instructions of the 8080 and many more. It has all the registers of the 8080 plus a number of additional registers. We will cover only those aspects of the Z80 that are found in the 8080 and 8085 at this time.

Accumulator

The 8080/8085/Z80 chips have one 8-bit accumulator. It operates as described in the New Concepts section of this chapter. Its operation is shown in Fig. 15-28. The Z80 also has a second *alternate* accumulator.

General-Purpose Registers

The 8080/8085/Z80 chips have an abundance of general-purpose registers. These registers are arranged in pairs. Notice the arrangement of one of these pairs in Fig. 15-29.

In this figure, 8 bits of data can go into and out of either register B or C. Or, 16 bits can go into and out of the pair, at which point they act as one 16-bit register.

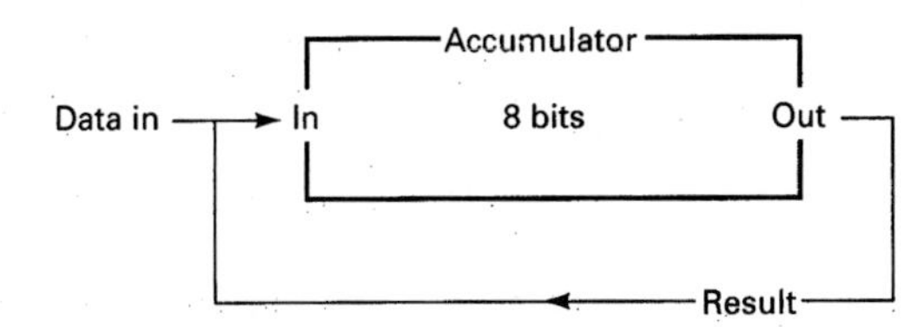

Fig. 15-28 8080/8085/Z80 accumulator model.

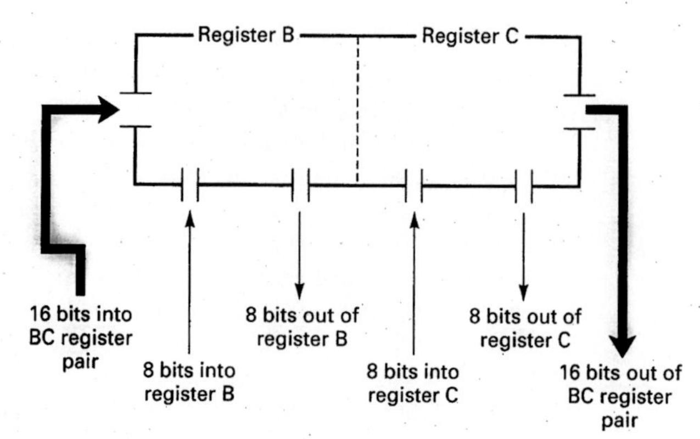

Fig. 15-29 Model of 8080/8085/Z80 general-purpose registers.

There are three sets of these general-purpose register pairs. They are the BC pair, the DE pair, and the HL pair. The letters B, C, D, and E are assigned to stand for each register. The letters H and L stand for high and low. The HL register pair is usually used for a different purpose than the other two pairs. We will discuss that purpose more in a later chapter.

Each of these registers has a mate, or "alternate," register in the Z80.

Program Counter

The 8080/8085/Z80 chips each have a 16-bit program counter which operates as described in the New Concepts section of this chapter. This program counter, as is the case with the 6502 family and the 6800 family, is divided into two halves for some operations. The upper byte or 8 bits are called the PC_H (for *program counter high*), and the lower byte is called the PC_L (for *program counter low*). See Fig. 15-30.

Most of the time the program counter operates as one 16-bit counter, but there are times, particularly when subroutines are involved, when division into 2 bytes is necessary. The display for the program counter will appear as four hexadecimal digits as shown in the figure.

Index Register(s)

The 8080 and 8085 have no index registers. The Z80 has two—an X index register and a Y index register. The index registers in the Z80 are each 16 bits wide.

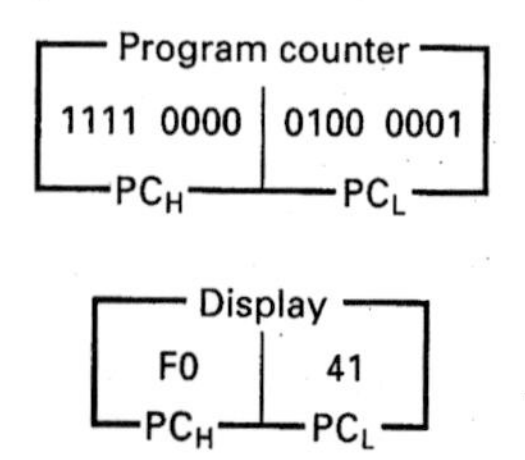

Fig. 15-30 Sixteen-bit 8080/8085/Z80 program counter and display.

Status Register

The status register in the 8080 and 8085 contains five flags in an 8-bit register. See Fig. 15-31.

The parity flag involves a topic which has not been discussed yet. *Parity* refers to the number of 1s in a binary number. *Even parity* exists when there is an even number of 1s. For example, the binary number 0110 000 has even parity because it has two 1s, and 2 is an even number. *Odd parity* exists when there is an odd number of 1s. For example, the binary number 0111 0000 has odd parity because there are three 1s, and 3 is an odd number. It is sometimes useful to keep track of parity for error-checking routines and in data communications. If the parity is even, the parity flag becomes set (1); if parity is odd, it clears (0).

The Z80 has the same five flags as the 8080 and 8085, and in the same positions, plus one additional flag. See Fig. 15-32.

The half-carry flag in the Z80 has exactly the same function as the auxiliary carry in the 8085/8080.

The parity flag in the Z80 has a dual role—that of parity checking and that of warning the programmer of 2's-complement overflow. Also, the Z80 has a negative or sign flag (the 8080 and 8085 do not have one) which operates as described in the New Concepts section of this chapter.

Stack and Stack Pointer

The 8080, 8085, and Z80 each have a stack with a 16-bit stack pointer which operates as described in the New Concepts section of this chapter.

Complete Model

Let's look at a complete model of the 8080/8085/Z80 family of microprocessors. Refer to Fig. 15-33 at this time.

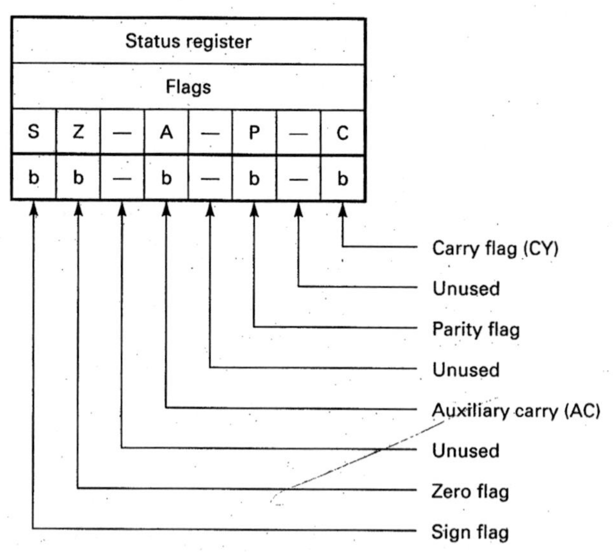

Fig. 15-31 8080/8085 status register. (b's represent bits.)

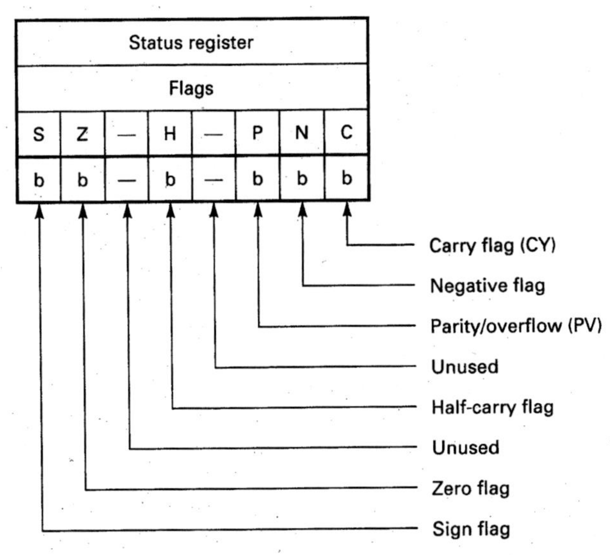

Fig. 15-32 Z80 status register. (b's represent bits.)

A couple of points concerning differences between the 8080/8085 and the Z80 should be noted. Figure 15-33 is a model of the 8080/8085. The Z80 has an additional set of alternate registers and two index registers which are not shown in the model. The status register in the Z80 has an additional flag called the *negative flag*. And the auxiliary carry flag in the 8080/8085 is usually called the half-carry flag in the Z80.

In our model we will not show the binary numbers that are actually in each register or location but rather the hexadecimal numbers which appear in the display of microprocessor trainers. The exception is the status register in which both binary and hexadecimal are shown. The small

Addresses	Memory
0000	hh
0001	hh
0002	hh
0003	hh
0004	hh
0005	hh
0006	hh
0007	hh
0008	hh
0009	hh
000A	hh

	Accumulator hh
Register B hh	Register C hh
Register D hh	Register E hh
Register H hh	Register L hh
SP_H—Stack pointer—SP_L hh	hh
PC_H—Program counter—PC_L hh	hh
	Status register S Z — A — P — C b b — b — b — b h | h

Fig. 15-33 Complete 8080/8085 and Z80 (8080 subset) programming model.

h's and b's represent the data that would be in each register or memory location. Each "h" stands for one hexademical digit or nibble, which is to say 4 bits. Each "b" stands for 1 bit. When we use this model in later chapters, we will place actual values in place of the h's and b's.

There is one point of significant difference between the 8080/8085/Z80 family and the 6502 or 6800 family. In the case of the 6502 and 6800 microprocessors, the registers and accumulators are completely independent of one another. In the 8080/8085/Z80 family, the six registers, namely B and C, D and E, and H and L, can operate as six independent 8-bit registers or as three 16-bit register pairs. This allows single operations to be performed on 16-bit data words.

15-6 8086/8088 FAMILY

In this section we will examine the 8086 and 8088 microprocessors from Intel. The 8088 is the microprocessor used in the popular IBM PCs, XTs, and compatibles. The 80286 used in ATs and the 80386 can also be used with this text.

Since the 8086/8088 chips are the successors of the 8085, they are similar to it but have many additional registers and capabilities.

Accumulator(s)

The 8086/8088 has an accumulator (shown in Fig. 15-34) which is 16 bits wide and is called AX. The upper 8 bits is called AH (*accumulator high*), and the lower 8 bits is called AL (*accumulator low*).

General-Purpose Registers

The 8086/8088 has three 16-bit or six 8-bit general-purpose registers (besides the accumulator). These are shown in Fig. 15-34 and are called the BX, CX, and DX registers. Each can be divided into an upper and lower byte called BH, BL, CH, CL, DH, and DL, respectively. Also note in the figure that A stands for accumulator, B for base, C

Accumulator AX	
AH hh	AL hh
Base BX	
BH hh	BL hh
Count CX	
CH hh	CL hh
Data DX	
DH hh	DL hh

Fig. 15-34 8086/8088 accumulator and general-purpose registers.

for count, and D for data. This can help you remember the main functions of each register.

Instruction Pointer

Instead of a program counter, the 8086/8088 has an *instruction pointer* which does what the program counter does in the 8-bit microprocessors. The instruction pointer is 16 bits wide.

Index Registers

The 8086/8088 has several index registers and pointers including the base pointer, source index, and destination index. All are 16 bits wide. These are not used alone, as with the 8-bit chips, but are used in combination with registers called *segment registers*. Figure 15-35 is a model of the 8086/8088 pointers and index registers.

Stack and Stack Pointer

The 8086/8088 stack is a standard memory stack (as are all the 8-bit microprocessors we've covered). The 8086/8088, however, can have a *very* large stack, up to 64K (65,536 bytes). The location of the top-of-the-stack is calculated by using both the stack pointer and the stack segment.

Status Register

The status register containing the 8086/8088 flags is 16 bits wide, although not all 16 bits are used. This register, shown in Fig. 15-36, has a lower byte (8 bits) which is exactly the same as the 8-bit 8085 microprocessor's status register. It has the same flags in the same positions. The upper byte has four flags which the 8085 does not have.

The first flag is the *trap* flag, which controls a single-step mode of operation.

Source index hhhh
Destination index hhhh
Stack pointer hhhh
Base pointer hhhh

Fig. 15-35 8086/8088 index registers and pointers.

Flags: New	Flags: 8085-like
— — — — O D I T	S Z — A — P — C
— — — — b b b b	b b — b — b — b
h ¦ h	h ¦ h

Fig. 15-36 8086/8088 flag register. (b's represent bits; h's represent hex digits.)

The *interrupt enable* flag controls the interrupt request pin on the microprocessor chip.

The *direction* flag controls whether the source index and destination index increment or decrement during string operations.

Finally, the *overflow* flag alerts the programmer to the existence of an arithmetic overflow when set. This is a condition in which the legal range for signed binary numbers of a particular word size has been exceeded.

Segment Registers

The 8086/8088 microprocessor has several other registers which do not exist on the 8-bit chips. These are the *segment registers*. We'll explain very briefly how they are used at this time.

All the pointers and index registers in the 8086/8088 chips are 16 bits wide; 2^{16} is 65,536 (64K) bytes. The address bus, however, is 20 bits wide. We can have memory locations extending up to 2^{20} or 1,048,576 (1 mega-) bytes. None of the pointers, including the instruction pointer, would be able to point to this wide of a range of addresses. To solve this problem, segment registers are used. Their contents are combined with the contents of the various pointers and index registers to form an address which is 20 bits wide. Exactly how this is done will be explained in a later chapter.

Complete Model

Figure 15-37 is a complete model of the 8086/8088 microprocessors.

In the model shown in Fig. 15-37 the placeholders for each binary digit are not shown. Rather, the hexadecimal digits that would be seen on a computer or trainer are indicated. The exception is the status register, in which both binary and hexadecimal placeholders are shown. The small h's and b's represent the data that would be in each register or memory location. Each "h" stands for one hexadecimal digit or nibble, which is 4 bits. Each "b" stands for 1 bit. When we use this model in later chapters, we will place actual values in place of the h's and b's.

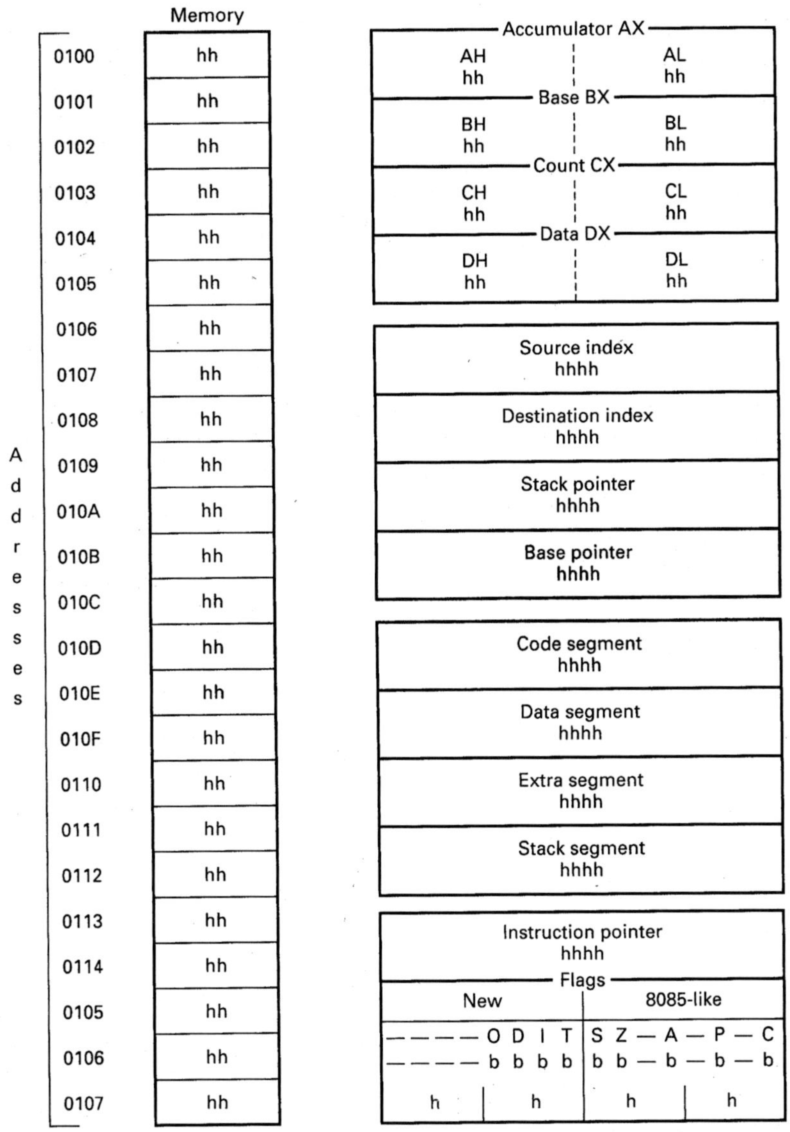

Addresses	Memory
0100	hh
0101	hh
0102	hh
0103	hh
0104	hh
0105	hh
0106	hh
0107	hh
0108	hh
0109	hh
010A	hh
010B	hh
010C	hh
010D	hh
010E	hh
010F	hh
0110	hh
0111	hh
0112	hh
0113	hh
0114	hh
0105	hh
0106	hh
0107	hh

Register	High	Low
Accumulator AX	AH hh	AL hh
Base BX	BH hh	BL hh
Count CX	CH hh	CL hh
Data DX	DH hh	DL hh

Register
Source index hhhh
Destination index hhhh
Stack pointer hhhh
Base pointer hhhh

Register
Code segment hhhh
Data segment hhhh
Extra segment hhhh
Stack segment hhhh

Register
Instruction pointer hhhh

Flags

New								8085-like							
—	—	—	—	O	D	I	T	S	Z	—	A	—	P	—	C
—	—	—	—	b	b	b	b	b	b	—	b	—	b	—	b
h				h				h				h			

Fig. 15-37 Complete 8086/8088 microprocessor programming model.

GLOSSARY

accumulator A register in a microprocessor which can not only store a byte or word of data but can have its contents operated on, with the result of that operation going back into the accumulator, replacing the previous value.

address Binary numbers which are assigned to consecutive memory locations. Specific memory locations are accessed through their addresses.

address bus A set of conductors upon which binary addresses travel to memory chips.

data bus A set of conductors which carry binary data to and from the microprocessor, memory, and I/O devices.

fetching The act of going to memory to get an instruction which is to be decoded and executed.

flag One of the bits in the status register. (See status register.)

general-purpose registers Locations which can store a byte or word of data similar to RAM but which are inside the microprocessor itself. Certain operations can usually be performed on the contents of registers.

index register A register which can be incremented and decremented and whose primary function is to point to data (often used in tables).

program counter A special-purpose register whose purpose is to keep track of the next instruction to be fetched from memory.

RAM An acronym for random-access memory. This type of memory loses its data when power is removed.

ROM An acronym for read-only memory. This type of memory does not lose its data when power is removed.

stack An area (usually in RAM) which holds vital information during subroutines and interrupts. It can also be used by the programmer as a LIFO (last-in-first-out) data storage area.

status register (condition code register) A special register whose individual bits show the status of certain conditions or the results of certain operations.

SELF-TESTING REVIEW

Read each of the following and provide the missing words. Answers appear at the beginning of the next question.

1. ________ is the type of memory which can have its contents changed thousands of times per second.
2. (*RAM*) The ________ of a memory location is similar to the address of your home and the ________ inside the memory location is similar to the beds, chairs, dishes, and so on, in your home.
3. (*address, data*) The ________ of a memory location is necessary to specify which of many locations is to be written to or read from.
4. (*address*) The address bus is usually ________ (unidirectional, bidirectional).
5. (*unidirectional*) The data bus is usually ________.
6. (*bidirectional*) Each different bit position in binary numbers represents a certain power of ________.
7. (2) Probably the most used register in a microprocessor is the ________.
8. (*accumulator*) A register which helps microprocessors to work with tables of data is the ________.
9. (*index register*) When a flag has a ________ in it, this indicates that the condition which the flag tests has *not* come true.
10. (*0*) When a flag has a ________ in it, this indicates that the condition which the flag tests *has* come true. (*1*)

PROBLEMS

General

15-1. By what means is one memory location differentiated from another?

15-2. Using decimal numbers, how many combinations can be represented by using only five digits?

15-3. Using binary numbers, how many combinations can be represented by using only 20 bits?

15-4. If we had $20{,}000_{10}$ memory locations, what would be the least number of address lines needed to describe each location? (Hint: Change 20,000 to binary or hex and determine the number of bits needed.)

15-5. What register can have its contents altered in the greatest variety of ways and is the real "workhorse" in the microprocessor?

15-6. In simplest terms, what are general-purpose registers?

15-7. What advantage do registers have over RAM?

15-8. What has the sole purpose of keeping track of the next instruction to be fetched?

15-9. In what register are the flags located?

15-10. What has happened if the zero flag has a 1 in it?

15-11. Which flag will be set if a carry from bit 7 of the accumulator is produced during an arithmetic operation?

15-12. Which flag is primarily used with binary-coded decimal numbers?

15-13. When normal stack instructions are used, can a number be pulled from somewhere in the middle of the stack?

15-14. What is taking a number from the top of the stack called?

15-15. If an 8-bit register contained the binary number 1101 1110, what hexadecimal number would appear as the display or readout for that register?

15-16. What are the binary contents of a register whose hexadecimal display reads 2A?

15-17. What would the hexadecimal display of a 16-bit register with $1100\ 0101\ 1000\ 0001_2$ as its contents read?

6502 Family

15-18. How many general-purpose registers does the 6502 have?

15-19. How wide are the index registers in the 6502?

15-20. What flag, when set, tells the 6502 to assume that binary-coded decimal (BCD) numbers are being used?

15-21. What is the maximum size of the 6502 stack?

6800 Family

15-22. How many accumulators does the 6800 have?

15-23. How wide is the 6800 program counter?

15-24. How many memory locations can the 6800 program counter reference or point to?

15-25. What are the 2 most significant bits in the 6800 condition code register used for?

8080/8085/Z80 Family

15-26. How many 8-bit general-purpose registers does the 8085 have?

15-27. How many index registers does the 8085 have?

15-28. How wide is the 8085 stack pointer?

8086/8088 Family

15-29. Describe how the 8088 accumulator is labeled and arranged.

15-30. How many 8-bit general-purpose registers does the 8088 have?

15-31. In the 8088 what has the same function as the program counter in the 8-bit microprocessors?

15-32. What 8-bit microprocessor is the lower byte of the 8088 flag register patterned after?

15-33. How large can the 8088 stack be?

DATA TRANSFER INSTRUCTIONS

New Concepts

So far we've been able to get an overview of computers, computer architecture, microprocessor architecture, programming, languages, flowcharting, and hardware. Now let's take a closer look at some of these areas.

Instruction Sets

The commands that microprocessors understand are called *instructions*, and the complete "vocabulary" of each chip is called its *instruction set*.

We will be studying the 6502, 6800/6808, 8080/8085/Z80, and 8086/8088 microprocessor families and each family's instruction set. We will deviate from this plan in two respects.

Rather than study the entire Z80 instruction set, we will study only those instructions which are common to the 8080 and 8085. (The Z80 has many instructions which neither the 8080 nor the 8085 understands. However, the Z80 understands *all* the instructions of the other two chips with only two exceptions.)

Also, we will not study the entire 8086/8088 instruction set but will omit the loop and string instructions since they have no counterpart in the 8-bit microprocessors.

Organization of This Text

You may find it helpful to know how this programming portion of the text was developed.

We are ready to begin learning about microprocessor instructions. The instructions being discussed in each chapter, the sequence in which the instructions are being presented, the sequence of the chapters, and the instruction categories have all been carefully planned.

As mentioned before, this text centers around the most popular general-purpose 8-bit microprocessors (the 6502 family, the 6800/6808 family, and the 8080/8085/Z80 family) and the 16-bit 8086/8088 family. During the preparation of this text, the instruction sets of each of these microprocessors were carefully analyzed, and it was found that each chip's instructions fell into natural groups. After each instruction was placed into its natural category, it was possible to identify those categories which were common to every microprocessor family. Those instructions which did not fall naturally into one of these common groups were placed in the group in which they most nearly fit. In short, a consistent and uniform method of classifying instructions was applied to each microprocessor family. In the tables section of this book (Part 4) you will find the complete instruction set of each chip broken down into these groups or categories.

Next, the chapters were planned to reflect these same groups. Thus, rather than trying to make the microprocessors fit the scheme of this text, the text was designed around the natural characteristics of the microprocessors. Each chip's instruction set has been broken down into the same categories as the others, and the appendixes and chapters treat each chip family equally.

Organization within Each Chapter

Most chapters start with a New Concepts section (which is where we are now). The discussion here is general—that is, it can be applied equally well to all microprocessor families and does not focus on any one family. Then, after this general discussion, the remainder of the chapter is divided into family-specific sections.

For example, if you are using the 6808 microprocessor, you would read the New Concepts section and then go immediately to the 6800/6808 Family section. There, specific information will be given to help you apply the principles discussed in the New Concepts section to the 6800/6808 microprocessors.

Now let's look at our first instruction category.

16-1 CPU CONTROL INSTRUCTIONS

The easiest instruction to learn about is an instruction which does nothing, and surprisingly, there is such an instruction. Let's look at it.

The No Operation Instruction

The *no operation* instruction does exactly that: It does nothing. This is a waste of time, and wasting time is what this instruction does best.

A microprocessor is quite fast, in some situations too fast. We can give it a certain number of these no operation instructions to stall it until a certain amount of time passes.

The no operation instruction has another use—that of filling space in the program. When writing programs, we must sometimes insert additional instructions into the middle of a program to alter the way it works or to fix a problem.

If you use one of the simpler monitors (instead of an assembler, or a monitor with an insert feature), it may not have a feature which will let you insert instructions into the middle of a program you have entered. When this happens, you must rewrite every part of the program beginning from the point at which the inserted instruction must be placed, to the end. By adding some no operation instructions at various locations in the program when you first write it, some spaces will have been created where new instructions can go. The new instructions can simply take the place of the no operation instructions.

The Halt Instruction

Called *wait*, *halt*, or *break* (depending on the microprocessor), this instruction has the obvious purpose of stopping the microprocessor. There is no *go* instruction—we'll see how that is done shortly—but there must be a way to stop the program. In some microprocessor families this is not the *only* function of this instruction, but this is all we need to be concerned with at this time.

16-2 DATA TRANSFER INSTRUCTIONS

This category of instructions has the job of transferring or moving data from one place to another. Before studying these instructions, we need to consider a basic concept.

Physical Places

Sometimes people think that when we speak of moving data from one place to another within a microprocessor, we are referring only to the "net effect" of the transfer, and that nothing actually moved.

If this were so, the operation of a microprocessor would resemble what happens when you go to the bank and transfer money from your savings account into your checking account. Though the net effect of the transfer is to decrease the amount of money in the savings account and to increase the amount in the checking account, you know that no one in the bank actually picked up the money in the savings account and placed it in another spot where your checking account was. It all happened "on paper."

This is not the case with microprocessors. The accumulators, general-purpose registers, program counter, index registers, and so on, are all real places. While it is true that tiny numbers don't move around inside the chip, the voltages representing these numbers can be made to appear in various places, so for all practical purposes the numbers themselves move.

If you experience difficulty visualizing what a program does, it may help to write down the contents of each register and/or memory location. Then as each location is changed by the program, change it on your paper. We will use this technique in many of the figures.

Where Data Is Transferred

Data is moved between registers or between registers and memory. The number of possible combinations depends on the microprocessor and how many registers it has. Figure 16-1 shows some typical possibilities.

How Data Is Transferred

Different microprocessor instruction sets use different terms to represent the act of transferring data. "Move," "load," "store," and "transfer" are all common terms.

Though we will use the term "moving," and even though thinking of it in that way will work as you become proficient, in the beginning a distinction has to be made. *When a*

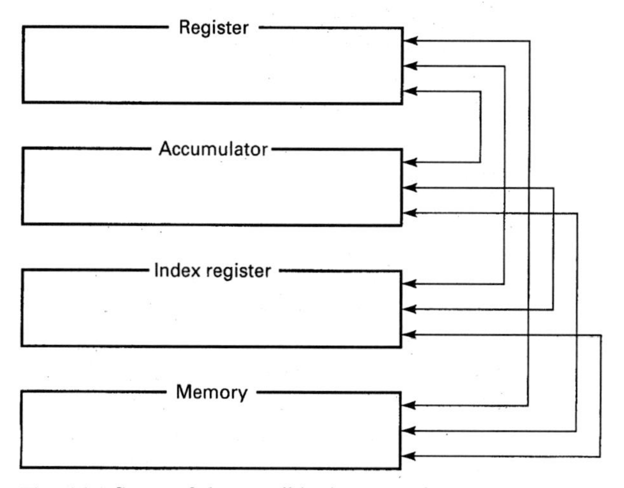

Fig. 16-1 Some of the possible data transfer combinations.

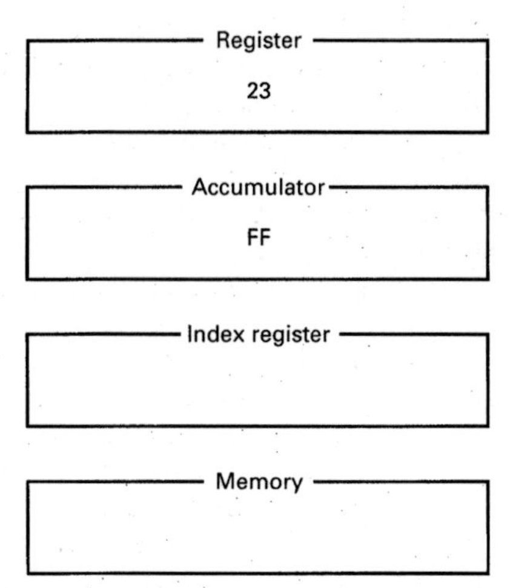

Fig. 16-2 An example of a transfer instruction.

move, load, transfer, or store instruction is executed, a duplicate of the data is actually being placed in the target register or destination.

If you were to move your car from one parking spot to another in a parking lot, your car would no longer be in its original place. This is true moving. This is *not* what happens in a microprocessor. If, however, you photocopy an important document, place the copy in a filing cabinet, and keep the original, you have not actually moved the *document* to the filing cabinet, but rather you have moved a *copy* of the document. This *is* what happens in a microprocessor.

An Example of a Transfer Instruction

Look at Fig. 16-2.

Suppose we wanted to transfer the FF in the accumulator to the register, which now contains 23. We would write a program which instructs the microprocessor to transfer the contents of the accumulator to the register. The result of this action is shown in Fig. 16-3.

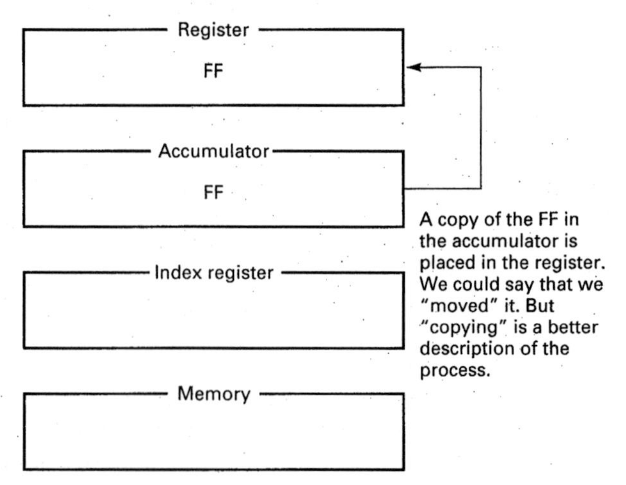

Fig. 16-3 An example of a transfer instruction.

Notice that the original FF in the accumulator is still there. We simply made a copy of it and placed the copy in the register. The original contents of the register are lost.

Now go to the section of this chapter which discusses your particular microprocessor family.

Specific Microprocessor Families

16-3 6502 FAMILY

Let's see how the ideas which were introduced in the New Concepts section apply to the 6502 microprocessor family.

CPU Control Instructions

The 6502 family has a *no operation* instruction which uses the mnemonic NOP. Refer to the Expanded Table of 6502 Instructions Listed by Category in Part 4 of this text.

Look at the NOP instruction, which is the very first instruction in this table. In the third column, the Boolean/Arithmetic Operation column, we see that this instruction does "nothing," just as we said it would. Also notice the hexadecimal number under the Op (op code) column, in this case EA. This is the actual hexadecimal code for NOP. Don't worry about the rest of the NOP information at this time.

The 6502 family doesn't have an actual halt instruction, but the instruction which serves its purpose is the **BReaK** instruction. Refer to the table again. Notice that the **BReaK** instruction uses the mnemonic BRK and has an op code of 00.

Data Transfer Instructions

Look under the **BReaK** instruction and you will see the beginning of the Data Transfer Instructions section of the table. In this section you will see a list of all of the different types of data transfer instructions available in the 6502 family. (To those with previous microprocessor experience: You may notice that we have excluded transfer instructions involving the stack. This is intentional. They have been included in the Stack Instructions category.)

Direction of Data Transfer

Let's look at the data transfer instructions more closely. The first instruction listed is the **LoaD A**ccumulator instruction. The boldfaced letters show where the LDA mnemonic came from. The third column shows the Boolean/Arithmetic Operation. This is a concise and graphic way to state exactly what this instruction does. It shows M, which stands for memory, moving toward A, which stands for the accumulator. To put it another way, the contents of a certain memory location are being transferred into the accumulator.

Recall from the New Concepts section that moving or transferring is actually more like making a copy of what's in a particular location and placing the copy in the destination.

Referring to the Expanded Table of 6502 Instructions, notice that the second and third instructions, LDX and LDY, are similar to the LDA. The difference is that they copy the contents of a particular memory location and place it in either the X register or the Y register instead of the accumulator.

It may help to have a mental picture of our programming model of the 6502, shown in Fig. 16-4, as we discuss these instructions.

We have talked about moving or copying the contents of some particular memory location to the accumulator, the X register, or the Y register. Now let's consider doing the reverse.

Look at the fourth, fifth, and sixth instructions in the table. They are STA, STX, and STY, that is, *Store the contents of the accumulator in a memory location, store the contents of the X register in a memory location*, and *store the contents of the Y register in a memory location*, respectively. The *store* instructions are just the reverse of the *load* instructions. (See the Boolean/Arithmetic Operation column.)

Now, continue referring to both the table and Fig. 16-4. The next two instructions (TAX and TXA) allow you to transfer the contents of the accumulator and X register between each other. The last two instructions (TAY and TYA) allow you to transfer the contents of the accumulator and the Y register between each other.

Op Codes

Does your computer or microprocessor trainer understand the words "load accumulator"? No. Does it understand the mnemonic LDA? No, but if you are using an assembler, the assembler translates the mnemonics into binary numbers which it does understand. (If you use a hexadecimal keypad or type in hex numbers, you do not have an assembler.) The point here is that the microprocessor inside your computer does not understand English words like "load" or mnemonics like LDA.

If you are using an assembler, the assembler program is translating the mnemonics, which the microprocessor does not understand, into something it does understand. What does the microprocessor understand? Binary numbers. In our case we will enter them as their equivalent hexadecimal value and let the monitor or assembler translate that into binary. For our purposes, at least at this point, we'll say that the microprocessor understands hexadecimal. (The monitor is part of the *firmware* built into your microprocessor trainer.)

Refer to the Expanded Table of 6502 Instructions. If we wanted to tell the microprocessor to load the accumulator from memory (the first data transfer instruction, LDA) the microprocessor chip would actually need the hex code in the seventh column over, the Op code column (Op for short). We would place the hex number A9, AD, A5, A1, B1, B5, BD, or B9, depending on which variation of the instruction we wanted to use, in the computer's memory as the first instruction to execute.

Let's try another example. What if you wanted to have the microprocessor store the contents of the Y register in memory? What would be the hex number the microprocessor would need to understand what you wanted to do? (You should have said either 8C or 84 or 94 from the STY instruction.)

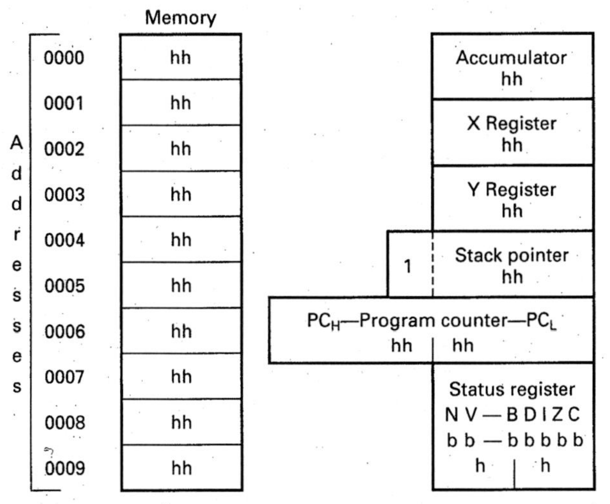

Fig. 16-4 Complete 6502 programming model.

Sample 6502 Program

Program Objective

Let's create a program which will

1. Place the number 11 in the accumulator.
2. Stop.

Creating the Program

Refer to the Data Transfer Instructions section of the Expanded Table. Do you see an instruction which could be used to place a number in the accumulator? Look in the Boolean/Arithmetic Operation column. You need an instruction which has an arrow pointing to the accumulator. There are three such instructions—LDA, TXA, and TYA. Since we don't want to involve the X register or Y register, LDA will be our choice.

The next step is to determine which of the LDA instructions to use. There are eight. The key to this decision is in the Address Mode column. The LDA instruction which has Immediate in the address column is the one we want.

Addr	Obj	Assembler	Comment
0000	A9	LDA #$11	Load the accumulator with the number (11) immediately following the LDA# op code (A9)
0001	11		
0002	00	BRK	Halt

Fig. 16-5 Sample program. (*Note:* The addresses should be an area where user programs can be placed. If 0000 is not such a place on your system, then you will need to change these addresses.)

Immediate addressing tells the microprocessor that the data it needs will be coming *immediately* after the op code. We will learn more about addressing modes in the next chapter.

Finally, you want the program to stop. The instruction which does this is in the CPU Control Instructions section of the Expanded Table. The BRK instruction is the obvious choice.

Entering the Program

The completed program is shown in Fig. 16-5. We'll see how to enter it into your microprocessor first by using an assembler and then without an assembler.

Note that the column labeled Obj contains the actual 6502 op codes while the Assembler column contains the mnemonic and data in a format similar to that which is used by an assembler.

Refer to the LDA instruction in the Expanded Table. To the right of the word Immediate, you see LDA #$dd. This is in the Assembler Notation column and describes how many assemblers require that you type this instruction. With eight different **L**oa**D A**ccumulator instructions, the assembler must know which one you want. The format of the information after the LDA is how the various forms of the command are differentiated. The # means that the data to be used is coming *immediately* after the command itself. The $ indicates that it is a hexadecimal number. The dd simply stands for two hexadecimal digits of data. (Each d stands for one nibble or 4 bits.)

It is important to remember that we are talking about a typical assembler format; however, there is no absolute standard that must be followed. Refer to the manual which came with your assembler, or ask your instructor for information about your assembler's format.

We are going to enter this program into memory starting at location 0000 (hexadecimal). If the trainer you are using does not allow programs to be placed in these memory locations, refer to your manual and substitute addresses which are valid for your trainer or computer for those shown in Fig. 16-5.

If you are using an assembler, please enter the program at this time. It will look similar to what is shown in Fig. 16-6.

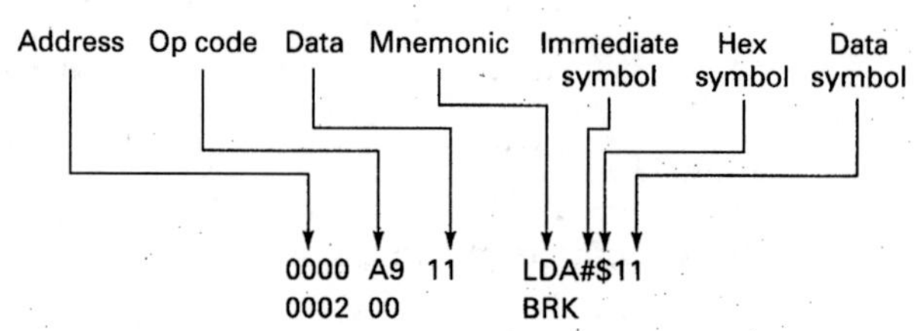

Fig. 16-6 Disassembly of the sample program. (The mnemonic and the data to the right of the mnemonic are all that's typed in during assembly.)

Now place 0s in the accumulator, the X register, and the Y register so that you will know what numbers are in each register before the program is run. Refer to Fig. 16-7 to see what memory and the registers should look like.

If you are *not* using an assembler, you must look up the op codes by hand in the Expanded Table. This is called *hand-assembly*. Let's go through the necessary steps for hand-assembly.

To the right of the LDA #$dd, in the op code (op for short) column you will see the hexadecimal number A9. This is the 6502 op code, which stands for *Load the accumulator with the number immediately following this op code*. Set your trainer so that the memory address at which the next instruction will be loaded is someplace within the area allowed for user programs. We chose 0000, but you

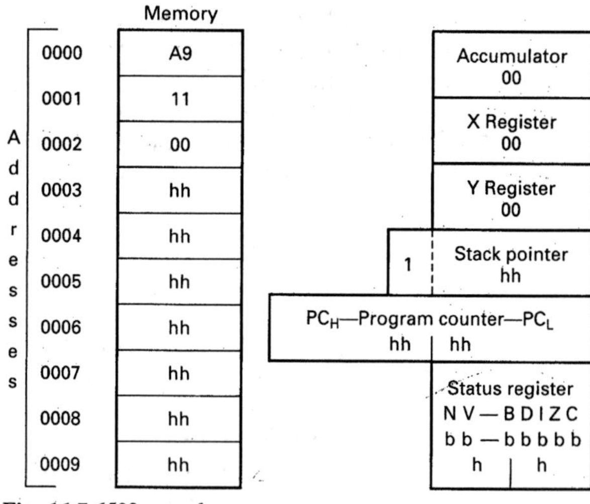

Fig. 16-7 6502 sample program.

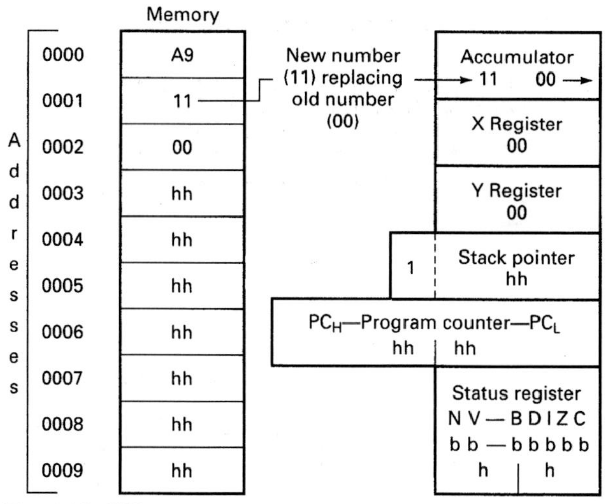

Fig. 16-8 6502 sample program.

may need to use another location. Enter the number A9 into the first available memory location. Since this was a *load accumulator immediate* instruction, the microprocessor will expect the next address, which *immediately* follows the op code, to contain the number which is to be placed in the accumulator. Therefore enter 11 next. In the third address enter 00, which is the op code for the BRK instruction.

Enter 0s into the accumulator, X register, and Y register at this time so that you will know the condition of these registers before the program is run.

If you check your registers and memory, you should see what is shown in Fig. 16-7 (although you may have placed the program at a different memory location). The h's and b's represent hex and binary digits which we are not concerned with at this time.

Running the Program

Let's use Fig. 16-8 during our analysis of program operation. The first op code is A9, which means *Load the accumulator with the contents of the next memory location*, or more properly, *Place a copy of the contents of the next memory location in the accumulator*. As you see, the number 11 is replacing 00 in the accumulator. The program then continues to the next instruction op code, 00, which stands for BREAK, and stops.

Checking the Results of Program (Analysis)

After running the program, you should have 00 in the X register, 00 in the Y register, and 11 in the accumulator. The program does what we designed it to do.

Here's one for you to try.

EXAMPLE 16-1

Manually place 00s in the accumulator, the X register, and the Y register. Next, write a program which will

1. Place the hex number EE in the accumulator.
2. Transfer (copy) the contents of the accumulator (A) into the X register (X).
3. Transfer (copy) the contents of the accumulator (A) into the Y register (Y).
4. Stop.

SOLUTION

Figure 16-9 shows the completed program. Figure 16-10 shows memory and the registers and what happens during program execution.

16-4 6800/6808 FAMILY

Let's see how the ideas which were introduced in the New Concepts section apply to the 6800/6808 microprocessor family.

CPU Control Instructions

The 6800/6808 family has a *no operation* instruction which uses the mnemonic NOP. Refer to the Expanded Table of 6800 Instructions Listed by Category in Part 4 of this text.

In the third column, called the Boolean/Arithmetic Operation column, we see that this instruction does "nothing," just as we said it would. Also notice the hexadecimal number under the op (op code) column, in this case 01. This is the actual hex code for NOP.

The 6800 family doesn't have an actual halt instruction, but the instruction which serves its purpose is the **WAI**t for Interrupt instruction. (Bold type and capital letters

Addr	Obj	Assembler	Comment
0000	A9	LDA #$EE	Copy the hex number EE into the accumulator (A)
0001	EE		
0002	AA	TAX	Transfer the contents of A into X
0003	A8	TAY	Transfer the contents of A into Y
0004	00	BRK	Stop

Fig. 16-9 Example 16-1 program listing.

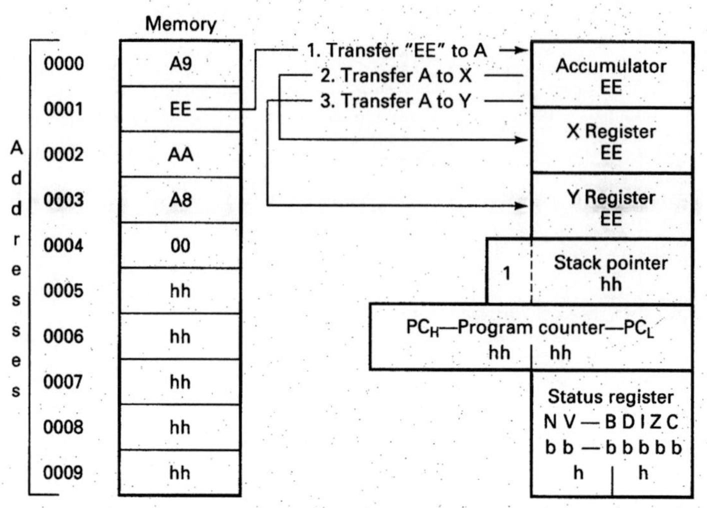

Fig. 16-10 Example 16-1 program analysis.

identify the mnemonic.) Refer to the Expanded Table of 6800 Instructions. Notice that the wait for interrupt instruction uses the mnemonic WAI and has an op code of 3E.

Data Transfer Instructions

Look in the Expanded Table at the next entry underneath the WAI instruction. This is the first entry in the Data Transfer Instructions section, which is a list of all of the different types of data transfer instructions available in the 6800/6808 family. (To those with previous microprocessor experience: You may notice that we have excluded transfer instructions involving the stack. This is intentional. They have been included in the Stack Instructions category.)

Direction of Data Transfer

Let's look at this Data Transfer section a little more closely. The first instruction listed is the **L**oa**D** **A**ccumulator **A** instruction. The boldfaced letters show where the LDAA mnemonic came from. The third column shows the Boolean/Arithmetic Operation. This is a concise and graphic way to state exactly what this instruction does. It shows M, which stands for memory, moving toward A, which stands for the accumulator. To put it another way, the contents of a certain memory location are being transferred into the accumulator.

Recall from the New Concepts section that moving or transferring is actually more like making a copy of what's in a particular location and placing the copy in the destination.

Referring to the table, notice that the second (**L**oa**D** Accumulator **B**) and seventh (**L**oa**D** **X** register) instructions are similar to the first (LDAA). The difference is that they copy the contents of a particular memory location and place it either in accumulator B or in the X register instead of accumulator A.

It may help to have a mental picture of our programming model of the 6800, shown in Fig. 16-11, as we discuss these instructions.

We have talked about moving or copying the contents of some particular memory location to accumulator A, accumulator B, or the X register. Now let's consider doing the reverse.

Look at the third, fourth, and eighth instructions in the Expanded Table. They are STAA, STAB, and STX, which is to say, *store the contents of accumulator A in a memory location, store the contents of accumulator B in a memory location*, and *store the contents of the X register in a memory location*, respectively. The STORE instructions are just the reverse of the LOAD instructions. (Note the Boolean/Arithmetic Operation column.)

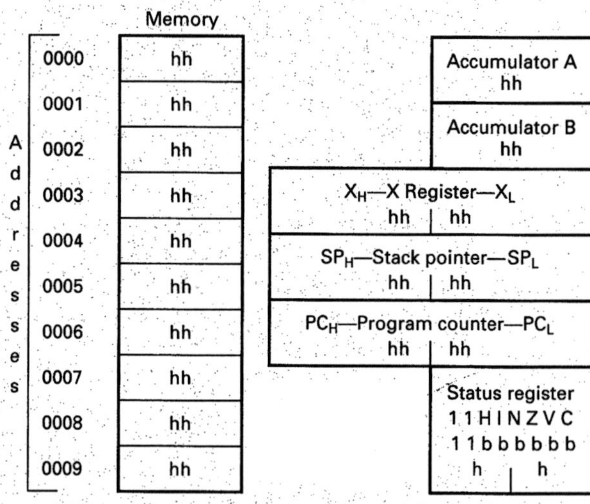

Fig. 16-11 Complete 6800/6808 programming model.

Continue referring to both the Expanded Table and Fig. 16-11. Instructions 5 and 6 in the Expanded Table (TAB and TBA) allow you to transfer the contents of accumulator A and accumulator B between each other.

The last three instructions (CLR, CLRA, and CLRB) simply transfer or place the number zero in accumulator A or B or in a memory location.

Op Codes

Does your computer or microprocessor trainer understand the words "load accumulator A"? No. Does it understand the mnemonic LDAA? If you are using an assembler, the assembler translates the mnemonic into binary numbers, which it does understand. (If you can type the mnemonic LDAA into your computer or trainer, you have an assembler. If instead you must use a hexadecimal keypad or type in hex numbers, you do not have an assembler.) The point here is that the microprocessor inside your computer does not understand English words like "load" or mnemonics like LDAA.

If you are using an assembler, the assembler program is translating the mnemonics, which the microprocessor does not understand, into something it does understand. What does the microprocessor understand? Binary numbers. In our case we will enter them as their equivalent hexadecimal value and let the monitor or assembler translate that into binary. For our purposes, at least at this point, we'll say that the microprocessor understands hexadecimal. (The monitor is part of the *firmware* built into your microprocessor trainer.)

Look again at the Expanded Table. If we wanted to tell the microprocessor to load the accumulator from memory (the first data transfer instruction, LDAA), the microprocessor chip would actually need the hex code in the seventh column over, the op code column (op for short). We would place the hex number 86, 96, A6, or B6 (depending on which variation of the instruction we wanted to use) in the computer's memory as the first instruction to execute. (We'll talk more about these variations later.)

Let's look at another example. What if you wanted to have the microprocessor store the contents of the X register in memory? What would be the hex number the microprocessor would need to understand what you wanted to do? You should have said either DF or EF or FF from the STX instruction.

Sample 6800/6808 Program

Program Objective

Let's create a program which will

1. Place the number 11 in the accumulator.
2. Stop.

Creating the Program

Refer to the Data Transfer Instructions section of the Expanded Table. Do you see an instruction which could be used to place a number in the accumulator? Look in the Boolean/Arithmetic Operation column. You need an instruction which has an arrow pointing to the accumulator. There are three such instructions—LDAA, TBA, and CLRA. Since we don't want to involve accumulator B, and since we don't want to clear accumulator A, LDAA will be our choice.

The next step is to determine which LDAA instruction to use. There are four. The key to this decision is in the Address Mode column. The LDAA instruction which has Immediate in the address column is the one we want. *Immediate addressing* tells the microprocessor that the data it needs will be coming *immediately* after the op code. We will learn more about addressing modes in the next chapter.

Finally, you want the program to stop. The instruction which does this is in the CPU Control Instructions section of the Expanded Table. The WAI instruction is the correct choice.

Entering the Program

The completed program is shown in Fig. 16-12. We'll see how to enter it into your microprocessor first by using an assembler and then without an assembler.

Note that the column labeled Obj contains the actual 6800 op codes, and the Assembler column contains the mnemonic and data in a format similar to that used by an assembler.

Addr	Obj	Assembler	Comment
0000	86	LDAA #$11	Load the accumulator with the number (11) immediately following the LDAA# op code (86)
0001	11		
0002	3E	WAI	Halt

Fig. 16-12 Sample program. (*Note:* The addresses should be an area where user programs can be placed. If 0000 is not such a place on your system, then you will need to change these addresses.)

Refer to the LDAA instruction in the Expanded Table. To the right of the word Immediate you see LDAA #$dd. This is in the Assembler Notation column and describes how many assemblers require that you type this instruction. With four different **LoaD** Accumulator **A** instructions, the assembler must know which one you want. The format of the information after the LDAA is how the different forms of the command are differentiated. The # means that the data to be used is coming *immediately* after the command itself. The $ indicates that it is a hexadecimal number. The dd simply stands for two hexadecimal digits of data. (Each d stands for one nibble or 4 bits.)

It is important to remember that we are talking about a typical assembler format; however, there is no absolute standard that must be followed. Refer to the manual which came with your assembler or ask your instructor for information about your assembler's format.

We are going to enter this program into memory starting at location 0000 (hexadecimal). If the trainer you are using does not allow programs to be placed in these memory locations, refer to your manual and substitute valid addresses in place of those shown in Fig. 16-12.

If you are using an assembler, please enter the program now. It will look similar to what is shown in Fig. 16-13.

Also place 0s in accumulator A, accumulator B, and the X (index) register so that you will know what numbers are in each register before you run the program. Refer to Fig. 16-14 to see what the memory and registers should look like.

If you are *not* using an assembler, you must look up the op codes by hand in the Expanded Table. This is called *hand-assembly*. Let's go through the necessary steps for hand-assembly.

To the right of the LDAA #$dd, in the op code (op for short) column you will see the hexadecimal number 86. This is the 6800/6808 op code, which stands for *Load accumulator A with the number immediately following this op code*. Set your trainer so that the memory address where the next instruction will be loaded is someplace within the area allowed for user programs. We chose 0000, but you may need to use another location. Enter the number 86 into the first available memory location. Since this was a *Load Accumulator A Immediate* instruction, the microprocessor will expect the next address, which *immediately* follows the op code, to contain the number which is to be placed in accumulator A. Therefore enter 11 next. In the third address enter 3E, which is the op code for the WAI instruction.

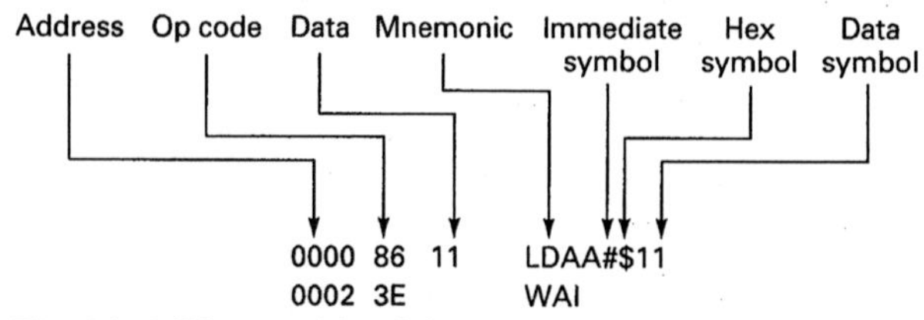

Fig. 16-13 Disassembly of the sample program.

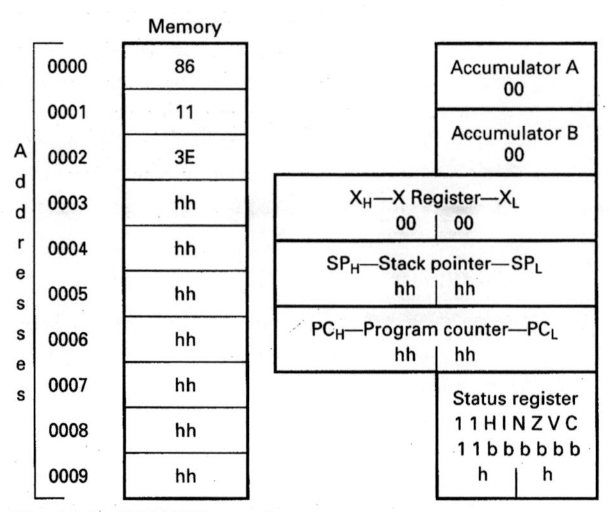

Fig. 16-14 6800/6808 sample program.

Enter 0s into accumulator A, accumulator B, and the X (index) register now so that you will know the condition of these registers before the program is run.

If you check your registers and memory, you should see what is shown in Fig. 16-14 (although you may have placed the program at a different memory location). The h's and b's represent hex and binary digits which we are not concerned with now.

Running the Program

Let's use Fig. 16-15 during our analysis of program operation.

The first op code is 86, which means, *Load accumulator A with the contents of the next memory location*, or more properly, *Place a copy of the contents of the next memory location in accumulator A*. As you see, the number 11 is

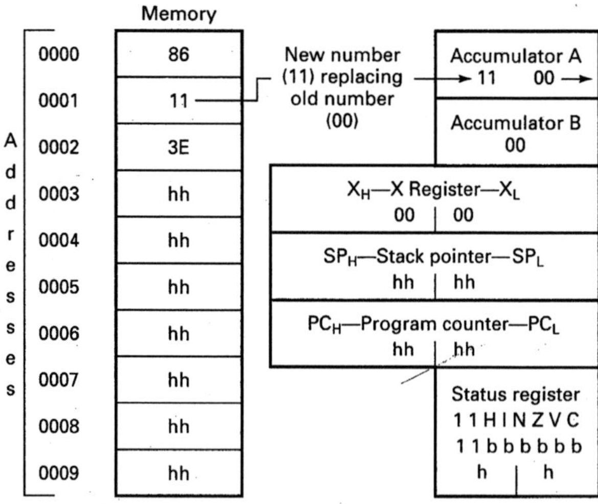

Fig. 16-15 6800/6808 sample program.

Addr	Obj	Assembler	Comment
0000	86	LDAA #$EE	Load accumulator A with the hex number immediately following the LDAA# op code (86)
0001	EE		
0002	16	TAB	Transfer the contents of A into B
0003	3E	WAI	Stop

Fig. 16-16 Example 16-2 program.

replacing 00 in the accumulator. The program then continues to the next instruction op code, 3E, which stands for **WAI**, and stops.

Checking the Results of Program (Analysis)

After running the program, you should have 00 in accumulator B and the X (index) register and 11 in accumulator A. The program does what we designed it to do.

Here's one for you to try.

EXAMPLE 16-2

First manually place 00s in accumulator A, accumulator B, and the X register. Then write a program which will

1. Load accumulator A with the hex number EE.
2. Transfer a copy of the contents of the accumulator A into accumulator B.
3. Stop.

SOLUTION

Figure 16-16 shows the completed program. Figure 16-17 shows the memory and registers and what happens during program execution.

16-5 8080/8085/Z80 FAMILY

Let's see how the ideas which were introduced in the New Concepts section apply to the 8080/8085/Z80 microprocessor family.

CPU Control Instructions

The 8080/8085/Z80 family has a *no operation* instruction which uses the mnemonic NOP. Refer to the Expanded Table of 8085/8080 and Z80 (8080 Subset) Instructions Listed by Category in Part 4 of this text.

In the ninth column, called the Boolean/Arithmetic Operation column, we see that this instruction does "nothing," as we said it would. Also notice the hexadecimal number under the op (op code) column, in this case 00. This is the actual hex code for NOP.

The 8080/8085/Z80 family has an actual halt instruction. Refer to the Expanded Table again. Notice that the halt instruction uses the mnemonic HLT [Z80 = HALT] and has an op code of 76.

Data Transfer Instructions

Refer to the Expanded Table. Underneath the halt instruction you will see the MOV A,A [Z80 = LD A,A] instruction

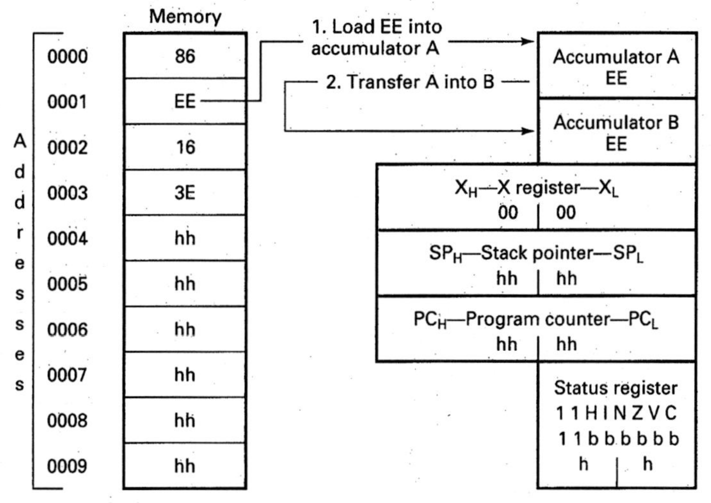

Fig. 16-17 Example 16-2 program analysis.

at the beginning of the Data Transfer Instructions section. This section is a list of all of the different types of data transfer instructions available in the 8080/8085/Z80 family. (To those with previous microprocessor experience: You may notice that we have excluded transfer instructions involving the stack. This is intentional. They have been included in the Stack Instructions category.)

Direction of Data Transfer

Let's look at the data transfer section a little more closely. The second instruction listed is the **MOV**e data to **A** from **B** instruction. The boldfaced letters help show where the MOV A,B mnemonic came from. (If you are using the Z80 microprocessor, it is the **L**oa**D** data into **A** from **B** instruction. The boldfaced letters show where the LD A,B mnemonic came from.) The ninth column shows the Boolean/Arithmetic Operation. This is a concise and graphic way to state exactly what this instruction does. It shows B, which stands for register B, moving toward A, which stands for the accumulator. To put it another way, the contents of register B are being transferred into the accumulator.

Recall from the New Concepts section that moving or transferring is actually more like making a copy of what's in a particular location and placing the copy in the destination.

It may help to have a mental picture of our programming model of the 8085/8080/Z80, shown in Fig. 16-18, as we discuss these instructions.

There are many directions in which data could be transferred with an accumulator, six registers, and memory. This can be seen in the Expanded Table. The first eight instructions transfer the contents of a register or memory location into the accumulator. (This can be seen in the Operation column and the Boolean/Arithmetic Operation column.) The second group of eight instructions copy the contents of the accumulator, one of the registers, or memory into register B. The third group of eight transfer data into register C. The fourth group into D. The fifth into E. The sixth into H. The seventh into L. And the eighth into a memory location. This makes 8 × 8 or 64 instructions just to do simple data transfers between registers.

The next group of eight instructions consists of the *Move Immediate* instructions. They move a specified number directly into a register or memory.

We will leave it to you to glance at the rest of the data transfer instructions in the Expanded Table.

If you have used the 6502 family or 6800/6808 family chips before (especially the 6502 family) and are now studying the 8085/Z80 family for the first time, you may be surprised by the great number of different instructions this family has. This is offset, however, by the relatively few addressing modes available and the simplicity this can offer the programmer. (The 6502 family, by contrast, has very few different instructions but has a large number of addressing modes for an 8-bit chip from its era.)

Op Codes

Does your computer or microprocessor trainer understand the statement "Move data to A from B"? No. Does it understand the mnemonic MOV A,B? If you are using an assembler, the assembler translates the mnemonic into binary numbers, which it does understand. (If you can type the mnemonic MOV A,B into your computer or trainer, you have an assembler. If instead you must see a hexadecimal keypad or type in hex numbers, you do not have an assembler.) The point here is that the microprocessor inside your computer does not understand English words like "Move" or mnemonics like MOV A,B.

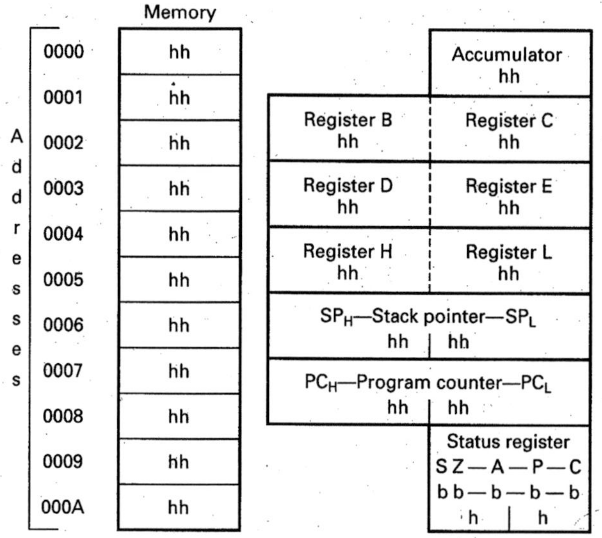

Fig. 16-18 Complete 8080/8085 and Z80 (8080 subset) programming model.

If you are using an assembler, the assembler program is translating the mnemonics, which the microprocessor does not understand, into something it does understand. What does the microprocessor understand? Binary numbers. In our case we will enter them as their equivalent hexadecimal value and let the monitor or assembler translate that into binary. For our purposes, at least at this point, we'll say that the microprocessor understands hexadecimal. (The monitor is part of the *firmware* built into your microprocessor trainer.)

Look again at the Data Transfer section of the table. If we wanted to tell the microprocessor to load the accumulator from register B (the second data transfer instruction, MOV A,B *[LD A,B]*, the microprocessor chip would actually need the hex code in the eighth column over, the op code column (op for short). We would place the hex number 78 in the computer's memory as the first instruction to execute.

Let's look at another example. What if you wanted to have the microprocessor copy the contents of the C register into the accumulator? What would be the hex number the microprocessor would need to understand what you wanted to do? You should have said 79 from the MOV A,C *[LD A,C]* instruction.

Sample 8085/Z80 Program

(*Note:* Since we are simultaneously covering the 8085 and Z80 microprocessors, we will give the 8085 mnemonic first, followed by the Z80 mnemonic in *italic* print and enclosed by square brackets, for example, MVI A,dd *[LD A,dd]*.)

Program Objective

Let's create a program which will

1. Place the number 11 in the accumulator.
2. Stop.

Creating the Program

Refer to the Data Transfer Instructions section of the Expanded Table. Do you see an instruction which could be used to place a number in the accumulator?

[*Note:* You may want to use the Mini Table of 8085/Z80 (8080 Subset) Instructions listed by Category at this time. There are so many 8085/Z80 data transfer instructions that it may prove to be a bit time-consuming to page through the Expanded Table.]

Look in the Boolean/Arithmetic Operation column (simply labeled Operation in the Mini Table). You need an instruction which has an arrow pointing to the accumulator (indicated by an A). There are 12 such instructions; using 8085 mnemonics, they are MOV A,A; MOV A,B; MOV A,C; MOV A,D; MOV A,E: MOV A,H; MOV A,L; MOV A,M; MVI A,dd; LDAX B; LDAX D; and LDA aaaa. *[Using Z80 mnemonics, they are LD A,A; LD A,B; LD A,C; LD A,D; LD A,E; LD A,H; LD A,L; LD A, (HL); LD A,dd; LD A, (BC); LD A, (DE); and LD A, (aaaa).]*

The next step is to determine which one of these instructions to use. The key to this decision can be found in the Operation or (Boolean/Arithmetic Operation) column. The data transfer instruction we want is one which will take a number (which we will place *immediately* after the instruction op code) and will transfer it into the accumulator.

The first eight instructions mentioned above take a number which is already in one of the seven 8085/Z80 registers or memory and place it in the accumulator. This is not what we want. The last three instructions take a number or data byte from a memory location and place it in the accumulator. This is not what we want either. The MVI A,dd (MoVe Immediate **dd** to A) *[Z80 = LD A, dd (LoaD **dd** into A)]* instruction takes the number *immediately* following the move instruction and places it in the accumulator. This is what we want since it allows us to specify the number 11 right after the op code for the move instruction.

Finally, you want the program to stop. The instruction which does this is in the CPU Control Instructions section. The halt instruction is the obvious choice.

Entering the Program

The completed program is shown in Fig. 16-19. We'll see how to enter it into your microprocessor first using an assembler and then without an assembler.

Note that the column labeled Obj contains the actual 8085 and Z80 op codes, and the Assembler column contains the mnemonic and data in a format similar to that used by an assembler.

Refer to the MVI A,dd *[LD A,dd]* instruction in the Mini Table. These mnemonics are used by assemblers, which means that you must type the instruction using this format. To the right of the mnemonic, in the Op column, is the op code for that particular instruction. The 8085 and Z80 microprocessors use the same op codes: Only the mnemonics are different. The dd simply stands for two hexadecimal digits of data. (Each d stands for one nibble or 4 bits.)

We are going to enter this program into memory starting at location 0000 (hexadecimal). If the trainer you are using does not allow programs to be placed in these memory locations, refer to your manual to determine where programs can be placed in memory and substitute those addresses.

If you are using an assembler, please enter the program now. It will look similar to what is shown in Fig. 16-20.

Also place 0s in the accumulator and all the general-purpose registers (registers B, C, D, E, H, and L) so that you will know what numbers are in each register before you run the program. Refer to Fig. 16-21 to see what the memory and registers should look like.

If you are not using an assembler, you must look up the op codes by hand in either the Expanded Table or the Mini Table. This is called *hand assembly*. Let's go through the necessary steps for hand assembly.

8085 mnemonics

Addr	Obj	Assembler	Comment
0000	3E	MVI A, 11	Load the accumulator with the number (11) immediately following the MVI op code (3E)
0001	11		
0002	76	HALT	Halt

Z80 mnemonics

Addr	Obj	Assembler	Comment
0000	3E	LD A, 11	Load the accumulator with the number (11) immediately following the LD A,dd op code (3E)
0001	11		
0002	76	HALT	Halt

Fig. 16-19 Sample program. (*Note:* The addresses should be an area where user programs can be placed. If 0000 is not such a place on your system, then you will need to change these addresses.)

If you look up the MVI A,dd *[LD A,dd]* mnemonic in either the Expanded Table or the Mini Table (for the 8080/8085/Z80), you will see the hex number 3E in the Op column. This is the op code which stands for, "**M**o**V**e the number **I**mmediately following this op code into the **A**ccumulator." *["**L**oa**D** the number following this op code into the **A**ccumulator."]* Set your trainer so that the memory address where the next instruction will be loaded is someplace within the area allowed for user programs. We chose 0000, but you may need to use another location. Enter the hex number 3E into the first available memory location. Since this was a **M**o**V**e **I**mmediate to **A**ccumulator *[**L**oa**D** **A**ccumulator]* instruction, the microprocessor will expect the next address, which *immediately* follows the op code, to contain the number which is to be placed in the accumulator. Therefore enter 11 next. In the third address enter 76, which is the op code for the halt instruction.

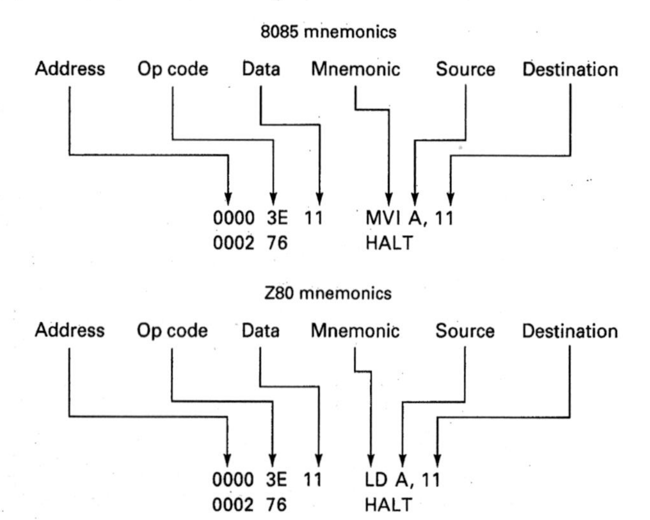

Fig. 16-20 Disassembly of the sample program.

Enter 0s into the accumulator and all the general-purpose registers at this time so that you will know the conditions of these registers before the program is run.

If you check your registers and memory, you should see what is shown in Fig. 16-21 (although you may have placed the program at a different memory location). The h's and b's represent hex and binary digits which we are not concerned with at this time.

Running the Program

Let's use Fig. 16-22 during our analysis of program operation.

The first op code is 3E, which means, *Load the accumulator with the contents of the next memory location*, or

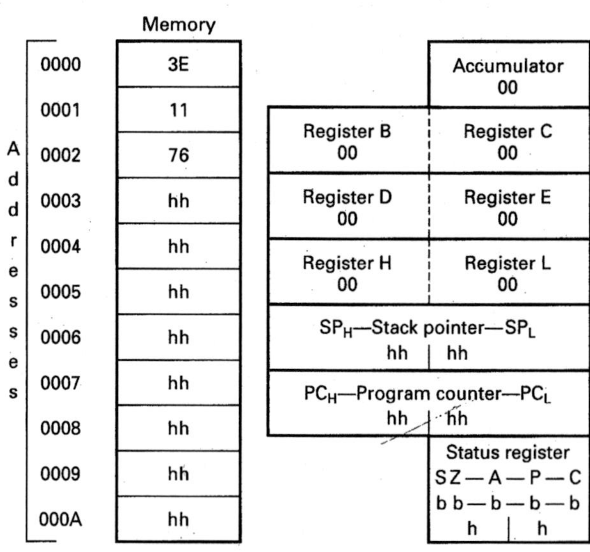

Fig. 16-21 8085/Z80 sample program.

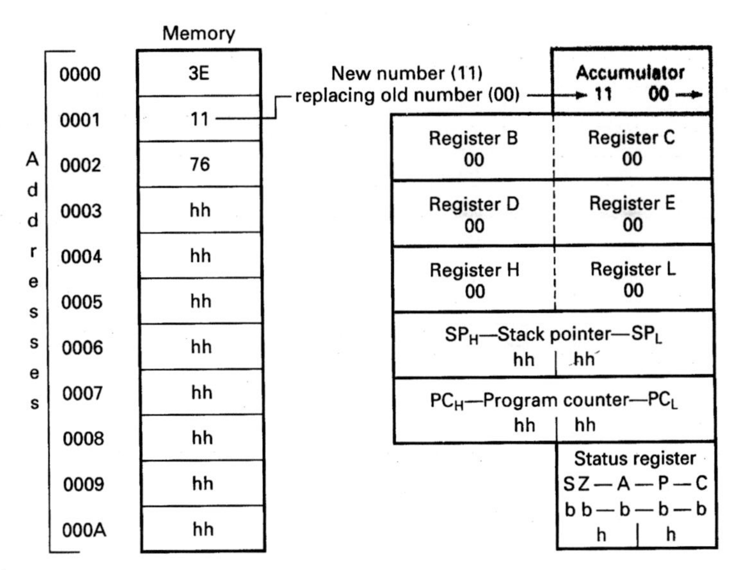

Fig. 16-22 8085/Z80 sample program.

more properly, *Place a* copy *of the contents of the next memory location in the accumulator*. As you see, the number 11 is replacing 00 in the accumulator. The program then continues to the next instruction op code, 76, which stands for halt, and stops.

Checking the Results of Program (Analysis)

After running the program, you should have 00 in all the general-purpose registers and 11 in the accumulator. The program does what we designed it to do.

Here's one for you to try.

EXAMPLE 16-3

First manually place 00s in the accumulator and all general-purpose registers. Then write a program which will

1. Place the hex number EE in the accumulator.
2. Move (copy) the contents of the accumulator (A) into register B.
3. Move (copy) the contents of the accumulator (A) into register C.
4. Stop.

SOLUTION

Figure 16-23 shows the completed program in both 8085 and Z80 mnemonics. Figure 16-24 shows the memory and registers and what happens during program execution.

16-6 8086/8088 FAMILY

We will approach the 16-bit 8086/8088 microprocessor a little differently than we did the 8-bit microprocessors. The 8-bit sections are designed to fit the needs of a person using op code charts and hand assembly in the earlier chapters and an assembler in the later chapters.

In the 16-bit section we assume that you are using the DOS DEBUG utility in the earlier chapters. DEBUG is readily available to all who use MS-DOS–type machines, and it is less sophisticated than assemblers, which keeps you closer to the hardware during the early part of the learning process.

In later chapters we will use both an assembler and DEBUG in figures and in answers to chapter questions. This will allow you to explore the advantages of a full-featured assembler and to continue to use DEBUG if you wish.

One final point should be kept in mind. This text is designed to make the learning process as simple as possible for the beginner. A 16-bit chip like the 8086/8088 is quite complex for the beginner. Therefore we do not attempt to cover every aspect of this chip.

CPU Control Instructions

The 8086/8088 has a no operation (NOP) instruction which works as described in the New Concepts section of this chapter. A brief description of the NOP instruction can be found in the CPU Control Instructions section of the Expanded Table of 8086/8088 Instruction Listed by Category in Part 4 of this text. The NOP has an op code of 90 and affects no flags.

The 8086/8088 has a halt instruction which functions as described in the New Concepts section. A description of this instruction appears in the CPU Control Instructions section of the 8086/8088 instruction set. Its mnemonic is HLT, and its op code is F4.

8085 mnemonics

Addr	Obj	Assembler	Comment
0000	3E	MVI A,EE	Place the hex number EE in the accumulator (A)
0001	EE		
0002	47	MOV B,A	Copy into register B the contents of A
0003	4F	MOV C,A	Copy into register C the contents of A
0004	76	HALT	Stop

Z80 mnemonics

Addr	Obj	Assembler	Comment
0000	3E	LD A,EE	Place the hex number EE in the accumulator (A)
0001	EE		
0002	47	LD B,A	Copy into register B the contents of A
0003	4F	LD C,A	Copy into register C the contents of A
0004	76	HALT	Stop

Fig. 16-23 Example 16-3 program.

Data Transfer Instructions

The 8086/8088 has eight instructions which we have placed in the Data Transfer Instructions section. While the Expanded Table of 8086/8088 Instructions Listed by Category lists all eight of these instructions, the most versatile and by far the most useful for the beginner is the **MOVe** instruction.

A copy of our programming model for the 8086/8088 appears in Fig. 16-25.

Direction of Data Transfer

A move can be *from* (source) a register, memory, or an immediate number *to* (destination) a register or memory. While *either* the source or the destination can be a memory location, *both* cannot be memory locations in the same instruction. The source and destination must both be either 8 bits wide or 16 bits wide; you can't mix data widths in the same instruction. And finally, you can't move from one segment register to another.

As you have seen from the programming model, the 8086/8088 has several 8-bit and 16-bit registers. This causes the number of move combinations between registers alone to number in the hundreds. A few examples are

MOV	AL,DL	AL ← DL
MOV	BH,BL	BH ← BL
MOV	AX,DX	AX ← DX
MOV	SP,BP	SP ← BP
MOV	SI,DI	SI ← DI
MOV	BX,DS	BX ← DS
MOV	AL,76	AL ← 76

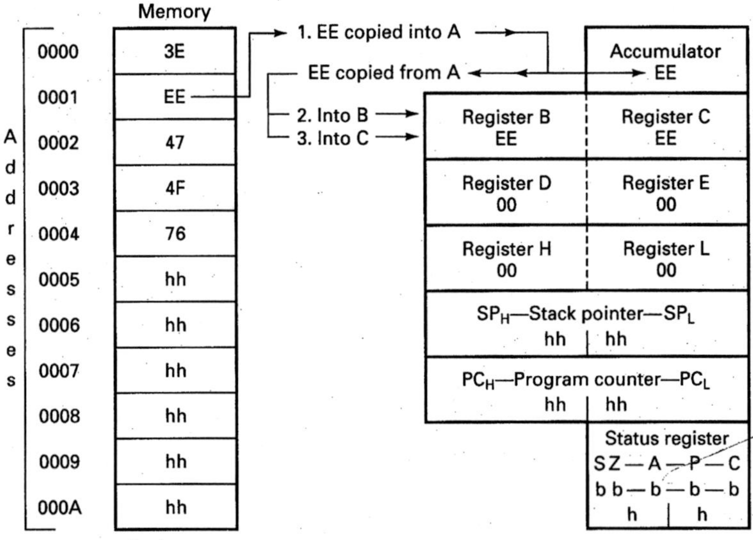

Fig. 16-24 Example 16-3 program analysis.

Addresses	Memory
0100	hh
0101	hh
0102	hh
0103	hh
0104	hh
0105	hh
0106	hh
0107	hh
0108	hh
0109	hh
010A	hh
010B	hh
010C	hh
010D	hh
010E	hh
010F	hh
0110	hh
0111	hh
0112	hh
0113	hh
0114	hh
0115	hh
0116	hh
0117	hh

Register	High	Low
Accumulator AX	AH hh	AL hh
Base BX	BH hh	BL hh
Count CX	CH hh	CL hh
Data DX	DH hh	DL hh

Source index	hhhh
Destination index	hhhh
Stack pointer	hhhh
Base pointer	hhhh
Code segment	hhhh
Data segment	hhhh
Extra segment	hhhh
Stack segment	hhhh
Instruction pointer	hhhh

Flags

New								8085-like							
—	—	—	—	O	D	I	T	S	Z	—	A	—	P	—	C
—	—	—	—	b	b	b	b	b	b	—	b	—	b	—	b
h				h				h				h			

Fig. 16-25 8080/8086 programming model.

MOV	AX,89E3	AX ← 89E3
MOV	[1234],AX	memory location 1234 ← AX
MOV	BL,[4456]	BL ← memory location 4456
MOV	DX,[BX+DI]	DX ← memory location found by adding the contents of BX and DI
MOV	AX,[BX+DI+0200]	AX ← memory location pointed to by the sum of the contents of BX, the contents of DI, and the hex number 200_{16}

The left column shows the instruction exactly as it appears when disassembled by DEBUG. The right column indicates where the data comes from and where it goes.

Sample 8086/8088 Program

Figure 16-26 shows a sequence of commands that will demonstrate a simple **MOV**e instruction and give you practice entering programs into DEBUG.

First, we started DEBUG by typing

C>debug

at the DOS prompt as shown. DEBUG responded with a

-

which indicates it is waiting for a command.

```
C>DEBUG
-r
AX=0000  BX=0000  CX=0000  DX=0000  SP=6D5E  BP=0000  SI=0000  DI=0000
DS=992A  ES=992A  SS=992A  CS=992A  IP=0100   NV UP EI PL NZ NA PO NC
992A:0100 7420          JZ      0122
-a
992A:0100 mov al,dl
992A:0102
-u 100 101
992A:0100 88D0          MOV     AL,DL
-r
AX=0000  BX=0000  CX=0000  DX=0000  SP=6D5E  BP=0000  SI=0000  DI=0000
DS=992A  ES=992A  SS=992A  CS=992A  IP=0100   NV UP EI PL NZ NA PO NC
992A:0100 88D0          MOV     AL,DL
-rdx
DX 0000
:00f3
-r
AX=0000  BX=0000  CX=0000  DX=00F3  SP=6D5E  BP=0000  SI=0000  DI=0000
DS=992A  ES=992A  SS=992A  CS=992A  IP=0100   NV UP EI PL NZ NA PO NC
992A:0100 88D0          MOV     AL,DL
-t
AX=00F3  BX=0000  CX=0000  DX=00F3  SP=6D5E  BP=0000  SI=0000  DI=0000
DS=992A  ES=992A  SS=992A  CS=992A  IP=0102   NV UP EI PL NZ NA PO NC
992A:0102 65            DB      65
-q

C>
```

Fig. 16-26 MOVe instruction (DEBUG screens).

Next we typed an "r," which stands for register. This causes DEBUG to display the values of all registers as shown in Fig. 16-27.

We will now duplicate (several times) that portion of Fig. 16-26 (in bold type) which shows the values in various registers. You should compare these sections (as we progress through each figure) to our 8086/8088 programming model in Fig. 16-25.

The current values of the general-purpose registers are shown in bold type in Fig. 16-28.

The values of the stack pointer, base pointer, source index, and destination index are shown in bold type in Fig. 16-29.

The values in the segment registers are shown in bold in Fig. 16-30.

The value of the instruction pointer and the current status of the flags are shown in bold in Fig. 16-31.

Finally, the address, op code, and assembler notation for the next instruction which is to be executed are shown in bold type in Fig. 16-32.

The area shown in bold type in Fig. 16-33 illustrates how we then typed an "a," which is the DEBUG assemble command, at the DEBUG prompt.

-a <ENTER>

```
-r
AX=0000  BX=0000  CX=0000  DX=0000  SP=6D5E  BP=0000  SI=0000  DI=0000
DS=992A  ES=992A  SS=992A  CS=992A  IP=0100   NV UP EI PL NZ NA PO NC
992A:0100 7420          JZ      0122
-
```

Fig. 16-27 DEBUG screens (cont.).

```
-r
AX=0000  BX=0000  CX=0000  DX=0000  SP=6D5E  BP=0000  SI=0000  DI=0000
DS=992A  ES=992A  SS=992A  CS=992A  IP=0100   NV UP EI PL NZ NA PO NC
992A:0100 7420          JZ      0122
-
```

Fig. 16-28 DEBUG screens (cont.).

```
-r
AX=0000  BX=0000  CX=0000  DX=0000  SP=6D5E  BP=0000  SI=0000  DI=0000
DS=992A  ES=992A  SS=992A  CS=992A  IP=0100   NV UP EI PL NZ NA PO NC
992A:0100 7420          JZ      0122
-
```

Fig. 16-29 DEBUG screens (cont.).

```
-r
AX=0000  BX=0000  CX=0000  DX=0000  SP=6D5E  BP=0000  SI=0000  DI=0000
DS=992A  ES=992A  SS=992A  CS=992A  IP=0100   NV UP EI PL NZ NA PO NC
992A:0100 7420          JZ      0122
-
```

Fig. 16-30 DEBUG screens (cont.).

```
-r
AX=0000  BX=0000  CX=0000  DX=0000  SP=6D5E  BP=0000  SI=0000  DI=0000
DS=992A  ES=992A  SS=992A  CS=992A  IP=0100   NV UP EI PL NZ NA PO NC
992A:0100 7420          JZ      0122
```

Fig. 16-31 DEBUG screens (cont.).

```
-r
AX=0000  BX=0000  CX=0000  DX=0000  SP=6D5E  BP=0000  SI=0000  DI=0000
DS=992A  ES=992A  SS=992A  CS=992A  IP=0100   NV UP EI PL NZ NA PO NC
992A:0100 7420          JZ      0122
-
```

Fig. 16-32 DEBUG screens (cont.).

```
C>DEBUG
-r
AX=0000  BX=0000  CX=0000  DX=0000  SP=6D5E  BP=0000  SI=0000  DI=0000
DS=992A  ES=992A  SS=992A  CS=992A  IP=0100   NV UP EI PL NZ NA PO NC
992A:0100 7420          JZ      0122
-a
992A:0100 mov al,dl
992A:0102
-u 100 101
992A:0100 88D0          MOV     AL,DL
-r
AX=0000  BX=0000  CX=0000  DX=0000  SP=6D5E  BP=0000  SI=0000  DI=0000
DS=992A  ES=992A  SS=992A  CS=992A  IP=0100   NV UP EI PL NZ NA PO NC
992A:0100 88D0          MOV     AL,DL
-rdx
DX 0000
:00f3
-r
AX=0000  BX=0000  CX=0000  DX=0000  SP=6D5E  BP=0000  SI=0000  DI=0000
DS=992A  ES=992A  SS=992A  CS=992A  IP=0100   NV UP EI PL NZ NA PO NC
992A:0100 88D0          MOV     AL,DL
-t

AX=00F3  BX=0000  CX=0000  DX=00F3  SP=6D5E  BP=0000  SI=0000  DI=0000
DS=992A  ES=992A  SS=992A  CS=992A  IP=0102   NV UP EI PL NZ NA PO NC
992A:0102 65            DB      65
-q

C>
```

Fig. 16-33 DEBUG screens (cont.).

DEBUG then responded with

992A:0100

which is the address at which our program will start. The 992A is the memory segment, and 0100 is the memory location within that segment. If you try this program on your computer, your segment will probably not be the same as ours. This is normal and will not affect the results of the program.

We then typed

mov al,dl <ENTER>

and DEBUG responded with

992A:0102

which is the address of the next available memory location. We then pressed <ENTER> to terminate assembly, and DEBUG waited for our next command.

We told DEBUG to create or *assemble* the machine code for the MOV AL,DL instruction. Then we wanted to check to see that this is what DEBUG did. We wanted to *disassemble* the machine code. The DEBUG command for this is "u," which stands for unassemble (DEBUG's name for disassemble). The next command in our program is

-u 100 101

which tells DEBUG to unassemble memory locations 100_{16}–101_{16} within the current code segment. DEBUG responded with

992A:0100 88D0 MOV AL,DL

992A is the current code segment. 0100 is the memory location of the first byte of this instruction. 88D0 is the machine code for MOV AL,DL, which was the assembly-language instruction we typed in.

We typed the register command, and DEBUG again displayed the current status of all registers. DEBUG's response is shown in Fig. 16-34.

When DEBUG displays the registers, it also displays the instruction which it finds at the memory location pointed to by the instruction pointer in the current code segment. These appear in bold type in Fig. 16-34. Our MOV AL,DL instruction appears in the assembly-language section.

Since our instruction said to move the contents of register DL to register AL, we needed to place some value in register DL. Notice that at this point AX, BX, CX, and DX all contained 0000. Even if the contents of DL were copied to AL, we wouldn't see any difference. We needed to place some value in DL which we could observe.

The area in bold type in Fig. 16-35 shows our next command

-rdx

which told DEBUG we wanted to change the value in register DX. DEBUG responded with

DX 0000
:

which was the current contents of register DX. The cursor waited after the colon. If we had typed in a value, that value would have been placed in the DX register. If we had pressed the <ENTER> key, the value in DX would not have changed.

```
-r
AX=0000  BX=0000  CX=0000  DX=0000  SP=6D5E  BP=0000  SI=0000  DI=0000
DS=992A  ES=992A  SS=992A  CS=992A  IP=0100   NV UP EI PL NZ NA PO NC
992A:0100 88D0          MOV     AL,DL
```

Fig. 16-34 DEBUG screens (cont.).

```
C>DEBUG
-r
AX=0000  BX=0000  CX=0000  DX=0000  SP=6D5E  BP=0000  SI=0000  DI=0000
DS=992A  ES=992A  SS=992A  CS=992A  IP=0100   NV UP EI PL NZ NA PO NC
992A:0100 7420          JZ      0122
-a
992A:0100 mov al,dl
992A:0102
-u 100 101
992A:0100 88D0          MOV     AL,DL
-r
AX=0000  BX=0000  CX=0000  DX=0000  SP=6D5E  BP=0000  SI=0000  DI=0000
DS=992A  ES=992A  SS=992A  CS=992A  IP=0100   NV UP EI PL NZ NA PO NC
992A:0100 88D0          MOV     AL,DL
-rdx
DX 0000
:00f3
-r
AX=0000  BX=0000  CX=0000  DX=00F3  SP=6D5E  BP=0000  SI=0000  DI=0000
DS=992A  ES=992A  SS=992A  CS=992A  IP=0100   NV UP EI PL NZ NA PO NC
992A:0100 88D0          MOV     AL,DL
-t

AX=00F3  BX=0000  CX=0000  DX=00F3  SP=6D5E  BP=0000  SI=0000  DI=0000
DS=992A  ES=992A  SS=992A  CS=992A  IP=0102   NV UP EI PL NZ NA PO NC
992A:0102 65            DB      65
-q

C>
```

Fig. 16-35 DEBUG screens (cont.).

We wanted to place a new number in DL. However, we could not single out the low byte of the DX register, so we simply placed 0s in the high byte and our number in the low byte. We typed that number (00f3) and pressed <ENTER>.

:00f3 <ENTER>

Figure 16-36 shows how we again used the register command ("r").

Notice that the value in register DX has been changed to the value we typed in.

Running the Program

Next we wanted the computer to execute the MOV AL,DL instruction. However, we did not want it to continue any further than that. Even though we had not entered any other instruction into the computer, there were others. When we turned the computer on, each unused memory location contained some number, even if it was 00_{16}. Most of these "random" numbers were actually the op code for some instruction. We didn't want these "random" instructions to execute.

DEBUG has a command called trace which executes the next instruction (the one displayed at the bottom of the register display) and then stops and automatically displays the contents of the registers for viewing. This is what we did in Fig. 16-37.

Notice that the value in DX has been *copied* into register AX. Notice also that the **I**nstruction **P**ointer has been incremented to the position of the next instruction in memory which is displayed at the bottom of the register display (in bold type).

Checking the Results

Figure 16-38 shows the operation of the program by using our programming model to illustrate the movement of F3 from one register to the other.

Our program worked. In the future we will not discuss each 8086/8088 program in such detail, but we have done so here to give you an idea of how to monitor the execution of a program. We have also introduced you to some DEBUG commands. Remember that the DEBUG commands—*assemble, unassemble, trace, register,* and *quit*—are not assembly-language instructions but are commands to the DEBUG utility, which helps you to enter, modify, and execute assembly-language instructions.

Finally, you may want to exit from the DEBUG program. That command is simply the **q**uit command, which is entered with the letter q. You will then be returned to the DOS prompt.

EXAMPLE 16-4

Place the number FE in register DH. Place the number 12 in DL. Then write a program that will

1. Copy DH to AH.
2. Copy DL to BH.

Use the trace command to execute the program and follow its operation.

```
-r
AX=0000  BX=0000  CX=0000  DX=00F3  SP=6D5E  BP=0000  SI=0000  DI=0000
DS=992A  ES=992A  SS=992A  CS=992A  IP=0100   NV UP EI PL NZ NA PO NC
992A:0100 88D0          MOV     AL,DL
```

Fig. 16-36 DEBUG screens (cont.).

```
-t

AX=00F3  BX=0000  CX=0000  DX=00F3  SP=6D5E  BP=0000  SI=0000  DI=0000
DS=992A  ES=992A  SS=992A  CS=992A  IP=0102   NV UP EI PL NZ NA PO NC
992A:0102 65            DB      65
```

Fig. 16-37 DEBUG screens (cont.).

	Memory
0100	88
0101	D0
0102	hh
0103	hh
0104	hh
0105	hh

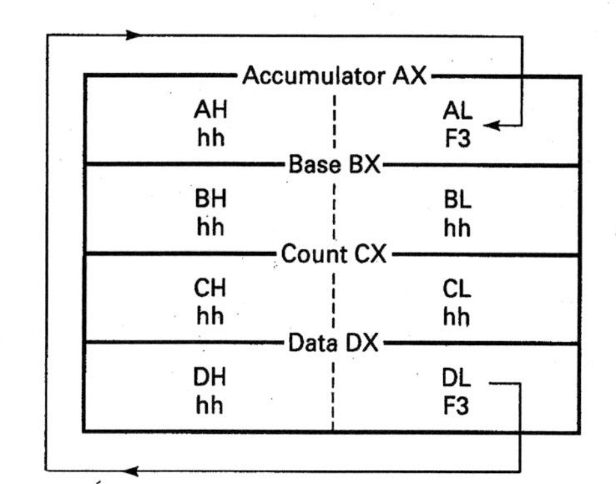

Fig. 16-38 MOVe instruction (programming model).

```
C>DEBUG
-r
AX=0000  BX=0000  CX=0000  DX=0000  SP=404E  BP=0000  SI=0000  DI=0000
DS=9BFB  ES=9BFB  SS=9BFB  CS=9BFB  IP=0100   NV UP EI PL NZ NA PO NC
9BFB:0100 A3FF72        MOV     [72FF],AX                        DS:72FF=FF1F
-rdx
DX 0000
:fe12
-r
AX=0000  BX=0000  CX=0000  DX=FE12  SP=404E  BP=0000  SI=0000  DI=0000
DS=9BFB  ES=9BFB  SS=9BFB  CS=9BFB  IP=0100   NV UP EI PL NZ NA PO NC
9BFB:0100 A3FF72        MOV     [72FF],AX                        DS:72FF=FF1F
-a
9BFB:0100 mov ah,dh
9BFB:0102 mov bh,dl
9BFB:0104
-r
AX=0000  BX=0000  CX=0000  DX=FE12  SP=404E  BP=0000  SI=0000  DI=0000
DS=9BFB  ES=9BFB  SS=9BFB  CS=9BFB  IP=0100   NV UP EI PL NZ NA PO NC
9BFB:0100 88F4          MOV     AH,DH
-t

AX=FE00  BX=0000  CX=0000  DX=FE12  SP=404E  BP=0000  SI=0000  DI=0000
DS=9BFB  ES=9BFB  SS=9BFB  CS=9BFB  IP=0102   NV UP EI PL NZ NA PO NC
9BFB:0102 88D7          MOV     BH,DL
-t

AX=FE00  BX=1200  CX=0000  DX=FE12  SP=404E  BP=0000  SI=0000  DI=0000
DS=9BFB  ES=9BFB  SS=9BFB  CS=9BFB  IP=0104   NV UP EI PL NZ NA PO NC
9BFB:0104 3C3A          CMP     AL,3A
-
```

Fig. 16-39 Example 16-4 (DEBUG screens).

SOLUTION

Figure 16-39 shows the process of changing the contents of the registers, entering the assembly-language instructions, and tracing program execution. Especially notice the areas in bold type. (They, of course, will not appear in bold on your computer screen.)

Figure 16-40 shows the same program, illustrating the movement of the data with our programming model.

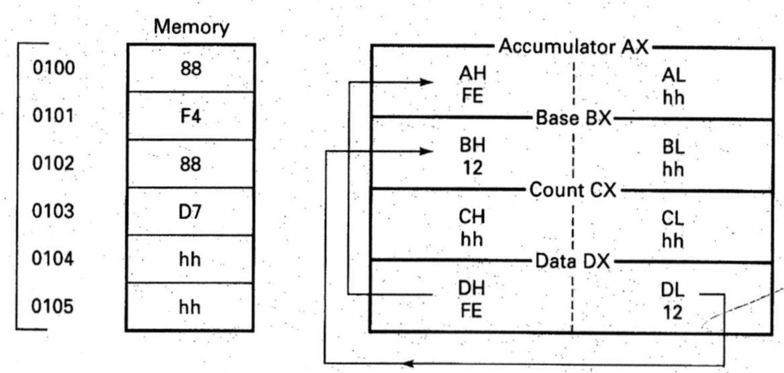

Fig. 16-40 Example 16-4 (programming model).

SELF-TESTING REVIEW

Read each of the following and provide the missing words. Answers appear at the beginning of the next question.

1. The various instructions which form the instruction set of most microprocessors fall into natural _________ or groups.
2. (*categories*) The _________ and _________ are the microprocessor chips found in IBM PC compatibles.
3. (*8086, 8088*) A technique which is sometimes helpful when analyzing a program involves _________ the contents of each register or memory location and updating each as it changes in the program.
4. (*writing*) When we talk of moving, loading, transferring, or storing data, while working with the microprocessors in this text, are we referring to *moving* in the sense that the data no longer exists in its original location?
5. (*No*) When we talk about moving, loading, transferring, or storing data, we are actually _________ the data.
6. (*copying*) If you can type mnemonics into your computer or trainer, it must have an _________.
7. (*assembler*) What do microprocessors understand?
8. (*binary numbers*) An assembler translates mnemonics into _________.
(*binary numbers*)

PROBLEMS

General

16-1. What does the NOP (no operation) instruction do?

16-2. What are two purposes of the NOP instruction?

16-3. If you move, load, or transfer the contents of the accumulator to a general-purpose register, what is left in the accumulator?

6502 Family

16-4. What is the op code for the NOP instruction?

16-5. What is the op code for the **BReaK** instruction?

16-6. What is the op code for the TAX (Transfer Accumulator to **X** register) instruction?

16-7. What does the TYA instruction do?

16-8. What does the mnemonic STX stand for?

16-9. Which instruction would you use to copy the contents of the Y register into a memory location?

16-10. Write a program which will
  a. Place the number 45_{16} in the accumulator.
  b. Transfer the contents of the accumulator to the X register.
  c. Stop.

6800/6808 Family

16-11. What is the op code for the NOP instruction?

16-12. What is the op code for the WAI instruction?

16-13. What is the op code for the TAB (Transfer accumulator **A** to accumulator **B**) instruction?

16-14. What does the TBA instruction do?

16-15. What does the mnemonic CLRA stand for?

16-16. Which instruction would you use to copy the contents of accumulator B into a memory location?

16-17. Write a program which will
  a. Place the number 89_{16} in accumulator B.
  b. Copy the contents of accumulator B to accumulator A.
  c. Stop.

8080/8085/Z80 Family

16-18. What is the op code for the NOP instruction?

16-19. What is the op code for the HALT instruction?

16-20. What is the op code for the Mov A,D *[LD A,D]* instruction?

16-21. What does the MOV B,C *[LD B,C]* instruction do?

16-22. What does the mnemonic MVI A,dd *[LD A,dd]* stand for?

16-23. Which instruction would you use to store the contents of the accumulator in a memory location?

16-24. Write a program which will
  a. Place the number 78_{16} immediately into the accumulator.
  b. Copy the contents of the accumulator into register C.
  c. Stop.

8086/8088 Family

16-25. What is the DOS utility which we are using in this text to do assembly, disassembly, running, and debugging of 8086/8088 assembly-language programs?

16-26. What three areas can serve as a source for the 8088 **MOV**e instruction?

16-27. What are the two areas which can serve as destinations for the 8088 MOV instruction?

16-28. Which area *cannot* be both a source and a destination at the same time?

16-29. What is the source of a MOV AL,DL instruction?

16-30. What is the destination of a MOV AL,76 instruction?

16-31. Does the instruction MOV B,[4456] move the number 4456 or the contents of memory location 4456_{16} to register B?

16-32. What does the DEBUG command "r" stand for and what does it do?

16-33. What does the DEBUG command "a" stand for and what does it do?

16-34. What does the DEBUG command "u" stand for and what does it do?

16-35. What does the DEBUG trace command do?

16-36. What is the DEBUG quit command?

16-37. Using DEBUG, write an 8086/8088 assembly program which will

a. Place the number 89_{16} into the register BL.

b. Copy the contents of BL into CL.
(*Note:* Use DEBUG's trace command to execute the program to see if it works.)

ADDRESSING MODES—I

New Concepts

In this chapter we will study the simplest of the different addressing modes. This will provide a foundation for the next couple of chapters. In a later chapter we will look at the more complex addressing modes. First we need to learn what an addressing mode is.

17-1 WHAT IS AN ADDRESSING MODE?

In an earlier chapter we used the system of addressing homes as a way to describe memory addressing. Let's use the same idea to describe addressing modes.

If you are moving and want to describe to the movers how to get to your new home so that they can deliver your belongings, you would give them the name of the state, city, street, and house number.

But what if you were moving to an apartment in Canada? In that case you would give them the name of the country, province, city, street, apartment complex, and apartment number.

Or what if you were moving to a backwoods cabin for a summer in the wilderness? You would give them the name of the state, county, county road, the direction and number of miles to travel on that county road, and finally landmarks to help them find the cabin. (OK, you probably won't have a truck moving all of your belongings to a wilderness cabin, but the analogy worked well up to that point.)

You can see that we need more than one way to "address" or describe a location because not every method works in every circumstance. This is what addressing modes are about.

How you describe a location you want to transfer a number to can depend on several factors. Remember that while the addressing mode which should be used is very apparent in some cases, in other cases choosing the best addressing mode requires skill that must be developed over time.

17-2 THE PAGING CONCEPT

Before we go any further into the subject of addressing modes, we need to look at the concept of paging. *Paging* is the concept of dividing memory into blocks of 256 bytes each. Each block is called a *page*. We have to look at how we count in hexadecimal to see why this number was chosen.

The number 256 was chosen because that is how far you can count using only two hex digits. Actually, FF, which is the highest two-digit hex number, is 255 (decimal), but if you count 00, you have 256 different numbers, or in this case, memory locations.

Counting from 00_{16} to FF_{16} using four hex digits looks like this:

```
0000
0001
0002
 :
 :
00FE
00FF
```

Notice that the left two digits are always 0. The range of hex numbers from 00 to FF is called *page 00* (sometimes called the *zero page*).

The next number after 00FF is 0100. Let's continue counting from there:

```
0100
0101
0102
 :
 :
01FE
01FF
```

Notice that the left two digits are 01. This is called *page one*.

The next number in the sequence is 0200. This is the beginning of page two. Page two ends with 02FF, after which comes 0300, the beginning of page three.

This process continues up to $FFFF_{16}$. There are 256 of these pages, with 256 bytes per page.

The addressing modes of the 8085 do not reference page numbers; however, the 6800/6808 and 6502 do have addressing modes that depend upon the concept of paging.

17-3 BASIC ADDRESSING MODES

We are now going to study the four most basic addressing modes. As you read about each mode, first and foremost try to understand the concept. The actual name of the addressing mode may be different for the microprocessor which you are using. After you read about these four modes, go to the section which covers your particular microprocessor for specific details.

Implied Addressing

In *implied addressing*, sometimes called *inherent addressing*, no address is necessary because the location is implied in the instruction itself. It is the simplest of all addressing modes. You used this mode in Chap. 16, which discussed the CPU control instructions. Remember the NOP (no operation) instruction? Do we have to tell it where to do nothing? No. No addressing is necessary.

Another example would be the case of a microprocessor which has only one accumulator and a certain index register. The 6502, for example, has an instruction called

TAX

which means

Transfer **A**ccumulator to **X** register

There is only one accumulator, and the specified register is the X register. The microprocessor knows exactly where the accumulator and the X register are, so we say that the address is *implied* in the instruction itself. The data will be transferred from the accumulator to the X register.

Register (Accumulator) Addressing

Register addressing, sometimes called *accumulator addressing*, involves only internal registers or an accumulator(s) and no external RAM. For example, the 8085 microprocessor has an instruction called

MOV A,B

which means

MOVe data to **A** from **B**

Since the data is being moved from one register to another, no other address information is needed. The names of the registers are enough.

It should be noted that with some microprocessors it is not clear whether this is considered to be a separate addressing mode or a special subtype of the implied addressing mode. See your particular microprocessor section for details.

Immediate Addressing

Immediate addressing is a mode in which the number or data to be operated on or moved is in the memory location *immediately* following the instruction op code. For example, the 6800/6808 microprocessor has an instruction called

LDAA #$dd

which means

Loa**D** **A**ccumulator **A** with the two hexadecimal ($) digits of data (dd) immediately (#) following this op code

In the computer's memory there will be the hex number 86, which is the op code for the LDAA immediate instruction, followed immediately by the two hex digits which we have called dd (since we don't know what their actual value is right now).

Direct Addressing

Direct addressing uses an op code followed by a 1- or 2-byte memory address where the data which is to be used can be found. The data is outside of the microprocessor itself, in one of the many thousands of memory locations. The Z80 has an instruction called

LD A, (aaaa)

which means

Loa**D** the **A**ccumulator with the data found at memory location (aaaa)

Here of course the aaaa is four hex digits, which makes a 16-bit address. The microprocessor will go to memory address aaaa and place a copy of the contents of that address in the accumulator.

Keep in mind that some microprocessors do not call this "direct" addressing and that some have more than one form of this addressing mode.

Specific Microprocessor Families

Go to the section which discusses the microprocessor you are using.

17-4 6502 FAMILY

The 6502 uses the implied and immediate modes as described in the New Concepts section of this chapter. The register mode and direct mode are a little different.

Implied Addressing

For an example of implied addressing refer to the Data Transfer Instructions part of the 6502 instruction set in the Expanded Table of 6502 Instructions Listed by Category.

Find the TAX instruction, which is an example of implied addressing as noted in the Address Mode column. Notice that it **T**ransfers the contents of the **A**ccumulator to the **X** register as indicated in the Operation and Boolean/Arithmetic Operation columns. Of course, no other information is needed since both of these locations are inside the microprocessor itself.

Register (Accumulator) Addressing

The 6502 doesn't use register addressing as a dominant addressing mode like the 8080/8085 does. It does use it in four instances, however, and calls it *accumulator* addressing.

For example, the 6502 instruction

ASL

which stands for **A**rithmetic **S**hift **L**eft, shifts every bit in the accumulator to the left one place. The operand is in the accumulator.

There are only four 6502 instructions which use the register or accumulator addressing mode: they are ASL A (**A**rithmetic **S**hift **L**eft **A**ccumulator), LSR A (**L**ogical **S**hift **R**ight **A**ccumulator), ROL A (**RO**tate **L**eft **A**ccumulator), and ROR A (**RO**tate **R**ight **A**ccumulator). All these instructions can be found in the Rotate and Shift Instructions section of the Expanded Table of 6502 Instructions Listed by Category.

Immediate Addressing

Now let's look at an example of immediate addressing. In the Data Transfer section of the 6502 instruction set, notice the first form of the LDA instruction. It uses immediate addressing, which is what the # in the Assembler Notation column stands for. The $ means that the number is hexadecimal (not decimal). The dd stands for two hex digits such as 35 or E2. To load the accumulator with the hex number E2, you would type

LDA #$E2

Direct Addressing

The 6502 has two different types of direct addressing. One is called zero page addressing, and the other absolute addressing.

Zero page addressing is direct addressing in which the target address is in page zero of memory, somewhere in the first 256 bytes of memory, between 0000_{16} and $00FF_{16}$. Since the first two hex digits of any address in zero page are 00s, the 00s can be omitted, making it possible to describe the address with only 1 byte.

Absolute addressing is a form of direct addressing in which the target address can be anywhere from 0000_{16} to $FFFF_{16}$. This requires four hex digits, which is a 2-byte address.

Referring again to the LDA instruction, the third form down is the zero page addressing form of the instruction. Notice that the assembler notation form appears as

LDA $aa

The two lowercase a's indicate a two-digit hex address. The second form of the LDA instruction is the absolute addressing form. The assembler notation in this case appears as

LDA $aaaa

which means that the address consists of four hex digits (2 bytes).

(*Note:* The 6502 microprocessor expresses addresses in reverse low-byte/high-byte order!)

6502 Summary

Some examples are

```
NOP         ← Implied addressing
ASL         ← Register (accumulator) addressing
LDA #$35    ← Immediate addressing
LDA $1E     ← Direct (zero page) addressing
LDA $123D   ← Direct (absolute) addressing
```

17-5 6800/6808 FAMILY

The 6800/6808 uses the implied and immediate modes as described in the New Concepts section. The register and direct modes are a little different.

Implied Addressing

For an example of implied addressing, refer to the Data Transfer Instructions part of the Expanded Table of 6800 Instructions Listed by Category.

Find the TAB instruction, which is an example of implied addressing as noted in the Address Mode column. Notice that it transfers the contents of accumulator A to accumulator B as indicated in the Operation and Boolean/Arithmetic Operation columns. No other information is needed since both of these locations are inside of the microprocessor itself.

Register (Accumulator) Addressing

The 6800/6808 doesn't use register addressing as a dominant addressing mode the way the 8080/8085 does. Technically, it does use it, however, and calls it *accumulator* addressing. Since it is often considered a special form of implied addressing by many who use the 6800/6808, it has not been included in the Address Mode column of the instruction sheets but rather falls under the title of Implied addressing.

Immediate Addressing

Now let's look at an example of immediate addressing. In the Data Transfer section of the 6800/6808 instruction set notice the first form of the LDAA instruction. It uses immediate addressing, which is what the # in the Assembler Notation column stands for. The $ means that the number is hexadecimal (not decimal). The dd stands for two hex digits such as E2. The instruction which would **L**oa**D** accumulator **A** with the value E2 would appear as

LDAA #$E2

Direct Addressing

The 6800/6808 has two different types of direct addressing. One is called direct addressing, and the other extended addressing.

Direct addressing is a form of direct addressing in which the target address is in page zero of memory—that is, somewhere in the first 256 bytes of memory between 0000_{16} and $00FF_{16}$. Since the first two hex digits of any address in this range are 00, the 00s can be omitted, making it possible to designate the address with only 1 byte.

Extended addressing is a form of direct addressing in which the target address can be anywhere from 0000_{16} to $FFFF_{16}$. This requires four hex digits, which is a 2-byte address.

Referring to the LDAA instruction, notice that the second form down is the direct addressing form of the instruction. The assembler notation appears as

LDAA $aa

which means there are only two hex address digits, indicated by aa (a stands for address). The fourth LDAA form is the extended addressing form of the instruction. The assembler notation in this case appears as

LDAA $aaaa

which means that there are four hex address digits (2 bytes).

6800/6808 Summary

Some examples are

TAB	← Implied addressing
TAB	← Register (accumulator) addressing
LDAA #$35	← Immediate addressing
LDAA $1E	← Direct addressing
LDAA $123D	← Direct (extended) addressing

17-6 8080/8085/Z80 FAMILY

The 8080/8085/Z80 uses the implied, immediate, register, and direct addressing modes as described in the New Concepts section of this chapter.

Note that the Z80 has all of the addressing modes that the 8080/8085 has, plus a number of addressing modes that the 8080/8085 does not have. We do not include these additional modes of the Z80 in either the text or the instruction set tables in this book. Refer to one of the many books available about the Z80 to learn about these other modes.

We need to bring your attention to a sometimes confusing fact about the 8080/8085/Z80 mnemonics. Look at the Data Transfer Instructions section of the Expanded Table of 8080/8085/Z80 (8080 subset) Instructions Listed by Category. Now look at the MOV A,B *[Z80 = LD A,B]* instruction (the second instruction in this section). Notice in the Boolean/Arithmetic Operation column that the data is moving from B toward A. This means that the mnemonic places the destination register before the source register. This is true of the entire 8080/8085/Z80 instruction set. The MOV A,B instruction is moving data *to* A *from* B. (*Note:* The 6502 and 6800/6808 are just the reverse.)

Implied Addressing

An example of implied addressing can be seen in the CPU Control Instructions section of the 8080/8085/Z80 instruction set. The NOP instruction uses implied addressing since no address is necessary. In the Flag Instructions section you can see another example. The STC (**S**e**T C**arry flag) instruction uses implied addressing. The carry flag is inside the 8080/8085/Z80 microprocessor. Therefore no other address information is needed.

Register (Accumulator) Addressing

This form of addressing is called *register addressing* with the 8080/8085 (in contrast to the term *accumulator addressing* used by the 6502 and 6808). The 8080/8085/Z80 uses this form of addressing very frequently. In fact, if you browse through the Data Transfer Instructions section of the Expanded Table of 8085/8080 and Z80 (8080 Subset) Instructions Listed by Category, you will find that most of these instructions use this form of addressing.

For example, the instruction MOV A,B *[Z80 = LD A,B]* moves or makes a copy of the data in the B register and places it in the A register. (We normally call this the accumulator.) Since external memory is not utilized, and both the source of the data and its destination are inside the microprocessor, this information is sufficient.

Immediate Addressing

The 8080/8085/Z80 microprocessors use the immediate mode as described in the New Concepts section at the beginning of the chapter.

To see an example of this mode, scan through the 8080/8085/Z80 instruction set in the Data Transfer Section until you come to the MVI A,dd *[Z80 = LD A,dd]* instruction (the 64th instruction in that section). You'll notice in the Address Mode column that this is labeled as using the immediate addressing mode. This means that the op code for this instruction (3E) would be followed immediately by the two hex digits we want moved.

If the Hex number C8 was the value we wanted to load into the accumulator, the 8080/8085 assembly-language notation would appear as

MVI A,C8 *[LD A,C8]*

The second instruction, in brackets and in italics, is the Z80 form.

Direct Addressing

The direct addressing mode as implemented in the 8080/8085/Z80 microprocessors works as described in the New Concepts section of this chapter.

The 8080/8085/Z80 has only one form of direct addressing. (The 6502 and the 6800/6808 have two forms of this addressing mode.)

To find an example of this mode, scan through the Data Transfer Instructions section of the 8080/8085/Z80 instruction set until you find the LDA aaaa *[LD A,(aaaa)]* instruction (the 78th instruction in this section). The op code for this instruction is 3A. It uses 3 bytes of memory. The 1st byte will be the op code, 3A. The 2d and 3d bytes will be the address of the memory location where the data can be found.

(*Note:* The 8080/8085/Z80 microprocessors express addresses in reverse low-byte/high-byte order!) If we wanted to load the accumulator from memory location 1234, the 3 bytes of object code would be

3A 34 12

in the op code/high-byte/low-byte sequence.

The assembly-language notation for this instruction would appear as

LDA 1234 *[LD A, (1234)]*

8080/8085/Z80 Summary

Some examples are

NOP	← Implied addressing
MOV A,B *[LD A,B]*	← Register addressing
MVI A,C8 *[LD A,C8]*	← Immediate addressing
LDA 1234 *[LD A, (1234)]*	← Direct addressing

17-7 8086/8088 FAMILY

Most of the 8086/8088 instructions are implemented as described in the New Concepts section of this chapter.

We need to bring to your attention a sometimes confusing fact about 8086/8088 mnemonics. The 8086/8088 mnemonics place the destination register before the source register. This is true of the entire 8086/8088 instruction set. The MOV AL,BL instruction is moving data *to* AL *from* BL. (*Note:* This is similar to the 8080/8085/Z80 microprocessors.)

Implied Addressing

Implied addressing works on the 8086/8088 microprocessors as described in the New Concepts section of this chapter. Two examples are HLT (halt) and NOP (no operation).

Register Addressing

Register addressing also works as described in the New Concepts section of this chapter. Since the 8086/8088 chips have eight 8-bit (or four 16-bit) general-purpose registers in addition to a number of other special-purpose registers, there are *hundreds* of move combinations. Let's look at one of them.

The instruction which moves the contents of the CX register into the BX register looks like this:

MOV BX,CX

Again you should notice that where the data is going to (BX) is written *first*, and where the data is coming from (CX) is written *last*.

Since only registers are involved, all of which are inside the microprocessor, no other information is needed by the microprocessor.

Immediate Addressing

Immediate addressing on the 8086/8088 is as described in the New Concepts section of this chapter. For example, the instruction MOV AL,37 would place the hexadecimal number 37 in the AL register.

Memory Segmentation

Before we can discuss direct addressing, we need to look at a feature of the 8086/8088 microprocessors which does not exist in any of the 8-bit microprocessors used in this book. That feature is memory segmentation.

Earlier in this chapter we discussed the paging concept. Segmentation is an extension of that concept. The 8-bit microprocessors use 16-bit addresses. That gives them a range from 0000_{16} to $FFFF_{16}$. In decimal that is 65,535, which gives us a total of 65,536 different memory locations counting location 0000_{16}. Another way to express this is as 64 kilobytes, or 64K. Notice that the addresses from 0000_{16} to $FFFF_{16}$ use four hex digits. The two right-most digits express which byte is being referred to. The two left-most digits express which page the bytes are in. There are 256 bytes per page and 256 pages from 0000_{16} to $FFFF_{16}$.

The 8086/8088 chips use a larger 20-bit address instead of the 16-bit address used by the 8-bit chips. Twenty bits is five hexadecimal digits. This provides a range from 00000_{16} to $FFFFF_{16}$. In decimal this is 1,048,575, which gives us 1,048,576 memory locations (since we can count 00000_{16}), or 1 megabyte of memory.

A segment is a 64K block of memory; thus there could be as many as 16 nonoverlapping segments in 1M (megabyte) of memory. Unlike a memory page, however, a segment is not bound to a certain location. The only requirement is that a segment must start on a 16-byte memory boundary. Segments can be nonoverlapping, they can partially overlap, or they can be superimposed with one exactly on top of the other. The 8086/8088 has four segment registers and so can manage four different segments at a time.

Direct addressing uses not only the address specified in the instruction but also the address in one of the segment registers. In the case of move instructions, the data segment register is used. The process involves adding the address you have specified to the address in the data segment register after shifting the data segment register to the left one hexadecimal digit. For example, if you said

MOV DL,[0100]

and if the data segment register contained 2000, the address would be calculated in the following manner.

2000	data segment register (shifted left)
+ 0100	address
20100	effective address

Notice that the contents of the data segment register have been shifted to the left one place. (You can think of it as adding a 0 to the right side of the data segment register.) So the MOV DL,[0100] instruction places a copy of the data found at memory location 20100_{16} (*not* location 0100_{16}) in the DL register.

We generally won't be concerned with segment registers in this text since our programs are simple and very small. All the segment registers will be the same, so the offset (the address of the instruction pointer) will be all we must pay attention to.

Direct Addressing

Except for memory segmentation, direct addressing on the 8086/8088 is quite like that used on the 8-bit chips. When we use the term *direct addressing* in reference to the 8086/8088, we are referring to the direct form of addressing used when manipulating data. (See the following topic, Program Direct Addressing, for the other use of direct addressing.)

For example, if the data segment register contains the number 0723, and the instruction

MOV BL,[0100]

is encountered, the contents of memory location 07330_{16} (07230 + 0100 = 07330) would be copied into the BL register.

Program Direct Addressing

Program direct addressing is no different from direct addressing: It is simply direct addressing used for a different purpose.

Program direct addressing is used with JMP and CALL instructions. These instructions direct the "flow" of the program. They are not used to manipulate data. Which instruction or subroutine is to be executed next can be altered with the JMP and CALL instructions.

For example, the instruction

JMP 100

tells the microprocessor to execute the instruction found at location 0100 (hex) in the program segment. This is an example of program direct addressing.

The offset (100 in the above example) is added to the *code segment* register rather than the data segment register. Remember that the contents of the code segment register, like the data segment register, are shifted one hexadecimal place to the left before being added to the offset.

8086/8088 Summary

Some examples are

NOP ← Implied addressing
MOV BX,CX ← Register addressing
MOV AL,37 ← Immediate addressing
MOV BL,[0100] ← Direct addressing
JMP 100 ← Direct (program direct) addressing

ARITHMETIC AND FLAGS

In this chapter we will study the arithmetic instructions of each of our microprocessor families. We will also look at the closely related topic of flags, at how they react to arithmetic instructions, and at the instructions which control them.

New Concepts

There are several main topics in this chapter. We will learn about (and review) the number systems microprocessors use. We will study addition and subtraction (as well as multiplication and division on the 16-bit 8086/8088). And finally we will study the flags which are affected by these arithmetic operations and how to alter the condition of those flags.

18-1 MICROPROCESSORS AND NUMBERS

We must first look at the kind of numbers a microprocessor performs arithmetic operations on. You have already studied much of this in earlier chapters.

Binary and Hexadecimal Numbers

We introduced binary numbers in Chap. 1. If that was the first time you had ever seen numbers in another base system, the whole subject may have been a bit confusing. It all becomes quite natural, though, with time and experience.

At this point there are a couple of very important skills which you must have. You should be able to look at an 8-bit binary number and know the decimal value of each of the bit's positions. This is illustrated in Fig. 18-1.

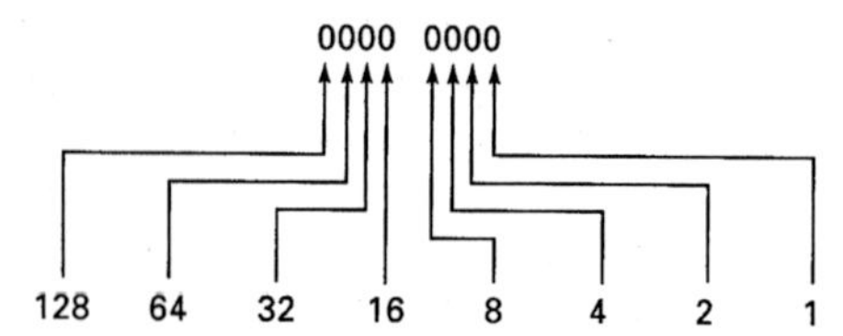

Fig. 18-1 Decimal values of each bit of an 8-bit binary number.

You should also be able to add the decimal values of each binary digit to determine the decimal value of the complete binary number. See Chap. 1 if you have forgotten how to do this.

Another skill which was stressed in Chap. 1 is now necessary if you are to work with microprocessors effectively. This is the ability to recognize any 4-bit binary number, its hexadecimal equivalent, and its decimal equivalent. The table which illustrates this appeared in Chap. 1 as Table 1-4 and is repeated here as Fig. 18-2.

If you are unsure about any of these concepts, review Chap. 1.

Binary-Coded Decimal Numbers

Binary-coded decimal numbers are just that: They are decimal numbers that happen to have each digit represented by its 4-bit binary equivalent. For example

$$0100_2 = 4_{10} \quad \text{and} \quad 0001_2 = 1_{10}$$

Hexadecimal	Binary	Decimal
0	0000	0
1	0001	1
2	0010	2
3	0011	3
4	0100	4
5	0101	5
6	0110	6
7	0111	7
8	1000	8
9	1001	9
A	1010	10
B	1011	11
C	1100	12
D	1101	13
E	1110	14
F	1111	15

Fig. 18-2 Hexadecimal-binary-decimal conversion chart.

Therefore the BCD (Binary Coded Decimal) equivalent of the decimal number 41 is

0100 0001

Each nibble (group of 4 bits) stands for one decimal digit. The number as a whole is still a decimal number, however.

ASCII

ASCII code is different from decimal, binary, hexadecimal, and BCD in that it is not a number system but rather a way to represent various symbols with different patterns of 1s and 0s. Each pattern of 1s and 0s stands for a different letter of the alphabet (uppercase or lowercase), digit, punctuation mark, or other useful character.

We use number systems to count and to perform mathematical computations. We don't use ASCII for these purposes. We use ASCII code to represent characters used in normal written communication.

Do not try to memorize the ASCII code. Using charts when needed will suffice. If a large amount of data is necessary, we usually have some device, primarily the standard computer keyboard, to create these ASCII characters. A table (Table 1-6) showing the ASCII code appears in Chap. 1.

Microprocessors and Number Conversions

Microprocessors "think" in binary numbers. They use binary numbers for calculations and logical operations. Since binary numbers can be displayed as hexadecimal numbers with fewer digits, we often display binary numbers as their hexadecimal equivalents when people must enter or interpret those numbers.

The BCD numbers are used in certain situations to aid the people who must read them. For this reason some microprocessors have instructions which can convert answers resulting from binary mathematical operations to binary-coded decimal numbers. We will look at these operations later in this chapter.

Bit Positions

Sometimes students are confused when people talk about a certain "bit." There are two ways to describe a particular bit: by the binary power of 2 reflected in its position and by its location, from right to left. Look at Fig. 18-3.

You will see both methods used in the workplace and in other textbooks, so you should become comfortable with each.

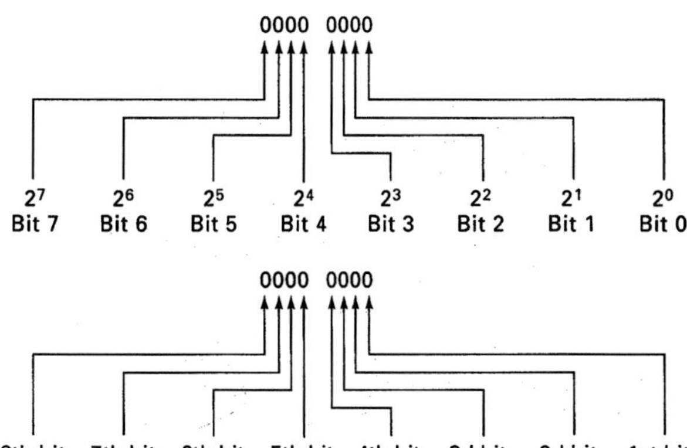

Fig. 18-3 Two methods for describing bit positions.

18-2 ARITHMETIC INSTRUCTIONS

We will now review basic binary math and look at typical microprocessor instructions which perform mathematical computations. Remember that we are now discussing techniques and instructions which are common to most microprocessors. We will study instructions specific to each microprocessor family in its appropriate section later in this chapter.

Addition

Each microprocessor family included in this text has at least one addition instruction. Most have more than one.

When adding binary numbers the microprocessor produces two types of information: (1) the sum of the two numbers (answer), (2) and information indicating whether there were carries in certain columns.

If you don't remember how to add binary numbers, you may want to review Chap. 6 now. There are really only five binary addition combinations to remember:

(1)	(2)	(3)	(4)	(5)
0	0	1	1	1
+0	+1	+0	+1	1
0	1	1	10	+1
				11

The first three combinations produce the same answer as they do in the decimal number base system. Combination #4 is simply saying that 1 + 1 = 2, except that the 2 is binary ($10_2 = 2_{10}$). You should say combination #4 to yourself as, "1 plus 1 equals 0, carry 1." Likewise, the fifth combination is saying that 1 + 1 + 1 = 3, except that the 3 is binary ($11_2 = 3_{10}$). You should express combination #5 as, "1 plus 1 plus 1 equals 1, carry 1." The last two combinations are the only new ones that you should memorize, since they are the only two that are different from our decimal number system.

To continue our review, let's see how to add several columns. It is common (and very practical) to show 8-bit binary numbers in two groups of four (as 2 nibbles). Refer to Fig. 18-4.

As you study Fig. 18-4, you will see that each of the

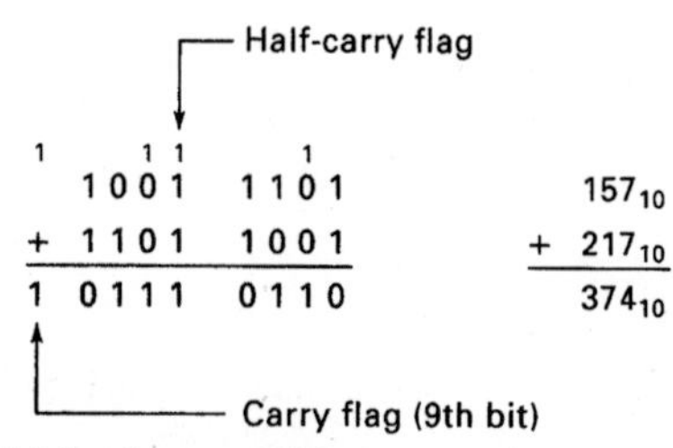

Fig. 18-4 Multi-column addition.

individual additions in each column is one of the five combinations we presented a moment ago.

Now let's continue using this example as we talk about two other closely related subtopics.

Carry Flag

The first flag we'll study is the carry flag. The *carry flag,* during addition, lets us know that the 8-bit sum is not the complete answer. If the carry flag is set (has a value of 1), it indicates that a 9th bit was produced.

Let's look again at Fig. 18-4. Notice the sum shown in the decimal version of the example. The decimal answer is 374. Now look at the binary version of the example. If you were to use only the right-most 8 bits (the 8 least significant bits), the sum would appear to be 118_{10} ($0111\ 0110_2 = 118_{10}$), which is not the correct sum. The 9th bit, which appears at the far left (the most significant bit), would not appear in an 8-bit accumulator. The 9th bit would exist in the carry flag (so to speak). The 1 in the carry flag would indicate a carry from column 8 to column 9. Again, we cannot see a 9th bit since the accumulator only holds 8 bits. (If you are using a 16-bit microprocessor, the function of the carry flag is the same as that described above except that it indicates the presence of a 17th bit, which will not fit into a 16-bit accumulator.

Substraction also affects the carry flag. We will discuss that a little later in this chapter.

Half-Carry Flag

Some (but not all) of our microprocessors have a half-carry flag. A *half-carry flag* indicates that a carry has occurred from the 4th-bit column to the 5th-bit column. The half-carry has been marked in Fig. 18-4.

Overflow Flag

The *overflow flag* alerts the programmer to a condition that is similar to, but not the same as, that to which the carry flag alerts the programmer. All our featured microprocessor families have an overflow flag except the 8080/8085. To understand what the overflow flag does, we need to take a closer look at 2's-complement arithmetic and signed binary numbers.

Each of our microprocessor families has one or more accumulators. All are 8-bit accumulators except the 8086/8088, which has a 16-bit accumulator. Let's focus our discussion on the 8-bit microprocessors.

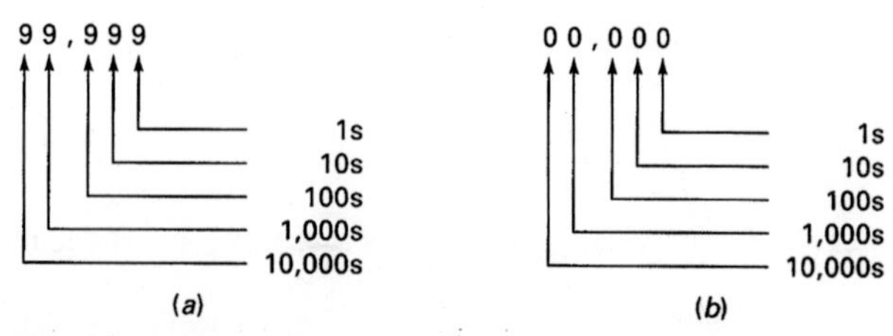

Fig. 18-5 (*a*) Automobile odometer. (*b*) Automobile odometer reset.

If we do not expect to ever need negative numbers in a particular application, we can let the binary range of 0000 0000 to 1111 1111 represent decimal numbers 0 to 255. These are called *unsigned binary numbers.* However, if we need to represent negative numbers, we must use the 2's-complement form of the numbers we wish to make negative. When we allow both positive and negative numbers, we are using *signed binary numbers.*

We introduced 2's-complement numbers in Chap. 6. The concept was compared to that of the odometer on a car. Remember that the accumulator, like the odometer of a car, can contain only a certain number of digits. Most cars display 5 digits plus 10ths of a mile. If we disregard the 10ths digit, we have just 5 places. Of course, the highest number which can be represented is 99,999 miles. There aren't enough digits to show 100,000 miles. The 1 is lost, and only the 00,000 remains. The odometer has reset. Figure 18-5 illustrates this.

The accumulator of a microprocessor has this same limitation. If you continuously increment an 8-bit accumulator, you will eventually reach a maximum number beyond which the accumulator would have to have another digit. Figure 18-6 illustrates this. The accumulator, like the odometer, will reset to zero if it is incremented one more time.

When working with 2's-complement binary numbers, we assume that the accumulator can also be rolled backward, so to speak, to represent negative numbers. One less than zero is 11111111_2, which would be equal to -1_{10}. One less than that would be 11111110_2, which would be equal to -2_{10}. This process would continue as shown in Fig. 18-7.

As Fig. 18-7 illustrates, -128_{10} is as far as we can go

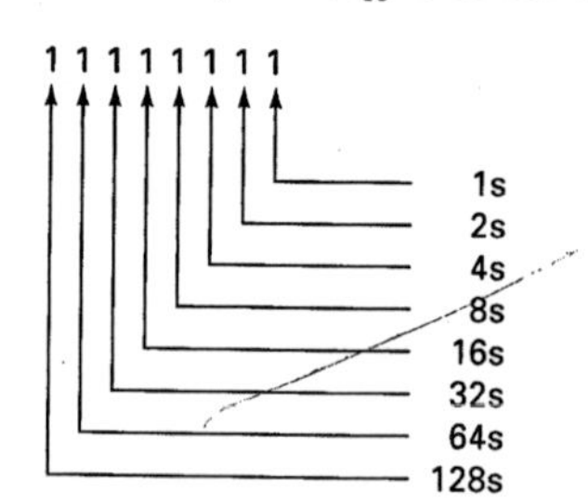

Fig. 18-6 Eight-bit accumulator.

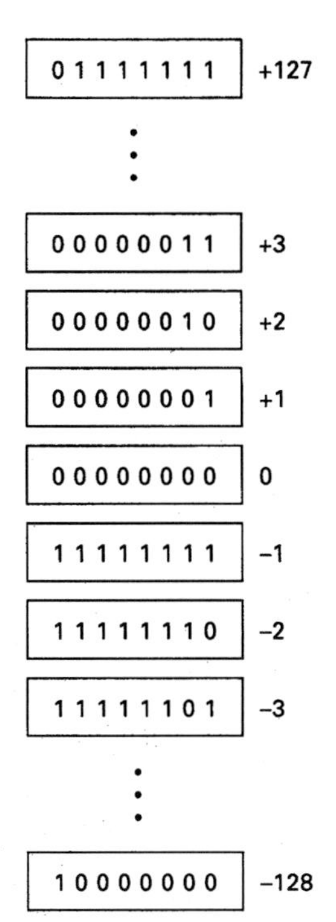

Fig. 18-7 Eight-bit 2's-complement range.

on the negative end. The reason for this is that one less than 10000000_2 is 01111111_2, which, if you look at the top of Fig. 18-7, is already being used as the equivalent of $+127_{10}$. When working with 8-bit 2's-complement numbers, we regard all numbers which have a 1 as the MSB (most significant bit) as negative. Numbers with a 0 in the MSB are positive. This means that the range for 8-bit 2's-complement binary numbers is $+127_{10}$ to -128_{10} inclusive.

Let's review a little. If we are using all 8 bits to represent numbers from 00_{10} to 255_{10}, we refer to these numbers as unsigned binary numbers. If we are using the MSB to signify whether a number is positive or negative, we have a range of -128_{10} to $+127_{10}$. These are called signed binary numbers.

There is a simple procedure by which you can determine how to form a negative binary (or hexadecimal) number. First, write the binary equivalent of the positive form of the number. For example

$$10_{10} = 0000\ 1010_2 = 0A_{16}$$

Now invert each bit of the binary number.

0000 1010 becomes 1111 0101

Then add 1.

$$\begin{array}{r} 1111\ 0101 \\ +\qquad\quad 1 \\ \hline 1111\ 0110 \end{array}$$

Therefore

$$-10_{10} = 1111\ 0110_2 = F6_{16}$$

Notice that the MSB of the binary number is 1, as we said it would be.

To determine what value a negative-signed binary number represents, reverse the above process. If you had the binary number

1111 0110 (the number created a moment ago)

invert each bit

0000 1001

and then add 1.

$$\begin{array}{r} {\scriptstyle 1} \\ 0000\ 1001 \\ +\qquad\quad 1 \\ \hline 0000\ 1010 \end{array}$$

Notice that we now have the binary number for 10_{10}. (A small 1 indicates a carry.) We have found that the binary number 1111 0110 is the signed binary number for -10_{10}.

The question now is how to interpret certain numbers. For example,

$$\begin{array}{r} 125_{10} \\ +\ 50_{10} \\ \hline 175_{10} \end{array} \qquad \begin{array}{r} 0111\ 1101_2 \\ +0011\ 0010_2 \\ \hline 1010\ 1111_2 \end{array} \qquad \begin{array}{r} 125_{10} \\ +\ 50_{10} \\ \hline -81_{10} \end{array}$$

We know that $125_{10} + 50_{10} = 175_{10}$. As you will notice in this example, however, the binary number for 175_{10} (which is $1010\ 1111_2$) is also the binary number for -81_{10}. So if we didn't know what two numbers this was the sum of, how would we know how to interpret this answer? If we simply found the binary number $1010\ 1111_2$ in a register, how would we know if it was meant to be $+175_{10}$ or -81_{10}? The answer is that we wouldn't. (The number -81_{10} is, of course, the wrong answer. We will deal with that part of the problem in just a bit.) We must know whether we are using unsigned binary numbers or signed binary numbers before we see the answer. It is simply a matter of agreement beforehand.

We have been preparing to explain the purpose of the overflow flag. We are now ready. The previous example, which produced a sum of $+175_{10}$ ($1010\ 1111_2$), would have set the overflow flag in an 8-bit microprocessor. The overflow flag tells the programmer that the last answer produced was outside the range of $+127_{10}$ to -128_{10} ($0111\ 1111_2$ to $1000\ 0000_2$ or $7F_{16}$ to 80_{16}). If the programmer understood this answer to represent an unsigned binary number, he or she would ignore the flag. If, however, this was intended to be a signed binary number, the programmer would know that this answer, if taken as a signed binary number, is incorrect because it has exceeded the range for 8-bit signed binary numbers.

The range for unsigned 16-bit binary numbers is 0_{10} (0000 0000 0000 0000_2 or 0000_{16}) to $65{,}535_{10}$ (1111 1111 1111 1111_2 or $FFFF_{16}$). The range for signed 16-bit binary numbers is $+\ 32{,}767_{10}$ (0111 1111 1111 1111_2 or $7FFF_{16}$) to $-32{,}768_{10}$ (1000 0000 0000 0000_2 or 8000_{16}).

Addition-with-Carry

The previous section on addition discussed the carry flag. The carry flag signals the programmer that the result of an operation has exceeded 8 bits.

The carry flag has another use, though. The carry from the 8th bit to the 9th bit (which is what the carry represents) can be used during multiple-precision arithmetic. We use multiple-precision arithmetic when the accumulator cannot accept numbers large enough for the desired operation.

Multiple-Precision Binary Numbers

Until now we have assumed that any numbers we want to add would occupy only 1 byte of memory. This is called a *single-precision* number. One-byte unsigned numbers can range from 0 to 255. Two-byte unsigned numbers can range from 0 to 65,535. These are called *double-precision* binary numbers. Three-byte unsigned binary numbers can range from 0 to 16,777,215. These are called *triple-precision* numbers.

When we construct a double-precision number, we use the same techniques to determine its value as when we work with a single-precision number. Recall from earlier chapters that each binary position has a value and that each value is twice as large as the value to its right. If you have a calculator which will calculate powers of a number, it is quite easy to determine the value of a double-precision binary number. Refer to Fig. 18-8.

You see that the least significant bit (LSB) has a value of 2^0. This is equal to the number 1. (If you try this on a scientific calculator, it should give you that answer.) To determine the value of a double-precision number, add the value of each position which has a 1 in it. This will give you its decimal value. For example, to calculate the value of the binary number

0100 0001 0000 0010

you would enter

$$2^{14} + 2^8 + 2^1 = 16{,}642_{10}$$

into your calculator to get the above answer.

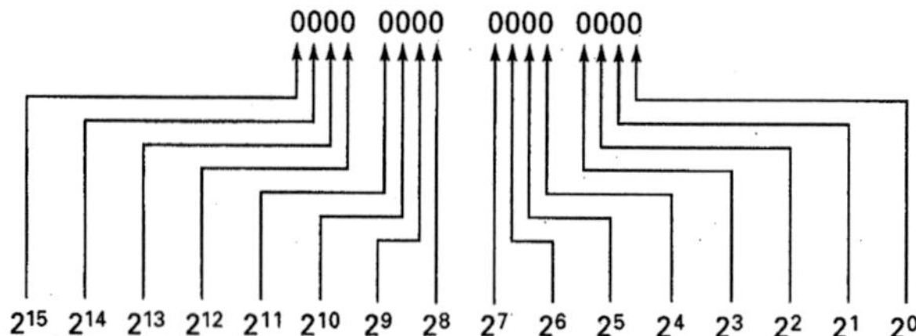

Fig. 18-8 Powers of 2 for a double-precision binary number.

Add-with-Carry

Let's step through a double-precision addition problem. Remember that we will be using the carry flag. Figure 18-9 shows an example.

The least significant bytes (LSBs) are on the right. They occupy the positions which have the least value. The most significant bytes are on the left. They occupy the positions which have the most value.

As you can see, several carries occur in this example. We are interested in the carry from the LSB to the MSB. That carry would actually be held in the carry flag of the microprocessor.

A typical microprocessor program to add these two binary numbers (using English phrases instead of microprocessor instructions) would appear as follows:

CLEAR CARRY FLAG
LOAD ACCUMULATOR WITH LSB OF ADDEND
ADD THE LSB OF THE AUGEND TO THE ACCUMULATOR
STORE THE LSB OF THE SUM IN MEMORY
LOAD THE ACCUMULATOR WITH THE MSB OF THE ADDEND
ADD-WITH-CARRY THE MSB OF THE AUGEND TO THE ACCUMULATOR
STORE THE MSB OF THE SUM IN MEMORY

Notice that we simply add the LSB of each number, but we *add-with-carry* the MSB of each number. When the microprocessor sees the add-with-carry instruction, it actually adds three numbers. It adds the addend (MSB), augend (MSB), and the carry flag. This brings the carry from the LSB into the MSB.

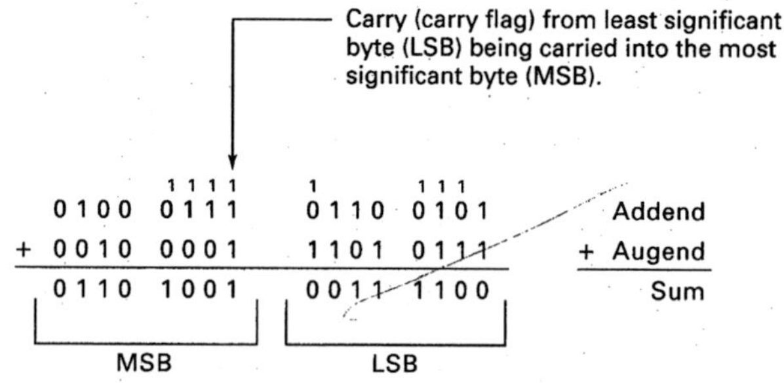

Fig. 18-9 Double-precision addition-with-carry.

Subtraction

Each of the microprocessor families included in this text has at least one subtraction instruction. Most have more than one.

When subtracting binary numbers, the microprocessor produces two types of information: (1) The difference between the two numbers (answer) and (2) whether there were borrows in certain columns.

If you don't remember how to subtract binary numbers, you may want to review Chap. 6 now. There are really only four binary combinations you need to remember:

(1)	(2)	(3)	(4)
0	1	1	$^{1}0$
−0	−0	−1	−1
0	1	0	1

The first three combinations produce the same answer as they do in the decimal-number base system. Combination #4 requires a borrow, which is shown by the small 1 set as a superscript. You cannot have 0 and subtract 1 from it. If you can borrow a 1 from the next-higher column, the subtraction becomes possible. If there is a higher column from which to borrow, this combination is really $2_{10} - 1_{10} = 1_{10}$. That is, 10_2 is created after the borrow occurs, and now the top number is larger than the bottom number. The carry flag is used if there is no higher column from which to borrow. You might say that it now becomes a "borrow" flag.

The last combination is the only new one that you will need to memorize since it is the only one that is different from our decimal number system.

The above discussion appeared in Chap. 6 and has been reviewed here for your convenience.

To continue our review, let's see how to subtract several columns. As in addition, it is common (and very practical) to show 8-bit binary numbers in two groups of four (as two nibbles). Refer to Fig. 18-10.

As you study Fig. 18-10, you will see that each individual subtraction in each column is one of the four combinations we presented a moment ago. When a borrow occurs, we have shown the borrowed 1 as a superscript 1 next to the 0 which needed it. The 1 that was borrowed from is crossed off, and its new value, 0, is shown above it.

Negative (Sign) Flag

The *negative flag*, sometimes called the *sign flag*, tells us whether the number in the accumulator is a positive or negative number. Since the most signficant bit of the accumulator is the sign bit (when using signed binary numbers), the negative flag simply reflects the status of that bit. If the most significant bit is 0, the negative flag is 0, and this is a positive number. If the most significant bit is 1, the negative flag will be 1, and this is a negative number.

While the negative flag always indicates the status of bit 7 of the accumulator, it is up to the programmer to determine whether the number is to be interpreted as a signed or unsigned binary number.

Figure 18-11 illustrates how the negative flag works.

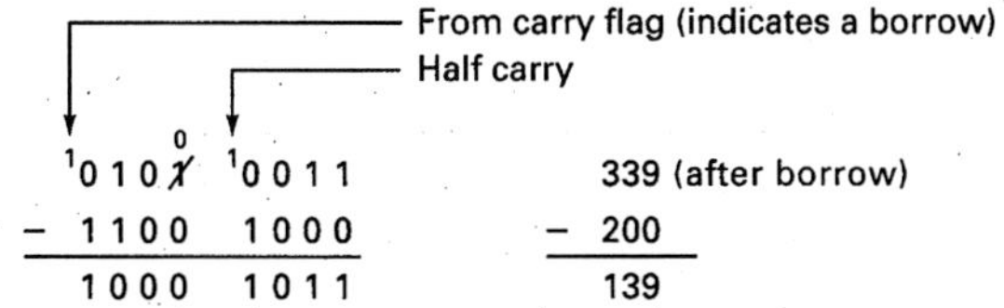

Fig. 18-10 Subtraction of binary numbers.

Zero Flag

The *zero flag* shows that the last operation produced a result of 0. This does not apply just to the accumulator but can apply to other registers as well. This is especially helpful when repeatedly decrementing (reducing by 1) an index register to determine the number of times a loop has executed. Knowing when a register has reached 0 is also useful when branching to other parts of a program and when determining whether or not to activate (call or enter) certain subroutines.

The one unusual feature of the zero flag is that it contains a 1 when the result is 0, and the flag is 0 when the result is anything other than 0. While this may appear confusing at first, it becomes second nature as you gain experience with microprocessors.

The idea here is that a 0 says that something is false or has not occurred. A 0 says, "No, this number was *not* the number zero."

A 1 says that something is true or has occurred. A 1 says, "Yes, this number *is* the number zero."

Subtraction-with-Carry (Borrow)

The same carry flag that informs us that an addition problem produced a sum which carried a 1 into the 9th bit also tells us something about subtraction problems. Now it tells us that to produce the answer (difference) the microprocessor had to borrow a 1 from a 9th bit. This occurs when the top number (minuend) is smaller than the bottom number (subtrahend).

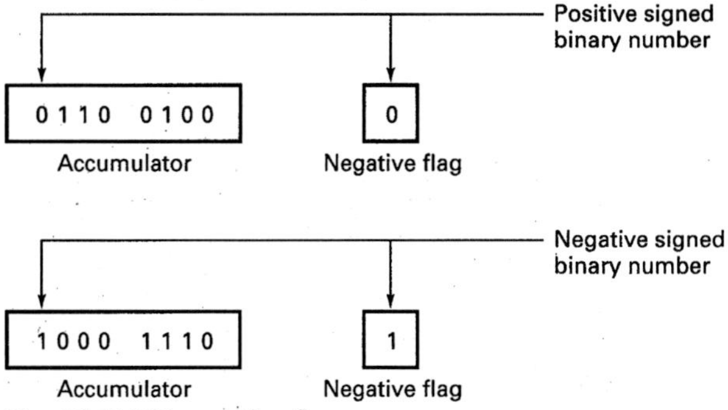

Fig. 18-11 The negative flag.

Refer again to Fig. 18-10. Notice that a borrow was required from a column that doesn't actually exist. There is no 9th column. The carry flag acts as that column. It tells us that a borrow from this "imaginary" column was necessary.

Most microprocessors set the carry flag (make it a 1) when a borrow is necessary (1 = true). The exception to this is the 6502 microprocessor. It clears the carry flag, as though the borrow actually came from the flag itself. In the 6502 you must set the carry flag before you start a subtraction problem so that, if a borrow is necessary, a 1 will be present.

Some microprocessors also monitor the 4th bit during subtraction. This is the half-carry flag which was mentioned earlier in this chapter.

Multiplication and Division

The 8-bit microprocessors featured in this text do not have multiplication or division instructions (the 6809, a relative of the 6800 and 6808, does have a **MUL**tiply instruction). However, the 16-bit 8086/8088 has both multiply and divide instructions, which will be discussed in the 8086/8088 section of this chapter.

There are several software algorithms for both multiplication and division which work well with the 8-bit microprocessors.

18-3 FLAG INSTRUCTIONS

Each of our microprocessors has instructions to alter the state of its flags. Which of their flags and how many of their flags can be directly altered vary.

The 8080/8085 has the fewest instructions for setting and clearing flags. The 6502, 8086/8088, and 6800/6808 all have the ability to set and clear many of their flags directly. The 6800/6808 has an instruction which makes it possible to move the status of all the flags into accumulator A and to copy the contents of accumulator A into the flag register. All our microprocessors except the 6800/6808 have the ability to push all the flags onto the stack and retrieve them from the stack. The 6800/6808 can accomplish the same task by transferring the flags to accumulator A and then to the stack in a two-step process.

We'll discuss the specific uses for each flag instruction in the Specific Microprocessor Families section of this chapter. The uses for flag instructions can be generalized, however. We use the flags primarily during arithmetic operations and for control of loops, branches, and subroutines.

Since we use the flags to give us information about the outcome of arithmetic operations, we often need to set or clear flags before these math operations so that we are certain of their exact condition before the operation begins.

We use flags to determine whether or not certain loops should be repeated, whether branches into other parts of the program should be taken, and whether certain subroutines should be called. Flags are used to make decisions about which microprocessor instructions should be executed next. This is the same as saying that the flags are used by the program to make decisions. For these reasons we may want to set or clear certain flags before or after certain instructions are executed.

Specific Microprocessor Families

Let's study the arithmetic and flag instructions for each of our microprocessor families. We'll be using short routines to study operations for which each microprocessor has specific instructions. We will not develop long routines to facilitate arithmetic operations which are not inherent to each microprocessor family. This will help you to become familiar with your microprocessor's basic arithmetic and flag instructions.

18-4 6502 FAMILY

The 6502 probably has the fewest different arithmetic instructions of any of our microprocessor families. However, by conscientiously setting and clearing the appropriate flags before arithmetic operations, this chip performs math operations adequately.

Arithmetic Instructions

The 6502 does not have normal *add* and *subtract* instructions. It has only *add-with-carry* and *subtract-with-carry*. Both of these instructions use the value in the accumulator as one of their operands with another value which can be an immediate value, or a value in memory, in addition to the value in the carry flag. The value in memory can be addressed any one of seven different ways. Let's see how to use these instructions.

Addition-with-Carry

Let's start with a very simply addition program. Figure 18-12 illustrates this type of program.

Notice first that we have used the CLC (**CL**ear **C**arry) instruction before we even loaded the accumulator with our first operand. This is necessary when using the 6502 microprocessor. If the carry flag is set from a previous operation, the ADC (**AD**d-with-**C**arry) instruction will add

$$
\begin{array}{r}
^{1\,1} \\
0100\ \ 1001 \\
+\ 0001\ \ 1110 \\
\hline
0110\ \ 0111
\end{array}
\qquad
\begin{array}{r}
49_{16} \\
+\ 1E_{16} \\
\hline
67_{16}
\end{array}
\qquad
\begin{array}{r}
73_{10} \\
+\ 30_{10} \\
\hline
103_{10}
\end{array}
$$

Addr	Obj	Assembler	Comment
0340	18	CLC	Prepare for addition problem
0341	A9	LDA #$49	Load accumulator with first number (49)
0342	49		
0343	69	ADC #$1E	Add 1E to the number in the accumulator and place the answer in the accumulator
0344	1E		
0345	00	BRK	Stop

Fig. 18-12 Simple 6502 addition problem.

the 1 in the carry flag to the answer and will cause the answer to be incorrect (it will be 1 greater than the correct result).

Pay particular attention to the accumulator and the processor status register. Notice their contents both before and after you run the program. (You may want to write down their values before and after so that you can study their behavior.) You will find that the accumulator will have the number 67_{16} in it (which is the correct answer) and that only the BRK (**BReaK**) flag will be set.

Let's look at the processor status register a little more closely. Refer to Fig. 18-13 now.

Examining the flags from right to left, let's consider each and why it was or was not set during the last problem.

The carry flag would have been set if a carry from the 8th bit to the 9th bit (which doesn't exist, so it goes into the carry flag) had occurred, but none did.

The zero flag would have been set if the answer had been 0, but it wasn't.

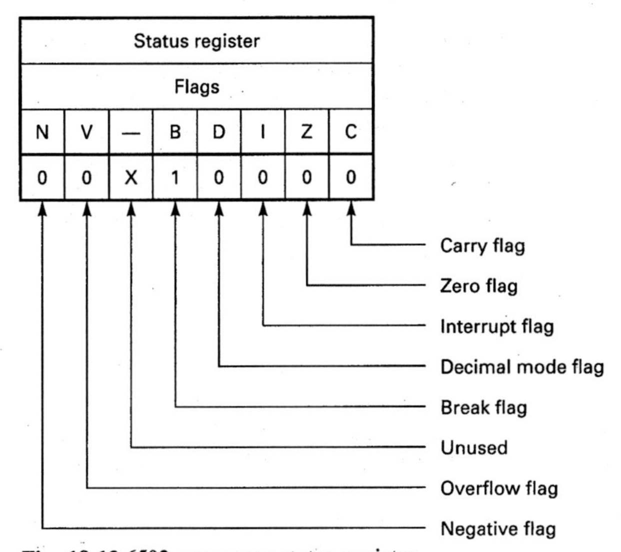

Fig. 18-13 6502 processor status register.

Don't worry about the interrupt flag since we haven't introduced this subject yet.

We dealt with the two operands as though they were hexadecimal numbers so we didn't set the decimal flag.

The break flag was set because we used the break instruction to stop the program.

The status of the unused flag doesn't matter.

We did not exceed the range of decimal +127 to −128 (hexadecimal 7F to 80); therefore the overflow flag was not set.

Finally, we did not have a 1 in the 8th bit of the accumulator so the answer could not have been negative; therefore the negative flag was not set.

The Negative Flag

Let's look at a problem which produces a negative answer. Refer to Fig. 18-14 now.

Notice that this is exactly the same problem that was used in Fig. 18-12 except that we have changed the first operand, which used to be 49_{16} into $C9_{16}$, which is the decimal number -55_{10}, if we consider these numbers to be signed binary numbers. We know that $-55_{10} + 30_{10} = -25_{10}$. Since this is a negative answer, we know that the negative flag should be set after the program is run.

Load the program and run it. Again write down the contents of the accumulator and processor status register before and after running the program so that you can compare them. After the program is run, the accumulator should contain the value $E7_{16}$. The processor status register should contain B0.

Let's examine the status register again. The binary value for $B0_{16}$ is $1011\ 0000_2$. If you put those bits into the appropriate positions in the status register as shown in Fig. 18-15, you will see that 3 bits or flags are set.

The break flag is again set because we used the break instruction to stop the program. We do not care about the status of the unused bit.

```
  1 1
  1100  1001
+ 0001  1110
  ----------
  1110  0111
```

$$\begin{array}{r} C9_{16} \\ +\ 1E_{16} \\ \hline E7_{16} \end{array} \qquad \begin{array}{r} -55_{10} \\ +\ 30_{10} \\ \hline -25_{10} \end{array}$$

Addr	Obj	Assembler	Comment
0340	18	CLC	Prepare for addition problem
0341	A9	LDA #$C9	Load accumulator with first number (C9)
0342	C9		
0343	69	ADC #$1E	Add 1E to the number in the accumulator and place the answer in the accumulator
0344	1E		
0345	00	BRK	Stop

Fig. 18-14 Simple 6502 addition problem with negative answer.

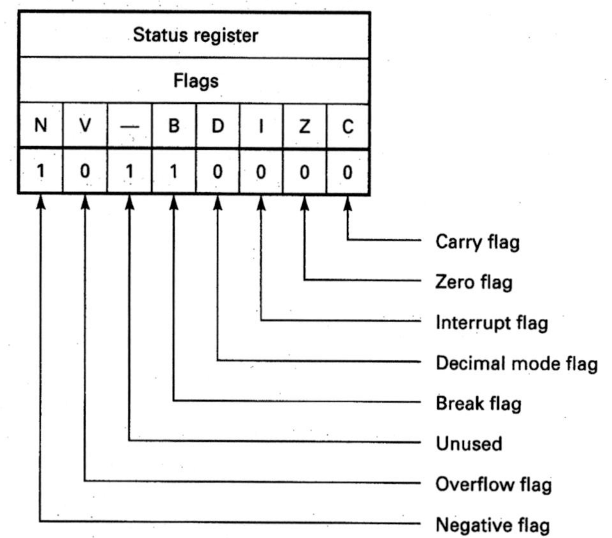

Fig. 18-15 6502 status register after an addition problem which produces a negative answer.

The negative flag is now set, however. This is what we expected to see. The sum of the addition problem was -25_{10} ($E7_{16}$). If we assume that our numbers are signed binary numbers, then any number that has a 1 in the 8th bit is negative. $E7_{16}$ has a 1 in the 8th bit. The negative flag simply reflects the state of the 8th bit.

The Zero Flag

Now let's change the program so that we get a sum of 0. Then we can see how the flags react to this situation.

Figure 18-16 shows the problem and the program to solve the problem.

We are again assuming that our numbers are signed binary numbers. The problem is $C9_{16} + 37_{16} = 00_{16}$, which is $-55_{10} + 55_{10} = 0_{10}$. You should go through the binary addition now before you run the program. Notice both the answer and any carries.

Write down the contents of the accumulator and the processor status register before and after running the program. You will notice that we are using the same program as in the last problem but have again changed one of the operands.

```
1 1111  1111
  1100  1001
+ 0011  0111
  ----------
1 0000  0000
```

$$\begin{array}{r} C9_{16} \\ +\ 37_{16} \\ \hline 00_{16} \end{array} \qquad \begin{array}{r} -55_{10} \\ +\ 55_{10} \\ \hline 0_{10} \end{array}$$

Addr	Obj	Assembler	Comment
0340	18	CLC	Prepare for addition problem
0341	A9	LDA #$C9	Load accumulator with first number (C9)
0342	C9		
0343	69	ADC #$37	Add 37 to the number in the accumulator and place the answer in the accumulator
0344	37		
0345	00	BRK	Stop

Fig. 18-16 Simple 6502 addition problem which produces a sum of 0.

Now enter and run the program. The accumulator should contain 00_{16}, and the status register should contain 33. If you place the bits of the status register in their proper places as shown in Fig. 18-17, you will see how the flags have responded to this problem.

Notice that the break flag and unused flag have again been set as before. The value of the unused flag has no meaning, and the break flag simply shows that we used a break to stop the program.

The zero flag is set, as we supposed it would be. The carry flag is also set. Notice in the binary addition that a carry did indeed occur from the 8th to a nonexistent 9th bit (which the carry flag acts as).

Status register							
Flags							
N	V	—	B	D	I	Z	C
0	0	1	1	0	0	1	1

Fig. 18-17 6502 status register after an addition problem which produces a sum of 0.

The Overflow Flag

When the overflow flag is set, it tells us that if the numbers which were just added or subtracted are signed binary numbers, then the valid range for such numbers has been exceeded and the result is incorrect. The valid range for 8-bit microprocessors, which the 6502 is, is +127 to −128. Let's change our problem to create an overflow.

Figure 18-18 shows our problem and program. Notice in this problem that we are assuming that all values are to be interpreted as signed binary values.

The problem shown here is $123_{10} + 111_{10} =$ ______. First go through the binary addition and enter the program. Then write down the values in the accumulator and processor status register, run the program, and finally write down the ending values of the accumulator and status register.

Status register							
Flags							
N	V	—	B	D	I	Z	C
1	1	1	1	0	0	0	0

Fig. 18-19 6502 status register after an addition problem which creates an overflow.

Figure 18-19 shows what the value in the status register should be.

You should have a sum of EA_{16} in the accumulator and $F0_{16}$ in the status register. EA_{16} is the correct sum if you are using *unsigned* binary numbers! If you interpret EA_{16} as a signed binary number, it has a value of -22_{10}. This is *not* the correct answer. We have exceeded our valid range for signed binary numbers.

The status register has a value of F0. This means that in addition to the unused flag and the break flag, both the overflow and the negative flags have been set.

It makes sense for the negative flag to be set because the 8th bit of the accumulator is set. This indicates a negative number if the value is a signed binary number.

The overflow flag is set because we have exceeded our range of $7F_{16}$ (127_{10}) to 80_{16} (-128_{10}), giving an incorrect result.

The Decimal Flag

Because of differences in the way binary and decimal numbers round, and because numeric output to humans is usually decimal, it is sometimes better to actually do arithmetic calculations by using decimal numbers rather than binary numbers. Actually, true decimal numbers are not used. Rather, a mixture of binary and decimal, called binary-coded decimal, is used. (The method used to create BCD numbers is covered in Chap. 1 and they have been discussed subsequently. You should review that section of

```
  1111  111
  0111  1011        7B16      12310
+ 0110  1111     +  6F16    + 11110
  1110  1010        EA16      23410
```

Addr	Obj	Assembler	Comment
0340	18	CLC	Prepare for addition problem
0341	A9	LDA #$7B	Load accumulator with first number (7B)
0342	7B		
0343	69	ADC #$6F	Add 6F to the number in the accumulator and place the answer in the accumulator
0344	6F		
0345	00	BRK	Stop

Fig. 18-18 Simple 6502 addition problem which produces an overflow.

Chap. 1 now if you are unsure of what BCD numbers are or how they are formed.)

One of the problems encountered when using BCD numbers is that, as the binary nibbles are added, invalid results are sometimes obtained.

Most microprocessors have an instruction called *decimal adjust* (or something similar). This instruction changes the number in the accumulator to what it would be if the last two numbers operated on had been BCD numbers instead of binary numbers. The 6502 handles this a little differently. It requires that you set a flag designed just for this purpose and enter a "decimal mode," so to speak. When the decimal flag is set, all operands are assumed to be packed BCD numbers.

Let's look at an example. In Fig. 18-20 we have compared a decimal addition problem to the binary version of the same problem.

First notice the difference between BCD and binary addition. BCD addition is not the same as binary addition. BCD is *decimal* addition using four binary digits to represent each decimal digit.

The program shown in Fig. 18-20 will help you understand the difference between binary and BCD addition (and subtraction). This program does the addition problem twice,

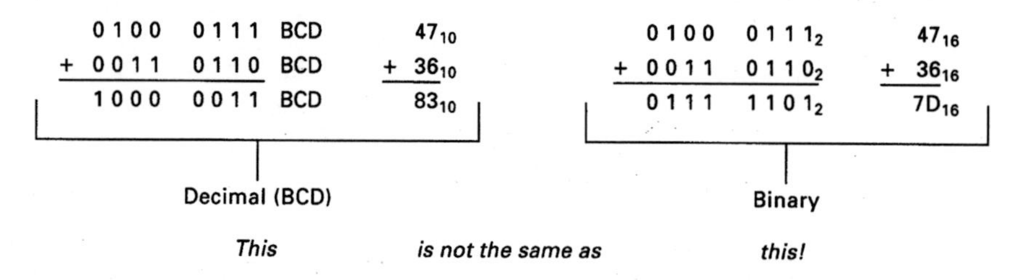

Addr	Obj	Assembler	Comment
0340	D8	CLD	Prepare to do ***binary*** addition
0341	18	CLC	
0342	A9	LDA #$47	This is being interpreted as a binary number
0343	A7		
0344	69	ADC #$36	This also is being considered a binary number
0345	36		
0346	8D	STA $03A0	We'll store the ***binary*** answer in memory location 03A0
0347	A0		
0348	03		
0349	08	PHP	Put the flags on the stack
034A	68	PLA	Transfer flags from stack to accumulator
034B	8D	STA $03A1	We'll store the status of the flags from the ***binary*** addition in the memory location immediately following the ***binary*** sum, which is location 03A1
034C	A1		
034D	03		
034E	F8	SED	Prepare for ***decimal*** addition
034F	18	CLC	
0350	A9	LDA #$47	This number is being interpreted as a ***decimal*** number
0351	47		
0352	69	ADC #$36	This number likewise is being considered a ***decimal*** number
0353	36		
0354	8D	STA $03A2	We'll store the ***decimal*** answer in memory location 03A2
0355	A2		
0356	03		
0357	08	PHP	Put the flags on the stack
0358	68	PLA	Transfer the flags to the accumulator
0359	8D	STA $03A3	We'll store the status of the flags resulting from this ***decimal*** addition in the memory location immediately following the ***decimal*** sum, that is, location 03A3
035A	A3		
035B	03		
035C	00	BRK	Stop

Fig. 18-20 Binary vs. BCD addition.

once using binary numbers and once using BCD numbers. The result of the binary addition is stored in memory location $03A0_{16}$, and the resulting flags in location $03A1_{16}$. The result of the BCD addition is stored in location $03A2_{16}$, and the resulting flags in location $03A3_{16}$. Enter and run this program to see what results you get. (Don't be concerned about the reference to the stack in the program. We'll study the stack in a later chapter. For now just think of it as a temporary storage area.) When we ran the program we found the following:

location $03A0_{16}$ = binary sum = 7D
location $03A1_{16}$ = binary flags = 30
location $03A2_{16}$ = BCD sum = 83
location $03A3_{16}$ = BCD flags = F8

The status of the binary flags indicates only that the break instruction had been used to stop the program. No other flags were set.

The status of the flags after the BCD addition indicates that the decimal flag was set. (We set this flag to get into the "decimal mode.") The negative flag was set but has no valid meaning. It was simply following the state of the 8th bit of the accumulator. The overflow flag was set, but it also has no valid meaning in BCD arithmetic.

Subtraction-with-Carry

Subtraction-with-carry is the opposite of addition-with-carry. As in addition, there is no simple subtract instruction, only subtract-with-carry.

The 6502 handles borrows differently from the way most other microprocessors do. Most microprocessors set the carry flag if either a carry or a borrow occurs. The 6502 sets the flag if a carry occurs and clears the flag if a borrow occurs. It is important to remember that *the carry flag must be set before a subtraction problem (or the first section of a multiple-precision subtraction problem) so that if a borrow is needed, it can clear the carry flag,* which then indicates that the borrow has occurred. *If the carry flag is not set before starting the subtraction, the answer will be incorrect. (It will be 1 less than the correct result.)*

Figure 18-21 illustrates the correct way to write a program to do single-precision subtraction.

You should assemble and run this program. When we did, we found that the result in the accumulator was FF. We also found that the overflow and negative flags had been set. The negative flag was set because the 8th bit of the answer is a 1, which indicates a negative-signed binary number. The overflow flag was set because $7F_{16} = 127_{10}$, and $80_{16} = -128_{10}$; therefore

$$\begin{array}{r} 127 \\ --128 \\ \hline 255 \end{array}$$

and 255_{10} is outside the valid range for 8-bit signed binary numbers. (The valid range is $+127_{10}$ to -128_{10}.)

18-5 6800/6808 FAMILY

The 6800/6808 has a variety of *add* and *subtract* instructions which can use either of its two accumulators and can address memory locations in several ways. The 6800/6808 can also add and subtract binary-coded decimal (BCD) numbers.

Arithmetic Instructions

The 6800/6808 has *add, subtract, add-with-carry, subtract-with-carry, add accumulator A to accumulator B, subtract accumulator B from accumulator A,* and *decimal adjust accumulator A* instructions. These instructions use the value in one of the accumulators as one of their operands and another value which can be an immediate value or a value in memory. Let's see how to use these instructions.

Addition

Let's start with a very simple addition program. Figure 18-22 illustrates this type of program.

Pay particular attention to the accumulator and the condition code register (status register). Notice their contents both before and after you run the program. (You may want to write down their values before and after so you can study

Addr	Obj	Assembler	Comment
0340	38	SEC	*Remember this step!*
0341	A9	LDA #$7F	
0342	7F		
0343	E9	SBC #$80	
0344	80		
0345	00	BRK	

Fig. 18-21 Subtraction-with-carry.

$$\begin{array}{rr} & ^{1\,1} \\ & 0100\ \ 1001 \\ + & 0001\ \ 1110 \\ \hline & 0110\ \ 0111 \end{array} \qquad \begin{array}{rr} & 49_{16} \\ + & 1E_{16} \\ \hline & 67_{16} \end{array} \qquad \begin{array}{rr} & 73_{10} \\ + & 30_{10} \\ \hline & 103_{10} \end{array}$$

Addr	Obj	Assembler	Comment
0000	86	LDAA #$49	Load accumulator with first number (49)
0001	49		
0002	8B	ADDA #$1E	Add 1E to the number in the accumulator and place the answer in the accumulator
0003	1E		
0004	3E	WAI	Stop

Fig. 18-22 Simple 6800/6808 addition problem.

their behavior.) You will find that the accumulator will have the number 67_{16} in it (which is the correct answer) and that only the half-carry flag will be set.

Let's look at the status register a little more closely. Refer to Fig. 18-23 now.

Examining the flags from right to left, let's consider each and why it was or was not set.

The carry flag would have been set if a carry from the 8th bit to the 9th bit (which doesn't exist, so it goes into the carry flag) had occurred, but none did.

We did not exceed the range of $+127_{10}$ to -128_{10} (hexadecimal 7F to 80); therefore the overflow flag was not set.

The zero flag would have been set if the answer had been zero, but it wasn't.

We did not have a 1 in the 8th bit of the accumulator so the answer could not have been negative; therefore the negative flag was not set.

Don't worry about the interrupt flag since we haven't introduced this subject yet.

The half-carry flag was set because we had a carry from the 4th bit to the 5th bit. (Information about the half-carry is useful when dealing with BCD numbers.)

The status of the unused flags doesn't matter.

The Negative Flag

Now let's look at a problem that produces a negative answer. See Fig. 18-24.

Notice that this is exactly the same problem as the last one except that we have changed the first operand, which was 49_{16}, into $C9_{16}$, which is the number -55_{10} if we consider these numbers to be signed binary numbers. We know that $-55_{10} + 30_{10} = -25_{10}$. Since this is a negative answer, we know that the negative flag should be set after the program is run.

Write down the contents of the accumulator and processor status register before running the program so that you know what the initial conditions are. Now load the program and run it. After you run the program, the accumulator should

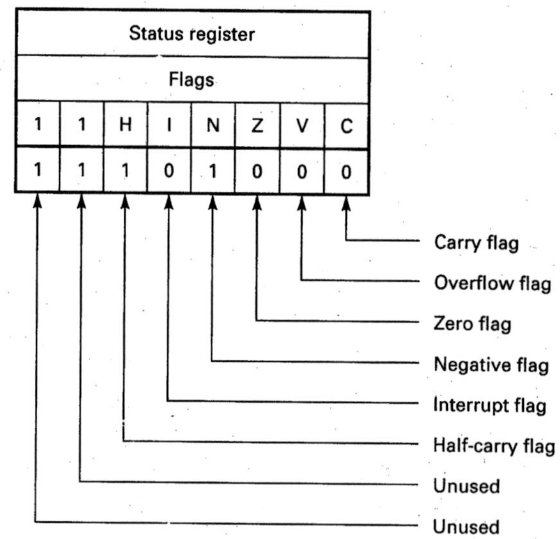

Fig. 18-23 6800/6808 status register.

$$
\begin{array}{rr}
 & {\scriptstyle 1\,1} \\
 & 1100\ \ 1001 \\
+ & 0001\ \ 1110 \\
\hline
 & 1110\ \ 0111
\end{array}
\qquad
\begin{array}{rr}
 & C9_{16} \\
+ & 1E_{16} \\
\hline
 & E7_{16}
\end{array}
\qquad
\begin{array}{rr}
 & -55_{10} \\
+ & 30_{10} \\
\hline
 & -25_{10}
\end{array}
$$

Addr	Obj	Assembler	Comment
0000	86	LDAA #$C9	Load accumulator with first number (C9)
0001	C9		
0002	8B	ADDA #$1E	Add 1E to the number in the accumulator and place the answer in the accumulator
0003	1E		
0004	3E	WAI	Stop

Fig. 18-24 Simple 6800/6808 addition problem with negative answer.

contain the value $E7_{16}$. The status register should contain XX101000.

Let's examine the status register again. If you put the bits into their appropriate positions in the status register as shown in Fig. 18-25, you will see that 2 bits or flags are set.

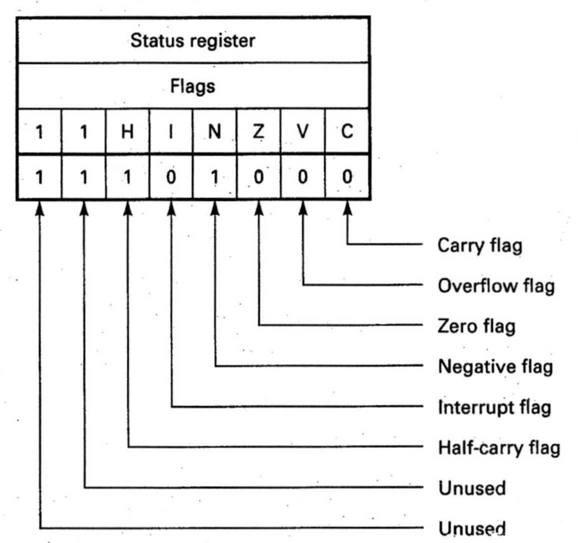

Fig. 18-25 6800/6808 status register after an addition problem which produces a negative answer.

The half-carry flag is again set because we had a carry from the 4th to the 5th bit of the result. The difference is that the negative flag is now set. This is what we expected to see. The sum of the addition problem was -25_{10} ($E7_{16}$). If we assume that our numbers are signed binary numbers, then any number that has a 1 in the 8th bit is negative. $E7_{16}$ has a 1 in the 8th bit.

The Zero Flag

Now let's change the program slightly so that we get a sum of 0. Then we can see how the flags react to this situation.

Figure 18-26 shows the problem and the program to solve the problem.

We are again assuming that our numbers are signed binary numbers. The problem is $C9_{16} + 37_{16} = 00_{16}$ ($-55_{10} + 55_{10} = 0_{10}$). You should go through the binary addition of these two numbers now before you run the program. Notice both the answer and the carries.

Again write down the contents of the accumulator and the status register before and after running the program. You will notice that we are using the same program as the ast example but have changed one of the operands.

Now enter and run the program. The accumulator should contain 00_{16}, and the status register should contain $XX100101_2$. If you place the bits of the status register value in their

$$
\begin{array}{rr}
 & {\scriptstyle 1\ \ 1111\ \ 111\ } \\
 & 1100\ \ 1001 \\
+ & 0011\ \ 0111 \\
\hline
 & 1\ \ 0000\ \ 0000
\end{array}
\qquad
\begin{array}{rr}
 & C9_{16} \\
+ & 37_{16} \\
\hline
 & 00_{16}
\end{array}
\qquad
\begin{array}{rr}
 & -55_{10} \\
+ & 55_{10} \\
\hline
 & 0_{10}
\end{array}
$$

Addr	Obj	Assembler	Comment
0000	86	LDAA #$C9	Load accumulator with first number (C9)
0001	C9		
0002	8B	ADDA #$37	Add 37 to the number in the accumulator and place the answer in the accumulator
0003	37		
0004	3E	WAI	Stop

Fig. 18-26 Simple 6800/6808 addition problem which produces a sum of 0.

Status register							
Flags							
1	1	H	I	N	Z	V	C
1	1	1	0	0	1	0	1

Fig. 18-27 6800/6808 status register after an addition problem which produces a sum of 0.

proper places as shown in Fig. 18-27, you will see how the flags have responded to this problem.

Notice that the half-carry flag has again been set. The zero flag is set, as we supposed it would be. The carry flag is also set. Notice in the binary addition that a carry did indeed occur from the 8th to a nonexistent 9th bit (which the carry flag acts as).

The Overflow Flag

When the overflow flag is set, it tells us that if the numbers which the microprocessor just added or subtracted are signed binary numbers, the valid range for such numbers has been exceeded and the result is incorrect. The valid range for 8-bit microprocessors is +127 to −128. Let's change our problem to create an overflow.

Figure 18-28 shows our problem and program. Note that in this problem we are assuming that all values are to be interpreted as *signed binary* values.

This problem is $123_{10} + 111_{10} =$ ______. First go through the binary addition and enter the program. Then write down the values of the accumulator and status register, run the program, and finally write down the final values of the accumulator and status register.

Figure 18-29 shows what the value in the status register should be.

You should have a sum of EA_{16} in the accumulator and $XX101010_2$ in the status register. EA_{16} is the correct sum if you are using *unsigned* binary numbers! If you interpret EA_{16} as a signed binary number, it has a value of -22_{10}.

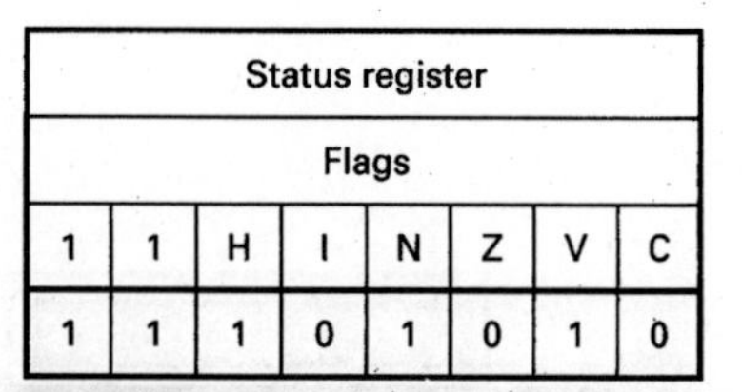

Status register							
Flags							
1	1	H	I	N	Z	V	C
1	1	1	0	1	0	1	0

Fig. 18-29 6800/6808 status register after an addition problem which creates an overflow.

This is *not* the correct answer. We have exceeded our valid range for signed binary numbers.

The status register has a value of XX101010. This means that in addition to the half-carry flag, both the overflow and the negative flags have been set.

It makes sense for the negative flag to be set because the 8th bit of the accumulator is set. This indicates a negative number if the value is a signed binary number.

The overflow flag is set because we have exceeded our range of $7F_{16}$ (127_{10}) to 80_{16} (-128_{10}), and the result is incorrect.

Decimal Addition

Because of differences in the way binary and decimal numbers round, and because numeric output to humans is usually decimal, it is sometimes helpful to actually do arithmetic calculations by using decimal numbers rather than binary numbers. Actually, true decimal numbers are not used. Rather a mixture of binary and decimal, called binary-coded decimal, is used. (The method used to create BCD numbers is covered in Chap. 1, and they have been discussed subsequently. You should review that section of Chap. 1 now if you are at all unsure of what BCD numbers are or how they are formed.)

One of the problems encountered in using BCD numbers is that, as the binary nibbles are added, invalid results are sometimes obtained.

Most microprocessors have an instruction called *decimal adjust* (or something similar). This instruction changes the

```
  1111  111
  0111  1011         7B16        12310
+ 0110  1111     +   6F16      + 11110
  ---------          ----        -----
  1110  1010         EA16        23410
```

Addr	Obj	Assembler	Comment
0000	86	LDAA #$7B	Load accumulator with first number (7B)
0001	7B		
0002	8B	ADDA #$6F	Add 6F to the number in the accumulator and place the answer in the accumulator
0003	6F		
0004	3E	WAI	Stop

Fig. 18-28 Simple 6800/6808 addition problem which produces an overflow.

number in the accumulator to what it would be if the last two numbers operated on were packed BCD (binary-coded decimal) numbers instead of binary numbers.

Let's look at an example. Figure 18-30 compares a decimal addition problem to the binary version of the same problem.

Notice first the difference between BCD and binary addition. BCD addition is not at all the same as binary addition. BCD is *decimal* addition using four binary digits to represent each decimal digit.

The program shown in Fig. 18-30 will help you understand the difference between binary and BCD addition (and subtraction). This program does the addition problem twice, once using binary numbers and once using BCD numbers. The result of the binary addition is stored in memory location $A0_{16}$, and the resulting flags in location $A1_{16}$. The result of the BCD addition is stored in location $A2_{16}$, and the resulting flags in location $A3_{16}$. Enter and run this program to see what results you obtain. When we ran the program, we found the following:

location $A0_{16}$ = binary sum = 7D
location $A1_{16}$ = binary flags = 000000

location $A2_{16}$ = BCD sum = 83
location $A3_{16}$ = BCD flags = 001000

The status of the binary flags indicates that no flags were set as a result of the binary addition. After the BCD addition, the negative flag was set but has no valid meaning. It is simply following the state of the 8th bit of the accumulator.

Subtraction

Subtraction is the opposite of addition. All the flags operate the same except the carry flag. After subtraction, the carry flag indicates whether or not a borrow has occurred. You can think of it as a "borrow" flag. A 1 in the carry flag position indicates that a borrow from the nonexistent 9th bit was required to do the subtraction. A 0 indicates that no borrow from the 9th bit was required.

Figure 18-31 illustrates how to write a program to do single-precision subtraction.

You should assemble and run this program. When we did, we found that the result in the accumulator was FF.

0100	0111	BCD	47_{10}
+ 0011	0110	BCD	+ 36_{10}
1000	0011	BCD	83_{10}

Decimal (BCD)

0100	0111_2	47_{16}
+ 0011	0110_2	+ 36_{16}
0111	1101_2	$7D_{16}$

Binary

This *is not the same as* *this!*

Addr	Obj	Assembler	Comment
0000	86	LDAA #$47	This is being interpreted as a binary number
0001	47		
0002	8B	ADDA #$36	This also is being considered a binary number
0003	36		
0004	97	STAA #A0	We'll store the ***binary*** answer in memory location 03A0
0005	A0		
0006	07	TPA	Transfer flags to accumulator
0007	97	STAA #A1	We'll store the status of the flags from the ***binary*** addition in memory location A1
0008	A1		
0009	86	LDAA #$47	This number is being interpreted as a ***decimal*** number
000A	47		
000B	8B	ADDA #$36	This number likewise is being considered a ***decimal*** number
000C	36		
000D	19	DAA	Make the answer decimal
000E	97	STAA $A2	We'll store the ***decimal*** answer in memory location 03A2
000F	A2		
0010	07	TPA	Transfer the flags to the accumulator
0011	97	STAA $A3	We'll store the status of the flags resulting from this ***decimal*** addition in memory location A3
0012	A3		
0013	3E	WAI	Stop

Fig. 18-30 Binary vs. BCD addition.

Addr	Obj	Assembler	Comment
0000	86	LDAA #$7F	
0001	7F		
0002	80	SUBA #$80	
0003	80		
0004	3E	WAI	

Fig. 18-31 Subtraction.

We also found that the overflow, negative, and carry flags had been set. The negative flag was set because the 8th bit of the answer is a 1, which indicates a negative-signed binary number. The overflow flag was set because $7F_{16} = 127_{10}$, and $80_{16} = -128_{10}$; therefore

$$\begin{array}{r} 127 \\ -(-128) \\ \hline 255 \end{array}$$

and 255_{10} is outside the valid range for 8-bit signed binary numbers (the valid range is $+127_{10}$ to -128_{10}). The carry flag was set because a borrow from a 9th bit was needed to complete the subtraction.

18-6 8080/8085/Z80 FAMILY

The 8080/8085/Z80 family has a variety of *add* and *subtract* instructions. The 8080/8085/Z80 can also work with binary-coded decimal (BCD) numbers.

Arithmetic Instructions

The 8080/8085/Z80 family has *add, subtract, add-with-carry, subtract-with-borrow, immediate mode* and *decimal adjust accumulator A* instructions. These instructions use the value in the accumulator as one of their operands and another value in one of the other registers as the other operand. Let's see how to use these instructions.

Addition

Let's start with a very simple addition program. Figure 18-32 illustrates this type of program.

Pay particular attention to the accumulator and the status register. Notice their contents both before and after you run the program. (You may want to write down their values before and after so that you can study their behavior.) You will find that the accumulator will have the number 67_{16} in it (which is the correct answer) and that only the half-carry flag will be set.

Let's look at the status register a little more closely. Refer to Fig. 18-33.

$$\begin{array}{rr} & {}^{1\,1} \\ & 0100\ \ 1001 \\ + & 0001\ \ 1110 \\ \hline & 0110\ \ 0111 \end{array} \qquad \begin{array}{rr} & 49_{16} \\ + & 1E_{16} \\ \hline & 67_{16} \end{array} \qquad \begin{array}{rr} & 73_{10} \\ + & 30_{10} \\ \hline & 103_{10} \end{array}$$

Addr	Obj	Assembler	Comment
1800	3E	MVI A,49	Load accumulator with first number (49)
1801	49		
1802	C6	ADI 1E	Add 1E to the number in the accumulator and place the answer in the accumulator
1803	1E		
1804	76	HALT	Stop

(8080/8085 mnemonics)

Addr	Obj	Assembler	Comment
1800	3E	LD A,49	Load accumulator with first number (49)
1801	49		
1802	C6	ADD A,1E	Add 1E to the number in the accumulator and place the answer in the accumulator
1803	1E		
1804	76	HALT	Stop

(Z80 mnemonics)

Fig. 18-32 Simple 8080/8085/Z80 addition problem.

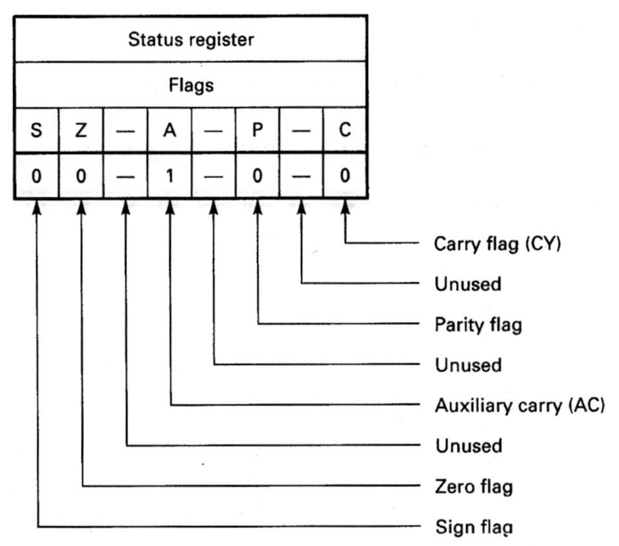

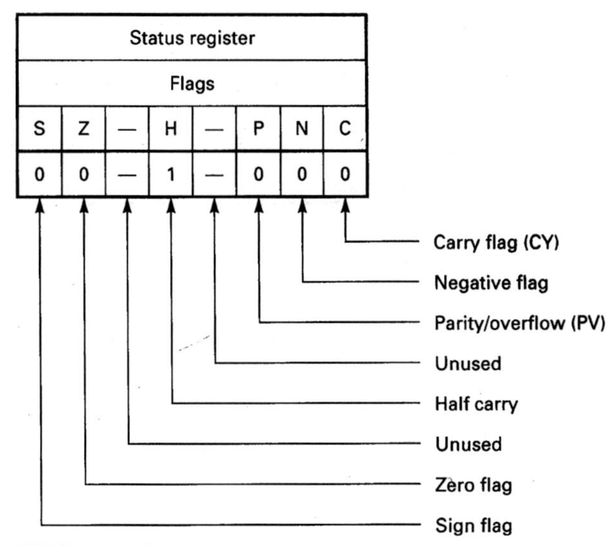

Fig. 18-33 8080/8085/Z80 status registers after addition problem.

Examining the flags from right to left, let's consider each and why it was or was not set.

The carry flag would have been set if a carry from the 8th bit to the 9th bit (which doesn't exist, so it goes into the carry flag) had occurred, but none did.

(*Note to Z80 users: Ignore the negative flag.*)

The parity flag was not set because the answer 0110 0111_2 has an odd number of 1s. That is to say it has odd parity, which is indicated by a 0. (*Note to Z80 users: We did not exceed the range of decimal +127 to −128—hexadecimal 7F to 80; therefore the parity/overflow flag was not set.*)

The microprocessor set the auxiliary carry (half-carry) flag because we had a carry from the 4th bit to the 5th bit. (Information about the half-carry is useful when dealing with BCD numbers.)

The zero flag would have been set if the answer had been zero, but it wasn't.

We did not have a 1 in the 8th bit of the accumulator so the answer could not have been negative; therefore the sign flag was not set.

The status of the unused flags doesn't matter.

The Sign Flag

Let's look at a problem that produces a negative answer. See Fig. 18-34.

Notice that this is the same problem as the last one except that we have changed the first operand. It used to be 49_{16}, but it is now $C9_{16}$, which is the decimal number -55_{10} if we consider these numbers to be signed binary numbers. We know that $-55_{10} + 30_{10} = -25_{10}$. Since this is a negative answer, we know that the sign flag should be set after the program is run.

Write down the contents of the accumulator and status register before running the program so that you know the initial conditions. Load the program and run it. After the program is run, the accumulator should contain the value $E7_{16}$. The status register should contain 10-1-1-0 *[Z80 = 10-1-000]*.

Let's examine the status register. If you put the status register bits into the appropriate positions in the status register as shown in Fig. 18-35, you will see what the bits indicate.

The auxiliary-carry *[half-carry]* is set again because we had a carry from the 4th to the 5th bit of the result.

The difference this time is that the sign flag is now set. This is what we expected to see. The sum of the addition problem was -25_{10} ($E7_{16}$). If we assume our numbers are signed binary numbers, then any number that has a 1 in the 8th bit is negative. $E7_{16}$ has a 1 in the 8th bit. The sign flag simply reflects the state of the 8th bit.

[*Note to 8085 users:* Your parity flag is 1 this time because the answer ($E7_{16}$) has an even number of 1s in it and even parity is indicated by a 1 in the parity flag. *Note to Z80 users: Your parity/overflow flag is 0 just like last time because the answer did not exceed the range from* $+127_{10}$ *to* -128_{10}.]

The Zero Flag

Now let's change the program slightly so that we get a sum of 0. That way we can see how the flags react to this situation.

Figure 18-36 shows the problem and the program to solve the problem.

We are again assuming that our numbers are signed binary numbers. The problem is $C9_{16} + 37_{16} = 00_{16}$

```
          1 1
          1100   1001        C9₁₆        -55₁₀
        + 0001   1110      + 1E₁₆      + 30₁₀
          1110   0111        E7₁₆        -25₁₀
```

Addr	Obj	Assembler	Comment
1800	3E	MVI A,C9	Load accumulator with first number (C9)
1801	C9		
1802	C6	ADI 1E	Add 1E to the number in the accumulator and place the answer in the accumulator
1803	1E		
1804	76	HALT	Stop

(8080/8085 mnemonics)

Addr	Obj	Assembler	Comment
1800	3E	LD A,C9	Load accumulator with first number (C9)
1801	C9		
1802	C6	ADD A,1E	Add 1E to the number in the accumulator and place the answer in the accumulator
1803	1E		
1804	76	HALT	Stop

(Z80 mnemonics)

Fig. 18-34 Simple 8080/8085/Z80 addition problem with negative answer.

$(-55_{10} + 55_{10} = 0_{10})$. You should go through the binary addition of these two numbers now before you run the program. Notice both the answer and the carries.

Again write down the contents of the accumulator and the status register before and after running the program. You will notice that we are using the same program but have changed one of the operands.

Now enter and run the program. The accumulator should contain 00_{16} and the status register should contain 01-1-1-1 *[Z80 = 01-1-001]*. If you place the bits of the status register value in their proper places as shown in Fig. 18-37, you will see how the flags have responded to this problem.

Notice that the half-carry flag has again been set.

The zero flag is set, as we supposed it would be.

The carry flag is also set. Notice in the binary addition

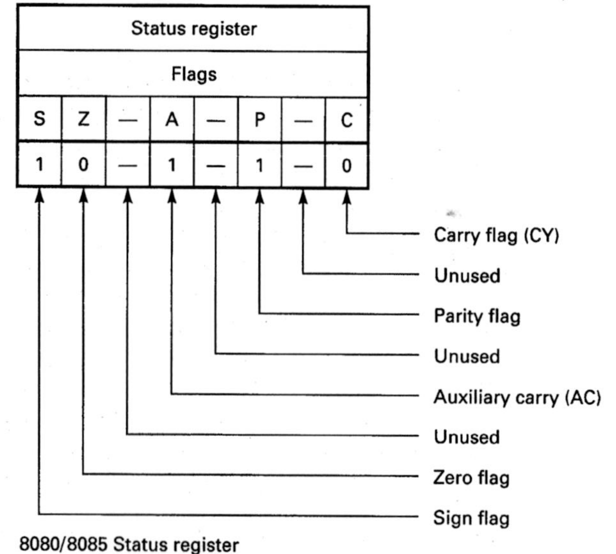

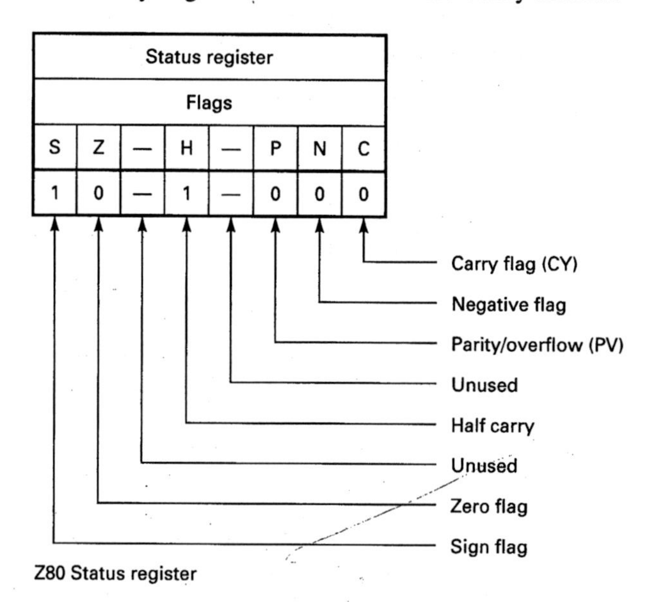

Fig. 18-35 8080/8085/Z80 status registers after an addition problem which produces a negative answer.

1 1111 111			
1100 1001	$C9_{16}$	-55_{10}
+ 0011 0111	+ 37_{16}	+ 55_{10}
1 0000 0000	00_{16}	0_{10}

Addr	Obj	Assembler	Comment
1800	3E	MVI A,C9	Load accumulator with first number (C9)
1801	C9		
1802	C6	ADI 37	Add 37 to the number in the accumulator and place the answer in the accumulator
1803	37		
1804	76	HALT	Stop

(8080/8085 mnemonics)

Addr	Obj	Assembler	Comment
1800	3E	LD A,C9	Load accumulator with first number (C9)
1801	C9		
1802	C6	ADD A,37	Add 37 to the number in the accumulator and place the answer in the accumulator
1803	37		
1804	76	HALT	Stop

(Z80 mnemonics)

Fig. 18-36 Simple 8080/8085/Z80 addition problem which produces a sum of 0.

that a carry did indeed occur from the 8th bit to a nonexistent 9th bit (which the carry flag acts as).

The *8085 parity flag* is set indicating an even number of 1s. (*Note to Z80 users: Your parity/overflow flag is cleared indicating you have not exceeded the range for 8-bit signed binary numbers, from* $+127_{10}$ *to* -128_{10}.)

The Parity Flag *[Z80: Parity/Overflow Flag]*

The 8080/8085 and Z80 microprocessors differ slightly in the function of this flag. Let's look at the 8080/8085 first.

Status register							
Flags							
S	Z	—	A	—	P	—	C
0	1	—	1	—	1	—	1

8080/8085 Status register

Status register							
Flags							
S	Z	—	H	—	P	N	C
0	1	—	1	—	0	0	1

Z80 Status register

Fig. 18-37 8080/8085/Z80 status registers after an addition problem which produces a sum of 0.

The 8080/8085 microprocessors have a parity flag which simply tells us how many 1s are in the accumulator after an arithmetic or a logic operation. Even parity exists when an even number of 1s are in the accumulator. Odd parity exists when an odd number of 1s exist in the accumulator. Even parity is shown by a 1 in the parity flag, and odd parity by a zero in the parity flag.

The Z80 has a combination parity/overflow flag. During logic operations it indicates parity as just described for the 8080/8085. During arithmetic operations, however, it acts as an overflow flag.

When an overflow flag is set, it tells us that if the numbers which were just added or subtracted are signed binary numbers, then the valid range for such numbers has been exceeded and the result is incorrect. The valid range for 8-bit microprocessors is $+127$ to -128. Let's change our problem to create an overflow.

Figure 18-38 shows our problem and program. In this problem it is important to note that we are assuming that all values are to be interpreted as *signed binary* values.

This problem is $+123_{10} + 111_{10} =$ ______. First go through the binary addition and enter the program. Then write down the values in the accumulator and status register, run the program, and finally write down the values of the accumulator and status register after the program has run.

Figure 18-39 shows what the value in the status register should be.

You should have a sum of EA_{16} in the accumulator and 10-1-0-0 *[Z80: 10-1-100]* in the status register. EA_{16} is the correct sum if you are using *unsigned* binary numbers! If

$$\begin{array}{r} 1111\ \ 111\ \ \\ 0111\ \ 1011 \\ +\ 0110\ \ 1111 \\ \hline 1110\ \ 1010 \end{array} \qquad \begin{array}{r} 7B_{16} \\ +\ 6F_{16} \\ \hline EA_{16} \end{array} \qquad \begin{array}{r} 123_{10} \\ +111_{10} \\ \hline 234_{10} \end{array}$$

Addr	Obj	Assembler	Comment
1800	3E	MVI A,7B	Load accumulator with first number (7B)
1801	7B		
1802	C6	ADI 6F	Add 6F to the number in the accumulator and place the answer in the accumulator
1803	6F		
1804	76	HALT	Stop

(8080/8085 mnemonics)

Addr	Obj	Assembler	Comment
1800	3E	LD A,7B	Load accumulator with first number (7B)
1801	7B		
1802	C6	ADD A,6F	Add 6F to the number in the accumulator and place the answer in the accumulator
1803	6F		
1804	76	HALT	Stop

(Z80 mnemonics)

Fig. 18-38 Simple 8080/8085/Z80 addition problem which produces an overflow.

you interpret EA_{16} as a signed binary number, it has a value of -22_{10}. This is *not* the correct answer. We have exceeded our valid range for signed binary numbers.

The status registers of both the 8080/8085 and the Z80 microprocessors have a 1 in the half-carry flag as before. Now however, both also have a sign flag that is set. It makes sense for the sign flag to be set because the 8th bit of the accumulator is set. This indicates a negative number if the value is a signed binary number.

Status register							
Flags							
S	Z	—	A	—	P	—	C
1	0	—	1	—	0	—	0

8080/8085 Status register

Status register							
Flags							
S	Z	—	H	—	P	N	C
1	0	—	1	—	1	0	0

Z80 Status register

Fig. 18-39 8080/8085/Z80 status registers after an addition problem which creates an overflow.

The parity flag of the 8080/8085 is 0 because the answer (EA_{16}) contains five 1s and 5 is an odd number. However, the parity/overflow flag of the Z80 acts as an overflow flag during an arithmetic instruction and is 1 because we have exceeded our range of $7F_{16}$ (127_{10}) to 80_{16} (-128_{10}) for 8-bit signed binary numbers, and the result is therefore incorrect.

Decimal Addition

Because of differences in the way binary and decimal numbers round, and because numeric output to humans is usually decimal, it is sometimes useful to do arithmetic calculations by using decimal numbers rather than binary numbers. Actually, true decimal numbers are not used. Rather a mixture of binary and decimal, called binary-coded decimal is used. (The method used to create BCD numbers is covered in Chap. 1, and they have been discussed subsequently. You should review that section of Chap. 1 now if you are unsure of what BCD numbers are or how they are formed.)

One of the problems involved in using BCD numbers is that as the binary nibbles are added, invalid results are sometimes obtained.

Most microprocessors have an instruction called *decimal adjust* (or something similar). This instruction changes the number in the accumulator to what it would be if the last two numbers operated on had been packed BCD numbers instead of binary numbers.

Let's look at an example. Figure 18-40 compares a decimal addition problem to the binary version of the same problem.

Notice first the difference between BCD and binary addition. BCD addition is not at all the same as binary addition. BCD is *decimal* addition using 4 bits to represent each decimal digit.

The program shown in Fig. 18-40 will help you understand the difference between binary and BCD addition (and subtraction). This program does the addition problem twice, once using binary numbers and once using BCD numbers. The result of the binary addition is stored in memory location $18A0_{16}$, and the resulting flags in location $18A1_{16}$. The result of the BCD addition is stored in location $18A2_{16}$, and the resulting flags in location $18A3_{16}$. Enter and run this program to see what results you get. When we ran the program, we found the following:

location $18A0_{16}$ = binary sum = 7D
location $18A1_{16}$ = binary flags = 00-0-1-0
[Z80:00-0-000]
location $18A2_{16}$ = BCD sum = 83
location $18A3_{16}$ = BCD flags = 10-1-0-0
[Z80: 10-1-000]

The status of the flags after the binary addition indicates that no flags were set (except the 8080/8085 parity flag indicating even parity).

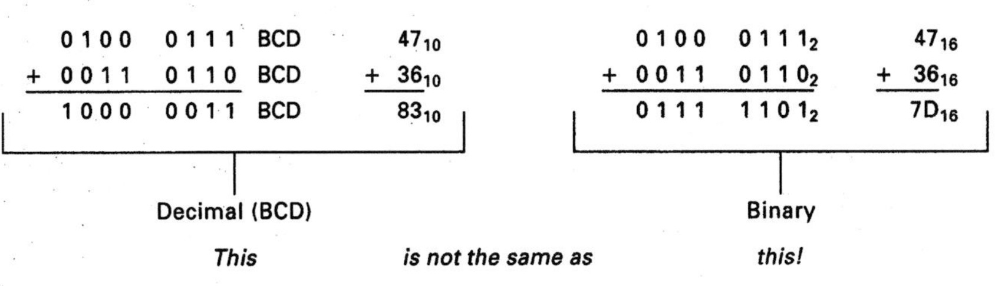

Addr	Obj	Assembler	Comment
1800	3E	MVI A,47	This is being interpreted as a binary number
1801	47		
1802	C6	ADI 36	This also is being considered a binary number
1803	36		
1804	32	STA 18A0	We'll store the ***binary*** answer in memory location 18A0
1805	A0		
1806	18		
1807	F5	PUSH PSW	Put the flags and accumulator in stack
1808	C1	POP B	Retrieve flags and accumulator into register B and C
1809	79	MOV A,C	Move the flags from register C to the accumulator
180A	32	STA 18A1	We'll store the status of the flags from the ***binary*** addition in memory location 18A1
180B	A1		
180C	18		
180D	3E	MVI A,47	This is being interpreted as a ***decimal*** number
180E	47		
180F	C6	ADI 36	This also is being considered a ***decimal*** number
1810	36		
1811	27	DAA	Convert the answer to ***decimal***
1812	32	STA 18A2	We'll store the ***decimal*** answer in memory location 18A2
1813	A2		
1814	18		
1815	F5	PUSH PSW	Put the flags and accumulator in stack
1816	C1	POP B	Retrieve flags and accumulator into registers B and C
1817	79	MOV A,C	Move the flags from register C to the accumulator
1818	32	STA 18A3	We'll store the status of the flags from the ***binary*** addition in memory location 18A3
1819	A3		
181A	18		
181B	76	HALT	Stop

(8080/8085 mnemonics)

Fig. 18-40 Binary vs. BCD addition. (Continued on next page.)

Addr	Obj	Assembler	Comment
1800	3E	LD A,47	This is being interpreted as a binary number
1801	47		
1802	C6	ADD A,36	This also is being considered a binary number
1803	36		
1804	32	LD (18A0),A	We'll store the ***binary*** answer in memory location 18A0
1805	A0		
1806	18		
1807	F5	PUSH AF	Put the flags and the accumulator in stack
1808	C1	POP BC	Retrieve flags and accumulator into registers B and C
1809	79	LD A,C	Move the flags from register C to the accumulator
180A	32	LD (18A1),A	We'll store the status of the flags from the ***binary*** addition in memory location 18A1
180B	A1		
180C	18		
180D	3E	LD A,47	This is being interpreted as a ***decimal*** number
180E	47		
180F	C6	ADD A,36	This also is being considered a ***decimal*** number
1810	36		
1811	27	DAA	Convert the answer to ***decimal***
1812	32	LD (18A2),A	We'll store the ***decimal*** answer in memory location 18A2
1813	A2		
1814	18		
1815	F5	PUSH AF	Put the flags and accumulator in stack
1816	C1	POP BC	Retrieve flags and accumulator into registers B and C
1817	79	LD A,C	Move the flags from register C to the accumulator
1818	32	LD (18A3),A	We'll store the status of the flags from the ***binary*** addition in memory location 18A3
1819	A3		
181A	18		
181B	76	HALT	Stop

(Z80 mnemonics)

Fig. 18-40 (Continued)

After the BCD addition, the sign flag was set, but it has no valid meaning. It simply follows the state of the 8th bit of the accumulator.

Several points must be kept in mind when doing decimal addition and subtraction on the 8080/8085 and Z80 microprocessors.

With the 8080/8085 microprocessors, the DAA instruction only works after addition. Also, the DAA instruction works only with the accumulator.

With the Z80 microprocessor, the DAA instruction can be used after either addition or subtraction. This is made possible by the addition of the negative flag which the 8080/8085 does not have. This flag simply keeps track of whether an addition or subtraction was just performed. This flag is used in combination with the half-carry flag to correct the BCD answers.

Subtraction

Subtraction is the opposite of addition. All the flags operate the same except the carry flag. After subtraction, the carry flag indicates whether a borrow has occurred. You can think of it as a "borrow" flag. A 1 in the carry flag position indicates that a borrow from the nonexistent 9th bit was required to do the subtraction. A 0 indicates that no borrow from the 9th bit was required.

Figure 18-41 illustrates how to write a program to do single-precision subtraction.

You should assemble and run this program. When we did, we found that the result in the accumulator was FF. We also found that the overflow (*Z80*), sign, and carry flags had been set. The sign flag was set because the 8th bit of the answer is a 1 which indicates a negative-signed binary number. The overflow flag was set because $7F_{16} = 127_{10}$ and $80_{16} = -128_{10}$; therefore

$$\begin{array}{r} 127 \\ -(-128) \\ \hline 255 \end{array}$$

and 255_{10} is outside the valid range for 8-bit signed binary numbers. (The valid range is $+127_{10}$ to -128_{10}.) The

Addr	Obj	Assembler	Comment
1800	3E	MVI A,7F	
1801	7F		
1802	D6	SUI 80	
1803	80		
1804	76	HALT	

(8080/8085 mnemonics)

Addr	Obj	Assembler	Comment
1800	3E	LD A,7F	
1801	7F		
1802	D6	SUB A,80	
1803	80		
1804	76	HALT	

(Z80 mnemonics)

Fig. 18-41 Subtraction.

carry flag was set because a borrow from a 9th bit was needed to complete the subtraction.

18-7 8086/8088 FAMILY

The 8086/8088 has a variety of arithmetic instructions and various support instructions. The 8086/8088 can also work with ASCII and binary-coded decimal (BCD) numbers.

Arithmetic Instructions

The 8086/8088 has *add, subtract, add-with-carry, subtract-with-borrow, ASCII adjust, multiply, divide, integer multiply, integer divide,* and *conversion* instructions. These instructions use a value in one of the registers, memory, or an immediate number as their operands. Let's see how to use these instructions.

DEBUG Revisited

In just a moment we are going to begin studying some sample arithmetic programs for the 8086/8088 microprocessor. However, we must first learn more about the DEBUG utility.

Until now, we have assembled each program with DEBUG and then executed the program by using the *trace* command. Trace executes one instruction, displays the contents of the registers and flags, and then stops. This works well when the program is only a few lines long or when you must carefully observe the effect each instruction has on the registers. It is very slow, however.

DEBUG has another command which executes an entire program without stopping until the end. This is the g (go) command. Of course, the computer has to know where to start. If you just use the g command the computer assumes that it should start program execution at the memory location indicated by the instruction pointer (IP). If that is not where you want to start, such as when you want to execute a program for the second time, you have two ways to specify where to start. One way is to change the instruction pointer with the r (register) command. This is accomplished as follows:

```
–rip
IP 0100
:0100
–
```

You should start your assembly-language programs at or after address 0100H. The other way is to specify a starting point as part of the g (go) command. To start at memory location 0100H, for example, you would type

```
g=0100
```

Execution would start at address 0100 even though the instruction pointer might not contain that address.

When you use the g (go) command, the computer also has to know where to stop. You might think that the HLT (HaLT) instruction would work just fine. When you are using DEBUG, however, a different instruction is needed to stop the program. You are using DEBUG to control the computer. When your assembly-language routine is finished, control of the computer must be returned to DEBUG. DOS (the computer's disk operating system) has a routine which will do this. This routine is accessed by executing the

INT 20 instruction. For example, the arithmetic program we're going to study shortly looks like this:

```
MOV AL,49
ADD AL,1E
INT 20
```

Notice the use of INT 20 to stop program execution. There are a number of these DOS functions which handle the computer's housekeeping chores.

The g command is faster than individual t (trace) commands, and we can tell the computer where to start and stop, but it has one major disadvantage. When you use the r (register) command to view the registers after the program has run, they will have the same values they had in them before the program was run. This doesn't give you a chance to study the registers and flags to learn about how the program works.

The solution to this problem is breakpoints. A *breakpoint* is an address where you want program execution to stop. A breakpoint is specified as part of the g command. The difference between using a breakpoint to stop the program and INT 20 is that, when the breakpoint is reached, all the registers will be automatically displayed and their contents will not have been returned to their previous values. This allows you to see what all the registers and flags look like at that exact point in the program. For example:

```
g 0104
```

tells DEBUG to start program execution at the address indicated by the instruction pointer and to stop at address 0104. Notice that the instruction at address 0104 will not be executed. Instructions or data at address 0103 will be the last that the program will use. After the program stops at address 0104, the contents of the registers and flags will be automatically displayed.

The starting and stopping points for program execution can be combined into one command. For example:

```
g=0100 0104
```

will cause program execution to start at address 0100 and to stop at address 0104. The contents of the registers and flags will be automatically displayed.

When you run programs using a breakpoint, you need to remember that the instruction pointer will not be reset. Therefore, you'll have to change it back to the program's starting point if you wish to run the program more than once or specify the starting point in the g command as just shown.

If you wish, the t (trace) command can still be used to execute instructions one at a time.

Now let's try running a short program which will show you how to use these DEBUG commands and will allow you to learn about the 8086/8088 ADD instruction.

Addition

Let's start with a very simple addition program. Figure 18-42 illustrates this type of program.

We have shown the program twice: the first time using the g command without a breakpoint, showing that the registers will in fact be the same as before the program was run, and the second time using the g command with a breakpoint, showing that the contents of the registers will reflect how the program alters them. From this point on in this text we will use a breakpoint to stop program execution. Being able to see how a program affects the registers is important since our primary purpose is to explain what the program has accomplished and how it functions by studying the registers and flags after it has run.

Pay particular attention to AL and the flags. Notice their contents both before and after the program is run. You will find that the accumulator will have the number 67_{16} in it (which is the correct answer) and that only the auxiliary flag will be set.

Let's look at the flags a little more closely. Examining the flags from right to left, let's consider each and why it was or was not set. Refer to the bottom portion of Fig. 18-42.

There was no carry (NC) because no carry from the 8th bit to the 9th bit (which doesn't exist so it goes into the carry flag) occurred.

The parity was odd (PO) because the answer, $0110\,0111_2$, has an odd number of 1s.

There was an auxiliary carry (AC) because we had a carry from the 4th bit to the 5th bit. (Information about the half-carry is useful when dealing with BCD numbers.)

There was no zero (NZ) because the answer wasn't 0.

The answer was positive or ''plus'' (PL) because we did not have a 1 in the 8th bit of the accumulator so the answer could not have been negative.

Don't worry about the enable interrupt (EI) or auto-increment (UP) flags for the moment.

There was no overflow (NV) because we did not exceed our range for valid 8-bit signed binary numbers $+127$ to -128.

In fact, if you compare the state of the flags before the program was run with the state of the flags after it was run, only one of them changed. That was the auxiliary carry (AC) flag.

The Sign Flag

Let's look at a problem that produces a negative answer. See Fig. 18-43 at this time.

Notice that this is exactly the same problem as the last one except that we have changed the first operand, which used to be 49_{16} into $C9_{16}$ (-55_{10} if we consider these numbers to be signed binary numbers). We know that $-55_{10} + 30_{10} = -25_{10}$. Since this is a negative answer, we know that the sign flag should be set after the program is run.

```
  1 1
  0100  1001        49₁₆       73₁₀
+ 0001  1110      + 1E₁₆     + 30₁₀
  0110  0111        67₁₆      103₁₀
```

```
B>DEBUG
-r
AX=0000  BX=0000  CX=0000  DX=0000  SP=FFEE  BP=0000  SI=0000  DI=0000
DS=8FFD  ES=8FFD  SS=8FFD  CS=8FFD  IP=0100   NV UP EI PL NZ NA PO NC
8FFD:0100 7420          JZ      0122
-
-a
8FFD:0100 mov al,49
8FFD:0102 add al,1e
8FFD:0104 int 20
8FFD:0106
-
-u 0100 0105
8FFD:0100 B049          MOV     AL,49
8FFD:0102 041E          ADD     AL,1E
8FFD:0104 CD20          INT     20
-
-g

Program terminated normally
-r
AX=0000  BX=0000  CX=0000  DX=0000  SP=FFEE  BP=0000  SI=0000  DI=0000
DS=8FFD  ES=8FFD  SS=8FFD  CS=8FFD  IP=0100   NV UP EI PL NZ NA PO NC
8FFD:0100 B049          MOV     AL,49
-
```

Program assembled, unassembled, and executed without using a breakpoint to stop program (notice that registers have returned to their previous state)

```
B>DEBUG
-r
AX=0000  BX=0000  CX=0000  DX=0000  SP=FFEE  BP=0000  SI=0000  DI=0000
DS=8FFD  ES=8FFD  SS=8FFD  CS=8FFD  IP=0100   NV UP EI PL NZ NA PO NC
8FFD:0100 7420          JZ      0122
-
-a
8FFD:0100 mov al,49
8FFD:0102 add al,1e
8FFD:0104 int 20
8FFD:0106
-
-u 0100 0105
8FFD:0100 B049          MOV     AL,49
8FFD:0102 041E          ADD     AL,1E
8FFD:0104 CD20          INT     20
-
-g 0104

AX=0067  BX=0000  CX=0000  DX=0000  SP=FFEE  BP=0000  SI=0000  DI=0000
DS=8FFD  ES=8FFD  SS=8FFD  CS=8FFD  IP=0104   NV UP EI PL NZ AC PO NC
8FFD:0104 CD20          INT     20
```

Program assembled, unassembled, and executed using a breakpoint to stop and display contents of registers and flags (notice that registers have *not* returned to their previous state)

Fig. 18-42 Simple 8086/8088 addition problem.

```
  1 1
  1100  1001        C9₁₆      -55₁₀
+ 0001  1110      + 1E₁₆    + 30₁₀
  1110  0111        E7₁₆      -25₁₀
```

```
-r
AX=0000  BX=0000  CX=0000  DX=0000  SP=FFEE  BP=0000  SI=0000  DI=0000
DS=8FFD  ES=8FFD  SS=8FFD  CS=8FFD  IP=0100   NV UP EI PL NZ NA PO NC
8FFD:0100 B0C9            MOV     AL,C9
-a
8FFD:0100 mov al,c9
8FFD:0102 add al,1e
8FFD:0104 int 20
8FFD:0106
-
-u 0100 0104
8FFD:0100 B0C9            MOV     AL,C9
8FFD:0102 041E            ADD     AL,1E
8FFD:0104 CD20            INT     20
-
-g 104

AX=00E7  BX=0000  CX=0000  DX=0000  SP=FFEE  BP=0000  SI=0000  DI=0000
DS=8FFD  ES=8FFD  SS=8FFD  CS=8FFD  IP=0104   NV UP EI NG NZ AC PE NC
8FFD:0104 CD20            INT     20
```

Fig. 18-43 Simple 8086/8088 addition problem with a negative answer.

Assemble the program and run it. Observe the contents of AL and the status register before and after running the program so that you can compare them. After the program is run, AL should contain the value $E7_{16}$. The status register shows that only two flags have changed in exactly the same way as in the last example. The parity flag indicates even parity, and we have an auxiliary carry, just like the last example.

The difference this time is that we have a negative (NG) answer. This is what we expected to see. The sum of the addition problem was -25_{10} ($E7_{16}$). If we assume that our numbers are 8-bit signed binary numbers, then any number that has a 1 in the 8th bit is negative. $E7_{16}$ has a 1 in the 8th bit. The sign flag simply reflects the state of the most significant bit (8th or 16th depending on whether we are using 8-bit or 16-bit numbers).

The Zero Flag

Now let's change the program slightly so that we obtain a sum of 0. Then we can see how the flags react to this situation.

Figure 18-44 shows the problem and the program to solve the problem.

We are again assuming that our numbers are signed binary numbers. The problem is $C9_{16} + 37_{16} = 00_{16}$, which is $-55_{10} + 55_{10} = 0_{10}$. You should go through the binary addition of these two numbers now before you run the program. Notice both the answer and the carries. Notice also that we are using the same program as in the last example but have changed one of the operands.

We have a carry out of the most-significant bit (CY), we have a half-carry (AC), we have even parity (PE), and of course the zero flag indicates that our answer was in fact 0 (ZR). You should be able to look at the problem itself and at the flags before and after the program was run and be able to see why the flags have responded the way they have.

The Parity Flag

The 8086/8088 microprocessor has a parity flag which simply tells us how many 1s are in the accumulator after an arithmetic or logic operation. Even parity exists when an even number of 1s are in the accumulator. Odd parity exists when an odd number of 1s exist in the accumulator. Even parity (PE) and odd parity (PO) are indicated in the flags section of the DEBUG display.

Overflow Flag

When the overflow flag is set, it tells us that if the numbers which were just added or subtracted are signed binary numbers, then the valid range for such numbers has been exceeded and the result is incorrect. The valid range for 8-bit calculations is $+127$ to -128. The valid range for 16-bit calculations is $+32{,}767$ to $-32{,}768$. Let's modify our problem to create an overflow.

Figure 18-45 shows our problem and program. Remember that we are assuming that all values are to be interpreted as 8-bit *signed binary* values.

This problem is $123_{10} + 111_{10} =$ ______. First go through the binary addition and enter the program. Then write down the values you think will be found in AL and

```
 1  1111   1111
    1100   1001        C9₁₆       -55₁₀
 +  0011   0111      + 37₁₆     +  55₁₀
 1  0000   0000        00₁₆         0₁₀
```

```
-r
AX=0000  BX=0000  CX=0000  DX=0000  SP=FFEE  BP=0000  SI=0000  DI=0000
DS=8FFD  ES=8FFD  SS=8FFD  CS=8FFD  IP=0100   NV UP EI PL NZ NA PO NC
8FFD:0100 7420          JZ      0122
-
-a
8FFD:0100 mov al,c9
8FFD:0102 add al,37
8FFD:0104 int 20
8FFD:0106
-
-u 0100 0104
8FFD:0100 B0C9          MOV     AL,C9
8FFD:0102 0437          ADD     AL,37
8FFD:0104 CD20          INT     20
-
-g 104

AX=0000  BX=0000  CX=0000  DX=0000  SP=FFEE  BP=0000  SI=0000  DI=0000
DS=8FFD  ES=8FFD  SS=8FFD  CS=8FFD  IP=0104   NV UP EI PL ZR AC PE CY
8FFD:0104 CD20          INT     20
-
```

Fig. 18-44 Simple 8086/8088 addition problem which produces a sum of 0.

the status register, run the program, and finally note the final values of AL and the status register.

You should have a sum of EA_{16} in the accumulator and should find that there has been an auxiliary carry (AC), that the sign bit indicates that this is a negative number (NG), and that there has been an overflow (OV). EA_{16} is the correct sum *if you are using unsigned binary numbers!* If you interpret EA_{16} as a signed binary number, it has a value of -22_{10}. This is *not* the correct answer. We have exceeded our valid range for 8-bit signed binary numbers.

```
 1111   1111
 0111   1011        7B₁₆       123₁₀
+0110   1111      + 6F₁₆      +111₁₀
 1110   1010        EA₁₆       234₁₀
```

```
-r
AX=0000  BX=0000  CX=0000  DX=0000  SP=FFEE  BP=0000  SI=0000  DI=0000
DS=8FFD  ES=8FFD  SS=8FFD  CS=8FFD  IP=0100   NV UP EI PL NZ NA PO NC
8FFD:0100 7420          JZ      0122
-
-a
8FFD:0100 mov al,7b
8FFD:0102 add al,6f
8FFD:0104 int 20
8FFD:0106
-
-u 0100 0104
8FFD:0100 B07B          MOV     AL,7B
8FFD:0102 046F          ADD     AL,6F
8FFD:0104 CD20          INT     20
-
-g 104

AX=00EA  BX=0000  CX=0000  DX=0000  SP=FFEE  BP=0000  SI=0000  DI=0000
DS=8FFD  ES=8FFD  SS=8FFD  CS=8FFD  IP=0104   OV UP EI NG NZ AC PO NC
8FFD:0104 CD20          INT     20
-
```

Fig. 18-45 Simple 8086/8088 addition problem which produces an overflow.

0100	0111	BCD	47_{10}	0100	0111_2		47_{16}
+ 0011	0110	BCD	+ 36_{10}	+ 0011	0110_2		+ 36_{16}
1000	0011	BCD	83_{10}	0111	1101_2		$7D_{16}$

Decimal (BCD) — Binary

This *is not the same as* *this!*

```
-r
AX=0000  BX=0000  CX=0000  DX=0000  SP=FFEE  BP=0000  SI=0000  DI=0000
DS=8FFD  ES=8FFD  SS=8FFD  CS=8FFD  IP=0100   NV UP EI PL NZ NA PO NC
8FFD:0100 7420          JZ      0122
-
-a
8FFD:0100 mov al,47          ;1st operand (binary)
8FFD:0102 add al,36          ;add 2d, put sum in al (binary)
8FFD:0104 mov [01A0],al      ;store sum
8FFD:0107 pushf              ;copy flags
8FFD:0108 pop bx             ;retrieve flags
8FFD:0109 mov [01A1],bx      ;store flags
8FFD:010D mov al,47          ;1st operand
8FFD:010F add al,36          ;add 2d, put sum in al (binary)
8FFD:0111 daa                ;convert sum to BCD
8FFD:0112 mov [01A3],al      ;store BCD sum
8FFD:0115 pushf              ;copy flags
8FFD:0116 pop bx             ;retrieve flags
8FFD:0117 mov [01A4],bx      ;store flags
8FFD:011B int 20             ;return to DEBUG
8FFD:011D
-
-g 011b

AX=0083  BX=FA92  CX=0000  DX=0000  SP=FFEE  BP=0000  SI=0000  DI=0000
DS=8FFD  ES=8FFD  SS=8FFD  CS=8FFD  IP=011B   OV UP EI NG NZ AC PO NC
8FFD:011B CD20          INT     20
-
-d 01A0 01AF
8FFD:01A0  7D 06 F2 83 92 FA 79 6E-74 61 78 20 65 72 72 6F   }.....yntax erro
```

Fig. 18-46 Binary vs. BCD addition.

Decimal Addition

Because of differences in the way binary and decimal numbers round, and because numeric output to humans is usually decimal, it is sometimes helpful to do arithmetic calculations by using decimal numbers rather than binary numbers. Actually, true decimal numbers are not used. Rather a mixture of binary and decimal, called binary-coded decimal, is used. (The method used to create BCD numbers is covered in Chap. 1, and they have been discussed subsequently. You should review that section of Chap. 1 now if you are unsure of what BCD numbers are or how they are formed.)

One of the problems involved in using BCD numbers is that as the binary nibbles are added invalid results are sometimes obtained.

Most microprocessors have an instruction called *decimal adjust* (or something similar). This instruction changes the number in the accumulator to what it would be if the last two numbers operated on had been packed BCD numbers instead of binary numbers.

Let's look at an example. Figure 18-46 is a decimal addition problem which is compared to the binary version of the same problem.

Notice first the difference between BCD and binary addition. BCD addition is not at all the same as binary addition. BCD is *decimal* addition using 4 bits to represent each decimal digit.

The program shown in Fig. 18-46 will help you understand the difference between binary and BCD addition (and subtraction). This program does the addition problem twice, once using binary numbers and once using BCD numbers. The result of the binary addition is stored in memory location $01A0_{16}$, and the resulting flags in locations $01A1_{16}$ and $01A2_{16}$. The result of the BCD addition is stored in location $01A3_{16}$, and the resulting flags in locations $01A4_{16}$ and $01A5_{16}$. Assemble and run this program to see whether you obtain the same results. When we ran the program we found the following:

location $01A0_{16}$ = binary sum = 7D
location $01A1_{16}$ = binary flags (low byte) = 06

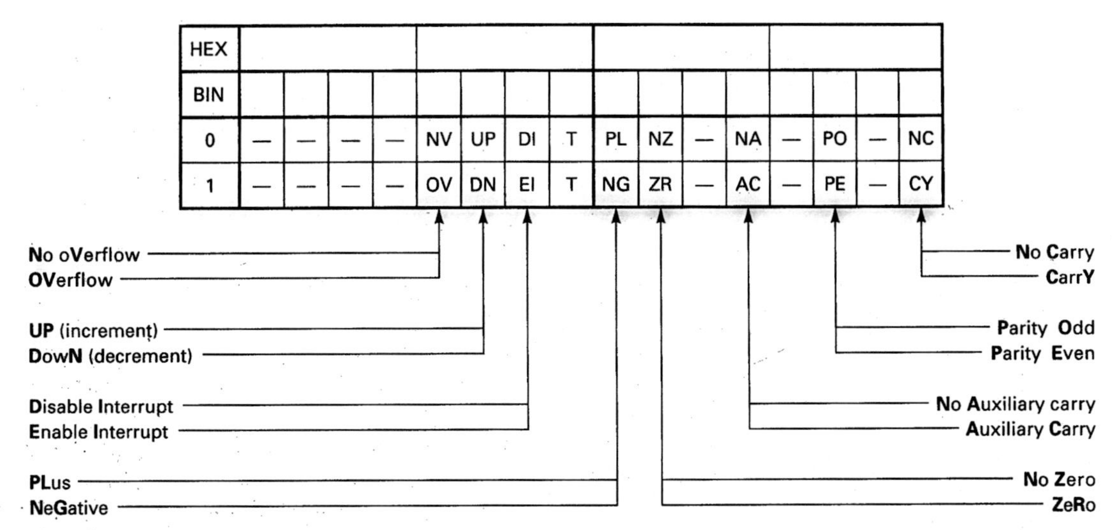

Fig. 18-47 8086/8088 flag chart.

location $01A2_{16}$ = binary flags (high byte) = F2
location $01A3_{16}$ = BCD sum = 83
location $01A4_{16}$ = BCD flags (low byte) = 92
location $01A5_{16}$ = BCD flags (high byte) = FA

Figure 18-47 will help you understand what the stored flag values mean.

When you store the value of the flag register (status register), you can place the hexadecimal values on the chart in Fig. 18-47. You can then convert the hexadecimal values to binary values and look in the 0 row or 1 row to see what conditions the flags indicate existed at a certain point in the program.

In this example we have placed the values of the flags after the binary addition in Fig. 18-48.

First notice that we have reversed the order of the hexadecimal values for the flags. The PUSHF instruction pushes the current value of the flags onto the stack. The POP BX then retrieves that value into the BX register. At this point the values are still in their correct order. In fact, if you will refer to Fig. 18-46, those areas have been printed in bold to illustrate this fact. Notice that BX contains $FA92_{16}$. Look at memory locations $01A4_{16}$ and $01A5_{16}$. Notice that those 2 bytes have been reversed. PUSH and POP instructions do not reverse the bytes. However, MOV instructions do. The MOV instruction places the data into memory in a low-byte/high-byte order, which has the effect of reversing the bytes when the memory locations are examined.

The binary addition problem produced an answer of $7D_{16}$, as we expected. There were no carries or overflows, and we have even parity. These conditions are shown in bold type in Fig. 18-48.

The BCD addition produced a sum of 83_{BCD}, as we thought it would. Don't be concerned about the flags at this point. Simply notice that they are different. They reflect

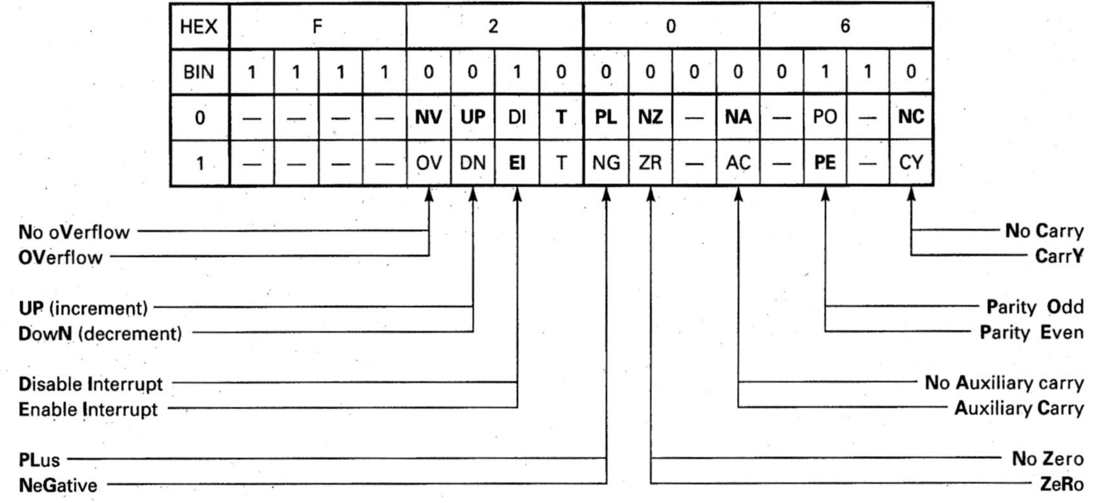

Fig. 18-48 Conditions after the binary addition.

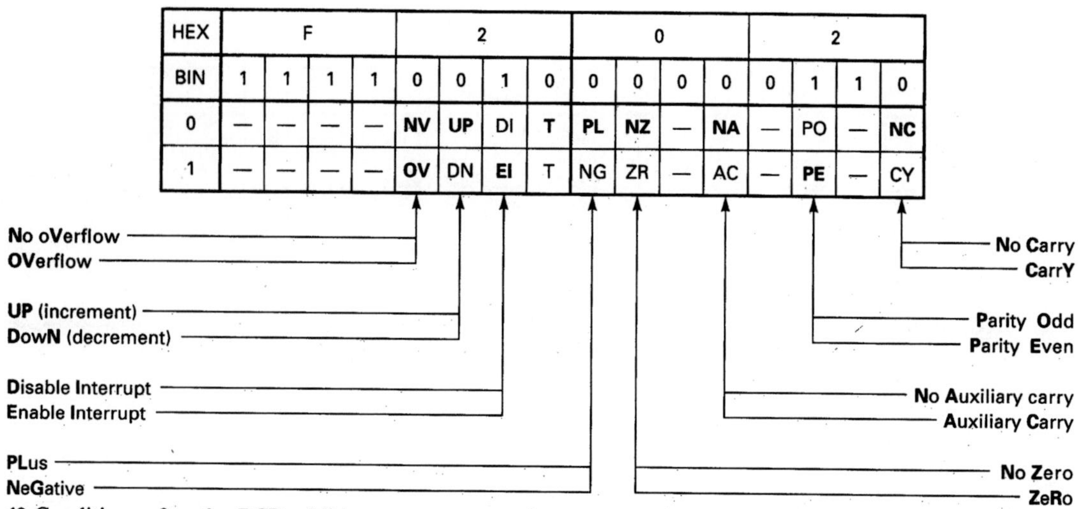

Fig. 18-49 Conditions after the BCD addition.

conditions that result from the conversion from binary to BCD (see Fig. 18-49.)

Subtraction

Subtraction is the opposite of addition. All the flags operate the same except the carry flag. After subtraction, the carry flag indicates whether a borrow has occurred or not. You can think of it as a "borrow" flag. A 1 in the carry flag position indicates that a borrow from a nonexistent bit was required to do the subtraction. A 0 indicates that no borrow was required.

Figure 18-50 illustrates how to write a program to do single-precision subtraction.

You should assemble and run this program. When we did, we found that the result in AL was FF. We also found that there was an overflow, the answer was negative, and there was a carry. The answer was negative because the 8th bit of the answer is a 1, which indicates an 8-bit negative signed binary number. There was an overflow because $7F_{16} = 127_{10}$ and $80_{16} = -128_{10}$; therefore

$$\begin{array}{r} 127 \\ -(-128) \\ \hline 255 \end{array}$$

and 255_{10} is outside the valid range for 8-bit signed binary numbers. (The valid range is $+127_{10}$ to -128_{10}.) There was a carry because a borrow from a 9th bit was needed to complete the subtraction.

Multiplication

The 8-bit microprocessors featured in this text do not have a multiply instruction. To multiply, the programmer must use many instructions to accomplish what the 8086/8088 does with just one instruction.

There are several ways the 8086/8088 can multiply. It can multiply signed binary numbers by using the **I**nteger **MUL**tiply (IMUL) instruction. It can also multiply unsigned binary numbers by using the **MUL**tiply (MUL) instruction.

Whether signed or unsigned, the 8086/8088 can multiply two 8-bit binary numbers to produce a 16-bit answer or two 16-bit binary numbers to produce a 32-bit answer.

```
-r
AX=0000  BX=0000  CX=0000  DX=0000  SP=379E  BP=0000  SI=0000  DI=0000
DS=9C86  ES=9C86  SS=9C86  CS=9C86  IP=0100   NV UP EI PL NZ NA PO NC
9C86:0100 8BFE          MOV     DI,SI
-
-a
9C86:0100 mov al,7f         ;load first operand
9C86:0102 sub al,80         ;subtract second operand
9C86:0104 int 20            ;return control to DEBUG
9C86:0106
-
-g 0104

AX=00FF  BX=0000  CX=0000  DX=0000  SP=379E  BP=0000  SI=0000  DI=0000
DS=9C86  ES=9C86  SS=9C86  CS=9C86  IP=0104   OV UP EI NG NZ NA PE CY
9C86:0104 CD20          INT     20
-
```

Fig. 18-50 Subtraction.

```
  1E
× FC
1DBB

B>DEBUG
-r
AX=0000  BX=0000  CX=0000  DX=0000  SP=379E  BP=0000  SI=0000  DI=0000
DS=9C86  ES=9C86  SS=9C86  CS=9C86  IP=0100   NV UP EI PL NZ NA PO NC
9C86:0100 8BFE          MOV     DI,SI
-
-a
9C86:0100 mov al,1E         ;first operand
9C86:0102 mov bl,FC         ;second operand (no immediate mode allowed)
9C86:0104 mul bl            ;mul automatically uses value value in al or ax
9C86:0106 int 20            ;return control to DEBUG
9C86:0108
-
-g 106

AX=1D88  BX=00FC  CX=0000  DX=0000  SP=379E  BP=0000  SI=0000  DI=0000
DS=9C86  ES=9C86  SS=9C86  CS=9C86  IP=0106   OV UP EI PL NZ NA PO CY
9C86:0106 CD20          INT     20
-
```

Fig. 18-51 Eight-bit multiplication on the 8086/8088.

If two 8-bit numbers are to be multiplied, one of them must be placed in AL. The other can be in a register or memory location. *Immediate mode multiplication is not allowed.* That is, you cannot do this:

```
mov al,1E
mul al,FC
int 20
```

You cannot specify a number to be multiplied by the number in AL in the instruction itself. You must move it to a register or memory location.

The problem 1E × FC and the program to solve it are shown in Fig. 18-51.

Notice that we moved the first number into AL and then the second into BL. We then only needed to say

mul bl

because the microprocessor assumes that the first number is in AL. The answer is placed in AX.

The only two flags that have any meaning after a MUL or IMUL instruction are the overflow and carry flags. If the upper byte of the answer (AH) is 00, then both of these flags will be cleared. Any other result in AH causes both of these flags to be set. Since the value in AH in our example is not 0, both the overflow and carry flags are set after the program is run.

Figure 18-52 illustrates a 16-bit multiplication problem.

```
    FFE2
×   12D3
12D0CB46

B>DEBUG
-r
AX=0000  BX=0000  CX=0000  DX=0000  SP=440E  BP=0000  SI=0000  DI=0000
DS=9BBF  ES=9BBF  SS=9BBF  CS=9BBF  IP=0100   NV UP EI PL NZ NA PO NC
9BBF:0100 C3            RET
-
-a
9BBF:0100 mov ax,FFE2      ;first 16-bit operand
9BBF:0103 mov bx,12D3      ;second 16-bit operand
9BBF:0106 mul bx           ;multiply ax by bx
9BBF:0108 int 20           ;return control to DEBUG
9BBF:010A
-
-g 108

AX=CB46  BX=12D3  CX=0000  DX=12D0  SP=440E  BP=0000  SI=0000  DI=0000
DS=9BBF  ES=9BBF  SS=9BBF  CS=9BBF  IP=0108   OV UP EI PL NZ NA PO CY
9BBF:0108 CD20          INT     20
-
-
```

Fig. 18-52 Sixteen-bit multiplication on the 8086/8088.

```
       58 remainder 2
FB ) 564A
```

```
B>DEBUG
-r
AX=0000  BX=0000  CX=0000  DX=0000  SP=45DE  BP=0000  SI=0000  DI=0000
DS=9BA2  ES=9BA2  SS=9BA2  CS=9BA2  IP=0100   NV UP EI PL NZ NA PO NC
9BA2:0100 2126A105      AND     [05A1],SP                      DS:05A1=E903
-
-a
9BA2:0100 mov ax,564A       ;dividend (16-bits)
9BA2:0103 mov bl,FB         ;divisor (8-bits)
9BA2:0105 div bl            ;divide ax by bl
9BA2:0107 int 20            ;return control to DEBUG
9BA2:0109
-
-g 107

AX=0258  BX=00FB  CX=0000  DX=0000  SP=45DE  BP=0000  SI=0000  DI=0000
DS=9BA2  ES=9BA2  SS=9BA2  CS=9BA2  IP=0107   NV UP EI PL NZ AC PO CY
9BA2:0107 CD20          INT     20
-
```

Fig. 18-53 A 16-bit number divided by an 8-bit number using the 8086/8088 DIV instruction.

The process is similar to that used in 8-bit multiplication. You use 16-bit registers instead of 8-bit, and the answer is 32-bits wide! The upper 2 bytes (16 bits) are found in DX, and the lower 2 bytes are found in AX.

The flags respond as they do for 8-bit multiplication.

Division

We handle division in a way which is similar to, yet the opposite of, the way multiplication is handled.

When division is done, the dividend (number to be divided) must be twice as wide (16 or 32 bits) as the divisor (8 or 16 bits). Figure 18-53 illustrates how a 16-bit dividend is divided by an 8-bit divisor.

Notice how we again moved the operands into a register to prepare for the actual division. Our 16-bit dividend ($564A_{16}$) was placed in AX and the 8-bit divisor (FB_{16}) was placed in BL. Notice that we simply say

div bl

and the microprocessor assumes we are dividing BL into AX.

Now notice how the answer is displayed. The answer is 58_{16}, with a remainder of 2_{16}. The quotient appears in the lower half of AX (AL), and the remainder is in the upper half of AX (AH). This is where the answer to a problem which divides a 16-bit number by an 8-bit number is found.

Figure 18-54 illustrates how to divide a 32-bit binary number by a 16-bit binary number.

To perform this type of problem, you must place the most significant 16 bits of the dividend in register DX. Place the least significant 16 bits of the dividend in register AX. Then place the 16-bit divisor in BX or CX. After the division the answer (quotient) will be found in register AX, with the remainder in register DX.

GLOSSARY

ASCII American Standard Code for Information Interchange. A binary code in which letters of the alphabet, numbers, punctuation, and certain control characters are represented.

BCD (binary-coded decimal) Decimal numbers which replace each decimal digit with its 4-bit binary equivalent.

multiple-precision number A number which is composed of more than one binary word.

single-precision number A number which is composed of one binary word. In an 8-bit microprocessor this is an 8-bit number, and in a 16-bit microprocessor this is a 16-bit number.

```
      789A remainder 8
45CE ) 20E28DF4
```

```
B>DEBUG
-r
AX=0000  BX=0000  CX=0000  DX=0000  SP=4ECE  BP=0000  SI=0000  DI=0000
DS=9B13  ES=9B13  SS=9B13  CS=9B13  IP=0100   NV UP EI PL NZ NA PO NC
9B13:0100 7420          JZ      0122
-
-
-a
9B13:0100 mov dx,20E2      ;most significant word of dividend
9B13:0103 mov ax,8DF4      ;least significant word of dividend
9B13:0106 mov bx,45CE      ;divisor
9B13:0109 div bx           ;divide DXAX register pair by BX
9B13:010B int 20           ;return control to DEBUG
9B13:010D
-
-g 10b

AX=789A  BX=45CE  CX=0000  DX=0008  SP=4ECE  BP=0000  SI=0000  DI=0000
DS=9B13  ES=9B13  SS=9B13  CS=9B13  IP=010B   NV UP EI NG NZ AC PE CY
9B13:010B CD20          INT     20
-
```

Fig. 18-54 A 32-bit number divided by a 16-bit number using the 8086/8088 microprocessor.

SELF-TESTING REVIEW

Read each of the following and provide the missing words. Answers appear at the beginning of the next question.

1. Binary-coded decimal numbers are decimal numbers in which each digit is represented by its _________-_________ _________ equivalent.
2. (*4-bit binary*) What is the binary value for 10_{10}?
3. (*1010_2*) When 8-bit binary numbers are added, the carry flag indicates when a carry from the _________ bit to the _________ bit has occurred.
4. (*8th, 9th*) When 8-bit binary numbers are added, the half-carry flag indicates when a carry from the _________ bit to the _________ bit has occurred.
5. (*4th, 5th*) A number which can be represented by 1 byte is called a _________-precision number.
6. (*single*) After subtraction, the carry flag indicates whether or not a _________ has occurred.
7. (*borrow*) Do the 8-bit microprocessors featured in this text have multiply or divide instructions? (*No*)

PROBLEMS

General

18-1. What two types of information are generated by a microprocessor during addition?

18-2. What does $1_2 + 1_2 + 1_2 = ?$

18-3. What is

$$\begin{array}{r} 1010\ 1110_2 \\ +\ 0011\ 0111_2 \\ \hline \end{array}$$

18-4. What is

$$\begin{array}{r} 0111\ 1111\ 0110\ 1101 \\ +\ 0001\ 1000\ 1111\ 0110 \\ \hline \end{array}$$

18-5. When we are using all 8 bits to represent the numbers 1_{10} to 255_{10}, we refer to these as ______ binary numbers.

18-6. When we use 8-bit binary numbers to represent values from -128_{10} to $+127_{10}$, we refer to these as _________binary numbers.

18-7. Find the 8-bit signed binary value for -100_{10}.

18-8. What flag warns the programmer that the last answer produced exceeds the valid range for signed binary numbers?

18-9. What flag tells the programmer whether the number in the accumulator is positive or negative?

Specific Microprocessor Families

Solve the following problems using the microprocessor of your choice.

18-10. Write a program which will add the unsigned binary numbers 67_{16} and 23_{16}. Determine which flags are altered by the program and why.

18-11. Write a program which will subtract the signed binary number $4D_{16}$ from $7F_{16}$. Determine which flags are altered by the program and why.

18-12. Write a program which will add the *decimal* numbers 40_{10} and 52_{10}.

18-13. With your computer or microprocessor trainer store the unsigned binary numbers 67_{16} and 23_{16} in two consecutive memory locations. Now write a program which will find the sum of these two numbers and store the sum in a free memory location. (8086/8088 users: To store 67_{16} and 23_{16} in memory locations use the DEBUG e (enter) command. For example, typing

–e 0180 67 23 00 00

will enter 67, 23, 00, and 00 into memory locations 0180, 0181, 0182, and 0183, respectively.)

LOGICAL INSTRUCTIONS

This chapter discusses the logical instructions of our featured microprocessors. These instructions, along with the arithmetic and shift and rotate instructions, give us the ability to alter bits and bytes (data) in a predictable fashion.

You may wish to review logic gates before beginning this chapter. Microprocessors use logical instructions the way digital circuits use logic gates.

New Concepts

There are really only four basic logical functions: AND, OR, EXCLUSIVE-OR, and NOT. The NAND, NOR, EXCLUSIVE-NOR, and NEGate functions are simply extensions of the four basic functions.

We will look at each of the basic four plus a couple of other special instructions some of the microprocessors have. We will also discuss masking, a primary use of the logical instructions.

19-1 THE AND INSTRUCTION

When we AND 2 bits or conditions, we are saying that the output bit, or condition, is true only if both the input bits, or conditions, are true. For example, there will be a voltage at the output of a circuit only if there is voltage at both of its inputs. Or, a bit in memory will be 1 only if 2 other input bits are also 1. Or, a drill will begin to lower only if the workpiece has been secured and the worker's hands are away from the bit.

Input		Output
B	A	Y
0	0	0
0	1	0
1	0	0
1	1	1

Fig. 19-1 AND truth table.

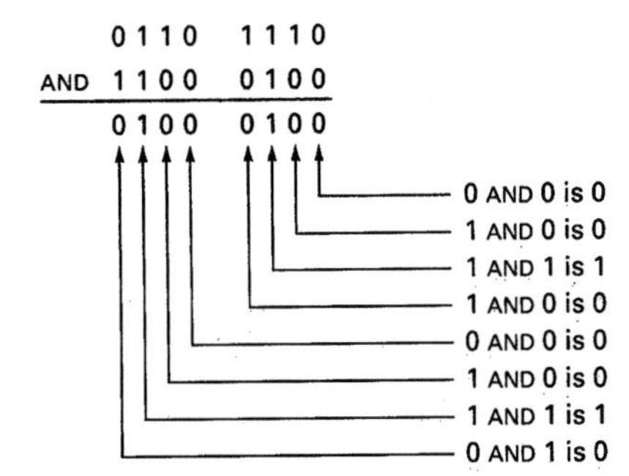

Fig. 19-2 ANDing 2 bytes together.

ANDing Bits

The truth table to AND 2 bits, or conditions, is shown in Fig. 19-1. Notice that the only way to get a 1 out is to put two 1s in.

ANDing Bytes

We can AND entire bytes, or words also. We simply apply the logic shown in the table to each bit. It's almost like turning a truth table on its side. For example, a problem in which we must AND 2 bytes is shown in Fig. 19-2.

Notice that we have applied the logic from the AND truth table to each bit. The only 1s in the answer are in columns where both the inputs are also 1.

EXAMPLE 19-1

Solve the following logical problem.

1011 1110 AND 0111 0001 is ???? ????

SOLUTION

```
     1011 1110
AND  0111 0001
     ---------
     0011 0000
```

Masking

A common use of the AND instruction is to AND bits or bytes with a mask. A *mask* allows us to change some bits in a certain way while allowing others to pass through unchanged. Look at the example shown in Fig. 19-3.

Notice that the upper nibble of the data byte passed through the 1s of the mask unchanged. However, every bit of the lower nibble passing through the 0s was cleared.

ANDing a mask to data can be viewed in either of two ways. You can say that selected data bits pass through unchanged while all others are cleared. Or, you can say that selected data bits are cleared while others pass through unaltered.

EXAMPLE 19-2

Devise a mask which when ANDed to an 8-bit data byte will clear all bits except the first 2 (2 least significant bits).

SOLUTION

0000 0011

For example:

```
     1111 1111   ←data
AND  0000 0011   ←mask
     ---------
     0000 0011
```

19-2 THE OR INSTRUCTION

When we OR 2 bits or conditions, we are saying that the output will be true (or 1) if either of the input bits or conditions is true (1) or if both of the input bits or conditions are true.

ORing Bits

The truth table to OR 2 bits or conditions is shown in Fig. 19-4.

Notice that you get a 1 out if any input is a 1. Or, to look at it another way, the only way to get a 0 out is to have 0s at both inputs.

```
     1 0 0 1   1 0 0 1  ←—— data
AND  1 1 1 1   0 0 0 0  ←—— mask
     -----------------
     1 0 0 1   0 0 0 0
```

Fig. 19-3 Using the AND instruction to mask bits.

Input		Output
B	A	Y
0	0	0
0	1	1
1	0	1
1	1	1

Fig. 19-4 OR truth table.

ORing Bytes

We can OR entire bytes, or words also. We simply apply the logic shown in the table to each bit. For example, the same problem used in the previous section, but now ORing the 2 bytes together, is shown in Fig. 19-5.

Notice that we have used the logic from the OR truth table and applied it to each bit. The only 0s in the answer are in columns where both the inputs are also 0.

EXAMPLE 19-3

Solve the following logical problem.

1011 1110 OR 0111 0001 is ???? ????

SOLUTION

```
    1011 1110
OR  0111 0001
    ---------
    1111 1111
```

Masking

A common use of the OR instruction is to OR bits or bytes with a mask. A *mask* allows some bits to pass through unchanged while others are changed in a certain way. Look at the example shown in Fig. 19-6.

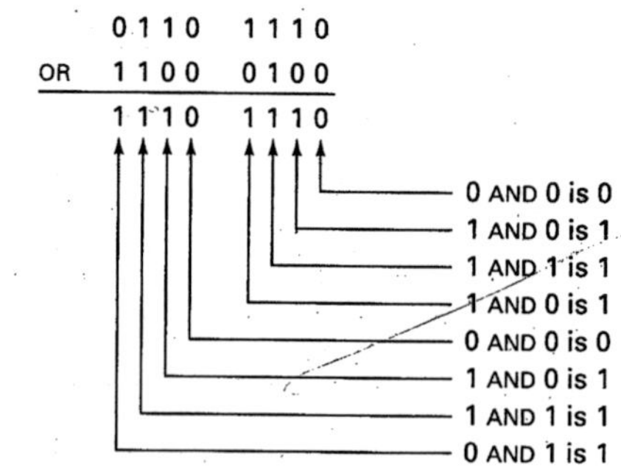

Fig. 19-5 ORing two bytes together.

```
      1001  1001  ←—— data
OR    1111  0000  ←—— mask
      1111  1001
```

Fig. 19-6 Using the OR instruction to mask bits.

Notice that the lower nibble of the data byte passing through the 0s of the mask was unchanged while every bit of the upper nibble passing through the 1s was set.

ORing a mask to data can be viewed in either of two ways. You can allow selected data bits to pass through unchanged while all others are set. Or, you can allow selected data bits to be set while all others pass through unaltered.

EXAMPLE 19-4

Devise a mask which when ORed to an 8-bit data byte will set all bits except the first 2 (2 least significant bits).

SOLUTION

1111 1100
For example:

```
      0000 0000   ← data
OR    1111 1100   ← mask
      1111 1100
```

19-3 THE EXCLUSIVE-OR (EOR, XOR) INSTRUCTION

When we EXCLUSIVELY OR (EOR, XOR) 2 bits or conditions, we are saying that the output bit or condition is true only if one or the other of the input bits or conditions is true, but not both. For example, there will be a voltage at the output of a circuit only if there is voltage at one or the other, but not both, of its inputs.

XORing Bits

The truth table to XOR 2 bits or conditions is shown in Fig. 19-7.

Input		Output
B	A	Y
0	0	0
0	1	1
1	0	1
1	1	0

Fig. 19-7 XOR truth table.

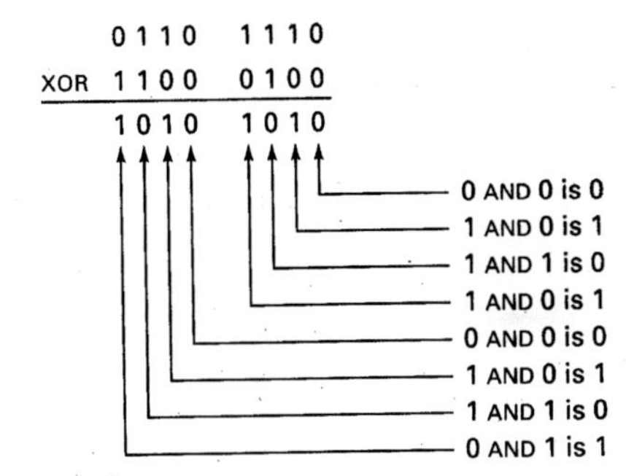

Fig. 19-8 XORing two bytes together.

Notice that the only way to get a 1 out is to have one, but not both, of the inputs be a 1.

XORing Bytes

We can XOR entire bytes, or words also. We simply apply the logic shown in the table to each bit. For example, the same problem shown in the previous two sections, but this time XORing the 2 bytes, is shown in Fig. 19-8.

Notice that we have used the logic from the XOR truth table and applied it to each bit. The only 1s in the answer are in columns where one but not both the inputs are 1.

EXAMPLE 19-5

Solve the following logical problem.

1011 1110 XOR 0111 0001 is ???? ????

SOLUTION

```
      1011 1110
XOR   0111 0001
      1100 1111
```

Masking

A common use of the XOR instruction is to XOR bits or bytes with a mask. A *mask* allows some bits to pass through unchanged while others are changed in a certain way. Look at the example shown in Fig. 19-9.

Notice that the lower nibble of the data byte passed through the 0s of the mask unchanged while every bit of the upper nibble passing through the 1s was inverted.

XORing a mask to data can be viewed in either of two ways. You can allow selected data bits to pass through

```
      1001  1001  ←—— data
XOR   1111  0000  ←—— mask
      0110  1001
```

Fig. 19-9 Using the XOR instruction to mask bits.

unchanged while all others are inverted. Or, you can allow selected data bits to be inverted while all others pass through unaltered.

EXAMPLE 19-6

Devise a mask which when XORed to an 8-bit data byte will invert all bits except the first 2 (2 least significant bits).

SOLUTION

1111 1100

For example:

```
    1111 1111    ←data
XOR 1111 1100    ←mask
    0000 0011
```

19-4 THE NOT INSTRUCTION

When we NOT or invert bits or conditions, we are saying that the output bit or condition is the opposite of the input bit or condition. For example, if there is a voltage at the input, there will not be one at the output; or if there is *no* voltage at the input, there *will* be a voltage at the output.

NOT-ing (Inverting) Bits

The truth table for the NOT function is shown in Fig. 19-10.

Input	Output
A	Y
0	1
1	0

Fig. 19-10 NOT truth table.

NOT-ing (Inverting) Bytes

We can NOT or invert entire bytes, or words also. We simply apply the logic shown in the table to each bit. An example of inverting or complementing a number is shown in Fig. 19-11.

NOT 1111 0000 is 0000 1111

Fig. 19-11 "NOT-ing" or inverting a binary number.

Notice that we have changed every 0 to a 1 and every 1 to a 0—that is, we have inverted every bit of the byte. This is the 1's complement of the number.

EXAMPLE 19-7

Solve the following logical problem.

NOT 1011 1110 is ???? ????

SOLUTION

0100 0001

19-5 THE NEG (NEGATE) INSTRUCTION

The NEGate instruction finds the 2's complement of a number. To find the 2's complement, we first find the 1's complement and then add 1. An example is shown in Fig. 19-12.

```
  1111  0000  ←—— number
  0000  1111  ←—— 1's complement
+          1  ←—— add 1
  0001  0000  ←—— 2's complement (original number NEGated)
```

Fig. 19-12 NEGating a number (2's complement).

Specific Microprocessor Families

Let's see how these instructions work in the different microprocessor families.

19-6 6502 FAMILY

The 6502 has three of the instructions discussed in the New Concepts section of this chapter plus one instruction not discussed there. These are the AND, OR, EOR, and BIT instructions. Let's look at each.

The AND Instruction

The 6502 AND instruction works exactly as described in the New Concepts section. If we use the same example we used in Fig. 19-3 in the New Concepts section, we will find that the 6502 does in fact AND bytes as discussed.

Figure 19-13 shows our original problem and solution plus a 6502 program which solves the problem. After running this program, you will find that the accumulator contains 90_{16}. This is exactly what we expected after 99_{16} was masked with FO_{16}.

```
      1001  1001  <---- data
AND   1111  0000  <---- mask
      ----------
      1001  0000
```

```
0340   A9 99    LDA #$99     ;load A with 1001 1001
0342   29 F0    AND #$F0     ;AND mask
0344   00       BRK          ;stop
```

Fig. 19-13 Using the 6502 AND instruction to mask bits.

```
      1001  1001  <---- data
OR    1111  0000  <---- mask
      ----------
      1111  1001
```

```
0340   A9 99    LDA #$99     ;load A with 1001 1001
0342   09 F0    ORA #$F0     ;OR mask
0344   00       BRK          ;stop
```

Fig. 19-14 Using the 6502 OR instruction to mask bits.

If you check the 6502 instruction set, you will find that the AND instruction affects the negative and zero flags. In this case the negative flag is set because the 8th bit of the accumulator is 1, indicating a 2's-complement negative number.

The OR Instruction

The 6502 OR instruction also works exactly as described in the New Concepts section. If we use the example from Fig. 19-6 in the New Concepts section, we find that the 6502 does OR bytes as discussed there.

Figure 19-14 shows our original problem and solution plus a 6502 program which solves the problem. After entering and running the program, you will find that the accumulator contains $F9_{16}$. This is the value we expected the accumulator to have.

The OR instruction also affects the negative and zero flags. You will find that the negative flag is again set because the 8th bit is 1, indicating a 2's-complement negative number.

The EOR Instruction

The 6502 EOR instruction also works as described in the New Concepts section. If we use the example from Fig. 19-9 in the New Concepts section, we'll find that the 6502 does EOR bytes as discussed there.

Figure 19-15 shows our original problem and solution plus a 6502 program which solves the problem. Enter and run the program. We expected the accumulator to have 69_{16} after EORing.

The EOR instruction also affects the negative and zero flags. This time neither is set; the result is neither negative nor zero.

The BIT Instruction

The BIT instruction was not described in the New Concepts section and is somewhat unusual. Refer to the BIT instruction in the Expanded Table of 6502 Instructions Listed by Category.

The BIT instruction ANDs a memory location with the accumulator. However, the result is not stored anywhere. Neither the accumulator nor the memory location is changed.

If the result of the AND is zero, the zero flag is set. If the result is not zero, the zero flag is not set.

The negative and overflow flags are affected in an unusual way. The status of the negative and overflow flags is not determined by the result of the AND process but rather is copied from bits 6 and 7 (7th and 8th bits) of the memory location.

A program which illustrates the operation of the BIT instruction is shown in Fig. 19-16.

The BIT instruction is useful when using the flags to control branching. You can alter the flags with a logical condition without actually changing the accumulator or memory location.

```
      1001  1001  <---- data
EOR   1111  0000  <---- mask
      ----------
      0110  1001
```

```
0340   A9 99    LDA #$99     ;load A with 1001 1001
0342   49 F0    EOR #$F0     ;EOR mask
0344   00       BRK          ;stop
```

Fig. 19-15 Using the 6502 EOR instruction to mask bits.

```
0340  A9 C0     LDA #$C0      ;load A with 1100 0000
0342  8D A0 03  STA #03A0     ;store 1100 0000 in location 03A0
0345  A9 00     LDA #$00      ;load A with 0000 0000
0347  2C A0 03  BIT $03A0     ;AND A (0000 0000) with 0340 (1100 0000)
034A  00        BRK           ;stop

After running the program:

negative flag = 1, overflow flag = 1, break flag = 1, zero flag = 1, accumulator = 00
```

Fig. 19-16 Using the 6502 BIT instruction.

19-7 6800/6808 FAMILY

The 6800/6808 has all the instructions discussed in the New Concepts Section plus one instruction not discussed there. These are the ANDA/ANDB, ORAA/ORAB, EORA/EORB, BITA/BITB, COM/COMA/COMB, and NEG/NEGA/NEGB instructions. Notice that each instruction has a mnemonic for each accumulator and that some (COM and NEG) have one for memory locations also. Let's look at each.

Clearing the Flags

The 6800/6808 examples which follow cover both the result of the logical operation and the condition of the flags. It is helpful to be able to clear the flags before the examples are run so that the previous condition of the flags is not confused with the effect the example had on the flags.

Place the following program in an area of memory you do not plan to use for the examples. Then run this program to clear both accumulators and all flags before running each example program.

```
xxxx 4F   CLRA   ;clear A
xxxx 5F   CLRB   ;clear B
xxxx 06   TAP    ;clear flags
xxxx 3E   WAI    ;stop
```

The ANDA/ANDB Instruction

The 6800/6808 AND instruction works exactly as described in the New Concepts section. If we use the same example we discussed in the New Concepts section (Fig. 19-3), we will find that the 6800/6808 does in fact AND bytes as discussed.

Figure 19-17 shows our original problem and solution plus a 6800/6808 program which solves the problem. If you will notice the condition of the accumulator and flags after running this program, you will find that the accumulator contains 90_{16} as we expected.

If you check the 6800/6808 instruction set, you will find that the AND instruction affects the negative and zero flags. (The overflow flag is always cleared.) In this case the negative flag is set because the 8th bit of the accumulator is 1, indicating a 2's-complement negative number. (It is assumed that the flags just discussed were cleared before the program was started.)

The ORAA/ORAB Instruction

The 6800/6808 ORAA/ORAB instruction also works exactly as described in the New Concepts section. We'll use the example found in Fig. 19-6 in the New Concepts section.

Figure 19-18 shows our original problem and solution plus a 6800/6808 program which solves the problem. After entering and running the program, you will find that the accumulator contains $F9_{16}$. This is what we expected.

The OR instruction also affects the negative and zero flags. (The overflow flag is always cleared.) The negative flag is again set because the 8th bit of A is 1, indicating a 2's-complement negative number.

The EORA/EORB Instruction

Let's look at the 6800/6808 EORA/EORB instruction. If we use the example from Fig. 19-9 in the New Concepts section, we will find that the 6800/6808 does EOR bytes as discussed.

Figure 19-19 shows our original problem and solution from Fig. 19-9 plus a 6800/6808 program which solves the problem. Enter and run the program. You will find that the accumulator contains 69_{16}.

The EOR instruction also affects the negative and zero flags. (The overflow flag is always cleared.) In this case neither was set; the result is neither negative nor zero.

```
      1001  1001  <---- data
  AND 1111  0000  <---- mask
      1001  0000

0000   86 99    LDAA #$99    ;load A with 1001 1001
0002   84 F0    ANDA #$F0    ;AND mask
0004   3E       WAI          ;stop
```

Fig. 19-17 Using the 6800/6808 AND instruction to mask bits.

```
           1001  1001  <---- data
        OR 1111  0000  <---- mask
           ---------
           1111  1001
```

```
0000  86 99    LDAA #$99      ;load A with 1001 1001
0002  8A FO    ORAA #$FO      ;OR mask
0004  3E       WAI            ;stop
```

Fig. 19-18 Using the 6800/6808 ORAA/ORAB instruction to mask bits.

```
           1001  1001  <---- data
       XOR 1111  0000  <---- mask
           ---------
           0110  1001
```

```
0000  86 99    LDAA #$99      ;load A with 1001 1001
0002  88 FO    EORA #$FO      ;EOR mask
0004  3E       WAI            ;stop
```

Fig. 19-19 Using the 6800/6808 EORA/EORB instruction to mask bits.

The BITA/BITB Instruction

The BIT instruction was not described in the New Concepts section. Refer to the BIT instruction in the Expanded Table of 6800/6808 Instructions Listed by Category.

The BIT instruction ANDs a memory location with one of the accumulators. However, the result is not stored anywhere. Neither the accumulator nor the memory location is changed.

If the result of the AND is zero, the zero flag is set. If the result is not zero, the zero flag is not set. If the result of the AND is a negative 2's-complement number, the negative flag is set. Regardless of the result, the overflow flag is cleared.

A program which illustrates the operation of the BIT instruction is shown in Fig. 19-20.

The BIT instruction is useful when the flags are used to control branching. You can alter the flags with a logical condition without actually changing the accumulator or memory location.

The COM/COMA/COMB Instruction

The complement instruction (COM/COMA/COMB) finds the 1's complement of each bit in the byte that's being complemented. That is, it inverts every bit in the byte. An example problem and a 6800/6808 program to solve the problem are shown in Fig. 19-21.

After running this program, you should find 55_{16} in A and the carry flag set.

Referring to the 6800/6808 instruction set, you will find that the COM instructions affect the negative and zero flags. In addition, they always clear the overflow flag and set the carry flag. In this example the negative flag is clear because the result (55_{16}) is not a negative number. Nor is it zero; therefore the zero flag is not set. The overflow flag is automatically cleared, and the carry flag automatically set.

The NEG/NEGA/NEGB Instruction

The NEG/NEGA/NEGB (negate) instructions are very similar to the COM/COMA/COMB instructions. The NEG instructions,

```
0000  86 FF    LDAA #$FF      ;load A with 1111 1111
0002  85 CO    BITA #$CO      ;AND A with 1100 0000
0004  3E       WAI            ;stop

        After running the program:

              A = FF          flags = 001000
```

Fig. 19-20 Using the 6800/6808 BITA instruction.

```
1010  1010  <---- original number (AA16)
0101  0101  <---- 1's complement of original number (5516)
```

```
0000  86 AA    LDAA #$AA      ;load A with 1010 1010
0002  43       COMA           ;invert all bits (0101 0101) (55h)
0003  3E       WAI            ;stop
```

Fig. 19-21 Using the 6800/6808 COM/COMA/COMB instructions.

```
         0101  1111  <—— original number (95₁₀)
         1010  0000  <—— 1's complement
       +          1  <—— plus 1
         ----------
         1010  0001  <—— 2's complement (-95₁₀)

0000  86 5F    LDAA#$5F     ;load A with 0101 1111
0002  40       NEGA         ;2's complement of A
0003  3E       WAI          ;stop
```

Fig. 19-22 Using the 6800/6808 NEG/NEGA/NEGB instructions.

however, find the 2's complement of a number instead of the 1's complement. Recall that the 2's complement is found by first finding the 1's complement and then adding 1.

Figure 19-22 shows an example problem and program using the negate instruction.

After running the program you will have $A1_{16}$ in the accumulator and the negative and carry flags set.

The negative flag is set because the 8th bit of A is set indicating a 2's-complement negative number.

Why the carry flag is set requires a little explanation. One way to look at a 2's-complement number is to view it as a 1's-complement number with 1 added to it. There is another point of view, however.

Remember how we described the creation of negative numbers as being like rotating an odometer backward? The original number used in this example is $0101\ 1111_2$, which is 95_{10}. If we rotate our odometer backward from 00 by 95 places, we will arrive at the binary number 1010 0001. Rotating the odometer backward from 00 is the same as subtracting from 00.

Now think about subtracting a number from 00. Would a borrow from the carry bit be required? Yes, because any number is larger than 0 and a borrow would be required to subtract it from 00. To subtract 95 from 00 requires a borrow, which is why the carry flag is set.

If you think about it, the carry flag would have been set regardless of what number we would have used. When you use the NEG instruction, the only time the carry flag won't be set is if you negate the number 00, because subtracting 00 from 00 does not require a borrow.

19-8 8080/8085/Z80 FAMILY

The 8080/8085/Z80 has four of the instructions discussed in the New Concepts section, although one has a different name. These are the AND (ANA [AND]), OR (ORA [OR]), XOR (XRA [XOR]), and NOT (CMA [CPL]) instructions. (Z80 mnemonics are shown in brackets.) Let's look at each.

The ANA [AND] Instruction

The 8080/8085/Z80 ANA [AND] instruction works as described in the New Concepts section. If we use the example from Fig. 19-3 in the New Concepts section, we will find that the 8080/8085/Z80 does in fact AND bytes as discussed.

Figure 19-23 shows our original problem and solution plus an 8080/8085/Z80 program which solves the problem. If you will notice the condition of the accumulator and flags after running this program, you will find that the accumulator has a 90_{16} in it as we expected. The sign, auxiliary carry, and parity flags will be set.

If you check the 8085/Z80 instruction set, you will find that the AND instruction affects the sign, zero, and parity

```
        1001  1001  <—— data
    AND 1111  0000  <—— mask
        ----------
        1001  0000

8085 program

1800  3E 99    MVI A,99     ;load A with 1001 1001
1802  06 F0    MVI B,F0     ;load B with mask (1111 0000)
1804  A0       ANA B        ;AND A with mask
1805  76       HLT          ;stop

Z80 program

1800  3E 99    LD A,99      ;load A with 1001 1001
1802  06 F0    LD B,F0      ;load B with mask (1111 0000)
1804  A0       AND B        ;AND A with mask
1805  76       HALT         ;stop
```

Fig. 19-23 Using the 8080/8085/Z80 ANA [AND] instruction to mask bits.

```
             1001   1001  <----- data
        OR   1111   0000  <----- mask
             ---------
             1111   1001
```

```
8085 program

1800   3E 99      MVI A,99        ;load A with number (1001 1001)
1802   06 F0      MVI B,F0        ;load B with mask (1111 0000)
1804   B0         ORA B           ;OR number and mask
1805   76         HLT             ;stop

Z80 program

1800   3E 99      LD A,99         ;load A with number (1001 1001)
1802   06 F0      LD B,F0         ;load B with mask (1111 0000)
1804   B0         OR B            ;OR number and mask
1805   76         HALT            ;stop
```

Fig. 19-24 Using the 8085/Z80 OR instruction to mask bits.

flags. The AND instruction *always* sets the auxiliary carry [half-carry] flag and always clears the carry flag. (*Note:* If you are using an 8080 microprocessor, the auxiliary flag works a little differently than it does in the 8085 and Z80. Check the Expanded Table.)

The sign flag is set because this is a negative number. The zero flag is clear because the result was not zero. The auxiliary flag is set because it is always set by this instruction. The parity flag is set because there are an even number of 1s. And the carry flag is clear because that flag is always cleared by the AND instruction.

The ORA [OR] Instruction

The 8085/Z80 OR instruction also works as described in the New Concepts section. We'll use the example from Fig. 19-6 in the New Concepts section.

Figure 19-24 shows our original problem and solution plus an 8085/Z80 program which solves the problem. After entering and running the program, you will find that the accumulator has a value of $F9_{16}$ and that the sign and parity flags have been set.

We expected the accumulator to have $F9_{16}$ after ORing. The OR instruction set the sign flag because $F9_{16}$ is a 2's-complement negative number. The parity flag is set because there are an even number of 1s in $F9_{16}$ ($1111\ 1001_2$). The zero flag is clear because the result ($F9_{16}$) is not zero. All other flags are automatically cleared by the OR instruction.

The XRA [XOR] Instruction

Let's look at the 8085/Z80 XOR instruction. If we use the example from Fig. 19-9 in the New Concepts section, we'll find that the 8085/Z80 does XOR bytes as discussed.

Figure 19-25 shows our original problem and solution plus an 8085/Z80 program which solves the problem. After entering and running the program, you will find that the accumulator contains 69_{16} and that only the parity flag is set. Examine the figure and the Expanded Table to find why this is so.

```
             1001   1001  <----- data
       XOR   1111   0000  <----- mask
             ---------
             0110   1001
```

```
8085 program

1800   3E 99      MVI A,99        ;load A with number (1001 1001)
1802   06 F0      MVI B,F0        ;load B with mask (1111 0000)
1804   A8         XRA B           ;XOR number with mask
1805   76         HLT             ;stop

Z80 program

1800   3E 99      LD A,99         ;load A with number (1001 1001)
1802   06 F0      LD B,F0         ;load B with mask (1111 0000)
1804   A8         XOR B           ;XOR number with mask
1805   76         HALT            ;stop
```

Fig. 19-25 Using the 8085/Z80 XOR instruction to mask bits.

```
                    NOT  1010 1010  is  0101 0101

8085 program

1800   3E AA      MVI A,AA          ;load A with 1010 1010
1802   2F         CMA               ;invert all bits (0101 0101) (55h)
1803   76         HLT               ;stop

Z80 program

1800   3E AA      LD A,AA           ;load A with 1010 1010
1802   2F         CPL               ;invert all bits (0101 0101) (55h)
1803   76         HALT              ;stop
```

Fig. 19-26 Using the 8085/Z80 complement instruction.

The CMA [CPL] Instruction

The complement instruction (CMA [CPL]) finds the 1's complement of each bit in the byte that's being complemented. That is, it inverts every bit in the byte. An example problem and an 8085/Z80 program to solve the problem are shown in Fig. 19-26.

After running this program, you should find the value 55_{16} in A. If you are using an 8085, you will find that none of the flags has been affected or changed by the CMA instruction. If you are using a Z80, you will find that the half-carry and parity flags have been set. The Z80 always sets these two flags after the CPL instruction.

19-9 8086/8088 FAMILY

The 8086/8088 has all the instructions discussed in the New Concepts section. These include the AND, OR, XOR, NOT, and NEG instructions. Let's look at each.

The AND Instruction

The 8086/8088 AND instruction works as described in the New Concepts section. If we use the example from Fig. 19-3 in the New Concepts section, we find that the 8086/8088 does in fact AND bytes as discussed.

Figure 19-27 shows our original problem and solution plus an 8086/8088 program which solves the problem. Notice the condition of the accumulator and flags before and after running this program.

After masking 99_{16} with FO_{16}, 90_{16} is exactly what we expected. If you check the 8086/8088 instruction set, you will find that the AND instruction affects the sign, zero, and parity flags. The overflow and carry flags are always cleared (NV, NC), and the auxiliary flag is undefined. In this case the sign flag is set (NG) because the 8th bit of the accumulator is 1, indicating a 2's-complement negative number.

The OR Instruction

The 8086/8088 OR instruction also works as described earlier in the New Concepts section. We'll use the example from Fig. 19-6 in the New Concepts section.

Figure 19-28 shows our original problem and solution plus an 8086/8088 program which solves the problem. After entering and running the program, you will find that AL has a value of $F9_{16}$ as we expected.

```
         1001  1001  <---- data
    AND  1111  0000  <---- mask
         ----------
         1001  0000

AX=0000  BX=0000  CX=0000  DX=0000  SP=F75E  BP=0000  SI=0000  DI=0000
DS=908A  ES=908A  SS=908A  CS=908A  IP=0100   NV UP EI PL NZ NA PO NC
908A:0100 B099          MOV     AL,99
-
-a 100
908A:0100 MOV AL,99           ;load A with 1001 1001
908A:0102 AND AL,FO           ;AND A with mask
908A:0104 INT 20              ;return control to DEBUG
908A:0106
-
-g 0104

AX=0090  BX=0000  CX=0000  DX=0000  SP=F75E  BP=0000  SI=0000  DI=0000
DS=908A  ES=908A  SS=908A  CS=908A  IP=0104   NV UP EI NG NZ NA PE NC
908A:0104 CD20          INT     20
-
```

Fig. 19-27 Using the 8086/8088 AND instruction to mask bits.

```
    1001  1001  <---- data
OR  1111  0000  <---- mask
    ----------
    1111  1001
```

```
-r
AX=0000  BX=0000  CX=0000  DX=0000  SP=F83E  BP=0000  SI=0000  DI=0000
DS=907C  ES=907C  SS=907C  CS=907C  IP=0100   NV UP EI PL NZ NA PO NC
907C:0100 B099          MOV     AL,99
-
-a
907C:0100 MOV AL,99         ;load A with number (1001 1001)
907C:0102 OR  AL,F0         ;OR number and mask
907C:0104 INT 20            ;stop
907C:0106
-
-g 104

AX=00F9  BX=0000  CX=0000  DX=0000  SP=F83E  BP=0000  SI=0000  DI=0000
DS=907C  ES=907C  SS=907C  CS=907C  IP=0104   NV UP EI NG NZ NA PE NC
907C:0104 CD20          INT     20
-
```

Fig. 19-28 Using the 8086/8088 OR instruction to mask bits.

The OR instruction also affects certain flags. The sign flag is set (NG) because this is a 2's-complement negative number. The overflow flag is cleared (NV) because the OR instruction always clears it. The carry flag is also cleared for the same reason (NC). We have even parity (PE), and the result is not zero (NZ).

The XOR Instruction

Let's look at the 8086/8088 XOR instruction using the example from Fig. 19-9 in the New Concepts section.

Figure 19-29 shows our original problem and solution plus an 8086/8088 program which solves the problem. After entering and running the program, you will find that AL has a value of 69_{16}. This is what we expected.

The XOR instruction affects the flags in the same way as the OR and AND instructions. Examine the flags that are affected by this instruction to see whether they responded as you expected.

The NOT Instruction

The invert instruction (NOT) finds the 1's complement of each bit in the byte that's being complemented. That is, it inverts every bit in the byte. An example problem and an 8086/8088 program to solve the problem are shown in Fig. 19-30.

After running this program, you should find the value 55_{16} in AL. And since this instruction does not affect *any*

```
     1001  1001  <---- data
XOR  1111  0000  <---- mask
     ----------
     0110  1001
```

```
-r
AX=0000  BX=0000  CX=0000  DX=0000  SP=F60E  BP=0000  SI=0000  DI=0000
DS=909F  ES=909F  SS=909F  CS=909F  IP=0100   NV UP EI PL NZ NA PO NC
909F:0100 7420          JZ      0122
-
-a
909F:0100 MOV AL,99         ;load A with number (1001 1001)
909F:0102 XOR AL,F0         ;XOR number with mask (1111 0000)
909F:0104 INT 20            ;return control to DEBUG
909F:0106
-
-g 104

AX=0069  BX=0000  CX=0000  DX=0000  SP=F60E  BP=0000  SI=0000  DI=0000
DS=909F  ES=909F  SS=909F  CS=909F  IP=0104   NV UP EI PL NZ NA PE NC
909F:0104 CD20          INT     20
-
```

Fig. 19-29 Using the 8086/8088 XOR instruction to mask bits.

NOT 1010 1010 is 0101 0101

```
-r
AX=0000  BX=0000  CX=0000  DX=0000  SP=F51E  BP=0000  SI=0000  DI=0000
DS=90AE  ES=90AE  SS=90AE  CS=90AE  IP=0100   NV UP EI PL NZ NA PO NC
90AE:0100 7420          JZ      0122
-
-a
90AE:0100 MOV AL,AA          ;load A with number (1010 1010)
90AE:0102 NOT AL             ;invert all bits of number (0101 0101) (55h)
90AE:0104 INT 20             ;return control to DEBUG
90AE:0106
-
-g 104

AX=0055  BX=0000  CX=0000  DX=0000  SP=F51E  BP=0000  SI=0000  DI=0000
DS=90AE  ES=90AE  SS=90AE  CS=90AE  IP=0104   NV UP EI PL NZ NA PO NC
90AE:0104 CD20          INT     20
-
```

Fig. 19-30 Using the 8086/8088 NOT instruction.

flags, you should find that every flag is exactly as it was before the instruction was executed.

The NEG Instruction

The NEG (negate) instruction is very similar to the NOT instruction. The NEG instruction, however, finds the 2's complement instead of the 1's complement. Recall that the 2's complement is found by first finding the 1's complement and then adding 1.

Figure 19-31 shows an example problem and program using the negate instruction. After running the program, you will have $A1_{16}$ in the accumulator.

Notice also that the sign flag is set (NG) as well as the carry flag (CY).

The negative flag is set because the 8th bit of AL is set indicating a 2's-complement negative number.

Why the carry flag is set requires a little explanation. One way to look at a 2's-complement number is to view it as a 1's-complement number with 1 added to it. There is another point of view, however.

Remember how we described the creation of negative numbers as being like rotating an odometer backward? The original number we used in this example is $0101\ 1111_2$, which is 95_{10}. If we rotate our odometer backward from 00 by 95 places, we will arrive at the binary number 1010 0001. Rotating the odometer backward from 00 is the same as subtracting from 00.

Now think about subtracting a number from 00. Would a borrow from the carry bit be required? Yes, because any

```
  0101 1111  ←—— original number (95₁₀)
  1010 0000  ←—— 1's complement
+         1  ←—— plus 1
  ---------
  1010 0001  ←—— 2's complement (–95₁₀)
```

```
-r
AX=0000  BX=0000  CX=0000  DX=0000  SP=F15E  BP=0000  SI=0000  DI=0000
DS=90EA  ES=90EA  SS=90EA  CS=90EA  IP=0100   NV UP EI PL NZ NA PO NC
90EA:0100 7420          JZ      0122
-
-a
90EA:0100 MOV AL,5F          ;load A with number (0101 1111)
90EA:0102 NEG AL             ;find 2's complement of number in AL
90EA:0104 INT 20             ;return control to DEBUG
90EA:0106
-
-g 104

AX=00A1  BX=0000  CX=0000  DX=0000  SP=F15E  BP=0000  SI=0000  DI=0000
DS=90EA  ES=90EA  SS=90EA  CS=90EA  IP=0104   NV UP EI NG NZ NA PO CY
90EA:0104 CD20          INT     20
-
```

Fig. 19-31 Using the 8086/8088 NEG instruction.

number is larger than 00 and a borrow would be required to subtract it from 00. To subtract 95 from 00 required a borrow, which is why the carry flag was set.

If you think about it, the carry flag would have been set regardless of what number we used. When you use the NEG instruction, the only time the carry flag won't be set is when you negate the number 00 itself, because subtracting 00 from 00 does not require a borrow.

SELF-TESTING REVIEW

Read each of the following and provide the missing words. Answers appear at the beginning of the next question.

1. Name the four basic logical instructions.

   _________ _________

   _________ _________

2. *(AND, OR, XOR, and NOT)* When we AND 2 bits, we are saying that the output bit will be 1 only if both inputs bits are _________.
3. *(1)* A mask allows us to change some bits in a byte while allowing others to pass through _________.
4. *(unchanged)* When we _________ two bits together, we are saying that the output bit will be a 1 if either or both of the input bits are 1.
5. *(OR)* When ORing bits, the only way to get a _______ out is to have both inputs be _________.
6. *(0, 0)* When using the XOR instruction, if both input bits are the same, the output bit will be a ___ (0, 1).
7. *(0)* When XORing bits, the only way to get a 1 out is for (both, either) _________ of the input bits to be a 1.
8. *(either)* When we NOT or invert bits, we are saying that the output bit is the _________ (same as, opposite of) the input bit.
9. *(opposite of)* To NEGate a number is to find the 2's complement of the number. This involves finding the _________ _________ and then adding _________.

   (1's complement, 1)

PROBLEMS

General

19-1.
```
    1011 1100
AND 0110 1010
```

19-2. Devise a mask which, used with the AND instruction, would allow all bits to pass through unaltered except the most significant. The most significant should be cleared.

19-3.
```
   0110 1110
OR 0011 0101
```

19-4. Devise a mask which, used with the OR instruction, would allow all bits to pass through unaltered except the most significant. The most significant should be set.

19-5.
```
    0101 0101
XOR 0011 1111
```

19-6. Devise a mask which, used with the XOR instruction, would invert all bits except the 2 most significant. The 2 most significant should pass through unaltered.

19-7. Invert the binary number 0111 1011.

19-8. Negate the number 0110 1110 (8-bit answer).

Specific Microprocessor Families

Solve the following problems by using the microprocessor of your choice.

19-9. Write and run a program which will place the binary number 1100 1001 in the accumulator and then AND it with the binary number 1011 1101.

19-10. Write and run a program which will place the number CC_{16} in the accumulator and then use the OR instruction to set every bit in the lower nibble of the accumulator while allowing every bit in the upper nibble to remain unchanged.

Advanced Problems

Solve the following problems using the microprocessor of your choice.

19-11. Write and run a program which will:

a. place 45_{16} in the accumulator.
b. add $2F_{16}$ to the number in the accumulator.
c. use a mask to invert every bit in the lower nibble of the sum yet not alter the upper nibble.
d. subtract $0001\ 1100_2$ from the last result.
e. create another mask (using the AND instruction) which will allow all bits of the last result to remain unchanged except the least significant 3 bits which should be cleared.

19-12. ASCII values for the digits 0 through 9 are shown below.

0	0011 0000
1	0011 0001
2	0011 0010
3	0011 0011
4	0011 0100
5	0011 0101
6	0011 0110
7	0011 0111
8	0011 1000
9	0011 1001

It may sometimes be desirable to change an ASCII number into its binary equivalent. For this problem, write and run a program which uses a mask to change the ASCII value for 5 into its binary equivalent.

SHIFT AND ROTATE INSTRUCTIONS

In this chapter we'll study two relatively straightforward concepts—*shifting* and *rotating*. Shifts and rotates can be used for parallel-to-serial data conversion, serial-to-parallel data conversion, multiplication, division, and other tasks.

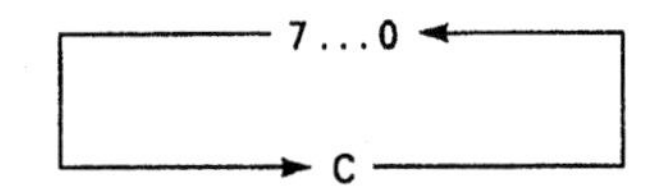

Fig. 20-1 Typical rotate left instruction.

New Concepts

The concepts of rotating and shifting are quite simple. Let's look at each in its "generic" form; then, as usual, we'll study each microprocessor family. The microprocessors' instructions which perform each of these functions differ only slightly.

20-1 ROTATING

Rotating bits is exactly what it sounds like—moving bits in a circle. Let's look at a typical rotate instruction to start our discussion. Figure 20-1 shows a typical *rotate left* instruction.

Figure 20-2 illustrates each step involved when a bit is rotated eight times. Figure 20-2 first shows an 8-bit accumulator and carry flag. The accumulator is loaded with the value 01_{16}, and the carry flag is cleared. Next, a sequence of eight rotate lefts are performed. Notice that the 1 just keeps moving 1 bit position each time.

Microprocessors can rotate toward the right or left. Some also have other forms of rotation in which the carry flag is involved in a slightly different way. We'll look at those in the Specific Microprocessor Families section.

20-2 SHIFTING

Shifting, like rotating, is exactly what it sounds like. And, like rotating, shifting can be toward the left or right. The 8080/8085 is the only microprocessor family being studied in this text which does not have shift instructions. The 8080/8085 has only rotate instructions.

Let's look first at the concept of shifting toward the left. Figure 20-3 illustrates what is known as a *logical shift left* or *arithmetic shift left*. Bits are shifted one at a time toward the left, with the bit in the 8th position (bit 7) being shifted into the carry flag.

Two things should be noticed which make this instruction different from the rotate instruction. First, the contents of the carry flag do not "wrap around" to bit 0; its contents are simply lost. Second, 0s are automatically shifted into bit 0 (least significant bit).

Look at Fig. 20-4 for an example of this type of shifting. We have loaded the value 99_{16} into the accumulator and have cleared the carry flag. Next we execute eight consecutive shifts. Notice that

1. 0s keep coming in from the left.
2. The bits in the accumulator keep shifting 1 bit to the left.
3. The bits shift from the most significant bit of the accumulator into the carry flag.
4. Bits shifting out of the carry flag are lost.

Shifts to the right are possible also. Figure 20-5 shows a typical *logical* shift to the right. This is basically the opposite of the shift left.

Figure 20-6 shows a typical *arithmetic* shift to the right. The *arithmetic shift right* instruction duplicates whatever was in the most significant bit and moves copies of it to the right with each shift.

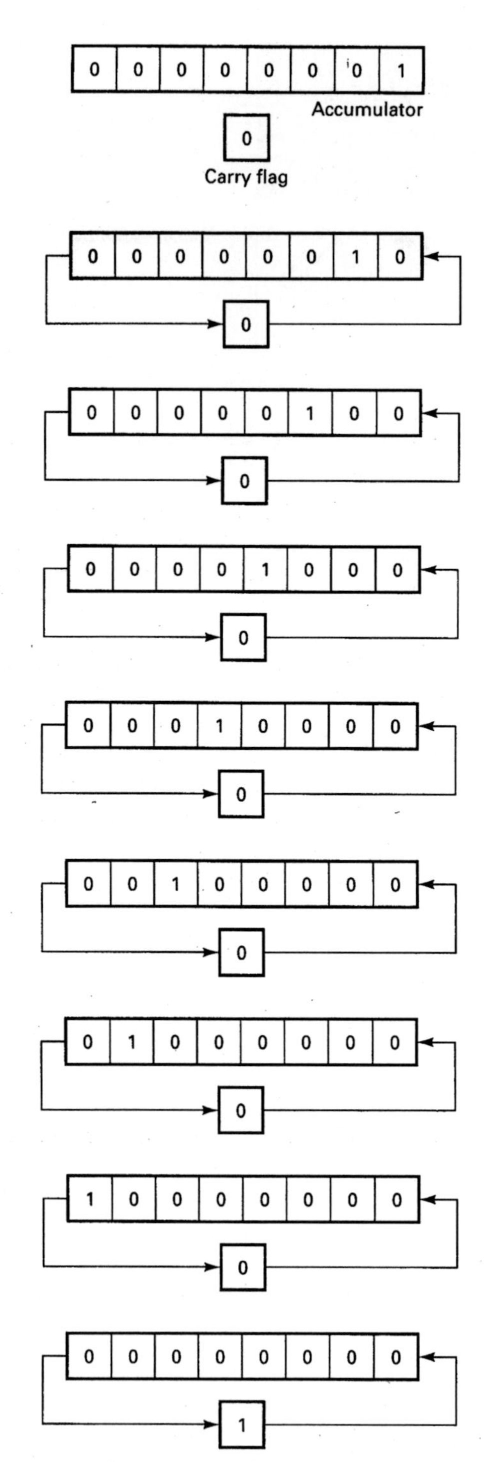

Fig. 20-2 Rotating left eight times.

20-3 AN EXAMPLE

Let's look at an example which uses the rotate instruction. It is often useful to be able to move a nibble of data from one part of a register to the other.

For example, let's say we wanted to clear every bit in

Fig. 20-3 Typical arithmetic shift left or logical shift left.

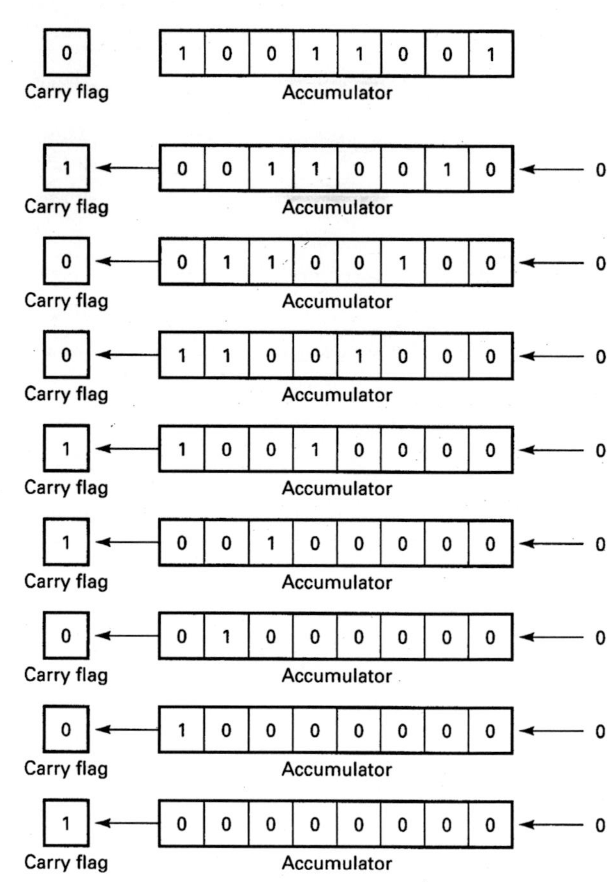

Fig. 20-4 Eight typical arithmetic shift left or logical shift left instructions.

0 → 7 . . . 0 → C

Fig. 20-5 Typical logical shift right.

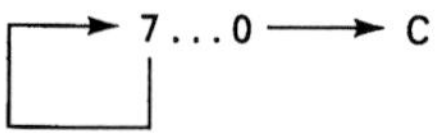

→ 7 . . . 0 → C

Fig. 20-6 Typical arithmetic shift right.

the upper nibble of the accumulator and then move every bit of the lower nibble into the upper nibble. There are no instructions for moving a nibble from one place to another. The rotate instruction can help accomplish this, though. Figure 20-7 shows our problem.

First, we'll use a mask to clear out the upper bit. This is shown in Fig. 20-8.

Next we'll clear the carry bit (since this bit will be rotated into the least significant bit of the lower nibble). Then we'll rotate toward the left four times. This is shown in Fig. 20-9.

If you compare the final value in Fig. 20-9 with our initial value in Fig. 20-9, you'll see that we have moved the lower nibble into the upper nibble, which is what we wanted to do.

Upper nibble	Lower nibble
1 1 0 0	1 1 0 1

Fig. 20-7 Situation in which we want to clear the upper nibble and then move every bit of the lower nibble into the upper nibble.

```
      1100  1101
AND   0000  1111
      ----------
      0000  1101
```

Fig. 20-8 Using the AND instruction to mask off the upper nibble.

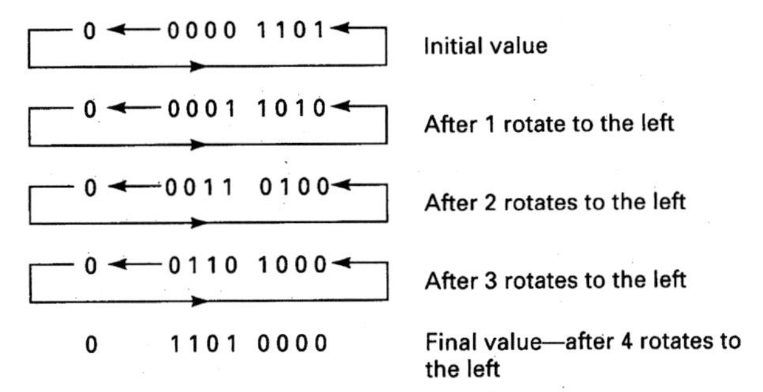

Fig. 20-9 Using the rotate through carry instruction to move the lower nibble into the upper nibble.

Specific Microprocessor Families

Let's study the shift and rotate instructions for each of our microprocessor families.

20-4 6502 FAMILY

The 6502 has two rotate instructions and two shift instructions. Let's look at them.

The ROL and ROR Instructions

The 6502 ROL (**RO**tate **L**eft) instruction works as described in the New Concepts section of this chapter and as shown in Fig. 20-10. Figure 20-10 is taken from the Rotate and Shift Instructions section of the Expanded Table of 6502 Instructions Listed by Category.

In Fig. 20-10 the "7 . . . 0" represents bits 0 through 7 of a byte. Here the "byte" is the value in the accumulator. The "C" represents the carry bit of the status register.

The ROL instruction causes each bit to move to the left one place. Bit 7 moves into the carry bit (flag), and the carry bit moves into bit 0.

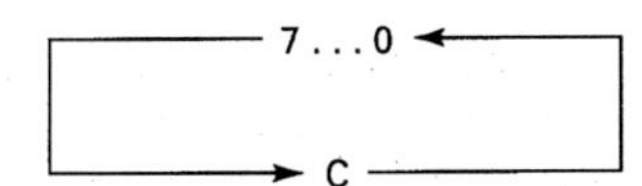

Fig. 20-10 6502 **RO**tate **L**eft instruction.

```
0001  0340           .org $0340
0002  0340           ;
0003  0340 18        CLC
0004  0341 A9 01     LDA #$01
0005  0343 2A        ROL A
0006  0344 2A        ROL A
0007  0345 2A        ROL A
0008  0346 2A        ROL A
0009  0347 2A        ROL A
0010  0348 2A        ROL A
0011  0349 2A        ROL A
0012  034A 2A        ROL A
0013  034B 00        BRK
0014  034C           ;
0015  034C           .end
```

Fig. 20-11 6502 program which rotates left eight times.

The ROR (**RO**tate **R**ight) instruction uses the same concept as the ROL instruction and affects flags in the same way. It simply rotates the bits in the opposite direction.

Figure 20-11 shows a program which clears the carry flag and rotates the accumulator toward the left eight times. If you have a monitor which can single-step ("walk") through the program, cause it to do so, and check the accumulator and carry flag after each step.

If you cannot single-step, then use a break (BRK) instruction after each ROL instruction so that you can observe the movement of the bits in the accumulator. After each BRK you will have to make your trainer or computer begin program execution again at the next ROL instruction to see the shifting action continue.

The ASL and LSR Instructions

The 6502 shift instructions also work as described in the New Concepts section of this chapter. The **A**rithmetic **S**hift **L**eft instruction is shown in Fig. 20-12.

C ←— 7...0 ←— 0

Fig. 20-12 6502 arithmetic shift left instruction.

The **L**ogical **S**hift **R**ight instruction is shown in Fig. 20-13. Both are quite simple.

0 —→ 7...0 —→ C

Fig. 20-13 6502 logical shift right instruction.

An Example

Let's look at the same example which was used in the New Concepts section. Remember, our objective was to clear the upper nibble and then to move the lower nibble of the accumulator into the upper nibble of the accumulator. Figure 20-14 shows our original problem.

Upper nibble	Lower nibble
1 1 0 0	1 1 0 1

Fig. 20-14 Situation in which we want to clear the upper nibble and then move every bit of the lower nibble into the upper nibble.

```
0001  0340            .org $0340
0002  0340            ;
0003  0340 29 0F      AND #$0F          ;mask off upper nibble
0004  0342 18         CLC               ;clear the carry flag
0005  0343 2A         ROL A             ;rotate left four times
0006  0344 2A         ROL A
0007  0345 2A         ROL A
0008  0346 2A         ROL A
0009  0347 00         BRK
0010  0348            ;
0011  0348            .end
```

Fig. 20-15 6502 program which clears the upper nibble of the accumulator and then moves the lower nibble into the upper nibble.

A 6502 program which can solve this problem is shown in Fig. 20-15. Manually place the initial value of CD_{16} in the accumulator before running the program. After the program is run, you should find the value $D0_{16}$ in the accumulator.

20-5 6800/6808 FAMILY

The 6800/6808 has two rotate instructions and three shift instructions.

The ROL/ROLA/ROLB and ROR/RORA/RORB Instructions

The 6800/6808 ROL/ROLA/ROLB instructions work as described in the New Concepts section and as shown in Fig. 20-16. Figure 20-16 is taken from the Rotate and Shift Instructions section of the Expanded Table of 6800/6808 Instructions Listed by Category.

In Fig. 20-16 the "7 . . . 0" represents bits 0 through 7 of a byte. In this case the "byte" is the value in a memory location, accumulator A, or accumulator B. The "C" represents the carry bit of the status register.

The ROL/ROLA/ROLB instructions cause each bit to move to the left one place. Bit 7 moves into the carry bit (flag), and the carry bit moves into bit 0.

The ROR/RORA/RORB (**RO**tate **R**ight) instructions use the same concept as the ROL/ROLA/ROLB instructions and affect flags in the same way. They simply rotate the bits in the opposite direction.

Figure 20-17 shows a program which clears the carry flag and rotates accumulator A toward the left eight times.

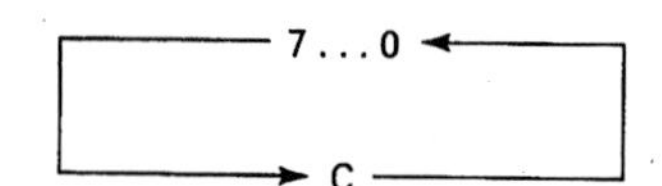

Fig. 20-16 6800/6808 ROL/ROLA/ROLB instructions.

```
0001  0000            .org $0000
0002  0000            ;
0003  0000 0C         CLC
0004  0001 86 01      LDAA #$01
0005  0003 49         ROLA
0006  0004 49         ROLA
0007  0005 49         ROLA
0008  0006 49         ROLA
0009  0007 49         ROLA
0010  0008 49         ROLA
0011  0009 49         ROLA
0012  000A 49         ROLA
0013  000B 3E         WAI
0014  000C            ;
0015  000C            .end
```

Fig. 20-17 6800/6808 program which rotates left eight times.

If you have a monitor which can single-step ("walk") through the program, cause it to do so and check the accumulator and carry flag after each step.

The ASL/ASLA/ASLB, ASR/ASRA/ASRB, and LSR/LSRA/LSRB Instructions

The 6800/6808 shift instructions also work as described in the New Concepts section of this chapter. The **A**rithmetic **S**hift **L**eft instruction is shown in Fig. 20-18.

The **A**rithmetic **S**hift **R**ight instruction is shown in Fig. 20-19. The **L**ogical **S**hift **R**ight instruction is shown in Fig. 20-20. All are quite simple.

C ← 7...0 ← 0

Fig. 20-18 6800/6808 arithmetic shift left instruction.

→ 7...0 → C

Fig. 20-19 6800/6808 arithmetic shift right instruction.

0 → 7...0 → C

Fig. 20-20 6800/6808 logical shift right instruction.

An Example

Let's look at the same example which was used in the New Concepts section. Remember, our objective was to clear the upper nibble and then to move the lower nibble of the accumulator into the upper nibble of the accumulator. Figure 20-21 shows our original problem.

Upper nibble	Lower nibble
1 1 0 0	1 1 0 1

Fig. 20-21 Situation in which we want to clear the upper nibble and then move every bit of the lower nibble into the upper nibble.

A 6800/6808 program which can solve this problem is shown in Fig. 20-22. Manually place the initial value of CD_{16} in the accumulator before running the program. After the program is run, you should find the value $D0_{16}$ in the accumulator.

```
0001  0000          .org $0000
0002  0000          ;
0003  0000 84 0F    ANDA #$0F      ;mask off upper nibble
0004  0002 0C       CLC            ;clear the carry flag
0005  0003 49       ROLA           ;rotate left four times
0006  0004 49       ROLA
0007  0005 49       ROLA
0008  0006 49       ROLA
0009  0007 3E       WAI
0010  0008          ;
0011  0008          .end
```

Fig. 20-22 6800/6808 program which moves the lower nibble of the accumulator into the upper nibble.

20-6 8080/8085/Z80 FAMILY

The 8080 and 8085 have four rotate instructions and no shift instructions. We will place the Z80 form of the instructions in square brackets. (The Z80 does have several multibyte shift instructions which we will not study at this time because the 8080 and 8085 do not share these instructions.)

The RAL [RLA] and RAR [RRA] Instructions

The 8080/8085/Z80 RAL [RLA] (**R**otate **A** **L**eft [**R**otate **L**eft **A**] instructions work as described in the New Concepts section and as shown in Fig. 20-23. Figure 20-23 is taken from the Rotate and Shift Instructions section of the Expanded Table of 8080/8085/Z80 Instructions Listed by Category.

In Fig. 20-23 the "7 . . . 0" represents bits 0 through 7 of a byte. In this case the "byte" is the value in the accumulator. The "C" represents the carry bit of the status register.

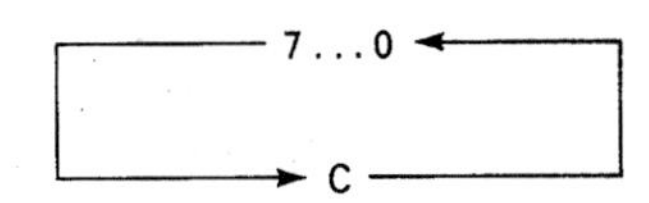

Fig. 20-23 The 8080/8085/Z80 RAL [RLA] instruction.

The RAL [RLA] instruction causes each bit to move to the left one place. Bit 7 moves into the carry bit (flag), and the carry bit moves into bit 0.

The RAR [RRA] instruction uses the same concept as the RAL [RLA] instruction and affects flags in the same way. It simply rotates the bits in the opposite direction.

Figure 20-24 shows a program which clears the carry flag and rotates the accumulator toward the left eight times. If you have a monitor which can single-step ("walk") through the program, cause it to do so and check the accumulator and carry flag after each step.

The RLC [RLCA] and RRC [RRCA] Instructions

The RLC [RLCA] (**R**otate **L**eft with **C**arry [**R**otate **L**eft with **C**arry **A**]) and RRC [RRCA] (**R**otate **R**ight with **C**arry [**R**otate **R**ight with **C**arry **A**]) instructions work just a little differently from the other rotate instructions we have discussed. The RLC [RLCA] instruction is shown in Fig. 20-25.

The RRC [RRCA] instruction is shown in Fig. 20-26.

In the case of the RLC [RLCA] instruction, all bits in the accumulator move toward the left. The bit rotating out of bit 7 goes into the carry flag *and* around into bit 0 of the accumulator.

In the case of the RRC [RRCA] instruction, all bits in the accumulator move toward the right. The bit rotating out of bit 0 goes into the carry flag *and* around into bit 7 of the accumulator.

An Example

Let's look at the same example which was used in the New Concepts section. Remember, our objective was to clear the upper nibble and then to move the lower nibble of the accumulator into the upper nibble of the accumulator. Figure 20-27 shows our original problem.

8080/8085 program

```
0001  1800          .org 1800h
0002  1800          ;
0003  1800 3E 01    MVI A,01H
0004  1802 17       RAL
0005  1803 17       RAL
0006  1804 17       RAL
0007  1805 17       RAL
0008  1806 17       RAL
0009  1807 17       RAL
0010  1808 17       RAL
0011  1809 17       RAL
0012  180A 76       HLT
0013  180B          ;
0014  180B          .end
```

Z80 program

```
0001  1800          .org 1800h
0002  1800          ;
0003  1800 3E 01    LD A,01H
0004  1802 17       RLA
0005  1803 17       RLA
0006  1804 17       RLA
0007  1805 17       RLA
0008  1806 17       RLA
0009  1807 17       RLA
0010  1808 17       RLA
0011  1809 17       RLA
0012  180A 76       HALT
0013  180B          ;
0014  180B          .end
```

Fig. 20-24 8080/8085 and Z80 programs which rotate left eight times.

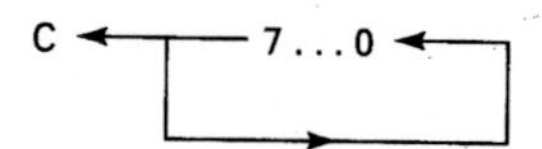

Fig. 20-25 8080/8085/Z80 RLC [RLCA] instruction.

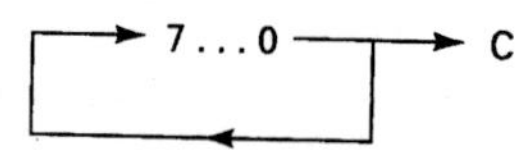

Fig. 20-26 8080/8085/Z80 RRC [RRCA] instruction.

Upper nibble	Lower nibble
1 1 0 0	1 1 0 1

Fig. 20-27 Situation in which we want to clear the upper nibble and then move every bit of the lower nibble into the upper nibble.

An 8080/8085/Z80 program which can solve this problem is shown in Fig. 20-28. Manually place the initial value of CD_{16} in the accumulator before running the program. After the program is run, you should find the value $D0_{16}$ in the accumulator.

20-7 8086/8088 FAMILY

The 8086/8088 has four rotate instructions and three shift instructions. They are discussed starting on the next page.

8080/8085 program

```
0001  1800          .org 1800h
0002  1800          ;
0003  1800 E6 0F    ANI 0FH     ;mask off upper nibble
0004  1802 37       STC         ;set the carry flag then
0005  1803 3F       CMC         ;  complement it
0006  1804 17       RAL         ;rotate left four times
0007  1805 17       RAL
0008  1806 17       RAL
0009  1807 17       RAL
0010  1808 76       HLT
0011  1809          ;
0012  1809          .end
```

Z80 program

```
0001  1800          .org 1800h
0002  1800          ;
0003  1800 E6 0F    AND 0FH     ;mask off upper nibble
0004  1802 37       SCF         ;set the carry flag then
0005  1803 3F       CCF         ;  complement it
0006  1804 17       RLA         ;rotate left four times
0007  1805 17       RLA
0008  1806 17       RLA
0009  1807 17       RLA
0010  1808 76       HALT
0011  1809          ;
0012  1809          .end
```

Fig. 20-28 8080/8085 and Z80 programs which clear the upper nibble of the accumulator and then move the lower nibble into the upper nibble.

The RCL and RCR Instructions

The RCL and RCR instructions work as described in the New Concepts section of this chapter and as shown in Fig. 20-29. Figure 20-29 is taken from the Rotate and Shift Instructions section of the Expanded Table of 8086/8088 Instructions Listed by Category.

In Fig. 20-29 the "MSB . . . LSB" represents bits 0 through 7 of a byte or bits 0 through 15 of a word. The "C" represents the carry bit of the status register.

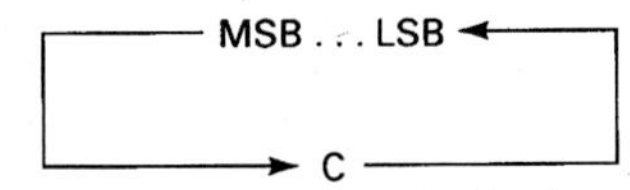

Fig. 20-29 The 8086/8088 RCL and RCR instructions.

The RCL instruction causes each bit to move to the left one place. The MSB moves into the carry bit (flag), and the carry bit moves into the LSB.

The RCR instruction uses the same concept as the RCL instruction and affects flags in the same way. It simply rotates the bits in the opposite direction.

Figure 20-30 shows a program which clears the carry flag and rotates AL toward the left eight times. We then single-step through the program. Follow each step and pay particular attention to AL and the carry flag.

```
C>DEBUG
-r
AX=0000  BX=0000  CX=0000  DX=0000  SP=FFEE  BP=0000  SI=0000  DI=0000
DS=775B  ES=775B  SS=775B  BS=775B  IP=0100   NV UP EI PL NZ NA PO NC
775B:0100 7420          JZ      0122
-
-a
775B:0100 CLC
775B:0101 MOV AL,01
775B:0103 RCL AL,1
775B:0105 RCL AL,1
775B:0107 RCL AL,1
775B:0109 RCL AL,1
775B:010B RCL AL,1
775B:010D RCL AL,1
775B:010F RCL AL,1
775B:0111 RCL AL,1
775B:0113 INT 20
775B:0115
-
-r
AX=0000  BX=0000  CX=0000  DX=0000  SP=FFEE  BP=0000  SI=0000  DI=0000
DS=775B  ES=775B  SS=775B  CS=775B  IP=0100   NV UP EI PL NZ NA PO NC
775B:0100 F8            CLC
-t

AX=0000  BX=0000  CX=0000  DX=0000  SP=FFEE  BP=0000  SI=0000  DI=0000
DS=775B  ES=775B  SS=775B  CS=775B  IP=0101   NV UP EI PL NZ NA PO NC
775B:0101 B001          MOV     AL,01
-t

AX=0001  BX=0000  CX=0000  DX=0000  SP=FFEE  BP=0000  SI=0000  DI=0000
DS=775B  ES=775B  SS=775B  CS=775B  IP=0103   NV UP EI PL NZ NA PO NC
775B:0103 D0D0          RCL     AL,1
-t

AX=0002  BX=0000  CX=0000  DX=0000  SP=FFEE  BP=0000  SI=0000  DI=0000
DS=775B  ES=775B  SS=775B  CS=775B  IP=0105   NV UP EI PL NZ NA PO NC
775B:0105 D0D0          RCL     AL,1
-t

AX=0004  BX=0000  CX=0000  DX=0000  SP=FFEE  BP=0000  SI=0000  DI=0000
DS=775B  ES=775B  SS=775B  CS=775B  IP=0107   NV UP EI PL NZ NA PO NC
775B:0107 D0D0          RCL     AL,1
-t

AX=0008  BX=0000  CX=0000  DX=0000  SP=FFEE  BP=0000  SI=0000  DI=0000
DS=775B  ES=775B  SS=775B  CS=775B  IP=0109   NV UP EI PL NZ NA PO NC
775B:0109 D0D0          RCL     AL,1
-t
```

Fig. 20-30 8086/8088 RCL instruction.

```
AX=0010  BX=0000  CX=0000  DX=0000  SP=FFEE  BP=0000  SI=0000  DI=0000
DS=775B  ES=775B  SS=775B  CS=775B  IP=010B   NV UP EI PL NZ NA PO NC
775B:010B D0D0          RCL     AL,1
-t

AX=0020  BX=0000  CX=0000  DX=0000  SP=FFEE  BP=0000  SI=0000  DI=0000
DS=775B  ES=775B  SS=775B  CS=775B  IP=010D   NV UP EI PL NZ NA PO NC
775B:010D D0D0          RCL     AL,1
-t

AX=0040  BX=0000  CX=0000  DX=0000  SP=FFEE  BP=0000  SI=0000  DI=0000
DS=775B  ES=775B  SS=775B  CS=775B  IP=010F   NV UP EI PL NZ NA PO NC
775B:010F D0D0          RCL     AL,1
-t

AX=0080  BX=0000  CX=0000  DX=0000  SP=FFEE  BP=0000  SI=0000  DI=0000
DS=775B  ES=775B  SS=775B  CS=775B  IP=0111   OV UP EI PL NZ NA PO NC
775B:0111 D0D0          RCL     AL,1
-t

AX=0000  BX=0000  CX=0000  DX=0000  SP=FFEE  BP=0000  SI=0000  DI=0000
DS=775B  ES=775B  SS=775B  CS=775B  IP=0113   OV UP EI PL NZ NA PO CY
775B:0113 CD20          INT     20
-t
```

Fig. 20-30 (cont.)

The ROL and ROR Instructions

The ROL (**RO**tate **L**eft) and ROR (**RO**tate **R**ight) instructions work just a little differently from the other rotate instructions we have discussed. The ROL instruction is shown in Fig. 20-31.

The ROR instruction is shown in Fig. 20-32.

The drawings shown here are slightly different from those shown in the instruction-set description, but if you'll look closely, you'll see that they are really the same.

In the case of the ROL instruction, all bits move toward the left. The bit rotating out of the MSB goes into the carry flag *and* around into the LSB.

In the case of the ROR instruction, all bits move toward the right. The bit rotating out of the LSB goes into the carry flag *and* around into the MSB.

Fig. 20-31 8086/8088 ROL instruction.

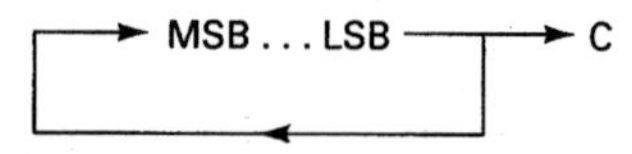

Fig. 20-32 8086/8088 ROR instruction.

The SAL/SHL, SAR, and SHR Instructions

The 8086/8088 shift instructions work as described in the New Concepts Section of this chapter. The Shift Arithmetic Left/**SH**ift logical **L**eft instruction is shown in Fig. 20-33.

The Shift Arithmetic **R**ight instruction is shown in Fig. 20-34. The **SH**ift logical **R**ight instruction is shown in Fig. 20-35. All are quite simple.

C ← MSB . . . LSB ← 0

Fig. 20-33 8086/8088 SAL/SHL instruction.

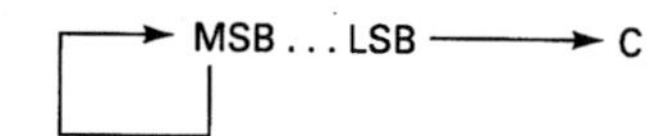

Fig. 20-34 8086/8088 shift arithmetic right instruction.

0 → MSB . . . LSB → C

Fig. 20-35 8086/8088 shift logical right instruction.

An Example

Let's look at the same example which was used in the New Concepts section. Remember, our objective was to clear the upper nibble and then to move the lower nibble of AL into the upper nibble of AL. Figure 20-36 shows our original problem.

An 8086/8088 program which can solve this problem is shown in Fig. 20-37. Manually place the initial value of CD_{16} in AL before running the program. After the program is run, you should find the value $D0_{16}$ in AL.

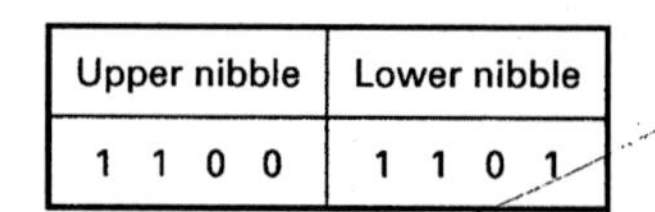

Fig. 20-36 Situation in which we want to clear the upper nibble and then move every bit of the lower nibble into the upper nibble.

```
C>DEBUG
-r
AX=0000  BX=0000  CX=0000  DX=0000  SP=FFEE  BP=0000  SI=0000  DI=0000
DS=77B0  ES=77B0  SS=77B0  BS=77B0  IP=0100   NV UP EI PL NZ NA PO NC
77B0:0100 240F          AND     AL,0F
-
-rax
AX 0000
:00cd
-
-a
77B0:0100 AND AL,0F           ;mask off upper nibble
77B0:0102 CLC                 ;clear the carry flag
77B0:0103 RCL AL,1            :rotate left four times
77B0:0105 RCL AL,1
77B0:0107 RCL AL,1
77B0:0109 RCL AL,1
77B0:010B INT 20
77B0:010D
-
-r
AX=00CD  BX=0000  CX=0000  DX=0000  SP=FFEE  BP=0000  SI=0000  DI=0000
DS=77B0  ES=77B0  SS=77B0  CS=77B0  IP=0100   NV UP EI PL NZ NA PO NC
77B0:0100 240F          AND     AL,0F
-
-g 010b

AX=00D0  BX=0000  CX=0000  DX=0000  SP=FFEE  BP=0000  SI=0000  DI=0000
DS=77B0  ES=77B0  SS=77B0  CS=77B0  IP=010B   OV UP EI PL NZ NA PO NC
77B0:010B CD20          INT     20
-
```

Fig. 20-37 8086/8088 program which clears the upper nibble of AL and then moves the lower nibble into the upper nibble.

SELF-TESTING REVIEW

Read each of the following and provide the missing words. Answers appear at the beginning of the next question.

1. Does rotating or shifting move the carry bit into one of the ends of the affected register? ________
2. *(Rotating)* Does rotating or shifting move 0s into one of the ends of the affected register? ________
3. *(Shifting)* Which of the following instructions duplicates the current value of the most significant bit and makes it the new value of the most significant bit? Rotate right, rotate left, logical shift right, logical shift left, arithmetic shift right, or arithmetic shift left?
(Arithmetic shift right)

PROBLEMS

Specific Microprocessor Families

Solve the following problems using the microprocessor of your choice.

20-1. Write a program which will place the number 34_{16} in the accumulator, clear the lower nibble (F), and then move the upper nibble (C) into the lower nibble by using a rotate instruction. (Write the program so that if the carry flag happens to be set (1) prior to running the program, it will *not* rotate the 1 from the carry flag into the upper nibble of the accumulator.)

20-2. The ASCII value for numbers is the same as the hex value for numbers except that the ASCII value has a 3 as a prefix. For example, the ASCII value for 0 is 30, the ASCII value for 1 is 31, the ASCII value for 2 is 32, the ASCII value for 3 is 33, and so on.

Write a program that will place the hex value 23 in the accumulator and will then take the upper nibble (2h), change it to its ASCII value (32h), and store it in a memory location. The program should then take the lower nibble (3h), change it to its ASCII value (33h), and store it in another memory location.

Restrictions: (1) You cannot use shift instructions (but you may use rotate instructions). (2) You must make the program so that it will work for any original value, not just 23h. (That value was picked randomly.)

Hints: (1) You should store the original value (23) in a memory location so that you can use it more than once. (2) You will need to use rotate instructions, masks, and arithmetic instructions. (3) You need to set aside three memory locations: one for the original value (23h), one for theASCII value for 2 (32h), and one for the ASCII value for 3 (33h).

20-3. Place the ASCII value for 8 (38h) in one memory location and the ASCII value for 9 (39h) in another location. Then write a program which will take these two ASCII values, convert them to their hex equivalents (8h and 9h), and combine them into a 1-byte, 2-digit, hex number (89h).

20-4. Since the value of a binary digit doubles in value each time it is moved to the left by one place, and becomes one-half of its previous value each time it is moved to the right one place, it is possible to multiply and divide by shifting/rotating.

Write a program which will load the value $1C_{16}$ into the accumulator and multiply it by 8 by shifting it.

6502, 6800/6808, and 8086/8088 users: You should use the arithmetic shift left type of instruction because it automatically shifts 0s into the least significant bit.

8086/8088 users: You have an actual multiply instruction but shouldn't use it for this program, since this chapter is intended to help you write programs using shift and rotate instructions.

Z80 users: You have an arithmetic shift left type of instruction, but you cannot use it here because it is not part of the 8080/8085 instruction subset. Use the following procedure for the 8080/8085.

8080/8085 users: You do not have any shift instructions; therefore, you should alternately clear the carry flag and rotate to achieve an effect similar to that of the arithmetic shift left instruction.

All users: There are other ways to multiply. This simply illustrates one way, and not necessarily the best or easiest for your particular microprocessor.

ADDRESSING MODES—II

New Concepts

In this chapter we'll study some of the more complex addressing modes. The different microprocessor families will show more variation at this point than they did in our earlier chapter on basic addressing modes.

The 6502 has more addressing modes than any other 8-bit microprocessor. Some are used quite often, but several are used with only a few instructions. Since the 6502 has no general-purpose registers and only one accumulator, it must use memory very often and is therefore said to have a *memory-intensive* architecture.

The 6800/6808 has a moderate number of different addressing modes, and students learning about it should not have difficulty. The 6800/6808 also lacks general-purpose registers but does have two accumulators. It is also considered to have a memory-intensive architecture.

The 8080/8085 has the fewest number of addressing modes of any of the 8-bit microprocessors. Students will find it easiest to learn in this respect. (The Z80 has more addressing modes, but those beyond the ones the 8080/8085 has will not be studied at this time.) The 8080/8085 has six general-purpose registers in addition to an accumulator and is therefore said to have a *register-intensive* architecture.

The 8086/8088, being a successor to and relative of the 8080/8085, has many general-purpose registers. Because it is a 16-bit microprocessor, it also has many addressing modes.

To summarize, the 6502 has 56 different instructions which use one or more of 13 addressing modes. When you combine the instructions and addressing modes, you produce 152 different op codes.

The 6800/6808 has 107 different instructions which use one or more of seven addressing modes. The 6800/6808 has 197 different op codes.

The 8080/8085 has 246 different instructions which have only one addressing mode each. There are five different addressing modes. This provides a total of 246 different op codes.

The 8086/8088 has 24 addressing modes (they are presented in 11 addressing-mode categories in this text) and approximately 91 different assembly-language instructions. This is just part of the picture, however.

Each 8086/8088 instruction can have many variations, the **MOV**e instruction probably being the best example. MOV is considered one assembly-language instruction; yet the 8086/8088 recognizes 28 different assembly-language forms of the MOV instruction (*move to a register, move immediate, move byte to memory, move word to register*, and so on). Each of the 28 assembly-language forms can have many different machine-level instructions which may be composed of up to 6 bytes (with eight 8-bit registers; the ability to move any one of them to any other produces 10s of different machine-level instructions just for moving 8-bit registers).

To put it simply, there are hundreds of variations of the MOV instruction alone. The possible variations of all 91 different assembly-language instructions number somewhere between 3,000 and 4,000.

How can anyone learn so many combinations? First, if you are using the 8086 or 8088, you will be concentrating on learning about the 91 different assembly-language instructions, not every possible variation. Second, once you learn any one instruction, MOV, for example, most of the variations will seem very natural. It's not like rote memorization.

Which microprocessor is easiest to learn? That's hard to say. They each have strengths and weaknesses. And which feature is a strength and which is a weakness depend on what you as the programmer want to do.

(*Note:* Do not try to memorize all of these addressing modes at this time. Read this chapter and then refer back to it as you need to in the chapters to come.)

(*Additional Note:* Reference will be made in this chapter to concepts and instructions which have not yet been covered. This is necessary to explain the various advanced addressing modes. This method of organizing the text has the great advantage of placing all necessary information regarding addressing modes in two easy-to-locate chapters.)

21-1 ADVANCED ADDRESSING MODES

Some addressing modes which will be described in this chapter use a multistep process to find the address of the data or the next instruction to be executed. There may be one or more intermediate addresses, but the final address at which the data or instruction is to be found will be referred to as the *effective address*.

There are three fundamental advanced addressing modes, although some microprocessors also feature variations of these three.

Relative Addressing

Relative addressing is a mode in which your destination is described relative to where you are now. You aren't directed to an absolute memory location but rather to an address higher or lower than where you are now.

This form of addressing is not used to describe where to find data but rather where the program should find its next instruction. But let's back up just a bit.

In an earlier chapter we described the program counter and its function (the 8086/8088 uses the term *instruction pointer* instead of program counter). It keeps track of the next memory location to be accessed. Normally the locations are taken in order. The microprocessor gets an instruction, goes to the next byte in memory to get the next instruction or data, then to the next, and so forth. Sometimes, however, we need to "jump" or "branch" to a different area in memory to get our next instruction, for example, when we want to repeat a section of the program. (This saves time compared to writing a portion of a program many times if it is to be executed many times.)

Relative addressing involves 2 bytes (on 8-bit microprocessors). The first is the op code for the jump or branch instruction. The second byte tells how far and in what direction the microprocessor should jump. The second byte is a signed binary number—that is, it can be positive or negative. If it's positive, the microprocessor jumps forward in memory (to a higher-numbered address). If it's negative, it jumps backward (to a lower-numbered address). There is a limit, however, to how far you can jump with this form of addressing. On 8-bit microprocessors the range is from -128_{10} to $+127_{10}$ bytes. On 16-bit microprocessors the range is from $-32{,}768_{10}$ to $+32{,}767_{10}$ bytes.

The next task is to determine exactly what point we start counting from. For example, if we tell the microprocessor to jump forward 10 memory locations, where do we start counting from? We must again look at the program counter.

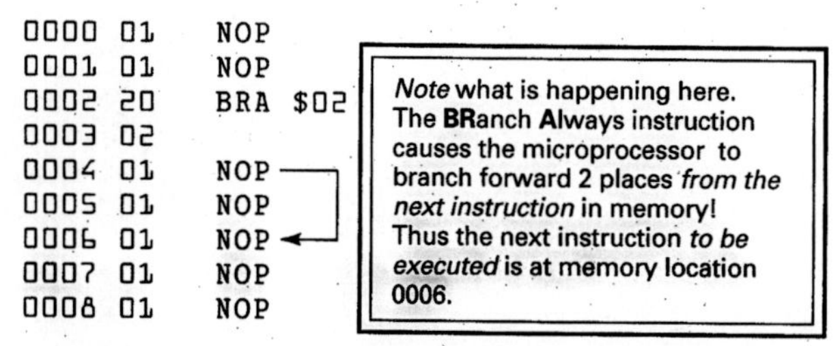

Fig. 21-1 An example of relative branching forward using the 6800/6808.

The program counter always points to the *next* memory location to be accessed. In the case of relative jumps, it points to the next instruction after the jump instruction.

We start counting from the memory location being pointed to by the program counter when the jump instruction is being executed. This memory location is *not* the location of the jump instruction itself, and it is *not* the byte after the jump instruction, but is the *next instruction* in memory, which is usually two memory locations *after* the jump instruction.

Let's look at an example. Refer to Fig. 21-1. The 6800/6808 has an instruction called BRA (**BR**anch **A**lways), which uses relative addressing.

The four-digit numbers in the left column are memory addresses. The two-digit numbers in the next column are op codes. The third column contains the assembly-language mnemonics. Memory location 0002 contains the op code 20, which is the op code for the BRA instruction. The next memory location, 0003, contains the number 02, which is the same 02 referred to in the BRA $02 instruction.

The NOPs are simply dummy instructions placed there, in this example, so that we have something to skip over when the branch is implemented. Again, memory address 0002 contains the op code for BRA, which is 20. Address 0003 contains the number of places we wish to move *relative to where the program counter will be while it's executing this instruction!* Since the program counter is always pointing to the next instruction in memory, it will contain 0004. $0004_{16} + 02_{16} = 0006_{16}$. This is the next instruction to be executed.

Now let's try branching backward. Figure 21-2 shows an example.

At this point a review of 2's-complement negative numbers may be in order. Remember the odometer? Let's look at it again, in decimal first.

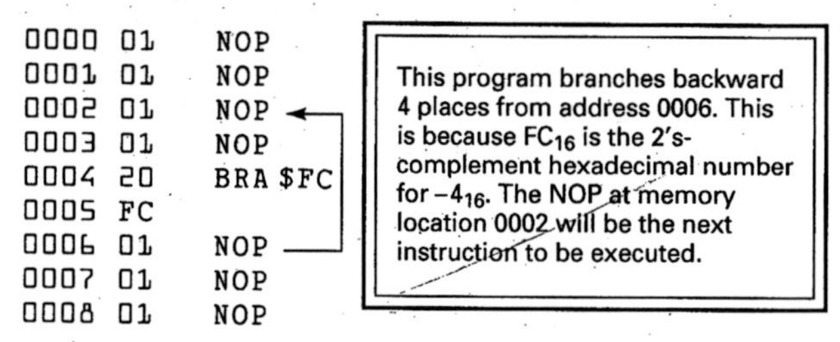

Fig. 21-2 An example of relative branching backward using the 6800/6808.

Negative 2's-Complement Numbers

Let's say you buy a brand-new car and the odometer reads 00,000. Now suppose your odometer rolls forward if the car drives forward, and rolls backward if the car drives backward. Let's drive backward from 00,000.

00,000
99,999
99,998
99,997
99,996

We could say that driving backward is like creating negative numbers: 99,999 is 1 mile less than 00,000. What's 1 less than 0? Minus one, of course. 99,998 is 2 miles less than 00,000. What's 2 less than 0? Minus two is. Let's look at some odometer readings from driving backward and their negative equivalents, along with some odometer readings from driving forward and their positive equivalents.

00,003	+3
00,002	+2
00,001	+1
00,000	0
99,999	−1
99,998	−2
99,997	−3
99,996	−4

Now let's show the same situation with a 1-byte hexadecimal odometer.

03	+3
02	+2
01	+1
00	0
FF	−1
FE	−2
FD	−3
FC	−4

Now look at Fig. 21-2 again. Do you see where the FC came from? It's −4.

What if you had to have a negative number like -40_{10}? Counting backward in hexadecimal would require too much time. There are several options. First, experiment with your calculator. Most scientific calculators now convert numbers back and forth between decimal, binary, octal, and hexadecimal. Many even do calculations in all number bases. Try entering -4_{10} and converting it to hexadecimal. If the calculator handles negative conversions, you'll get many F's and a C at the end. Simply ignore all the leading F's and use just the last two digits, the final FC.

If your calculator does conversions between decimal and hexadecimal but won't handle negative numbers, you can use another technique. A two-digit hexadecimal number is made up of 8 binary bits, each representing a power of 2. Find 2^8 and then subtract the number you wish to make negative. In the case of −4, for instance, take $2^8_{10} - 4_{10} = 252_{10}$. Now convert 252_{10} to hexadecimal; it should be FC. (To do the same thing with a 16-bit number, use 2^{16} instead of 2^8.)

Or, should no calculator be handy at the time, use the technique described in Chap. 6, that of taking the 2's complement of the number you wish to make negative. In the case of −4 it looks like this:

0000 0100	+4
1111 1011	1's complement (invert all bits)
+ 1	add 1
1111 1100	2's complement for −4
↓ ↓	
F C	converted to hexadecimal

Indirect Addressing

Indirect addressing is an addressing mode in which the *data* does not appear after the op code (as in immediate addressing), nor does *its memory location* appear after the op code (as in direct addressing), but rather a memory location follows the op code, and in this location is *another* address where the data may be found. It's like finding the address of an address. (*Indirect addressing* is indeed a fitting name.)

There are two basic types of indirect addressing: absolute indirect addressing and register indirect addressing. The 6502 uses absolute indirect addressing. The 8080/8085/Z80 uses register indirect addressing. The 8086/8088 uses register indirect addressing for data and program indirect addressing for jumps (which we'll study later). The 6800/6808 has no indirect addressing (indirect addressing was added to the 6809).

Let's look at an example of this addressing mode and then develop the topic further in the Specific Microprocessor Families section of this chapter. The 6502 has an instruction which looks like this

JMP ($aaaa)

which means **JuMP** indirect (indicated by the parentheses) to the address indicated by aaaa. If the address were 1000_{16}, it would be written as

JMP ($1000)

This tells us that at memory location 1000 and 1001 we can find the address the microprocessor should jump to. The address found at these two locations is loaded into the program counter. (It takes two locations because addresses in the 6502 are 16 bits wide but memory locations are only 8 bits wide.)

Indexed Addressing

Indexed addressing involves using a register called an *index register*, with a number called an *offset*, to calculate the address where the data is located. Let's look at an example using the 6800/6808.

One version of the 6800/6808's load accumulator A instruction looks like this

LDAA $ff,X

which means

> **LoaD** Accumulator **A** with the value in the memory location found by adding the contents of the X register to the hexadecimal offset ff.

For example, if the X register contains the number 1000_{16} and the instruction is written as

LDAA $22,X

we calculate the address where the data is located in this way

$$X + ff = \text{address}$$
$$1000_{16} + 22_{16} = 1022_{16}$$

The microprocessor then goes to address 1022 and places a copy of its contents in accumulator A.

You might be curious as to why we would want an addressing mode like this. One reason is its usefulness in accessing individual pieces of data in a data table. The index register can be incremented (increased by 1) or decremented (decreased by 1) easily, allowing the programmer to access each item in the table.

The 6502 microprocessor has two index registers, the X register and the Y register, and it has six different types of indexed addressing! The 6800/6808 has only one index register, the X register, with only one type of indexed addressing. The 8080/8085 has no index registers at all (the Z80 has two, X and Y) and has no indexed addressing mode. The 8086/8088 has two index registers, the source index and the destination index, and has several types of indexed addressing.

Specific Microprocessor Families

Go to the section which discusses your particular microprocessor.

21-2 6502 FAMILY

The 6502's numerous addressing modes make it unusual among 8-bit microprocessors. It has 13 different addressing modes. Allow us to offer a few words of encouragement at this time.

First, don't expect everything to make sense in the beginning. It takes time before all these new concepts become clear and you feel comfortable with them. Incidentally, the subject of addressing modes is the *only* difficult aspect of the 6502. In fact, the 6502 has the fewest different *instructions* of any of the 8-bit microprocessors—only 56 (the 6800/6808 has 107; the 8080/8085 has 246).

Relative Addressing

The relative addressing mode occurs in only one category of 6502 instruction, the Conditional Jump (Branch) category. Look at that section of the Expanded Table of 6502 Instructions Listed by Category. No other category uses this type of addressing, and this category uses no other type of addressing.

The subject of branching is coming in a later chapter, but it is necessary to discuss branching instructions for a moment to continue our coverage of the relative addressing mode.

The status register is where the 6502's flags are located. They keep track of certain events. If the result of the last calculation were 0, for instance, the zero flag bit would contain a 1. If we wanted to know whether the last result was a 0, we would check the zero flag. A 1 would mean yes, and a 0 would mean no. If we wanted the program to perform one action if the result of the last operation was a 0, and another if the result of the last operation was not a 0, we would write our program so that it would check the zero flag.

Let's look at the BEQ instruction. The assembler notation looks like this

BEQ $rr

which means

> **B**ranch rr bytes from where the program counter is now and do what it says to do there if the result of the last operation was **EQ**ual to 0.

You'll notice that the Operation column of the instruction table has a shorter version of that description.

Let's look at a program fragment. Refer to Fig. 21-3.

After the BEQ instruction and its operand in locations 0007 and 0008 have been fetched, the program counter will have already incremented to 0009, which is where we start counting for the branch (jump).

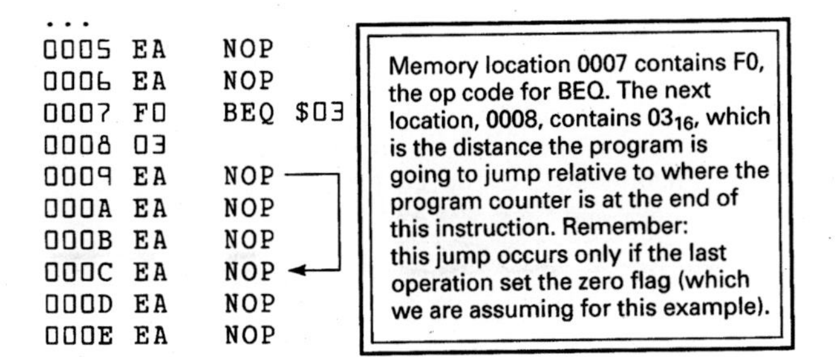

Fig. 21-3 6502 example of relative addressing. *Note:* The zero flag is assumed to be set from a previous operation.

Refer back to the New Concepts section of this chapter to see how a backward branch or jump would work and how to use 2's-complement negative numbers.

Indirect Addressing

There is only one 6502 instruction which uses the indirect addressing mode. That instruction is the JMP instruction, which is found in the Unconditional Jump Instructions category in the Expanded Table of 6502 Instructions Listed by Category.

This particular instruction can be used with two different addressing modes. In the absolute addressing mode, the microprocessor simply jumps to the specified address. When written this way

JMP $aaaa

it means

> **JuMP** to address $aaaa_{16}$ and continue program execution from that point.

In the indirect addressing mode, however, it would be written this way

JMP ($aaaa)

and would mean

> **JuMP** to the address which can be found at memory location aaaa and aaaa + 1.

```
0000 6C    JMP ($0004)
0001 04
0002 00
0003 EA    NOP
0004 1F<-----------low byte----|
0005 01<-----------high byte---|
...
011F next instruction    <-----
0120
...
```

This is where the effective address is being stored. 011F is placed in the program counter.

This is the location of the next instruction to be executed.

Fig. 21-4 Example of 6502 indirect addressing mode.

This would load the contents of memory location aaaa into the low byte of the program counter (PC_L). The contents of memory location aaaa + 1 would be loaded into the high byte of the program counter (PC_H). (This reverse low-byte/high-byte order is normal for the 6502.)

Let's look at an example. If you refer to Fig. 21-4, you will see that the instruction

JMP ($0004)

does not mean that address 0004 is where the program is supposed to jump to, but rather that location 0004 contains the address it's supposed to jump to.

Indexed Addressing

Indexed addressing is the subject of the remainder of this 6502 section. There are four basic indexed addressing modes, and two more which use a mixture of indexed and indirect addressing.

It should be noted that while the 6502 family has a great number of addressing modes which use the index registers, it is the only family which has index registers which are only 8-bits wide. The 6800/6808, Z80, and 8086/8088 all have 16-bit index registers. Keep this in mind if you use the 6502 in addition to one of the other microprocessors.

Zero Page,X and Absolute,X Addressing

You may remember from the New Concepts section of this chapter that the 6502 has two index registers, X and Y, and six different forms of indexed addressing. Here are the first two of the six forms. The difference between these two forms is the range of addresses possible.

These first two forms, and the next two, are so similar to the description in the New Concepts section that you will probably have little difficulty understanding them. If you don't remember how the indexed form of addressing works, go back and reread the description now.

Look in the Data Transfer Instructions category of the Expanded Table of 6502 Instructions Listed by Category. We will use the LDA instruction to illustrate the *zero page,X* and *absolute,X* addressing modes.

First notice the Assembler Notation column for the zero page,X and absolute,X forms of the LDA instruction. For these two the assembler notation is

LDA $ff,X ← zero page,X
LDA $ffff,X ← absolute,X

In both cases the offset (ff or ffff) is a hexadecimal number which is going to be added to the value in the X register. The sum of these two values provides the address of the data which is to be loaded into the accumulator.

For example, if the X register contained the hexadecimal number 10, the instruction

LDA $034E,X

would add those two values,

$$034E_{16} + 10_{16} = 035E_{16}$$

and place a copy of the contents of memory location $035E_{16}$ in the accumulator.

When zero page,X addressing is used, the offset (the number being added to the X register) is two hex digits wide and the X register is also two hex digits wide. Two hex digits can address memory locations only in page 0 (00_{16} to FF_{16}). When this addressing mode is used, it is assumed that the data is somewhere in page 0. *If the sum of the offset and the X register is greater than FF_{16} then the most significant digit is truncated and only the first two digits are used!* For example, if the X register contained FF, the instruction

LDA $04,X

would add the offset to the X register

$$04_{16} + FF_{16} = 103_{16} \quad \textit{(The 1 will be dropped.)}$$

so the data will be retrieved from location 03_{16}! Numbers larger than FF_{16} *wrap around* to the beginning of page 0.

When absolute,X addressing is used, the offset is a four-digit hexadecimal number ranging from 0000_{16} to $FFFF_{16}$. This allows the data to be located anywhere in the entire 6502 address range. If the sum of the offset and the X register exceeds $FFFF_{16}$, then the microprocessor again performs a *wraparound* back to 0000_{16}.

Zero Page,Y and Absolute,Y Addressing

Notice in the Data Transfer Instructions section of the Expanded Table of 6502 Instructions Listed by Category that the LDX instruction uses both *absolute,Y* and *zero page,Y* addressing. These work exactly the same as absolute,X and zero page,X, except that they use the Y register instead.

The absolute,X, absolute,Y, and zero page,X addressing modes are used by many 6502 instructions. Zero page,Y addressing is used by only two instructions, however—LDX and STX.

Indirect Indexed Addressing

Indirect indexed addressing, as the name implies, is a mixture of indirect addressing and indexed addressing. Notice that the word "indirect" is first, and the word "indexed" is next. In this form of addressing, the indirect part of the address calculation is accomplished first; then the indexing is taken into consideration.

Refer to this form of the LDA instruction in the Data Transfer Instructions section of the Expanded Table of 6502 Instructions Listed by Category. Remember the word order—*indirect*, then *indexed*; and notice the assembler notation—LDA ($aa),Y.

To understand the assembler notation for this form of addressing, it helps to remember one of the rules of algebra. In algebra, expressions are read from left to right, and when parentheses are encountered, they are read from the inside to the outside. Let's look at an example.

LDA ($aa),Y

The $aa stands for a two-digit hexadecimal address. Because only two digits are allowed, this address must be between 00_{16} and FF_{16}. At this address, and the one following it (aa and aa + 1), is a 16-bit address stored in reverse low-byte/high-byte order. This address is then added to the Y register to produce the actual (effective) address where the operand (data) is stored. Notice that we worked our way from left to right and from the inside toward the outside as we analyzed this instruction.

For example, let's say that

Y register = 10_{16}
memory location 2D = 00
memory location 2E = C0

If we write the instruction

LDA ($2D),Y

the microprocessor will look in addresses 2D and 2E and use their contents to form another address, C000. It will then take the number $C000_{16}$ and add it to the Y register:

$$C000_{16} + 10_{16} = C010_{16}$$

$C010_{16}$ is where the data is actually stored.

To summarize,

LDA ($aa),Y

means

> **LoaD** the Accumulator with the contents of an address formed by adding the contents of memory location aa and aa + 1 (low-byte/high-byte order) to the Y register.

Indexed Indirect Addressing

This form of addressing is also a mixture of indexed and indirect addressing, but it is the reverse of the previous indirect indexed addressing.

It will be helpful here, as in the previous explanation, to think of how algebraic expressions are written, from left to right and from the inside to the outside.

We will again use the LDA instruction. Look at the indexed indirect form of this instruction. In the Assembler Notation column it appears as

LDA ($ff,X)

In this form of addressing, the microprocessor takes the two-digit offset (ff_{16}) and then adds it to the value found in the X register. (If the sum of ff and X is greater than FF_{16}, the sum will be truncated so that only the two least significant digits remain.) The address formed by the sum of ff and the X register and the following address contain the effective address stored in reverse low-byte/high-byte order.

Let's try an example. If

$$\text{X register} = 10_{16}$$

and we write the instruction

LDA ($11,X)

then the microprocessor will add 11_{16} to the X register

$$11_{16} + 10_{16} = 21_{16}$$

creating the address 21_{16}. *However, this is not where the operand (data) is stored!* At addresses 21_{16} and 22_{16} the effective address is stored in reverse low-byte/high-byte order. So if

memory location 21 = 00
memory location 22 = C0

then the address $C000_{16}$ is created. *Memory address $C000_{16}$ does contain the operand!*

To summarize,

LDA ($ff,X)

means

> **LoaD** the Accumulator with the contents of the memory location *pointed to* by the contents of memory location ff + X and ff + X + 1.

21-3 6800/6808 FAMILY

The 6800/6808 microprocessor has only two addressing modes which must be covered in this chapter—relative addressing and indexed addressing. (The 6800/6808 has no form of indirect addressing.)

Relative Addressing

The 6800/6808 uses relative addressing with all of its branch instructions. These fall into three instruction categories, Unconditional Jump (Branch) Instructions, Conditional Jump (Branch) Instructions, and Subroutine Instructions. This form of addressing works exactly as described in the New Concepts section of this chapter. (In fact, the 6800/6808 was used as our example in that section.)

Let's go over this mode again by using the program fragment in Fig. 21-5.

Since 02_{16} is a positive number, we branch forward by that many spaces *starting with the memory location which will be pointed to by the program counter after the BRA instruction and its operand have been fetched.*

It is important to remember that the BRA operand is a 2's-complement signed binary number and thus can be either negative or positive within a range from $+127_{10}$ to -128_{10}. A negative number indicates a backward branch, and a positive number indicates a forward branch.

Indexed Addressing

The subject of indexed addressing, as discussed in the New Concepts section, was illustrated by using the 6800/6808. We present that information again here for your convenience.

```
0010 20    BRA $(02)
0011 02
0012 01    NOP---+
0013 01    NOP   |
0014 01    NOP<--+
0015 01    NOP
```

Fig. 21-5 An example of relative addressing.

One version of the 6800/6808's load accumulator A instruction looks like this

LDAA $ff,X

which means

> **LoaD** Accumulator **A** with the value in the memory location found by adding the contents of the X register to the hexadecimal offset ff.

For example, if the X register contained the number 1000_{16} and the instruction were written as

LDA $22,X

we would calculate the address where the data was located in this way:

$$X + ff = \text{address}$$
$$1000_{16} + 22_{16} = 1022_{16}$$

We would go to address 1022 and place a copy of its contents in accumulator A.

21-4 8080/8085/Z80 FAMILY

The 8080/8085 microprocessor is easier to learn in some respects than the other 8-bit microprocessors. One reason is that the 8080/8085 has the fewest number of addressing modes. And while the 8080/8085 has the most number of different instructions (246, in contrast to the 6502 with only 56 and the 6800/6808 with 107), each instruction works with only one addressing mode (in contrast to the 6502, which has some instructions which operate in as many as eight different addressing modes).

As we talk about the 8080/8085/Z80 family, you should remember that although the Z80 is treated as a part of the 8080/8085 family in this text, it is a significantly enhanced member of the 8080/8085 family. It has many multibyte instructions and several addressing modes which the 8080/8085 does not have. At this time we will cover only those aspects of the Z80 which it has in common with the 8080/8085.

Register Indirect Addressing

The only advanced addressing mode which the 8080/8085 has is register indirect addressing. Although indirect addressing was covered in the New Concepts section of this chapter, register indirect addressing was not covered since it is a variation of indirect addressing which, among the 8-bit microprocessors, is unique to this family.

Register indirect addressing uses the contents of a 16-bit register pair (most often the HL register pair) as a pointer for the operand.

For example, refer to the Data Transfer Instructions section of the Expanded Table of 8085/8080 and Z80 (8080 Subset) Instructions Listed by Category and look at the MOV A,M [Z80 = LD A,(HL)] instruction. (The MOV A,M instruction is the eighth instruction in this category.) The 8085 form is written

MOV A,M

which means

> **MOV**e to the **A**ccumulator the number found at the **M**emory location pointed to by the HL register pair.

The Z80 form is written

LD A,(HL)

which means

> **LoaD** the **A**ccumulator with the number found at the memory location pointed to (parens) by the **HL** register pair.

which says the same thing the 8085 form did but in different words.

To give an example, if

$$\text{register pair HL} = 1000_{16}$$

and you entered MOV A,M [Z80 LD A,(HL)] into your assembler, the microprocessor would go to memory location 1000_{16} and place a copy of its contents in the accumulator.

There are a few occasions when either the BC or the DE register pair is used instead of the HL pair. You may want to page through the Expanded Table of 8085/8080 and Z80 (8080 Subset) Instructions Listed by Category to see some of the instructions that use this addressing mode.

21-5 8086/8088 FAMILY

Because the 8086/8088 is a 16-bit microprocessor, it uses a greater number of addressing modes than the 8-bit microprocessors, and the modes are more complex. We covered the basic 8086/8088 addressing modes in a previous chapter and will try to give a simple, yet sufficiently complete description of each of the advanced modes at this time.

Register Relative Addressing

Register relative addressing uses two numbers, added together, to determine the address of the source. This form of addressing is especially useful in addressing arrays (tables of data).

Some examples of register relative addressing using the format used by DEBUG (an MS-DOS utility which helps to "debug" programs and includes an assembler and disassembler) are

```
MOV    AL,[BX+0100]
MOV    AX,[DI+0200]
MOV    [SI+0500],CL
MOV    [BP+20],BL
MOV    DI,[BX+0400]
```

Figure 21-6 illustrates how this form of addressing works. The instruction

```
MOV    AL,[BX+0100]
```

is used as an example. Notice first the brackets surrounding the BX+0100. This is required by DEBUG and indicates that the two numbers added together (the value in register BX + 0100_{16}) will point to the location of the data being moved to AL.

We can use the number in the source (0100) to indicate the location of the beginning of the table. The value in the register indicated in the source operand tells us which item in the table is the desired data item.

Notice in Fig. 21-6 that 0100 is the beginning of the table and that 03 (the value in BX) is the data item we need. We need the fourth item in the table starting at address 0100. The contents of memory location 0103 (E3) have been copied to register AL.

It is important to remember that we have added the displacement (0100) to the value in the indicated register (BX) to form an address (0103) *in the current data segment!*

Program Relative Addressing

Program relative addressing is used with JMP and CALL instructions. This mode specifies where the next program instruction is located without using absolute addressing. This allows you to write relocatable assembly-language programs.

Figure 21-7 shows an 8086/8088 instruction which is *not* using program relative addressing. (We'll show you program relative addressing in a moment.) This figure is using direct addressing. We have listed the same line of code three times.

The first line shows the code as it appeared on our computer after being disassembled by DEBUG.

The second line shows DEBUG's disassembly broken into its major components. The *address* is the address of the current memory location. We did not type the address; DEBUG picked that address for us. The *machine code* contains the actual bits which will tell the 8086/8088 what to do. The *assembly language* is what we typed in when using DEBUG.

The third line shows even greater detail. Notice that the code segment the program is to jump to (8888) and location within the segment (0100) are actually contained in the machine code (the bytes are reversed).

Figure 21-8 shows a JMP instruction written using DEBUG which does use program relative addressing, instead of direct addressing as in Fig. 21-7.

Line one shows the information as it appeared on our screen when disassembled by DEBUG.

Line two illustrates the major components of the disassembly. We typed in the assembly language, and DEBUG provided us with the machine code.

The third line shows the components in greater detail. The most interesting fact is that the address we specified as our target address is not the same as the address DEBUG generated. Let's see what DEBUG did.

The JMP op code, EB, is in memory location 0100 as indicated in the "location within segment" portion of the

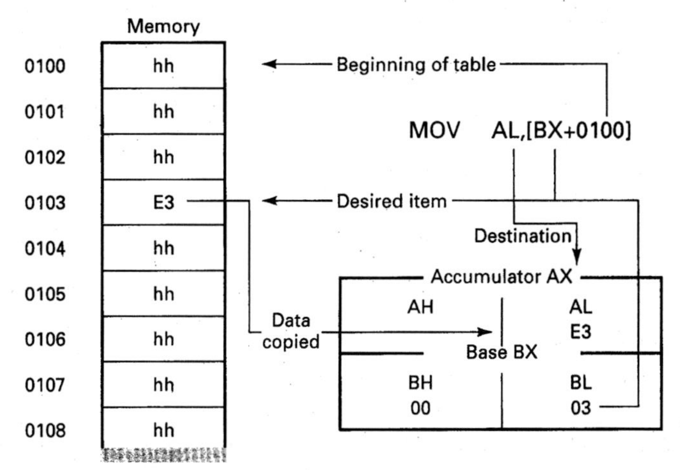

Fig. 21-6 Register relative addressing.

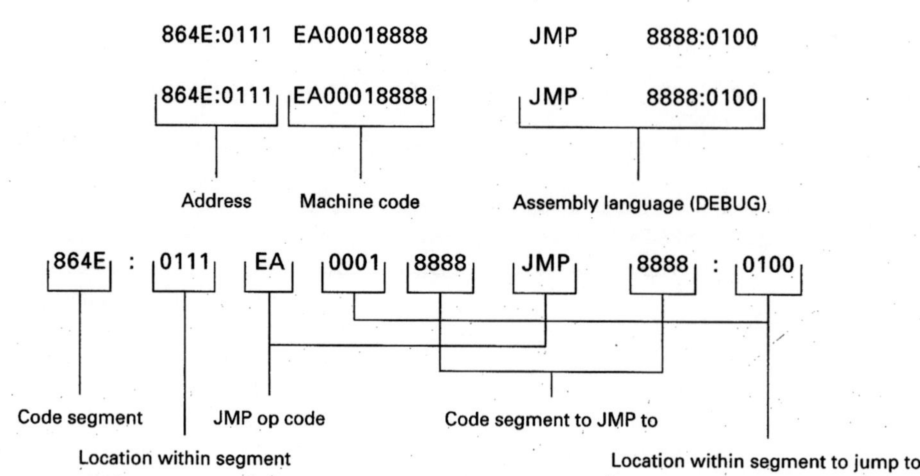

Fig. 21-7 Direct addressing.

line. That means the next *byte*, 0E, is in address 0101. (DEBUG does not show the 0101.) Therefore, the next *instruction* is at memory location 0102.

How far is it from memory location 0102 to our target address of 0110? Remember, these are *hexadecimal* numbers.

$$0110_{16} - 0102_{16} = E_{16}$$

To reach the target address of 0110, the microprocessor will have to jump forward a number of spaces from the point (the instruction) at which the instruction pointer is pointing when this instruction is executed; the number of spaces is E_{16}. The 0E in Fig. 21-8 was calculated by DEBUG as the position of our target *relative* to where the instruction pointer will be when this instruction is being executed.

Relative addressing tells the microprocessor how far to jump forward or backward from the instruction after the JMP instruction. The next instruction is used because the instruction pointer always points to the *next* instruction *to be* executed.

A positive relative address signifies a jump forward; a negative relative address signifies a jump backward.

Register Indirect Addressing

Register indirect addressing uses a register to point to a memory location rather than specifying that location directly. BX, BP, SI, and DI are used as pointers. All of them except BP point to locations in the *data segment*; BP points to a location in the *stack segment*. The registers can point to either the source or the destination operand.

An assembly-language instruction which uses indirect addressing is shown in Fig. 21-9.

The format of the instruction line in bold print in Fig. 21-9 is the format that DEBUG uses. (The code segment on your computer will probably not be the same as the one shown in Fig. 21-9.)

Most of the different components of the instruction line in bold have been identified in the figure. [BX] is labeled as the source. The brackets around BX indicate that the operand is not the contents of BX; rather the operand will be found at the address *pointed to* by BX.

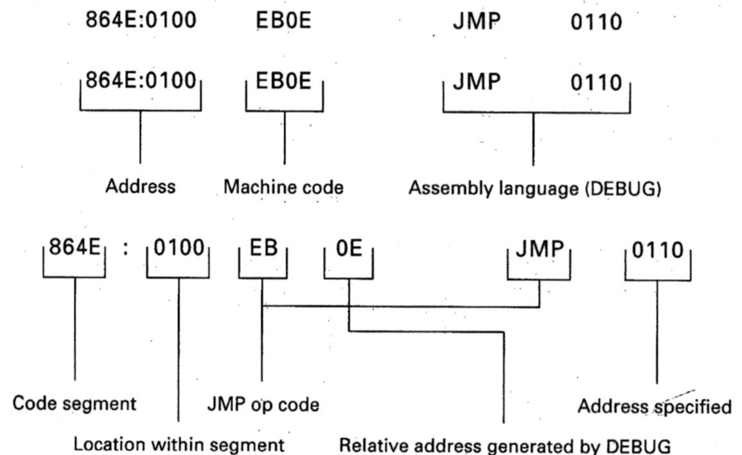

Fig. 21-8 Program relative addressing.

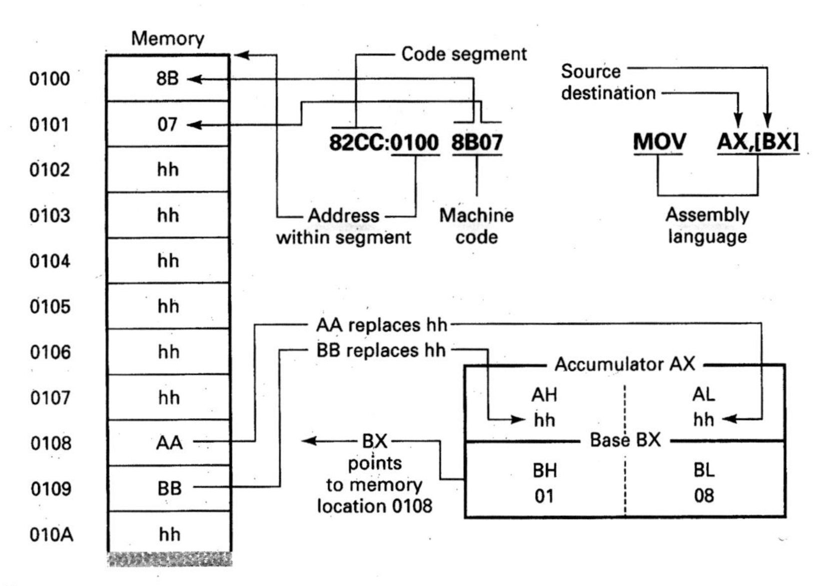

Fig. 21-9 Register indirect addressing.

If you look at the contents of BX, you will see the value 0108. That means that the actual operand is in memory location 0108. In this case we are moving a 16-bit word rather than an 8-bit byte. Since it takes two memory locations to hold a whole word, we will find the operand in locations 0108 and 0109. The 16-bit values in locations 0108 and 0109 are copied into AX, which is the destination.

Figure 21-10 is a screen dump of Fig. 21-9 obtained by using DEBUG.

In the first line

-d 100 10f

tells DEBUG to "dump" the contents of memory locations 0100_{16} through $010F_{16}$ to the screen so that we can see them. The hyphen halfway through the memory dump separates those 16 bytes into two sections to make the display easier to read. We have shown the contents of locations 0100 and 0101 in bold because they are the object code for the

MOV AX,[BX]

instruction. The contents of memory locations 0108 and 0109 are in bold because they are the locations being pointed to by register BX.

The -r tells DEBUG to display its registers. We have shown the contents of registers AX and BX in bold in this illustration because they are the two registers being referred to in this example.

The -t is the DEBUG trace command. This tells DEBUG to execute the next instruction and then stop. The next instruction is

MOV AX,[BX]

Notice the contents of register AX after the trace command. The contents of memory locations 0108 and 0109 have been copied to register AX as was illustrated in Fig. 21-9.

Take some time to compare Figs. 21-9 and 21-10. You may notice that the code segments in the two figures differ. That's because we created the figures on two different days, and the memory arrangement in our computer was not exactly the same both days. This is normal and something you should expect to see as you try these figures and

```
-d 100 10f
9029:0100  8B 07 00 00 00 00 00 00-AA BB 00 00 00 00 00 00   ................
-
-r
AX=0000  BX=0108  CX=0000  DX=0000  SP=FD6E  BP=0000  SI=0000  DI=0000
DS=9029  ES=9029  SS=9029  CS=9029  IP=0100   NV UP EI PL NZ NA PO NC
9029:0100 8B07          MOV     AX,[BX]                          DS:0108=BBAA
-
-t

AX=BBAA  BX=0108  CX=0000  DX=0000  SP=FD6E  BP=0000  SI=0000  DI=0000
DS=9029  ES=9029  SS=9029  CS=9029  IP=0102   NV UP EI PL NZ NA PO NC
9029:0102 0000          ADD     [BX+SI],AL                       DS:0108=AA
```

Fig. 21-10 DEBUG screen dump of Fig. 21-9.

examples on your computer. Everything will be the same except the code segment, and that will almost never match ours.

Again, in the case of register indirect addressing, at least one of the operands is in a memory location pointed to by the value in a register (BX, BP, SI, DI).

Program Indirect Addressing

Program indirect addressing is used by CALL and JMP instructions. It allows the memory location where the program is to fetch its next instruction to be stored in a register, in a memory location pointed to by a register, or in a memory location pointed to by a register with a displacement.

Normally instructions are stored in memory in sequential order, with the microprocessor fetching one after another. When a JMP instruction uses direct addressing, the address the microprocessor is to jump to is placed immediately after the jump instruction itself.

A CALL instruction causes the microprocessor to go to another area of memory where a subroutine is stored, execute the subroutine, and then return to where it left off before it began the subroutine. The CALL, like the JMP instruction, can use direct addressing and place the location of the subroutine immediately after the CALL instruction.

When either the CALL or the JMP instruction uses one of the 16-bit registers (AX, BX, CX, DX, SP, BP, SI, or DI), it means that the destination for the JMP or CALL is located in that register. For example

```
JMP    AX
```

instructs the microprocessor to look in register AX and jump to the location stored in AX. That is, AX ''points'' to the correct memory location.

When either the JMP or the CALL instruction uses a register placed inside brackets ([BX], [BP], [SI], or [DI]), it means that register contains an address, and that address contains another address, which is the actual destination for the JMP or CALL. For example,

```
JMP    [BX]
```

instructs the microprocessor to look in register BX. Let's say BX=0200. Next the microprocessor looks at address 0200 and 0201. There it will find another address which is its actual destination.

When either the JMP or the CALL instruction uses one of the registers with brackets ([BX], [BP], [SI], or [DI]) and a displacement, the microprocessor is instructed to add the displacement to the contents of the register, forming an address, and then to look at that address and get another address, which is the actual destination. For example

```
JMP    [BX+0100]
```

instructs the microprocessor to add 0100_{16} to the value in BX. Let's say that BX contains 0500_{16}.

$$0500_{16} + 0100_{16} = 0600_{16}$$

The microprocessor now looks in addresses 0600 and 0601 and gets another address. This is the destination address where the next instruction is to be fetched or the subroutine begins.

Base plus Index Addressing

Base plus index addressing also uses the concept of calculating the address where data is located rather than using direct addressing, which explicitly states where the data is located.

When *base plus index* addressing is used, the contents of one of the *base registers* (either BX or BP) and the contents of one of the *index registers* (either SI or DI) are added to calculate the address of the operand. For example,

```
MOV    AX,[BX+DI]
```

instructs the microprocessor to add the value in register BX to the value in register DI. This sum is the location of the data which is to be copied into register AX. This is illustrated in Fig. 21-11.

Base plus index addressing is useful for working with tables of data. The base register (BX or BP) can point to the beginning of the data table. The index register (SI or DI) can then point to the specific piece of data within the table. The program can then increment or decrement the index register to point to the next or preceding piece of data in the table.

Base Relative plus Index Addressing

Base relative plus index addressing combines the features of base plus index addressing and register relative addressing. Examples of base relative plus index addressing are

```
MOV    DX,[BX+SI+10]
MOV    [BX+DI+20],AX
```

In the first example, the microprocessor would add the values in registers BX and SI and the number 10_{16}. The sum is the memory location of the data which is to be copied into register DX.

In the second example, the microprocessor would copy the contents of register AX to a memory location whose address would be calculated by finding the sum of 20_{16}, the value in register BX, and the value in register DI.

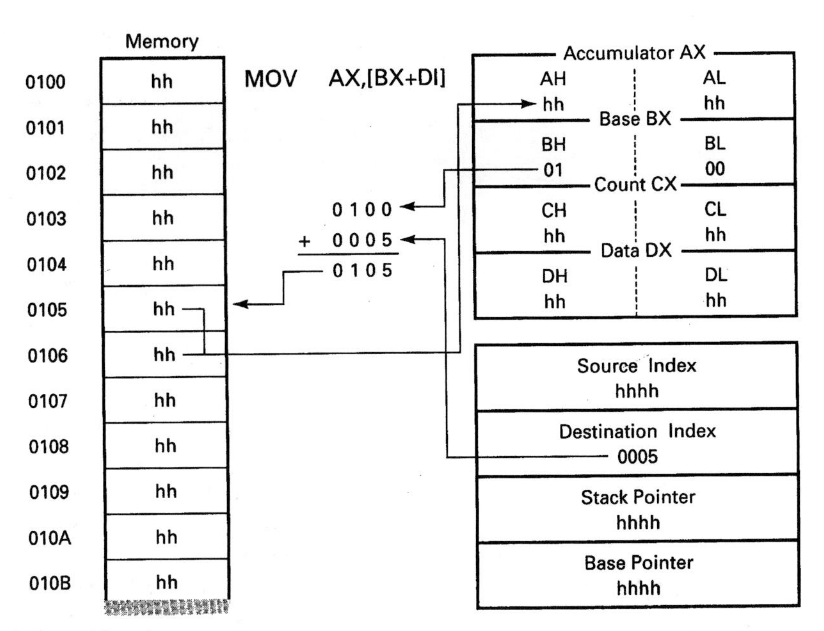

Fig. 21-11 Base plus index addressing.

This addressing mode is useful for working with two-dimensional data tables. The displacement (the number) can point to the beginning of the table, since this is the constant value. The base register (BX) can point to the first of the two dimensions (for example, a record in a file or an area in a data table). The index register (SI or DI) can then point to the specific memory location containing the desired data (for example, a field within a record within a file, or a specific piece of data in a data table). The program can then increment or decrement the base register to point to the next or previous record in the file and increment or decrement the index register to point to the next or previous field in the record.

BRANCHING AND LOOPS

In this chapter we'll study branching and loops. A branch instruction (or jump instruction) causes the program to "skip" forward or backward and to execute instructions from this new memory location.

A loop involves executing a series of microprocessor instructions and then branching backward to repeat the same set of instructions. This "loop" is finally broken, or exited from, when some condition is met.

The previous chapter introduced you to the remainder of the addressing modes (the more difficult ones) which had not been covered in the earlier chapter on addressing. From this point on we will use many of the different types of addressing modes available to each microprocessor. You should refer back to either of the chapters on addressing whenever necessary.

New Concepts

We'll study unconditional branching (or jumping) first; then we'll discuss the slightly more difficult subject of conditional branching. Later we'll look at loops and how to control them through the use of conditions and counters.

22-1 UNCONDITIONAL JUMPS

The simplest type of branch or jump is an unconditional one. This means that the program will jump to the indicated memory location every time this part of the program is run. The jump can be forward or backward.

With unconditional jumps, most of the microprocessors featured in this text use some form of direct or indirect addressing to indicate where the next instruction should be fetched from. The exceptions to this are the 6800/6808, which can also use relative addressing, and the 8086/8088, which also uses relative addressing, at least for jumps within a single memory segment.

To jump forward, you simply indicate the address of the next instruction to be executed. We'll look at exactly how the different addressing modes are used in the Specific Microprocessor Families section of this chapter.

22-2 CONDITIONAL BRANCHING

Conditional branching, like unconditional branching, causes program execution to continue with an instruction which is not the next instruction in memory. We either skip forward or backward from where we are now. Whether or not program execution does skip depends on a certain condition.

The microprocessor determines whether a condition is true or not true by the condition of the flags. To be able to predict whether or not a condition will be true when the microprocessor reaches the point at which the conditional branch occurs, one must know how the preceding instructions affect the flags. How each instruction affects each of the flags is shown in several of the instruction-set tables for each microprocessor.

When we branch *forward*, we have the effect of skipping over a certain number of instructions, if certain conditions exist, and not skipping over them if those conditions do not exist. Figure 22-1 shows a generic example of branching forward.

When we branch *backward*, the instructions between where we branched from and where we branched to are executed again. They could in fact be executed many times. This creates a loop which will not be exited from until some condition is met. Figure 22-2 shows a generic example of branching backward.

In Fig. 22-2, we are not branching backward from address 0009 because of the instruction at that memory location. Rather, we are branching backward because of the instruc-

```
0000    INSTRUCTION
0001    DATA
0002    INSTRUCTION
0003    DATA
0004    INSTRUCTION
0005    INSTRUCTION
0006    INSTRUCTION
0007    COND JUMP
0008    00A0
.
.
.
.
.
00A0    INSTRUCTION
00A1    DATA
00A2    INSTRUCTION
00A3    END
```

This area is skipped over if condition exits. If condition doesn't exist, this area is not skipped over.

Fig. 22-1 Example of generic forward conditional jump.

tion at location 0007 and the address at location 0008. The arrow is drawn from location 0009 because that will be the instruction pointed to by the program counter or instruction pointer when the branch occurs. Remember, the instruction pointer or program counter points to the *next* instruction to be executed, not the one currently being executed.

22-3 COMPARE AND TEST INSTRUCTIONS

Many (but not all) microprocessor instructions affect the flags. The flags then tell something about the results of the instruction. There are instructions, however, *compare* and *test* instructions, which actually do nothing except affect flags.

For example, the arithmetic instructions actually accomplish some task, such as adding, subtracting, multiplying, or dividing, and also affect the flags depending on the result of the operation. Compare and test instructions, however, compare a register or memory location to another, to zero, or AND two registers, without producing any result or changing any register or memory location—that is, no answer is produced. The flags, however, respond just as if an answer had been produced. A conditional branch instruction can then check the flags and determine whether a certain condition is true or false and then branch or not branch accordingly.

22-4 INCREMENT AND DECREMENT INSTRUCTIONS

Sometimes you may want to repeat a section of your program a certain number of times. A register or memory location is used to count how many times the section has been repeated. This register or memory location being used as a counter can either count up (increment) to a certain value or count down (decrement) to a certain value. Since it is easy to test for the occurrence of zero (just check the zero flag), counters often start at a certain number and decrement to zero. When the counter reaches zero, we know how many times that section of the program has repeated.

This technique produces a loop and uses conditional branching in a way that is similar to that discussed in the last section, although the intent is a little different. In the last section we were talking about situations when you want to branch if an operation produces a certain result. In this section we are discussing situations when we simply want something to be repeated a certain number of times.

22-5 NESTED LOOPS

It's possible to nest loops one inside the other. Figure 22-3 shows what this looks like.

The operand immediately following the conditional branch instruction may not be the actual address to branch to but rather the value needed by some other form of addressing such as relative addressing.

Remember also that we do not branch from the memory location containing the conditional branch instruction; nor do we branch from the next address which determines where we branch to, but from the instruction after that.

In Fig. 22-3 you can see that an inner loop will be repeated until the conditions necessary for the program to "drop through" the bottom of the loop exist, in which case the program may go back to the beginning of the outer loop, depending again on the conditions which exist.

```
0000    INSTRUCTION
0001    DATA
0002    INSTRUCTION
0003    DATA
0004    INSTRUCTION
0005    INSTRUCTION
0006    INSTRUCTION
0007    COND JUMP
0008    0000
0009    INSTRUCTION
```

This area is repeated if certain condition exists. This area is not repeated if condition does not exist.

Fig. 22-2 Example of generic backward conditional jump.

```
0000        INSTRUCTION                   <----------+
0001        data                                     |
0002        INSTRUCTION                              |
0003        data                                     |
0004        INSTRUCTION                              |
0005        data                                     |
0006        INSTRUCTION               <------+       |
0007        data                             |       |
0008        INSTRUCTION                      |       |
0009        data                             |       |
000A        CONDITIONAL BRANCH BACKWARDS     |       |
000B        0006                             |       |
000C        INSTRUCTION               -------+       |
000D        data                                     |
000E        INSTRUCTION                              |
000F        data                                     |
0010        CONDITIONAL BRANCH BACKWARDS             |
0011        0000                                     |
0012        INSTRUCTION               ---------------+
```

Fig. 22-3 Generic nested loops.

Specific Microprocessor Families

Let's look at each of our microprocessors' instructions to see how branching and loops are handled.

22-6 6502 FAMILY

The 6502 microprocessor family has a variety of instructions to handle unconditional jumps, conditional branching, comparing, incrementing, and decrementing. We'll look at several tasks and see how the 6502 microprocessor handles them.

You should enter each program into your computer or microprocessor trainer and single-step through it, watching the appropriate registers, memory locations, and flags to understand how each program works.

Unconditional Jumps

The forward unconditional jump using absolute addressing is easiest to understand. An example is shown in Fig. 22-4.

The program begins by loading the accumulator with FF_{16}. In a moment we are going to subtract another number from FF_{16}. First, however, we need to jump to the area of memory where the subtract instruction is. We have placed the subtract instruction several memory locations forward from this point to show, in a very simple manner, how the unconditional jump instruction operates.

The next instruction is our jump instruction. In the source code column of line 0004 the instruction

JMP MINUS

appears, which might be different from what you were expecting.

The instruction is saying to jump to a place called MINUS. To be able to jump to a place with a certain name is not a native ability of the 6502 microprocessor. Our assembler is making this possible. Line 0008 has the label *MINUS* in the label column. This is the place we want to jump to. Notice the address at the MINUS label. The address is 0348. Now look back at line 0004. In the op code column you see 4C, which is the op code for an unconditional jump. Then come the numbers 48 03. If you reverse those two sets of numbers, you have 0348. This is the memory location of the instruction labeled MINUS. If you use an assembler, you can use labels and the assembler will calculate the address for you. If you are hand-assembling these programs, you must enter the address as shown in the op code column, in the reverse low-byte/high-byte order. If you are using an assembler which does not allow labels, you will need to use the format shown in the 6502 tables. Namely,

JMP $0348

```
0001  0340                      .org $0340      ;beginning of code
0002  0340
0003  0340 A9 FF       START:   LDA #$FF        ;minuend
0004  0342 4C 48 03             JMP MINUS       ;forward unconditional jump
0005  0345 EA                   NOP
0006  0346 EA                   NOP             ;misc. instructions
0007  0347 EA                   NOP
0008  0348 38          MINUS:   SEC             ;prepare for subtraction
0009  0349 E9 EE                SBC #$EE        ;subtrahend
0010  034B 8D 4F 03             STA ANSWER      ;store difference
0011  034E 00                   BRK             ;stop
0012  034F
0013  034F 00          ANSWER   .db     $00     ;memory area for answer
0014  0350                                      ;  (initialized to 00)
0015  0350                      .end
```

Fig. 22-4 Forward unconditional jump with the 6502 microprocessor.

After the jump instruction are several NOPs which could be other instructions or just unused memory in a particular microprocessor system.

Line 0008 is the next instruction to be executed. It sets the carry flag in preparation for the subtraction instruction. In line 0009 we subtract EE_{16} from FF_{16} (in the accumulator). In line 0010 we store the result of our subtraction in a memory location called *ANSWER*. Look at line 0013, labeled ANSWER. In the op code column are the initials .db. They stand for *define byte*. We are telling the assembler to reserve a memory location, namely, a single byte of memory, with the name ANSWER. The assembler is initializing the memory location ANSWER with a value of 0. Our program can then put any other number we wish in that location.

Notice also that the memory location of ANSWER is $034F_{16}$. In the op code column of line 0010 we see 8D 4F 03. 8D is the op code for storing the value of the accumulator in a certain memory location. If you reverse the order of 4F 03, you have 034F, which is the memory location of ANSWER. Again, the assembler made life simpler by figuring out where the next available memory location would be and setting aside that location for the ANSWER.

Finally, in line 0011 the program stops.

You should enter this program and single-step through it, making sure that everything works as described.

Conditional Branches

Now let's see an example of conditional branching. Figure 22-5 shows such an example.

In this program we are going to do several things differently from the way they were done in the last program. First, we are using a conditional jump or branch rather than an unconditional one. Second, we are branching backward rather than forward. Third, we are creating a loop by branching backward and repeating a section of the program. Finally, we are using a register as a counter to control how many times the loop repeats.

In line 0003 we place the number 3_{16} in the X register. This register *controls* how many times we will branch backward. In line 0004 we clear the Y register making it 00_{16} so that it can be used to *count* how many times the loop repeats.

Line 0005 marks the beginning of the loop; we have named that location *REPEAT*. In this line we increment the Y register since we are beginning to pass through the loop, in this case for the first time. The Y register is keeping track of how many times the loop is passed through. Line 0006 represents the fact that there could be many instructions inside the loop which are going to be repeated.

Line 0007 decrements (reduces by 1) the X register. The X register keeps track of how many times through the loop are remaining.

Line 0008 is where we meet our conditional branch instruction. BNE means **B**ranch if **N**ot **E**qual. Your first thought might be, "Not equal to what?" If you check the Expanded Table for the 6502, you'll see it is **B**ranch if the last result is **N**ot **E**qual to 0.

All the conditional branch instructions are influenced by the most recent instruction that affected the flag they check. In this case the zero flag is checked. What was the last instruction which sets or clears the zero flag? The DEX (**DE**crement **X** register) instruction. If the X register were reduced to 0, the zero flag would be set. Has the X register been reduced to 0? On this first pass through the loop, it gets reduced from 3 to 2. No, the X register is not equal to 0.

The branch instruction says, "**B**ranch if the last result is **N**ot **E**qual to 0." Clearly this is true: the last result is not 0, so we branch. Branch to where? We branch to the memory location known as *REPEAT*. Notice that the location called REPEAT, in line 0005, is memory location 0344_{16}. Now look again at line 0008. D0 is the op code for the BNE instruction, and FB is where it is branching to. Is FB the memory location of REPEAT? No. The BNE instruction uses relative addressing. FB_{16} is a negative-signed binary number telling us how many places to move from where we are now. FB_{16} is -5_{10}. We must branch five memory location backward from memory location 0349_{16}.

It will be helpful to enter this program into your computer or microprocessor trainer and single-step through it. We've gone through the loop only once in our discussion here.

```
0001  0340                    .ORG $0340
0002  0340
0003  0340 A2 03    START:    LDX #$03      ;initialize X (repeats)
0004  0342 A0 00              LDY #$00      ;initialize Y
0005  0344 C8       REPEAT:   INY           ;times loop has repeated
0006  0345 EA                 NOP           ;misc instructions
0007  0346 CA                 DEX           ;decrement X
0008  0347 D0 FB              BNE REPEAT    ;if X not equal to 0 then
0009  0349                                  ;  branch back to start of
0010  0349                                  ;  loop
0011  0349 00                 BRK           ;stop
0012  034A
0013  034A                    .END
```

Fig. 22-5 A backward conditional jump creating a loop with the 6502 microprocessor.

Pay special attention to the X register, the Y register, and the zero flag.

Compare Instructions

The *compare* instructions allow us to compare the values in two registers and/or memory locations, and to set the flags accordingly, without changing either of the original values. The appropriate branch instruction can then cause program execution to continue at the desired location. The program in Fig. 22-6 will allow you to observe the compare instructions.

The program simply loads the value 05_{16} into the accumulator and compares the numbers 04_{16}, 06_{16}, and 05_{16} to it. If you will refer to the Expanded Table of 6502 Instructions and look in the Operation column, you will see what we mean by "compare."

To "compare" means to subtract the number you are "comparing" from the number being "compared to." For example, line 0004 of the program in Fig. 22-6 sets the flags as though 04_{16} had been subtracted from 05_{16}, without actually changing the value in the accumulator.

Lines 0005 and 0006 likewise subtract 06_{16} and 05_{16}, respectively, from the value in the accumulator without altering the accumulator.

A point needs to be made at this time about the carry flag in the 6502 microprocessor. Most microprocessors set a flag (value of 1) to say, "Yes, this condition exists." For example, setting the zero flag (value of 1) means, "Yes, the last value (or current value) is a zero." When a flag is reset (value of zero) it means "No, this condition does not exist."

The 6502 handles the carry flag in an unusual way. It is inverted. After addition this flag will appear as expected. A 1 means that a carry occurred, and a 0 means that a carry did not occur. After subtraction, however, a 1 means that a borrow did not occur, and a 0 means that a borrow did occur. Be careful to remember this exception when using 6502 compare instructions to prepare for branch instructions.

This program's only purpose is to allow you to see how the flags are affected by each compare instruction. Enter the program and single-step through it. Watch the flags after each instruction and make sure that you understand why they react the way they do.

An Example Program

We'll now look at an example program which uses a compare instruction, increment instructions, and a conditional branch instruction. This program looks at two numbers in memory, determines which is larger, and then places the larger value in a third memory location. It also uses a form of indexed addressing. Refer to Fig. 22-7 at this time.

After entering this program into your computer or trainer, but before running it, you must place values of your choice into the two memory locations indicated in the notes at the beginning of the program.

This program uses the X register to help point to the next memory location to load a number from or store a number in. The first instruction in line 0008 initializes the X register with a value of 00_{16}.

Memory location $03A0_{16}$ is the beginning of a series of memory locations which this program uses. A common way to address successive memory locations is to use some form of indexed addressing. Location $03A0_{16}$ is the beginning of the list, and the X register will point to each successive number in the list. In line 0009 we load the accumulator with the first number from the list. The memory location of this number is formed by adding $03A0_{16}$ to the value of the X register, which is 00_{16} at this moment, to form the address of the first number in the list, in location $03A0_{16}$.

In line 0010 we increment the X register to a value of 01_{16} so that it points to the next number.

In line 0011 we compare the value held in memory location $03A1_{16}$ to the value in the accumulator. If the value in the accumulator is larger, then no borrow will be needed to perform the comparison (which involves subtraction). Therefore the carry flag will be set.

We find in line 0012 that, if the carry flag is set, then we branch forward to line 0014. This will be the case if the value in the accumulator is the larger value. In line 0014 the X register is incremented so that it points to the last memory location. In line 0015 we store the value now in the accumulator in that final memory location.

If during the comparison in line 0011 the value in the accumulator is smaller, a borrow is required to perform the comparison (involving subtraction) and the carry flag is cleared. In line 0012 the carry flag is not set and the branch does not occur. Therefore, the next instruction in line 0013 is executed. This instruction loads the second number into

```
0001  0340                        .org $0340
0002  0340
0003  0340 A9 05      START:     LDA #$05        ;initial value
0004  0342 C9 04                 CMP #$04        ;compare each of these numbers
0005  0344 C9 06                 CMP #$06        ;  to A and set flags as though
0006  0346 C9 05                 CMP #$05        ;  each had been subtracted from A
0007  0348 00                    BRK
0008  0349
0009  0349                       .end
```

Fig. 22-6 Using the compare instruction.

```
0001  0000                   ;place a number in memory location $0340 and another in $03A1,
0002  0000                   ;  this program will determine which is larger and place
0003  0000                   ;  the larger in location $03A2  (Note: Do not use two
0004  0000                   ;  numbers which are equal.)
0005  0000
0006  0340                             .org $0340
0007  0340
0008  0340 A2 00             START:    LDX #$00          ;initialize X register
0009  0342 BD A0 03                    LDA $03A0,X       ;load A from mem 03A0 + 00 = 03A0
0010  0345 E8                          INX               ;point to next mem loc
0011  0346 DD A0 03                    CMP $03A0,X       ;compare data in mem 03A0 +
                                                           01 = 03A1 to A
0012  0349 B0 03                       BCS FOUND         ;if A is larger jump forward to Found;
0013  034B BD A0 03                    LDA $03A0,X       ;  otherwise load A from mem 03A0 +
                                                             01 = 03A1
0014  034E E8                FOUND:    INX               ;point to next mem loc
0015  034F 9D A0 03                    STA $03A0,X       ;store A in mem 03A0 + 02 = 03A2
0016  0352 00                          BRK               ;stop
0017  0353
0018  0353                             .end
```

Fig. 22-7 An example 6502 program.

the accumulator. Obviously, if the first number is not the larger, the second one must be. After loading the accumulator with the second number in line 0013, we continue in lines 0014 and 0015 to store that value in the third memory location.

This program will give you an idea how to use some of the new instructions in this chapter and how to use indexed addressing.

22-7 6800/6808 FAMILY

The 6800/6808 microprocessor family has a variety of instructions to handle unconditional jumps and branches, conditional branching, comparing, incrementing, and decrementing. We'll look at several tasks and see how the 6800/6808 microprocessor handles them.

You should enter each program into your computer or microprocessor trainer and single-step through it, watching the appropriate registers, memory locations, and flags to understand how each program works.

Unconditional Jumps

The forward unconditional jump using extended addressing is probably easiest to understand. An example is shown in Fig. 22-8.

(*Technical Note:* We have started this program at address 0100_{16} rather than our usual 0000_{16}. Addresses from 0000_{16} to $00FF_{16}$ form page 0 of memory. Some instructions can use direct addressing, if the desired location is on page 0, or extended addressing, if the desired location is on a memory page other than page 0. Our particular assembler had trouble handling forward references on page 0. Switching to a page other than page 0 provided a simple solution to this problem.)

The program begins by loading accumulator A with FF_{16}. In a moment we are gong to subtract another number from this one. First we need to jump to the area of memory where the subtract instruction is. We have placed the subtract instruction several memory locations forward from this point to show, in a very simple manner, how the unconditional jump instruction operates.

```
0001  0100                        .org $0100       ;beginning of code
0002  0100
0003  0100 86 FF       START:     LDAA #$FF        ;minuend
0004  0102 7E 01 08               JMP MINUS        ;forward unconditional jump
0005  0105 01                     NOP
0006  0106 01                     NOP              ;misc. instructions
0007  0107 01                     NOP
0008  0108 80 EE       MINUS:     SUBA #$EE        ;subtrahend
0009  010A B7 01 0E               STAA ANSWER      ;store difference
0010  010D 3E                     WAI              ;stop
0011  010E
0012  010E 00          ANSWER     .db      $00     ;memory area for answer
0013  010F                                         ;  (initialized to 00)
0014  010F                        .end
```

Fig. 22-8 Forward unconditional jump with the 6800/6808 microprocessor. (*Note* that address is $0100 rather than $0000. This prevents an assembler error caused by a forward reference to a label on zero page.)

The next instruction is our jump instruction. In the source code column of line 0004 the instruction

JMP MINUS

appears, which might be different than what you were expecting.

The instruction is saying to jump to a place called *MINUS*. To be able to jump to a place with a certain name is not a native ability of the 6800/6808 microprocessor. Our assembler is making this possible. Line 0008 has the label MINUS in the label column. This is the place we want to jump to. Notice the address at the MINUS label. The address is 0108. Now look back at line 0004. In the op code column you see 7E, which is the op code for an unconditional jump. Then come the numbers 01 08. This is the memory location of the instruction labeled MINUS. If you use an assembler, you can use labels and the assembler will calculate the address for you. If you are hand-assembling these programs, you must enter the address as shown in the op code column. If you are using an assembler which does not allow labels, you will need to use the format shown in the 6800/6808 instruction-set tables. Namely

JMP $0108

After the jump instruction are several NOPs which could be other instructions or just unused memory in a particular microprocessor system.

In line 0008 we subtract EE_{16} from FF_{16} (in accumulator A). In line 0009 we store the result of our subtraction in a memory location called *ANSWER*. Look at line 0012, labeled ANSWER. In the op code column are the initials .db. They stand for *define byte*. We are telling the assembler to reserve a memory location, namely, a single byte of memory, with the name ANSWER. The assembler is initializing the memory location ANSWER with a value of 0. Our program can then put any other number we wish in that location.

Notice also that the memory location of ANSWER is $010E_{16}$. In the op code column of line 0009 we see B7 01 0E. The op code for storing the value of the accumulator in a certain memory location is B7. 010E is the memory location of ANSWER. Again, the assembler made life simpler by figuring out where the next available memory location would be and setting aside that location for the ANSWER.

Finally, in line 0010 the program stops.

You should enter this program and single-step through it, making sure everything works as described.

Conditional Branches

Now let's see an example of conditional branching. Figure 22-9 shows such an example.

In this program we are going to do several things differently from the way they were done in the last program. First, we are using a conditional jump or branch rather than an unconditional one. Second, we are branching backward rather than forward. Third, we are creating a loop by branching backward and repeating a section of the program. Finally, we are using a register as a counter to control how many times the loop repeats.

In line 0003 we place the number 3_{16} in the X register. This register *controls* how many times we will branch backward. In line 0004 we clear accumulator B, making it 00_{16} so that it can be used to *count* how many times the loop repeats.

Line 0005 marks the beginning of the loop, and we have named that location *REPEAT*. In this line we increment accumulator B since we are beginning to pass through the loop, in this case for the first time. Accumulator B is keeping track of how many times the loop is passed through. Line 0006 represents the fact that there could be many instructions inside this loop which are going to be repeated.

Line 0007 decrements (reduces by 1) the X register. The X register keeps track of how many times to go through the loop remain.

Line 0008 is where we meet our conditional branch instruction. BNE means **B**ranch if **N**ot **E**qual. Your first thought might be, "Not equal to what?" If you check the Expanded Table for the 6800/6808, you'll see that it is **B**ranch if **N**ot **E**qual to 0.

```
0001  0000                      .ORG $0000
0002  0000
0003  0000 CE 00 03   START:    LDX #$0003       ;initialize X (repeats)
0004  0003 C6 00                LDAB #$00        ;initialize B
0005  0005 5C         REPEAT:   INCB             ;times loop has repeated
0006  0006 01                   NOP              ;misc. instructions
0007  0007 09                   DEX              ;decrement X
0008  0008 26 FB                BNE REPEAT       ;if X not equal to 0 then
0009  000A                                       ;  branch back to start of
0010  000A                                       ;  loop
0011  000A 3E                   WAI              ;stop
0012  000B
0013  000B                      .END
```

Fig. 22-9 A backward conditional jump creating a loop with the 6502 microprocessor.

All the conditional branch instructions are influenced by the most recent instruction that affected the flag they check. In this case the zero flag is checked. What was the last instruction which sets or clears the zero flag? The DEX (**DE**crement **X** register) instruction. If the X register was reduced to 0, the zero flag would be set. Has the X register been reduced to 0? On this first pass through the loop, it gets reduced from 3 to 2. No, the X register is not equal to 0.

The branch instruction says, "**B**ranch if **N**ot **E**qual to 0." Clearly this is true: the last result is not 0, so we branch. Branch to where? We branch to the memory location known as REPEAT. Notice that the location called REPEAT, in line 0005, is memory location 0005_{16}. Now look again at line 0008. The op code for the BNE instruction is 26, and FB is where it's branching to. Is FB the memory location of REPEAT? No. The BNE instruction uses relative addressing. FB_{16} is a negative-signed binary number telling us how many places to move from where we are now. FB_{16} is -5_{10}. We must branch five memory locations backward from memory location $000A_{16}$.

It will be helpful to enter this program into your computer or microprocessor trainer and single-step through it. Pay special attention to the X register, accumulator B, and the zero flag.

Compare Instructions

The *compare* instructions allow us to compare the values in two registers and/or memory locations and to set the flags accordingly without changing either of the original values. The appropriate branch instruction can then cause program execution to continue at the desired location. The program in Fig. 22-10 allows you to observe how the compare instructions work.

The program simply loads the value 05_{16} into accumulator A and compares the numbers 04_{16}, 06_{16}, and 05_{16} to it. If you refer to the Expanded Table of 6800/6808 Instructions and look in the Operation column, you will see what we mean by "compare."

To "compare" means to subtract the number you are "comparing" from the number being "compared to." For example, line 0004 of the program in Fig. 22-10 sets the flags as though 04_{16} had been subtracted from 05_{16}, in accumulator A, without actually changing the value in the accumulator.

Line 0005 and 0006 likewise subtract 06_{16} and 05_{16}, respectively, from the value in accumulator A without altering the accumulator.

This program's only purpose is to allow you to see how the flags are affected by each compare instruction. Enter the program and single-step through it. Watch the flags after each step and make sure that you understand why they react the way they do.

An Example Program

We'll now look at an example program which uses a compare instruction, increment instructions, and a conditional branch instruction. This program looks at two numbers in memory, determines which is larger, and then places the larger value in a third memory location. It also uses a form of indexed addressing. Refer to Fig. 22-11 at this time.

After entering this program into your computer or trainer but before running it, you must place values of your choice into the two memory locations indicated in the notes at the beginning of the program.

This program uses the X register to help point to the next memory location to load a number from or store a number in. The first instruction in line 0008 initializes the X register with a value of $01A0_{16}$.

Memory location $01A0_{16}$ is the beginning of a series of memory locations which this program uses. A common way to address successive memory locations is to use some form of indexed addressing. Location $01A0_{16}$ is the beginning of the list, and the X register will point to each successive number in the list. In line 0009 we load the accumulator with the first number from the list. The memory location of this number is formed by adding 00_{16} to the value in the X register, which is $01A0_{16}$, to form the address of the first number in the list, at location $01A0_{16}$.

In line 0010 we increment the X register to a value of $01A1_{16}$ so that it points to the next number.

In line 0011 we compare the value held in memory location $01A1_{16}$ to the value in accumulator A. If the value in the accumulator is larger, then no borrow will be needed to perform the comparison (which involves subtraction). Therefore the carry flag will be clear.

We find in line 0012 that, if the carry flag is clear, then we branch forward to line 0014. This will be the case if the value in the accumulator is the larger value. In line 0014 the X register is incremented, so it points to the last

```
0001  0000                  .org $0000
0002  0000
0003  0000 86 05    START:  LDAA #$05       ;initial value
0004  0002 81 04            CMPA #$04       ;compare each of these numbers
0005  0004 81 06            CMPA #$06       ;  to A and set flags as though
0006  0006 81 05            CMPA #$05       ;  each had been subtracted from A
0007  0008 3E               WAI
0008  0009
0009  0009                  .end
```

Fig. 22-10 Using the compare instruction.

```
0001  0000                ;place a number in memory location $01A0 and another in $01A1;
0002  0000                ;  this program will determine which is larger and place
0003  0000                ;  the larger in location $01A2  (Note: Do not use two
0004  0000                ;  numbers which are equal.)
0005  0000
0006  0100                       .org $0100
0007  0100
0008  0100 CE 01 A0    START:    LDX #$01A0         ;initialize X register
0009  0103 A6 00                 LDAA $00,X         ;load A from mem 01A0 + 00 = 01A0
0010  0105 08                    INX                ;point to next mem loc
0011  0106 A1 00                 CMPA $00,X         ;compare data in mem 01A0 + 00 =
                                                       01A1 to A
0012  0108 24 02                 BCC FOUND          ;if A is larger jump forward to Found;
0013  010A A6 00                 LDAA $00,X         ;  otherwise load A from mem 01A1 +
                                                       00 = 01A1
0014  010C 08          FOUND:    INX                ;point to next mem loc
0015  010D A7 00                 STAA $00,X         ;store A in mem 01A2 + 00 = 01A2
0016  010F 3E                    WAI                ;stop
0017  0110
0018  0110                       .end
```

Fig. 22-11 An example 6800/6808 program.

memory location. In line 0015 we store the value now in accumulator A in that final memory location.

If, during the comparison in line 0011 the value in the accumulator is smaller, a borrow is required to perform the comparison (involving subtraction) and the carry flag is set. In line 0012 the carry flag is not clear and the branch does not occur. Therefore, the next instruction in line 0013 is executed. This instruction loads the second number into the accumulator. Obviously, if the first number is not the larger, the second one must be. After loading accumulator A with the second number in line 0013, we continue in lines 0014 and 0015 to store that value in the third memory location.

This program will give you an idea how to use some of the new instructions in this chapter and how to use indexed addressing.

22-8 8080/8085/Z80 FAMILY

The 8080/8085/Z80 microprocessor family has a variety of instructions to handle unconditional jumps, conditional branching, comparing, incrementing, and decrementing. We'll look at several tasks and see how the 8080/8085/Z80 microprocessor handles them.

You should enter each program into your computer or microprocessor trainer and single-step through it, watching the appropriate registers, memory locations, and flags to understand how each program works.

Remember that we will show both 8080/8085 and Z80 programs in the figures and that in the text we will show 8080/8085 mnemonics first with Z80 mnemonics in brackets.

Unconditional Jumps

The forward unconditional jump using direct addressing is probably easiest to understand. An example is shown in Fig. 22-12.

The program begins by loading the accumulator with FF_{16}. In a moment we are going to subtract another number from this one. First we need to jump to the area of memory where the subtract instruction is. We have placed the subtract instruction several memory locations forward from this point to show, in a very simple manner, how the unconditional jump instruction operates.

The next instruction is our jump instruction. In the source code column of line 0004 the instruction

JMP MINUS [JP MINUS]

appears, which might be different than what you were expecting.

The instruction is saying to jump to a place called *MINUS*. To be able to jump to a place with a certain name is not a native ability of the 8080/8085/Z80 microprocessor. Our assembler is making this possible. Line 0008 has the label MINUS in the label column. This is the place we want to jump to. Notice the address at the MINUS label. The address is 1808. Now look back at line 0004. In the op code column you see C3, which is the op code for an unconditional jump. Then come the numbers 08 18. If you reverse these two sets of numbers, you have 1808. This is the memory location of the instruction labeled MINUS. If you use an assembler, you can use labels and the assembler will calculate the address for you. If you are hand-assembling these programs, you must enter the address as shown in the op code column, in the reverse low-byte/high-byte order. If you are using an assembler which does not allow labels, you will need to use the format shown in the 8080/8085/Z80 instruction-set tables. Namely

JMP aaaa [JP aaaa]

After the jump instruction are several NOPs which could be other instructions or just unused memory in a particular microprocessor system.

```
8080/8085 program

0001  1800                              .org 1800h       ;beginning of code
0002  1800
0003  1800 3E FF            START:   MVI A,0FFh       ;minuend
0004  1802 C3 08 18                  JMP MINUS        ;forward unconditional jump
0005  1805 00                        NOP
0006  1806 00                        NOP              ;misc. instructions
0007  1807 00                        NOP
0008  1808 D6 EE            MINUS:   SUI 0EEh         ;subtrahend
0009  180A 32 0E 18                  STA ANSWER       ;store difference
0010  180D 76                        HLT              ;stop
0011  180E
0012  180E 00               ANSWER   .db     00h      ;memory area for answer
0013  180F                                            ;  (initialized to 00)
0014  180F                           .end

Z80 program

0001  1800                           .org 1800h       ;beginning of code
0002  1800
0003  1800 3E FF            START:   LD A,0FFh        ;minuend
0004  1802 C3 08 18                  JP MINUS         ;forward unconditional jump
0005  1805 00                        NOP
0006  1806 00                        NOP              ;misc. instructions
0007  1807 00                        NOP
0008  1808 D6 EE            MINUS:   SUB 0EEh         ;subtrahend
0009  180A 32 0E 18                  LD (ANSWER),A    ;store difference
0010  180D 76                        HALT             ;stop
0011  180E
0012  180E 00               ANSWER   .db     00h      ;memory area for answer
0013  180F                                            ;  (initialized to 00)
0014  180F                           .end
```

Fig. 22-12 Forward unconditional jump with the 8080/8085/Z80 microprocessor.

In line 0008 we subtract EE_{16} from FF_{16} (in the accumulator). In line 0009 we store the result of our subtraction in a memory location called *ANSWER*. Look at line 0012, labeled ANSWER. In the op code column are the initials .db. They stand for *define byte*. We are telling the assembler to reserve a memory location, namely, a single byte of memory, with the name ANSWER. The assembler is initializing the memory location ANSWER with a value of 0. Our program can then put any other number we wish in that location.

Notice also that the memory location of ANSWER is $180E_{16}$. In the op code column of line 0009 we see 32 0E 18. The op code for storing the value of the accumulator in a certain memory location is 32. If you reverse 0E 18, you have 180E, which is the memory location of ANSWER. Again the assembler made life simpler by figuring out where the next available memory location would be and setting aside that location for the ANSWER.

Finally, in line 0010, the program stops.

You should enter this program and single-step through it, making sure that everything works as described.

Conditional Branches

Now let's see an example of conditional branching. Figure 22-13 shows such an example.

In this program we are going to do several things differently from the way they were done in the last program. First, we are using a conditional jump or branch rather than an unconditional one. Second, we are branching backward rather than forward. Third, we are creating a loop by branching backward and repeating a section of the program. Finally, we are using a register as a counter to control how many times the loop repeats.

In line 0003 we place the number 3_{16} in register B. This register *controls* how many times we will branch backward. In line 0004 we clear register C making it 00_{16} so that it can be used to *count* how many times the loop repeats.

Line 0005 marks the beginning of the loop, and we have named that location *REPEAT*. In this line we increment register C since we are beginning to pass through the loop, in this case for the first time. Register C is keeping track of how many times the loop is passed through. Line 0006 represents the fact that there could be many instructions inside this loop which are going to be repeated.

Line 0007 decrements (reduces by one) register B. Register B keeps track of how many times we have left to go through the loop.

Line 0008 is where we meet our conditional branch instruction. JNZ means **J**ump if **N**ot **Z**ero. [JP NZ means **JumP** if **N**ot **Z**ero.] Your first thought might be, "If what isn't zero?"

8080/8085 program

```
0001   1800                   .ORG 1800h
0002   1800
0003   1800 06 03    START:   MVI B,03h       ;initialize B (repeats)
0004   1802 0E 00             MVI C,00h       ;initialize C
0005   1804 0C       REPEAT:  INR C           ;times loop has repeated
0006   1805 00                NOP             ;misc instructions
0007   1806 05                DCR B           ;decrement B
0008   1807 C2 04 18          JNZ REPEAT      ;if B not equal to 0 then
0009   180A                                   ;  branch back to start of
0010   180A                                   ;  loop
0011   180A 76                HLT             ;stop
0012   180B
0013   180B                   .END
```

Z80 program

```
0001   1800                   .ORG 1800h
0002   1800
0003   1800 06 03    START:   LD B,03h        ;initialize B (repeats)
0004   1802 0E 00             LD C,00h        ;initialize C
0005   1804 0C       REPEAT:  INC C           ;times loop has repeated
0006   1805 00                NOP             ;misc instructions
0007   1806 05                DEC B           ;decrement B
0008   1807 C2 04 18          JP NZ,REPEAT    ;if B not equal to 0 then
0009   180A                                   ;  branch back to start of
0010   180A                                   ;  loop
0011   180A 76                HALT            ;stop
0012   180B
0013   180B                   .END
```

Fig. 22-13 A backward conditional jump creating a loop with the 8080/8085/Z80 microprocessor.

All the conditional branch instructions are influenced by the most recent instruction that affected the flag they check. In this case the zero flag is checked. What was the last instruction which sets or clears the zero flag? The DCR B (**DeCR**ement **B**) [DEC B (**DEC**rement **B**)] instruction. If register B were reduced to 0, the zero flag would be set. Has register B been reduced to zero? On this first pass through the loop, it gets reduced from 3 to 2. No, register B is not equal to 0.

The jump instruction says, "Jump if not zero." Clearly this is true: the last result is not 0, so we do jump. Jump to where? We jump to the memory location known as *REPEAT*. Notice that the location called REPEAT, in line 0005, is memory location 1804_{16}. Now look again at line 0008. C2 is the op code for the JNZ [JP NZ] instruction. If you reverse the two sets of numbers 04 18, you form 1804, which is the memory location of the REPEAT label.

It will be helpful to enter this program into your computer or microprocessor trainer and single-step through it. Pay special attention to register B, register C, and the zero flag.

Compare Instructions

The *compare* instructions allow us to compare the values in two registers and/or memory locations and to set the flags accordingly without changing either of the original values. The appropriate jump instruction can then cause program execution to continue at the desired location. The program in Fig. 22-14 allows you to experiment with the compare instructions.

This program loads the value 05_{16} into the accumulator and compares the numbers 04_{16}, 06_{16}, and 05_{16} to it. If you will refer to the Expanded Table of 8080/8085/Z80 Instructions and look in the Operation column, you will see what we mean by "compare."

To "compare" means to subtract the number you are comparing from the number being "compared to." For example, line 0004 of the program in Fig. 22-14 sets the flags as though 04_{16} had been subtracted from 05_{16}, without actually changing the value in the accumulator.

Lines 0005 and 0006 likewise subtract 06_{16} and 05_{16}, respectively, from the value in the accumulator without altering the accumulator.

This program's only purpose is to allow you to see how the flags are affected by each compare instruction. Enter the program and single-step through it. Watch the flags after each step and make sure you understand why they react the way they do.

An Example Program

We'll now look at an example program which uses a compare instruction, increment instructions, and a condi-

8080/8085 program

```
0001    1800                       .org 1800h
0002    1800
0003    1800 3E 05      START:     MVI A,05h      ;initial value
0004    1802 FE 04                 CPI 04h        ;compare each of these numbers
0005    1804 FE 06                 CPI 06h        ;  to A and set flags as though
0006    1806 FE 05                 CPI 05h        ;  each had been subtracted from A
0007    1808 76                    HLT
0008    1809
0009    1809                       .end
```

Z80 program

```
0001    1800                       .org 1800h
0002    1800
0003    1800 3E 05      START:     LD A,05h       ;initial value
0004    1802 FE 04                 CP 04h         ;compare each of these numbers
0005    1804 FE 06                 CP 06h         ;  to A and set flags as though
0006    1806 FE 05                 CP 05h         ;  each had been subtracted from A
0007    1808 76                    HALT
0008    1809
0009    1809                       .end
```

Fig. 22-14 Using the compare instruction.

tional branch instruction. This program looks at two numbers in memory, determines which is larger, and then places the larger value in a third memory location. It also uses register indirect addressing. Refer to Fig. 22-15 at this time.

After entering this program into your computer or trainer, but before running it, you must place values of your choice into the two memory locations indicated in the notes at the beginning of the program.

This program uses the HL register pair to help point to the next memory location to load a number from or store a number in. The first instruction in line 0008 initializes the HL register pair with a value of $18A0_{16}$.

Memory location $18A0_{16}$ is the beginning of a series of memory locations which this program uses. A common way to address successive memory locations is to use some form of indexed addressing. The 8080/8085 does not actually have an index register; however, the HL register pair can be used with register indirect addressing to accomplish much the same thing. Location $18A0_{16}$ is the beginning of the list, and the HL register pair will point to each successive number in the list. In line 0009 we load the accumulator with the first number from the list. The memory location of this number is pointed to by the value in the HL register pair.

In line 0010 we increment the HL register pair to a value of $18A1_{16}$ so that it points to the next number.

In line 0011 we compare the value held in memory location $18A1_{16}$ to the value in the accumulator. If the value in the accumulator is larger, then no borrow will be needed to perform the comparison (which involves subtraction). Therefore the carry flag will be clear.

We find in line 0012 that, if the carry flag is clear, then we branch forward to line 0014. This will be the case if the value in the accumulator is the larger value. In line 0014 the HL register pair is incremented so that it points to the last memory location. In line 0015 we store the value now in the accumulator in that final memory location.

If, during the comparison in line 0011 the value in the accumulator is smaller, a borrow is required to perform the comparison (involving subtraction) and the carry flag is set. In line 0012 the carry flag is not clear and the branch does not occur. Therefore the next instruction in line 0013 is executed. This instruction loads the second number into the accumulator. Obviously, if the first number is not the larger, the second one must be. After loading the accumulator with the second number in line 0013, we continue in lines 0014 and 0015 to store that value in the third memory location.

This program will give you an idea how to use some of the new instructions in this chapter and how to use register indirect addressing.

22-9 8086/8088 FAMILY

The 8086/8088 microprocessor family has a variety of instructions to handle unconditional jumps, conditional branching, comparing, incrementing, and decrementing. We'll look at several typical tasks and see how the 8086/8088 microprocessor handles them.

You should enter each program into your computer or microprocessor trainer and single-step through it, watching the appropriate registers, memory locations, and flags to understand how each program works.

8080/8085 program

```
0001  0000                  ;place a number in memory location 18A0h and another in 18A1h;
0002  0000                  ;  this program will determine which is larger and place
0003  0000                  ;  the larger in location 18A2h  (Note: Do not use two
0004  0000                  ;  numbers which are equal.)
0005  0000
0006  1800                          .org 1800h
0007  1800
0008  1800 21 A0 18         START:  LXI H,18A0h         ;initialize HL register
0009  1803 7E                       MOV A,M             ;load A from mem 18A0
0010  1804 23                       INX H               ;point to next mem loc
0011  1805 BE                       CMP M               ;compare data in mem 18A1 to A
0012  1806 D2 0A 18                 JNC FOUND           ;if A is larger jump forward to Found;
0013  1809 7E                       MOV A,M             ;  otherwise load A from mem 18A1
0014  180A 23               FOUND:  INX H               ;point to next mem loc
0015  180B 77                       MOV M,A             ;store A in mem 18A2
0016  180C 76                       HLT                 ;stop
0017  180D
0018  180D                          .end
```

Z80 program

```
0001  0000                  ;place a number in memory location 18A0h and another in 18A1h;
0002  0000                  ;  this program will determine which is larger and place
0003  0000                  ;  the larger in location  18A2h  (Note: Do not use two
0004  0000                  ;  numbers which are equal.)
0005  0000
0006  1800                          .org 1800h
0007  1800
0008  1800 21 A0 18         START:  LD HL,18A0h         ;initialize HL register
0009  1803 7E                       LD A,(HL)           ;load A from mem 18A0
0010  1804 23                       INC HL              ;point to next mem loc
0011  1805 BE                       CP (HL)             ;compare data in mem 18A1 to A
0012  1806 D2 0A 18                 JP NC,FOUND         ;if A is larger jump forward to Found;
0013  1809 7E                       LD A,(HL)           ;  otherwise load A from mem 18A1
0014  180A 23               FOUND:  INC HL              ;point to next mem loc
0015  180B 77                       LD (HL),A           ;store A in mem 18A2
0016  180C 76                       HALT                ;stop
0017  180D
0018  180D                          .end
```

Fig. 22-15 An example 8080/8085/Z80 program.

Using An Assembler

We need to explain a few things about using an assembler with the 8086/8088 microprocessor. Look at Fig. 22-16 for a moment. The

```
page ,132
```

command tells the assembler to create a list file (Fig. 22-16 is a list file) that is up to 132 columns wide. This gives us more room for the comments at the ends of the lines.

The top portion above the program, which reads

```
CODE    SEGMENT
        ASSUME CS:CODE, DS:CODE, SS:CODE
        ORG 100h
```

and the bottom portion, which reads

```
CODE    ENDS
        END     START
```

are required by the assembler. This information has to do with where in memory we want the program to be and how we want to handle memory segmentation. This model allows the program to be assembled and linked to form an .EXE file which can then be converted to a .COM file with the EXE2BIN DOS utility. A complete discussion of these concepts is beyond the scope of this text. If you will use this model, however, you will be able to use DEBUG to examine the file and use the trace command to single-step through it.

After you assemble and link the file, use the EXE2BIN

```
 1                                   page ,132
 2
 3 0000                              CODE     SEGMENT
 4                                            ASSUME CS:CODE, DS:CODE, SS:CODE
 5 0100                                       ORG 100h
 6
 7 0100   B0 FF                      START:   MOV AL,0FFh      ;minuend
 8 0102   EB 03                               JMP SHORT MINUS ;forward unconditional jump
 9 0104   90                                  NOP
10 0105   90                                  NOP              ;misc. instructions
11 0106   90                                  NOP
12 0107   2C EE                      MINUS:   SUB AL,0EEh      ;subtrahend
13 0109   A2 010E R                           MOV ANSWER,AL    ;store difference
14 010C   CD 20                               INT 20h          ;stop
15
16 010E   00                         ANSWER   DB      00h      ;memory area for answer
17                                                             ;  (initialized to 0)
18
19 010F                              CODE     ENDS
20
21                                            END     START
```

Fig. 22-16 Forward unconditional jump with the 8086/8088 microprocessor (using an assembler).

utility to change it to a .COM file. Then load the file (filename.ext) by typing

debug filename.ext

at the DOS prompt.

Unconditional Jumps

The forward unconditional jump using direct addressing is probably the easiest to understand. Look again at Fig. 22-16. The same program entered with DEBUG is shown in Fig. 22-17.

The program begins by loading AL with FF_{16}. In a moment we are going to subtract another number from this one. First we need to jump to the area of memory where the subtract instruction is. We have placed the subtract instruction several memory locations forward from this point to show, in a very simple manner, how the unconditional jump instruction operates.

```
C>DEBUG
-r
AX=0000  BX=0000  CX=0000  DX=0000  SP=FFEE  BP=0000  SI=0000  DI=0000
DS=3F3D  ES=3F3D  SS=3F3D  CS=3F3D  IP=0100   NV UP EI PL NZ NA PO NC
3F3D:0100 B0FF          MOV     AL,FF
-
-a
3F3D:0100 MOV AL,FF          ;minuend
3F3D:0102 JMP 0107           ;forward unconditional jump
3F3D:0104 NOP
3F3D:0105 NOP                ;misc. instructions
3F3D:0106 NOP
3F3D:0107 SUB AL,EE          ;subtrahend
3F3D:0109 MOV [010E],AL      ;store difference
3F3D:010C INT 20             ;stop
3F3D:010E
-
-
-u 100 10d
3F3D:0100 B0FF          MOV     AL,FF
3F3D:0102 EB03          JMP     0107
3F3D:0104 90            NOP
3F3D:0105 90            NOP
3F3D:0106 90            NOP
3F3D:0107 2CEE          SUB     AL,EE
3F3D:0109 A20E01        MOV     [010E],AL
3F3D:010C CD20          INT     20
-
```

Fig. 22-17 Forward unconditional jump with the 8086/8088 microprocessor (using DEBUG).

The next instruction is our jump instruction. In the source-code column of line 8 in Fig. 22-16 the instruction

JMP SHORT MINUS

appears, which might be different from what you were expecting.

The instruction is saying to jump to a placed called *MINUS*. To be able to jump to a place with a certain name is not a native ability of the 8086/8088 microprocessor. Our assembler is making this possible. Line 12 has the label MINUS in the label column. This is the place we want to jump to. Notice the address at the MINUS label. The address is 0107. Now look back at line 8. In the op code column you see EB, which is the op code for an unconditional jump. Then comes the number 03. This is the number of memory locations by which we must move forward from the instruction after the JMP instruction. Moving forward 03 places takes us to memory location 0107. This is the memory location of the instruction labeled MINUS. If you use an assembler, you can use labels and the assembler will calculate the relative address for you. The term *SHORT* tells the assembler that this place called MINUS is within 127 bytes of our current location.

If you are using DEBUG to assemble these programs, you must enter the program as shown in Fig. 22-17. Toward the top of Fig. 22-17 we simply say

JMP 0107

Notice further down in Fig. 22-17 where we disassembled the program that JMP 0107 disassembles to EB03. Our assembler and DEBUG produced the same code.

After the jump instruction are several NOPs which could be other instructions or just unused memory in a particular microprocessor system.

In line 12 of Fig. 22-16 we subtract EE_{16} from FF_{16} (in AL). In line 13 we store the result of our subtraction in a memory location called *ANSWER*. Look at line 16, labeled ANSWER. In the op code column are the initials DB. This stands for *define byte*. We are telling the assembler to reserve a memory location, namely, a single byte of memory, with the name ANSWER. The assembler is initializing the memory location ANSWER with a value of 0. Our program can then put any other number we wish in that location.

Notice also that the memory location of ANSWER is $010E_{16}$. In the op code column of line 13 we see A2 010E. A2 is the op code for storing the value of AL in a certain memory location. Again the assembler made life simpler by figuring out where the next available memory location would be and setting aside that location for the ANSWER.

If you used DEBUG as shown in Fig. 22-17, then you had to specify memory location 010E as shown.

Finally, in line 14 of Fig. 22-16, the program stops.

You should enter this program and single-step through it, making sure that everything works as described. This is shown in Fig. 22-18.

```
-r
AX=0000  BX=0000  CX=0000  DX=0000  SP=FFEE  BP=0000  SI=0000  DI=0000
DS=3F3D  ES=3F3D  SS=3F3D  CS=3F3D  IP=0100   NV UP EI PL NZ NA PO NC
3F3D:0100 B0FF          MOV     AL,FF
-
-t

AX=00FF  BX=0000  CX=0000  DX=0000  SP=FFEE  BP=0000  SI=0000  DI=0000
DS=3F3D  ES=3F3D  SS=3F3D  CS=3F3D  IP=0102   NV UP EI PL NZ NA PO NC
3F3D:0102 EB03          JMP     0107
-
AX=00FF  BX=0000  CX=0000  DX=0000  SP=FFEE  BP=0000  SI=0000  DI=0000
DS=3F3D  ES=3F3D  SS=3F3D  CS=3F3D  IP=0107   NV UP EI PL NZ NA PO NC
3F3D:0107 2CEE          SUB     AL,EE
-t

AX=0011  BX=0000  CX=0000  DX=0000  SP=FFEE  BP=0000  SI=0000  DI=0000
DS=3F3D  ES=3F3D  SS=3F3D  CS=3F3D  IP=0109   NV UP EI PL NZ NA PE NC
3F3D:0109 A20E01        MOV     [010E],AL                     DS:010E=83
-t

AX=0011  BX=0000  CX=0000  DX=0000  SP=FFEE  BP=0000  SI=0000  DI=0000
DS=3F3D  ES=3F3D  SS=3F3D  CS=3F3D  IP=010C   NV UP EI PL NZ NA PE NC
3F3D:010C CD20          INT     20
-
-d 0100 010F
3F3D:0100   B0 FF EB 03 90 90 90 2C-EE A2 0E 01 CD 20 11 3F   ...........,.......?
-
-
```

Fig. 22-18 Forward unconditional jump with the 8086/8088 microprocessor (single-stepping with the Trace command).

```
1                                         page ,132
2
3 0000                                    CODE    SEGMENT
4                                                 ASSUME CS:CODE, DS:CODE, SS:CODE
5 0100                                            ORG 100h
6
7 0100  B1 03                             START:  MOV CL,03h        ;initialize CL (repeats)
8 0102  B5 00                                     MOV CH,00h        ;initialize CH
9 0104  FE C5                             REPEAT: INC CH            ;times loop has repeated
10 0106  90                                       NOP               ;misc. instructions
11 0107  FE C9                                    DEC CL            ;decrement CL
12 0109  75 F9                                    JNZ REPEAT        ;if CL not equal to 0 then
13                                                                  ;  branch back to start of
14                                                                  ;  loop
15 010B  CD 20                                    INT 20h           ;stop
16
17 010D                                   CODE    ENDS
18
19                                                END     START
```

Fig. 22-19 A backward conditional jump creating a loop with the 8086/8088 microprocessor (using an assembler).

Conditional Branches

Now let's see an example of conditional branching. Figure 22-19 shows such an example using an assembler.

Figure 22-20 shows the same program using DEBUG.

In this program we are going to do several things differently from the way they were done in the last program. First, we are using a conditional jump or branch rather than an unconditional one. Second, we are branching backward rather than forward. Third, we are creating a loop by branching backward and repeating a section of the program. Finally, we are using a register as a counter to control how many times the loop repeats.

In line 7 of Fig. 22-19 we place the number 3_{16} in CL. This register *controls* how many times we will branch backward. In line 8 we clear CH, making it 00_{16} so that it can be used to *count* how many times the loop repeats.

Line 9 marks the beginning of the loop, and we have named that location *REPEAT*. In this line we increment CH since we are beginning to pass through the loop, in this case for the first time. Register CH is keeping track of how many times the loop is passed through. Line 10 represents the fact that there could be many instructions inside this loop which are going to be repeated.

Line 11 decrements (reduced by 1) register CL. Register CL keeps track of how many times we have left to go through the loop.

Line 12 is where we meet our conditional branch instruction. JNZ means **J**ump if **N**ot **Z**ero. Your first thought might be, "If what isn't zero?"

All the conditional branch instructions are influenced by the most recent instruction that affected the flag they check. In this case the zero flag is checked. What was the last instruction which sets or clears the zero flag? The DEC CL

```
-a 100
77B3:0100 MOV CL,03          ;initialize CL (repeats)
77B3:0102 MOV CH,00          ;initialize CH
77B3:0104 INC CH             ;times loop has repeated
77B3:0106 NOP                ;misc instructions
77B3:0107 DEC CL             ;decrement CL
77B3:0109 JNZ 0104           ;if CL not equal to 0 then
77B3:010B                    ;  branch back to start of
77B3:010B                    ;  loop
77B3:010B INT 20             ;stop
77B3:010D
-
-u 100 10c
77B3:0100 B103          MOV     CL,03
77B3:0102 B500          MOV     CH,00
77B3:0104 FEC5          INC     CH
77B3:0106 90            NOP
77B3:0107 FEC9          DEC     CL
77B3:0109 75F9          JNZ     0104
77B3:010B CD20          INT     20
-
```

Fig. 22-20 A backward conditional jump creating a loop with the 8086/8088 microprocessor (using DEBUG).

```
 1                          page ,132
 2
 3                          ;place a number in memory location DATA and another in DATA+1;
 4                          ;  this program will determine which is larger and place
 5                          ;  the larger in location DATA+2  (Note: Do not use two
 6                          ;  numbers which are equal.)
 7
 8 0000                     CODE     SEGMENT
 9                                   ASSUME CS:CODE, DS:CODE, SS:CODE
10 0100                              ORG 0100h
11
12 0100   BB 0000           START:   MOV BX,00h              ;initialize BX register
13 0103   8A 87 0119 R               MOV AL,[DATA + BX]      ;move byte to AL from mem loc DATA
14 0107   43                         INC BX                  ;point to next mem loc (DATA + 1)
15 0108   3A 87 0119 R               CMP AL,[DATA + BX]      ;compare byte in mem DATA + 1 to AL
16 010C   77 04                      JA FOUND                ;if AL is larger jump forward to Found
17 010E   8A 87 0119 R               MOV AL,[DATA + BX]      ;  otherwise move byte to AL from mem
                                                                DATA + 1
18 0112   43                FOUND:   INC BX                  ;point to next mem loc (DATA + 2)
19 0113   88 87 0119 R               MOV [DATA + BX],AL      ;move byte in AL to mem DATA + 2
20 0117   CD 20                      INT 20h                 ;stop
21
22 0119   05 04 00          DATA     DB      05h,04h,00h     ;you can use different values for the
23                                                           ;  first two numbers
24
25 011C                     CODE     ENDS
26
27                                   END     START
```

Fig. 22-23 An example 8086/8088 program (using an assembler).

```
C>DEBUG
-r
AX=0000  BX=0000  CX=0000  DX=0000  SP=FFEE  BP=0000  SI=0000  DI=0000
DS=3F3D  ES=3F3D  SS=3F3D  CS=3F3D  IP=0100   NV UP EI PL NZ NA PO NC
3F3D:0100 BB0000        MOV     BX,0000
-
-a
3F3D:0100 MOV     BX,0000                ;initialize BX register
3F3D:0103 MOV     AL,[BX+0119]           ;move byte to AL from mem loc 0119 + 0
3F3D:0107 INC     BX                     ;point to next mem loc 0119 + 1
3F3D:0108 CMP     AL,[BX+0119]           ;compare byte in mem 0119 + 1 to AL
3F3D:010C JA      0112                   ;if AL is larger jump forward to 0112,
3F3D:010E MOV     AL,[BX+0119]           ;  otherwise move byte to AL from 0119 + 1
3F3D:0112 INC     BX                     ;point to next mem loc 0119 + 2
3F3D:0113 MOV     [BX+0119],AL           ;move byte in AL to mem 0119 + 2
3F3D:0117 INT     20                     ;stop
3F3D:0119
-
-
-u 0100 0118
3F3D:0100 BB0000        MOV     BX,0000
3F3D:0103 8A871901      MOV     AL,[BX+0119]
3F3D:0107 43            INC     BX
3F3D:0108 3A871901      CMP     AL,[BX+0119]
3F3D:010C 7704          JA      0112
3F3D:010E 8A871901      MOV     AL,[BX+0119]
3F3D:0112 43            INC     BX
3F3D:0113 88871901      MOV     [BX+0119],AL
3F3D:0117 CD20          INT     20
```

Fig. 22-24 An example 8086/8088 program (using DEBUG).

```
-
-e 0119
3F3D:0119  5E.05    F6.04    8B.00    07.00
-
-d 0110 011f
3F3D:0110  19 01 43 88 87 19 01 CD-20 05 04 00 00 89 46 EE   ..C..... .....F.
-


-r
AX=0000  BX=0000  CX=0000  DX=0000  SP=FFEE  BP=0000  SI=0000  DI=0000
DS=3F3D  ES=3F3D  SS=3F3D  CS=3F3D  IP=0100   NV UP EI PL NZ NA PO NC
3F3D:0100 BB0000        MOV     BX,0000
-t

AX=0000  BX=0000  CX=0000  DX=0000  SP=FFEE  BP=0000  SI=0000  DI=0000
DS=3F3D  ES=3F3D  SS=3F3D  CS=3F3D  IP=0103   NV UP EI PL NZ NA PO NC
3F3D:0103 8A871901      MOV     AL,[BX+0119]                   DS:0119=05
-t

AX=0005  BX=0000  CX=0000  DX=0000  SP=FFEE  BP=0000  SI=0000  DI=0000
DS=3F3D  ES=3F3D  SS=3F3D  CS=3F3D  IP=0107   NV UP EI PL NZ NA PO NC
3F3D:0107 43            INC     BX
-t

AX=0005  BX=0001  CX=0000  DX=0000  SP=FFEE  BP=0000  SI=0000  DI=0000
DS=3F3D  ES=3F3D  SS=3F3D  CS=3F3D  IP=0108   NV UP EI PL NZ NA PO NC
3F3D:0108 3A871901      CMP     AL,[BX+0119]                   DS:011A=04
-t

AX=0005  BX=0001  CX=0000  DX=0000  SP=FFEE  BP=0000  SI=0000  DI=0000
DS=3F3D  ES=3F3D  SS=3F3D  CS=3F3D  IP=010C   NV UP EI PL NZ NA PO NC
3F3D:010C 7704          JA      0112
-t

AX=0005  BX=0001  CX=0000  DX=0000  SP=FFEE  BP=0000  SI=0000  DI=0000
DS=3F3D  ES=3F3D  SS=3F3D  CS=3F3D  IP=0112   NV UP EI PL NZ NA PO NC
3F3D:0112 43            INC     BX
-t

AX=0005  BX=0002  CX=0000  DX=0000  SP=FFEE  BP=0000  SI=0000  DI=0000
DS=3F3D  ES=3F3D  SS=3F3D  CS=3F3D  IP=0113   NV UP EI PL NZ NA PO NC
3F3D:0113 88871901      MOV     [BX+0119],AL                   DS:011B=00
-t

AX=0005  BX=0002  CX=0000  DX=0000  SP=FFEE  BP=0000  SI=0000  DI=0000
DS=3F3D  ES=3F3D  SS=3F3D  CS=3F3D  IP=0117   NV UP EI PL NZ NA PO NC
3F3D:0117 CD20          INT     20
-
-
-d 0110 011f
3F3D:0110  19 01 43 88 87 19 01 CD-20 05 04 05 00 89 46 EE   ..C..... .....F.
-
-
```

Fig. 22-24 (cont.)

GLOSSARY

decrement To decrease. Most microprocessors decrement registers or memory locations by 1.

increment To increase. Most microprocessors increment registers or memory locations by 1.

loop A group of instructions which can be executed more than once. The program "falls through" the loop when some condition exists or when the loop has been executed a predetermined number of times.

nest To fit one inside another. Loops can be nested by having one small loop executing within a larger loop.

SELF-TESTING REVIEW

Read each of the following and provide the missing words. Answers appear at the beginning of the next question.

1. Branches or jumps can be made to execute all the time or only when certain conditions exist. That is, branches and loops can be ________ or ________.
2. *(conditional, unconditional)* When a program branches backward and repeats a group of instructions, it is called a ________.
3. *(loop)* Compare instructions generally (though not always) set and clear the microprocessor's flags as though ________ had occurred. *(subtraction)*

PROBLEMS

Solve the following problems by using the microprocessor of your choice.

You may have some difficulty with the following two problems; therefore only two are given. As you begin each problem, do not immediately think of which microprocessor instructions to use. Instead, think about the problem itself and visualize what the memory locations will contain. Think of how to move the data between registers and memory locations to solve the problem, and *then* think about what instructions can be used to accomplish the moves.

22-1. Write a program which will use the first number in a list of unsigned binary numbers as a reference, will compare that number to each of the following numbers in the list, and will then stop when it finds the first number in the list which is smaller than or equal to the reference number. Finally, the program should store that first number which was smaller or equal to the reference number in a memory location called ANSWER.

(*Important:* The numbers in the list must be considered unsigned binary numbers. At least *one number* in the list *must* be smaller than or equal to the reference number. All the numbers *may* be smaller or equal to the reference. The program will be most interesting if more than one, but not all, the numbers are smaller than or equal to the reference.)

(*Note:* You will need to enter the list of numbers before running the program. *The list must have a minimum of two numbers* and can have as many additional numbers as you wish. We have started the list of numbers at memory location \$03A0 for the 6502, \$01A0 for the 6800/6808, and at 18A0h for the 8080/8085/Z80, and at a location labeled LIST for the 8086/8088.)

22-2. Write a program which will look at a list of numbers which you will store in memory. The end of this list will be indicated by the number 00. The number 00 cannot be used anywhere in the list except to mark its end. Write the program so that it will add each pair of consecutive numbers. That is, if the list contained the numbers 06_{16}, $2E_{16}$, 36_{16}, 42_{16}, and 00_{16}, it would perform the following additions:

$$06_{16} + 2E_{16} = 34_{16}$$
$$2E_{16} + 36_{16} = 64_{16}$$
$$36_{16} + 42_{16} = 78_{16}$$

The program should not add the 00_{16} to the preceding number since 00_{16} is not one of the numbers in the list but indicates the end of the list.

When the program adds the first two numbers, it should place their sum in a memory location called *LRGST* (largest). As it adds each of the following pairs, it should compare their sum with the number in LRGST. If the new sum is larger than the number in LRGST, then the new largest number should be placed in LRGST. Thus, after the program has added all the pairs together, LRGST will contain the largest sum that was created. All numbers should be considered unsigned binary numbers.

(*Note:* The list must contain at least one number, with the number 00 following it to indicate the end of the list. In this case no sum should appear in LRGST because there can be no sum with a list of only one number. The list can contain any number of numbers beyond one.)

(*Note:* We have used the numbers $2E_{16}$, $3C_{16}$, $1B_{16}$, 46_{16}, and 00_{16} to end the list, in that order, in the answer key. You should try altering your list to make sure it works under various circumstances.)

SUBROUTINE AND STACK INSTRUCTIONS

At this point we have covered most of the instruction set of each of the microprocessors featured in this text. Two final topics, however, the stack and subroutines, may be the most important ones. Without subroutines, programs written for these microprocessors would be unmanageable. Subroutines are used when there are tasks which must be executed or used many times. The subroutine provides a way to write a program segment which can handle a specific task and be reused.

The stack is important because it supports subroutines by storing information the microprocessor needs when it tries to return from a subroutine.

New Concepts

This chapter deals with subroutines and with the stack, especially as the stack relates to subroutines. The use of the stack in passing parameters between subroutines or in mixed-language programs is beyond the scope of this text and is not discussed.

We discussed the stack in Chap. 15. We'll review a portion of that chapter here.

23-1 STACK AND STACK POINTER

The stack, in the case of the microprocessors used in this text, is located in RAM. Refer to Fig. 23-1.

The structure of the stack is a *first-in-last-out* (FILO) type of structure. Unlike main memory, where you can access any data item in any order, the stack is designed so that you can access only the top of the stack. If you want to place data in the stack, it must go on top, and if you wish to remove data from the stack, it must be on top before it can be removed.

Let's see how the situation in Fig. 23-1 has come to be. To do that, refer to Fig. 23-2. Data item #1 is the first item we wish to place on the stack.

At this time the stack pointer is "pointing" to memory location 0008; therefore, data item #1 will be placed in the stack at that memory location. Putting a piece of data in the stack is called *pushing* data onto the stack. It is as though the data is being pushed in from the top. Now look at Fig. 23-3.

We have pushed data item #1 onto the stack, and the stack pointer has been decremented or decreased by 1,

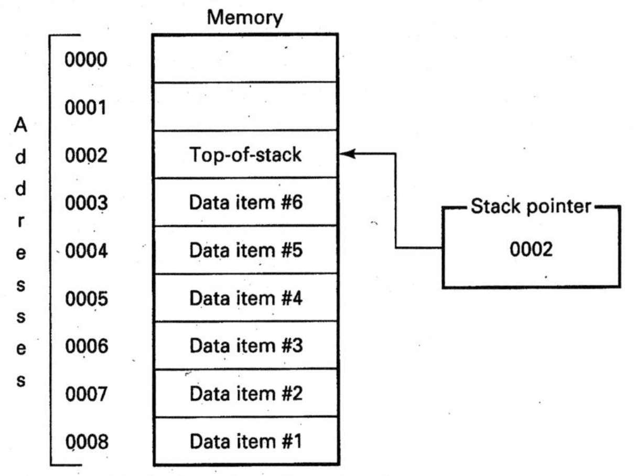

Fig. 23-1 Typical stack and stack pointer.

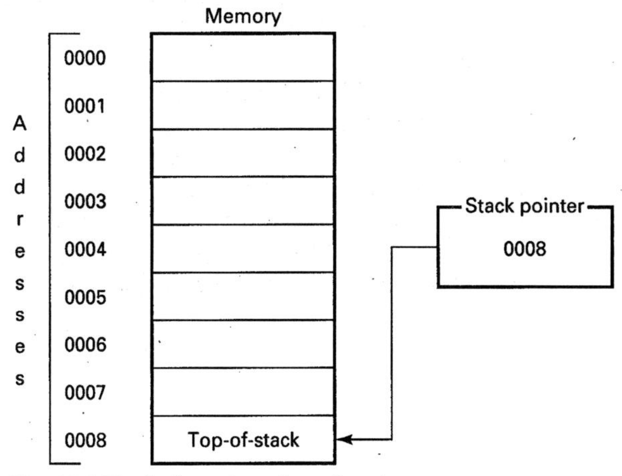

Fig. 23-2 Typical stack and stack pointer.

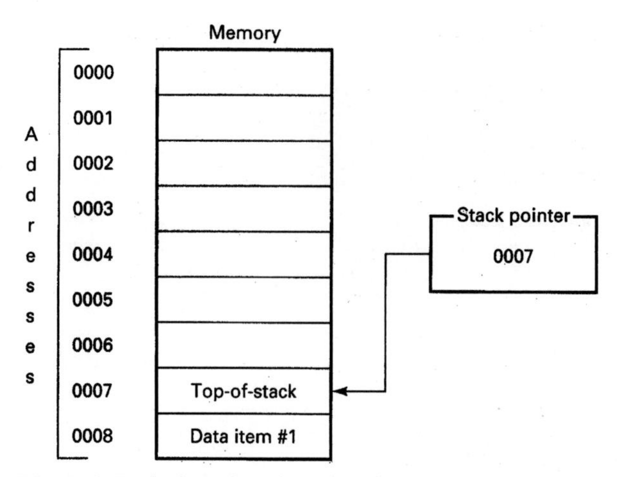

Fig. 23-3 Typical stack and stack pointer.

which means that it is now pointing to memory location 0007. Now 0007 is the top-of-the-stack. Now let's push data item #2 onto the stack. The stack will appear as it does in Fig. 23-4.

When data item #2 was *pushed* onto the stack, it went into the location which was being pointed to by the stack pointer, which was 0007. The stack pointer was then decremented to 0006. This process will be repeated until the stack appears as it did in Fig. 23-1.

At some point we will need this data in the stack, so we will remove it from the top-of-the-stack. This is called *popping* or *pulling* the data from the stack. We simply reverse the whole process. As each data item is removed, the stack pointer will drop, which in this case means that it will increment or point to the next-greater memory address.

23-2 BRANCHING VERSUS SUBROUTINES

In Chap. 22, where branching was discussed, we saw that branching causes program execution to *jump* or *branch* to another section of the program. This may be an unconditional jump or a conditional jump. In either case the instructions immediately following the jump instruction may not be executed. If we branch to another section of the program, it is because we don't want to execute the instructions immediately following the branch instructions.

Subroutines also allow us to jump to another section of the program to execute instructions there. Subroutines differ from jumps or branches, however, in that the instructions which immediately follow the subroutine instruction are executed later. (The act of starting to execute a subroutine is referred to as *jumping to a subroutine* if you are using a 6502 or 6800/6808 microprocessor. It is referred to as *calling a subroutine* if you are using an 8080/8085/Z80 or 8086/8088 microprocessor.)

After the microprocessor jumps to a subroutine or calls a subroutine, the instructions in the subroutine begin to execute. At the end of the subroutine is an instruction called the *return* instruction. The return instruction is usually the last instruction in the subroutine; it tells the microprocessor to go back to the place in the program where it was when the subroutine was called and to pick up where it left off. This is shown in Fig. 23-5.

It is also possible for a subroutine to call another subroutine. These *nested* subroutines then sort of "unwind" and return in the reverse order relative to that in which they were called. This is illustrated in Fig. 23-6.

23-3 HOW DO SUBROUTINES RETURN?

The ability of a subroutine to return to the exact location it came from, especially when nested several layers deep, raises the question of how it knows where to return to.

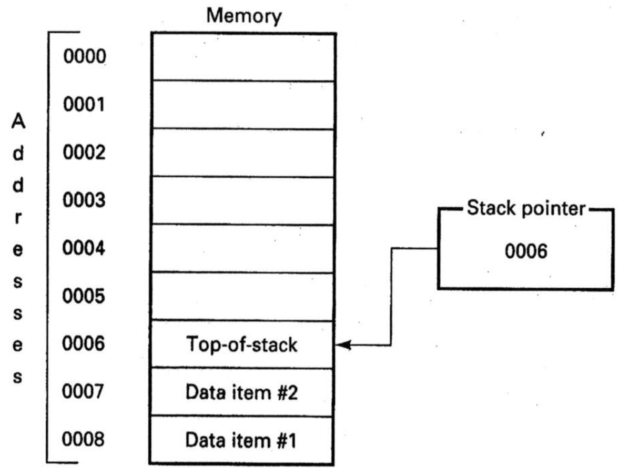

Fig. 23-4 Typical stack and stack pointer.

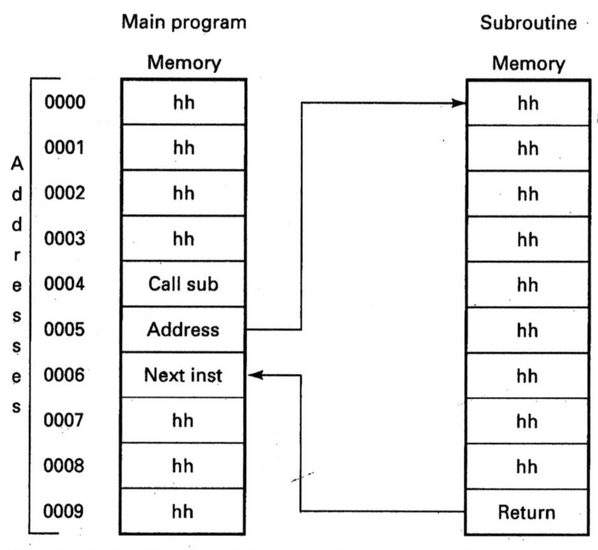

Fig. 23-5 "Calling" or "jumping" to a subroutine.

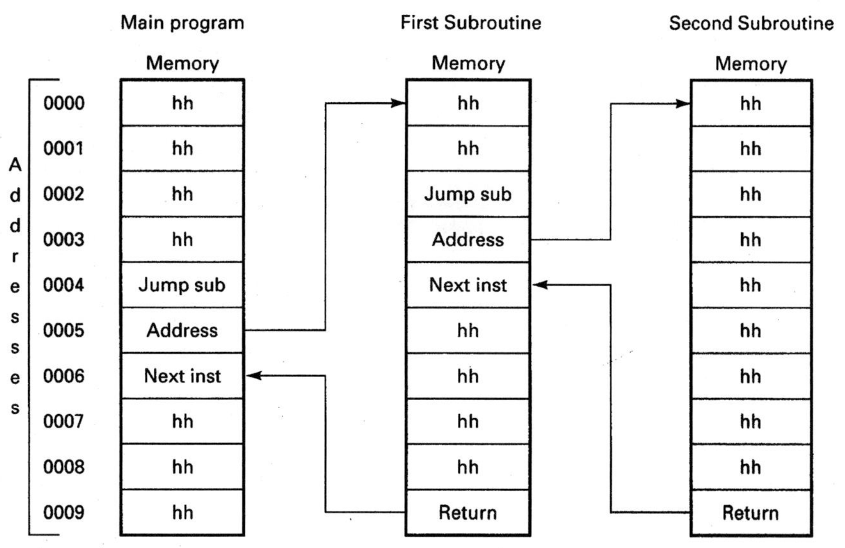

Fig. 23-6 Nested subroutines.

That is, how does it know where it came from? The answer lies in what happens just before the microprocessor leaves the main program, or current subroutine, to go to the subroutine being called.

The microprocessor must know two things before a subroutine can be called or jumped to. First, it must know where it's going, and second, it must know how to get back.

The instruction *jump to subroutine* or *call subroutine* contains the address of the desired subroutine. This may be in the form of an absolute address or an offset of some sort. This is the destination.

The program counter (8086/8088 instruction pointer) contains the address of the next instruction to be executed. This is the point to which the microprocessor needs to return. Refer to Fig. 23-7.

When the subroutine is called, the contents of the program counter are pushed onto the stack. This requires more than one push, since in the case of the 8-bit microprocessors the stack is only 8 bits wide but the program counter is 16 bits wide. (The 8088 stores not only the instruction pointer but may also store the code segment, depending on the type of call—*near* or *far*.)

After the program counter (instruction pointer) is pushed onto the stack, the address of the subroutine which is being called or jumped to is placed in the program counter (instruction pointer), and program execution begins at this new address.

Execution now continues in the subroutine until a return instruction is encountered. Refer to Fig. 23-8.

At this point, the address of the next instruction which was to be executed after the subroutine jump or call, which

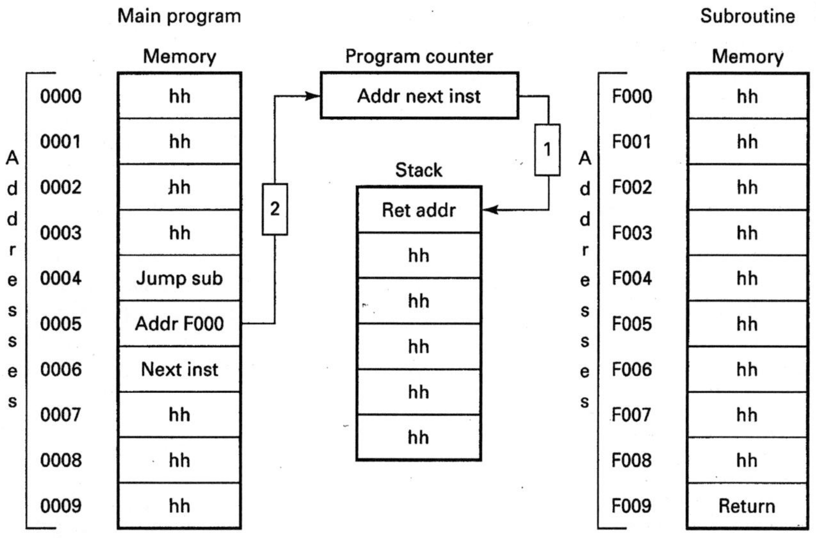

Fig. 23-7 Calling a subroutine.

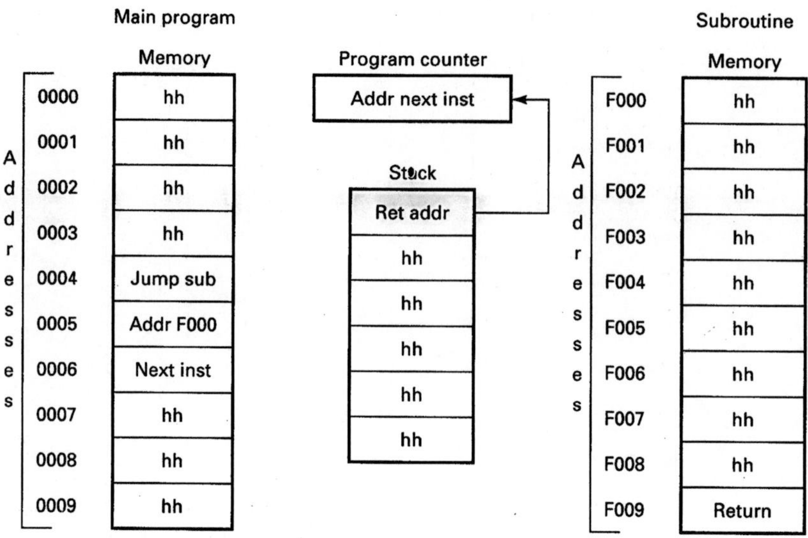

Fig. 23-8 Returning from a subroutine.

has been stored on the stack, is pulled or popped from the stack and placed in the program counter (instruction pointer). Execution then proceeds from that point forward in the main program.

To summarize:

1. The call or jump to subroutine instruction is encountered.
2. The program counter (instruction pointer) is already pointing to the next instruction to be executed (in this section of the program code).
3. The contents of the program counter (instruction pointer) are pushed onto the stack.
4. The address of the subroutine is placed in the program counter (instruction pointer).
5. Program execution now begins in the subroutine.
6. When a return instruction is encountered, the return address, which has been previously stored in the stack, is pulled from the stack and placed in the program counter (instruction pointer).
7. Program execution continues from where it left off before the subroutine was called or jumped to.

23-4 PUSHING AND POPPING REGISTERS

When a subroutine is called or jumped to, the use and operation of the stack are automatic. You don't have to tell the microprocessor to store the return address on the stack. It is done automatically.

In addition to the automatic use of the stack in subroutine calls, the stack can be used directly by the programmer for other purposes. Although each microprocessor is different, in general, you can push onto the stack, and pull from the stack, the contents of some or most of the microprocessor's registers. This is often used to pass values from the main program to subroutines and back, or from subroutine to subroutine. These values are sometimes referred to as *parameters*. The use of the stack in parameter passing, however, is beyond the scope of this text.

Specific Microprocessor Families

Let's look at each of our featured microprocessors. We will not go into great detail about what each microprocessor does automatically before and after a subroutine is called. Rather, we will give examples which show how to call a subroutine and how to nest subroutines.

23-5 6502 FAMILY

The 6502 microprocessor works as described in the New Concepts section of this chapter. There is one point worth noting, however.

The stack pointer of the 6502 is a little different from that of the other microprocessors featured in this text. The changeable portion of the stack pointer is only 8 bits wide (all the others are 16 bits wide) and a 9th bit is always set to 1. This means that the location of the stack must lie in the range from address 0100 to 01FF. This is shown in Fig. 23-9.

Setting the Stack Pointer

Our first example program illustrates how to set the stack pointer to a desired address and then call a subroutine. It

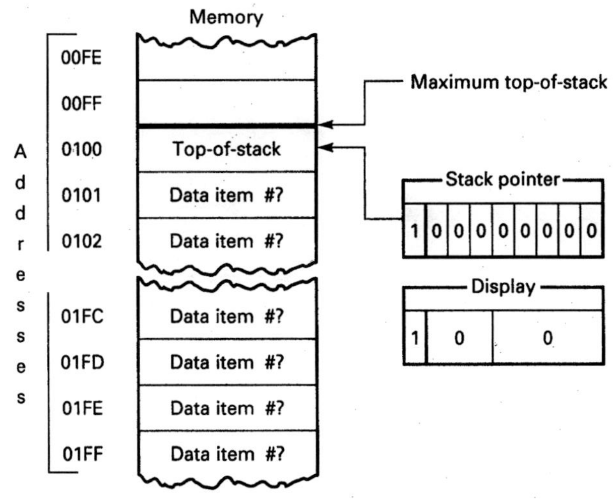

Fig. 23-9 6502 family stack and stack pointer.

is important to note that, with the simple programs we have used throughout this text, setting the stack pointer is normally not required. The microprocessor trainer or computer you are working with will have an operating system that will set the stack pointer to a logical address based on available memory.

Figure 23-10 contains our example program. It sets the stack pointer to a desired address and then calls a subroutine. The subroutine does not actually do anything. It gives you a chance to single-step through a program and watch the stack pointer and program counter.

Calling More than One Subroutine (Not Nested)

Our next example program is shown in Fig. 23-11.

The two subroutines shown here occur one after the other. They are *not* nested. You should single-step through this program and watch the stack pointer and program counter. This is important because the next program will also contain two subroutines, but they *will* be nested. We want you to see the difference between the two.

Again, these first programs do not do anything. Just observe the behavior of the program counter and the stack pointer.

Nesting Subroutines

The program shown in Fig. 23-12 also has two subroutines. They *are* nested, however.

Single-step through this program and watch the stack pointer and the program counter carefully. Notice how they act differently from the way they did in the last program. When you are inside the second subroutine, the stack is holding the return addresses for both subroutines. That's why it decrements further.

Pushing Registers

The example program shown in Fig. 23-13 shows how to use the stack to move information from one register to another.

The program pushes the flags onto the stack and then pulls them from off the stack into the accumulator. The

```
0001  0340                    .ORG $0340
0002  0340                    ;
0003  0340 A2 F9     START:   LDX #$F9     ;load number for stack pointer
0004  0342 9A                 TXS          ;load stack pointer
0005  0343 EA                 NOP          ;misc instructions
0006  0344 20 48 03           JSR SUBRTN   ;jump to subroutine (watch stack pointer)
0007  0347 00                 BRK          ;stop
0008  0348 EA        SUBRTN:  NOP          ;misc instructions
0009  0349 60                 RTS          ;return from subroutine
0010  034A                    ;
0011  034A                    .END
```

Fig. 23-10 6502 program loading stack pointer and calling a subroutine.

```
0001  0340                    .ORG $0340
0002  0340
0003  0340 EA        START:   NOP
0004  0341 20 49 03           JSR RTNE_1   ;
0005  0344 EA                 NOP          ;
0006  0345 20 4B 03           JSR RTNE_2   ;
0007  0348 00                 BRK          ;
0008  0349 EA        RTNE_1:  NOP          ;
0009  034A 60                 RTS          ;
0010  034B EA        RTNE_2:  NOP          ;
0011  034C 60                 RTS          ;
0012  034D
0013  034D                    .END
```

Watch the stack pointer as each subroutine is "called" or "jumped to," and as execution returns from each back to the main program. These subroutines are *not* nested.

Fig. 23-11 6502 program with two subroutines *not* nested.

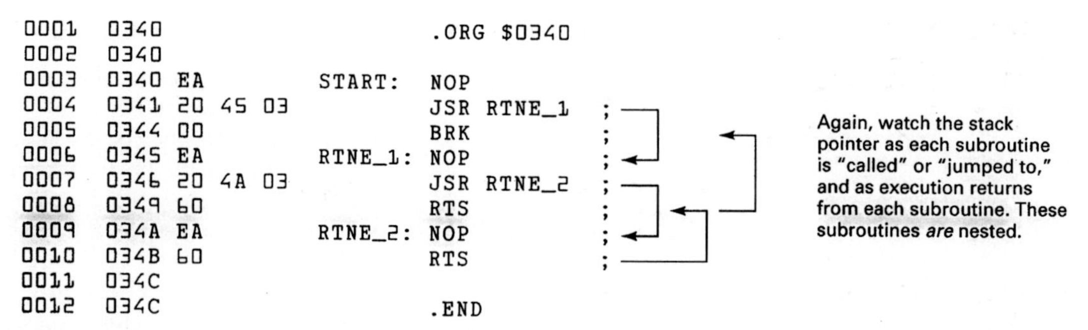

```
0001  0340                        .ORG $0340
0002  0340
0003  0340 EA          START:     NOP
0004  0341 20 45 03               JSR RTNE_1   ;
0005  0344 00                     BRK          ;
0006  0345 EA          RTNE_1:    NOP          ;
0007  0346 20 4A 03               JSR RTNE_2   ;
0008  0349 60                     RTS          ;
0009  034A EA          RTNE_2:    NOP          ;
0010  034B 60                     RTS          ;
0011  034C
0012  034C                        .END
```


Fig. 23-12 6502 program with two *nested* subroutines.

```
0001  0340                   .ORG $0340
0002  0340
0003  0340 08       START:   PHP          :push flags then decrement stack pointer
0004  0341 68                PLA          ;pull then increment stack pointer
0005  0342 00                BRK          ;stop
0006  0343
0007  0343                   .END
```

Fig. 23-13 6502 program which pushes a register.

bits of the accumulator, which represent the status of the flags, can now be examined by the program or stored in memory.

A Useful Program Containing a Subroutine

Let's take a look at the program shown in Fig. 23-14.

This program's purpose is as follows:

> This program will read a list of five signed binary numbers. As it reads each number, it will determine whether that number is positive or 0 or negative. If it is positive or 0, it will do nothing with the number. If the number is negative, a subroutine will be entered. This subroutine will find the absolute value of the number (that is, it will make the negative number positive). It will then write this positive number into memory in place of the original negative number. (We used the decimal numbers 3, −4, −2, 0, and 5.) (*Note:* If the microprocessor being used here has a negate instruction, that instruction will not be used.)

Enter this program into your microprocessor trainer or computer and single-step through it. Study the program and make sure that you understand its operation.

```
0001  0340                             .org $0340
0002  0340
0003  0340 A2 00          START:       LDX #$00          ;address of beginning of list
0004  0342 A0 06                       LDY #$06          ;counter
0005  0344 88             GETNUM:      DEY               ;decrement counter
0006  0345 F0 19                       BEQ DONE          ;if no items left end program
0007  0347 BD 61 03                    LDA $LIST,X       ;load number from list
0008  034A C9 00                       CMP #$00          ;is it positive/zero or negative?
0009  034C 10 05                       BPL NEXT          ;if positive get next number now
0010  034E F0 03                       BEQ NEXT          ;if zero get next number now
0011  0350 20 57 03                    JSR NEGNUM        ;if negative call subroutine
0012  0353 E8             NEXT:        INX               ;point to next number in list
0013  0354 4C 44 03                    JMP GETNUM        ;branch back to beginning
0014  0357 49 FF          NEGNUM:      EOR #$FF          ;invert all bits of negative number
0015  0359 18                          CLC               ;prepare for addition
0016  035A 69 01                       ADC #$01          ;add 1 to inverted bits
0017  035C 9D 61 03                    STA $LIST,X       ;write absolute value over
                                                           old negative value
0018  035F 60                          RTS               ;return
0019  0360 00             DONE:        BRK               ;stop
0020  0361
0021  0361 03FCFE0005     LIST:        .db     3, -4, -2, 0, 5    ;list of 5 numbers
0022  0366
0023  0366                             .end
```

Fig. 23-14 A useful 6502 program which contains a subroutine.

```
0001  0100                     .ORG $0100
0002  0100                     ;
0003  0100 8E 01 FF   START:   LDS #$01FF   ;load stack pointer
0004  0103 01                  NOP          ;misc instructions
0005  0104 BD 01 08            JSR SUBRTN   ;jump to subroutine (watch stack pointer)
0006  0107 3E                  WAI          ;stop
0007  0108 01         SUBRTN:  NOP          ;misc instructions
0008  0109 39                  RTS          ;return from subroutine
0009  010A                     ;
0010  010A                     .END
```

Fig. 23-15 6800/6808 program loading stack pointer and calling a subroutine.

23-6 6800/6808 FAMILY

The 6800/6808 microprocessor works as described in the New Concepts section of this chapter. We'll look at several sample programs which you can enter into your microprocessor trainer or computer and examine.

Setting the Stack Pointer

Our first example program illustrates how to set the stack pointer to a desired address and then call a subroutine. It is important to note that, with the simple programs we have used throughout this text, setting the stack pointer is normally not required. The microprocessor trainer or computer you are working with will have an operating system that will set the stack pointer to a logical address based on available memory.

Figure 23-15 contains our example program. It sets the stack pointer to a desired address and then calls a subroutine. The subroutine does not actually do anything. It gives you a chance to single-step through a program and watch the stack pointer and program counter.

Calling More than One Subroutine (Not Nested)

Our next example program is shown in Fig. 23-16.

The two subroutines shown here occur one after the other. They are *not* nested. You should single-step through this program and watch the stack pointer and program counter. This is important because the next program will also contain two subroutines, but they *will* be nested. We want you to see the difference between the two.

Again, these first programs do not do anything. Just observe the behavior of the program counter and the stack pointer.

Nesting Subroutines

The program shown in Fig. 23-17 also has two subroutines. They *are* nested, however.

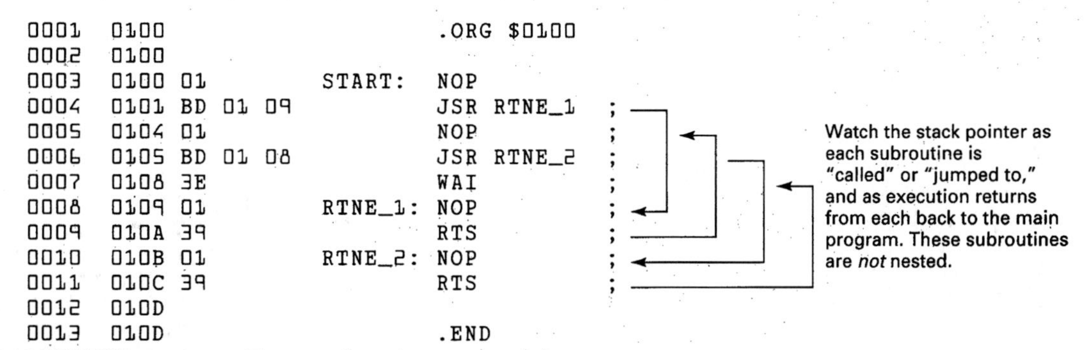

```
0001  0100                     .ORG $0100
0002  0100
0003  0100 01         START:   NOP
0004  0101 BD 01 09            JSR RTNE_1   ;
0005  0104 01                  NOP          ;
0006  0105 BD 01 0B            JSR RTNE_2   ;
0007  0108 3E                  WAI          ;
0008  0109 01         RTNE_1:  NOP          ;
0009  010A 39                  RTS          ;
0010  010B 01         RTNE_2:  NOP          ;
0011  010C 39                  RTS          ;
0012  010D
0013  010D                     .END
```


Fig. 23-16 6800/6808 program with two subroutines *not* nested.

```
0001  0100                     .ORG $0100
0002  0100
0003  0100 01         START:   NOP
0004  0101 BD 01 05            JSR RTNE_1   ;
0005  0104 3E                  WAI          ;
0006  0105 01         RTNE_1:  NOP          ;
0007  0106 BD 01 0A            JSR RTNE_2   ;
0008  0109 39                  RTS          ;
0009  010A 01         RTNE_2:  NOP          ;
0010  010B 39                  RTS          ;
0011  010C
0012  010C                     .END
```

Again, watch the stack pointer as each subroutine is "called" or "jumped to," and as execution returns from each subroutine. These subroutines *are* nested.

Fig. 23-17 6800/6808 program with two *nested* subroutines.

```
0001  0100                      .ORG $0100
0002  0100
0003  0100 86 12      START:    LDAA #$12    ;load values into
0004  0102 C6 34                LDAB #$34    ;  registers
0005  0104 36                   PSHA         ;push then decrement stack pointer
0006  0105 37                   PSHB         ;push then decrement stack pointer again
0007  0106 32                   PULA         ;pull then increment stack pointer
0008  0107 33                   PULB         ;pull then increment stack pointer again
0009  0108 3E                   WAI          ;stop
0010  0109
0011  0109                      .END
```

Fig. 23-18 6800/6808 program which pushes a register.

Single-step through this program and watch the stack pointer and the program counter carefully. Notice how they act differently from the way they did in the last program. When you are inside the second subroutine, the stack is holding the return addresses for both subroutines. That's why it decrements further.

Pushing Registers

The example program shown in Fig. 23-18 shows how to use the stack to move information from one register to another.

The program loads accumulators A and B with a value, pushes A and B onto the stack, and then pulls them from the stack in reverse order. This places the data that was in A in B and the data that was in B in A.

A Useful Program Containing a Subroutine

Let's take a look at the program shown in Fig. 23-19. This program's purpose is as follows:

> This program will read a list of five signed binary numbers. As it reads each number, it will determine whether that number is positive or 0 or negative. If it is positive or 0, it will do nothing with the number. If the number is negative, a subroutine will be entered. This subroutine will find the absolute value of the number (that is, it will make the negative number positive). It will then write this positive number into memory in place of the original negative number. (We used the decimal numbers 3, −4, −2, 0, and 5.) (*Note:* If the microprocessor being used here has a negate instruction, that instruction will not be used.)

Enter this program into your microprocessor trainer or computer and single-step through it. Study the program and make sure that you understand its operation.

```
0001  0100                           .org $0100
0002  0100
0003  0100 CE 01 1B      START:      LDX #$LIST       ;address of beginning of list
0004  0103 C6 06                     LDAB #$06        ;counter
0005  0105 5A            GETNUM:     DECB             ;decrement counter
0006  0106 27 12                     BEQ DONE         ;if no items left end program
0007  0108 A6 00                     LDAA $00,X       ;load number from list
0008  010A 81 00                     CMPA #$00        ;is it positive/zero or negative?
0009  010C 2C 03                     BGE NEXT         ;if positive get next number now
0010  010E BD 01 14                  JSR NEGNUM       ;if negative call subroutine
0011  0111 08            NEXT:       INX              ;point to next number in list
0012  0112 20 F1                     BRA GETNUM       ;branch back to beginning
0013  0114 43            NEGNUM:     COMA             ;invert all bits of negative number
0014  0115 8B 01                     ADDA #$01        ;add 1 to inverted bits
0015  0117 A7 00                     STAA $00,X       ;write absolute value over
                                                        old negative value
0016  0119 39                        RTS              ;return
0017  011A 3E            DONE:       WAI              ;stop
0018  011B
0019  011B 03FCFE0005    LIST:       .db      3, -4, -2, 0, 5      ;list of 5 numbers
0020  0120
0021  0120                           .end
```

Fig. 23-19 A useful 6800/6808 program which contains a subroutine.

23-7 8080/8085/Z80 FAMILY

The 8080/8085/Z80 microprocessor works as described in the New Concepts section of this chapter. We'll look at several sample programs which you can enter into your microprocessor trainer or computer and examine.

The 8080/8085/Z80 microprocessors do have two features

that the other microprocessors featured in this text don't have: They have the ability to perform conditional subroutine calls and to perform conditional returns from subroutines. All the other microprocessors featured in this text have only unconditional calls and unconditional returns.

Setting the Stack Pointer

Our first example program illustrates how to set the stack pointer to a desired address and then call a subroutine. It is important to note that, with the simple programs we have used throughout this text, setting the stack pointer is normally not required. The microprocessor trainer or computer you are working with will have an operating system that will set the stack pointer to a logical address based on available memory.

Figure 23-20 contains our example program. It sets the stack pointer to a desired address and then calls a subroutine. The subroutine does not actually do anything. It gives you a chance to single-step through a program and watch the stack pointer and program counter.

Calling More than One Subroutine (Not Nested)

Our next example program is shown in Fig. 23-21.

The two subroutines shown here occur one after the other. They are *not* nested. You should single-step through this program and watch the stack pointer and program counter. This is important because the next program will also contain two subroutines, but they *will* be nested. We want you to see the difference between the two.

```
8080/8085 program

0001  1800                     .ORG 1800h
0002  1800                     ;
0003  1800 31 9E 1F  START:    LXI SP, 1F9Eh    ;load stack pointer
0004  1803 00                  NOP              ;misc instructions
0005  1804 CD 08 18            CALL SUBRTN      ;call subroutine (watch stack pointer)
0006  1807 76                  HLT              ;stop
0007  1808 00        SUBRTN:   NOP              ;misc instructions
0008  1809 C9                  RET              ;return from subroutine
0009  180A                     ;
0010  180A                     .END

Z80 program

0001  1800                     .ORG 1800h
0002  1800                     ;
0003  1800 31 9E 1F  START:    LD SP,1F9Eh      ;load stack pointer
0004  1803 00                  NOP              ;misc instructions
0005  1804 CD 08 18            CALL SUBRTN      ;call subroutine (watch stack pointer)
0006  1807 76                  HALT             ;stop
0007  1808 00        SUBRTN:   NOP              ;misc instructions
0008  1809 C9                  RET              ;return from subroutine
0009  180A                     ;
0010  180A                     .END
```

Fig. 23-20 8080/8085/Z80 program loading stack pointer and calling a subroutine.

Again, these first programs do not do anything. Just observe the behavior of the program counter and the stack pointer.

Nesting Subroutines

The program shown in Fig. 23-22 also has two subroutines. They *are* nested, however.

Single-step through this program and watch the stack pointer and the program counter carefully. Notice how they act differently from the way they did in the last program. When you are inside the second subroutine, the stack is holding the return addresses for both subroutines. That's why it decrements further.

Pushing Registers

The example program shown in Fig. 23-23 shows how to use the stack to move information from one register to another.

The program loads register pairs BC and DE with a value, pushes BC and DE onto the stack, and then pulls them from the stack in reverse order. This places the data that was in BC in DE, and the data that was in DE in BC.

A Useful Program Containing a Subroutine

Let's take a look at the program shown in Fig. 23-24.

This program's purpose is as follows:

> This program will read a list of five signed binary numbers. As it reads each number, it will determine

```
8080/8085 program

0001   1800                               .ORG 1800h
0002   1800
0003   1800 00             START:   NOP
0004   1801 CD 09 18                CALL RTNE_1
0005   1804 00                      NOP
0006   1805 CD 0B 18                CALL RTNE_2
0007   1808 76                      HLT
0008   1809 00             RTNE_1:  NOP
0009   180A C9                      RET
0010   180B 00             RTNE_2:  NOP
0011   180C C9                      RET
0012   180D
0013   180D                         .END
```

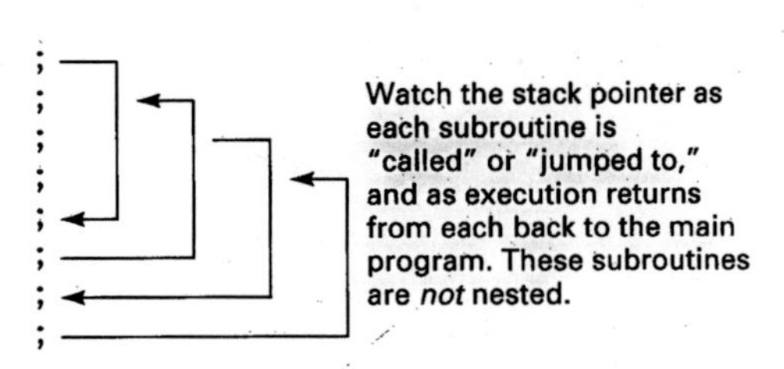


```
Z80 program

0001   1800                               .ORG 1800h
0002   1800
0003   1800 00             START:   NOP
0004   1801 CD 09 18                CALL RTNE_1
0005   1804 00                      NOP
0006   1805 CD 0B 18                CALL RTNE_2
0007   1808 76                      HLT
0008   1809 00             RTNE_1:  NOP
0009   180A C9                      RET
0010   180B 00             RTNE_2:  NOP
0011   180C C9                      RET
0012   180D
0013   180D                         .END
```

Watch the stack pointer as each subroutine is "called" or "jumped to," and as execution returns from each back to the main program. These subroutines are *not* nested.

Fig. 23-21 8080/8085/Z80 program with two subroutines *not* nested.

```
8080/8085 program

0001   1800                               .ORG 1800h
0002   1800
0003   1800 00             START:   NOP
0004   1801 CD 05 18                CALL RTNE_1
0005   1804 76                      HLT
0006   1805 00             RTNE_1:  NOP
0007   1806 CD 0A 18                CALL RTNE_2
0008   1809 C9                      RET
0009   180A 00             RTNE_2:  NOP
0010   180B C9                      RET
0011   180C
0012   180C                         .END
```

Again, watch the stack pointer as each subroutine is "called" or "jumped to," and as execution returns from each subroutine. These subroutines *are* nested.

```
Z80 program

0001   1800                               .ORG 1800h
0002   1800
0003   1800 00             START:   NOP
0004   1801 CD 05 18                CALL RTNE_1
0005   1804 76                      HALT
0006   1805 00             RTNE_1:  NOP
0007   1806 CD 0A 18                CALL RTNE_2
0008   1809 C9                      RET
0009   180A 00             RTNE_2:  NOP
0010   180B C9                      RET
0011   180C
0012   180C                         .END
```

Again, watch the stack pointer as each subroutine is "called" or "jumped to," and as execution returns from each subroutine. These subroutines *are* nested.

Fig. 23-22 8080/8085/Z80 program with two *nested* subroutines.

```
8080/8085 program

0001  1800                       .ORG 1800h
0002  1800
0003  1800 01 34 12   START:     LXI B,1234h    ;load values into
0004  1803 11 78 56              LXI D,5678h    ;  registers
0005  1806 C5                    PUSH B         ;push then decrement stack pointer
0006  1807 D5                    PUSH D         ;push then decrement stack pointer again
0007  1808 C1                    POP B          ;pull then increment stack pointer
0008  1809 D1                    POP D          ;pull then increment stack pointer again
0009  180A 76                    HLT            ;stop
0010  180B
0011  180B                       .END


Z80 program

0001  1800                       .ORG 1800h
0002  1800
0003  1800 01 34 12   START:     LD BC,1234h    ;load values into
0004  1803 11 78 56              LD DE,5678h    ;  registers
0005  1806 C5                    PUSH BC        ;push then decrement stack pointer
0006  1807 D5                    PUSH DE        ;push then decrement stack pointer again
0007  1808 C1                    POP BC         ;pull then increment stack pointer
0008  1809 D1                    POP DE         ;pull then increment stack pointer again
0009  180A 76                    HALT           ;stop
0010  180B
0011  180B                       .END
```

Fig. 23-23 8080/8085/Z80 program which pushes a register.

whether that number is positive or 0 or negative. If it is positive or 0, it will do nothing with the number. If the number is negative, a subroutine will be entered. This subroutine will find the absolute value of the number (that is, it will make the negative number positive). It will then write this positive number into memory in place of the original negative number. (We used the decimal numbers 3, −4, −2, 0, and 5.) (*Note:* If the microprocessor being used here has a negate instruction, that instruction will not be used.)

Enter this program into your microprocessor trainer or computer and single-step through it. Study the program and make sure that you understand its operation.

23-8 8086/8088 FAMILY

The 8086/8088 microprocessor works as described in the New Concepts section of this chapter. The 8086/8088 can have a very large stack, up to 64K (65,536 bytes). The location of the top-of-the-stack is calculated by using both the stack pointer and the stack segment.

We'll look at several sample programs which you can enter into your microprocessor trainer or computer and examine.

Setting the Stack Pointer

Our first example program illustrates how to set the stack pointer to a desired address and then call a subroutine. It is important to note that, with the simple programs we have used throughout this text, setting the stack pointer is normally not required. The microprocessor trainer or computer you are working with will have an operating system that will set the stack pointer to a logical address based on available memory.

Figure 23-25 contains our example program. It sets the stack pointer to a desired address and then calls a subroutine. The subroutine does not actually do anything. It gives you a chance to single-step through a program and watch the stack pointer and program counter.

Calling More than One Subroutine (Not Nested)

Our next example program is shown in Fig. 23-26.

The two subroutines shown here occur one after the other. They are *not* nested. You should single-step through this program and watch the stack pointer and program counter. This is important because the next program will also contain two subroutines, but they *will* be nested. We want you to see the difference between the two.

Again, these first programs do not do anything. Just observe the behavior of the program counter and the stack pointer.

Nesting Subroutines

The program shown in Fig. 23-27 also has two subroutines. They *are* nested, however.

```
8080/8085 program

0001  1800                     .org 1800h
0002  1800
0003  1800 21 1C 18   START:   LXI H,LIST     ;address of beginning of list
0004  1803 06 06               MVI B,06h      ;counter
0005  1805 05         GETNUM:  DCR B          ;decrement counter
0006  1806 CA 1B 18            JZ DONE        ;if no items left end program
0007  1809 7E                  MOV A,M        ;load number from list
0008  180A FE 00               CPI 00h        ;is it positive/zero or negative?
0009  180C F2 12 18            JP NEXT        ;if positive get next number now
0010  180F CD 16 18            CALL NEGNUM    ;if negative call subroutine
0011  1812 23         NEXT:    INX H          ;point to next number in list
0012  1813 C3 05 18            JMP GETNUM     ;branch back to beginning
0013  1816 2F         NEGNUM:  CMA            ;invert all bits of negative number
0014  1817 C6 01               ADI 01h        ;add 1 to inverted bits
0015  1819 77                  MOV M,A        ;write absolute value over old negative valu
0016  181A C9                  RET            ;return
0017  181B 76         DONE:    HLT            ;stop
0018  181C
0019  181C 03FCFE0005 LIST:    .db    3, -4, -2, 0, 5     ;list of 5 numbers
0020  1821
0021  1821                     .end
```

```
Z80 program

0001  1800                     .org 1800h
0002  1800
0003  1800 21 1C 18   START:   LD HL,LIST     ;address of beginning of list
0004  1803 06 06               LD B,06h       ;counter
0005  1805 05         GETNUM:  DEC B          ;decrement counter
0006  1806 CA 1B 18            JP Z,DONE      ;if no items left end program
0007  1809 7E                  LD A,(HL)      ;load number from list
0008  180A FE 00               CP 00h         ;is it positive/zero or negative?
0009  180C F2 12 18            JP P,NEXT      ;if positive get next number now
0010  180F CD 16 18            CALL NEGNUM    ;if negative call subroutine
0011  1812 23         NEXT:    INC HL         ;point to next number in list
0012  1813 C3 05 18            JP GETNUM      ;branch back to beginning
0013  1816 2F         NEGNUM:  CPL            ;invert all bits of negative number
0014  1817 C6 01               ADD A,01h      ;add 1 to inverted bits
0015  1819 77                  LD (HL),A      ;write absolute value over old negative valu
0016  181A C9                  RET            ;return
0017  181B 76         DONE:    HALT           ;stop
0018  181C
0019  181C 03FCFE0005 LIST:    .db    3, -4, -2, 0, 5     ;list of 5 numbers
0020  1821
0021  1821                     .end
```

Fig. 23-24 A useful 8080/8085/Z80 program which contains a subroutine.

Single-step through this program and watch the stack pointer and the program counter carefully. Notice how they act differently from the way they did in the last program. When you are inside the second subroutine, the stack is holding the return addresses for both subroutines. That's why it decrements further.

Pushing Registers

The example program shown in Fig. 23-28 shows how to use the stack to move information from one register to another.

The program loads registers AX and BX with a value, pushes AX and BX onto the stack, then pulls them from the stack in reverse order. This places the data that was in AX in BX, and the data that was in BX in AX.

A Useful Program Containing a Subroutine

Let's take a look at the program shown in Fig. 23-29.

This program's purpose is as follows:

> This program will read a list of five signed binary numbers. As it reads each number, it will determine

8086/8088 program (with assembler)

```
 1                      page ,132
 2
 3 0000                 CODE     SEGMENT
 4                               ASSUME CS:CODE, DS:CODE, SS:CODE
 5 0100                          ORG 0100h
 6
 7 0100  BC FFF9        START:   MOV SP,0FFF9h          ;load stack pointer
 8 0103  90                      NOP                    ;misc instructions
 9 0104  E8 0109 R               CALL SHORT SUBRTN      ;call subroutine (watch stack pointer)
10 0107  CD 20                   INT 20h                ;stop
11 0109  90             SUBRTN:  NOP                    ;misc instructions
12 010A  C3                      RET                    ;return from subroutine
13
14 010B                 CODE     ENDS
15
16                               END     START
```

8086/8088 program (with DEBUG)

```
MOV     SP,FFF9     ;load stack pointer
NOP                 ;misc instructions
CALL    0109        ;call subroutine (watch stack pointer)
INT     20          ;stop
NOP                 ;misc instructions
RET                 ;return from subroutine
```

Fig. 23-25 8086/8088 program loading stack pointer and calling a subroutine.

8086/8088 program (with assembler)

```
 1                      page ,132
 2
 3 0000                 CODE     SEGMENT
 4                               ASSUME CS:CODE, DS:CODE, SS:CODE
 5 0100                          ORG 0100h
 6
 7 0100  90             START:   NOP
 8 0101  E8 010A R               CALL SHORT RTNE_1      ;
 9 0104  90                      NOP                    ;
10 0105  E8 010C R               CALL SHORT RTNE_2      ;
11 0108  CD 20                   INT 20h                ;
12 010A  90             RTNE_1:  NOP                    ;
13 010B  C3                      RET                    ;
14 010C  90             RTNE_2:  NOP                    ;
15 010D  C3                      RET                    ;
16
17 010E                 CODE     ENDS
18
19                               END     START
```

Watch the stack pointer as each subroutine is "called" or "jumped to," and as execution returns from each back to the main program. These subroutines are *not* nested.

8086/8088 program (with DEBUG)

```
NOP
CALL    010A    ;
NOP             ;
CALL    010C    ;
INT     20      ;
NOP             ;
RET             ;
NOP             ;
RET             ;
```

Watch the stack pointer as each subroutine is "called" or "jumped to," and as execution returns from each back to the main program. These subroutines are *not* nested.

Fig. 23-26 8086/8088 program with two subroutines *not* nested.

8086/8088 program (with assembler)

```
 1                         page ,132
 2
 3 0000                    CODE     SEGMENT
 4                                  ASSUME CS:CODE, DS:CODE, SS:CODE
 5 0100                             ORG 0100h
 6
 7 0100  90                START:   NOP
 8 0101  E8 0106 R                  CALL SHORT RTNE_1   ;
 9 0104  CD 20                      INT 20h             ;
10 0106  90                RTNE_1:  NOP                 ;
11 0107  E8 010B R                  CALL SHORT RTNE_2   ;
12 010A  C3                         RET                 ;
13 010B  90                RTNE_2:  NOP                 ;
14 010C  C3                         RET                 ;
15
16 010D                    CODE     ENDS
17
18                                  END      START
```

Again, watch the stack pointer as each subroutine is "called" or "jumped to," and as execution returns from each subroutine. These subroutines *are* nested.

8086/8088 program (with DEBUG)

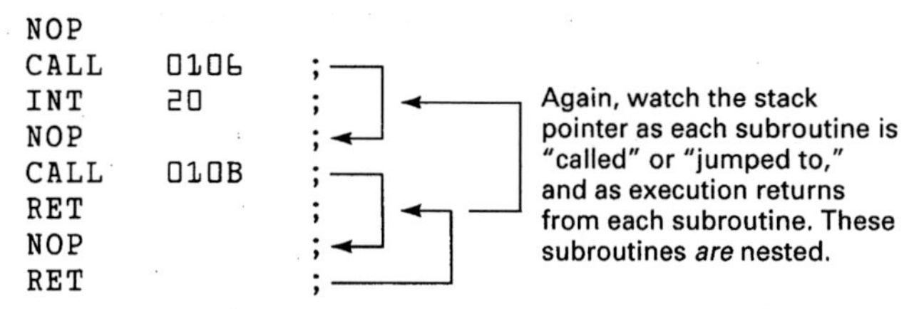


```
NOP
CALL    0106     ;
INT     20       ;
NOP              ;
CALL    010B     ;
RET              ;
NOP              ;
RET              ;
```

Fig. 23-27 8086/8088 program with two *nested* subroutines.

8086/8088 program (with assembler)

```
 1                         page ,132
 2
 3 0000                    CODE     SEGMENT
 4                                  ASSUME CS:CODE, DS:CODE, SS:CODE
 5 0100                             ORG 0100h
 6
 7 0100  B8 1234           START:   MOV AX,1234h      ;load values into
 8 0103  BB 5678                    MOV BX,5678h      ;  registers
 9 0106  50                         PUSH AX           ;push then decrement stack pointer
10 0107  53                         PUSH BX           ;push then decrement stack pointer again
11 0108  58                         POP AX            ;pop then increment stack pointer
12 0109  5B                         POP BX            ;pop then increment stack pointer again
13 010A  CD 20                      INT 20h           ;stop
14
15 010C                    CODE     ENDS
16
17                                  END    START
```

8086/8088 program (with DEBUG)

```
MOV      AX,1234     ;load values into
MOV      BX,5678     ;  registers
PUSH     AX          ;push then decrement stack pointer
PUSH     BX          ;push then decrement stack pointer again
POP      AX          ;pop then increment stack pointer
POP      BX          ;pop then increment stack pointer again
INT      20          ;stop
```

Fig. 23-28 8086/8088 program which pushes a register.

8086/8088 program (with assembler)

```
 1                          page ,132
 2
 3 0000                     CODE    SEGMENT
 4                                  ASSUME CS:CODE, DS:CODE, SS:CODE
 5 0100                             ORG 0100h
 6
 7 0100   BB 0000           START:  MOV BX,00h                    ;address of beginning of list
 8 0103   B1 06                     MOV CL,06h                    ;counter
 9 0105   FE C9             GETNUM: DEC CL                        ;decrement counter
10 0107   74 17                     JZ DONE                       ;if no items left end program
11 0109   8A 87 0122 R              MOV AL,[BYTE PTR LIST + BX]   ;load number from list
12 010D   3C 00                     CMP AL,00h                    ;is it positive/zero or negative?
13 010F   7D 03                     JGE NEXT                      ;if positive get next number now
14 0111   E8 0117 R                 CALL NEGNUM                   ;if negative call subroutine
15 0114   43                NEXT:   INC BX                        ;point to next number in list
16 0115   EB EE                     JMP GETNUM                    ;branch back to beginning
17 0117   F6 D0             NEGNUM: NOT AL                        ;invert all bits of negative number
18 0119   04 01                     ADD AL,01h                    ;add 1 to inverted bits
19 011B   88 87 0122 R              MOV [BYTE PTR LIST + BX],AL   ;write absolute value over old
                                                                    negative value
20 011F   C3                        RET                           ;return
21 0120   CD 20             DONE:   INT 20h                       ;stop
22
23 0122   03 FC FE 00 05    LIST:   db      3, -4, -2, 0, 5       ;list of 5 numbers
24
25
26 0127                     CODE    ENDS
27
28                                  END     START
```

8086/8088 program (with DEBUG)

```
a 0100
MOV     BX,0000         ;address of beginning of list
MOV     CL,06           ;counter
DEC     CL              ;decrement counter
JZ      0120            ;if no items left end program
MOV     AL,[BX+0122]    ;load number from list
CMP     AL,00           ;is it positive/zero or negative?
JGE     0114            ;if positive get next number now
CALL    0117            ;if negative call subroutine
INC     BX              ;point to next number in list
JMP     0105            ;branch back to beginning
NOT     AL              ;invert all bits of negative number
ADD     AL,01           ;add 1 to inverted bits
MOV     [BX+0122],AL    ;write absolute value over old negative value
RET                     ;return
INT     20              ;stop

e 0122  03 FC FE 00 05
```

Fig. 23-29 A useful 8086/8088 program which contains a subroutine.

whether that number is positive or 0 or negative. If it is positive or 0, it will do nothing with the number. If the number is negative, a subroutine will be entered. This subroutine will find the absolute value of the number (that is, it will make the negative number positive). It will then write this positive number into memory in place of the original negative number. (We used the decimal numbers 3, −4, −2, 0, and 5.) (*Note:* If the microprocessor being used here has a negate instruction, that instruction will not be used.)

Enter this program into your microprocessor trainer or computer and single-step through it. Study the program and make sure that you understand its operation.

SELF-TESTING REVIEW

Read each of the following and provide the missing words. Answers appear at the beginning of the next question.

1. ________ are used when there are common tasks which must be executed or used many times.
2. *(Subroutines)* The structure of the stack is a ________ type of structure.
3. *(FILO)* The act of putting a piece of data on the top of the stack is called ________ the data onto the stack.
4. *(pushing)* The act of removing a piece of data from the top of the stack is called ________ or ________ the data from the stack.
5. *(pulling, popping)* The instruction that is usually the last instruction in a subroutine, and that tells the microprocessor to go back to the place where it was before the subroutine was called, is the ________ instruction.
6. *(return)* In general, the programmer can push onto and pull from the stack one or more of the microprocessor's ________.

(registers)

PROBLEMS

Solve the following problem using the microprocessor of your choice. This will be the longest program you have written thus far. Therefore, this chapter has only this one program for you to write. The program can be considered correct only if it causes the correct values to be placed in the counter variables *and* alters the original list correctly.

23-1. A 1-byte unsigned number can range from 00 to FF. Each number in this range has a corresponding ASCII value. The primary categories within the ASCII table are shown below. (The characters from 80–FF are not actually official ASCII characters but are used to form the extended IBM character set.)

00 – 1F	various control characters
20 – 2F	punctuation marks
30 – 39	numbers
3A–40	punctuation marks
41 – 5A	uppercase letters
5B–60	punctuation marks
61 – 7A	lowercase letters
7B–7F	punctuation marks
80 – FF	foreign letters, boxes, math symbols, miscellaneous

Write a program in which the main part of the program examines consecutive bytes from a list which ends with the number FF. This main program section then determines which category each value in the list is from. Different subroutines will then be called, depending on which category a value belongs to.

If the value represents a lowercase letter, a subroutine called LOWER will increment a memory location called NUM_LW, which indicates the number of lowercase letters found.

If the value represents an uppercase letter, a subroutine called UPPER will increment a memory location called NUM_UP, which indicates the number of uppercase letters found.

If the value represents a number, a subroutine called NUM will change the number to its corresponding binary value. (The ASCII value for a number and the binary value for that number are not the same.) The subroutine will then store the binary value in the list in place of the original ASCII value and then increment a memory location called NUM_N, which indicates the number of numbers found.

If the number represents a control character, the program will do nothing.

If the value represents a punctuation mark, a subroutine called PUNCT will increment a memory location called NUM_P, which indicates the number of punctuation marks found.

If the value represents one of the special characters in the range from 80 to FF, a subroutine called SPECL will change the uppermost bit of the number from a 1 to a 0. This change will cause the value to fit into one of the previously mentioned categories. The subroutine SPECL will then return to the main program, which is to be arranged in such a way that this converted value will be evaluated a second time to determine its new category and have the appropriate subroutine called.

Place the following hexadecimal values in the list: 00, 1F, 20, 2F, 30, 39, 3A, 40, 41, 5A, 5B, 60, 61, 7A, 7B, 7F, 80, and FF. (FF is not actually a value to be evaluated but marks the end of the list.)

PART 4

MICROPROCESSOR INSTRUCTION SET TABLES

EXPANDED TABLE OF 8085/8080 AND Z80 (8080 SUBSET) INSTRUCTIONS LISTED BY CATEGORY

Micro	Mnemonic	Operation	8085>SZ-A-P-C Z80>SZ-H-PNC	T	#	Address Mode	Op	Boolean/Arith. Operation	Notes

CPU Control Instructions

Micro	Mnemonic	Operation	8085>SZ-A-P-C Z80>SZ-H-PNC	T	#	Address Mode	Op	Boolean/Arith. Operation	Notes
8085	NOP	No OPeration	xx-x-x-x	4	1	Implied	00	nothing	**Can be used to create time delays or leave extra spaces for instructions to be inserted at a later time.**
Z80	NOP	No OPeration	xx-x-xxx						
8085	HLT	**HALT**	xx-x-x-x	5	1	Implied	76	stop processing	(8080 = 7 states)
Z80	HALT	**HALT**	xx-x-xxx	4					

Data Transfer Instructions

Micro	Mnemonic	Operation	8085>SZ-A-P-C Z80>SZ-H-PNC	T	#	Address Mode	Op	Boolean/Arith. Operation	Notes
8085	MOV A,A	MOVe data to **A** from **A**	xx-x-x-x	4	1	Register	7F	A ← A	(8080 = 5 T states)
Z80	LD A,A	LoaD data into **A** from **A**	xx-x-xxx						
8085	MOV A,B	MOVe data to **A** from **B**	xx-x-x-x	4	1	Register	78	A ← B	(8080 = 5 T states)
Z80	LD A,B	LoaD data into **A** from **B**	xx-x-xxx						
8085	MOV A,C	MOVe data to **A** from **C**	xx-x-x-x	4	1	Register	79	A ← C	(8080 = 5 T states)
Z80	LD A,C	LoaD data into **A** from **C**	xx-x-xxx						
8085	MOV A,D	MOVe data to **A** from **D**	xx-x-x-x	4	1	Register	7A	A ← D	(8080 = 5 T states)
Z80	LD A,D	LoaD data into **A** from **D**	xx-x-xxx						
8085	MOV A,E	MOVe data to **A** from **E**	xx-x-x-x	4	1	Register	7B	A ← E	(8080 = 5 T states)
Z80	LD A,E	LoaD data into **A** from E	xx-x-xxx						

EXPANDED TABLE OF 8085/8080 AND Z80 (8080 SUBSET) INSTRUCTIONS LISTED BY CATEGORY (*Continued*)

Micro	Mnemonic	Operation	8085>SZ-A-P-C Z80>SZ-H-PNC	T	#	Address Mode	Op	Boolean/Arith. Operation	Notes
8085	MOV A,H	MOVe data to A from H	xx-x-x-x	4	1	Register	7C	A ← H	(8080 = 5 T states)
Z80	LD A,H	LoaD data into A from H	xx-x-xxx						
8085	MOV A,L	MOVe data to A from L	xx-x-x-x	4	1	Register	7D	A ← L	(8080 = 5 T states)
Z80	LD A,L	LoaD data into A from L	xx-x-xxx						
8085	MOV A,M	MOVe data to A from M	xx-x-x-x	7	1	Reg Ind	7E	A ← M_{HL}	The data byte found at the memory location pointed to by the HL register pair is copied into the accumulator.
Z80	LD A,(HL)	LoaD data into A from (HL)	xx-x-xxx						
8085	MOV B,A	MOVe data to B from A	xx-x-x-x	4	1	Register	47	B ← A	(8080 = 5 T states)
Z80	LD B,A	LoaD data into B from A	xx-x-xxx						
8085	MOV B,B	MOVe data to B from B	xx-x-x-x	4	1	Register	40	B ← B	(8080 = 5 T states)
Z80	LD B,B	LoaD data into B from B	xx-x-xxx						
8085	MOV B,C	MOVe data to B from C	xx-x-x-x	4	1	Register	41	B ← C	(8080 = 5 T states)
Z80	LD B,C	LoaD data into B from C	xx-x-xxx						
8085	MOV B,D	MOVe data to B from D	xx-x-x-x	4	1	Register	42	B ← D	(8080 = 5 T states)
Z80	LD B,D	LoaD data into B from D	xx-x-xxx						
8085	MOV B,E	MOVe data to B from E	xx-x-x-x	4	1	Register	43	B ← E	(8080 = 5 T states)
Z80	LD B,E	LoaD data into B from E	xx-x-xxx						
8085	MOV B,H	MOVe data to B from H	xx-x-x-x	4	1	Register	44	B ← H	(8080 = 5 T states)
Z80	LD B,H	LoaD data into B from H	xx-x-xxx						

Micro	Mnemonic	Operation	8085 > SZ-A-P-C Z80 > SZ-H-PNC	T	#	Address Mode	Op	Boolean/Arith. Operation	Notes
8085	MOV B,L	MOVe data to B from L	xx-x-x-x	4	1	Register	45	B ← L	(8080 = 5 T states)
Z80	LD B,L	LoaD data into B from L	xx-x-xxx						
8085	MOV B,M	MOVe data to B from M_{HL}	xx-x-x-x	7	1	Reg Ind	46	B ← M_{HL}	The data byte found at the memory location pointed to by the HL register pair is copied into register B.
Z80	LD B,(HL)	LoaD data into B from (HL)	xx-x-xxx						
8085	MOV C,A	MOVe data to C from A	xx-x-x-x	4	1	Register	4F	C ← A	(8080 = 5 T states)
Z80	LD C,A	LoaD data into C from A	xx-x-xxx						
8085	MOV C,B	MOVe data to C from B	xx-x-x-x	4	1	Register	48	C ← B	(8080 = 5 T states)
Z80	LD C,B	LoaD data into C from B	xx-x-xxx						
8085	MOV C,C	MOVe data to C from C	xx-x-x-x	4	1	Register	49	C ← C	(8080 = 5 T states)
Z80	LD C,C	LoaD data into C from C	xx-x-xxx						
8085	MOV C,D	MOVe data to C from D	xx-x-x-x	4	1	Register	4A	C ← D	(8080 = 5 T states)
Z80	LD C,D	LoaD data into C from D	xx-x-xxx						
8085	MOV C,E	MOVe data to C from E	xx-x-x-x	4	1	Register	4B	C ← E	(8080 = 5 T states)
Z80	LD C,E	LoaD data into C from E	xx-x-xxx						
8085	MOV C,H	MOVe data to C from H	xx-x-x-x	4	1	Register	4C	C ← H	(8080 = 5 T states)
Z80	LD C,H	LoaD data into C from H	xx-x-xxx						
8085	MOV C,L	MOVe data to C from L	xx-x-x-x	4	1	Register	4D	C ← L	(8080 = 5 T states)
Z80	LD C,L	LoaD data into C from L	xx-x-xxx						

EXPANDED TABLE OF 8085/8080 AND Z80 (8080 SUBSET) INSTRUCTIONS LISTED BY CATEGORY (*Continued*)

Micro	Mnemonic	Operation	8085 > SZ-A-P-C Z80 > SZ-H-PNC	T	#	Address Mode	Op	Boolean/Arith. Operation	Notes
8085 Z80	MOV C,M LD C,(HL)	MOVe data to C from M_{HL} LoaD data into C from (HL)	xx-x-x-x xx-x-xxx	7	1	Reg Ind	4E	C ← M_{HL}	The data byte found at the memory location pointed to by the HL register pair is copied into register C.
8085 Z80	MOV D,A LD D,A	MOVe data to D from A LoaD data into D from A	xx-x-x-x xx-x-xxx	4	1	Register	57	D ← A	(8080 = 5 T states)
8085 Z80	MOV D,B LD D,B	MOVe data to D from B LoaD data into D from B	xx-x-x-x xx-x-xxx	4	1	Register	50	D ← B	(8080 = 5 T states)
8085 Z80	MOV D,C LD D,C	MOVe data to D from C LoaD data into D from C	xx-x-x-x xx-x-xxx	4	1	Register	51	D ← C	(8080 = 5 T states)
8085 Z80	MOV D,D LD D,D	MOVe data to D from D LoaD data into D from D	xx-x-x-x xx-x-xxx	4	1	Register	52	D ← D	(8080 = 5 T states)
8085 Z80	MOV D,E LD D,E	MOVe data to D from E LoaD data into D from E	xx-x-x-x xx-x-xxx	4	1	Register	53	D ← E	(8080 = 5 T states)
8085 Z80	MOV D,H LD D,H	MOVe data to D from H LoaD data into D from H	xx-x-x-x xx-x-xxx	4	1	Register	54	D ← H	(8080 = 5 T states)
8085 Z80	MOV D,L LD D,L	MOVe data to D from L LoaD data into D from L	xx-x-x-x xx-x-xxx	4	1	Register	55	D ← L	(8080 = 5 T states)
8085 Z80	MOV D,M LD D,(HL)	MOVe data to D from M_{HL} LoaD data into D from (HL)	xx-x-x-x xx-x-xxx	7	1	Reg Ind	56	D ← M_{HL}	The data byte found at the memory location pointed to by the HL register pair is copied into register D.

Micro	Mnemonic	Operation	8085>SZ-A-P-C Z80>SZ-H-PNC	T	#	Address Mode	Op	Boolean/Arith. Operation	Notes
8085	MOV E,A	MOVe data to E from A	xx-x-x-x	4	1	Register	5F	E ← A	(8080 = 5 T states)
Z80	LD E,A	LoaD data into E from A	xx-x-xxx						
8085	MOV E,B	MOVe data to E from B	xx-x-x-x	4	1	Register	58	E ← B	(8080 = 5 T states)
Z80	LD E,B	LoaD data into E from B	xx-x-xxx						
8085	MOV E,C	MOVe data to E from C	xx-x-x-x	4	1	Register	59	E ← C	(8080 = 5 T states)
Z80	LD E,C	LoaD data into E from C	xx-x-xxx						
8085	MOV E,D	MOVe data to E from D	xx-x-x-x	4	1	Register	5A	E ← D	(8080 = 5 T states)
Z80	LD E,D	LoaD data into E from D	xx-x-xxx						
8085	MOV E,E	MOVe data to E from E	xx-x-x-x	4	1	Register	5B	E ← E	(8080 = 5 T states)
Z80	LD E,E	LoaD data into E from E	xx-x-xxx						
8085	MOV E,H	MOVe data to E from H	xx-x-x-x	4	1	Register	5C	E ← H	(8080 = 5 T states)
Z80	LD E,H	LoaD data into E from H	xx-x-xxx						
8085	MOV E,L	MOVe data to E from L	xx-x-x-x	4	1	Register	5D	E ← L	(8080 = 5 T states)
Z80	LD E,L	LoaD data into E from L	xx-x-xxx						
8085	MOV E,M	MOVe data to E from M_{HL}	xx-x-x-x	7	1	Reg Ind	5E	E ← M_{HL}	The data byte found at the memory location pointed to by the HL register pair is copied into register E.
Z80	LD E,(HL)	LoaD data into E from (HL)	xx-x-xxx						
8085	MOV H,A	MOVe data to H from A	xx-x-x-x	4	1	Register	67	H ← A	(8080 = 5 T states)
Z80	LD H,A	LoaD data into H from A	xx-x-xxx						

EXPANDED TABLE OF 8085/8080 AND Z80 (8080 SUBSET) INSTRUCTIONS LISTED BY CATEGORY (*Continued*)

Micro	Mnemonic	Operation	8085>SZ-A-P-C Z80>SZ-H-PNC	T	#	Address Mode	Op	Boolean/Arith. Operation	Notes
8085	MOV H,B	MOVe data to H from B	xx-x-x-x	4	1	Register	60	H ← B	(8080 = 5 T states)
Z80	LD H,B	LoaD data into H from B	xx-x-xxx						
8085	MOV H,C	MOVe data to H from C	xx-x-x-x	4	1	Register	61	H ← C	(8080 = 5 T states)
Z80	LD H,C	LoaD data into H from C	xx-x-xxx						
8085	MOV H,D	MOVe data to H from D	xx-x-x-x	4	1	Register	62	H ← D	(8080 = 5 T states)
Z80	LD H,D	LoaD data into H from D	xx-x-xxx						
8085	MOV H,E	MOVe data to H from E	xx-x-x-x	4	1	Register	63	H ← E	(8080 = 5 T states)
Z80	LD H,E	LoaD data into H from E	xx-x-xxx						
8085	MOV H,H	MOVe data to H from H	xx-x-x-x	4	1	Register	64	H ← H	(8080 = 5 T states)
Z80	LD H,H	LoaD data into H from H	xx-x-xxx						
8085	MOV H,L	MOVe data to H from L	xx-x-x-x	4	1	Register	65	H ← L	(8080 = 5 T states)
Z80	LD H,L	LoaD data into H from L	xx-x-xxx						
8085	MOV H,M	MOVe data to H from M_{HL}	xx-x-x-x	7	1	Reg Ind	66	H ← M_{HL}	The data byte found at the memory location pointed to by the HL register pair is copied into register H.
Z80	LD H,(HL)	LoaD data into H from (HL)	xx-x-xxx						
8085	MOV L,A	MOVe data to L from A	xx-x-x-x	4	1	Register	6F	L ← A	(8080 = 5 T states)
Z80	LD L,A	LoaD data into L from A	xx-x-xxx						
8085	MOV L,B	MOVe data to L from B	xx-x-x-x	4	1	Register	68	L ← B	(8080 = 5 T states)
Z80	LD L,B	LoaD data into L from B	xx-x-xxx						

Micro	Mnemonic	Operation	8085 > SZ-A-P-C Z80 > SZ-H-PNC	T	#	Address Mode	Op	Boolean/Arith. Operation	Notes
8085 Z80	MOV L,C LD L,C	**MOVe data to L from C** **LoaD data into L from C**	xx-x-x-x xx-x-xxx	4	1	Register	69	L ← C	(8080 = 5 T states)
8085 Z80	MOV L,D LD L,D	**MOVe data to L from D** **LoaD data into L from D**	xx-x-x-x xx-x-xxx	4	1	Register	6A	L ← D	(8080 = 5 T states)
8085 Z80	MOV L,E LD L,E	**MOVe data to L from E** **LoaD data into L from E**	xx-x-x-x xx-x-xxx	4	1	Register	6B	L ← E	(8080 = 5 T states)
8085 Z80	MOV L,H LD L,H	**MOVe data to L from H** **LoaD data into L from H**	xx-x-x-x xx-x-xxx	4	1	Register	6C	L ← H	(8080 = 5 T states)
8085 Z80	MOV L,L LD L,L	**MOVe data to L from L** **LoaD data into L from L**	xx-x-x-x xx-x-xxx	4	1	Register	6D	L ← L	(8080 = 5 T states)
8085 Z80	MOV L,M LD L,(HL)	**MOVe data to L from M_{HL}** **LoaD data into L from (HL)**	xx-x-x-x xx-x-xxx	7	1	Reg Ind	6E	L ← M_{HL}	The data byte found at the memory location pointed to by the HL register pair is copied into register L.
8085 Z80	MOV M,A LD (HL),A	**MOVe data to M_{HL} from A** **LoaD data into (HL) from A**	xx-x-x-x xx-x-xxx	7	1	Reg Ind	77	M_{HL} ← A	The data in the accumulator is copied into the memory location pointed to by the HL register pair.
8085 Z80	MOV M,B LD (HL),B	**MOVe data to M_{HL} from B** **LoaD data into (HL) from B**	xx-x-x-x xx-x-xxx	7	1	Reg Ind	70	M_{HL} ← B	The data in register B is copied into the memory location pointed to by the HL register pair.
8085 Z80	MOV M,C LD (HL),C	**MOVe data to M_{HL} from C** **LoaD data into (HL) from C**	xx-x-x-x xx-x-xxx	7	1	Reg Ind	71	M_{HL} ← C	The data in register C is copied into the memory location pointed to by the HL register pair.

EXPANDED TABLE OF 8085/8080 AND Z80 (8080 SUBSET) INSTRUCTIONS LISTED BY CATEGORY (*Continued*)

Micro	Mnemonic	Operation	8085>SZ-A-P-C Z80>SZ-H-PNC	T	#	Address Mode	Op	Boolean/Arith. Operation	Notes
8085 Z80	MOV M,D LD (HL),D	MOVe data to M_{HL} from D LoaD data into (HL) from D	xx-x-x-x xx-x-xxx	7	1	Reg Ind	72	$M_{HL} \leftarrow D$	The data in register D is copied into the memory location pointed to by the HL register pair.
8085 Z80	MOV M,E LD (HL),E	MOVe data to M_{HL} from E LoaD data into (HL) from E	xx-x-x-x xx-x-xxx	7	1	Reg Ind	73	$M_{HL} \leftarrow E$	The data in register E is copied into the memory location pointed to by the HL register pair.
8085 Z80	MOV M,H LD (HL),H	MOVe data to M_{HL} from H LoaD data into (HL) from H	xx-x-x-x xx-x-xxx	7	1	Reg Ind	74	$M_{HL} \leftarrow H$	The data in register H is copied into the memory location pointed to by the HL register pair.
8085 Z80	MOV M,L LD (HL),L	MOVe data to M_{HL} from L LoaD data into (HL) from L	xx-x-x-x xx-x-xxx	7	1	Reg Ind	75	$M_{HL} \leftarrow L$	The data in register L is copied into the memory location pointed to by the HL register pair.
8085 Z80	MVI A,dd LD A,dd	MoVe Immediate dd to A LoaD dd into A	xx-x-x-x xx-x-xxx	7	2	Immed	3E	A ← dd	The data byte immediately following the op code is copied into the accumulator.
8085 Z80	MVI B,dd LD B,dd	MoVe Immediate dd to B LoaD dd into B	xx-x-x-x xx-x-xxx	7	2	Immed	06	B ← dd	The data byte immediately following the op code is copied into register B.
8085 Z80	MVI C,dd LD C,dd	MoVe Immediate dd to C LoaD dd into C	xx-x-x-x xx-x-xxx	7	2	Immed	0E	C ← dd	The data byte immediately following the op code is copied into register C.
8085 Z80	MVI D,dd LD D,dd	MoVe Immediate dd to D LoaD dd into D	xx-x-x-x xx-x-xxx	7	2	Immed	16	D ← dd	The data byte immediately following the op code is copied into register D.
8085 Z80	MVI E,dd LD E,dd	MoVe Immediate dd to E LoaD dd into E	xx-x-x-x xx-x-xxx	7	2	Immed	1E	E ← dd	The data byte immediately following the op code is copied into register E.
8085 Z80	MVI H,dd LD H,dd	MoVe Immediate dd to H LoaD dd into H	xx-x-x-x xx-x-xxx	7	2	Immed	26	H ← dd	The data byte immediately following the op code is copied into register H.

Micro	Mnemonic	Operation	8085 > SZ-A-P-C Z80 > SZ-H-PNC	T	#	Address Mode	Op	Boolean/Arith. Operation	Notes
8085 Z80	MVI L,dd LD L,dd	MoVe Immediate dd to L LoaD dd into L	xx-x-x-x xx-x-xxx	7	2	Immed	2E	L ← dd	The data byte immediately following the op code is copied into register L.
8085 Z80	MVI M,dd LD (HL),dd	MoVe Immediate dd to M_{HL} LoaD dd into (HL)	xx-x-x-x xx-x-xxx	10	2	Immed/ Reg Ind	36	M_{HL} ← dd	The data byte immediately following the op code is copied into the memory location pointed to by the HL register pair.
8085 Z80	LXI B,dddd LD BC,dddd	Load eXtended Immediate dddd into register pair BC LoaD dddd into register pair BC	xx-x-x-x xx-x-xxx	10	3	Immed	01	BC ← dddd	Copy bytes 3 and 2 of the instruction into registers B and C respectively.
8085 Z80	LXI D,dddd LD DE,dddd	Load eXtended Immediate dddd into register pair DE LoaD dddd into register pair DE	xx-x-x-x xx-x-xxx	10	3	Immed	11	DE ← dddd	Copy bytes 3 and 2 of the instruction into registers D and E respectively.
8085 Z80	LXI H,dddd LD HL,dddd	Load eXtended Immediate dddd into register pair HL LoaD dddd into register pair HL	xx-x-x-x xx-x-xxx	10	3	Immed	21	HL ← dddd	Copy bytes 3 and 2 of the instruction into registers H and L respectively.
8085 Z80	LDAX B LD A,(BC)	LoaD Accumulator eXtended with data from mem loc BC LoaD Accumulator with data from mem loc (BC)	xx-x-x-x xx-x-xxx	7	1	Reg Ind	0A	A ← M_{BC}	Copy the data byte found at the memory location pointed to by the BC register pair into the accumulator.
8085 Z80	LDAX D LD A,(DE)	LoaD Accumulator eXtended with data from mem loc DE LoaD Accumulator with data from mem loc (DE)	xx-x-x-x xx-x-xxx	7	1	Reg Ind	1A	A ← M_{DE}	Copy the data byte found at the memory location pointed to by the DE register pair into the accumulator.
8085 Z80	LHLD aaaa LD HL,(aaaa)	Load HL Direct with data starting at aaaa LoaD HL with data starting at (aaaa)	xx-x-x-x xx-x-xxx	16	3	Direct	2A	L ← M_{aaaa} H ← M_{aaaa+1}	Copy the data byte found at memory location aaaa into the L register and the data byte found at the next memory location (aaaa+1) into the H register.

EXPANDED TABLE OF 8085/8080 AND Z80 (8080 SUBSET) INSTRUCTIONS LISTED BY CATEGORY *(Continued)*

Micro	Mnemonic	Operation	8085 > SZ-A-P-C Z80 > SZ-H-PNC	T	#	Address Mode	Op	Boolean/Arith. Operation	Notes
8085	LDA aaaa	LoaD Accumulator with data from mem loc aaaa	xx-x-x-x	13	3	Direct	3A	$A \leftarrow M_{aaaa}$	Copy the contents of memory location aaaa into the Accumulator.
Z80	LD A,(aaaa)	LoaD Accumulator with data from mem loc (aaaa)	xx-x-xxx						
8085	STA aaaa	STore Accumulator in mem loc aaaa	xx-x-x-x	13	3	Direct	32	$M_{aaaa} \leftarrow A$	Copy the contents of the accumulator into memory location aaaa.
Z80	LD (aaaa),A	LoaD mem loc (aaaa) with the contents of the Accumulator	xx-x-xxx						
8085	STAX B	STore Accumulator eXtended at mem loc BC	xx-x-x-x	7	1	Reg Ind	02	$M_{BC} \leftarrow A$	Copy the contents of the accumulator into the memory location pointed to by the BC register pair.
Z80	LD (BC),A	LoaD mem loc (BC) with the contents of the Accumulator	xx-x-xxx						
8085	STAX D	STore Accumulator eXtended at mem loc DE	xx-x-x-x	7	1	Reg Ind	12	$M_{DE} \leftarrow A$	Copy the contents of the accumulator into the memory location pointed to by the DE register pair.
Z80	LD (DE),A	LoaD mem loc (DE) with the contents of the Accumulator	xx-x-xxx						
8085	SHLD aaaa	Store HL Direct at mem loc aaaa	xx-x-x-x	16	3	Direct	22	$M_{aaaa} \leftarrow L$ $M_{aaaa+1} \leftarrow H$	Copy the contents of register L into memory location aaaa and the contents of register H into the next (aaaa+1) memory location.
Z80	LD (aaaa),HL	LoaD mem loc starting at (aaaa) with contents of HL)	xx-x-xxx						
8085	XCHG	eXCHanGe DE with HL	xx-x-x-x	4	1	Register	EB	$DE \leftrightarrow HL$	Exchange the contents of the DE and HL register pairs.
Z80	EX DE,HL	EXchange DE with HL	xx-x-xxx						

Flag Instructions

Micro	Mnemonic	Operation	8085 > SZ-A-P-C Z80 > SZ-H-PNC	T	#	Address Mode	Op	Boolean/Arith. Operation	Notes
8085	STC	SeT Carry flag	xx-x-x-1	4	1	Implied	37	$C \leftarrow 1$	The carry flag is normally designated as "CY" for the 8080/8085.
Z80	SCF	Set Carry Flag	xx-x-xx1						

Micro	Mnemonic	Operation	8085 > SZ-A-P-C Z80 > SZ-H-PNC	T	#	Address Mode	Op	Boolean/Arith. Operation	Notes
8085	CMC	CoMplement Carry flag	xx-x-x-C	4	1	Implied	3F	$C \leftarrow \overline{C}$	The carry flag is normally designated as "CY" for the 8080/8085.
Z80	CCF	Complement Carry Flag	xx-x-xxC						

Arithmetic Instructions

Micro	Mnemonic	Operation	Flags	T	#	Address Mode	Op	Boolean/Arith. Operation	Notes
8085	ADD A	**ADD A** to A	SZ-A-P-C	4	1	Register	87	$A \leftarrow A + A$	
Z80	ADD A,A	**ADD A** to **A**	SZ-H-P0C						
8085	ADD B	**ADD B** to A	SZ-A-P-C	4	1	Register	80	$A \leftarrow A + B$	
Z80	ADD A,B	**ADD B** to **A**	SZ-H-P0C						
8085	ADD C	**ADD C** to A	SZ-A-P-C	4	1	Register	81	$A \leftarrow A + C$	
Z80	ADD A,C	**ADD C** to **A**	SZ-H-P0C						
8085	ADD D	**ADD D** to A	SZ-A-P-C	4	1	Register	82	$A \leftarrow A + D$	
Z80	ADD A,D	**ADD D** to **A**	SZ-H-P0C						
8085	ADD E	**ADD E** to A	SZ-A-P-C	4	1	Register	83	$A \leftarrow A + E$	
Z80	ADD A,E	**ADD E** to **A**	SZ-H-P0C						
8085	ADD H	**ADD H** to A	SZ-A-P-C	4	1	Register	84	$A \leftarrow A + H$	
Z80	ADD A,H	**ADD H** to **A**	SZ-H-P0C						
8085	ADD L	**ADD L** to A	SZ-A-P-C	4	1	Register	85	$A \leftarrow A + L$	
Z80	ADD A,L	**ADD L** to **A**	SZ-H-P0C						
8085	ADD M	**ADD** M_{HL} to A	SZ-A-P-C	7	1	Reg Ind	86	$A \leftarrow A + M_{HL}$	Add the data byte whose memory location is pointed to by the HL register pair to the accumulator and store the results in the accumulator.
Z80	ADD A,(HL)	**ADD (HL)** to **A**	SZ-H-P0C						
8085	ADC A	AdD with Carry **A** to A	SZ-A-P-C	4	1	Register	8F	$A \leftarrow A + A + C$	The carry flag is usually designated by "CY" for the 8080/8085.
Z80	ADC A,A	AdD with Carry **A** to **A**	SZ-H-P0C						

EXPANDED TABLE OF 8085/8080 AND Z80 (8080 SUBSET) INSTRUCTIONS LISTED BY CATEGORY *(Continued)*

Micro	Mnemonic	Operation	8085>SZ-A-P-C Z80>SZ-H-PNC	T	#	Address Mode	Op	Boolean/Arith. Operation	Notes
8085 Z80	ADC B ADC A,B	AdD with Carry B to A AdD with Carry B to A	SZ-A-P-C SZ-H-P0C	4	1	Register	88	A ← A + B + C	The carry flag is usually designated by "CY" for the 8080/8085.
8085 Z80	ADC C ADC A,C	AdD with Carry C to A AdD with Carry C to A	SZ-A-P-C SZ-H-P0C	4	1	Register	89	A ← A + C + C	The carry flag is usually designated by "CY" for the 8080/8085.
8085 Z80	ADC D ADC A,D	AdD with Carry D to A AdD with Carry D to A	SZ-A-P-C SZ-H-P0C	4	1	Register	8A	A ← A + D + C	The carry flag is usually designated by "CY" for the 8080/8085.
8085 Z80	ADC E ADC A,E	AdD with Carry E to A AdD with Carry E to A	SZ-A-P-C SZ-H-P0C	4	1	Register	8B	A ← A + E + C	The carry flag is usually designated by "CY" for the 8080/8085.
8085 Z80	ADC H ADC A,H	AdD with Carry H to A AdD with Carry H to A	SZ-A-P-C SZ-H-P0C	4	1	Register	8C	A ← A + H + C	The carry flag is usually designated by "CY" for the 8080/8085.
8085 Z80	ADC L ADC A,L	AdD with Carry L to A AdD with Carry L to A	SZ-A-P-C SZ-H-P0C	4	1	Register	8D	A ← A + L + C	The carry flag is usually designated by "CY" for the 8080/8085.
8085 Z80	ADC M ADC A,(HL)	AdD with Carry M_{HL} to A AdD with Carry (HL) to A	SZ-A-P-C SZ-H-P0C	7	1	Reg Ind	8E	A ← A + M_{HL} + C	Add to the accumulator both the contents of the memory location pointed to by the HL register pair, and the carry flag, and then place this result in the accumulator.
8085 Z80	SUB A SUB A	SUBtract A from A SUBtract A from A	SZ-A-P-C SZ-H-P1C	4	1	Register	97	A ← A - A	
8085 Z80	SUB B SUB B	SUBtract B from A SUBtract B from A	SZ-A-P-C SZ-H-P1C	4	1	Register	90	A ← A - B	
8085 Z80	SUB C SUB C	SUBtract C from A SUBtract C from A	SZ-A-P-C SZ-H-P1C	4	1	Register	91	A ← A - C	

Micro	Mnemonic	Operation	8085 > SZ-A-P-C Z80 > SZ-H-PNC	T	#	Address Mode	Op	Boolean/Arith. Operation	Notes
8085	SUB D	SUBtract D from A	SZ-A-P-C	4	1	Register	92	A ← A - D	
Z80	SUB D	SUBtract D from A	SZ-H-P1C						
8085	SUB E	SUBtract E from A	SZ-A-P-C	4	1	Register	93	A ← A - E	
Z80	SUB E	SUBtract E from A	SZ-H-P1C						
8085	SUB H	SUBtract H from A	SZ-A-P-C	4	1	Register	94	A ← A - H	
Z80	SUB H	SUBtract H from A	SZ-H-P1C						
8085	SUB L	SUBtract L from A	SZ-A-P-C	4	1	Register	95	A ← A - L	
Z80	SUB L	SUBtract L from A	SZ-H-P1C						
8085	SUB M	SUBtract M_{HL} from A	SZ-A-P-C	7	1	Reg Ind	96	A ← A - M_{HL}	Subtract the contents of the memory location pointed to by the HL register pair from the contents of the accumulator.
Z80	SUB (HL)	SUBtract (HL) from A	SZ-H-P1C						
8085	SBB A	SuBtract with Borrow A from A	SZ-A-P-C	4	1	Register	9F	A ← A - A - C	
Z80	SBC A,A	SuBtract with Carry A from A	SZ-H-P1C						
8085	SBB B	SuBtract with Borrow B from A	SZ-A-P-C	4	1	Register	98	A ← A - B - C	
Z80	SBC A,B	SuBtract with Carry B from A	SZ-H-P1C						
8085	SBB C	SuBtract with Borrow C from A	SZ-A-P-C	4	1	Register	99	A ← A - C - C	
Z80	SBC A,C	SuBtract with Carry C from A	SZ-H-P1C						
8085	SBB D	SuBtract with Borrow D from A	SZ-A-P-C	4	1	Register	9A	A ← A - D - C	
Z80	SBC A,D	SuBtract with Carry D from A	SZ-H-P1C						
8085	SBB E	SuBtract with Borrow E from A	SZ-A-P-C	4	1	Register	9B	A ← A - E - C	
Z80	SBC A,E	SuBtract with Carry E from A	SZ-H-P1C						

EXPANDED TABLE OF 8085/8080 AND Z80 (8080 SUBSET) INSTRUCTIONS LISTED BY CATEGORY *(Continued)*

Micro	Mnemonic	Operation	8085 > SZ-A-P-C Z80 > SZ-H-PNC	T	#	Address Mode	Op	Boolean/Arith. Operation	Notes
8085	SBB H	SuBtract with Borrow H from A	SZ-A-P-C	4	1	Register	9C	A ← A - H - C	
Z80	SBC A,H	SuBtract with Carry H from A	SZ-H-P1C						
8085	SBB L	SuBtract with Borrow L from A	SZ-A-P-C	4	1	Register	9D	A ← A - L - C	
Z80	SBC A,L	SuBtract with Carry L from A	SZ-H-P1C						
8085	SBB M	SuBtract with Borrow M_{HL} from A	SZ-A-P-C	7	1	Reg Ind	9E	A ← A - M_{HL} - C	Subtract from the contents of the accumulator both the carry flag and the contents of the memory location pointed to by the HL register pair.
Z80	SBC A,(HL)	SuBtract with Carry (HL) from A	SZ-H-P1C						
8085	DAD B	Double AdD BC to HL	xx-x-x-C	10	1	Register	09	HL ← HL + BC	
Z80	ADD HL,BC	ADD BC to HL	xx-x-x0C	11					
8085	DAD D	Double AdD DE to HL	xx-x-x-C	10	1	Register	19	HL ← HL + DE	
Z80	ADD HL,DE	ADD DE to HL	xx-x-x0C	11					
8085	DAD H	Double AdD HL to HL	xx-x-x-C	10	1	Register	29	HL ← HL + HL	
Z80	ADD HL,HL	ADD HL to HL	xx-x-x0C	11					
8085	ADI dd	AdD Immediate dd to A	SZ-A-P-C	7	2	Immed	C6	A ← A + dd	
Z80	ADD A,dd	ADD dd to A	SZ-H-P0C						
8085	ACI dd	AdD with Carry Immediate dd to A	SZ-A-P-C	7	2	Immed	CE	A ← A + dd + C	
Z80	ADC A,dd	AdD with Carry dd to A	SZ-H-P0C						
8085	SUI dd	SUbtract Immediate dd from A	SZ-A-P-C	7	2	Immed	D6	A ← A - dd	
Z80	SUB dd	SUBtract dd from A	SZ-H-P1C						

Micro	Mnemonic	Operation	8085 > SZ-A-P-C Z80 > SZ-H-PNC	T	#	Address Mode	Op	Boolean/Arith. Operation	Notes
8085 Z80	SBI dd SBC A,dd	Subtract with Borrow Immediate dd from A SuBtract with Carry dd from A	SZ-A-P-C SZ-H-P1C	7	2	Immed	DE	A ← A - dd - C	
8085 Z80	DAA DAA	Decimal Adjust A Decimal Adjust A	SZ-A-P-C SZ-H-PxC	4	1	Implied	27	A ← BCD (A)	The 8-bit contents of the accumulator are adjusted to form two 4-bit binary-coded-decimal (BCD) digits.

Logical Instructions

Micro	Mnemonic	Operation	Flags	T	#	Address Mode	Op	Boolean/Arith. Operation	Notes
8085 Z80	ANA A AND A	ANd A with A AND A with A	SZ-A-P-0 SZ-1-P00	4	1	Register	A7	A ← A AND A	(8085) A flag=1 (8080) A=ORing of bit 3 of the operands
8085 Z80	ANA B AND B	ANd A with B AND B with A	SZ-A-P-0 SZ-1-P00	4	1	Register	A0	A ← A AND B	(8085) A flag=1 (8080) A flag=ORing of bit 3 of the operands
8085 Z80	ANA C AND C	ANd A with C AND C with A	SZ-A-P-0 SZ-1-P00	4	1	Register	A1	A ← A AND C	(8085) A flag=1 (8080) A flag=ORing of bit 3 of the operands
8085 Z80	ANA D AND D	ANd A with D AND D with A	SZ-A-P-0 SZ-1-P00	4	1	Register	A2	A ← A AND D	(8085) A flag=1 (8080) A flag=ORing of bit 3 of the operands
8085 Z80	ANA E AND E	ANd A with E AND E with A	SZ-A-P-0 SZ-1-P00	4	1	Register	A3	A ← A AND E	(8085) A flag=1 (8080) A flag=ORing of bit 3 of the operands
8085 Z80	ANA H AND H	ANd A with H AND H with A	SZ-A-P-0 SZ-1-P00	4	1	Register	A4	A ← A AND H	(8085) A flag=1 (8080) A flag=ORing of bit 3 of the operands
8085 Z80	ANA L AND L	ANd A with L AND L with A	SZ-A-P-0 SZ-1-P00	4	1	Register	A5	A ← A AND L	(8085) A flag=1 (8080) A flag=ORing of bit 3 of the operands
8085 Z80	ANA M AND (HL)	ANd A with M_{HL} AND (HL) with A	SZ-A-P-0 SZ-1-P00	7	1	Reg Ind	A6	A ← A AND M_{HL}	(8085) A flag=1 (8080) A flag=ORing of bit 3 of the operands

EXPANDED TABLE OF 8085/8080 AND Z80 (8080 SUBSET) INSTRUCTIONS LISTED BY CATEGORY (*Continued*)

Micro	Mnemonic	Operation	8085>SZ-A-P-C Z80>SZ-H-PNC	T	#	Address Mode	Op	Boolean/Arith. Operation	Notes
8085	XRA A	eXclusively OR A with A	SZ-0-P-0	4	1	Register	AF	A ← A XOR A	
Z80	XOR A	eXclusively OR A with A	SZ-0-P00						
8085	XRA B	eXclusively OR A with B	SZ-0-P-0	4	1	Register	A8	A ← A XOR B	
Z80	XOR B	eXclusively OR A with B	SZ-0-P00						
8085	XRA C	eXclusively OR A with C	SZ-0-P-0	4	1	Register	A9	A ← A XOR C	
Z80	XOR C	eXclusively OR A with C	SZ-0-P00						
8085	XRA D	eXclusively OR A with D	SZ-0-P-0	4	1	Register	AA	A ← A XOR D	
Z80	XOR D	eXclusively OR A with D	SZ-0-P00						
8085	XRA E	eXclusively OR A with E	SZ-0-P-0	4	1	Register	AB	A ← A XOR E	
Z80	XOR E	eXclusively OR A with E	SZ-0-P00						
8085	XRA H	eXclusively OR A with H	SZ-0-P-0	4	1	Register	AC	A ← A XOR H	
Z80	XOR H	eXclusively OR A with H	SZ-0-P00						
8085	XRA L	eXclusively OR A with L	SZ-0-P-0	4	1	Register	AD	A ← A XOR L	
Z80	XOR L	eXclusively OR A with L	SZ-0-P00						
8085	XRA M	eXclusively OR A with M_{HL}	SZ-0-P-0	7	1	Reg Ind	AE	A ← A XOR M_{HL}	Exclusively OR the contents of the accumulator with the contents of the memory location pointed to by the HL register pair.
Z80	XOR (HL)	eXclusively OR A with (HL)	SZ-0-P00						
8085	ORA A	OR A with A	SZ-0-P-0	4	1	Register	B7	A ← A OR A	
Z80	OR A	OR A with A	SZ-0-P00						

Micro	Mnemonic	Operation	8085>SZ-A-P-C Z80>SZ-H-PNC	T	#	Address Mode	Op	Boolean/Arith. Operation	Notes
8085	ORA B	**OR A with B**	SZ-0-P-0	4	1	Register	B0	A ← A OR B	
Z80	OR B	**OR A with B**	SZ-0-P00						
8085	ORA C	**OR A with C**	SZ-0-P-0	4	1	Register	B1	A ← A OR C	
Z80	OR C	**OR A with C**	SZ-0-P00						
8085	ORA D	**OR A with D**	SZ-0-P-0	4	1	Register	B2	A ← A OR D	
Z80	OR D	**OR A with D**	SZ-0-P00						
8085	ORA E	**OR A with E**	SZ-0-P-0	4	1	Register	B3	A ← A OR E	
Z80	OR E	**OR A with E**	SZ-0-P00						
8085	ORA H	**OR A with H**	SZ-0-P-0	4	1	Register	B4	A ← A OR H	
Z80	OR H	**OR A with H**	SZ-0-P00						
8085	ORA L	**OR A with L**	SZ-0-P-0	4	1	Register	B5	A ← A OR L	
Z80	OR L	**OR A with L**	SZ-0-P00						
8085	ORA M	**OR A with** M_{HL}	SZ-0-P-0	7	1	Reg Ind	B6	A ← A OR M_{HL}	OR the contents of the accumulator with the contents of the memory location pointed to by the HL register pair.
Z80	OR (HL)	**OR A with (HL)**	SZ-0-P00						
8085	ANI dd	**AN**d **I**mmediate dd with A	SZ-A-P-0	7	2	Immed	E6	A ← A AND dd	(8085) A flag = 1 (8080) A flag = ORing of bit 3 of operands
Z80	AND dd	**AND** dd with A	SZ-1-P00						
8085	XRI dd	e**X**clusively **OR** **I**mmediate dd with A	SZ-0-P-0	7	2	Immed	EE	A ← A XOR dd	
Z80	XOR dd	e**X**clusively **OR** dd with A	SZ-0-P00						
8085	ORI dd	**OR** **I**mmediate dd with A	SZ-0-P-0	7	2	Immed	F6	A ← A OR dd	
Z80	OR dd	**OR** dd with A	SZ-0-P00						
8085	CMA	**C**o**M**plement **A**	xx-x-x-x	4	1	Implied	2F	A ← $\overline{A}$	Invert every bit in the accumulator. Form the 1's complement.
Z80	CPL	**C**om**PL**ement A	xx-1-x1x						

EXPANDED TABLE OF 8085/8080 AND Z80 (8080 SUBSET) INSTRUCTIONS LISTED BY CATEGORY (*Continued*)

Micro	Mnemonic	Operation	8085 > SZ-A-P-C Z80 > SZ-H-PNC	T	#	Address Mode	Op	Boolean/Arith. Operation	Notes
		Rotate and Shift Instructions							
8085	RLC	Rotate Left with Carry	xx-x-x-C	4	1	Implied	07	$C \leftarrow A_7 \ldots A_0 \leftarrow$ ($A_7 \rightarrow A_0$)	
Z80	RLCA	Rotate Left with Carry A	xx-0-x0C						
8085	RRC	Rotate Right with with Carry	xx-x-x-C	4	1	Implied	0F	$\rightarrow A_7 \ldots A_0 \rightarrow C$ ($A_0 \rightarrow A_7$)	
Z80	RRCA	Rotate Right with Carry A	xx-0-x0C						
8085	RAL	Rotate A Left	xx-x-x-C	4	1	Implied	17	$C \leftarrow A_7 \ldots A_0 \leftarrow$ ($C \rightarrow A_0$)	
Z80	RLA	Rotate Left A	xx-0-x0C						
8085	RAR	Rotate A Right	xx-x-x-C	4	1	Implied	1F	$\rightarrow A_7 \ldots A_0 \rightarrow C$ ($C \rightarrow A_7$)	
Z80	RRA	Rotate Right A	xx-0-x0C						
		Increment and Decrement Instructions							
8085	INR A	INcRement A	SZ-A-P-x	4	1	Register	3C	A ← A + 1	(8080 = 5 states)
Z80	INC A	INCrement A	SZ-H-P0x						
8085	INR B	INcRement B	SZ-A-P-x	4	1	Register	04	B ← B + 1	(8080 = 5 states)
Z80	INC B	INCrement B	SZ-H-P0x						
8085	INR C	INcRement C	SZ-A-P-x	4	1	Register	0C	C ← C + 1	(8080 = 5 states)
Z80	INC C	INCrement C	SZ-H-P0x						
8085	INR D	INcRement D	SZ-A-P-x	4	1	Register	14	D ← D + 1	(8080 = 5 states)
Z80	INC D	INCrement D	SZ-H-P0x						
8085	INR E	INcRement E	SZ-A-P-x	4	1	Register	1C	E ← E + 1	(8080 = 5 states)
Z80	INC E	INCrement E	SZ-H-P0x						

Micro	Mnemonic	Operation	8085>SZ-A-P-C Z80>SZ-H-PNC	T	#	Address Mode	Op	Boolean/Arith. Operation	Notes
8085	INR H	INcRement H	SZ-A-P-x	4	1	Register	24	H ← H + 1	(8080 = 5 states)
Z80	INC H	INCrement H	SZ-H-P0x						
8085	INR L	INcRement L	SZ-A-P-x	4	1	Register	2C	L ← L + 1	(8080 = 5 states)
Z80	INC L	INCrement L	SZ-H-P0x						
8085	INR M	INcRement M_{HL}	SZ-A-P-x	10	1	Reg Ind	34	$M_{HL} \leftarrow M_{HL} + 1$	
Z80	INC (HL)	INCrement (HL)	SZ-H-P0x	11					
8085	INX B	INcrement eXtended B	xx-x-x-x	6	1	Register	03	BC ← BC + 1	(8080 = 5 states)
Z80	INC BC	INCrement reg pair BC	xx-x-xxx						
8085	INX D	INcrement eXtended D	xx-x-x-x	6	1	Register	13	DE ← DE + 1	(8080 = 5 states)
Z80	INC DE	INCrement reg pair DE	xx-x-xxx						
8085	INX H	INcrement eXtended H	xx-x-x-x	6	1	Register	23	HL ← HL + 1	(8080 = 5 states)
Z80	INC HL	INCrement reg pair HL	xx-x-xxx						
8085	DCR A	DeCRement register A	SZ-A-P-x	4	1	Register	3D	A ← A - 1	(8080 = 5 states)
Z80	DEC A	DECrement register A	SZ-H-P1x						
8085	DCR B	DeCRement register B	SZ-A-P-x	4	1	Register	05	B ← B - 1	(8080 = 5 states)
Z80	DEC B	DECrement register B	SZ-H-P1x						
8085	DCR C	DeCRement register C	SZ-A-P-x	4	1	Register	0D	C ← C - 1	(8080 = 5 states)
Z80	DEC C	DECrement register C	SZ-H-P1x						
8085	DCR D	DeCRement register D	SZ-A-P-x	4	1	Register	15	D ← D - 1	(8080 = 5 states)
Z80	DEC D	DECrement register D	SZ-H-P1x						
8085	DCR E	DeCRement register E	SZ-A-P-x	4	1	Register	1D	E ← E - 1	(8080 = 5 states)
Z80	DEC E	DECrement register E	SZ-H-P1x						

EXPANDED TABLE OF 8085/8080 AND Z80 (8080 SUBSET) INSTRUCTIONS LISTED BY CATEGORY (*Continued*)

Micro	Mnemonic	Operation	8085>SZ-A-P-C Z80>SZ-H-PNC	T	#	Address Mode	Op	Boolean/Arith. Operation	Notes
8085	DCR H	DeCRement register H	SZ-A-P-x	4	1	Register	25	H ← H - 1	(8080 = 5 states)
Z80	DEC H	DECrement register H	SZ-H-P1x						
8085	DCR L	DeCRement register L	SZ-A-P-x	4	1	Register	2D	L ← L - 1	(8080 = 5 states)
Z80	DEC L	DECrement register L	SZ-H-P1x						
8085	DCR M	DeCRement M_{HL}	SZ-A-P-x	10	1	Reg Ind	35	M_{HL} ← M_{HL} - 1	
Z80	DEC (HL)	DECrement (HL)	SZ-H-P1x						
8085	DCX B	DeCrement eXtended register pair BC	xx-x-x-x	6	1	Register	0B	BC ← BC - 1	(8080 = 5 states)
Z80	DEC BC	DECrement register pair BC	xx-x-xxx						
8085	DCX D	DeCrement eXtended register pair DE	xx-x-x-x	6	1	Register	1B	DE ← DE - 1	(8080 = 5 states)
Z80	DEC DE	DECrement register pair DE	xx-x-xxx						
8085	DCX H	DeCrement eXtended register pair HL	xx-x-x-x	6	1	Register	2B	HL ← HL - 1	(8080 = 5 states)
Z80	DEC HL	DECrement register pair HL	xx-x-xxx						

Unconditional Jump Instructions

Micro	Mnemonic	Operation	8085>SZ-A-P-C Z80>SZ-H-PNC	T	#	Address Mode	Op	Boolean/Arith. Operation	Notes
8085	JMP aaaa	JuMP to mem loc aaaa	xx-x-x-x	10	3	Direct	C3	PC ← aaaa	
Z80	JP aaaa	JumP to mem loc aaaa	xx-x-xxx						
8085	PCHL	transfer to the Program Counter HL	xx-x-x-x	6	1	Register	E9	PC_H ← H PC_L ← L	(8080 = 5 states) Transfer the contents of register H to the high-order byte of the program counter and the contents of register L to the low-order byte of the program counter.
Z80	JP (HL)	JumP to (HL)	xx-x-xxx	4					

Micro	Mnemonic	Operation	8085 > SZ-A-P-C Z80 > SZ-H-PNC	T	#	Address Mode	Op	Boolean/Arith. Operation	Notes

Test (Compare) Instructions

Micro	Mnemonic	Operation	8085 > SZ-A-P-C Z80 > SZ-H-PNC	T	#	Address Mode	Op	Boolean/Arith. Operation	Notes
8085 Z80	CMP A CP A	CoMPare A to A ComPare A to A	SZ-A-P-C SZ-H-P1C	4	1	Register	BF	A - A	If A = A then the Z flag = 1. If A < A then the C flag = 1.
8085 Z80	CMP B CP B	CoMPare B to A ComPare B to A	SZ-A-P-C SZ-H-P1C	4	1	Register	B8	A - B	If A = B then the Z flag = 1. If A < B then the C flag = 1.
8085 Z80	CMP C CP C	CoMPare C to A ComPare C to A	SZ-A-P-C SZ-H-P1C	4	1	Register	B9	A - C	If A = C then the Z flag = 1. If A < C then the C flag = 1.
8085 Z80	CMP D CP D	CoMPare D to A ComPare D to A	SZ-A-P-C SZ-H-P1C	4	1	Register	BA	A - D	If A = D then the Z flag = 1. If A < D then the C flag = 1.
8085 Z80	CMP E CP E	CoMPare E to A ComPare E to A	SZ-A-P-C SZ-H-P1C	4	1	Register	BB	A - E	If A = E then the Z flag = 1. If A < E then the C flag = 1.
8085 Z80	CMP H CP H	CoMPare H to A ComPare H to A	SZ-A-P-C SZ-H-P1C	4	1	Register	BC	A - H	If A = H then the Z flag = 1. If A < H then the C flag = 1.
8085 Z80	CMP L CP L	CoMPare L to A ComPare L to A	SZ-A-P-C SZ-H-P1C	4	1	Register	BD	A - L	If A = L then the Z flag = 1. If A < L then the C flag = 1.
8085 Z80	CMP M CP (HL)	CoMPare M_{HL} to A ComPare (HL) to A	SZ-A-P-C SZ-H-P1C	7	1	Reg Ind	BE	A - M_{HL}	If A = M_{HL} then the Z flag = 1. If A < M_{HL} then the C flag = 1.
8085 Z80	CPI dd CP dd	ComPare Immediate dd to A ComPare dd to A	SZ-A-P-C SZ-H-P1C	7	2	Immed	FE	A - dd	If A = dd then the Z flag = 1. If A < dd then the C flag = 1.

EXPANDED TABLE OF 8085/8080 AND Z80 (8080 SUBSET) INSTRUCTIONS LISTED BY CATEGORY (*Continued*)

Micro	Mnemonic	Operation	8085>SZ-A-P-C Z80>SZ-H-PNC	T	#	Address Mode	Op	Boolean/Arith. Operation	Notes
		Conditional Jump (Branch) Instructions							
8085	JNZ aaaa	Jump if Not Zero to aaaa	xx-x-x-x	7/10	3	Direct	C2	PC ← aaaa if Z = 0	(8080 = 10 states) PC_L ← byte 2 PC_H ← byte 3
Z80	JP NZ,aaaa	JumP if Not Zero to aaaa	xx-x-xxx	10					
8085	JZ aaaa	Jump if Zero to aaaa	xx-x-x-x	7/10	3	Direct	CA	PC ← aaaa if Z = 1	(8080 = 10 states) PC_L ← byte 2 PC_H ← byte 3
Z80	JP Z,aaaa	JumP if Zero to aaaa	xx-x-xxx	10					
8085	JNC aaaa	Jump if No Carry to aaaa	xx-x-x-x	7/10	3	Direct	D2	PC ← aaaa if C = 0	(8080 = 10 states) PC_L ← byte 2 PC_H ← byte 3
Z80	JP NC,aaaa	JumP if No Carry to aaaa	xx-x-xxx	10					
8085	JC aaaa	Jump if Carry to aaaa	xx-x-x-x	7/10	3	Direct	DA	PC ← aaaa if C = 1	(8080 = 10 states) PC_L ← byte 2 PC_H ← byte 3
Z80	JP C,aaaa	JumP if Carry to aaaa	xx-x-xxx	10					
8085	JPO aaaa	Jump if Parity Odd to aaaa	xx-x-x-x	7/10	3	Direct	E2	PC ← aaaa if P = 0	(8080 = 10 states) PC_L ← byte 2 PC_H ← byte 3
Z80	JP PO,aaaa	JumP if Parity Odd to aaaa	xx-x-xxx	10					
8085	JPE aaaa	Jump if Parity Even to aaaa	xx-x-x-x	7/10	3	Direct	EA	PC ← aaaa if P = 1	(8080 = 10 states) PC_L ← byte 2 PC_H ← byte 3
Z80	JP PE,aaaa	JumP if Parity Even to aaaa	xx-x-xxx	10					
8085	JP aaaa	Jump if Plus to aaaa	xx-x-x-x	7/10	3	Direct	F2	PC ← aaaa if S = 0	(8080 = 10 states) PC_L ← byte 2 PC_H ← byte 3
Z80	JP P,aaaa	JumP if Plus to aaaa	xx-x-xxx	10					
8085	JM aaaa	Jump if Minus to aaaa	xx-x-x-x	7/10	3	Direct	FA	PC ← aaaa if S = 1	(8080 = 10 states) PC_L ← byte 2 PC_H ← byte 3
Z80	JP M,aaaa	JumP if Minus to aaaa	xx-x-xxx	10					
		Subroutine Instructions							
8085	CALL aaaa	CALL subroutine at aaaa	xx-x-x-x	18	3	Direct/ Reg Ind	CD	S ← PC_H S ← PC_L PC ← aaaa	(8080 = 17 states) The stack pointer is decremented as each new byte is pushed onto the stack. PC_H ← byte 3 PC_L ← byte 2
Z80	CALL aaaa	CALL subroutine at aaaa	xx-x-xxx	17					

Micro	Mnemonic	Operation	8085 > SZ-A-P-C Z80 > SZ-H-PNC	T	#	Address Mode	Op	Boolean/Arith. Operation	Notes
8085	CNZ aaaa	Call if Not Zero	xx-x-x-x	9/18				if Z = 0	(8080 = 11/17 states)
		subroutine at aaaa			3	Direct/	C4	$S \leftarrow PC_H$	The stack pointer is
Z80	CALL NZ,aaaa	CALL if Not Zero	xx-x-xxx	10/17		Reg Ind		$S \leftarrow PC_L$	decremented as each new byte
		subroutine at aaaa						$PC \leftarrow aaaa$	is pushed onto the stack.
									$PC_H \leftarrow$ byte 3
									$PC_L \leftarrow$ byte 2
8085	CZ aaaa	Call if Zero	xx-x-x-x	9/18				if Z = 1	(8080 = 11/17 states)
		subroutine at aaaa			3	Direct/	CC	$S \leftarrow PC_H$	The stack pointer is
Z80	CALL Z,aaaa	CALL if Zero	xx-x-xxx	10/17		Reg Ind		$S \leftarrow PC_L$	decremented as each new byte
		subroutine at aaaa						$PC \leftarrow aaaa$	is pushed onto the stack.
									$PC_H \leftarrow$ byte 3
									$PC_L \leftarrow$ byte 2
8085	CNC aaaa	Call if No Carry	xx-x-x-x	9/18				if C = 0	(8080 = 11/17 states)
		subroutine at aaaa			3	Direct/	D4	$S \leftarrow PC_H$	The stack pointer is
Z80	CALL NC,aaaa	CALL if No Carry	xx-x-xxx	10/17		Reg Ind		$S \leftarrow PC_L$	decremented as each new byte
		subroutine at aaaa						$PC \leftarrow aaaa$	is pushed onto the stack.
									$PC_H \leftarrow$ byte 3
									$PC_L \leftarrow$ byte 2
8085	CC aaaa	Call if Carry	xx-x-x-x	9/18				if C = 1	(8080 = 11/17 states)
		subroutine at aaaa			3	Direct/	DC	$S \leftarrow PC_H$	The stack pointer is
Z80	CALL C,aaaa	CALL if Carry	xx-x-xxx	10/17		Reg Ind		$S \leftarrow PC_L$	decremented as each new byte
		subroutine at aaaa						$PC \leftarrow aaaa$	is pushed onto the stack.
									$PC_H \leftarrow$ byte 3
									$PC_L \leftarrow$ byte 2
8085	CPO aaaa	Call if Parity Odd	xx-x-x-x	9/18				if P = 0	(8080 = 11/17 states)
		subroutine at aaaa			3	Direct/	E4	$S \leftarrow PC_H$	The stack pointer is
Z80	CALL PO,aaaa	CALL if Parity Odd	xx-x-xxx	10/17		Reg Ind		$S \leftarrow PC_L$	decremented as each new byte
		subroutine at aaaa						$PC \leftarrow aaaa$	is pushed onto the stack.
									$PC_H \leftarrow$ byte 3
									$PC_L \leftarrow$ byte 2
8085	CPE aaaa	Call if Parity Even	xx-x-x-x	9/18				if P = 1	(8080 = 11/17 states)
		subroutine at aaaa			3	Direct/	EC	$S \leftarrow PC_H$	The stack pointer is
Z80	CALL PE,aaaa	CALL if Parity Even	xx-x-xxx	10/17		Reg Ind		$S \leftarrow PC_L$	decremented as each new byte
		subroutine at aaaa						$PC \leftarrow aaaa$	is pushed onto the stack.
									$PC_H \leftarrow$ byte 3
									$PC_L \leftarrow$ byte 2
8085	CP aaaa	Call if Plus	xx-x-x-x	9/18				if S = 0	(8080 = 11/17 states)
		subroutine at aaaa			3	Direct/	F4	$S \leftarrow PC_H$	The stack pointer is
Z80	CALL P,aaaa	CALL if Plus	xx-x-xxx	10/17		Reg Ind		$S \leftarrow PC_L$	decremented as each new byte
		subroutine at aaaa						$PC \leftarrow aaaa$	is pushed onto the stack.
									$PC_H \leftarrow$ byte 3
									$PC_L \leftarrow$ byte 2

EXPANDED TABLE OF 8085/8080 AND Z80 (8080 SUBSET) INSTRUCTIONS LISTED BY CATEGORY (*Continued*)

Micro	Mnemonic	Operation	8085>SZ-A-P-C Z80>SZ-H-PNC	T	#	Address Mode	Op	Boolean/Arith. Operation	Notes
8085	CM aaaa	Call if Minus subroutine at aaaa	xx-x-x-x	9/18	3	Direct/ Reg Ind	FC	if S = 1 S ← PC_H S ← PC_L PC ← aaaa	(8080 = 11/17 states) The stack pointer is decremented as each new byte is pushed onto the stack. PC_H ← byte 3 PC_L ← byte 2
Z80	CALL M,aaaa	CALL if Minus subroutine at aaaa	xx-x-xxx	10/17					
8085	RET	RETurn	xx-x-x-x	10	1	Reg Ind	C9	PC_L ← S PC_H ← S	The stack pointer is incremented as each byte is popped from the stack.
Z80	RET	RETurn	xx-x-xxx						
8085	RNZ	Return if Not Zero	xx-x-x-x	6/12	1	Reg Ind	C0	if Z = 0 PC_L ← S PC_H ← S	(8080 = 5/11 states) The stack pointer is incremented as each byte is popped from the stack.
Z80	RET NZ	RETurn if Not Zero	xx-x-xxx	5/10					
8085	RZ	Return if Zero	xx-x-x-x	6/12	1	Reg Ind	C8	if Z = 1 PC_L ← S PC_H ← S	(8080 = 5/11 states) The stack pointer is incremented as each byte is popped from the stack.
Z80	RET Z	RETurn if Zero	xx-x-xxx	5/10					
8085	RNC	Return if No Carry	xx-x-x-x	6/12	1	Reg Ind	D0	if C = 0 PC_L ← S PC_H ← S	(8080 = 5/11 states) The stack pointer is incremented as each byte is popped from the stack.
Z80	RET NC	RETurn if No Carry	xx-x-xxx	5/10					
8085	RC	Return if Carry	xx-x-x-x	6/12	1	Reg Ind	D8	if C = 1 PC_L ← S PC_H ← S	(8080 = 5/11 states) The stack pointer is incremented as each byte is popped from the stack.
Z80	RET C	RETurn if Carry	xx-x-xxx	5/10					
8085	RPO	Return if Parity Odd	xx-x-x-x	6/12	1	Reg Ind	E0	if P = 0 PC_L ← S PC_H ← S	(8080 = 5/11 states) The stack pointer is incremented as each byte is popped from the stack.
Z80	RET PO	RETurn if Parity Odd	xx-x-xxx	5/10					
8085	RPE	Return if Parity Even	xx-x-x-x	6/12	1	Reg Ind	E8	if P = 1 PC_L ← S PC_H ← S	(8080 = 5/11 states) The stack pointer is incremented as each byte is popped from the stack.
Z80	RET PE	RETurn if Parity Even	xx-x-xxx	5/10					

Micro	Mnemonic	Operation	8085>SZ-A-P-C Z80>SZ-H-PNC	T	#	Address Mode	Op	Boolean/Arith. Operation	Notes
8085	RP	Return if Plus	xx-x-x-x	6/12				if S = 0	(8080 = 5/11 states)
					1	Reg Ind	F0	$PC_L \leftarrow S$	The stack pointer is
Z80	RET P	RETurn if Plus	xx-x-xxx	5/10				$PC_H \leftarrow S$	incremented as each byte is popped from the stack.
8085	RM	Return if Minus	xx-x-x-x	6/12				if S = 1	(8080 = 5/11 states)
					1	Reg Ind	F8	$PC_L \leftarrow S$	The stack pointer is
Z80	RET M	RETurn if Minus	xx-x-xxx	5/10				$PC_H \leftarrow S$	incremented as each byte is popped from the stack.
8085	RST 0	ReStarT 0	xx-x-x-x	12				$S \leftarrow PC_H$	(8080 = 11 states)
					1	Reg Ind	C7	$S \leftarrow PC_L$	The stack pointer is
Z80	RST 00H	ReStarT 00H	xx-x-xxx	11				PC ← 0000H	decremented as each new byte is pushed onto the stack.
8085	RST 1	ReStarT 1	xx-x-x-x	12				$S \leftarrow PC_H$	(8080 = 11 states)
					1	Reg Ind	CF	$S \leftarrow PC_L$	The stack pointer is
Z80	RST 08H	ReStarT 08H	xx-x-xxx	11				PC ← 0008H	decremented as each new byte is pushed onto the stack.
8085	RST 2	ReStarT 2	xx-x-x-x	12				$S \leftarrow PC_H$	(8080 = 11 states)
					1	Reg Ind	D7	$S \leftarrow PC_L$	The stack pointer is
Z80	RST 10H	ReStarT 10H	xx-x-xxx	11				PC ← 0010H	decremented as each new byte is pushed onto the stack.
8085	RST 3	ReStarT 3	xx-x-x-x	12				$S \leftarrow PC_H$	(8080 = 11 states)
					1	Reg Ind	DF	$S \leftarrow PC_L$	The stack pointer is
Z80	RST 18H	ReStarT 18H	xx-x-xxx	11				PC ← 0018H	decremented as each new byte is pushed onto the stack.
8085	RST 4	ReStarT 4	xx-x-x-x	12				$S \leftarrow PC_H$	(8080 = 11 states)
					1	Reg Ind	E7	$S \leftarrow PC_L$	The stack pointer is
Z80	RST 20H	ReStarT 20H	xx-x-xxx	11				PC ← 0020H	decremented as each new byte is pushed onto the stack.
8085	RST 5	ReStarT 5	xx-x-x-x	12				$S \leftarrow PC_H$	(8080 = 11 states)
					1	Reg Ind	EF	$S \leftarrow PC_L$	The stack pointer is
Z80	RST 28H	ReStarT 28H	xx-x-xxx	11				PC ← 0028H	decremented as each new byte is pushed onto the stack.
8085	RST 6	ReStarT 6	xx-x-x-x	12				$S \leftarrow PC_H$	(8080 = 11 states)
					1	Reg Ind	F7	$S \leftarrow PC_L$	The stack pointer is
Z80	RST 30H	ReStarT 30H	xx-x-xxx	11				PC ← 0030H	decremented as each new byte is pushed onto the stack.

EXPANDED TABLE OF 8085/8080 AND Z80 (8080 SUBSET) INSTRUCTIONS LISTED BY CATEGORY *(Continued)*

Micro	Mnemonic	Operation	8085>SZ-A-P-C Z80>SZ-H-PNC	T	#	Address Mode	Op	Boolean/Arith. Operation	Notes
8085	RST 7	ReStarT 7	xx-x-x-x	12	1	Reg Ind	FF	S ← PC_H S ← PC_L PC ← 0038H	(8080 = 11 states) The stack pointer is decremented as each new byte is pushed onto the stack.
Z80	RST 38H	ReStarT 38H	xx-x-xxx	11					

Stack Instructions

Micro	Mnemonic	Operation	8085>SZ-A-P-C Z80>SZ-H-PNC	T	#	Address Mode	Op	Boolean/Arith. Operation	Notes
8085	LXI SP,dddd	Load eXtended Immediate dddd into the Stack Pointer	xx-x-x-x	10	3	Immed	31	SP ← dddd	Copy bytes 3 and 2 of the instruction into the stack pointer.
Z80	LD SP,dddd	LoaD dddd into the Stack Pointer	xx-x-xxx						
8085	DAD SP	Double AdD SP to HL	xx-x-x-C	10	1	Register	39	HL ← HL + SP	
Z80	ADD HL,SP	ADD SP to HL	xx-x-x0C	11					
8085	INX SP	INcrement eXtended Stack Pointer	xx-x-x-x	6	1	Register	33	SP ← SP + 1	(8080 = 5 states)
Z80	INC SP	INCrement Stack Pointer	xx-x-xxx						
8085	DCX SP	DeCrement eXtended Stack Pointer	xx-x-x-x	6	1	Register	3B	SP ← SP - 1	(8080 = 5 states)
Z80	DEC SP	DECrement Stack Pointer	xx-x-xxx						
8085	PUSH B	PUSH reg pair BC	xx-x-x-x	12	1	Reg Ind	C5	S ← B S ← C	(8080 = 11 states) The stack pointer is decremented as each new byte is pushed onto the stack.
Z80	PUSH BC	PUSH reg pair BC	xx-x-xxx	11					
8085	PUSH D	PUSH reg pair DE	xx-x-x-x	12	1	Reg Ind	D5	S ← D S ← E	(8080 = 11 states) The stack pointer is decremented as each new byte is pushed onto the stack.
Z80	PUSH DE	PUSH reg pair DE	xx-x-xxx	11					
8085	PUSH H	PUSH reg pair HL	xx-x-x-x	12	1	Reg Ind	E5	S ← H S ← L	(8080 = 11 states) The stack pointer is decremented as each new byte is pushed onto the stack.
Z80	PUSH HL	PUSH reg pair HL	xx-x-xxx	11					

Micro	Mnemonic	Operation	8085>SZ-A-P-C Z80>SZ-H-PNC	T	#	Address Mode	Op	Boolean/Arith. Operation	Notes
8085	PUSH PSW	PUSH Processor Status Word	xx-x-x-x	12	1	Reg Ind	F5	S ← A	(8080 = 11 states) The stack pointer is decremented as each new byte is pushed onto the stack. The "flags" byte is assembled in the normal order of the flags (8080/8085 = SZ-A-P-C and Z80 = SZ-H-PNC) for that microprocessor.
Z80	PUSH AF	PUSH Accumulator and Flags	xx-x-xxx	11				S ← flags	
8085	POP B	POP reg pair BC	xx-x-x-x	10	1	Reg Ind	C1	C ← S	The stack pointer is incremented as each byte is popped from the stack.
Z80	POP BC	POP reg pair BC	xx-x-xxx					B ← S	
8085	POP D	POP reg pair DE	xx-x-x-x	10	1	Reg Ind	D1	E ← S	The stack pointer is incremented as each byte is popped from the stack.
Z80	POP DE	POP reg pair DE	xx-x-xxx					D ← S	
8085	POP H	POP reg pair HL	xx-x-x-x	10	1	Reg Ind	E1	L ← S	The stack pointer is incremented as each byte is popped from the stack.
Z80	POP HL	POP reg pair HL	xx-x-xxx					H ← S	
8085	POP PSW	POP Processor Status Word	SZ-A-P-C	10	1	Reg Ind	F1	flags ← S	The stack pointer is incremented as each byte is popped from the stack.
Z80	POP AF	POP Accumulator and Flag	SZ-H-PNC					A ← S	
8085	XTHL	eXchange top of sTack with reg pair HL	xx-x-x-x	16	1	Reg Ind	E3	L ↔ S	(8080 = 18 states) Stack pointer does not change
Z80	EX (SP),HL	EXchange $M_{(SP)}$ with reg pair HL	xx-x-xxx	19				H ↔ $S_{(next)}$	
8085	SPHL	move into SP the contents of reg pair HL	xx-x-x-x	6	1	Register	F9	SP ← HL	(8080 = 5 states)
Z80	LD SP,HL	LoaD into SP the contents of reg pair HL	xx-x-xxx						

Interrupt Instructions

Micro	Mnemonic	Operation	8085>SZ-A-P-C Z80>SZ-H-PNC	T	#	Address Mode	Op	Boolean/Arith. Operation	Notes
8085	DI	Disable Interrupts	xx-x-x-x	4	1	Implied	F3	IFF ← 0	
Z80	DI	Disable Interrupts	xx-x-xxx						

EXPANDED TABLE OF 8085/8080 AND Z80 (8080 SUBSET) INSTRUCTIONS LISTED BY CATEGORY (*Continued*)

Micro	Mnemonic	Operation	8085 > SZ-A-P-C Z80 > SZ-H-PNC	T	#	Address Mode	Op	Boolean/Arith. Operation	Notes
8085	EI	Enable Interrupts	xx-x-x-x	4	1	Implied	FB	IFF ← 1	
Z80	EI	Enable Interrupts	xx-x-xxx						
8085	RIM	(not covered here - see note at end of table)							
8085	SIM	(not covered here - see note at end of table)							

Input-Output Instructions

Micro	Mnemonic	Operation	8085 > SZ-A-P-C Z80 > SZ-H-PNC	T	#	Address Mode	Op	Boolean/Arith. Operation	Notes
8085	OUT dd	OUTput to port dd contents of A	xx-x-x-x	10	2	Direct	D3	dd port ← A	The contents of the accumulator are sent to a specified output port.
Z80	OUT dd,A	OUTput to port dd contents of A	xx-x-xxx	11					
8085	IN dd	INput into A one byte from port dd	xx-x-x-x	10	2	Direct	DB	A ← dd port (byte)	One byte from the specified port is copied into the accumulator.
Z80	IN A,dd	INput into A one byte from port dd	xx-x-xxx	11					

Address Modes

Implied
Register
Immediate
Direct
Register Indirect (Reg Ind)

Abbreviations and Explanations

PSW = program status word (flags)
S = stack
SP = stack pointer
PC = program counter
IFF = interrupt enable flip-flop
A = accumulator
B,C,D,E,H,L = registers
$_L$ = low-order byte
$_H$ = high-order byte
A_7-A_0 = accumulator bits 0 through 7

d = data (a single hex digit)
dd = data (two hex digits - 1 byte)
dddd = data (four hex digits - 2 bytes)

a = address (a single hex digit)
aa = address (two hex digits - 1 byte)
aaaa = address (four hex digits - 2 bytes)

Flags

If one of the flag letter designations is in the column for that particular flag it indicates that the flag is affected by this operation and could be set or cleared depending on the result of the operation. One of the following could also appear in a flag column:

- \- = no flag is represented by this column, a blank bit in the status register
- x = flag not affected by this operation
- 1 = flag always set by this operation
- 0 = flag always cleared by this operation

<u>8085</u>

S = sign flag
Z = zero flag
A = auxiliary carry flag (usually labeled "AC")
P = parity flag
C = carry flag (usually labeled "CY")

Z80

S = sign
Z = zero flag
H = half carry flag
P = parity/overflow flag (usually labeled "P/V")
N = negative flag
C = carry flag

Symbols in the Page Heading

T = T states
# = number of bytes

Special Notes

States = When two numbers appear in the "States" column separated by a slash, the lower number indicates the number of states if the condition is false and the operation does not occur, and the larger number indicates the number of states if the condition is true and the operation does occur.

8080 = The 8080 behaves the same as the 8085 unless special information is provided in the "Notes" column for the 8080.

RIM & SIM = These two instructions related to interrupts are not covered in this table. They apply only to the 8085 (neither is available in either the 8080 or Z80).

Addressing Modes - A Summary

Implied: These instructions contain the source and destination of the data by implication.

Register: In this mode the operand and its source are specified and data is operated on in the registers only.

Immediate: The data to be operated on follows the instruction op code in memory; that is, it is the next byte in memory after the instruction.

Direct: The full address of the location of the operand in contained in bytes 2 and 3, that is, the next two bytes in memory after the instruction. The low-order byte comes first, and the high-order second.

Register Indirect (Reg Ind): In this addressing mode several steps are involved. Included in the instruction is a register pair; the contents of that register pair contains the address where that operand may be found, not the operand itself.

MINI TABLE OF 8085/8080 AND Z80 (8080 SUBSET) INSTRUCTIONS LISTED BY CATEGORY

CPU Control Instructions

8085	Z80	Op	Operation
NOP	NOP	00	Nothing happens
HLT	HALT	76	Stop processing

Data Transfer Instructions

8085	Z80	Op	Operation
MOV A,A	LD A,A	7F	A ← A
MOV A,B	LD A,B	78	A ← B
MOV A,C	LD A,C	79	A ← C
MOV A,D	LD A,D	7A	A ← D
MOV A,E	LD A,E	7B	A ← E
MOV A,H	LD A,H	7C	A ← H
MOV A,L	LD A,L	7D	A ← L
MOV A,M	LD A,(HL)	7E	A ← M_{HL}
MOV B,A	LD B,A	47	B ← A
MOV B,B	LD B,B	40	B ← B
MOV B,C	LD B,C	41	B ← C
MOV B,D	LD B,D	42	B ← D
MOV B,E	LD B,E	43	B ← E
MOV B,H	LD B,H	44	B ← H
MOV B,L	LD B,L	45	B ← L
MOV B,M	LD B,(HL)	46	B ← M_{HL}
MOV C,A	LD C,A	4F	C ← A
MOV C,B	LD C,B	48	C ← B
MOV C,C	LD C,C	49	C ← C
MOV C,D	LD C,D	4A	C ← D
MOV C,E	LD C,E	4B	C ← E
MOV C,H	LD C,H	4C	C ← H
MOV C,L	LD C,L	4D	C ← L

8085	Z80	Op	Operation
MOV C,M	LD C,(HL)	4E	C ← M_{HL}
MOV D,A	LD D,A	57	D ← A
MOV D,B	LD D,B	50	D ← B
MOV D,C	LD D,C	51	D ← C
MOV D,D	LD D,D	52	D ← D
MOV D,E	LD D,E	53	D ← E
MOV D,H	LD D,H	54	D ← H
MOV D,L	LD D,L	55	D ← L
MOV D,M	LD D,(HL)	56	D ← M_{HL}
MOV E,A	LD E,A	5F	E ← A
MOV E,B	LD E,B	58	E ← B
MOV E,C	LD E,C	59	E ← C
MOV E,D	LD E,D	5A	E ← D
MOV E,E	LD E,E	5B	E ← E
MOV E,H	LD E,H	5C	E ← H
MOV E,L	LD E,L	5D	E ← L
MOV E,M	LD E,(HL)	5E	E ← M_{HL}
MOV H,A	LD H,A	67	H ← A
MOV H,B	LD H,B	60	H ← B
MOV H,C	LD H,C	61	H ← C
MOV H,D	LD H,D	62	H ← D
MOV H,E	LD H,E	63	H ← E
MOV H,H	LD H,H	64	H ← H
MOV H,L	LD H,L	65	H ← L
MOV H,M	LD H,(HL)	66	H ← M_{HL}
MOV L,A	LD L,A	6F	L ← A
MOV L,B	LD L,B	68	L ← B
MOV L,C	LD L,C	69	L ← C

8085	Z80	Op	Operation
MOV L,D	LD L,D	6A	$L \leftarrow D$
MOV L,E	LD L,E	6B	$L \leftarrow E$
MOV L,H	LD L,H	6C	$L \leftarrow H$
MOV L,L	LD L,L	6D	$L \leftarrow L$
MOV L,M	LD L,(HL)	6E	$L \leftarrow M_{HL}$
MOV M,A	LD (HL),A	77	$M_{HL} \leftarrow A$
MOV M,B	LD (HL),B	70	$M_{HL} \leftarrow B$
MOV M,C	LD (HL),C	71	$M_{HL} \leftarrow C$
MOV M,D	LD (HL),D	72	$M_{HL} \leftarrow D$
MOV M,E	LD (HL),E	73	$M_{HL} \leftarrow E$
MOV M,H	LD (HL),H	74	$M_{HL} \leftarrow H$
MOV M,L	LD (HL),L	75	$M_{HL} \leftarrow L$
MVI A,dd	LD A,dd	3E	$A \leftarrow dd$
MVI B,dd	LD B,dd	06	$B \leftarrow dd$
MVI C,dd	LD C,dd	0E	$C \leftarrow dd$
MVI D,dd	LD D,dd	16	$D \leftarrow dd$
MVI E,dd	LD E,dd	1E	$E \leftarrow dd$
MVI H,dd	LD H,dd	26	$H \leftarrow dd$
MVI L,dd	LD L,dd	2E	$L \leftarrow dd$
MVI M,dd	LD (HL),dd	36	$M_{HL} \leftarrow dd$
LXI B,dddd	LD BC,dddd	01	$BC \leftarrow dddd$
LXI D,dddd	LD DE,dddd	11	$DE \leftarrow dddd$
LXI H,dddd	LD HL,dddd	21	$HL \leftarrow dddd$
LDAX B	LD A,(BC)	0A	$A \leftarrow M_{BC}$
LDAX D	LD A,(DE)	1A	$A \leftarrow M_{DE}$
LHLD aaaa	LD HL,(aaaa)	2A	$L \leftarrow M_{aaaa}$ $H \leftarrow M_{aaaa+1}$
LDA aaaa	LD A,(aaaa)	3A	$A \leftarrow M_{aaaa}$
STA aaaa	LD (aaaa),A	32	$M_{aaaa} \leftarrow A$
STAX B	LD (BC),A	02	$M_{BC} \leftarrow A$

8085	Z80	Op	Operation
STAX D	LD (DE),A	12	$M_{DE} \leftarrow A$
SHLD aaaa	LD (aaaa),HL	22	$M_{aaaa} \leftarrow L$ $M_{aaaa+1} \leftarrow H$
XCHG	EX DE,HL	EB	$DE \leftrightarrow HL$

Flag Instructions

8085	Z80	Op	Operation
STC	SCF	37	$C \leftarrow 1$
CMC	CCF	3F	$C \leftarrow \overline{C}$

Arithmetic Instructions

8085	Z80	Op	Operation
ADD A	ADD A,A	87	$A \leftarrow A + A$
ADD B	ADD A,B	80	$A \leftarrow A + B$
ADD C	ADD A,C	81	$A \leftarrow A + C$
ADD D	ADD A,D	82	$A \leftarrow A + D$
ADD E	ADD A,E	83	$A \leftarrow A + E$
ADD H	ADD A,H	84	$A \leftarrow A + H$
ADD L	ADD A,L	85	$A \leftarrow A + L$
ADD M	ADD A,(HL)	86	$A \leftarrow A + M_{HL}$
ADC A	ADC A,A	8F	$A \leftarrow A + A + C$
ADC B	ADC A,B	88	$A \leftarrow A + B + C$
ADC C	ADC A,C	89	$A \leftarrow A + C + C$
ADC D	ADC A,D	8A	$A \leftarrow A + D + C$
ADC E	ADC A,E	8B	$A \leftarrow A + E + C$
ADC H	ADC A,H	8C	$A \leftarrow A + H + C$
ADC L	ADC A,L	8D	$A \leftarrow A + L + C$
ADC M	ADC A,(HL)	8E	$A \leftarrow A + M_{HL} + C$
SUB A	SUB A	97	$A \leftarrow A - A$
SUB B	SUB B	90	$A \leftarrow A - B$
SUB C	SUB C	91	$A \leftarrow A - C$

MINI TABLE OF 8085/8080 AND Z80 (8080 SUBSET) INSTRUCTIONS LISTED BY CATEGORY (*Continued*)

8085	Z80	Op	Operation
SUB D	SUB D	92	A ← A - D
SUB E	SUB E	93	A ← A - E
SUB H	SUB H	94	A ← A - H
SUB L	SUB L	95	A ← A - L
SUB M	SUB (HL)	96	A ← A - M_{HL}
SBB A	SBC A,A	9F	A ← A - A - C
SBB B	SBC A,B	98	A ← A - B - C
SBB C	SBC A,C	99	A ← A - C - C
SBB D	SBC A,D	9A	A ← A - D - C
SBB E	SBC A,E	9B	A ← A - E - C
SBB H	SBC A,H	9C	A ← A - H - C
SBB L	SBC A,L	9D	A ← A - L - C
SBB M	SBC A,(HL)	9E	A ← A - M_{HL} - C
DAD B	ADD HL,BC	09	HL ← HL + BC
DAD D	ADD HL,DE	19	HL ← HL + DE
DAD H	ADD HL,HL	29	HL ← HL + HL
ADI dd	ADD A,dd	C6	A ← A + dd
ACI dd	ADC A,dd	CE	A ← A + dd + C
SUI dd	SUB dd	D6	A ← A - dd
SBI dd	SBC A,dd	DE	A ← A - dd - C
DAA	DAA	27	A ← BCD (A)

Logical Instructions

8085	Z80	Op	Operation
ANA A	AND A	A7	A ← A AND A
ANA B	AND B	A0	A ← A AND B
ANA C	AND C	A1	A ← A AND C
ANA D	AND D	A2	A ← A AND D
ANA E	AND E	A3	A ← A AND E
ANA H	AND H	A4	A ← A AND H
ANA L	AND L	A5	A ← A AND L
ANA M	AND (HL)	A6	A ← A AND M_{HL}
XRA A	XOR A	AF	A ← A XOR A
XRA B	XOR B	A8	A ← A XOR B
XRA C	XOR C	A9	A ← A XOR C
XRA D	XOR D	AA	A ← A XOR D
XRA E	XOR E	AB	A ← A XOR E
XRA H	XOR H	AC	A ← A XOR H
XRA L	XOR L	AD	A ← A XOR L
XRA M	XOR (HL)	AE	A ← A XOR M_{HL}
ORA A	OR A	B7	A ← A OR A
ORA B	OR B	B0	A ← A OR B
ORA C	OR C	B1	A ← A OR C
ORA D	OR D	B2	A ← A OR D
ORA E	OR E	B3	A ← A OR E
ORA H	OR H	B4	A ← A OR H
ORA L	OR L	B5	A ← A OR L
ORA M	OR (HL)	B6	A ← A OR M_{HL}
ANI dd	AND dd	E6	A ← A AND dd
XRI dd	XOR dd	EE	A ← A XOR dd
ORI dd	OR dd	F6	A ← A OR dd
CMA	CPL	2F	A ← $\overline{A}$

Rotate and Shift Instructions

8085	Z80	Op	Operation
RLC	RLCA	07	C ← $A_7 \ldots A_0$ ← (rotate left, A_7 to A_0)
RRC	RRCA	0F	→ $A_7 \ldots A_0$ → C (rotate right, A_0 to A_7)

8085	Z80	Op	Operation
RAL	RLA	17	┌C←$A_7 \ldots A_0$←┐ (rotate left through carry)
RAR	RRA	1F	┌→$A_7 \ldots A_0$→C┐ (rotate right through carry)

Increment and Decrement Instructions

8085	Z80	Op	Operation
INR A	INC A	3C	A ← A + 1
INR B	INC B	04	B ← B + 1
INR C	INC C	0C	C ← C + 1
INR D	INC D	14	D ← D + 1
INR E	INC E	1C	E ← E + 1
INR H	INC H	24	H ← H + 1
INR L	INC L	2C	L ← L + 1
INR M	INC (HL)	34	$M_{HL} \leftarrow M_{HL} + 1$
INX B	INC BC	03	BC ← BC + 1
INX D	INC DE	13	DE ← DE + 1
INX H	INC HL	23	HL ← HL + 1
DCR A	DEC A	3D	A ← A - 1
DCR B	DEC B	05	B ← B - 1
DCR C	DEC C	0D	C ← C - 1
DCR D	DEC D	15	D ← D - 1
DCR E	DEC E	1D	E ← E - 1
DCR H	DEC H	25	H ← H - 1
DCR L	DEC L	2D	L ← L - 1
DCR M	DEC (HL)	35	$M_{HL} \leftarrow M_{HL} - 1$
DCX B	DEC BC	0B	BC ← BC - 1
DCX D	DEC DE	1B	DE ← DE - 1
DCX H	DEC HL	2B	HL ← HL - 1

Unconditional Jump Instructions

8085	Z80	Op	Operation
JMP aaaa	JP aaaa	C3	PC ← aaaa
PCHL	JP (HL)	E9	$PC_H \leftarrow H$ $PC_L \leftarrow L$

Test (Compare) Instructions

8085	Z80	Op	Operation
CMP A	CP A	BF	A - A
CMP B	CP B	B8	A - B
CMP C	CP C	B9	A - C
CMP D	CP D	BA	A - D
CMP E	CP E	BB	A - E
CMP H	CP H	BC	A - H
CMP L	CP L	BD	A - L
CMP M	CP (HL)	BE	$A - M_{HL}$
CPI dd	CP dd	FE	A - dd

Conditional Jump (Branch) Instructions

8085	Z80	Op	Operation
JNZ aaaa	JP NZ,aaaa	C2	PC ← aaaa If Z = 0
JZ aaaa	JP Z,aaaa	CA	PC ← aaaa If Z = 1
JNC aaaa	JP NC,aaaa	D2	PC ← aaaa If C = 0
JC aaaa	JP C,aaaa	DA	PC ← aaaa If C = 1
JPO aaaa	JP PO,aaaa	E2	PC ← aaaa If P = 0
JPE aaaa	JP PE,aaaa	EA	PC ← aaaa If P = 1
JP aaaa	JP P,aaaa	F2	PC ← aaaa If S = 0

MINI TABLE OF 8085/8080 AND Z80 (8080 SUBSET) INSTRUCTIONS LISTED BY CATEGORY (*Continued*)

8085	Z80	Op	Operation
JM aaaa	JP M,aaaa	FA	$PC \leftarrow aaaa$ If $S = 1$

Subroutine Instructions

8085	Z80	Op	Operation
CALL aaaa	CALL aaaa	CD	$S \leftarrow PC_H$ $S \leftarrow PC_L$ $PC \leftarrow aaaa$
CNZ aaaa	CALL NZ,aaaa	C4	If $Z = 0$ $S \leftarrow PC_H$ $S \leftarrow PC_L$ $PC \leftarrow aaaa$
CZ aaaa	CALL Z,aaaa	CC	If $Z = 1$ $S \leftarrow PC_H$ $S \leftarrow PC_L$ $PC \leftarrow aaaa$
CNC aaaa	CALL NC,aaaa	D4	If $C = 0$ $S \leftarrow PC_H$ $S \leftarrow PC_L$ $PC \leftarrow aaaa$
CC aaaa	CALL C,aaaa	DC	If $C = 1$ $S \leftarrow PC_H$ $S \leftarrow PC_L$ $PC \leftarrow aaaa$
CPO aaaa	CALL PO,aaaa	E4	If $P = 0$ $S \leftarrow PC_H$ $S \leftarrow PC_L$ $PC \leftarrow aaaa$
CPE aaaa	CALL PE,aaaa	EC	If $P = 1$ $S \leftarrow PC_H$ $S \leftarrow PC_L$ $PC \leftarrow aaaa$
CP aaaa	CALL P,aaaa	F4	If $S = 0$ $S \leftarrow PC_H$ $S \leftarrow PC_L$ $PC \leftarrow aaaa$
CM aaaa	CALL M,aaaa	FC	If $S = 1$ $S \leftarrow PC_H$ $S \leftarrow PC_L$ $PC \leftarrow aaaa$
RET	RET	C9	$PC_L \leftarrow S$ $PC_H \leftarrow S$
RNZ	RET NZ	C0	If $Z = 0$ $PC_L \leftarrow S$ $PC_H \leftarrow S$
RZ	RET Z	C8	If $Z = 1$ $PC_L \leftarrow S$ $PC_H \leftarrow S$
RNC	RET NC	D0	If $C = 0$ $PC_L \leftarrow S$ $PC_H \leftarrow S$
RC	RET C	D8	If $C = 1$ $PC_L \leftarrow S$ $PC_H \leftarrow S$
RPO	RET PO	E0	If $P = 0$ $PC_L \leftarrow S$ $PC_H \leftarrow S$
RPE	RET PE	E8	If $P = 1$ $PC_L \leftarrow S$ $PC_H \leftarrow S$
RP	RET P	F0	If $S = 0$ $PC_L \leftarrow S$ $PC_H \leftarrow S$
RM	RET M	F8	If $S = 1$ $PC_L \leftarrow S$ $PC_H \leftarrow S$
RST 0	RST 00H	C7	$S \leftarrow PC_H$ $S \leftarrow PC_L$ $PC \leftarrow 0000H$
RST 1	RST 08H	CF	$S \leftarrow PC_H$ $S \leftarrow PC_L$ $PC \leftarrow 0008H$
RST 2	RST 10H	D7	$S \leftarrow PC_H$ $S \leftarrow PC_L$ $PC \leftarrow 0010H$
RST 3	RST 18H	DF	$S \leftarrow PC_H$ $S \leftarrow PC_L$ $PC \leftarrow 0018H$
RST 4	RST 20H	E7	$S \leftarrow PC_H$ $S \leftarrow PC_L$ $PC \leftarrow 0020H$
RST 5	RST 28H	EF	$S \leftarrow PC_H$ $S \leftarrow PC_L$ $PC \leftarrow 0028H$

8085	Z80	Op	Operation
RST 6	RST 30H	F7	S ← PC_H S ← PC_L PC ← 0030H
RST 7	RST 38H	FF	S ← PC_H S ← PC_L PC ← 0038H

<u>Stack Instructions</u>

8085	Z80	Op	Operation
LXI SP,dddd	LD SP,dddd	31	SP ← dddd
DAD SP	ADD HL,SP	39	HL ← HL + SP
INX SP	INC SP	33	SP ← SP + 1
DCX SP	DEC SP	3B	SP ← SP - 1
PUSH B	PUSH BC	C5	S ← B S ← C
PUSH D	PUSH DE	D5	S ← D S ← E
PUSH H	PUSH HL	E5	S ← H S ← L
PUSH PSW	PUSH AF	F5	S ← A S ← flags
POP B	POP BC	C1	C ← S B ← S
POP D	POP DE	D1	E ← S D ← S
POP H	POP HL	E1	L ← S H ← S
POP PSW	POP AF	F1	flags ← S A ← S
XTHL	EX (SP),HL	E3	L ↔ S H ↔ S (next)
SPHL	LD SP,HL	F9	SP ← HL

<u>Interrupt Instructions</u>

8085	Z80	Op	Operation
DI	DI	F3	IFF ← 0
EI	EI	FB	IFF ← 1

<u>Input-Output Instructions</u>

8085	Z80	Op	Operation
OUT dd	OUT dd,A	D3	dd port ← A
IN dd	IN A,dd	DB	A ← dd port (byte)

CONDENSED TABLE OF 8085/8080 AND Z80 (8080 SUBSET) INSTRUCTIONS LISTED BY CATEGORY

CPU Control Instructions

8085	Z80	Op
NOP	NOP	00
HLT	HALT	76

Data Transfer Instructions

8085	Z80	Op
MOV A,A	LD A,A	7F
MOV A,B	LD A,B	78
MOV A,C	LD A,C	79
MOV A,D	LD A,D	7A
MOV A,E	LD A,E	7B
MOV A,H	LD A,H	7C
MOV A,L	LD A,L	7D
MOV A,M	LD A,(HL)	7E
MOV B,A	LD B,A	47
MOV B,B	LD B,B	40
MOV B,C	LD B,C	41
MOV B,D	LD B,D	42
MOV B,E	LD B,E	43
MOV B,H	LD B,H	44
MOV B,L	LD B,L	45
MOV B,M	LD B,(HL)	46
MOV C,A	LD C,A	4F
MOV C,B	LD C,B	48
MOV C,C	LD C,C	49
MOV C,D	LD C,D	4A
MOV C,E	LD C,E	4B
MOV C,H	LD C,H	4C
MOV C,L	LD C,L	4D
MOV C,M	LD C,(HL)	4E
MOV D,A	LD D,A	57
MOV D,B	LD D,B	50
MOV D,C	LD D,C	51
MOV D,D	LD D,D	52
MOV D,E	LD D,E	53
MOV D,H	LD D,H	54
MOV D,L	LD D,L	55
MOV D,M	LD D,(HL)	56
MOV E,A	LD E,A	5F
MOV E,B	LD E,B	58
MOV E,C	LD E,C	59
MOV E,D	LD E,D	5A
MOV E,E	LD E,E	5B

CONDENSED TABLE OF 8085/8080 AND Z80 (8080 SUBSET) INSTRUCTIONS LISTED BY CATEGORY (*Continued*)

8085	Z80	Op
MOV E,H	LD E,H	5C
MOV E,L	LD E,L	5D
MOV E,M	LD E,(HL)	5E
MOV H,A	LD H,A	67
MOV H,B	LD H,B	60
MOV H,C	LD H,C	61
MOV H,D	LD H,D	62
MOV H,E	LD H,E	63
MOV H,H	LD H,H	64
MOV H,L	LD H,L	65
MOV H,M	LD H,(HL)	66
MOV L,A	LD L,A	6F
MOV L,B	LD L,B	68
MOV L,C	LD L,C	69
MOV L,D	LD L,D	6A
MOV L,E	LD L,E	6B
MOV L,H	LD L,H	6C
MOV L,L	LD L,L	6D
MOV L,M	LD L,(HL)	6E
MOV M,A	LD (HL),A	77
MOV M,B	LD (HL),B	70
MOV M,C	LD (HL),C	71
MOV M,D	LD (HL),D	72
MOV M,E	LD (HL),E	73
MOV M,H	LD (HL),H	74
MOV M,L	LD (HL),L	75
MVI A,dd	LD A,dd	3E
MVI B,dd	LD B,dd	06
MVI C,dd	LD C,dd	0E
MVI D,dd	LD D,dd	16
MVI E,dd	LD E,dd	1E
MVI H,dd	LD H,dd	26
MVI L,dd	LD L,dd	2E
MVI M,dd	LD (HL),dd	36
LXI B,dddd	LD BC,dddd	01
LXI D,dddd	LD DE,dddd	11
LXI H,dddd	LD HL,dddd	21
LDAX B	LD A,(BC)	0A
LDAX D	LD A,(DE)	1A
LHLD aaaa	LD HL,(aaaa)	2A
LDA aaaa	LD A,(aaaa)	3A
STA aaaa	LD (aaaa),A	32
STAX B	LD (BC),A	02
STAX D	LD (DE),A	12
SHLD aaaa	LD (aaaa),HL	22
XCHG	EX DE,HL	EB

Flag Instructions

8085	Z80	Op
STC	SCF	37
CMC	CCF	3F

Arithmetic Instructions

8085	Z80	Op
ADD A	ADD A,A	87
ADD B	ADD A,B	80
ADD C	ADD A,C	81
ADD D	ADD A,D	82
ADD E	ADD A,E	83
ADD H	ADD A,H	84
ADD L	ADD A,L	85
ADD M	ADD A,(HL)	86
ADC A	ADC A,A	8F
ADC B	ADC A,B	88
ADC C	ADC A,C	89
ADC D	ADC A,D	8A
ADC E	ADC A,E	8B
ADC H	ADC A,H	8C
ADC L	ADC A,L	8D
ADC M	ADC A,(HL)	8E
SUB A	SUB A	97
SUB B	SUB B	90
SUB C	SUB C	91
SUB D	SUB D	92
SUB E	SUB E	93
SUB H	SUB H	94
SUB L	SUB L	95
SUB M	SUB (HL)	96
SBB A	SBC A,A	9F
SBB B	SBC A,B	98
SBB C	SBC A,C	99
SBB D	SBC A,D	9A
SBB E	SBC A,E	9B
SBB H	SBC A,H	9C
SBB L	SBC A,L	9D
SBB M	SBC A,(HL)	9E
DAD B	ADD HL,BC	09
DAD D	ADD HL,DE	19
DAD H	ADD HL,HL	29
ADI dd	ADD A,dd	C6
ACI dd	ADC A,dd	CE
SUI dd	SUB dd	D6
SBI dd	SBC A,dd	DE
DAA	DAA	27

Logical Instructions

8085	Z80	Op
ANA A	AND A	A7
ANA B	AND B	A0
ANA C	AND C	A1
ANA D	AND D	A2
ANA E	AND E	A3
ANA H	AND H	A4
ANA L	AND L	A5
ANA M	AND (HL)	A6
XRA A	XOR A	AF
XRA B	XOR B	A8
XRA C	XOR C	A9
XRA D	XOR D	AA
XRA E	XOR E	AB
XRA H	XOR H	AC
XRA L	XOR L	AD
XRA M	XOR (HL)	AE
ORA A	OR A	B7
ORA B	OR B	B0
ORA C	OR C	B1
ORA D	OR D	B2
ORA E	OR E	B3
ORA H	OR H	B4
ORA L	OR L	B5
ORA M	OR (HL)	B6
ANI dd	AND dd	E6
XRI dd	XOR dd	EE
ORI dd	OR dd	F6
CMA	CPL	2F

Rotate and Shift Instructions

8085	Z80	Op
RLC	RLCA	07
RRC	RRCA	0F
RAL	RLA	17
RAR	RRA	1F

Increment and Decrement Instructions

8085	Z80	Op
INR A	INC A	3C
INR B	INC B	04
INR C	INC C	0C
INR D	INC D	14
INR E	INC E	1C
INR H	INC H	24
INR L	INC L	2C
INR M	INC (HL)	34
INX B	INC BC	03
INX D	INC DE	13
INX H	INC HL	23
DCR A	DEC A	3D
DCR B	DEC B	05
DCR C	DEC C	0D
DCR D	DEC D	15
DCR E	DEC E	1D
DCR H	DEC H	25

8085	Z80	Op
DCR L	DEC L	2D
DCR M	DEC (HL)	35
DCX B	DEC BC	0B
DCX D	DEC DE	1B
DCX H	DEC HL	2B

Unconditional Jump Instructions

8085	Z80	Op
JMP aaaa	JP aaaa	C3
PCHL	JP (HL)	E9

Test (Compare) Instructions

8085	Z80	Op
CMP A	CP A	BF
CMP B	CP B	B8
CMP C	CP C	B9
CMP D	CP D	BA
CMP E	CP E	BB
CMP H	CP H	BC
CMP L	CP L	BD
CMP M	CP (HL)	BE
CPI dd	CP dd	FE

Conditional Jump (Branch) Instructions

8085	Z80	Op
JNZ aaaa	JP NZ,aaaa	C2
JZ aaaa	JP Z,aaaa	CA
JNC aaaa	JP NC,aaaa	D2
JC aaaa	JP C,aaaa	DA
JPO aaaa	JP PO,aaaa	E2
JPE aaaa	JP PE,aaaa	EA
JP aaaa	JP P,aaaa	F2
JM aaaa	JP M,aaaa	FA

Subroutine Instructions

8085	Z80	Op
CALL aaaa	CALL aaaa	CD
CNZ aaaa	CALL NZ,aaaa	C4
CZ aaaa	CALL Z,aaaa	CC
CNC aaaa	CALL NC,aaaa	D4
CC aaaa	CALL C,aaaa	DC
CPO aaaa	CALL PO,aaaa	E4
CPE aaaa	CALL PE,aaaa	EC
CP aaaa	CALL P,aaaa	F4
CM aaaa	CALL M,aaaa	FC
RET	RET	C9
RNZ	RET NZ	C0
RZ	RET Z	C8
RNC	RET NC	D0
RC	RET C	D8
RPO	RET PO	E0
RPE	RET PE	E8
RP	RET P	F0
RM	RET M	F8
RST 0	RST 00H	C7
RST 1	RST 08H	CF
RST 2	RST 10H	D7
RST 3	RST 18H	DF
RST 4	RST 20H	E7
RST 5	RST 28H	EF
RST 6	RST 30H	F7
RST 7	RST 38H	FF

Stack Instructions

8085	Z80	Op
LXI SP,dddd	LD SP,dddd	31
DAD SP	ADD HL,SP	39
INX SP	INC SP	33
DCX SP	DEC SP	3B
PUSH B	PUSH BC	C5
PUSH D	PUSH DE	D5
PUSH H	PUSH HL	E5
PUSH PSW	PUSH AF	F5
POP B	POP BC	C1
POP D	POP DE	D1
POP H	POP HL	E1
POP PSW	POP AF	F1
XTHL	EX (SP),HL	E3
SPHL	LD SP,HL	F9

Interrupt Instructions

8085	Z80	Op
DI	DI	F3
EI	EI	FB

Input-Output Instructions

8085	Z80	Op
OUT dd	OUT dd,A	D3
IN dd	IN A,dd	DB

CONDENSED TABLE OF 8085/8080 AND Z80 (8080 SUBSET) INSTRUCTIONS LISTED BY OP CODE

Op	8080/8085	Z80
00	NOP	NOP
01	LXI B,dddd	LD BC,dddd
02	STAX B	LD (BC),A
03	INX B	INC BC
04	INR B	INC B
05	DCR B	DEC B
06	MVI B,dd	LD B,dd
07	RLC	RLCA
09	DAD B	ADD HL,BC
0A	LDAX B	LD A,(BC)
0B	DCX B	DEC BC
0C	INR C	INC C
0D	DCR C	DEC C
0E	MVI C,dd	LD C,dd
0F	RRC	RRCA
11	LXI D,dddd	LD DE,dddd
12	STAX D	LD (DE),A
13	INX D	INC DE
14	INR D	INC D
15	DCR D	DEC D
16	MVI D,dd	LD D,dd
17	RAL	RLA
19	DAD D	ADD HL,DE
1A	LDAX D	LD A,(DE)
1B	DCX D	DEC DE
1C	INR E	INC E
1D	DCR E	DEC E
1E	MVI E,dd	LD E,dd
1F	RAR	RRA
21	LXI H,dddd	LD HL,dddd
22	SHLD aaaa	LD (aaaa),HL
23	INX H	INC HL
24	INR H	INC H

CONDENSED TABLE OF 8085/8080 AND Z80 (8080 SUBSET) INSTRUCTIONS LISTED BY OP CODE (*Continued*)

Op	8080/8085	Z80
25	DCR H	DEC H
26	MVI H,dd	LD H,dd
27	DAA	DAA
29	DAD H	ADD HL,HL
2A	LHLD aaaa	LD HL,(aaaa)
2B	DCX H	DEC HL
2C	INR L	INC L
2D	DCR L	DEC L
2E	MVI L,dd	LD L,dd
2F	CMA	CPL
31	LXI SP,dddd	LD SP,dddd
32	STA aaaa	LD (aaaa),A
33	INX SP	INC SP
34	INR M	INC (HL)
35	DCR M	DEC (HL)
36	MVI M,dd	LD (HL),dd
37	STC	SCF
39	DAD SP	ADD HL,SP
3A	LDA aaaa	LD A,(aaaa)
3B	DCX SP	DEC SP
3C	INR A	INC A
3D	DCR A	DEC A
3E	MVI A,dd	LD A,dd
3F	CMC	CCF
40	MOV B,B	LD B,B
41	MOV B,C	LD B,C
42	MOV B,D	LD B,D
43	MOV B,E	LD B,E
44	MOV B,H	LD B,H
45	MOV B,L	LD B,L
46	MOV B,M	LD B,(HL)
47	MOV B,A	LD B,A
48	MOV C,B	LD C,B
49	MOV C,C	LD C,C
4A	MOV C,D	LD C,D
4B	MOV C,E	LD C,E
4C	MOV C,H	LD C,H
4D	MOV C,L	LD C,L
4E	MOV C,M	LD C,(HL)
4F	MOV C,A	LD C,A
50	MOV D,B	LD D,B
51	MOV D,C	LD D,C
52	MOV D,D	LD D,D
53	MOV D,E	LD D,E
54	MOV D,H	LD D,H
55	MOV D,L	LD D,L
56	MOV D,M	LD D,(HL)
57	MOV D,A	LD D,A
58	MOV E,B	LD E,B
59	MOV E,C	LD E,C
5A	MOV E,D	LD E,D
5B	MOV E,E	LD E,E
5C	MOV E,H	LD E,H
5D	MOV E,L	LD E,L
5E	MOV E,M	LD E,(HL)

Op	8080/8085	Z80
5F	MOV E,A	LD E,A
60	MOV H,B	LD H,B
61	MOV H,C	LD H,C
62	MOV H,D	LD H,D
63	MOV H,E	LD H,E
64	MOV H,H	LD H,H
65	MOV H,L	LD H,L
66	MOV H,M	LD H,(HL)
67	MOV H,A	LD H,A
68	MOV L,B	LD L,B
69	MOV L,C	LD L,C
6A	MOV L,D	LD L,D
6B	MOV L,E	LD L,E
6C	MOV L,H	LD L,H
6D	MOV L,L	LD L,L
6E	MOV L,M	LD L,(HL)
6F	MOV L,A	LD L,A
70	MOV M,B	LD (HL),B
71	MOV M,C	LD (HL),C
72	MOV M,D	LD (HL),D
73	MOV M,E	LD (HL),E
74	MOV M,H	LD (HL),H
75	MOV M,L	LD (HL),L
76	HLT	HALT
77	MOV M,A	LD (HL),A
78	MOV A,B	LD A,B
79	MOV A,C	LD A,C
7A	MOV A,D	LD A,D
7B	MOV A,E	LD A,E
7C	MOV A,H	LD A,H
7D	MOV A,L	LD A,L
7E	MOV A,M	LD A,(HL)
7F	MOV A,A	LD A,A
80	ADD B	ADD A,B
81	ADD C	ADD A,C
82	ADD D	ADD A,D
83	ADD E	ADD A,E
84	ADD H	ADD A,H
85	ADD L	ADD A,L
86	ADD M	ADD A,(HL)
87	ADD A	ADD A,A
88	ADC B	ADC A,B
89	ADC C	ADC A,C
8A	ADC D	ADC A,D
8B	ADC E	ADC A,E
8C	ADC H	ADC A,H
8D	ADC L	ADC A,L
8E	ADC M	ADC A,(HL)
8F	ADC A	ADC A,A
90	SUB B	SUB B
91	SUB C	SUB C
92	SUB D	SUB D
93	SUB E	SUB E
94	SUB H	SUB H
95	SUB L	SUB L

Op	8080/8085	Z80
96	SUB M	SUB (HL)
97	SUB A	SUB A
98	SBB B	SBC A,B
99	SBB C	SBC A,C
9A	SBB D	SBC A,D
9B	SBB E	SBC A,E
9C	SBB H	SBC A,H
9D	SBB L	SBC A,L
9E	SBB M	SBC A,(HL)
9F	SBB A	SBC A,A
A0	ANA B	AND B
A1	ANA C	AND C
A2	ANA D	AND D
A3	ANA E	AND E
A4	ANA H	AND H
A5	ANA L	AND L
A6	ANA M	AND (HL)
A7	ANA A	AND A
A8	XRA B	XOR B
A9	XRA C	XOR C
AA	XRA D	XOR D
AB	XRA E	XOR E
AC	XRA H	XOR H
AD	XRA L	XOR L
AE	XRA M	XOR (HL)
AF	XRA A	XOR A
B0	ORA B	OR B
B1	ORA C	OR C
B2	ORA D	OR D
B3	ORA E	OR E
B4	ORA H	OR H
B5	ORA L	OR L
B6	ORA M	OR (HL)
B7	ORA A	OR A
B8	CMP B	CP B
B9	CMP C	CP C
BA	CMP D	CP D
BB	CMP E	CP E
BC	CMP H	CP H
BD	CMP L	CP L
BE	CMP M	CP (HL)
BF	CMP A	CP A
C0	RNZ	RET NZ
C1	POP B	POP BC
C2	JNZ aaaa	JP NZ,aaaa
C3	JMP aaaa	JP aaaa
C4	CNZ aaaa	CALL NZ,aaaa
C5	PUSH B	PUSH BC
C6	ADI dd	ADD A,dd
C7	RST 0	RST 00H
C8	RZ	RET Z
C9	RET	RET
CA	JZ aaaa	JP Z,aaaa
CC	CZ aaaa	CALL Z,aaaa
CD	CALL aaaa	CALL aaaa

Op	8080/8085	Z80
CE	ACI dd	ADC A,dd
CF	RST 1	RST 08H
D0	RNC	RET NC
D1	POP D	POP DE
D2	JNC aaaa	JP NC,aaaa
D3	OUT dd	OUT dd,A
D4	CNC aaaa	CALL NC,aaaa
D5	PUSH D	PUSH DE
D6	SUI dd	SUB dd
D7	RST 2	RST 10H
D8	RC	RET C
DA	JC aaaa	JP C,aaaa
DB	IN dd	IN A,dd
DC	CC aaaa	CALL C,aaaa
DE	SBI dd	SBC A,dd
DF	RST 3	RST 18H
E0	RPO	RET PO
E1	POP H	POP HL
E2	JPO aaaa	JP PO,aaaa
E3	XTHL	EX (SP),HL
E4	CPO aaaa	CALL PO,aaaa
E5	PUSH H	PUSH HL
E6	ANI dd	AND dd
E7	RST 4	RST 20H
E8	RPE	RET PE
E9	PCHL	JP (HL)
EA	JPE aaaa	JP PE,aaaa
EB	XCHG	EX DE,HL
EC	CPE aaaa	CALL PE,aaaa
EE	XRI dd	XOR dd
EF	RST 5	RST 28H
F0	RP	RET P
F1	POP PSW	POP AF
F2	JP aaaa	JP P,aaaa
F3	DI	DI
F4	CP aaaa	CALL P,aaaa
F5	PUSH PSW	PUSH AF
F6	ORI dd	OR dd
F7	RST 6	RST 30H
F8	RM	RET M
F9	SPHL	LD SP,HL
FA	JM aaaa	JP M,aaaa
FB	EI	EI
FC	CM aaaa	CALL M,aaaa
FE	CPI dd	CP dd
FF	RST 7	RST 38H

CONDENSED TABLE OF 8085/8080 AND Z80 (8080 SUBSET) INSTRUCTIONS LISTED ALPHABETICALLY BY 8085/8080 MNEMONIC

8085	Z80	Op
ACI dd	ADC A,dd	CE
ADC A	ADC A,A	8F
ADC B	ADC A,B	88
ADC C	ADC A,C	89
ADC D	ADC A,D	8A
ADC E	ADC A,E	8B
ADC H	ADC A,H	8C
ADC L	ADC A,L	8D
ADC M	ADC A,(HL)	8E
ADD A	ADD A,A	87
ADD B	ADD A,B	80
ADD C	ADD A,C	81
ADD D	ADD A,D	82
ADD E	ADD A,E	83
ADD H	ADD A,H	84
ADD L	ADD A,L	85
ADD M	ADD A,(HL)	86
ADI dd	ADD A,dd	C6
ANA A	AND A	A7
ANA B	AND B	A0
ANA C	AND C	A1
ANA D	AND D	A2
ANA E	AND E	A3
ANA H	AND H	A4
ANA L	AND L	A5
ANA M	AND (HL)	A6
ANI dd	AND dd	E6
CALL aaaa	CALL aaaa	CD
CC aaaa	CALL C,aaaa	DC
CM aaaa	CALL M,aaaa	FC
CMA	CPL	2F
CMC	CCF	3F
CMP A	CP A	BF
CMP B	CP B	B8
CMP C	CP C	B9
CMP D	CP D	BA
CMP E	CP E	BB
CMP H	CP H	BC
CMP L	CP L	BD
CMP M	CP (HL)	BE
CNC aaaa	CALL NC,aaaa	D4
CNZ aaaa	CALL NZ,aaaa	C4
CP aaaa	CALL P,aaaa	F4
CPE aaaa	CALL PE,aaaa	EC
CPI dd	CP dd	FE
CPO aaaa	CALL PO,aaaa	E4
CZ aaaa	CALL Z,aaaa	CC
DAA	DAA	27
DAD B	ADD HL,BC	09
DAD D	ADD HL,DE	19
DAD H	ADD HL,HL	29
DAD SP	ADD HL,SP	39
DCR A	DEC A	3D
DCR B	DEC B	05
DCR C	DEC C	0D
DCR D	DEC D	15
DCR E	DEC E	1D
DCR H	DEC H	25
DCR L	DEC L	2D
DCR M	DEC (HL)	35
DCX B	DEC BC	0B
DCX D	DEC DE	1B
DCX H	DEC HL	2B
DCX SP	DEC SP	3B
DI	DI	F3
EI	EI	FB
HLT	HALT	76
IN dd	IN A,dd	DB
INR A	INC A	3C
INR B	INC B	04
INR C	INC C	0C
INR D	INC D	14
INR E	INC E	1C
INR H	INC H	24
INR L	INC L	2C
INR M	INC (HL)	34
INX B	INC BC	03
INX D	INC DE	13
INX H	INC HL	23
INX SP	INC SP	33
JC aaaa	JP C,aaaa	DA
JM aaaa	JP M,aaaa	FA
JMP aaaa	JP aaaa	C3
JNC aaaa	JP NC,aaaa	D2
JNZ aaaa	JP NZ,aaaa	C2
JP aaaa	JP P,aaaa	F2
JPE aaaa	JP PE,aaaa	EA
JPO aaaa	JP PO,aaaa	E2
JZ aaaa	JP Z,aaaa	CA
LDA aaaa	LD A,(aaaa)	3A
LDAX B	LD A,(BC)	0A
LDAX D	LD A,(DE)	1A
LHLD aaaa	LD HL,(aaaa)	2A
LXI B,dddd	LD BC,dddd	01
LXI D,dddd	LD DE,dddd	11
LXI H,dddd	LD HL,dddd	21
LXI SP,dddd	LD SP,dddd	31
MOV A,A	LD A,A	7F
MOV A,B	LD A,B	78
MOV A,C	LD A,C	79
MOV A,D	LD A,D	7A
MOV A,E	LD A,E	7B

CONDENSED TABLE OF 8085/8080 AND Z80 (8080 SUBSET) INSTRUCTIONS LISTED ALPHABETICALLY BY 8085/8080 MNEMONIC ***(Continued)***

8085	Z80	Op	8085	Z80	Op	8085	Z80	Op
MOV A,H	LD A,H	7C	MOV L,H	LD L,H	6C	RPE	RET PE	E8
MOV A,L	LD A,L	7D	MOV L,L	LD L,L	6D	RPO	RET PO	E0
MOV A,M	LD A,(HL)	7E	MOV L,M	LD L,(HL)	6E	RRC	RRCA	0F
MOV B,A	LD B,A	47	MOV M,A	LD (HL),A	77	RST 0	RST 00H	C7
MOV B,B	LD B,B	40	MOV M,B	LD (HL),B	70	RST 1	RST 08H	CF
MOV B,C	LD B,C	41	MOV M,C	LD (HL),C	71	RST 2	RST 10H	D7
MOV B,D	LD B,D	42	MOV M,D	LD (HL),D	72	RST 3	RST 18H	DF
MOV B,E	LD B,E	43	MOV M,E	LD (HL),E	73	RST 4	RST 20H	E7
MOV B,H	LD B,H	44	MOV M,H	LD (HL),H	74	RST 5	RST 28H	EF
MOV B,L	LD B,L	45	MOV M,L	LD (HL),L	75	RST 6	RST 30H	F7
MOV B,M	LD B,(HL)	46	MVI A,dd	LD A,dd	3E	RST 7	RST 38H	FF
MOV C,A	LD C,A	4F	MVI B,dd	LD B,dd	06	RZ	RET Z	C8
MOV C,B	LD C,B	48	MVI C,dd	LD C,dd	0E	SBB A	SBC A,A	9F
MOV C,C	LD C,C	49	MVI D,dd	LD D,dd	16	SBB B	SBC A,B	98
MOV C,D	LD C,D	4A	MVI E,dd	LD E,dd	1E	SBB C	SBC A,C	99
MOV C,E	LD C,E	4B	MVI H,dd	LD H,dd	26	SBB D	SBC A,D	9A
MOV C,H	LD C,H	4C	MVI L,dd	LD L,dd	2E	SBB E	SBC A,E	9B
MOV C,L	LD C,L	4D	MVI M,dd	LD (HL),dd	36	SBB H	SBC A,H	9C
MOV C,M	LD C,(HL)	4E	NOP	NOP	00	SBB L	SBC A,L	9D
MOV D,A	LD D,A	57	ORA A	OR A	B7	SBB M	SBC A,(HL)	9E
MOV D,B	LD D,B	50	ORA B	OR B	B0	SBI dd	SBC A,dd	DE
MOV D,C	LD D,C	51	ORA C	OR C	B1	SHLD aaaa	LD (aaaa),HL	22
MOV D,D	LD D,D	52	ORA D	OR D	B2	SPHL	LD SP,HL	F9
MOV D,E	LD D,E	53	ORA E	OR E	B3	STA aaaa	LD (aaaa),A	32
MOV D,H	LD D,H	54	ORA H	OR H	B4	STAX B	LD (BC),A	02
MOV D,L	LD D,L	55	ORA L	OR L	B5	STAX D	LD (DE),A	12
MOV D,M	LD D,(HL)	56	ORA M	OR (HL)	B6	STC	SCF	37
MOV E,A	LD E,A	5F	ORI dd	OR dd	F6	SUB A	SUB A	97
MOV E,B	LD E,B	58	OUT dd	OUT dd,A	D3	SUB B	SUB B	90
MOV E,C	LD E,C	59	PCHL	JP (HL)	E9	SUB C	SUB C	91
MOV E,D	LD E,D	5A	POP B	POP BC	C1	SUB D	SUB D	92
MOV E,E	LD E,E	5B	POP D	POP DE	D1	SUB E	SUB E	93
MOV E,H	LD E,H	5C	POP H	POP HL	E1	SUB H	SUB H	94
MOV E,L	LD E,L	5D	POP PSW	POP AF	F1	SUB L	SUB L	95
MOV E,M	LD E,(HL)	5E	PUSH B	PUSH BC	C5	SUB M	SUB (HL)	96
MOV H,A	LD H,A	67	PUSH D	PUSH DE	D5	SUI dd	SUB dd	D6
MOV H,B	LD H,B	60	PUSH H	PUSH HL	E5	XCHG	EX DE,HL	EB
MOV H,C	LD H,C	61	PUSH PSW	PUSH AF	F5	XRA A	XOR A	AF
MOV H,D	LD H,D	62	RAL	RLA	17	XRA B	XOR B	A8
MOV H,E	LD H,E	63	RAR	RRA	1F	XRA C	XOR C	A9
MOV H,H	LD H,H	64	RC	RET C	D8	XRA D	XOR D	AA
MOV H,L	LD H,L	65	RET	RET	C9	XRA E	XOR E	AB
MOV H,M	LD H,(HL)	66	RLC	RLCA	07	XRA H	XOR H	AC
MOV L,A	LD L,A	6F	RM	RET M	F8	XRA L	XOR L	AD
MOV L,B	LD L,B	68	RNC	RET NC	D0	XRA M	XOR (HL)	AE
MOV L,C	LD L,C	69	RNZ	RET NZ	C0	XRI dd	XOR dd	EE
MOV L,D	LD L,D	6A	RP	RET P	F0	XTHL	EX (SP),HL	E3
MOV L,E	LD L,E	6B						

CONDENSED TABLE OF 8085/8080 AND Z80 (8080 SUBSET) INSTRUCTIONS LISTED ALPHABETICALLY BY Z80 MNEMONIC

Z80	8080/8085	Op
ADC A,(HL)	ADC M	8E
ADC A,A	ADC A	8F
ADC A,B	ADC B	88
ADC A,C	ADC C	89
ADC A,D	ADC D	8A
ADC A,dd	ACI dd	CE
ADC A,E	ADC E	8B
ADC A,H	ADC H	8C
ADC A,L	ADC L	8D
ADD A,(HL)	ADD M	86
ADD A,A	ADD A	87
ADD A,B	ADD B	80
ADD A,C	ADD C	81
ADD A,D	ADD D	82
ADD A,dd	ADI dd	C6
ADD A,E	ADD E	83
ADD A,H	ADD H	84
ADD A,L	ADD L	85
ADD HL,BC	DAD B	09
ADD HL,DE	DAD D	19
ADD HL,HL	DAD H	29
ADD HL,SP	DAD SP	39
AND (HL)	ANA M	A6
AND A	ANA A	A7
AND B	ANA B	A0
AND C	ANA C	A1
AND D	ANA D	A2
AND dd	ANI dd	E6
AND E	ANA E	A3
AND H	ANA H	A4
AND L	ANA L	A5
CALL aaaa	CALL aaaa	CD
CALL C,aaaa	CC aaaa	DC
CALL M,aaaa	CM aaaa	FC
CALL NC,aaaa	CNC aaaa	D4
CALL NZ,aaaa	CNZ aaaa	C4
CALL P,aaaa	CP aaaa	F4
CALL PE,aaaa	CPE aaaa	EC
CALL PO,aaaa	CPO aaaa	E4
CALL Z,aaaa	CZ aaaa	CC
CCF	CMC	3F
CP (HL)	CMP M	BE
CP A	CMP A	BF
CP B	CMP B	B8
CP C	CMP C	B9
CP D	CMP D	BA
CP dd	CPI dd	FE
CP E	CMP E	BB
CP H	CMP H	BC
CP L	CMP L	BD
CPL	CMA	2F
DAA	DAA	27
DEC (HL)	DCR M	35
DEC A	DCR A	3D
DEC B	DCR B	05
DEC BC	DCX B	0B
DEC C	DCR C	0D
DEC D	DCR D	15
DEC DE	DCX D	1B
DEC E	DCR E	1D
DEC H	DCR H	25
DEC HL	DCX H	2B
DEC L	DCR L	2D
DEC SP	DCX SP	3B
DI	DI	F3
EI	EI	FB
EX (SP),HL	XTHL	E3
EX DE,HL	XCHG	EB
HALT	HLT	76
IN A,dd	IN dd	DB
INC (HL)	INR M	34
INC A	INR A	3C
INC B	INR B	04
INC BC	INX B	03
INC C	INR C	0C
INC D	INR D	14
INC DE	INX D	13
INC E	INR E	1C
INC H	INR H	24
INC HL	INX H	23
INC L	INR L	2C
INC SP	INX SP	33
JP (HL)	PCHL	E9
JP aaaa	JMP aaaa	C3
JP C,aaaa	JC aaaa	DA
JP M,aaaa	JM aaaa	FA
JP NC,aaaa	JNC aaaa	D2
JP NZ,aaaa	JNZ aaaa	C2
JP P,aaaa	JP aaaa	F2
JP PE,aaaa	JPE aaaa	EA
JP PO,aaaa	JPO aaaa	E2
JP Z,aaaa	JZ aaaa	CA
LD (aaaa),A	STA aaaa	32
LD (aaaa),HL	SHLD aaaa	22
LD (BC),A	STAX B	02
LD (DE),A	STAX D	12
LD (HL),A	MOV M,A	77
LD (HL),B	MOV M,B	70
LD (HL),C	MOV M,C	71
LD (HL),D	MOV M,D	72
LD (HL),dd	MVI M,dd	36
LD (HL),E	MOV M,E	73
LD (HL),H	MOV M,H	74
LD (HL),L	MOV M,L	75
LD A,(aaaa)	LDA aaaa	3A
LD A,(BC)	LDAX B	0A
LD A,(DE)	LDAX D	1A
LD A,(HL)	MOV A,M	7E
LD A,A	MOV A,A	7F
LD A,B	MOV A,B	78
LD A,C	MOV A,C	79
LD A,D	MOV A,D	7A
LD A,dd	MVI A,dd	3E
LD A,E	MOV A,E	7B
LD A,H	MOV A,H	7C
LD A,L	MOV A,L	7D
LD B,(HL)	MOV B,M	46
LD B,A	MOV B,A	47
LD B,B	MOV B,B	40
LD B,C	MOV B,C	41
LD BC,dddd	LXI B,dddd	01
LD B,D	MOV B,D	42
LD B,dd	MVI B,dd	06
LD B,E	MOV B,E	43
LD B,H	MOV B,H	44
LD B,L	MOV B,L	45
LD C,(HL)	MOV C,M	4E
LD C,A	MOV C,A	4F
LD C,B	MOV C,B	48
LD C,C	MOV C,C	49
LD C,D	MOV C,D	4A
LD C,dd	MVI C,dd	0E
LD C,E	MOV C,E	4B
LD C,H	MOV C,H	4C
LD C,L	MOV C,L	4D
LD D,(HL)	MOV D,M	56
LD D,A	MOV D,A	57
LD D,B	MOV D,B	50
LD D,C	MOV D,C	51
LD D,D	MOV D,D	52
LD D,dd	MVI D,dd	16
LD D,E	MOV D,E	53
LD DE,dddd	LXI D,dddd	11
LD D,H	MOV D,H	54
LD D,L	MOV D,L	55
LD E,(HL)	MOV E,M	5E
LD E,A	MOV E,A	5F
LD E,B	MOV E,B	58
LD E,C	MOV E,C	59
LD E,D	MOV E,D	5A
LD E,dd	MVI E,dd	1E
LD E,E	MOV E,E	5B
LD E,H	MOV E,H	5C
LD E,L	MOV E,L	5D
LD H,(HL)	MOV H,M	66
LD H,A	MOV H,A	67
LD H,B	MOV H,B	60
LD H,C	MOV H,C	61
LD H,D	MOV H,D	62
LD H,dd	MVI H,dd	26
LD H,E	MOV H,E	63
LD H,H	MOV H,H	64
LD H,L	MOV H,L	65
LD HL,(aaaa)	LHLD aaaa	2A
LD HL,dddd	LXI H,dddd	21

CONDENSED TABLE OF 8085/8080 AND Z80 (8080 SUBSET) INSTRUCTIONS LISTED ALPHABETICALLY BY Z80 MNEMONIC (*Continued*)

Z80	8080/8085	Op	Z80	8080/8085	Op	Z80	8080/8085	Op
LD L,(HL)	MOV L,M	6E	PUSH BC	PUSH B	C5	SBC A,B	SBB B	98
LD L,A	MOV L,A	6F	PUSH DE	PUSH D	D5	SBC A,C	SBB C	99
LD L,B	MOV L,B	68	PUSH HL	PUSH H	E5	SBC A,D	SBB D	9A
LD L,C	MOV L,C	69	RET	RET	C9	SBC A,dd	SBI dd	DE
LD L,D	MOV L,D	6A	RET C	RC	D8	SBC A,E	SBB E	9B
LD L,dd	MVI L,dd	2E	RET M	RM	F8	SBC A,H	SBB H	9C
LD L,E	MOV L,E	6B	RET NC	RNC	D0	SBC A,L	SBB L	9D
LD L,H	MOV L,H	6C	RET NZ	RNZ	C0	SCF	STC	37
LD L,L	MOV L,L	6D	RET P	RP	F0	SUB (HL)	SUB M	96
LD SP,dddd	LXI SP,dddd	31	RET PE	RPE	E8	SUB A	SUB A	97
LD SP,HL	SPHL	F9	RET PO	RPO	E0	SUB dd	SUI dd	D6
NOP	NOP	00	RET Z	RZ	C8	SUB B	SUB B	90
OR (HL)	ORA M	B6	RLA	RAL	17	SUB C	SUB C	91
OR A	ORA A	B7	RLCA	RLC	07	SUB D	SUB D	92
OR B	ORA B	B0	RRA	RAR	1F	SUB E	SUB E	93
OR C	ORA C	B1	RRCA	RRC	0F	SUB H	SUB H	94
OR D	ORA D	B2	RST 00H	RST 0	C7	SUB L	SUB L	95
OR dd	ORI dd	F6	RST 08H	RST 1	CF	XOR (HL)	XRA M	AE
OR E	ORA E	B3	RST 10H	RST 2	D7	XOR A	XRA A	AF
OR H	ORA H	B4	RST 18H	RST 3	DF	XOR B	XRA B	A8
OR L	ORA L	B5	RST 20H	RST 4	E7	XOR C	XRA C	A9
OUT dd,A	OUT dd	D3	RST 28H	RST 5	EF	XOR D	XRA D	AA
POP AF	POP PSW	F1	RST 30H	RST 6	F7	XOR dd	XRI dd	EE
POP BC	POP B	C1	RST 38H	RST 7	FF	XOR E	XRA E	AB
POP DE	POP D	D1	SBC A,(HL)	SBB M	9E	XOR H	XRA H	AC
POP HL	POP H	E1	SBC A,A	SBB A	9F	XOR L	XRA L	AD
PUSH AF	PUSH PSW	F5						

EXPANDED TABLE OF 6800 INSTRUCTIONS LISTED BY CATEGORY

Mnemonic	Operation	Boolean/Arith. Operation	Flags HINZVC	Address Mode	Assembler Notation	Op	~	#	Notes
				CPU Control Instructions					
NOP	No OPeration	Nothing	xxxxxx	Implied	NOP	01	2	1	Only the program counter is incremented. No operation occurs.
WAI	WAIt for interrupt	PC + 1 → PC PC_L → S PC_H → S X_L → S X_H → S A → S B → S CCR → S	xIxxxx	Implied	WAI	3E	9	1	After those actions shown in the "Boolean/Arithmetic Operation" column take place, the current program is suspended. If I=0 and the Interrupt Request line is taken low then I=1 and the microprocessor will begin to execute a program whose address is found in memory locations FFF8 and FFF9.

Mnemonic	Operation	Boolean/Arith. Operation	Flags HINZVC	Address Mode	Assembler Notation	Op	~	#	Notes

Data Transfer Instructions

Mnemonic	Operation	Boolean/Arith. Operation	Flags HINZVC	Address Mode	Assembler Notation	Op	~	#	Notes
LDAA	LoaD Accumulator A	M → A	xxNZ0x	Immediate	LDAA #$dd	86	2	2	
				Direct	LDAA $aa	96	3	2	
				Indexed	LDAA $ff,X	A6	5	2	
				Extended	LDAA $aaaa	B6	4	3	
LDAB	LoaD Accumulator B	M → B	xxNZ0x	Immediate	LDAB #$dd	C6	2	2	
				Direct	LDAB $aa	D6	2	2	
				Indexed	LDAB $ff,X	E6	5	2	
				Extended	LDAB $aaaa	F6	4	3	
STAA	STore Accumulator A	A → M	xxNZ0x	Direct	STAA $aa	97	4	2	
				Indexed	STAA $ff,X	A7	6	2	
				Extended	STAA $aaaa	B7	5	3	
STAB	STore Accumulator B	B → M	xxNZ0x	Direct	STAB $aa	D7	4	2	
				Indexed	STAB $ff,X	E7	6	2	
				Extended	STAB $aaaa	F7	5	3	
TAB	Transfer A to B	A → B	xxNZ0x	Implied	TAB	16	2	1	
TBA	Transfer B to A	B → A	xxNZ0x	Implied	TBA	17	2	1	
LDX	LoaD X register	$M \rightarrow X_H$	xxNZ0x	Immediate	LDX #$dddd	CE	3	3	
		$(M + 1) \rightarrow X_L$		Direct	LDX $aa	DE	4	2	
				Indexed	LDX $ff,X	EE	6	2	
				Extended	LDX $aaaa	FE	5	3	
STX	STore X register	$X_H \rightarrow M$	xxNZ0x	Direct	STX $aa	DF	5	2	
		$X_L \rightarrow (M + 1)$		Indexed	STX $ff,X	EF	7	2	
				Extended	STX $aaaa	FF	6	3	
CLR	CLeaR memory location	00 → M	xx0100	Indexed	CLR $ff,X	6F	7	2	
				Extended	CLR $aaaa	7F	6	3	
CLRA	CLeaR accumulator A	00 → A	xx0100	Implied	CLRA	4F	2	1	
CLRB	CLeaR accumulator B	00 → B	xx0100	Implied	CLRB	5F	2	1	

Flag Instructions

Mnemonic	Operation	Boolean/Arith. Operation	Flags HINZVC	Address Mode	Assembler Notation	Op	~	#	Notes
CLC	CLear Carry flag	0 → C	xxxxx0	Implied	CLC	0C	2	1	

EXPANDED TABLE OF 6800 INSTRUCTIONS LISTED BY CATEGORY (*Continued*)

Mne-monic	Operation	Boolean/Arith. Operation	Flags HINZVC	Address Mode	Assembler Notation	Op	~	#	Notes
CLI	CLear Interrupt flag	0 → I	x0xxxx	Implied	CLI	0E	2	1	
CLV	CLear oVerflow flag	0 → V	xxxxVx	Implied	CLV	0A	2	1	
SEC	SEt Carry flag	1 → C	xxxxx1	Implied	SEC	0D	2	1	
SEI	SEt Interrupt flag	1 → I	x1xxxx	Implied	SEI	0F	2	1	
SEV	SEt oVerflow flag	1 → V	xxxx1x	Implied	SEV	0B	2	1	
TAP	Transfer Accumulator A to Processor condition code register	A → CCR	HINZVC	Implied	TAP	06	2	1	
TPA	Transfer Processor condition code register to accumulator A	CCR → A	xxxxxx	Implied	TPA	07	2	1	

Arithmetic Instructions

Mne-monic	Operation	Boolean/Arith. Operation	Flags HINZVC	Address Mode	Assembler Notation	Op	~	#	Notes
ADDA	ADD accumulator A to memory location	A + M → A	HxNZVC	Immediate	ADDA #$dd	8B	2	2	
				Direct	ADDA $aa	9B	3	2	
				Indexed	ADDA $ff,X	AB	5	2	
				Extended	ADDA $aaaa	BB	4	3	
ADDB	ADD accumulator B to memory location	B + M → B	HxNZVC	Immediate	ADDB #$dd	CB	2	2	
				Direct	ADDB $aa	DB	3	2	
				Indexed	ADDB $ff,X	EB	5	2	
				Extended	ADDB $aaaa	FB	4	3	
ABA	Add accumulator B to accumulator A	A + B → A	HxNZVC	Implied	ABA	1B	2	1	
ADCA	AdD with Carry accumulator A to memory location	A + M + C → A	HxNZVC	Immediate	ADCA #$dd	89	2	2	
				Direct	ADCA $aa	99	3	2	
				Indexed	ADCA $ff,X	A9	5	2	
				Extended	ADCA $aaaa	B9	4	3	
ADCB	AdD with Carry accumulator B to memory location	B + M + C → B	HxNZVC	Immediate	ADCB #$dd	C9	2	2	
				Direct	ADCB $aa	D9	3	2	
				Indexed	ADCB $ff,X	E9	5	2	
				Extended	ADCB $aaaa	F9	4	3	

Mnemonic	Operation	Boolean/Arith. Operation	Flags HINZVC	Address Mode	Assembler Notation	Op	~	#	Notes
SUBA	SUBtract memory location from accumulator A	A - M → A	xxNZVC	Immediate	SUBA #$dd	80	2	2	
				Direct	SUBA $aa	90	3	2	
				Indexed	SUBA $ff,X	A0	5	2	
				Extended	SUBA $aaaa	B0	4	3	
SUBB	SUBtract memory location from accumulator B	B - M → B	xxNZVC	Immediate	SUBB #$dd	C0	2	2	
				Direct	SUBB $aa	D0	3	2	
				Indexed	SUBB $ff,X	E0	5	2	
				Extended	SUBB $aaaa	F0	4	3	
SBA	Subtract accumulator B from accumulator A	A - B → A	xxNZVC	Implied	SBA	10	2	1	
SBCA	SuBtract with Carry memory location from accumulator A	A - M - C → A	xxNZVC	Immediate	SBCA #$dd	82	2	2	
				Direct	SBCA $aa	92	3	2	
				Indexed	SBCA $ff,X	A2	5	2	
				Extended	SBCA $aaaa	B2	4	3	
SBCB	SuBtract with Carry memory location from accumulator B	B - M - C → B	xxNZVC	Immediate	SBCB #$dd	C2	2	2	
				Direct	SBCB $aa	D2	3	2	
				Indexed	SBCB $ff,X	E2	5	2	
				Extended	SBCB $aaaa	F2	4	3	
DAA	Decimal Adjust accumulator A	(converts binary number into BCD number)	xxNZVC	Implied	DAA	19	2	1	Converts the number in A to the BCD number it would be if the last two operands had been BCD numbers.

Logical Instructions

Mnemonic	Operation	Boolean/Arith. Operation	Flags HINZVC	Address Mode	Assembler Notation	Op	~	#	Notes
ANDA	AND accumulator A with memory location	A AND M → A	xxNZ0x	Immediate	ANDA #$dd	84	2	2	
				Direct	ANDA $aa	94	3	2	
				Indexed	ANDA $ff,X	A4	5	2	
				Extended	ANDA $aaaa	B4	4	3	
ANDB	AND accumulator B with memory location	B AND M → B	xxNZ0x	Immediate	ANDB #$dd	C4	2	2	
				Direct	ANDB $aa	D4	3	2	
				Indexed	ANDB $ff,X	E4	5	2	
				Extended	ANDB $aaaa	F4	4	3	
ORAA	OR Accumulator A with memory location	A OR M → A	xxNZ0x	Immediate	ORAA #$dd	8A	2	2	
				Direct	ORAA $aa	9A	3	2	
				Indexed	ORAA $ff,X	AA	5	2	
				Extended	ORAA $aaaa	BA	4	3	
ORAB	OR Accumulator B with memory location	B OR M → B	xxNZ0x	Immediate	ORAB #$dd	CA	2	2	
				Direct	ORAB $aa	DA	3	2	
				Indexed	ORAB $ff,X	EA	5	2	
				Extended	ORAB $aaaa	FA	4	3	

EXPANDED TABLE OF 6800 INSTRUCTIONS LISTED BY CATEGORY (*Continued*)

Mnemonic	Operation	Boolean/Arith. Operation	Flags HINZVC	Address Mode	Assembler Notation	Op	~	#	Notes
EORA	Exclusively **OR** accumulator **A** with memory location	A EOR M → A	xxNZ0x	Immediate	EORA #$dd	88	2	2	
				Direct	EORA $aa	98	3	2	
				Indexed	EORA $ff,X	A8	5	2	
				Extended	EORA $aaaa	B8	4	3	
EORB	Exclusively **OR** accumulator **A** with memory location	B EOR M → B	xxNZ0x	Immediate	EORB #$dd	C8	2	2	
				Direct	EORB $aa	D8	3	2	
				Indexed	EORB $ff,X	E8	5	2	
				Extended	EORB $aaaa	F8	4	3	
BITA	**BIT** test accumulator **A**	A AND M	xxNZ0x	Immediate	BITA #$dd	85	2	2	Accumulator A and a memory location are ANDed but neither is changed. However, flags N and Z are affected accordingly.
				Direct	BITA $aa	95	3	2	
				Indexed	BITA $ff,X	A5	5	2	
				Extended	BITA $aaaa	B5	4	3	
BITB	**BIT** test accumulator **B**	B AND M	xxNZ0x	Immediate	BITB #$dd	C5	2	2	Accumulator B and a memory location are ANDed but neither is changed. However, flags N and Z are affected accordingly.
				Direct	BITB $aa	D5	3	2	
				Indexed	BITB $ff,X	E5	5	2	
				Extended	BITB $aaaa	F5	4	3	
COM	**COM**plement memory location (1's complement)	$\overline{M} \rightarrow M$	xxNZ01	Indexed	COM $ff,X	63	7	2	
				Extended	COM $aaaa	73	6	2	
COMA	**COM**plement accumulator **A** (1's complement)	$\overline{A} \rightarrow A$	xxNZ01	Implied	COMA	43	2	1	
COMB	**COM**plement accumulator **B** (1's complement)	$\overline{B} \rightarrow B$	xxNZ01	Implied	COMB	53	2	1	
NEG	**NEG**ate memory location (2's complement)	00 - M → M	xxNZVC	Indexed	NEG $ff,X	60	7	2	Affects the carry flag as if the memory location had been subtracted from zero.
				Extended	NEG $aaaa	70	6	3	
NEGA	**NEG**ate accumulator **A** (2's complement)	00 - A → A	xxNZVC	Implied	NEGA	40	2	1	Affects the carry flag as if accumulator A had been subtracted from zero.
NEGB	**NEG**ate accumulator **B** (2's complement)	00 - B → B	xxNZVC	Implied	NEGB	50	2	1	Affects the carry flag as if accumulator B had been subtracted from zero.

Rotate and Shift Instructions

Mnemonic	Operation	Boolean/Arith. Operation	Flags HINZVC	Address Mode	Assembler Notation	Op	~	#	Notes
ROL	**RO**tate memory location **L**eft	$M_7 \ldots M_0$ ← C ← M_7 (rotate through carry)	xxNZVC	Indexed	ROL $ff,X	69	7	2	
				Extended	ROL $aaaa	79	6	3	

Mnemonic	Operation	Boolean/Arith. Operation	Flags HINZVC	Address Mode	Assembler Notation	Op	~	#	Notes
ROLA	ROtate to the Left accumulator A	$C \leftarrow A_7 \ldots A_0 \leftarrow C$ (rotate left through C)	xxNZVC	Implied	ROLA	49	2	1	
ROLB	ROtate to the Left accumulator B	$C \leftarrow B_7 \ldots B_0 \leftarrow C$ (rotate left through C)	xxNZVC	Implied	ROLB	59	2	1	
ROR	ROtate memory location Right	$C \rightarrow M_7 \ldots M_0 \rightarrow C$ (rotate right through C)	xxNZVC	Indexed	ROR $ff,X	66	7	2	
				Extended	ROR $aaaa	76	6	3	
RORA	ROtate to the Right accumulator A	$C \rightarrow A_7 \ldots A_0 \rightarrow C$ (rotate right through C)	xxNZVC	Implied	RORA	46	2	1	
RORB	ROtate to the Right accumulator B	$C \rightarrow B_7 \ldots B_0 \rightarrow C$ (rotate right through C)	xxNZVC	Implied	RORB	56	2	1	
ASL	Arithmetic Shift Left memory location	$C \leftarrow M_7 \ldots M_0 \leftarrow 0$	xxNZVC	Indexed	ASL $ff,X	68	7	2	
				Extended	ASL $aaaa	78	6	3	
ASLA	Arithmetic Shift Left accumulator A	$C \leftarrow A_7 \ldots A_0 \leftarrow 0$	xxNZVC	Implied	ASLA	48	2	1	
ASLB	Arithmetic Shift Left accumulator B	$C \leftarrow B_7 \ldots B_0 \leftarrow 0$	xxNZVC	Implied	ASLB	58	2	1	
ASR	Arithmetic Shift Right memory location	$M_7 \ldots M_0 \rightarrow C$ (M_7 fed back)	xxNZVC	Indexed	ASR $ff,X	67	7	2	
				Extended	ASR $aaaa	77	6	3	
ASRA	Arithmetic Shift Right accumulator A	$A_7 \ldots A_0 \rightarrow C$ (A_7 fed back)	xxNZVC	Implied	ASRA	47	2	1	
ASRB	Arithmetic Shift Right accumulator B	$B_7 \ldots B_0 \rightarrow C$ (B_7 fed back)	xxNZVC	Implied	ASRB	57	2	1	
LSR	Logical Shift Right memory location	$0 \rightarrow M_7 \ldots M_0 \rightarrow C$	xx0ZVC	Indexed	LSR $ff,X	64	7	2	
				Extended	LSR $aaaa	74	6	3	
LSRA	Logical Shift Right accumulator A	$0 \rightarrow A_7 \ldots A_0 \rightarrow C$	xx0ZVC	Implied	LSRA	44	2	1	
LSRB	Logical Shift Right accumulator B	$0 \rightarrow B_7 \ldots B_0 \rightarrow C$	xx0ZVC	Implied	LSRB	54	2	1	

Mne-monic	Operation	Boolean/Arith. Operation	Flags HINZVC	Address Mode	Assembler Notation	Op	~	#	Notes
				Increment and Decrement Instructions					
INC	INCrement memory location	M + 1 → M	xxNZVx	Indexed	INC $ff,X	6C	7	2	
				Extended	INC $aaaa	7C	6	3	
INCA	INCrement accum-ulator A	A + 1 → A	xxNZVx	Implied	INCA	4C	2	1	
INCB	INCrement accum-ulator B	B + 1 → B	xxNZVx	Implied	INCB	5C	2	1	
DEC	DECrement memory location	M - 1 → M	xxNZVx	Indexed	DEC $ff,X	6A	7	2	
				Extended	DEC $aaaa	7A	6	3	
DECA	DECrement accum-ulator A	A - 1 → A	xxNZVx	Implied	DECA	4A	2	1	
DECB	DECrement accum-ulator B	B - 1 → B	xxNZVx	Implied	DECB	5A	2	1	
INX	INcrement X (index) register	X + 1 → X	xxxZxx	Implied	INX	08	4	1	
DEX	DEcrement X (index) register	X - 1 → X	XXXzXX	Implied	DEX	09	4	1	
				Unconditional Jump Instructions					
JMP	JuMP to memory location	X + ff → PC (indexed) aaaa → PC (extended)	xxxxxx	Indexed	JMP $ff,X	6E	4	2	
				Extended	JMP $aaaa	7E	3	3	
BRA	BRanch Always to memory loc-ation	PC + 2 + rr → PC	xxxxxx	Relative	BRA $rr	20	4	2	
				Test (Compare) Instructions					
CMPA	CoMPare memory location to accumulator A	A - M	xxNZVC	Immediate	CMPA #$dd	81	2	2	
				Direct	CMPA $aa	91	3	2	
				Indexed	CMPA $ff,X	A1	5	2	
				Extended	CMPA $aaaa	B1	4	3	
CMPB	CoMPare memory location to accumulator B	B - M	xxNZVC	Immediate	CMPB #$dd	C1	2	2	
				Direct	CMPB $aa	D1	3	2	
				Indexed	CMPB $ff,X	E1	5	2	
				Extended	CMPB $aaaa	F1	4	3	

Mne-monic	Operation	Boolean/Arith. Operation	Flags HINZVC	Address Mode	Assembler Notation	Op	~	#	Notes
CBA	Compare accumulator B to accumulator A	A - B	xxNZVC	Implied	CBA	11	2	1	
CPX	ComPare memory location to X (index) register	X_H - M X_L - (M+1)	xxNZVx	Immediate Direct Indexed Extended	CPX #$dddd CPX $aa CPX $ff,X CPX $aaaa	8C 9C AC BC	3 4 6 5	3 2 2 3	
TST	TEsT memory location for zero or minus	M - 00	xxNZ00	Indexed Extended	TST $ff,X TST $aaaa	6D 7D	7 6	2 3	
TSTA	TEsT accumulator A for zero or minus	A - 00	xxNZ00	Implied	TSTA	4D	2	1	
TSTB	TEsT accumulator B for zero or minus	B - 00	xxNZ00	Implied	TSTB	5D	2	1	

Conditional Jump (Branch) Instructions

Mne-monic	Operation	Boolean/Arith. Operation	Flags HINZVC	Address Mode	Assembler Notation	Op	~	#	Notes
BCC	Branch if Carry Clear	PC + 2 + rr → PC if C=0	xxxxxx	Relative	BCC $rr	24	4	2	
BCS	Branch if Carry Set	PC + 2 + rr → PC if C=1	xxxxxx	Relative	BCS $rr	25	4	2	
BEQ	Branch if result of last operation was EQual to zero	PC + 2 + rr → PC if Z=1	xxxxxx	Relative	BEQ $rr	27	4	2	
BGE	Branch if Greater than or Equal to zero	PC + 2 + rr → PC if N EOR V = 0	xxxxxx	Relative	BGE $rr	2C	4	2	This branch occurs after the instructions CBA, CMP, SBA, or SUB if the 2's-complement minuend is greater than or equal to the 2's-complement subtrahend creating an answer which is greater than or equal to zero.
BGT	Branch if Greater Than zero	PC + 2 + rr → PC if Z AND (N EOR V) = 0	xxxxxx	Relative	BGT $rr	2E	4	2	This branch occurs after the instructions CBA, CMP, SBA, or SUB if the 2's-complement minuend is greater than the 2's-complement subtrahend, creating an answer which is greater than zero.

EXPANDED TABLE OF 6800 INSTRUCTIONS LISTED BY CATEGORY (*Continued*)

Mne-monic	Operation	Boolean/Arith. Operation	Flags HINZVC	Address Mode	Assembler Notation	Op	~	#	Notes
BHI	Branch if HIgher	PC + 2 + rr → PC if C AND Z = 0	xxxxxx	Relative	BHI $rr	22	4	2	This branch occurs after the instructions CBA, CMP, SBA, or SUB if the unsigned binary minuend is greater than the unsigned binary subtrahend.
BLE	Branch if Less than or Equal to zero	PC + 2 + rr → PC if Z AND (N EOR V) = 1	xxxxxx	Relative	BLE $rr	2F	4	2	This branch occurs after the instructions CBA, CMP, SBA, or SUB if the 2's-complement minuend is less than or equal to the 2's-complement subtrahend, creating an answer which is less than or equal to zero.
BLS	Branch if Lower or the Same	PC + 2 + rr → PC if C OR Z = 1	xxxxxx	Relative	BLS $rr	23	4	2	This branch occurs after the instructions CBA, CMP, SBA, or SUB if the unsigned binary minuend is less than or equal to the unsigned binary subtrahend.
BLT	Branch if Less Than zero	PC + 2 + rr → PC if N EOR V = 1	xxxxxx	Relative	BLT $rr	2D	4	2	This branch occurs after the instructions CBA, CMP, SBA, or SUB if the 2's-complement minuend is less than the 2's-complement subtrahend, creating an answer which is less than zero.
BMI	Branch is MInus	PC + 2 + rr → PC if N=1	xxxxxx	Relative	BMI $rr	2B	4	2	
BNE	Branch if Not Equal to zero	PC + 2 + rr → PC if Z=1	xxxxxx	Relative	BNE $rr	26	4	2	
BVC	Branch if oVerflow Clear	PC + 2 + rr → PC if V=0	xxxxxx	Relative	BVC $rr	28	4	2	
BVS	Branch if oVerflow Set	PC + 2 + rr → PC if V=1	xxxxxx	Relative	BVS $rr	29	4	2	
BPL	Branch if PLus	PC + 2 + rr → PC if N=0	xxxxxx	Relative	BPL $rr	2A	4	2	

Mnemonic	Operation	Boolean/Arith. Operation	Flags HINZVC	Address Mode	Assembler Notation	Op	~	#	Notes
				Subroutine Instructions					
JSR	Jump SubRoutine	PC + 2 → PC PC_L → S PC_H → S SP - 2 → SP (ff+X) → PC	xxxxxx	Indexed	JSR $ff,X	AD	8	2	The program counter is incremented by 2 (Indexed) or 3 (Extended) and the program counter is pushed onto the stack 1 byte at a time. At the memory location indicated by the addressing mode will be found the address of the first instruction of the subroutine. This address is placed in the program counter.
		PC + 3 → PC PC_L → S PC_H → S SP - 2 → SP (aaaa) → PC		Extended	JSR $aaaa	BD	9	3	
RTS	ReTurn from Subroutine	S → PC_H S → PC_L SP + 2 → SP	xxxxxx	Implied	RTS	39	5	1	The address of the next instruction in the main program after the last JSR is loaded from the stack into the program counter 1 byte at a time.
BSR	Branch to SubRoutine	PC + 2 → PC PC_L → S PC_H → S SP - 2 → SP PC + rr → PC	xxxxxx	Relative	BSR $rr	8D	8	2	The program counter is incremented by 2 and pushed onto the stack 1 byte at a time. The memory location of the next instruction is then calculate by adding the 2's-complement binary number rr to the program counter. This instruction differs from JSR in the form of addressing it uses.
				Stack Instructions					
LDS	LoaD Stack pointer	M → SP_H (M + 1) → SP_L	xxNZ0x	Immediate Direct Indexed Extended	LDS #$dddd LDS $aa LDS $ff,X LDS $aaaa	8E 9E AE BE	3 4 6 5	3 2 2 3	
STS	STore Stack pointer	SP_H → M SP_L → (M + 1)	xxNZ0x	Direct Indexed Extended	STS $aa STS $ff,X STS $aaaa	9F AF BF	5 7 6	2 2 3	
PSHA	PuSH accumulator A onto the stack	A → S SP - 1 → SP	xxxxxx	Implied	PSHA	36	4	1	Whenever A or B is pushed onto the stack the stack pointer is decremented by 1. When the contents of the stack are placed in A or B the stack pointer is incremented by 1.
PSHB	PuSH accumulator B onto the stack	B → S SP - 1 → SP	xxxxxx	Implied	PSHB	37	4	1	
PULA	PUlL accumulator A from the stack	S → A SP + 1 → SP	xxxxxx	Implied	PULA	32	4	1	
PULB	PUlL accumulator B from the stack	S → B SP + 1 → SP	xxxxxx	Implied	PULB	33	4	1	

Mnemonic	Operation	Boolean/Arith. Operation	Flags HINZVC	Address Mode	Assembler Notation	Op	~	#	Notes
DES	DEcrement Stack pointer	SP - 1 → SP	xxxxxx	Implied	DES	34	4	1	
INS	INcrement Stack pointer	SP + 1 → SP	xxxxxx	Implied	INS	31	4	1	
TXS	Transfer X (index) register to Stack pointer	X - 1 → SP	xxxxxx	Implied	TXS	35	4	1	
TSX	Transfer Stack pointer to the X (index) register	SP + 1 → X	xxxxxx	Implied	TSX	30	4	1	
			Interrupt Instructions						
RTI	ReTurn from Interrupt	S → CCR S → B S → A S → X_H S → X_L S → PC_H S → PC_L	HINZVC	Implied	RTI	3B	10	1	
SWI	SoftWare Interrupt	PC + 1 → PC PC_L → S PC_H → S X_L → S X_H → S A → S B → S CCR → S	x1xxxx	Implied	SWI	3F	12	1	After the actions shown in the "Boolean/Arithmetic Operation" column take place, the microprocessor will begin to execute a program whose address is found in memory locations FFFA and FFFB.
			Input-Output Instructions						
none									The 6800/6808 has no special input and output instructions but rather memory-maps these operations.

Notes

Addressing Modes	Assembler Notation
Immediate	Mnemonic #$dd
Direct	Mnemonic $aa
Indexed	Mnemonic $ff,X
Extended	Mnemonic $aaaa
Implied	Mnemonic
Relative	Mnemonic $rr

Abbreviations and Explanations

a = address (one hex digit)
d = data (one hex digit)
f = offset (one hex digit) to be added to the X register (ff is positive -- $00-$ff which is decimal 0-255)
r = relative displacement (one hex digit) to be added to the program counter (rr is 2's-complement number and thus can be positive or negative, -128 to +127)
$ = indicates a hexadecimal number
# = indicates the data follows immediately after the instruction

$_L$ = low byte (lower byte of a two byte number)
$_H$ = high byte (upper byte of a two byte number)

Flags

H = instruction affects the half carry-flag
I = instruction affects the interrupt flag
N = instruction affects the negative flag
Z = instruction affects the zero flag
V = instruction affects the overflow flag
C = instruction affects the carry flag
0 = instruction always clears affected flag
1 = instruction always sets affected flag
x = flag not affected by instruction

CCR = condition code register (flags)
S = stack
SP = stack pointer
PC = program counter
() = contents of the memory location in the parenthesis

$M_7...M_0$ = memory bits 0-7 of a particular memory location
$A_7...A_0$ = bits 0-7 of accumulator a
$B_7...B_0$ = bits 0-7 of accumulator b

X = Index register

0 = One zero bit.
00 = One zero byte.

Symbols in the Page Heading

~ = clock cycles
# = # of bytes used by instruction (and following address or data if used)

Addressing Modes - Summary

Immediate (Mnemonic #$dd): In this addressing mode, the operand (data or number that something is being done to) is contained in the memory location(s) immediately following the instruction.

Direct (Mnemonic $aa): Direct addressing places the address of the operand in the byte following the instruction.

Indexed (Mnemonic $ff,X): This mode involves a couple of steps. First, the number ff (which is the byte after the instruction) is added to the value in the X register. The number ff is an 8-bit number which can only be positive (0-255 decimal). Then the operand is fetched from this newly formed address.

Extended (Mnemonic $aaaa): Extended addressing is the same as Direct except that a wider range is possible. The first byte is the instruction as in Direct addressing. The second and third bytes then form a 16-bit address where the operand can be found.

Implied (Mnemonic): When the operand is within the microprocessor itself implied addressing is used. In these cases the location of the operand is contained within the instruction itself. CLRA (CLeaR accumulator A) is an example of implied addressing.

Relative (Mnemonic $rr): Relative addressing is used exclusively with the branch and jump instructions. The byte following the instruction is an 8-bit 2's-complement number (+127 to -128) which is added to the contents of the program counter. This then is the address of the next instruction. The location of the next instruction is being indicated relative to the current location in memory (the current contents of the program counter).

SHORT TABLE OF 6800 INSTRUCTIONS LISTED ALPHABETICALLY

Mnemonic	Operation	Assembler Notation	Op
ABA	Add accumulator B to accumulator A	ABA	1B
ADCA	AdD with Carry accumulator A to memory location	ADCA #$dd	89
		ADCA $aa	99
		ADCA $ff,X	A9
		ADCA $aaaa	B9
ADCB	AdD with Carry accumulator B to memory location	ADCB #$dd	C9
		ADCB $aa	D9
		ADCB $ff,X	E9
		ADCB $aaaa	F9
ADDA	ADD accumulator A to memory location	ADDA #$dd	8B
		ADDA $aa	9B
		ADDA $ff,X	AB
		ADDA $aaaa	BB
ADDB	ADD accumulator B to memory location	ADDB #$dd	CB
		ADDB $aa	DB
		ADDB $ff,X	EB
		ADDB $aaaa	FB
ANDA	AND accumulator A with memory location	ANDA #$dd	84
		ANDA $aa	94
		ANDA $ff,X	A4
		ANDA $aaaa	B4
ANDB	AND accumulator B with memory location	ANDB #$dd	C4
		ANDB $aa	D4
		ANDB $ff,X	E4
		ANDB $aaaa	F4
ASL	Arithmetic Shift Left memory location	ASL $ff,X	68
		ASL $aaaa	78
ASLA	Arithmetic Shift Left accumulator A	ASLA	48
ASLB	Arithmetic Shift Left accumulator B	ASLB	58
ASR	Arithmetic Shift Right memory location	ASR $ff,X	67
		ASR $aaaa	77
ASRA	Arithmetic Shift Right accumulator A	ASRA	47
ASRB	Arithmetic Shift Right accumulator B	ASRB	57
BCC	Branch if Carry Clear	BCC $rr	24
BCS	Branch if Carry Set	BCS $rr	25
BEQ	Branch if result of last operation was EQual to zero	BEQ $rr	27
BGE	Branch if Greater than or Equal to zero	BGE $rr	2C
BGT	Branch if Greater Than zero	BGT $rr	2E
BHI	Branch if HIgher	BHI $rr	22
BITA	BIT test accumulator A	BITA #$dd	85
		BITA $aa	95
		BITA $ff,X	A5
		BITA $aaaa	B5
BITB	BIT test accumulator B	BITB #$dd	C5
		BITB $aa	D5
		BITB $ff,X	E5
		BITB $aaaa	F5
BLE	Branch if Less then or Equal to zero	BLE $rr	2F
BLS	Branch if Lower or the Same	BLS $rr	23
BLT	Branch if Less Than zero	BLT $rr	2D
BMI	Branch is MInus	BMI $rr	2B
BNE	Branch if Not Equal to zero	BNE $rr	26
BPL	Branch if PLus	BPL $rr	2A
BRA	BRanch Always to memory location	BRA $rr	20
BSR	Branch to SubRoutine	BSR $rr	8D
BVC	Branch if oVerflow Clear	BVC $rr	28
BVS	Branch if oVerflow Set	BVS $rr	29

Mnemonic	Operation	Assembler Notation	Op
CBA	Compare accumulator B to accumulator A	CBA	11
CLC	CLear Carry flag	CLC	0C
CLI	CLear Interrupt flag	CLI	0E
CLR	CLeaR memory location	CLR $ff,X CLR $aaaa	6F 7F
CLRA	CLeaR accumulator A	CLRA	4F
CLRB	CLeaR accumulator B	CLRB	5F
CLV	CLear oVerflow flag	CLV	0A
CMPA	CoMPare memory location to accumulator A	CMPA #$dd CMPA $aa CMPA $ff,X CMPA $aaaa	81 91 A1 B1
CMPB	CoMPare memory location to accumulator B	CMPB #$dd CMPB $aa CMPB $ff,X CMPB $aaaa	C1 D1 E1 F1
COM	COMplement memory location (1's complement)	COM $ff,X COM $aaaa	63 73
COMA	COMplement accumulator A (1's complement)	COMA	43
COMB	COMplement accumulator B (1's complement)	COMB	53
CPX	ComPare memory location to X (index) register	CPX #$dd CPX $aa CPX $ff,X CPX $aaaa	8C 9C AC BC
DAA	Decimal Adjust accumulator A	DAA	19
DEC	DECrement memory location	DEC $ff,X DEC $aaaa	6A 7A
DECA	DECrement accumulator A	DECA	4A
DECB	DECrement accumulator B	DECB	5A
DES	DEcrement Stack pointer	DES	34

Mnemonic	Operation	Assembler Notation	Op
DEX	DEcrement X (index) register	DEX	09
EORA	Exclusively OR accumulator A with memory location	EORA #$dd EORA $aa EORA $ff,X EORA $aaaa	88 98 A8 B8
EORB	Exclusively OR accumulator A with memory location	EORB #$dd EORB $aa EORB $ff,X EORB $aaaa	C8 D8 E8 F8
INC	INCrement memory location	INC $ff,X INC $aaaa	6C 7C
INCA	INCrement accumulator A	INCA	4C
INCB	INCrement accumulator B	INCB	5C
INS	INcrement Stack pointer	INS	31
INX	INcrement X (index) register	INX	08
JMP	JuMP to memory location	JMP $ff,X JMP $aaaa	6E 7E
JSR	Jump SubRoutine	JSR $ff,X JSR $aaaa	AD BD
LDAA	LoaD Accumulator A	LDAA #$dd LDAA $aa LDAA $ff,X LDAA $aaaa	86 96 A6 B6
LDAB	LoaD Accumulator B	LDAB #$dd LDAB $aa LDAB $ff,X LDAB $aaaa	C6 D6 E6 F6
LDS	LoaD Stack pointer	LDS #$dddd LDS $aa LDS $ff,X LDS $aaaa	8E 9E AE BE
LDX	LoaD X register	LDX #$dd LDX $aa LDX $ff,X LDX $aaaa	CE DE EE FE
LSR	Logical Shift Right memory location	LSR $ff,X LSR $aaaa	64 74

SHORT TABLE OF 6800 INSTRUCTIONS LISTED ALPHABETICALLY (*Continued*)

Mnemonic	Operation	Assembler Notation	Op
LSRA	Logical Shift Right accumulator A	LSRA	44
LSRB	Logical Shift Right accumulator B	LSRB	54
NEG	NEGate memory location (2's complement)	NEG $ff,X	60
		NEG $aaaa	70
NEGA	NEGate accumulator A (2's complement)	NEGA	40
NEGB	NEGate accumulator B (2's complement)	NEGB	50
NOP	No OPeration	NOP	01
ORAA	OR Accumulator A with memory location	ORAA #$dd	8A
		ORAA $aa	9A
		ORAA $ff,X	AA
		ORAA $aaaa	BA
ORAB	OR Accumulator B with memory location	ORAB #$dd	CA
		ORAB $aa	DA
		ORAB $ff,X	EA
		ORAB $aaaa	FA
PSHA	PuSH accumulator A onto the stack	PSHA	36
PSHB	PuSH accumulator B onto the stack	PSHB	37
PULA	PUlL accumulator A from the stack	PULA	32
PULB	PUlL accumulator B from the stack	PULB	33
ROL	ROtate memory location Left	ROL $ff,X	69
		ROL $aaaa	79
ROLA	ROtate to the Left accumulator A	ROLA	49
ROLB	ROtate to the Left accumulator B	ROLB	59
ROR	ROtate memory location Right	ROR $ff,X	66
		ROR $aaaa	76
RORA	ROtate to the Right accumulator A	RORA	46
RORB	ROtate to the Right accumulator B	RORB	56
RTI	ReTurn from Interrupt	RTI	3B
RTS	ReTurn from Subroutine	RTS	39
SBA	Subtract accumulator B from accumulator A	SBA	10
SBCA	SuBtract with Carry memory location from accumulator A	SBCA #$dd	82
		SBCA $aa	92
		SBCA $ff,X	A2
		SBCA $aaaa	B2
SBCB	SuBtract with Carry memory location from accumulator B	SBCB #$dd	C2
		SBCB $aa	D2
		SBCB $ff,X	E2
		SBCB $aaaa	F2
SEC	SEt Carry flag	SEC	0D
SEI	SEt Interrupt flag	SEI	0F
SEV	SEt oVerflow flag	SEV	0B
STAA	STore Accumulator A	STAA $aa	97
		STAA $ff,X	A7
		STAA $aaaa	B7
STAB	STore Accumulator B	STAB $aa	D7
		STAB $ff,X	E7
		STAB $aaaa	F7
STS	STore Stack pointer	STS $aa	9F
		STS $ff,X	AF
		STS $aaaa	BF
STX	STore X register	STX $aa	DF
		STX $ff,X	EF
		STX $aaaa	FF
SUBA	SUBtract memory location from accumulator A	SUBA #$dd	80
		SUBA $aa	90
		SUBA $ff,X	A0
		SUBA $aaaa	B0
SUBB	SUBtract memory location from accumulator B	SUBB #$dd	C0
		SUBB $aa	D0
		SUBB $ff,X	E0
		SUBB $aaaa	F0
SWI	SoftWare Interrupt	SWI	3F

Mnemonic	Operation	Assembler Notation	Op
TAB	Transfer A to B	TAB	16
TAP	Transfer Accumulator A to Processor condition code register	TAP	06
TBA	Transfer B to A	TBA	17
TPA	Transfer Processor condition code register to accumulator A	TPA	07
TST	TEsT memory location for zero or minus	TST $ff,X TST $aaaa	6D 7D
TSTA	TEsT accumulator A for zero or minus	TSTA	4D
TSTB	TEsT accumulator B for zero or minus	TSTB	5D
TSX	Transfer Stack pointer to the X (index) register	TSX	30
TXS	Transfer X (index) register to Stack pointer	TXS	35
WAI	WAit for Interrupt	WAI	3E

SHORT TABLE OF 6800 INSTRUCTIONS LISTED BY CATEGORY

Assembler Notation	Op	Boolean/Arith Operation	Flags HINZVC
CPU Control Instructions			
NOP	01	nothing	xxxxxx
WAI	3E	PC + 1 → PC PC_L → S PC_H → S X_L → S X_H → S A → S B → S CCR → S	xIxxxx
Data Transfer Instructions			
LDAA #$dd LDAA $aa LDAA $ff,X LDAA $aaaa	86 96 A6 B6	M → A	xxNZ0x
LDAB #$dd LDAB $aa LDAB $ff,X LDAB $aaaa	C6 D6 E6 F6	M → B	xxNZ0x
STAA $aa STAA $ff,X STAA $aaaa	97 A7 B7	A → M	xxNZ0x
STAB $aa STAB $ff,X STAB $aaaa	D7 E7 F7	B → M	xxNZ0x
TAB	16	A → B	xxNZ0x
TBA	17	B → A	xxNZ0x
LDX #$dddd LDX $aa LDX $ff,X LDX $aaaa	CE DE EE FE	M → X_H (M + 1) → X_L	xxNZ0x
STX $aa STX $ff,X STX $aaaa	DF EF FF	X_H → M X_L → (M + 1)	xxNZ0x
CLR $ff,X CLR $aaaa	6F 7F	00 → M	xx0100
CLRA	4F	00 → A	xx0100
CLRB	5F	00 → B	xx0100
Flag Instructions			
CLC	0C	0 → C	xxxxx0
LI	0E	0 → I	x0xxxx
CLV	0A	0 → V	xxxxVx
SEC	0D	1 → C	xxxxx1
SEI	0F	1 → I	x1xxxx

SHORT TABLE OF 6800 INSTRUCTIONS LISTED BY CATEGORY (*Continued*)

Assembler Notation	Op	Boolean/Arith Operation	Flags HINZVC
SEV	0B	1 → V	xxxx1x
TAP	06	A → CCR	HINZVC
TPA	07	CCR → A	xxxxxx

Arithmetic Instructions

Assembler Notation	Op	Boolean/Arith Operation	Flags HINZVC
ADDA #$dd	8B	A + M → A	HxNZVC
ADDA $aa	9B		
ADDA $ff,X	AB		
ADDA $aaaa	BB		
ADDB #$dd	CB	B + M → B	HxNZVC
ADDB $aa	DB		
ADDB $ff,X	EB		
ADDB $aaaa	FB		
ABA	1B	A + B → A	HxNZVC
ADCA #$dd	89	A + M + C → A	HxNZVC
ADCA $aa	99		
ADCA $ff,X	A9		
ADCA $aaaa	B9		
ADCB #$dd	C9	B + M + C → B	HxNZVC
ADCB $aa	D9		
ADCB $ff,X	E9		
ADCB $aaaa	F9		
SUBA #$dd	80	A - M → A	xxNZVC
SUBA $aa	90		
SUBA $ff,X	A0		
SUBA $aaaa	B0		
SUBB #$dd	C0	B - M → B	xxNZVC
SUBB $aa	D0		
SUBB $ff,X	E0		
SUBB $aaaa	F0		
SBA	10	A - B → A	xxNZVC
SBCA #$dd	82	A - M - C → A	xxNZVC
SBCA $aa	92		
SBCA $ff,X	A2		
SBCA $aaaa	B2		
SBCB #$dd	C2	B - M - C → B	xxNZVC
SBCB $aa	D2		
SBCB $ff,X	E2		
SBCB $aaaa	F2		
DAA	19	(converts binary add. of BCD characters into BCD format)	xxNZVC

Logical Instructions

Assembler Notation	Op	Boolean/Arith Operation	Flags HINZVC
ANDA #$dd	84	A AND M → A	xxNZ0x
ANDA $aa	94		
ANDA $ff,X	A4		
ANDA $aaaa	B4		
ANDB #$dd	C4	B AND M → B	xxNZ0x
ANDB $aa	D4		
ANDB $ff,X	E4		
ANDB $aaaa	F4		
ORAA #$dd	8A	A OR M → A	xxNZ0x
ORAA $aa	9A		
ORAA $ff,X	AA		
ORAA $aaaa	BA		
ORAB #$dd	CA	B OR M → B	xxNZ0x
ORAB $aa	DA		
ORAB $ff,X	EA		
ORAB $aaaa	FA		
EORA #$dd	88	A EOR M → A	xxNZ0x
EORA $aa	98		
EORA $ff,X	A8		
EORA $aaaa	B8		
EORB #$dd	C8	B EOR M → B	xxNZ0x
EORB $aa	D8		
EORB $ff,X	E8		
EORB $aaaa	F8		
BITA #$dd	85	A AND M	xxNZ0x
BITA $aa	95		
BITA $ff,X	A5		
BITA $aaaa	B5		
BITB #$dd	C5	B AND M	xxNZ0x
BITB $aa	D5		
BITB $ff,X	E5		
BITB $aaaa	F5		
COM $ff,X	63	$\overline{M} \rightarrow M$	xxNZ01
COM $aaaa	73		
COMA	43	$\overline{A} \rightarrow A$	xxNZ01
COMB	53	$\overline{B} \rightarrow B$	xxNZ01

Assembler Notation	Op	Boolean/Arith Operation	Flags HINZVC
NEG $ff,X	60	00 - M → M	xxNZVC
NEG $aaaa	70		
NEGA	40	00 - A → A	xxNZVC
NEGB	50	00 - B → B	xxNZVC

Rotate and Shift Instructions

Assembler Notation	Op	Boolean/Arith Operation	Flags HINZVC
ROL $ff,X	69	$C \leftarrow M_7 \ldots M_0 \leftarrow C$	xxNZVC
ROL $aaaa	79		
ROLA	49	$C \leftarrow A_7 \ldots A_0 \leftarrow C$	xxNZVC
ROLB	59	$C \leftarrow B_7 \ldots B_0 \leftarrow C$	xxNZVC
ROR $ff,X	66	$C \rightarrow M_7 \ldots M_0 \rightarrow C$	xxNZVC
ROR $aaaa	76		
RORA	46	$C \rightarrow A_7 \ldots A_0 \rightarrow C$	xxNZVC
RORB	56	$C \rightarrow B_7 \ldots B_0 \rightarrow C$	xxNZVC
ASL $ff,X	68	$C \leftarrow M_7 \ldots M_0 \leftarrow 0$	xxNZVC
ASL $aaaa	78		
ASLA	48	$C \leftarrow A_7 \ldots A_0 \leftarrow 0$	xxNZVC
ASLB	58	$C \leftarrow B_7 \ldots B_0 \leftarrow 0$	xxNZVC
ASR $ff,X	67	$M_7 \rightarrow M_7 \ldots M_0 \rightarrow C$	xxNZVC
ASR $aaaa	77		
ASRA	47	$A_7 \rightarrow A_7 \ldots A_0 \rightarrow C$	xxNZVC
ASRB	57	$B_7 \rightarrow B_7 \ldots B_0 \rightarrow C$	xxNZVC
LSR $ff,X	64	$0 \rightarrow M_7 \ldots M_0 \rightarrow C$	xx0ZVC
LSR $aaaa	74		
LSRA	44	$0 \rightarrow A_7 \ldots A_0 \rightarrow C$	xx0ZVC
LSRB	54	$0 \rightarrow B_7 \ldots B_0 \rightarrow C$	xx0ZVC

Increment and Decrement Instructions

Assembler Notation	Op	Boolean/Arith Operation	Flags HINZVC
INC $ff,X	6C	M + 1 → M	xxNZVx
INC $aaaa	7C		
INCA	4C	A + 1 → A	xxNZVx
INCB	5C	B + 1 → B	xxNZVx
DEC $ff,X	6A	M - 1 → M	xxNZVx
DEC $aaaa	7A		
DECA	4A	A - 1 → A	xxNZVx
DECB	5A	B - 1 → B	xxNZVx
INX	08	X + 1 → X	xxxZxx
DEX	09	X - 1 → X	xxxZxx

Unconditional Jump Instructions

Assembler Notation	Op	Boolean/Arith Operation	Flags HINZVC
JMP $ff,X	6E	X + ff → PC (indexed)	xxxxxx
JMP $aaaa	7E	aaaa → PC (extended)	
BRA $rr	20	PC + 2 + rr → PC	xxxxxx

Test (Compare) Instructions

Assembler Notation	Op	Boolean/Arith Operation	Flags HINZVC
CMPA #$dd	81	A - M	xxNZVC
CMPA $aa	91		
CMPA $ff,X	A1		
CMPA $aaaa	B1		
CMPB #$dd	C1	B - M	xxNZVC
CMPB $aa	D1		
CMPB $ff,X	E1		
CMPB $aaaa	F1		
CBA	11	A - B	xxNZVC
CPX #$dddd	8C	X_H - M	xxNZVx
CPX $aa	9C	X_L - (M+1)	
CPX $ff,X	AC		
CPX $aaaa	BC		
TST $ff,X	6D	M - 00	xxNZ00
TST $aaaa	7D		
TSTA	4D	A - 00	xxNZ00
TSTB	5D	B - 00	xxNZ00

SHORT TABLE OF 6800 INSTRUCTIONS LISTED BY CATEGORY (*Continued*)

Assembler Notation	Op	Boolean/Arith Operation	Flags HINZVC
Conditional Jump (Branch) Instructions			
BCC $rr	24	PC + 2 + rr → PC If C=0	xxxxxx
BCS $rr	25	PC + 2 + rr → PC If C=1	xxxxxx
BEQ $rr	27	PC + 2 + rr → PC If Z=1	xxxxxx
BGE $rr	2C	PC + 2 + rr → PC If N EOR V = 0	xxxxxx
BGT $rr	2E	PC + 2 + rr → PC If Z AND (N EOR V) = 0	xxxxxx
BHI $rr	22	PC + 2 + rr → PC If C AND Z = 0	xxxxxx
BLE $rr	2F	PC + 2 + rr → PC If Z AND (N EOR V) = 1	xxxxxx
BLS $rr	23	PC + 2 + rr → PC If C OR Z = 1	xxxxxx
BLT $rr	2D	PC + 2 + rr → PC If N EOR V = 1	xxxxxx
BMI $rr	2B	PC + 2 + rr → PC If N=1	xxxxxx
BNE $rr	26	PC + 2 + rr → PC If Z=1	xxxxxx
BVC $rr	28	PC + 2 + rr → PC If V=0	xxxxxx
BVS $rr	29	PC + 2 + rr → PC If V=1	xxxxxx
BPL $rr	2A	PC + 2 + rr → PC If N=0	xxxxxx
Subroutine Instructions			
JSR $ff,X	AD	PC + 2 → PC PC_L → S PC_H → S SP - 2 → SP (ff+X) → PC	xxxxxx
JSR $aaaa	BD	PC + 3 → PC PC_L → S PC_H → S SP - 2 → SP (aaaa) → PC	
RTS	39	S → PC_H S → PC_L SP + 2 → SP	xxxxxx
BSR $rr	8D	PC + 2 → PC PC_L → S PC_H → S SP - 2 → SP PC + rr → PC	xxxxxx
Stack Instructions			
LDS #$dddd	8E	M → SP_H (M + 1) → SP_L	xxNZ0x
LDS $aa	9E		
LDS $ff,X	AE		
LDS $aaaa	BE		
STS $aa	9F	SP_H → M SP_L → (M + 1)	xxNZ0x
STS $ff,X	AF		
STS $aaaa	BF		
PSHA	36	A → S SP - 1 → SP	xxxxxx
PSHB	37	B → S SP - 1 → SP	xxxxxx
PULA	32	S → A SP + 1 → SP	xxxxxx
PULB	33	S → B SP + 1 → SP	xxxxxx
DES	34	SP - 1 → SP	xxxxxx
INS	31	SP + 1 → SP	xxxxxx

Assembler Notation	Op	Boolean/Arith Operation	Flags HINZVC
TXS	35	X - 1 → SP	xxxxxx
TSX	30	SP + 1 → X	xxxxxx

Interrupt Instructions

Assembler Notation	Op	Boolean/Arith Operation	Flags HINZVC
RTI	3B	S → CCR S → B S → A S → X_H S → X_L S → PC_H S → PC_L	HINZVC

Assembler Notation	Op	Boolean/Arith Operation	Flags HINZVC
SWI	3F	PC + 1 → PC PC_L → S PC_H → S X_L → S X_H → S A → S B → S CCR → S	x1xxxx

Input-Output Instructions

none

CONDENSED TABLE OF 6800 INSTRUCTIONS LISTED BY CATEGORY

Assembler	Op
CPU Control Instructions	
NOP	01
WAI	3E
Data Transfer Instructions	
LDAA #$dd	86
LDAA $aa	96
LDAA $ff,X	A6
LDAA $aaaa	B6
LDAB #$dd	C6
LDAB $aa	D6
LDAB $ff,X	E6
LDAB $aaaa	F6
STAA $aa	97
STAA $ff,X	A7
STAA $aaaa	B7
STAB $aa	D7
STAB $ff,X	E7
STAB $aaaa	F7
TAB	16
TBA	17
LDX #$dddd	CE
LDX $aa	DE
LDX $ff,X	EE
LDX $aaaa	FE
STX $aa	DF
STX $ff,X	EF
STX $aaaa	FF
CLR $ff,X	6F
CLR $aaaa	7F
CLRA	4F
CLRB	5F
Flag Instructions	
CLC	0C
LI	0E
CLV	0A
SEC	0D
SEI	0F
SEV	0B
TAP	06
TPA	07
Arithmetic Instructions	
ADDA #$dd	8B
ADDA $aa	9B
ADDA $ff,X	AB
ADDA $aaaa	BB
ADDB #$dd	CB
ADDB $aa	DB
ADDB $ff,X	EB
ADDB $aaaa	FB
ABA	1B
ADCA #$dd	89
ADCA $aa	99
ADCA $ff,X	A9
ADCA $aaaa	B9
ADCB #$dd	C9
ADCB $aa	D9
ADCB $ff,X	E9
ADCB $aaaa	F9
SUBA #$dd	80
SUBA $aa	90
SUBA $ff,X	A0
SUBA $aaaa	B0
SUBB #$dd	C0
SUBB $aa	D0
SUBB $ff,X	E0
SUBB $aaaa	F0
SBA	10
SBCA #$dd	82
SBCA $aa	92
SBCA $ff,X	A2
SBCA $aaaa	B2
SBCB #$dd	C2
SBCB $aa	D2
SBCB $ff,X	E2
SBCB $aaaa	F2
DAA	19
Logical Instructions	
ANDA #$dd	84
ANDA $aa	94
ANDA $ff,X	A4
ANDA $aaaa	B4
ANDB #$dd	C4
ANDB $aa	D4
ANDB $ff,X	E4
ANDB $aaaa	F4
ORAA #$dd	8A
ORAA $aa	9A
ORAA $ff,X	AA
ORAA $aaaa	BA
ORAB #$dd	CA
ORAB $aa	DA
ORAB $ff,X	EA
ORAB $aaaa	FA
EORA #$dd	88
EORA $aa	98
EORA $ff,X	A8
EORA $aaaa	B8
EORB #$dd	C8
EORB $aa	D8
EORB $ff,X	E8
EORB $aaaa	F8

CONDENSED TABLE OF 6800 INSTRUCTIONS LISTED BY CATEGORY *(Continued)*

Assembler	Op
BITA #$dd	85
BITA $aa	95
BITA $ff,X	A5
BITA $aaaa	B5
BITB #$dd	C5
BITB $aa	D5
BITB $ff,X	E5
BITB $aaaa	F5
COM $ff,X	63
COM $aaaa	73
COMA	43
COMB	53
NEG $ff,X	60
NEG $aaaa	70
NEGA	40
NEGB	50
Rotate and Shift Instructions	
ROL $ff,X	69
ROL $aaaa	79
ROLA	49
ROLB	59
ROR $ff,X	66
ROR $aaaa	76
RORA	46
RORB	56
ASL $ff,X	68
ASL $aaaa	78
ASLA	48
ASLB	58
ASR $ff,X	67
ASR $aaaa	77

Assembler	Op
ASRA	47
ASRB	57
LSR $ff,X	64
LSR $aaaa	74
LSRA	44
LSRB	54
Increment and Decrement Instructions	
INC $ff,X	6C
INC $aaaa	7C
INCA	4C
INCB	5C
DEC $ff,X	6A
DEC $aaaa	7A
DECA	4A
DECB	5A
INX	08
DEX	09
Unconditional Jump Instructions	
JMP $ff,X	6E
JMP $aaaa	7E
BRA $rr	20
Test (Compare) Instructions	
CMPA #$dd	81
CMPA $aa	91
CMPA $ff,X	A1
CMPA $aaaa	B1

Assembler	Op
CMPB #$dd	C1
CMPB $aa	D1
CMPB $ff,X	E1
CMPB $aaaa	F1
CBA	11
CPX #$dddd	8C
CPX $aa	9C
CPX $ff,X	AC
CPX $aaaa	BC
TST $ff,X	6D
TST $aaaa	7D
TSTA	4D
TSTB	5D
Conditional Jump (Branch) Instructions	
BCC $rr	24
BCS $rr	25
BEQ $rr	27
BGE $rr	2C
BGT $rr	2E
BHI $rr	22
BLE $rr	2F
BLS $rr	23
BLT $rr	2D
BMI $rr	2B
BNE $rr	26
BVC $rr	28
BVS $rr	29
BPL $rr	2A
Subroutine Instructions	
JSR $ff,X	AD
JSR $aaaa	BD

Assembler	Op
RTS	39
BSR $rr	8D
Stack Instructions	
LDS #$dddd	8E
LDS $aa	9E
LDS $ff,X	AE
LDS $aaaa	BE
STS $aa	9F
STS $ff,X	AF
STS $aaaa	BF
PSHA	36
PSHB	37
PULA	32
PULB	33
DES	34
INS	31
TXS	35
TSX	30
Interrupt Instructions	
RTI	3B
SWI	3F
Input-Output Instructions	
none	

CONDENSED TABLE OF 6800 INSTRUCTIONS LISTED ALPHABETICALLY

Assembler	Op	Assembler	Op	Assembler	Op	Assembler	Op
ABA	1B	BMI $rr	2B	INS	31	RORB	56
ADCA $aa	99	BNE $rr	26	INX	08	RTI	3B
ADCA $aaaa	B9	BPL $rr	2A	JMP $aaaa	7E	RTS	39
ADCA $ff,X	A9	BRA $rr	20	JMP $ff,X	6E	SBA	10
ADCA #$dd	89	BSR $rr	8D	JSR $aaaa	BD	SBCA $aa	92
ADCB $aa	D9	BVC $rr	28	JSR $ff,X	AD	SBCA $aaaa	B2
ADCB $aaaa	F9	BVS $rr	29	LDAA $aa	96	SBCA $ff,X	A2
ADCB $ff,X	E9	CBA	11	LDAA $aaaa	B6	SBCA #$dd	82
ADCB #$dd	C9	CLC	0C	LDAA $ff,X	A6	SBCB $aa	D2
ADDA $aa	9B	CLI	0E	LDAA #$dd	86	SBCB $aaaa	F2
ADDA $aaaa	BB	CLR $aaaa	7F	LDAB $aa	D6	SBCB $ff,X	E2
ADDA $ff,X	AB	CLR $ff,X	6F	LDAB $aaaa	F6	SBCB #$dd	C2
ADDA #$dd	8B	CLRA	4F	LDAB $ff,X	E6	SEC	0D
ADDB $aa	DB	CLRB	5F	LDAB #$dd	C6	SEI	0F
ADDB $aaaa	FB	CLV	0A	LDS $aa	9E	SEV	0B
ADDB $ff,X	EB	CMPA $aa	91	LDS $aaaa	BE	STAA $aa	97
ADDB #$dd	CB	CMPA $aaaa	B1	LDS $ff,X	AE	STAA $aaaa	B7
ANDA $aa	94	CMPA $ff,X	A1	LDS #$dddd	8E	STAA $ff,X	A7
ANDA $aaaa	B4	CMPA #$dd	81	LDX $aa	DE	STAB $aa	D7
ANDA $ff,X	A4	CMPB $aa	D1	LDX $aaaa	FE	STAB $aaaa	F7
ANDA #$dd	84	CMPB $aaaa	F1	LDX $ff,X	EE	STAB $ff,X	E7
ANDB $aa	D4	CMPB $ff,X	E1	LDX #$dd	CE	STS $aa	9F
ANDB $aaaa	F4	CMPB #$dd	C1	LSR $aaaa	74	STS $aaaa	BF
ANDB $ff,X	E4	COM $aaaa	73	LSR $ff,X	64	STS $ff,X	AF
ANDB #$dd	C4	COM $ff,X	63	LSRA	44	STX $aa	DF
ASL $aaaa	78	COMA	43	LSRB	54	STX $aaaa	FF
ASL $ff,X	68	COMB	53	NEG $aaaa	70	STX $ff,X	EF
ASLA	48	CPX $aa	9C	NEG $ff,X	60	SUBA $aa	90
ASLB	58	CPX $aaaa	BC	NEGA	40	SUBA $aaaa	B0
ASR $aaaa	77	CPX $ff,X	AC	NEGB	50	SUBA $ff,X	A0
ASR $ff,X	67	CPX #$dd	8C	NOP	01	SUBA #$dd	80
ASRA	47	DAA	19	ORAA $aa	9A	SUBB $aa	D0
ASRB	57	DEC $aaaa	7A	ORAA $aaaa	BA	SUBB $aaaa	F0
BCC $rr	24	DEC $ff,X	6A	ORAA $ff,X	AA	SUBB $ff,X	E0
BCS $rr	25	DECA	4A	ORAA #$dd	8A	SUBB #$dd	C0
BEQ $rr	27	DECB	5A	ORAB $aa	DA	SWI	3F
BGE $rr	2C	DES	34	ORAB $aaaa	FA	TAB	16
BGT $rr	2E	DEX	09	ORAB $ff,X	EA	TAP	06
BHI $rr	22	EORA $aa	98	ORAB #$dd	CA	TBA	17
BITA $aa	95	EORA $aaaa	B8	PSHA	36	TPA	07
BITA $aaaa	B5	EORA $ff,X	A8	PSHB	37	TST $aaaa	7D
BITA $ff,X	A5	EORA #$dd	88	PULA	32	TST $ff,X	6D
BITA #$dd	85	EORB $aa	D8	PULB	33	TSTA	4D
BITB $aa	D5	EORB $aaaa	F8	ROL $aaaa	79	TSTB	5D
BITB $aaaa	F5	EORB $ff,X	E8	ROL $ff,X	69	TSX	30
BITB $ff,X	E5	EORB #$dd	C8	ROLA	49	TXS	35
BITB #$dd	C5	INC $aaaa	7C	ROLB	59	WAI	3E
BLE $rr	2F	INC $ff,X	6C	ROR $aaaa	76		
BLS $rr	23	INCA	4C	ROR $ff,X	66		
BLT $rr	2D	INCB	5C	RORA	46		

CONDENSED TABLE OF 6800 INSTRUCTIONS LISTED BY OP CODE

Op	Assembler	Op	Assembler	Op	Assembler	Op	Assembler
01	NOP	49	ROLA	8C	CPX #$dd	C4	ANDB #$dd
06	TAP	4A	DECA	8D	BSR $rr	C5	BITB #$dd
07	TPA	4C	INCA	8E	LDS #$dddd	C6	LDAB #$dd
08	INX	4D	TSTA	90	SUBA $aa	C8	EORB #$dd
09	DEX	4F	CLRA	91	CMPA $aa	C9	ADCB #$dd
0A	CLV	50	NEGB	92	SBCA $aa	CA	ORAB #$dd
0B	SEV	53	COMB	94	ANDA $aa	CB	ADDB #$dd
0C	CLC	54	LSRB	95	BITA $aa	CE	LDX #$dd
0D	SEC	56	RORB	96	LDAA $aa	D0	SUBB $aa
0E	CLI	57	ASRB	97	STAA $aa	D1	CMPB $aa
0F	SEI	58	ASLB	98	EORA $aa	D2	SBCB $aa
10	SBA	59	ROLB	99	ADCA $aa	D4	ANDB $aa
11	CBA	5A	DECB	9A	ORAA $aa	D5	BITB $aa
16	TAB	5C	INCB	9B	ADDA $aa	D6	LDAB $aa
17	TBA	5D	TSTB	9C	CPX $aa	D7	STAB $aa
19	DAA	5F	CLRB	9E	LDS $aa	D8	EORB $aa
1B	ABA	60	NEG $ff,X	9F	STS $aa	D9	ADCB $aa
20	BRA $rr	63	COM $ff,X	A0	SUBA $ff,X	DA	ORAB $aa
22	BHI $rr	64	LSR $ff,X	A1	CMPA $ff,X	DB	ADDB $aa
23	BLS $rr	66	ROR $ff,X	A2	SBCA $ff,X	DE	LDX $aa
24	BCC $rr	67	ASR $ff,X	A4	ANDA $ff,X	DF	STX $aa
25	BCS $rr	68	ASL $ff,X	A5	BITA $ff,X	E0	SUBB $ff,X
26	BNE $rr	69	ROL $ff,X	A6	LDAA $ff,X	E1	CMPB $ff,X
27	BEQ $rr	6A	DEC $ff,X	A7	STAA $ff,X	E2	SBCB $ff,X
28	BVC $rr	6C	INC $ff,X	A8	EORA $ff,X	E4	ANDB $ff,X
29	BVS $rr	6D	TST $ff,X	A9	ADCA $ff,X	E5	BITB $ff,X
2A	BPL $rr	6E	JMP $ff,X	AA	ORAA $ff,X	E6	LDAB $ff,X
2B	BMI $rr	6F	CLR $ff,X	AB	ADDA $ff,X	E7	STAB $ff,X
2C	BGE $rr	70	NEG $aaaa	AC	CPX $ff,X	E8	EORB $ff,X
2D	BLT $rr	73	COM $aaaa	AD	JSR $ff,X	E9	ADCB $ff,X
2E	BGT $rr	74	LSR $aaaa	AE	LDS $ff,X	EA	ORAB $ff,X
2F	BLE $rr	76	ROR $aaaa	AF	STS $ff,X	EB	ADDB $ff,X
30	TSX	77	ASR $aaaa	B0	SUBA $aaaa	EE	LDX $ff,X
31	INS	78	ASL $aaaa	B1	CMPA $aaaa	EF	STX $ff,X
32	PULA	79	ROL $aaaa	B2	SBCA $aaaa	F0	SUBB $aaaa
33	PULB	7A	DEC $aaaa	B4	ANDA $aaaa	F1	CMPB $aaaa
34	DES	7C	INC $aaaa	B5	BITA $aaaa	F2	SBCB $aaaa
35	TXS	7D	TST $aaaa	B6	LDAA $aaaa	F4	ANDB $aaaa
36	PSHA	7E	JMP $aaaa	B7	STAA $aaaa	F5	BITB $aaaa
37	PSHB	7F	CLR $aaaa	B8	EORA $aaaa	F6	LDAB $aaaa
39	RTS	80	SUBA #$dd	B9	ADCA $aaaa	F7	STAB $aaaa
3B	RTI	81	CMPA #$dd	BA	ORAA $aaaa	F8	EORB $aaaa
3E	WAI	82	SBCA #$dd	BB	ADDA $aaaa	F9	ADCB $aaaa
3F	SWI	84	ANDA #$dd	BC	CPX $aaaa	FA	ORAB $aaaa
40	NEGA	85	BITA #$dd	BD	JSR $aaaa	FB	ADDB $aaaa
43	COMA	86	LDAA #$dd	BE	LDS $aaaa	FE	LDX $aaaa
44	LSRA	88	EORA #$dd	BF	STS $aaaa	FF	STX $aaaa
46	RORA	89	ADCA #$dd	C0	SUBB #$dd		
47	ASRA	8A	ORAA #$dd	C1	CMPB #$dd		
48	ASLA	8B	ADDA #$dd	C2	SBCB #$dd		

CPU Control Instructions

ESC

ESCape

The ESC instruction allows the 8086/8088 to pass instructions to the 8087 math coprocessor. The instructions for the coprocessor appear as a 6-bit code embedded in the escape instruction. The 8086/8088 performs a NOP while the 8087 executes the instruction. [Flags affected - none]

HLT

HaLT

The HLT instruction causes the 8086/8088 to stop fetching and executing instructions and enter a halt state. To exit from the halt state the microprocessor must receive a hardware reset or interrupt signal. [Flags affected - none]

LOCK

LOCK

LOCK is a prefix which can be used in front of 8086/8088 instructions. It prevents any other processors from gaining access to the systems buses during the following instruction. [Flags affected - none]

NOP

No **OP**eration

The NOP instruction simply uses up three clock cycles during which nothing is done and no flags are affected. It is useful 1) in programs requiring time delays, and 2) as a means to hold space open in programs so instructions can be added at a later date. [Flags affected - none]

WAIT

WAIT

The WAIT instruction causes the 8086/8088 to enter a wait state or idle condition during which no further processing occurs (except valid interrupts) until a signal is received on the TEST pin. [Flags affected - none]

Data Transfer Instructions

LAHF

Load **AH** from **F**lag

The LAHF instruction copies the low-order byte of the flag (status) register to AH. The flags themselves are not affected. The low order byte of the 8086/8088 status register is the same as that of the 8085. This instruction is used primarily to translate 8085 software into 8086/8088 software. [Flags affected - none]

LDS

Load **D**ata **S**egment

The LDS instruction performs two distinct operations. First it loads two consecutive bytes of memory into one of the 16-bit general, index, or pointer registers. Then it loads the next two consecutive bytes of memory into the 16-bit DS register.

For example, if DI=1000 then:

LDS BX,[DI]

copies the contents of memory locations 1000 and 1001 of the data segment into register BX and the contents of memory locations 1002 and 1003 of the data segment into register DS.

[Flags affected - none]

LEA

Load Effective Address

The LEA instruction loads one of the 16-bit general, index, or pointer registers from another register or memory.

Example:

LEA CX,[SI]

copies the number (address) in the SI register to the CX register.

[Flags affected - none]

LES

Load Extra Segment

The LES instruction performs two distinct operations. First it loads two consecutive bytes of memory into one of the 16-bit general, index, or pointer registers. Then it loads the next two consecutive bytes of memory into the 16-bit ES register.

For example, if DI=1000 then:

LES BX,[DI]

copies the contents of memory locations 1000 and 1001 of the data segment into register BX and the contents of memory locations 1002 and 1003 of the data segment into register ES. [Flags affected - none]

MOV

MOVe

The MOV instruction copies the contents of a register, memory location, or immediate number to a register or memory location. The source and destination must both be of the same length and both cannot be memory locations. [Flags affected - none]

SAHF

Store AH in Flags

The SAHF instruction copies AH to the low-order byte of the flag (status) register. The low-order byte of the 8086/8088 status register is the same as that of the 8085. This instruction is used primarily to translate 8085 software into 8086/8088 software. After this instruction is executed SF, ZF, AF, PF, and CF will correspond to bits 7, 6, 4, 2, and 1 of AH respectively. [Flags affected - SF, ZF, AF, PF, CF]

XCHG

eXCHanGe

The XCHG instruction exchanges the contents of two registers or a register and a memory location. Segment registers cannot be used nor can two memory locations. The source and destination must be of the same length. [Flags affected - none]

XLAT

trans(X)LATe

The XLAT instruction is used to look up values in a table. First the location of the beginning of the table must be loaded into the BX register. Then the relative location within the table of the desired value must be placed in the AL register. When the XLAT instruction is executed the value of BX is added to AL to form an address. The contents of that address then replaces the former value in AL. This instruction can be used to translate ASCII values into EBCDIC values for example. [Flags affected - none]

Flag Instructions

CLC

CLear Carry flag

The CLC instruction places a zero (0) in the carry flag bit of the status register. [Flags affected - CF=0]

CLD

CLear Direction flag (auto-increment)

The CLD instruction places a zero (0) in the direction flag bit of the status register. When this flag is cleared (0), SI and DI will automatically increment when certain string instructions are executed. [Flags affected - DF=0]

CLI

CLear Interrupt-enable flag

The CLI instruction places a zero (0) in the interrupt-enable flag bit of the status register. When this flag is cleared (0) the 8086/8088 will not respond to interrupt signals on the INTR pin. Signals on the NMI pin are not affected however. [Flags affected - IF=0]

CMC

CoMplement Carry flag

The CMC instruction inverts the carry flag bit of the status register. If the CF is 0, it will be changed to a 1. If it is a 1, it will be changed to 0. [Flags affected - CF]

STC

SeT Carry flag

The STC instruction places a one (1) in the carry flag bit of the status register. [Flags affected - CF=1]

STD

SeT Direction flag (auto-decrement)

The STD instruction places a one (1) in the direction flag bit of the status register. When this flag is set (1), SI and DI will automatically decrement when certain string instructions are executed. [Flags affected - DF=1]

STI

SeT Interrupt enable flag

The STI instruction places a one (1) in the interrupt-enable flag bit of the status register. When this flag is set (1) the 8086/8088 will respond to interrupt signals on the INTR pin. [Flags affected - IF=1]

Arithmetic Instructions

AAA

ASCII Adjust for Addition

The AAA instruction can be used after addition to adjust or alter the number in AL to what it would be if the last two operands were ASCII numbers. AH will be cleared. [Flags affected - AF, CF, OF (undefined), SF (undefined), ZF (undefined), PF (undefined)]

AAD

ASCII Adjust for Division

The AAD instruction is used before division by a single-digit, unpacked, BCD number. First you must have an unpacked, two-digit, BCD number in AX. The AAD instruction can then be used to adjust that number. This adjustment must occur **before** any division can take place. The adjustment changes the two-digit, unpacked, BCD number in AX into its equivalent binary number in AL. AH is changed to 00h. Next, AX can be divided by an 8-bit, single-digit, unpacked, BCD number. The binary quotient will be in AL with the binary remainder in AH. **Note: To use this instruction with ASCII numbers the "3" in the upper nibble must be masked out of the numbers first.** [Flags affected - SF, ZF, PF, OF (undefined), AF (undefined), CF (undefined)]

AAM

ASCII Adjust for Multiplication

The AAM instruction adjusts the product after multiplication of two, unpacked, single-digit, BCD numbers. To use this instruction you must have two single-digit, unpacked, BCD numbers. One must be in AL and the other in a register or memory location. After you multiply the two single-digit, unpacked, BCD numbers the binary answer will be in AL. The AAM instruction will convert it to its unpacked BCD equivalent. **Note: To use this instruction with ASCII numbers you must first mask the "3" in the upper nibble.** [Flags affected - SF, ZF, PF, AF (undefined), OF (undefined), CF (undefined)]

AAS

ASCII Adjust for Subtraction

The AAS instruction can be used after subtraction to adjust or alter the number in AL to what it would be if the last two operands were ASCII numbers. AH will be cleared. [Flags affected - AF, CF, OF (undefined), SF (undefined), ZF (undefined), PF (undefined)]

ADC

AdD with Carry

The ADC instruction works the same as the ADD instruction except that it adds the value in the carry flag (CF) to the sum of the two operands. [Flags affected - CF, PF, AF, ZF, SF, OF]

ADD

ADD

The ADD instruction adds a binary number in a source register, memory location, or immediate number to a destination binary number in a register or memory location. The result is placed in the destination location. The source and destination are assumed to be binary, both must be of the same size (byte or word), and both cannot be memory locations. [Flags affected - CF, PF, AF, ZF, SF, OF]

CBW

Convert Byte to Word

The CBW instruction takes bit 7 (the highest-order bit) of AL and duplicates it in every bit of AH. This converts an 8-bit signed-binary number in AL into a 16-bit signed-binary number in AX. This must be done before division (IDIV) involving two 8-bit **signed-binary** numbers to convert the dividend (in AL) into its 16-bit form (in AX). (For unsigned-binary numbers place 00H in AH.) It can also be used before integer multiplication (IMUL) involving an 8-bit operand and a 16-bit operand. The 8-bit operand can be converted to a 16-bit operand before the IMUL instruction is executed. [Flags affected - none]

CWD

Convert Word to Double word

The CWD instruction is similar to the CBW instruction except that it converts 16-bit values into 32-bit values instead of 8-bit to 16-bit. It takes bit 15 (the highest-order bit) of AX and duplicates it in every bit of DX. This converts a 16-bit signed-binary number in AX into a 32-bit signed-binary number in DX:AX (high 16 bits in DX, low 16 bits in AX). This must be done before division involving two 16-bit numbers to convert the dividend (in AX) into its 32-bit form (in DX:AX). [Flags affected - none]

DAA

Decimal Adjust for Addition

The DAA instruction adjusts the contents of AL from a binary number to a packed BCD (binary coded decimal) number when used after addition. When addition is performed the operands are assumed to be binary numbers. If they were in fact packed BCD numbers then the DAA instruction would have to be used after the addition to correct the result. Note that DAA only works on AL so each byte of a multi-byte packed BCD number must be moved into AL, added, adjusted, and then the result moved back out to make room for the next byte. [Flags affected - SF, ZF, AF, PF, CF, OF (undefined)]

DAS

Decimal Adjust for Subtraction

The DAS instruction adjusts the contents of AL from a binary number to a packed BCD (binary-coded-decimal) number when used after subtraction. When subtraction is performed the operands are assumed to be binary numbers. If they were in fact packed BCD numbers then the DAS instruction would have to be used after the subtraction to correct the result. Note that DAS only works on AL so each byte of a multi-byte packed BCD number must be moved into AL, subtracted, adjusted, and then the result moved back out to make room for the next byte. [Flags affected - SF, ZF, AF, PF, CF, OF (undefined)]

DIV

DIVide (unsigned)

The DIV instruction can divide a 16-bit unsigned-binary number in AX by an 8-bit unsigned-binary number in a register or memory location. If you want to divide one 8-bit number by another you must first change the dividend in AL into a 16-bit number by placing 00H in AH. After execution the result (quotient) will be in AL and the remainder in AH.

DIV can also divide a 32-bit unsigned-binary number in DX:AX (high-order word in DX, low-order word in AX) by a 16-bit unsigned-binary number in a register or memory location. If you wish to divide one 16-bit number by another you must first convert the dividend in AX into a 32-bit number in DX:AX by placing 0000H in DX. The result (quotient) will be in AX and the remainder in DX. [Flags affected - OF (undefined), SF (undefined), ZF (undefined), AF (undefined), PF (undefined), CF (undefined)]

IDIV

Integer DIVision (signed)

The IDIV instruction can divide a 16-bit signed-binary number in AX by an 8-bit signed-binary number in a register or memory location. The result (quotient) will be in AL and the remainder in AH. It can also divide a 32-bit signed-binary number in DX:AX (high-order word in DX, low-order word in AX) by a 16-bit signed-binary number in a register or memory location. The result (quotient) will be in AX and the remainder in DX. **Important! - See CBW and CWD.** [Flags affected - OF (undefined), SF (undefined), ZF (undefined), AF (undefined), PF (undefined), CF (undefined)]

IMUL

Integer MULtiplication (signed)

The IMUL instruction multiplies a signed binary number in a register or memory location times a signed number in AL if 8-bit or AX if 16-bit. If two 8-bit numbers are multiplied then a 16-bit answer will be found in AX. If two 16-bit numbers are multiplied then a 32-bit answer will be found in DX:AX (high byte in DX, low byte in AX). To multiply an 8-bit signed binary number by a 16-bit signed-binary number see the CBW instruction. [Flags affected - OF, CF, SF (undefined), ZF (undefined), AF (undefined), PF (undefined)]

MUL

MULtiply (unsigned)

The MUL instruction multiplies an unsigned binary number in a register or memory location times an unsigned number in AL if 8-bit or AX if 16-bit. If two 8-bit numbers are multiplied then a 16-bit answer will be found in AX. If two 16-bit numbers are multiplied then a 32-bit answer will be found in DX:AX (high byte in DX, low byte in AX). [Flags affected - OF, CF, SF (undefined), ZF (undefined), AF (undefined), PF (undefined)]

SBB

SuBtract with Borrow

The SBB instruction is the same as the SUB instruction except that the value in the carry flag (CF) is also subtracted. That is, the source (second operand) and CF are both subtracted from the destination (first operand). The source and destination must both be either 8-bit or 16-bit. All values are assumed to be binary. [Flags affected - OF, SF, ZF, AF, PF, CF]

SUB

SUBtract

The SUB instruction subtracts the contents of a source (the second operand in 8086/8088 assembly language) register, memory location, or an immediate number from the contents of a destination (the first operand in 8086/8088 assembly language) register or memory location. The result is placed in the destination location. The source and destination must both be of the same size (byte or word) and both cannot be memory locations. [Flags affected - CF, PF, AF, ZF, SF, OF]

Logical Instructions

AND

logical AND

The AND instruction performs a logical AND of each bit of the source and destination operands. The source (second operand in 8086/8088 assembly language) can be an immediate number, register, or memory location. The destination can be a register or memory location. Both source and destination cannot be memory locations. Both operands can be 8-bit or both can be 16-bit. Neither can be a segment register. After execution the source is unchanged but the destination will contain the result of the ANDing operation. [Flags affected - OF=0, SF, ZF, PF, CF=0, AF (undefined)]

NEG

NEGate (2's complement)

The NEG instruction produces the 2's complement of a binary number. This can be done manually by inverting each bit then adding one (1). This instruction is also essentially the same as subtracting the number from zero. [Flags affected - OF, SF, ZF, AF, PF, CF]

NOT

NOT

The NOT instruction inverts every bit of the operand. The operand can be in a register or memory location. [Flags affected - none]

OR

OR

The OR instruction performs a logical OR of each bit of the source and destination operands. The source (second operand in 8086/8088 assembly language) can be an immediate number, register, or memory location. The destination can be a register or memory location. Both source and destination cannot be memory locations. Both operands can be 8-bit or both can be 16-bit. Neither can be a segment register. After execution the source is unchanged but the destination will contain the result of the ORing operation. [Flags affected - OF=0, SF, ZF, PF, CF=0, AF (undefined)]

XOR

eXclusive **OR**

The XOR instruction performs a logical XOR of each bit of the source and destination operands. The source (second operand in 8086/8088 assembly language) can be an immediate number, register, or memory location. The destination can be a register or memory location. Both source and destination cannot be memory locations. Both operands can be 8-bit or both can be 16-bit. Neither can be a segment register. After execution the source is unchanged but the destination will contain the result of the XORing operation. [Flags affected - OF=0, SF, ZF, PF, CF=0, AF (undefined)]

Rotate and Shift Instructions

RCL

Rotate through Carry to the Left

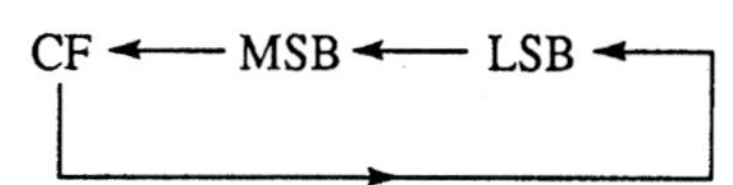

The RCL instruction rotates the bits of the destination as shown above. After an RCL instruction the destination will have rotated toward the left, the carry flag will hold the bit most recently rotated out of the MSB, and the LSB will hold the bit most recently rotated from the carry flag. The destination can be a register or memory location. If you want to rotate one bit position you specify a "1" in the instruction. If you want to rotate more than one bit position place the number of bits in the CL register and include that register in the instruction.

Examples:

RCL AX,1

rotates AX one bit position

RCL AX,CL

rotates AX the number of bit positions indicated by the number held in the CL register.

[Flags affected - OF, CF]

RCR

Rotate through **C**arry to the **R**ight

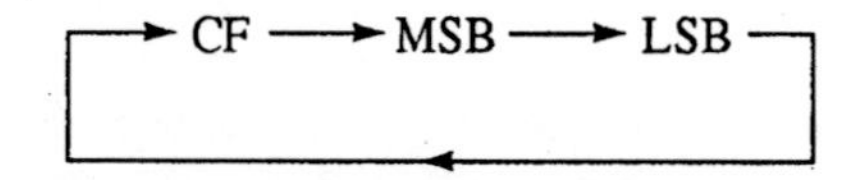

The RCR instruction rotates the bits of the destination as shown above. After an RCR instruction the destination will have rotated toward the right, the carry flag will hold the bit most recently rotated from the LSB, and the MSB will hold the bit most recently rotated from the carry flag. The destination can be a register or memory location. If you want to rotate one bit position you specify a "1" in the instruction. If you want to rotate more than one bit position place the number of bits in the CL register and include that register in the instruction.

Examples:

RCR AX,1

rotates AX one bit position

RCR AX,CL

rotates AX the number of bit positions indicated by the number held in the CL register.

[Flags affected - OF, CF]

ROL

ROtate **L**eft

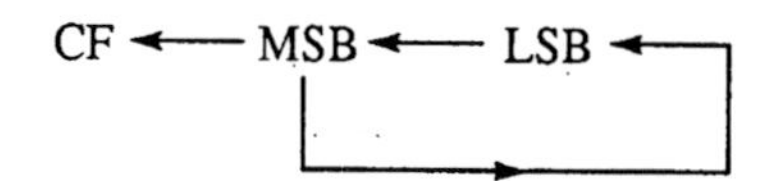

The ROL instruction rotates the bits of the destination as shown above. After an ROL instruction the destination will have rotated toward the left, and the carry flag and the LSB will both contain the same bit which was most recently rotated into them from the MSB. The destination can be a register or memory location. If you want to rotate one bit position you specify a "1" in the instruction. If you want to rotate more than one bit position place the number of bits in the CL register and include that register in the instruction.

Examples:

ROL AX,1

rotates AX one bit position

ROL AX,CL

rotates AX the number of bit positions indicated by the number held in the CL register.

[Flags affected - OF, CF]

ROR

ROtate Right

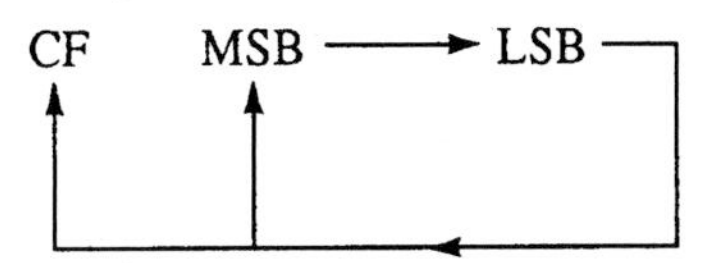

The ROR instruction rotates the bits of the destination as shown above. After an ROR instruction the destination will have rotated toward the right, and the carry flag and the MSB will both contain the same bit which was most recently rotated into them from the LSB. The destination can be a register or memory location. If you want to rotate one bit position you specify a "1" in the instruction. If you want to rotate more than one bit position place the number of bits in the CL register and include that register in the instruction.

Examples:

ROR AX,1
: rotates AX one bit position

ROR AX,CL
: rotates AX the number of bit positions indicated by the number held in the CL register.

[Flags affected - OF, CF]

SAL/SHL

Shift Arithmetic Left/SHift logical Left

CF ← MSB ← LSB ← 0

The SAL or SHL instruction shifts the bits of the destination as shown above. After an SAL/SHL instruction the destination will have shifted toward the left, the carry flag will contain the bit most recently shifted out of the MSB, and the LSB will contain a 0. The destination can be a register or memory location. If you want to rotate one bit position you specify a "1" in the instruction. If you want to rotate more than one bit position place the number of bits in the CL register and include that register in the instruction.

Examples:

SHL AX,1
: rotates AX one bit position

SHL AX,CL
: rotates AX the number of bit positions indicated by the number held in the CL register.

(DEBUG Note: DEBUG only allows the SHL mnemonic.)

[Flags affected - OF, SF, ZF, PF, CF, AF (undefined)]

SAR

Shift Arithmetic Right

```
┌──► MSB ──► LSB ──► CF
│     │
└─────┘
```

The SAR instruction shifts the bits of the destination as shown above. After an SAR instruction the destination will have shifted to the right, the MSB will contain what it did before the instruction (i.e., it duplicates itself and shifts a copy of itself to the right), and the carry flag will hold the bit most recently shifted out of the LSB. The destination can be a register or memory location. If you want to rotate one bit position you specify a "1" in the instruction. If you want to rotate more than one bit position place the number of bits in the CL register and include that register in the instruction.

Examples:

SAR AX,1

rotates AX one bit position

SAR AX,CL

rotates AX the number of bit positions indicated by the number held in the CL register.

[Flags affected - OF, SF, ZF, PF, CF, AF (undefined)]

SHR

SHift logical Right

0 ──► MSB ──► LSB ──► CF

The SHR instruction shifts the bits of the destination as shown above. After an SHR instruction the destination will have shifted toward the right, the MSB will contain a 0, and the carry flag will hold the bit most recently shifted in from the LSB. The destination can be a register or memory location. If you want to rotate one bit position you specify a "1" in the instruction. If you want to rotate more than one bit position place the number of bits in the CL register and include that register in the instruction.

Examples:

SHR AX,1

rotates AX one bit position

SHR AX,CL

rotates AX the number of bit positions indicated by the number held in the CL register.

[Flags affected - OF, SF, ZF, PF, CF, AF (undefined)]

Increment and Decrement Instructions

DEC

DECrement

The DEC instruction decreases the value in the destination by 1. The destination is assumed to be a binary number and can be a register (except a segment register) or memory location. It is worthwhile to note that the CF is not affected by this instruction. [Flags affected - OF, SF, ZF, AF, PF]

INC

INCrement

The INC instruction increases the value in the destination by 1. The destination is assumed to be a binary number and can be a register (except a segment register) or memory location. It is worthwhile to note that the CF is not affected by this instruction. [Flags affected - OF, SF, ZF, AF, PF]

Unconditional Jump Instructions

JMP

JuMP

JMP is an unconditional jump instruction which causes the 8086/8088 to continue executing instructions at some other place in the program. The jump can be classified as short, near, or far. The short and near instructions are relative to the current instruction pointer (IP) location. Since the IP always points to the next instruction to be executed you start counting forward or backward from the next instruction after the JMP instruction. A short jump can be up to a maximum of 127 memory bytes forward from the current IP position ($7E_{16}$ or $+127_{10}$) or up to 128 memory bytes backward from the current IP position (80_{16} or -128_{10}). A near jump can be anywhere within the current 64K code segment. The assembler will calculate this as being up to 32,767 bytes forward ($7FFF_{16}$ or $+32{,}767_{10}$) or 32,768 bytes backward (8000_{16} or $-32{,}768_{10}$) from the current IP position. A far jump can be anywhere in the 1-Mbyte addressing range of the 8086/8088. The far jump specifies both the desired code segment (CS) and the desired instruction pointer (IP). **DEBUG Note: When you want to JMP you do not need to be concerned about calculating the distance forward or backward from the current instruction pointer (IP) position. Simply specify the location you want to go to in the form**

JMP XXXX

where XXXX is the memory location (and therefore the desired instruction pointer value) for the short and near jumps and DEBUG will determine whether this is a forward or backward jump and will calculate the exact distance for you. Likewise if you want to use the value in a register as your destination simply specify that register and DEBUG will calculate the relative jump distance for you. In the case of the far jump specify the location you want to jump to in the form

JMP YYYY:XXXX

where YYYY is the code segment (CS) and XXXX is the instruction pointer (IP). [Flags affected - none]

Test (Compare) Instructions

CMP

CoMPare

The CMP instruction is used to compare two operands for the purpose of affecting flags according to the outcome. That is, the compare instruction subtracts the source operand (the second operand) from the destination (the first operand). Neither operand is changed; only the flags are affected. The source can be an immediate number, a register, or a memory location. The destination can be a register or memory location. Both operands cannot be memory locations. [Flags affected - OF, SF, ZF, AF, PF, CF]

TEST

TEST

The TEST instruction ANDs the source and destination operands but neither stores a result nor changes either operand. Rather, the flags are affected by the ANDing. This is useful before a conditional jump instruction. The source can be an immediate number, register, or memory location. The destination can be a register or memory location. Both operands cannot be memory locations. [Flags affected - OF=0, CF=0, SF, ZF, AF (undefined), PF (only lower 8 bits of destination)]

Conditional Jump (Branch) Instructions

JA/JNBE

Jump if Above/Jump if Not Below nor Equal

The JA/JNBE conditional jump instruction will cause program execution to transfer to another location in a range from +127 bytes to -128 bytes from the instruction following the jump instruction if CF=0 and if ZF=0 (both must be 0). If this condition is not true no jump occurs. When used after CMP, this instruction is referring to the unsigned values of the operands used by the CMP instruction. **DEBUG Note: Regardless of which mnemonic is used during assembly, DEBUG always disassembles this op code as JA.** [Flags affected - none]

JAE/JNB/JNC

Jump if Above or Equal/Jump if Not Below/Jump if No Carry

The JAE/JNB/JNC conditional jump instruction will cause program execution to transfer to another location in a range from +127 bytes to -128 bytes from the instruction following the jump instruction if CF=0. If this condition is not true no jump occurs. When used after CMP, this instruction is referring to the unsigned values of the operands used by the CMP instruction. **DEBUG Note: Regardless of which mnemonic is used during assembly, DEBUG always disassembles this op code as JNB.** [Flags affected - none]

JB/JNAE/JC

Jump if Below/Jump if Not Above nor Equal/Jump if Carry

The JB/JNAE/JC conditional jump instruction will cause program execution to transfer to another location in a range from +127 bytes to -128 bytes from the instruction following the jump instruction if CF=1. If this condition is not true no jump occurs. When used after CMP, this instruction is referring to the unsigned values of the operands used by the CMP instruction. **DEBUG Note: Regardless of which mnemonic is used during assembly, DEBUG always disassembles this op code as JB.** [Flags affected - none]

JBE/JNA

Jump if Below or Equal/Jump if Not Above

The JBE/JNA conditional jump instruction will cause program execution to transfer to another location in a range from +127 bytes to -128 bytes from the instruction following the jump instruction if CF=0 or ZF=1. If this condition is not true no jump occurs. When used after CMP, this instruction is referring to the unsigned values of the operands used by the CMP instruction. **DEBUG Note: Regardless of which mnemonic is used during assembly, DEBUG always disassembles this op code as JBE.** [Flags affected - none]

JCXZ

Jump if CX register is Zero

The JCXZ conditional jump instruction will cause program execution to transfer to another location in a range from +127 bytes to -128 bytes from the instruction following the jump instruction if the CX register is 0. If this condition is not true no jump occurs. [Flags affected - none]

JE/JZ

Jump if Equal to/Jump if Zero

The JE/JZ conditional jump instruction will cause program execution to transfer to another location in a range from +127 bytes to -128 bytes from the instruction following the jump instruction if ZF=1. If this condition is not true no jump occurs. When used after CMP, this instruction is referring to the values of the operands used by the CMP instruction. **DEBUG Note: Regardless of which mnemonic is used during assembly, DEBUG always disassembles this op code as JZ.** [Flags affected - none]

JG/JNLE

Jump if Greater/Jump if Not Less than nor Equal

The JG/JNLE conditional jump instruction will cause program execution to transfer to another location in a range from +127 bytes to -128 bytes from the instruction following the jump instruction if (SF XOR OF) OR ZF = 0. To say it another way, the jump occurs if the sign flag and the overflow flag are equal (both 0 or both 1) at the same time that the zero flag is 0. Only two combinations are possible. If SF=0, OF=0, and ZF=0 the jump occurs; or if SF=1, OF=1, and ZF=0 the jump also occurs. If this condition is not true no jump occurs. When used after CMP, this instruction is referring to the signed values of the operands used by the CMP instruction. **DEBUG Note: Regardless of which mnemonic is used during assembly, DEBUG always disassembles this op code as JG.** [Flags affected - none]

JGE/JNL

Jump if Greater than or Equal/Jump if Not Less

The JGE/JNL conditional jump instruction will cause program execution to transfer to another location in a range from +127 bytes to -128 bytes from the instruction following the jump instruction if SF=OF. If this condition is not true no jump occurs. When used after CMP, this instruction is referring to the signed values of the operands used by the CMP instruction. **DEBUG Note: Regardless of which mnemonic is used during assembly, DEBUG always disassembles this op code as JGE.** [Flags affected - none]

JL/JNGE

Jump if Less/Jump if Not Greater than nor Equal

The JGE/JNL conditional jump instruction will cause program execution to transfer to another location in a range from +127 bytes to -128 bytes from the instruction following the jump instruction if the SF does **not** equal the OF. If this condition is not true no jump occurs. When used after CMP, this instruction is referring to the signed values of the operands used by the CMP instruction. **DEBUG Note: Regardless of which mnemonic is used during assembly, DEBUG always disassembles this op code as JL.** [Flags affected - none]

JLE/JNG

Jump if Less than or Equal/Jump if Not Greater

The JLE/JNG conditional jump instruction will cause program execution to transfer to another location in a range from +127 bytes to -128 bytes from the instruction following the jump instruction if (SF XOR OF) OR ZF = 1. To say it another way, the jump occurs if the sign flag and the overflow flag are **not** equal, or if the zero flag is 0. Only two combinations do **not** produce the jump. If SF=0, OF=0, and ZF=0 then **no** jump occurs; or if SF=1, OF=1, and ZF=0 then **no** jump occurs. When used after CMP, this instruction is referring to the signed values of the operands used by the CMP instruction. **DEBUG Note: Regardless of which mnemonic is used during assembly, DEBUG always disassembles this op code as JLE.** [Flags affected - none]

JNE/JNZ

Jump if Not Equal to/Jump if Not Zero

The JNE/JNZ conditional jump instruction will cause program execution to transfer to another location in a range from +127 bytes to -128 bytes from the instruction following the jump instruction if ZF=0. If this condition is not true no jump occurs. When used after CMP, this instruction is referring to the values of the operands used by the CMP instruction. **DEBUG Note: Regardless of which mnemonic is used during assembly, DEBUG always disassembles this op code as JNZ.** [Flags affected - none]

JNO

Jump if Not Overflow

An overflow occurs when the result of a signed arithmetic operation is too large to fit in the register or memory location. The JNO conditional jump instruction will cause program execution to transfer to another location in a range from +127 bytes to -128 bytes from the instruction following the jump instruction if OF=0, that is, if an overflow has **not** occurred. If this condition is not true no jump occurs. [Flags affected - none]

JNP/JPO

Jump if Not Parity/Jump if Parity Odd

When the result of an operation which affects the parity flag has a result which has an odd number of 1s in it then the PF=0. The JNP/JPO conditional jump instruction will cause program execution to transfer to another location in a range from +127 bytes to -128 bytes from the instruction following the jump instruction if PF=0. If this condition is not true no jump occurs. **DEBUG Note: Regardless of which mnemonic is used during assembly, DEBUG always disassembles this op code as JPO.** [Flags affected - none]

JNS

Jump if Not Sign

The JNS conditional jump instruction will cause program execution to transfer to another location in a range from +127 bytes to -128 bytes from the instruction following the jump instruction if SF=0. If this condition is not true no jump occurs. Since a "0" in the sign flag occurs when the result of the last operation was a positive signed number, this instruction is essentially saying to jump if the last operation produced a positive signed result. [Flags affected - none]

JO

Jump if Overflow

An overflow occurs when the result of a signed arithmetic operation is too large to fit in the register or memory location. The JO conditional jump instruction will cause program execution to transfer to another location in a range from +127 bytes to -128 bytes from the instruction following the jump instruction if OF=1, that is, if an overflow has occurred. If this condition is not true no jump occurs. [Flags affected - none]

JP/JPE

Jump if Parity/Jump if Parity Even

When the result of an operation which affects the parity flag has a result which has an even number of 1s in it then the PF=1. The JP/JPE conditional jump instruction will cause program execution to transfer to another location in a range from +127 bytes to -128 bytes from the instruction following the jump instruction if PF=1. If this condition is not true no jump occurs. **DEBUG Note: Regardless of which mnemonic is used during assembly, DEBUG always disassembles this op code as JPE.** [Flags affected - none]

JS

Jump if Sign

The JS conditional jump instruction will cause program execution to transfer to another location in a range from +127 bytes to -128 bytes from the instruction following the jump instruction if SF=1. If this condition is not true no jump occurs. Since a "1" in the sign flag occurs when the result of the last operation was a negative signed number, this instruction is essentially saying to jump if the last operation produced a negative signed result. [Flags affected - none]

Subroutine Instructions

CALL

CALL procedure

The CALL instruction causes the 8086/8088 to leave its current location in the program and to begin executing a procedure (a small special purpose program or subroutine located in a different place in memory) and then automatically return after that procedure is finished. The call can be classified as near or far. The near instruction is relative to the current instruction pointer (IP) location. Since the IP always points to the next instruction to be executed you start counting forward or backward from the next instruction after the CALL instruction. A near call can be anywhere within the current 64K code segment. The assembler will calculate this as being up to 32,767 bytes forward ($7FFF_{16}$ or $+32,767_{10}$) or 32,768 bytes backward (8000_{16} or $-32,768_{10}$) from the current IP position. When a near call is executed the contents of the instruction pointer (IP) are pushed onto the stack so that the 8086/8088 will know where to return after the procedure has been finished. A far call can be anywhere in the 1-Mbyte addressing range of the 8086/8088. The far call specifies both the desired code segment (CS) and the desired instruction pointer

(IP). When a far call is executed the contents of both the instruction pointer (IP) and the code segment (CS) are pushed onto the stack so that the 8086/8088 will know where to return after the procedure has been finished. **DEBUG Note: When you want to CALL a procedure you do not need to be concerned about calculating the distance forward or backward from the current instruction pointer (IP) position. Simply specify the location of the procedure in the form**

CALL XXXX

where XXXX is the memory location (and therefore the desired instruction pointer value) for the near call and DEBUG will determine whether that location is forward or backward and will calculate the exact distance for you. Likewise if you want to use the value in a register as your destination simply specify that register and DEBUG will calculate the relative distance for you. In the case of a far call specify the location of the procedure in the form

CALL YYYY:XXXX

where YYYY is the code segment (CS) and XXXX is the instruction pointer (IP). (See also RETurn.) [Flags affected - none]

RET

RETurn from subroutine

The RET instruction is placed at the end of a procedure or subroutine. It marks the end of that procedure and causes the 8086/8088 to return to the instruction immediately following the CALL instruction which began this particular procedure. The 8086/8088 knows where to return because the CALL instruction pushed the contents of the instruction pointer (IP) onto the stack. The RET instruction pops the value of the IP from the stack and places it in the IP. In the case of a far call the return instruction pops both the IP value and the code segment (CS) value from the stack. **DEBUG Note: DEBUG accepts both RET and RETN as the mnemonics for a return from a near call. When disassembled both will appear as RET. To specify a return from a far call the mnemonic RETF must be used and it will be disassembled as RETF.** [Flags affected - none]

Stack Instructions

POP

POP from stack

The POP instruction copies the word at the top of the stack to the destination operand. The destination can be a general-purpose register, segment register, or two consecutive memory locations. (The CS register is illegal.) After the POP, the stack pointer (SP) is incremented by 2 to point to the new top-of-stack. [Flags affected - none]

POPF

POP Flags from stack

The POPF instruction copies the word at the top of the stack into the flag register, replacing the values of all flags. The stack pointer (SP) is then incremented by 2. (Using POPF and PUSHF provides a way to change the TF. There is no instruction for directly altering this flag.) [Flags affected - OF, DF, IF, TF, SF, ZF, AF, PF, CF]

PUSH

PUSH onto stack

The PUSH instruction decrements the stack pointer (SP) by 2 and then copies the source operand (word) to the new top-of-stack. The source can be a general-purpose register, segment register, or two consecutive memory locations. [Flags affected - none]

PUSHF

PUSH Flags onto stack

The PUSHF instruction decrements the stack pointer (SP) by 2 and then copies the flag register to the new top-of-stack. [Flags affected - none]

Interrupt Instructions

INT

INTerrupt

The INT instruction causes program execution to be transferred to a special type of routine whose address is pointed to by an interrupt vector. There are 256 interrupt vectors in memory locations 00000h to 003FFh. Each vector is 4 bytes in length and contains the address (CS:IP) of the routine which handles this particular type of interrupt. The INT operand is a decimal number from 0 through 255 which identifies which interrupt vector is to be used. The actual memory location of the interrupt is calculated by multiplying the operand by 4. That answer forms the decimal equivalent of the beginning of the four memory locations which hold the interrupt vector. When the INT instruction is executed the following occur:

1. The stack pointer is decremented by 2 and the flags are pushed onto the stack.
2. IF and TF are cleared.
3. The stack pointer is decremented by 2 and CS is pushed onto the stack.
4. The new CS is fetched from the interrupt vector and the interrupt vector + 1.
5. The stack pointer is decremented by 2 and IP is pushed onto the stack.
6. The new IP is fetched from the interrupt vector + 2 and the interrupt vector + 3.
7. Begin execution of the interrupt routine located at memory location CS:IP.

The routine will continue until a IRET instruction is encountered, at which point program execution will pick up where it left off immediately after the INT instruction. [Flags affected - IF and TF]

INTO

INTerrupt on **O**verflow

The INTO instruction initiates a software interrupt which is, in all respects, the same as that produced by the INT instruction except that the INTO instruction is conditional, and the operand cannot be specified but is automatically type 4. That is, the INTO instruction will branch to the interrupt routine only if OF=1 and there is no choice as to where the interrupt vector will come from. It will always be a type 4 interrupt which is held in the 4 bytes starting at memory location 10h. This instruction is most often used after arithmetic operations to handle any overflow conditions. See the discussion for the INT instruction for more details. [Flags affected - IF and TF]

IRET

Interrupt **RET**urn

The IRET instruction is used to return from an interrupt routine (whether a hardware or software interrupt). The IP, CS, and flags are all popped from the stack and program execution continues from the instruction immediately following the INT instruction. The IRET instruction has no operand. [Flags affected - OF, DF, IF, TF, SF, ZF, AF, PF, CF]

Input-Output Instructions

IN

INput

The IN instruction allows a byte or word to be acquired from an I/O device [source] and placed in AL (byte) or AX (word) [destination]. An I/O address [source operand] from 00h through FFh can be specified directly in the instruction. If an address larger than FFh is desired a 16-bit address can be placed in DX used as the source operand in the IN instruction. Only AX and AL can be used as destinations [destination operand] by the IN instruction.

Example:

IN AL,45 — copy a byte from I/O address 45h into AL

IN AX,78 — copy a word from I/O address 78h into AX

IN AL,DX — copy a byte from the I/O address pointed to by the contents of DX and place in AL

I/O port addresses F8h through FFh are reserved by Intel for future hardware and software products and should not be used for any other purpose. [Flags affected - none]

OUT

OUTput

The OUT instruction allows a byte or word to be sent from AL (byte) or AX (word) [source] to an I/O device [destination]. An I/O address [destination operand] from 00h through FFh can be specified directly in the instruction. If an address larger than FFh is desired a 16-bit address can be placed in DX used as the destination operand in the OUT instruction. Only AX and AL can be used as sources [source operand] by the OUT instruction.

Example:

OUT 45,AL — copy a byte from AL to I/O address 45h

OUT 78,AX copy a word from AX to I/O address 78h

OUT DX,AL copy a byte from AL to the I/O address pointed to by the contents of DX

I/O port addresses F8h through FFh are reserved by Intel for future hardware and software products and should not be used for any other purpose. [Flags affected - none]

String Instructions

CMPS/CMPSB/CMPSW **CoMpare Strings/CoMPare Strings Byte/CoMPare Strings Word**

The CMPS/CMPSB/CMPSW instruction is used to compare the contents of two memory bytes, two words, or two entire sections of memory. The SI (source index) is used to point to the source in the DS (data segment). The DI (destination index) is used to point to the destination in the ES (extra segment). The 8086/8088 makes the comparison by subtracting the destination from the source. Neither operand is changed by the comparison; only flags are affected. After the comparison DI and SI are automatically incremented (if DF=0) or decremented (if DF=1). The increment/decrement is 1 if the CMPB mnemonic is used or 2 if CMPW is used. The REP/REPE/REPZ and REPNE/REPNZ repeat prefixes can be used with this instruction to compare an entire section of memory. **DEBUG Note: Only the CMPSB and CMPSW mnemonics are accepted by DEBUG.** [Flags affected - OF, SF, ZF, AF, PF, CF]

LODS/LODSB/LODSW **LOaD String/LOaD String Byte/LOaD String Word**

The LODS/LODSB/LODSW instruction loads (copies) either a byte (LODSB) from the memory location pointed to by SI into AL, or a word (LODSW) from the memory location pointed to by SI into AX. SI is either automatically incremented by 1 (LODSB) or by 2 (LODSW) if DF=0, or SI is automatically decremented by 1 (LODSB) or by 2 (LODSW) if DF=1. The REP/REPE/REPZ and REPNE/REPNZ repeat prefixes can be used with this instruction. **DEBUG Note: DEBUG only accepts the LODSB and LODSW mnemonics.** [Flags affected - none].

MOVS/MOVSB/MOVSW **MOVe String/MOVe String Byte/MOVe String Word**

The MOVS/MOVSB/MOVSW instruction is used to transfer the contents of a block of memory to another area in memory. The SI (source index) is used to point to the source in the DS (data segment). The DI (destination index) is used to point to the destination in the ES (extra segment). After the move DI and SI are automatically incremented (if DF=0) or decremented (if DF=1). The increment/decrement is 1 if the MOVSB mnemonic is used or 2 if MOVSW is used. The REP/REPE/REPZ and REPNE/REPNZ repeat prefixes can be used with this instruction to move an entire section of memory. **DEBUG Note: Only the MOVSB and MOVSW mnemonics are accepted by DEBUG.** [Flags affected - none]

REP/REPE/REPZ

REPeat/**REP**eat if Equal/**REP**eat if Zero

REP/REPE/REPZ is a prefix which causes string instructions to be repeated the number of times indicated by the value in CX. Each time the string instruction is repeated CX is decremented by one. This continues 1) in the case of MOVS and STOS, **until** CX=0, or **2)** in the case of CMPS and SCAS, **until either** CX=0 **or** the compared bytes or words are **not** equal (ie. ZF=0). **DEBUG Note: REP, REPE, and REPZ are all mnemonics for the same op code and DEBUG disassembles all of them as REPZ.** [Flags affected - none]

REPNE/REPNZ

REPeat if Not Equal/**REP**eat if Not Zero

REPNE/REPNZ is a prefix which causes string instructions to be repeated the number of times indicated by the value in CX. Each time the string instruction is repeated CX is decremented by 1. This continues **1)** in the case of MOVS and STOS, **until** CX=0, or **2)** in the case of CMPS and SCAS, **until either** CX=0 **or** the compared bytes or words **are** equal (ie. ZF=1). **DEBUG Note: REPNE and REPNZ are mnemonics for the same op code and DEBUG disassembles all of them as REPNZ.** [Flags affected - none]

SCAS/SCASB/SCASW

SCAn String/**SCA**n String Byte/**SCA**n String Word

The SCAS/SCASB/SCASW instruction is used to check a string for the occurrence or non-occurrence of a particular byte or word. The instruction accomplishes this by subtracting the byte or word in the extra segment (ES) which is pointed to by DI from AL (if a byte) or AX (if a word). Neither the contents of the string nor those of AX/AL are changed; however the flags are affected by the operation. After the operation, DI is automatically incremented (if DF=0) or decremented (if DF=1). DI will be incremented or decremented by 1 for byte scans or by 2 for word scans. The REP/REPE/REPZ prefix can be used to scan for the non-occurrence of a byte or word. The REPNE/REPNZ prefix can be used to scan for the occurrence of a byte or word. **DEBUG Note: DEBUG only recognizes the SCASB and SCASW mnemonics.** [Flags affected - OF, SF, ZF, AF, PF, CF]

STOS/STOSB/STOSW

STOre String/**STO**re String Byte/**STO**re String Word

The STOS/STOSB/STOSW instruction copies a byte from AL or a word from AX to a memory location in the extra segment (ES) pointed to by DI. After the operation, DI is automatically incremented (if DF=0) or decremented (if DF=1). DI will be incremented or decremented by 1 for a byte store or by 2 for a word store. The REP/REPE/REPZ and REPNE/REPNZ repeat prefixes can be used with this instruction to store a certain value in a range of memory locations. **DEBUG Note: Only the STOSB and STOSW mnemonics are accepted by DEBUG.** [Flags affected - none]

Loop Instructions

LOOP

LOOP

The LOOP instruction provides a way to repeat a group of instructions the number of times indicated by the value in the CX register. The LOOP instruction unconditionally transfers program execution to a memory location in the range of -128 to +127 bytes from the address of the instruction immediately following the

LOOP instruction if CX > 0. Each time the LOOP instruction is executed CX is decremented by 1; then the value of CX is checked. If CX > 0, program execution will branch to the location indicated by the operand of the LOOP instruction. If CX = 0, the program does not branch and the instruction immediately following the LOOP instruction is executed next. As CX is decremented wraparound occurs from 0000h to FFFFh. [Flags affected - none]

LOOPE/LOOPZ

LOOP while **E**qual/**LOOP** while **Z**ero

The LOOPE/LOOPZ instruction provides a way to repeat a group of instructions the number of times indicated by the value in the CX register. The LOOPE/LOOPZ instruction transfers program execution to a memory location in the range of -128 to +127 bytes from the address of the instruction immediately following the LOOP instruction if CX > 0 **and** ZF=1. Each time the LOOP instruction is executed CX is decremented by 1; then the values of CX and ZF are checked. If CX > 0, program execution will branch to the location indicated by the operand of the LOOP instruction if ZF=1 also. If **either** CX = 0 **or** ZF=0, the program does not branch, and the instruction immediately following the LOOP instruction is executed next. As CX is decremented wraparound occurs from 0000h to FFFFh. [Flags affected - none]

LOOPNE/LOOPNZ

LOOP while **N**ot **E**qual/**LOOP** while **N**ot **Z**ero

The LOOPNE/LOOPNZ instruction provides a way to repeat a group of instructions the number of times indicated by the value in the CX register. The LOOPNE/LOOPNZ instruction transfers program execution to a memory location in the range of -128 to +127 bytes from the address of the instruction immediately following the LOOP instruction if CX > 0 **and** ZF=0. Each time the LOOP instruction is executed CX is decremented by 1; then the values of CX and ZF are checked. If CX > 0, program execution will branch to the location indicated by the operand of the LOOP instruction if ZF=0 also. If **either** CX = 0 **or** ZF=1, the program does not branch, and the instruction immediately following the LOOP instruction is executed next. As CX is decremented wraparound occurs from 0000h to FFFFh. [Flags affected - none]

CONDENSED TABLE OF 8086/8088 INSTRUCTIONS LISTED BY CATEGORY

CPU Control Instructions

ESC	**ESC**ape
HLT	**H**a**LT**
LOCK	**LOCK**
NOP	**N**o **OP**eration
WAIT	**WAIT**

Data Transfer Instructions

LAHF	**Load AH from Flag**
LDS	**Load Data Segment**
LEA	**Load Effective Address**
LES	**Load Extra Segment**
MOV	**MOVe**
SAHF	**Store AH in Flags**
XCHG	**eXCHanGe**
XLAT	**trans(X)LATe**

Flag Instructions

CLC	**CLear Carry flag**
CLD	**CLear Direction flag (auto-increment)**
CLI	**CLear Interrupt-enable flag**
CMC	**CoMplement Carry flag**
STC	**SeT Carry flag**
STD	**SeT Direction flag (auto-decrement)**
STI	**SeT Interrupt enable flag**

Arithmetic Instructions

AAA	**ASCII Adjust for Addition**
AAD	**ASCII Adjust for Division**
AAM	**ASCII Adjust for Multiplication**
AAS	**ASCII Adjust for Subtraction**
ADC	**AdD with Carry**
ADD	**ADD**
CBW	**Convert Byte to Word**
CWD	**Convert Word to Double word**
DAA	**Decimal Adjust for Addition**
DAS	**Decimal Adjust for Subtraction**
DIV	**DIVide (unsigned)**
IDIV	**Integer DIVision (signed)**
IMUL	**Integer MULtiplication (signed)**
MUL	**MULtiply (unsigned)**
SBB	**SuBtract with Borrow**
SUB	**SUBtract**

Logical Instructions

AND	**logical AND**
NEG	**NEGate (2's complement)**
NOT	**NOT**
OR	**OR**
XOR	**eXclusive OR**

Rotate and Shift Instructions

RCL	Rotate through Carry to the Left
RCR	Rotate through Carry to the Right
ROL	ROtate Left
ROR	ROtate Right
SAL/SHL	Shift Arithmetic Left/SHift logical LefT
SAR	Shift Arithmetic Right
SHR	SHift logical Right

Increment and Decrement Instructions

DEC	DECrement
INC	INCrement

Unconditional Jump Instructions

JMP	JuMP

Test (Compare) Instructions

CMP	CoMPare
TEST	TEST

Conditional Jump (Branch) Instructions

JA/JNBE	Jump if Above/Jump if Not Below nor Equal
JAE/JNB/JNC	Jump if Above or Equal/Jump if Not Below/Jump if No Carry
JB/JNAE/JC	Jump if Below/Jump if Not Above nor Equal/Jump if Carry
JBE/JNA	Jump if Below or Equal/Jump if Not Above
JCXZ	Jump if CX register is Zero
JE/JZ	Jump if Equal to/Jump if Zero
JG/JNLE	Jump if Greater/Jump if Not Less than nor Equal
JGE/JNL	Jump if Greater than or Equal/Jump if Not Less
JL/JNGE	Jump if Less/Jump if Not Greater than nor Equal
JLE/JNG	Jump if Less than or Equal/Jump if Not Greater
JNE/JNZ	Jump if Not Equal to/Jump if Not Zero
JNO	Jump if Not Overflow
JNP/JPO	Jump if Not Parity/Jump if Parity Odd
JNS	Jump if Not Sign
JO	Jump if Overflow
JP/JPE	Jump if Parity/Jump if Parity Even
JS	Jump if Sign

Subroutine Instructions

CALL	**CALL** procedure
RET	**RET**urn from subroutine

Stack Instructions

POP	**POP** from stack
POPF	**POP** **F**lags from stack
PUSH	**PUSH** onto stack
PUSHF	**PUSH** **F**lags onto stack

Interrupt Instructions

INT	**INT**errupt
INTO	**INT**errupt on **O**verflow
IRET	**I**nterrupt **RET**urn

Input-Output Instructions

IN	**IN**put
OUT	**OUT**put

String Instructions

CMPS/CMPSB/CMPSW	**C**o**M**pare **S**trings/**C**o**MP**are **S**trings **B**yte/**C**o**MP**are **S**trings **W**ord
LODS/LODSB/LODSW	**LO**a**D** **S**tring/**LO**a**D** **S**tring **B**yte/**LO**a**D** **S**tring **W**ord
MOVS/MOVSB/MOVSW	**MOV**e **S**tring/**MOV**e **S**tring **B**yte/**MOV**e **S**tring **W**ord
REP/REPE/REPZ	**REP**eat/**REP**eat if **E**qual/**REP**eat if **Z**ero
REPNE/REPNZ	**REP**eat if **N**ot **E**qual/**REP**eat if **N**ot **Z**ero
SCAS/SCASB/SCASW	**SCA**n **S**tring/**SCA**n **S**tring **B**yte/**SCA**n **S**tring **W**ord
STOS/STOSB/STOSW	**STO**re **S**tring/**STO**re **S**tring **B**yte/**STO**re **S**tring **W**ord

Loop Instructions

LOOP	**LOOP**
LOOPE/LOOPZ	**LOOP** while **E**qual/**LOOP** while **Z**ero
LOOPNE/LOOPNZ	**LOOP** while **N**ot **E**qual/**LOOP** while **N**ot **Z**ero

CONDENSED TABLE OF 8086/8088 INSTRUCTIONS LISTED ALPHABETICALLY

AAA	ASCII Adjust for Addition
AAD	ASCII Adjust for Division
AAM	ASCII Adjust for Multiplication
AAS	ASCII Adjust for Subtraction
ADC	AdD with Carry
ADD	ADD
AND	logical AND
CALL	CALL procedure
CBW	Convert Byte to Word
CLC	CLear Carry flag
CLD	CLear Direction flag (auto-increment)
CLI	CLear Interrupt-enable flag
CMC	CoMplement Carry flag
CMP	CoMPare
CMPS/CMPSB/CMPSW	CoMpare Strings/CoMPare Strings Byte/CoMPare Strings Word
CWD	Convert Word to Double word
DAA	Decimal Adjust for Addition
DAS	Decimal Adjust for Subtraction
DEC	DECrement
DIV	DIVide (unsigned)
ESC	ESCape
HLT	HaLT
IDIV	Integer DIVision (signed)
IMUL	Integer MULtiplication (signed)
IN	INput
INC	INCrement
INT	INTerrupt
INTO	INTerrupt on Overflow
IRET	Interrupt RETurn
JA/JNBE	Jump if Above/Jump if Not Below nor Equal
JAE/JNB/JNC	Jump if Above or Equal/Jump if Not Below/Jump if No Carry
JB/JNAE/JC	Jump if Below/Jump if Not Above nor Equal/Jump if Carry
JBE/JNA	Jump if Below or Equal/Jump if Not Above
JCXZ	Jump if CX register is Zero
JE/JZ	Jump if Equal to/Jump if Zero
JG/JNLE	Jump if Greater/Jump if Not Less than nor Equal
JGE/JNL	Jump if Greater than or Equal/Jump if Not Less
JL/JNGE	Jump if Less/Jump if Not Greater than nor Equal
JLE/JNG	Jump if Less than or Equal/Jump if Not Greater
JMP	JuMP unconditional
JNE/JNZ	Jump if Not Equal to/Jump if Not Zero
JNO	Jump if Not Overflow
JNP/JPO	Jump if Not Parity/Jump if Parity Odd
JNS	Jump if Not Sign
JO	Jump if Overflow
JP/JPE	Jump if Parity/Jump if Parity Even
JS	Jump if Sign
LAHF	Load AH from Flag
LDS	Load Data Segment

CONDENSED TABLE OF 8086/8088 INSTRUCTIONS LISTED ALPHABETICALLY (*Continued*)

LEA	**Load Effective Address**
LES	**Load Extra Segment**
LOCK	**LOCK**
LODS/LODSB/LODSW	**LOaD** String/**LOaD** String Byte/**LOaD** String Word
LOOP	**LOOP**
LOOPE/LOOPZ	**LOOP** while Equal/**LOOP** while Zero
LOOPNE/LOOPNZ	**LOOP** while Not Equal/**LOOP** while Not Zero
MOV	**MOVe**
MOVS/MOVSB/MOVSW	**MOVe** String/**MOVe** String Byte/**MOVe** String Word
MUL	**MULtiply** (unsigned)
NEG	**NEGate** (2's complement)
NOP	**No OPeration**
NOT	**NOT**
OR	**OR**
OUT	**OUTput**
POP	**POP** from stack
POPF	**POP Flags** from stack
PUSH	**PUSH** onto stack
PUSHF	**PUSH Flags** onto stack
RCL	**Rotate** through **Carry** to the **Left**
RCR	**Rotate** through **Carry** to the **Right**
REP/REPE/REPZ	**REPeat**/**REPeat** if **Equal**/**REPeat** if **Zero**
REPNE/REPNZ	**REPeat** if **Not Equal**/**REPeat** if **Not Zero**
RET	**RETurn** from subroutine
ROL	**ROtate Left**
ROR	**ROtate Right**
SAHF	**Store AH** in **Flags**
SAL/SHL	**Shift Arithmetic Left**/**SHift** logical **Left**
SAR	**Shift Arithmetic Right**
SBB	**SuBtract** with **Borrow**
SCAS/SCASB/SCASW	**SCAn String**/**SCAn String Byte**/**SCAn String Word**
SHR	**SHift** logical **Right**
STC	**SeT Carry** flag
STD	**SeT Direction** flag (auto-decrement)
STI	**SeT Interrupt** enable flag
STOS/STOSB/STOSW	**STOre String**/**STOre String Byte**/**STOre String Word**
SUB	**SUBtract**
TEST	**TEST**
WAIT	**WAIT**
XCHG	**eXCHanGe** (source with destination)
XLAT	trans(X)**LATe**
XOR	**eXclusive OR**

EXPANDED TABLE OF 6502 INSTRUCTIONS LISTED BY CATEGORY

Mne-monic	Operation	Boolean/Arith Operation	Flags NV-BDIZC	Address Mode	Assembler Notation	Op	~	#	Notes
				CPU Control Instructions					
NOP	No OPeration	Nothing	xx-xxxxx	Implied	NOP	EA	2	1	
BRK	BReaK (forced interrupt)	PC + 2 → S SP - 2 → SP PSR → S SP - 1 → S $FFFE → PC	xx-1x1xx	Implied	BRK	00	7	1	
				Data Transfer Instructions					
LDA	LoaD Accumulator	M → A	Nx-xxxZx	Immediate	LDA #$dd	A9	2	2	
				Absolute	LDA $aaaa	AD	4	3	
				Zero Page	LDA $aa	A5	3	2	
				Indxd Indct	LDA ($ff,X)	A1	6	2	
				Indct Indxd	LDA ($aa),Y	B1	5*	2	
				Zero page,X	LDA $ff,X	B5	4	2	
				Absolute,X	LDA $ffff,X	BD	4*	3	
				Absolute,Y	LDA $ffff,Y	B9	4*	3	
LDX	LoaD X register	M → X	Nx-xxxZx	Immediate	LDX #$dd	A2	2	2	
				Absolute	LDX $aaaa	AE	4	3	
				Zero page	LDX $aa	A6	3	2	
				Absolute,Y	LDX $ffff,Y	BE	4*	3	
				Zero page,Y	LDX $ff,Y	B6	4	2	
LDY	LoaD Y register	M → Y	Nx-xxxZx	Immediate	LDY #$dd	A0	2	2	
				Absolute	LDY $aaaa	AC	4	3	
				Zero page	LDY $aa	A4	3	2	
				Zero page,X	LDY $ff,X	B4	4	2	
				Absolute,X	LDY $ffff,X	BC	4*	3	
STA	STore Accumulator	A → M	xx-xxxxx	Absolute	STA $aaaa	8D	4	3	
				Zero page	STA $aa	85	3	2	
				Indxd Indct	STA ($ff,X)	81	6	2	
				Indct Indxd	STA ($aa),Y	91	6	2	
				Zero page,X	STA $ff,X	95	4	2	
				Absolute,X	STA $ffff,X	9D	5	3	
				Absolute,Y	STA $ffff,Y	99	5	3	
STX	STore X register	X → M	xx-xxxxx	Absolute	STX $aaaa	8E	4	3	
				Zero page	STX $aa	86	3	2	
				Zero page,Y	STX $ff,Y	96	4	2	
STY	STore Y register	Y → M	xx-xxxxx	Absolute	STY $aaaa	8C	4	3	
				Zero page	STY $aa	84	3	2	
				Zero page,X	STY $ff,X	94	4	2	
TAX	Transfer Accumulator to X register	A → X	Nx-xxxZx	Implied	TAX	AA	2	1	
TXA	Transfer X register to Accumulator	X → A	Nx-xxxZx	Implied	TXA	8A	2	1	

EXPANDED TABLE OF 6502 INSTRUCTIONS LISTED BY CATEGORY (*Continued*)

Mne-monic	Operation	Boolean/Arith Operation	Flags NV-BDIZC	Address Mode	Assembler Notation	Op	~	#	Notes
TAY	Transfer Accumulator to Y register	A → Y	Nx-xxxZx	Implied	TAY	A8	2	1	
TYA	Transfer Y register to Accumulator	Y → A	Nx-xxxZx	Implied	TYA	98	2	1	
				Flag Instructions					
CLC	CLear Carry flag	0 → C	xx-xxxx0	Implied	CLC	18	2	1	
CLD	CLear Decimal flag	0 → D	xx-x0xxx	Implied	CLD	D8	2	1	
CLI	CLear Interrupt flag	0 → I	xx-xx0xx	Implied	CLI	58	2	1	
CLV	CLear oVerflow flag	0 → V	x0-xxxxx	Implied	CLV	B8	2	1	
SEC	SEt Carry flag	1 → C	xx-xxxx1	Implied	SEC	38	2	1	
SED	SEt Decimal flag	1 → D	xx-x1xxx	Implied	SED	F8	2	1	
SEI	SEt Interrupt flag	1 → I	xx-xx1xx	Implied	SEI	78	2	1	
				Arithmetic Instructions					
ADC	AdD with Carry	A + M + C → A	NV-xxxZC	Immediate	ADC #$dd	69	2	2	The carry flag must be cleared before single-precision addition or before the first byte of multiple-precision addition.
				Absolute	ADC $aaaa	6D	4	3	
				Zero page	ADC $aa	65	3	2	
				Indxd Indct	ADC ($ff,X)	61	6	2	
				Indct Indxd	ADC ($aa),Y	71	5*	2	
				Zero page,X	ADC $ff,X	75	4	2	
				Absolute,X	ADC $ffff,X	7D	4*	3	
				Absolute,Y	ADC $ffff,Y	79	4*	3	
SBC	SuBtract with Carry	A - M - (1-C) → A	NV-xxxZC	Immediate	SBC #$dd	E9	2	2	The carry flag must be set before single-precision subtraction or before the first byte of multiple-precision subtraction.
				Absolute	SBC $aaaa	ED	4	3	
				Zero page	SBC $aa	E5	3	2	
		Note: (1-C) = Borrow		Indxd Indct	SBC ($ff,X)	E1	6	2	
				Indct Indxd	SBC ($aa),Y	F1	5*	2	
				Zero page,X	SBC $ff,X	F5	4	2	
				Absolute,X	SBC $ffff,X	FD	4*	3	The operation of the carry flag is inverted during subtraction.
				Absolute,Y	SBC $ffff,Y	F9	4*	3	
				Logical Instructions					
AND	logical AND	A AND M → A	Nx-xxxZx	Immediate	AND #$dd	29	2	2	
				Absolute	AND $aaaa	2D	4	3	
				Zero page	AND $aa	25	3	2	
				Indxd Indct	AND ($ff,X)	21	6	2	
				Indct Indxd	AND ($aa),Y	31	5	2	
				Zero page,X	AND $ff,X	35	4	2	
				Absolute,X	AND $ffff,X	3D	4*	3	
				Absolute,Y	AND $ffff,Y	39	4*	3	

Mnemonic	Operation	Boolean/Arith Operation	Flags NV-BDIZC	Address Mode	Assembler Notation	Op	~	#	Notes
EOR	Exclusive **OR**	A EOR M → A	Nx-xxxZx	Immediate	EOR #$dd	49	2	2	
				Absolute	EOR $aaaa	4D	4	3	
				Zero page	EOR $aa	45	3	2	
				Indxd Indct	EOR ($ff,X)	41	6	2	
				Indct Indxd	EOR ($aa),Y	51	5*	2	
				Zero page,X	EOR $ff,X	55	4	2	
				Absolute,X	EOR $ffff,X	5D	4*	3	
				Absolute,Y	EOR $ffff,Y	59	4*	3	
ORA	**OR** Accumulator	A OR M → A	Nx-xxxZx	Immediate	ORA #$dd	09	2	2	
				Absolute	ORA $aaaa	0D	4	3	
				Zero page	ORA $aa	05	3	2	
				Indxd Indct	ORA ($ff,X)	01	6	2	
				Indct Indxd	ORA ($aa),Y	11	5	2	
				Zero page,X	ORA $ff,X	15	4	2	
				Absolute,X	ORA $ffff,X	1D	4*	3	
				Absolute,Y	ORA $ffff,Y	19	4*	3	
BIT	test memory **BITs**	A AND M $M_7 \rightarrow N$ $M_6 \rightarrow V$	76-xxxZx	Absolute	BIT $aaaa	2C	4	3	Memory bits 7 and 6 are transferred into the N and V flags respectively.
				Zero page	BIT $aa	24	3	2	
				Rotate and Shift Instructions					
ASL	Arithmetic Shift Left	C ← 7...0 ← 0	Nx-xxxZC	Absolute	ASL $aaaa	0E	6	3	
				Zero page	ASL $aa	06	5	2	
				Accumulator	ASL A	0A	2	1	
				Zero page,X	ASL $ff,X	16	6	2	
				Absolute,X	ASL $ffff,X	1E	7	3	
LSR	Logical Shift Right	0 → 7...0 → C	0x-xxxZC	Absolute	LSR $aaaa	4E	6	3	
				Zero page	LSR $aa	46	5	2	
				Accumulator	LSR A	4A	2	1	
				Zero page,X	LSR $ff,X	56	6	2	
				Absolute,X	LSR $ffff,X	5E	7	3	
ROL	ROtate Left	7...0 ← C ← (bit 7)	Nx-xxxZC	Absolute	ROL $aaaa	2E	6	3	
				Zero page	ROL $aa	26	5	2	
				Accumulator	ROL A	2A	2	1	
				Zero page,X	ROL $ff,X	36	6	2	
				Absolute,X	ROL $ffff,X	3E	7	3	
ROR	ROtate Right	(bit 0) → C → 7...0	Nx-xxxZC	Absolute	ROR $aaaa	6E	6	3	
				Zero page	ROR $aa	66	5	2	
				Accumulator	ROR A	6A	2	1	
				Zero page,X	ROR $ff,X	76	6	2	
				Absolute,X	ROR $ffff,X	7E	7	3	
				Increment and Decrement Instructions					
INC	INCrement memory	M + 1 → M	Nx-xxxZx	Absolute	INC $aaaa	EE	6	3	
				Zero page	INC $aa	E6	5	2	
				Zero page,X	INC $ff,X	F6	6	2	
				Absolute,X	INC $ffff,X	FE	7	3	

EXPANDED TABLE OF 6502 INSTRUCTIONS LISTED BY CATEGORY (*Continued*)

Mne-monic	Operation	Boolean/Arith Operation	Flags NV-BDIZC	Address Mode	Assembler Notation	Op	~	#	Notes
INX	INcrement X register	X + 1 → X	Nx-xxxZx	Implied	INX	E8	2	1	
INY	INcrement Y register	Y + 1 → Y	Nx-xxxZx	Implied	INY	C8	2	1	
DEC	DECrement memory	M - 1 → M	Nx-xxxZx	Absolute	DEC $aaaa	CE	6	3	
				Zero page	DEC $aa	C6	5	2	
				Zero page,X	DEC $ff,X	D6	6	2	
				Absolute,X	DEC $ffff,X	DE	7	3	
DEX	DEcrement X register	X - 1 → X	Nx-xxxZx	Implied	DEX	CA	2	1	
DEY	DEcrement Y register	Y - 1 → Y	Nx-xxxZx	Implied	DEY	88	2	1	

Unconditional Jump Instructions

Mne-monic	Operation	Boolean/Arith Operation	Flags NV-BDIZC	Address Mode	Assembler Notation	Op	~	#	Notes
JMP	JuMP to new memory location	aaaa → PC {abs addressing}	xx-xxxxx	Absolute	JMP $aaaa	4C	3	3	In the indirect addressing mode, aaaa is not transferred into the PC but rather the contents of memory location aaaa and aaaa+1 are placed in the PC.
		(aaaa) → PC_L (aaaa + 1) → PC_H {indirect addressing}		Indirect	JMP ($aaaa)	6C	5	3	**Special Note:** Care should be used with this mode because of a bug in the 6502 chip family. If the indirect address is located at a page boundary {example, JMP ($5FFF)} an incorrect address will be generated.

Test (Compare) Instructions

Mne-monic	Operation	Boolean/Arith Operation	Flags NV-BDIZC	Address Mode	Assembler Notation	Op	~	#	Notes
CMP	CoMPare memory location to accumulator	A - M	Nx-xxxZC	Immediate	CMP #$dd	C9	2	2	
				Absolute	CMP $aaaa	CD	4	3	
				Zero page	CMP $aa	C5	3	2	
				Indxd Indct	CMP ($ff,X)	C1	6	2	
				Indct Indxd	CMP ($aa),Y	D1	5*	2	
				Zero page,X	CMP $ff,X	D5	4	2	
				Absolute,X	CMP $ffff,X	DD	4*	3	
				Absolute,Y	CMP $ffff,Y	D9	4*	3	
CPX	ComPare memory location to X register	X - M	Nx-xxxZC	Immediate	CPX #$dd	E0	2	2	
				Absolute	CPX $aaaa	EC	4	3	
				Zero page	CPX $aa	E4	3	2	
CPY	ComPare memory location to Y register	Y - M	Nx-xxxZC	Immediate	CPY #$dd	C0	2	2	
				Absolute	CPY $aaaa	CC	4	3	
				Zero page	CPY $aa	C4	3	2	

Mne-monic	Operation	Boolean/Arith Operation	Flags NV-BDIZC	Address Mode	Assembler Notation	Op	~	#	Notes
			Conditional Jump (Branch) Instructions						
BCC	Branch if Carry Clear	PC + rr → PC if C=0	xx-xxxxx	Relative	BCC $rr	90	2+2		
BCS	Branch if Carry Set	PC + rr → PC if C=1	xx-xxxxx	Relative	BCS $rr	B0	2+2		
BEQ	Branch if last result EQual to zero	PC + rr → PC if Z=1	xx-xxxxx	Relative	BEQ $rr	F0	2+2		
BNE	Branch if last result Not Equal to zero	PC + rr → PC if Z=0	xx-xxxxx	Relative	BNE $rr	D0	2+2		
BMI	Branch if last result a MInus (neg) number	PC + rr → PC if N=1	xx-xxxxx	Relative	BMI $rr	30	2+2		
BPL	Branch is last result a PLus (pos) number	PC + rr → PC if N=0	xx-xxxxx	Relative	BPL $rr	10	2+2		
BVC	Branch if oVerflow flag Clear	PC + rr → PC if V=0	xx-xxxxx	Relative	BVC $rr	50	2+2		
BVS	Branch if oVerflow flag Set	PC + rr → PC if V=1	xx-xxxxx	Relative	BVS $rr	70	2+2		
			Subroutine Instructions						
JSR	Jump to SubRoutine	PC + 2 → S aaaa → PC SP - 2 → SP	xx-xxxxx	Absolute	JSR $aaaa	20	6	3	
RTS	ReTurn from Subroutine	S (2 bytes) → PC PC + 1 → PC SP + 2 → SP	xx-xxxxx	Implied	RTS	60	6	1	
			Stack Instructions						
PHA	PusH Accumulator onto stack	A → S SP - 1 → SP	xx-xxxxx	Implied	PHA	48	3	1	
PLA	PulL Accumulator from stack	S → A SP + 1 → SP	Nx-xxxZx	Implied	PLA	68	4	1	
PHP	PusH Processor status register onto stack	PSR → S SP - 1 → SP	xx-xxxxx	Implied	PHP	08	3	1	

EXPANDED TABLE OF 6502 INSTRUCTIONS LISTED BY CATEGORY *(Continued)*

Mne-monic	Operation	Boolean/Arith Operation	Flags NV-BDIZC	Address Mode	Assembler Notation	Op	~	#	Notes
PLP	PulL Processor status register from stack	S → PSR SP + 1 → SP	NV-BDIZC	Implied	PLP	28	4	1	
TXS	Transfer X register into Stack pointer	X → SP	xx-xxxxx	Implied	TXS	9A	2	1	
TSX	Transfer Stack pointer into X register	SP → X	Nx-xxxZx	Implied	TSX	BA	2	1	
			Interrupt Instructions						
RTI	ReTurn from Interrupt	S → PSR SP + 1 → SP S (2 bytes) → PC SP + 2 → SP	NV-BDIZC	Implied	RTI	40	6	1	
			Input-Output Instructions						
none									The 6502 memory-maps all input and output rather than using special instructions.

Notes

Address Modes	Assembler Notation
Immediate	Mnemonic #$dd
Absolute	Mnemonic $aaaa
Zero page	Mnemonic $aa
Accumulator	Mnemonic A
Implied	Mnemonic
Indxd Indct	Mnemonic ($ff,X)
Indct Indxd	Mnemonic ($aa),Y
Zero page,X	Mnemonic $ff,X
Absolute,X	Mnemonic $ffff,X
Absolute,Y	Mnemonic $ffff,Y
Relative	Mnemonic $rr
Indirect	Mnemonic ($aaaa)
Zero page,Y	Mnemonic $ff,Y

Abbreviations and Explanations

Indxd Indct = Indexed Indirect
Indct Indxd = Indirect Indexed
a = address (one hex digit)
d = data (one hex digit)
f = address offset (one hex digit) ($ff is an unsigned binary number and is therefore positive)
r = relative address (one hex digit) ($rr is a 2's-complement signed binary number and can therefore be positive or negative)
* = add 1 cycle if page boundary crossed
+ = add 1 cycle if branch occurs; add 1 more cycle if branch crosses page
( ) =the contents of the address within parentheses form the actual address
7...0 = bits 0 through 7 of memory or the accumulator
M_7, M_6, etc. = Bits 7, 6, etc. of a memory location
$_L$ = low-order byte
$_H$ = high-order byte

PC = program counter
S = stack (contents of the top byte of the stack)
SP = stack pointer
PSR = processor status register (flags)
* = Add 1 cycle if crossing page boundary

Flags

0 = flag always cleared
1 = flag always set

x = flag not affected
N = negative flag
V = overflow flag
B = break flag
D = decimal flag
I = interrupt flag
Z = zero flag
C = carry flag

Symbols in the Page Heading

~ = clock cycles
# = # of bytes used by instruction (and following address or data if used)

Addressing Modes - Summary

Immediate (Mnemonic #$dd): The data to be operated on (#$dd) is in the next byte of memory after the instruction itself. Therefore no address is needed.

Absolute (Mnemonic $aaaa): The data to be operated on is found in the memory location indicated ($aaaa). This is a 2-byte address and can point to any place in the 6502's 64K (65,536 byte) addressing range.

Zero page (Mnemonic $aa): The data to be operated on is found in the memory location indicated ($aa). This is a 1-byte address and can point only to a place in page zero of memory. Page zero is address $00-$FF (decimal 0-255).

Accumulator (Mnemonic A): These are instructions which use implied addressing, where the data is already in the accumulator.

Implied (Mnemonic): These instructions indicate where the data is or will be within the instruction itself.

Indxd Indct (Mnemonic ($ff,X)): In this form of addressing, the operand (the number which is going to have something done to it) is found through a multistep process. First, the offset ($ff) is added to the X register to form an address (this address must be in page zero since both of these are 8-bit numbers). The microprocessor then gets the contents of this memory location and the following location to form another address where it will then find the data (operand).

Indct Indxd (Mnemonic ($aa),Y): This addressing mode is sometimes confused with the one above though it does work differently. First, the microprocessor goes to address $aa and the address immediately following $aa. It uses the contents of these two locations to form a 16-bit address to which the Y register is added. This then forms the actual address where the operand is located.

Zero page,X (Mnemonic $ff,X): In this form of addressing the number $ff is added to the X register to form a second address where the operand is located. Because both $ff and X are 8-bit binary numbers, the actual address must be in page zero. If the sum of these two numbers exceeds $FF (the end of page zero), any carry will be ignored and the address will "wrap around" to the beginning of page zero.

Absolute,X (Mnemonic $ffff,X): In this case, the 16-bit number $ffff is added to the X register to form the actual address. If this number exceeds hexadecimal $FFFF, the carry is ignored and the address "wraps around" to $0000 and continues from there.

Absolute,Y (Mnemonic $ffff,Y): This address mode works the same as Absolute,X except that the Y register is used instead.

Relative (Mnemonic $rr): $rr is a 2's-complement signed binary number; that is, it can be positive or negative. This number is added to the current contents of the program counter to determine the actual address. $rr is different from an offset ($ffff or $ff) because it is not added to another register but directly to the program counter itself. It directs the microprocessor relative to its current place in memory.

Indirect (Mnemonic ($aaaa)): In this mode, the contents of address $aaaa and the contents of the address immediately following it are used to form the actual address where the operand is to be found. (Only the JMP instruction uses this addressing mode.)

Zero page,Y (Mnemonic $ff,Y): This addressing mode is exactly like the "Zero page,X" mode except that register Y is used instead.

SHORT TABLE OF 6502 INSTRUCTIONS LISTED BY CATEGORY

Assembler Notation	Op	Boolean/Arith Operation	Flags NV-BDIZC
CPU Control Instructions			
NOP	EA	Nothing	xx-xxxxx
BRK	00	PC + 2 → S SP - 2 → SP PSR → S SP - 1 → S $FFFE → PC	xx-1x1xx
Data Transfer Instructions			
LDA #$dd	A9	M → A	Nx-xxxZx
LDA $aaaa	AD		
LDA $aa	A5		
LDA ($ff,X)	A1		
LDA ($aa),Y	B1		
LDA $ff,X	B5		
LDA $ffff,X	BD		
LDA $ffff,Y	B9		
LDX #$dd	A2	M → X	Nx-xxxZx
LDX $aaaa	AE		
LDX $aa	A6		
LDX $ffff,Y	BE		
LDX $ff,Y	B6		
LDY #$dd	A0	M → Y	Nx-xxxZx
LDY $aaaa	AC		
LDY $aa	A4		
LDY $ff,X	B4		
LDY $ffff,X	BC		
STA $aaaa	8D	A → M	xx-xxxxx
STA $aa	85		
STA ($ff,X)	81		
STA ($aa),Y	91		
STA $ff,X	95		
STA $ffff,X	9D		
STA $ffff,Y	99		
STX $aaaa	8E	X → M	xx-xxxxx
STX $aa	86		
STX $ff,Y	96		
STY $aaaa	8C	Y → M	xx-xxxxx
STY $aa	84		
STY $ff,X	94		
TAX	AA	A → X	Nx-xxxZx
TXA	8A	X → A	Nx-xxxZx
TAY	A8	A → Y	Nx-xxxZx
TYA	98	Y → A	Nx-xxxZx
Flag Instructions			
CLC	18	0 → C	xx-xxxx0
CLD	D8	0 → D	xx-x0xxx
CLI	58	0 → I	xx-xx0xx
CLV	B8	0 → V	x0-xxxxx
SEC	38	1 → C	xx-xxxx1
SED	F8	1 → D	xx-x1xxx
SEI	78	1 → I	xx-xx1xx
Arithmetic Instructions			
ADC #$dd	69	A + M + C → A	NV-xxxZC
ADC $aaaa	6D		
ADC $aa	65		
ADC ($ff,X)	61		
ADC ($aa),Y	71		
ADC $ff,X	75		
ADC $ffff,X	7D		
ADC $ffff,Y	79		
SBC #$dd	E9	A - M -	NV-xxxZC
SBC $aaaa	ED	(1-C) → A	
SBC $aa	E5		
SBC ($ff,X)	E1	Note: (1-C) =	
SBC ($aa),Y	F1	Borrow	
SBC $ff,X	F5		
SBC $ffff,X	FD		
SBC $ffff,Y	F9		
Logical Instructions			
AND #$dd	29	A AND M → A	Nx-xxxZx
AND $aaaa	2D		
AND $aa	25		
AND ($ff,X)	21		
AND ($aa),Y	31		
AND $ff,X	35		
AND $ffff,X	3D		
AND $ffff,Y	39		

Assembler Notation	Op	Boolean/Arith Operation	Flags NV-BDIZC
EOR #$dd	49	A EOR M → A	Nx-xxxZx
EOR $aaaa	4D		
EOR $aa	45		
EOR ($ff,X)	41		
EOR ($aa),Y	51		
EOR $ff,X	55		
EOR $ffff,X	5D		
EOR $ffff,Y	59		
ORA #$dd	09	A OR M → A	Nx-xxxZx
ORA $aaaa	0D		
ORA $aa	05		
ORA ($ff,X)	01		
ORA ($aa),Y	11		
ORA $ff,X	15		
ORA $ffff,X	1D		
ORA $ffff,Y	19		
BIT $aaaa	2C	A AND M	76-xxxZx
BIT $aa	24	$M_7 \rightarrow N$	
		$M_6 \rightarrow V$	

Rotate and Shift Instructions

Assembler Notation	Op	Boolean/Arith Operation	Flags NV-BDIZC
ASL $aaaa	0E	C ← 7...0 ← 0	Nx-xxxZC
ASL $aa	06		
ASL A	0A		
ASL $ff,X	16		
ASL $ffff,X	1E		
LSR $aaaa	4E	0 → 7...0 → C	0x-xxxZC
LSR $aa	46		
LSR A	4A		
LSR $ff,X	56		
LSR $ffff,X	5E		
ROL $aaaa	2E	7...0 ← ; → C	Nx-xxxZC
ROL $aa	26		
ROL A	2A		
ROL $ff,X	36		
ROL $ffff,X	3E		
ROR $aaaa	6E	→ 7...0 ; C ←	Nx-xxxZC
ROR $aa	66		
ROR A	6A		
ROR $ff,X	76		
ROR $ffff,X	7E		

Increment and Decrement Instructions

Assembler Notation	Op	Boolean/Arith Operation	Flags NV-BDIZC
INC $aaaa	EE	M + 1 → M	Nx-xxxZx
INC $aa	E6		
INC $ff,X	F6		
INC $ffff,X	FE		
INX	E8	X + 1 → X	Nx-xxxZx
INY	C8	Y + 1 → Y	Nx-xxxZx
DEC $aaaa	CE	M - 1 → M	Nx-xxxZx
DEC $aa	C6		
DEC $ff,X	D6		
DEC $ffff,X	DE		
DEX	CA	X - 1 → X	Nx-xxxZx
DEY	88	Y - 1 → Y	Nx-xxxZx

Unconditional Jump Instructions

Assembler Notation	Op	Boolean/Arith Operation	Flags NV-BDIZC
JMP $aaaa	4C	aaaa → PC {abs addressing}	xx-xxxxx
JMP ($aaaa)	6C	(aaaa) → PC_L (aaaa + 1) → PC_H {indirect addressing}	

Test (Compare) Instructions

Assembler Notation	Op	Boolean/Arith Operation	Flags NV-BDIZC
CMP #$dd	C9	A - M	Nx-xxxZC
CMP $aaaa	CD		
CMP $aa	C5		
CMP ($ff,X)	C1		
CMP ($aa),Y	D1		
CMP $ff,X	D5		
CMP $ffff,X	DD		
CMP $ffff,Y	D9		
CPX #$dd	E0	X - M	Nx-xxxZC
CPX $aaaa	EC		
CPX $aa	E4		
CPY #$dd	C0	Y - M	Nx-xxxZC
CPY $aaaa	CC		
CPY $aa	C4		

Conditional Jump (Branch) Instructions

Assembler Notation	Op	Boolean/Arith Operation	Flags NV-BDIZC
BCC $rr	90	PC + rr → PC if C=0	xx-xxxxx
BCS $rr	B0	PC + rr → PC if C=1	xx-xxxxx
BEQ $rr	F0	PC + rr → PC if Z=1	xx-xxxxx

SHORT TABLE OF 6502 INSTRUCTIONS LISTED BY CATEGORY (*Continued*)

Assembler Notation	Op	Boolean/Arith Operation	Flags NV-BDIZC
BNE $rr	D0	PC + rr → PC if Z=0	xx-xxxxx
BMI $rr	30	PC + rr → PC if N=1	xx-xxxxx
BPL $rr	10	PC + rr → PC if N=0	xx-xxxxx
BVC $rr	50	PC + rr → PC if V=0	xx-xxxxx
BVS $rr	70	PC + rr → PC if V=1	xx-xxxxx

Subroutine Instructions

Assembler Notation	Op	Boolean/Arith Operation	Flags NV-BDIZC
JSR $aaaa	20	PC + 2 → S aaaa → PC SP - 2 → SP	xx-xxxxx
RTS	60	S (2 bytes) → PC PC + 1 → PC SP + 2 → SP	xx-xxxxx

Stack Instructions

Assembler Notation	Op	Boolean/Arith Operation	Flags NV-BDIZC
PHA	48	A → S SP - 1 → SP	xx-xxxxx
PLA	68	S → A SP + 1 → SP	Nx-xxxZx
PHP	08	PSR → S SP - 1 → SP	xx-xxxxx
PLP	28	S → PSR SP + 1 → SP	NV-BDIZC
TXS	9A	X → SP	xx-xxxxx
TSX	BA	SP → X	Nx-xxxZx

Interrupt Instructions

Assembler Notation	Op	Boolean/Arith Operation	Flags NV-BDIZC
RTI	40	S → PSR SP + 1 → SP S (2 bytes) → PC SP + 2 → SP	NV-BDIZC

Input-Output Instructions

None

CONDENSED TABLE OF 6502 INSTRUCTIONS LISTED BY CATEGORY

CPU Control Instructions

NOP	EA
BRK	00

Data Transfer Instructions

LDA #$dd	A9
LDA $aaaa	AD
LDA $aa	A5
LDA ($ff,X)	A1
LDA ($aa),Y	B1
LDA $ff,X	B5
LDA $ffff,X	BD
LDA $ffff,Y	B9
LDX #$dd	A2
LDX $aaaa	AE
LDX $aa	A6
LDX $ffff,Y	BE
LDX $ff,Y	B6
LDY #$dd	A0
LDY $aaaa	AC
LDY $aa	A4
LDY $ff,X	B4
LDY $ffff,X	BC
STA $aaaa	8D
STA $aa	85
STA ($ff,X)	81
STA ($aa),Y	91
STA $ff,X	95
STA $ffff,X	9D
STA $ffff,Y	99
STX $aaaa	8E
STX $aa	86
STX $ff,Y	96
STY $aaaa	8C
STY $aa	84
STY $ff,X	94
TAX	AA
TXA	8A
TAY	A8
TYA	98

Flag Instructions

CLC	18
CLD	D8
CLI	58
CLV	B8
SEC	38
SED	F8
SEI	78

Arithmetic Instructions

ADC #$dd	69
ADC $aaaa	6D
ADC $aa	65
ADC ($ff,X)	61

CONDENSED TABLE OF 6502 INSTRUCTIONS LISTED BY CATEGORY (*Continued*)

Instruction	Opcode
ADC ($aa),Y	71
ADC $ff,X	75
ADC $ffff,X	7D
ADC $ffff,Y	79
SBC #$dd	E9
SBC $aaaa	ED
SBC $aa	E5
SBC ($ff,X)	E1
SBC ($aa),Y	F1
SBC $ff,X	F5
SBC $ffff,X	FD
SBC $ffff,Y	F9

Logical Instructions

Instruction	Opcode
AND #$dd	29
AND $aaaa	2D
AND $aa	25
AND ($ff,X)	21
AND ($aa),Y	31
AND $ff,X	35
AND $ffff,X	3D
AND $ffff,Y	39
EOR #$dd	49
EOR $aaaa	4D
EOR $aa	45
EOR ($ff,X)	41
EOR ($aa),Y	51
EOR $ff,X	55
EOR $ffff,X	5D
EOR $ffff,Y	59
ORA #$dd	09
ORA $aaaa	0D
ORA $aa	05
ORA ($ff,X)	01
ORA ($aa),Y	11
ORA $ff,X	15
ORA $ffff,X	1D
ORA $ffff,Y	19
BIT $aaaa	2C
BIT $aa	24

Rotate and Shift Instructions

Instruction	Opcode
ASL $aaaa	0E
ASL $aa	06
ASL A	0A
ASL $ff,X	16
ASL $ffff,X	1E
LSR $aaaa	4E
LSR $aa	46
LSR A	4A
LSR $ff,X	56
LSR $ffff,X	5E
ROL $aaaa	2E
ROL $aa	26
ROL A	2A
ROL $ff,X	36
ROL $ffff,X	3E
ROR $aaaa	6E
ROR $aa	66
ROR A	6A
ROR $ff,X	76
ROR $ffff,X	7E

Increment and Decrement Instructions

Instruction	Opcode
INC $aaaa	EE
INC $aa	E6
INC $ff,X	F6
INC $ffff,X	FE
INX	E8
INY	C8
DEC $aaaa	CE
DEC $aa	C6
DEC $ff,X	D6
DEC $ffff,X	DE
DEX	CA
DEY	88

Unconditional Jump Instructions

Instruction	Opcode
JMP $aaaa	4C
JMP ($aaaa)	6C

Test (Compare) Instructions

Instruction	Opcode
CMP #$dd	C9
CMP $aaaa	CD
CMP $aa	C5
CMP ($ff,X)	C1
CMP ($aa),Y	D1
CMP $ff,X	D5
CMP $ffff,X	DD
CMP $ffff,Y	D9
CPX #$dd	E0
CPX $aaaa	EC
CPX $aa	E4
CPY #$dd	C0
CPY $aaaa	CC
CPY $aa	C4

Conditional Jump (Branch) Instructions

Instruction	Opcode
BCC $rr	90
BCS $rr	B0
BEQ $rr	F0
BNE $rr	D0
BMI $rr	30
BPL $rr	10
BVC $rr	50
BVS $rr	70

Subroutine Instructions

Instruction	Opcode
JSR $aaaa	20
RTS	60

Stack Instructions

Instruction	Opcode
PHA	48
PLA	68
PHP	08
PLP	28
TXS	9A
TSX	BA

Interrupt Instructions

Instruction	Opcode
RTI	40

Input-Output Instructions

None

CONDENSED TABLE OF 6502 INSTRUCTIONS LISTED ALPHABETICALLY

Instruction	Opcode
ADC ($aa),Y	71
ADC ($ff,X)	61
ADC $aa	65
ADC $aaaa	6D
ADC $ffff,X	7D
ADC $ffff,Y	79
ADC $ff,X	75
ADC #$dd	69
AND ($aa),Y	31
AND ($ff,X)	21
AND $aa	25
AND $aaaa	2D
AND $ffff,X	3D
AND $ffff,Y	39
AND $ff,X	35
AND #$dd	29
ASL $aa	06
ASL $aaaa	0E
ASL $ffff,X	1E
ASL $ff,X	16
ASL A	0A
BCC $rr	90
BCS $rr	B0
BEQ $rr	F0
BIT $aa	24
BIT $aaaa	2C
BMI $rr	30
BNE $rr	D0
BPL $rr	10
BRK	00
BVC $rr	50
BVS $rr	70
CLC	18
CLD	D8
CLI	58
CLV	B8
CMP ($aa),Y	D1
CMP ($ff,X)	C1
CMP $aa	C5
CMP $aaaa	CD
CMP $ffff,X	DD
CMP $ffff,Y	D9
CMP $ff,X	D5
CMP #$dd	C9
CPX $aa	E4
CPX $aaaa	EC
CPX #$dd	E0
CPY $aa	C4
CPY $aaaa	CC
CPY #$dd	C0
DEC $aa	C6
DEC $aaaa	CE

CONDENSED TABLE OF 6502 INSTRUCTIONS LISTED ALPHABETICALLY (*Continued*)

Instruction	Op Code
DEC $ffff,X	DE
DEC $ff,X	D6
DEX	CA
DEY	88
EOR ($aa),Y	51
EOR ($ff,X)	41
EOR $aa	45
EOR $aaaa	4D
EOR $ffff,X	5D
EOR $ffff,Y	59
EOR $ff,X	55
EOR #$dd	49
INC $aa	E6
INC $aaaa	EE
INC $ffff,X	FE
INC $ff,X	F6
INX	E8
INY	C8
JMP ($aaaa)	6C
JMP $aaaa	4C
JSR $aaaa	20
LDA ($aa),Y	B1
LDA ($ff,X)	A1
LDA $aa	A5
LDA $aaaa	AD
LDA $ffff,X	BD
LDA $ffff,Y	B9
LDA $ff,X	B5
LDA #$dd	A9
LDX $aa	A6
LDX $aaaa	AE
LDX $ffff,Y	BE
LDX $ff,Y	B6
LDX #$dd	A2
LDY $aa	A4
LDY $aaaa	AC
LDY $ffff,X	BC
LDY $ff,X	B4
LDY #$dd	A0
LSR $aa	46
LSR $aaaa	4E
LSR $ffff,X	5E
LSR $ff,X	56
LSR A	4A
NOP	EA
ORA ($aa),Y	11
ORA ($ff,X)	01
ORA $aa	05
ORA $aaaa	0D
ORA $ffff,X	1D
ORA $ffff,Y	19
ORA $ff,X	15
ORA #$dd	09
PHA	48
PHP	08
PLA	68
PLP	28
ROL $aa	26
ROL $aaaa	2E
ROL $ffff,X	3E
ROL $ff,X	36
ROL A	2A
ROR $aa	66
ROR $aaaa	6E
ROR $ffff,X	7E
ROR $ff,X	76
ROR A	6A
RTI	40
RTS	60
SBC ($aa),Y	F1
SBC ($ff,X)	E1
SBC $aa	E5
SBC $aaaa	ED
SBC $ffff,X	FD
SBC $ffff,Y	F9
SBC $ff,X	F5
SBC #$dd	E9
SEC	38
SED	F8
SEI	78
STA ($aa),Y	91
STA ($ff,X)	81
STA $aa	85
STA $aaaa	8D
STA $ffff,X	9D
STA $ffff,Y	99
STA $ff,X	95
STX $aa	86
STX $aaaa	8E
STX $ff,Y	96
STY $aa	84
STY $aaaa	8C
STY $ff,X	94
TAX	AA
TAY	A8
TSX	BA
TXA	8A
TXS	9A
TYA	98

CONDENSED TABLE OF 6502 INSTRUCTIONS LISTED BY OP CODE

Op Code	Instruction
00	BRK
01	ORA ($ff,X)
05	ORA $aa
06	ASL $aa
08	PHP
09	ORA #$dd
0A	ASL A
0D	ORA $aaaa
0E	ASL $aaaa
10	BPL $rr
11	ORA ($aa),Y
15	ORA $ff,X
16	ASL $ff,X
18	CLC
19	ORA $ffff,Y
1D	ORA $ffff,X
1E	ASL $ffff,X
20	JSR $aaaa
21	AND ($ff,X)
24	BIT $aa
25	AND $aa
26	ROL $aa
28	PLP
29	AND #$dd
2A	ROL A
2C	BIT $aaaa
2D	AND $aaaa
2E	ROL $aaaa
30	BMI $rr
31	AND ($aa),Y
35	AND $ff,X
36	ROL $ff,X
38	SEC
39	AND $ffff,Y
3D	AND $ffff,X
3E	ROL $ffff,X
40	RTI
41	EOR ($ff,X)
45	EOR $aa
46	LSR $aa
48	PHA
49	EOR #$dd
4A	LSR A
4C	JMP $aaaa
4D	EOR $aaaa
4E	LSR $aaaa
50	BVC $rr
51	EOR ($aa),Y
55	EOR $ff,X
56	LSR $ff,X
58	CLI
59	EOR $ffff,Y
5D	EOR $ffff,X
5E	LSR $ffff,X
60	RTS
61	ADC ($ff,X)
65	ADC $aa
66	ROR $aa
68	PLA
69	ADC #$dd
6A	ROR A
6C	JMP ($aaaa)
6D	ADC $aaaa
6E	ROR $aaaa
70	BVS $rr
71	ADC ($aa),Y
75	ADC $ff,X
76	ROR $ff,X
78	SEI
79	ADC $ffff,Y
7D	ADC $ffff,X
7E	ROR $ffff,X
81	STA ($ff,X)
84	STY $aa
85	STA $aa
86	STX $aa
88	DEY
8A	TXA
8C	STY $aaaa
8D	STA $aaaa
8E	STX $aaaa
90	BCC $rr
91	STA ($aa),Y
94	STY $ff,X
95	STA $ff,X
96	STX $ff,Y
98	TYA
99	STA $ffff,Y
9A	TXS
9D	STA $ffff,X
A0	LDY #$dd
A1	LDA ($ff,X)
A2	LDX #$dd
A4	LDY $aa
A5	LDA $aa
A6	LDX $aa
A8	TAY
A9	LDA #$dd
AA	TAX
AC	LDY $aaaa
AD	LDA $aaaa
AE	LDX $aaaa
B0	BCS $rr
B1	LDA ($aa),Y
B4	LDY $ff,X
B5	LDA $ff,X
B6	LDX $ff,Y
B8	CLV
B9	LDA $ffff,Y
BA	TSX
BC	LDY $ffff,X
BD	LDA $ffff,X
BE	LDX $ffff,Y
C0	CPY #$dd
C1	CMP ($ff,X)
C4	CPY $aa

CONDENSED TABLE OF 6502 INSTRUCTIONS LISTED BY OP CODE (*Continued*)

Op Code	Instruction
C5	CMP $aa
C6	DEC $aa
C8	INY
C9	CMP #$dd
CA	DEX
CC	CPY $aaaa
CD	CMP $aaaa
CE	DEC $aaaa
D0	BNE $rr
D1	CMP ($aa),Y
D5	CMP $ff,X
D6	DEC $ff,X
D8	CLD
D9	CMP $ffff,Y
DD	CMP $ffff,X
DE	DEC $ffff,X
E0	CPX #$dd
E1	SBC ($ff,X)
E4	CPX $aa
E5	SBC $aa
E6	INC $aa
E8	INX
E9	SBC #$dd
EA	NOP
EC	CPX $aaaa
ED	SBC $aaaa
EE	INC $aaaa
F0	BEQ $rr
F1	SBC ($aa),Y
F5	SBC $ff,X
F6	INC $ff,X
F8	SED
F9	SBC $ffff,Y
FD	SBC $ffff,X
FE	INC $ffff,X

Appendixes

APPENDIX 1. THE ANALOG INTERFACE

The data in a microprocessor is in digital form. This differs from the outside world where data is in analog (continuous) form. To get digital data, we need to use an *analog-to-digital* (A/D) *converter;* it will convert analog voltage or current into an equivalent digital word.

Conversely, after a CPU has processed data, it is often necessary to convert the digital answer into an analog voltage or current. This conversion requires a *digital-to-analog* (D/A) converter.

The *analog interface* is the boundary where digital and analog meet, where the microcomputer connects to the outside world. At this interface, we find either an A/D converter (input side) or a D/A converter (output side). This chapter discusses some of the hardware and software found at the analog interface.

A1-1 OP-AMP BASICS

Let us briefly review the *operational amplifier* (op amp) because this device is used with D/A and A/D converters. We will zero in on the key features that make the op amp useful at the analog interface.

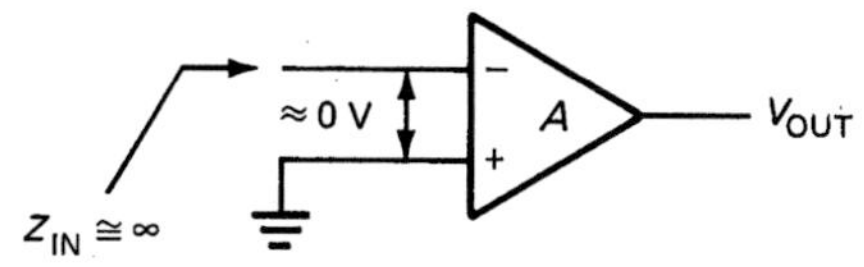

Fig. A1-1 Operational amplifier.

Virtual Ground

Figure A1-1 shows the symbol for an op amp. V_{OUT} is the output voltage with respect to ground. A is the open-loop voltage gain of the op amp, often more than 100,000. When connected as an inverter, the noninverting input (+ input) is grounded. The inverting input (− input) receives the signal voltage.

Because the voltage gain of an op amp is so large, the input voltage is in microvolts. To a first approximation, the input voltage may be treated as 0 V. Furthermore, the input impedance of the inverting input approaches infinity (sometimes FETs are used for the input stage, as in BIFET op amps). These key features, zero input voltage and infinite input impedance, make the inverting input a *virtual ground point.*

How is a virtual ground different from an ordinary ground? An ordinary ground has zero voltage while sinking any amount of current. A virtual ground, however, is a ground for voltage but not for current; it has zero voltage but can sink no current. In the discussion that follows, we will approximate the inverting input of an op amp as a virtual ground point: this means zero voltage and zero current.

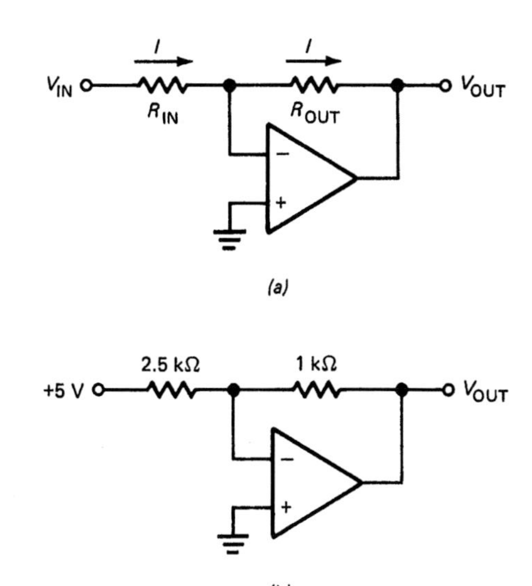

Fig. A1-2 Output current equals input current.

Output Voltage and Current

Figure A1-2*a* shows an inverting op amp with input and output resistors. V_{IN} is the input voltage with respect to ground, and V_{OUT} is the output voltage with respect to ground. Because of the high gain and input impedance, we can approximate the inverting input as a virtual ground point. Therefore, all the input voltage appears across the input resistor, which means that the input current is

$$I = \frac{V_{IN}}{R_{IN}} \qquad \text{(A1-1)}$$

Since none of the input current can enter the virtual ground point, it must pass through the output resistor. In other words, the output current equals the input current. And the output voltage is

$$V_{OUT} = -IR_{OUT} \qquad \text{(A1-2)}$$

The minus sign indicates phase inversion. If the input voltage is positive, the output voltage is negative.

As an example of calculating input current and output voltage, look at Fig. A1-2*b*. The input current is

$$I = \frac{5\ \text{V}}{2.5\ \text{k}\Omega} = 2\ \text{mA}$$

The output voltage is

$$V_{OUT} = -2\ \text{mA} \times 1\ \text{k}\Omega = -2\ \text{V}$$

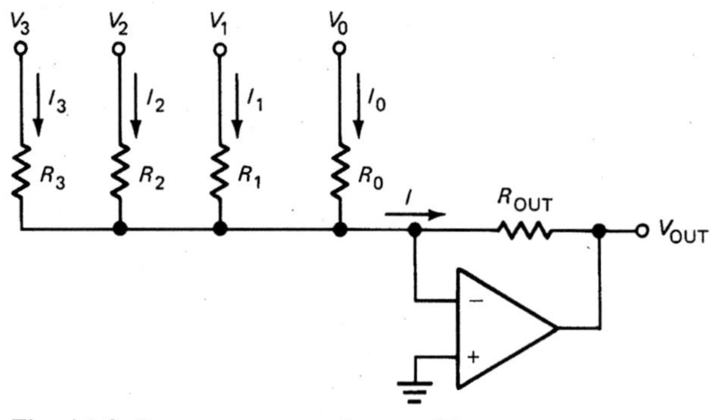

Fig. A1-3 Output current equals sum of input currents.

Summing Circuit

Figure A1-3 is an op-amp circuit whose output current is the *sum* of the input currents. Here is the proof. Because of the virtual ground point, each input voltage appears across its resistor. This means that the input currents are

$$I_3 = \frac{V_3}{R_3} \qquad I_2 = \frac{V_2}{R_2} \qquad I_1 = \frac{V_1}{R_1} \qquad I_0 = \frac{V_0}{R_0}$$

Kirchhoff's current law gives a total input current of

$$I = I_3 + I_2 + I_1 + I_0$$

Again, the virtual ground guarantees that all this input current goes through the output resistor. As before,

$$V_{OUT} = -IR_{OUT}$$

A1-2 A BASIC D/A CONVERTER

The op-amp summing circuit can be used to build a D/A converter by selecting input resistors that are weighted in binary progression. Figure A1-4 gives you the idea. V_{REF} is an accurate reference voltage, and the resistors are precision resistors to get accurate input currents. The switches can be open or closed. When all switches are open, all input currents are zero and the output current is zero.

All Bits High

When all switches are closed, the input currents are

$$I_3 = \frac{V_{REF}}{R} \qquad I_2 = \frac{V_{REF}}{2R} \qquad I_1 = \frac{V_{REF}}{4R} \qquad I_0 = \frac{V_{REF}}{8R}$$

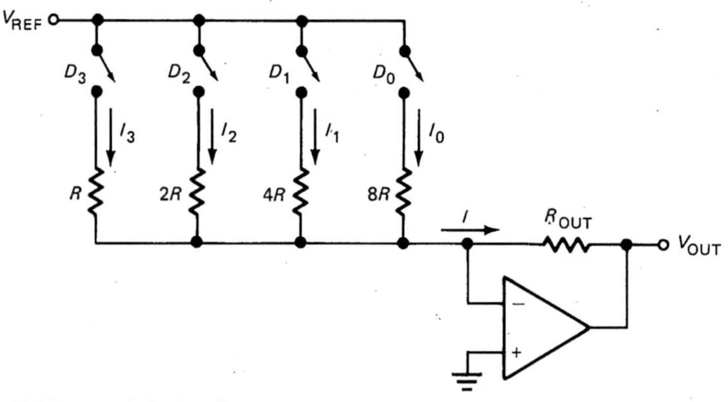

Fig. A1-4 D/A conversion with binary-weighted resistors.

The output current with all switches closed is the sum of all input currents and equals

$$I = \frac{V_{REF}}{R}(1 + 0.5 + 0.25 + 0.125) \quad \text{(A1-3)}$$

$$I = 1.875\frac{V_{REF}}{R}$$

By opening and closing switches we can produce 16 different output currents from 0 to $1.875V_{REF}/R$.

Any Digital Input

If 0 stands for an open switch and 1 for a closed switch, we can rewrite Eq. A1-3 as

$$I = \frac{V_{REF}}{R}(D_3 + 0.5D_2 + 0.25D_1 + 0.125D_0) \quad \text{(A1-4)}$$

In powers of 2,

$$I = \frac{V_{REF}}{R}(D_3 + 2^{-1}D_2 + 2^{-2}D_1 + 2^{-3}D_0) \quad \text{(A1-5)}$$

This says that the output current is the sum of binary-weighted input currents. In other words, we have a D/A converter. For instance, suppose $V_{REF} = 5$ V and $R = 5$ kΩ. Then the total output current varies from 0 to 1.875 mA, as shown in Table A1-1.

Current Switches

Figure A1-5 shows how we can transistorize the switching. Data bits D_3 through D_0 drive the bases of the transistors through the current-limiting resistors. When a bit is high, it produces enough base current to saturate its transistor. When a bit is low, the transistor is cut off. Since each transistor is saturated or cut off, it acts like a closed or open switch. (Base resistance is not critical; it need only be less than collector resistance multiplied by β_{dc}.)

TABLE A1-1. WEIGHTED D/A CONVERTER

D_3	D_2	D_1	D_0	Output current, mA	Fraction of maximum
0	0	0	0	0	0
0	0	0	1	0.125	$\frac{1}{15}$
0	0	1	0	0.25	$\frac{2}{15}$
0	0	1	1	0.375	$\frac{3}{15}$
0	1	0	0	0.5	$\frac{4}{15}$
0	1	0	1	0.625	$\frac{5}{15}$
0	1	1	0	0.75	$\frac{6}{15}$
0	1	1	1	0.875	$\frac{7}{15}$
1	0	0	0	1	$\frac{8}{15}$
1	0	0	1	1.125	$\frac{9}{15}$
1	0	1	0	1.25	$\frac{10}{15}$
1	0	1	1	1.375	$\frac{11}{15}$
1	1	0	0	1.5	$\frac{12}{15}$
1	1	0	1	1.625	$\frac{13}{15}$
1	1	1	0	1.75	$\frac{14}{15}$
1	1	1	1	1.875	$\frac{15}{15}$

If the lower 4 bits of an output port are connected to D_3 to D_0, the circuit of Fig. A1-5 will convert digital data to analog current. For instance, assume port 22H has been programmed as an output port in a minimum system. If the lower 4 bits of port 22H are connected to D_3 to D_0, this program segment will operate the D/A converter:

Label	Mnemonic	Comment
	MVI A,FFH	;Initialize accumulator
LOOP:	INR A	;Count up
	OUT 22H	;Output nibble
	JMP LOOP	;Get next nibble

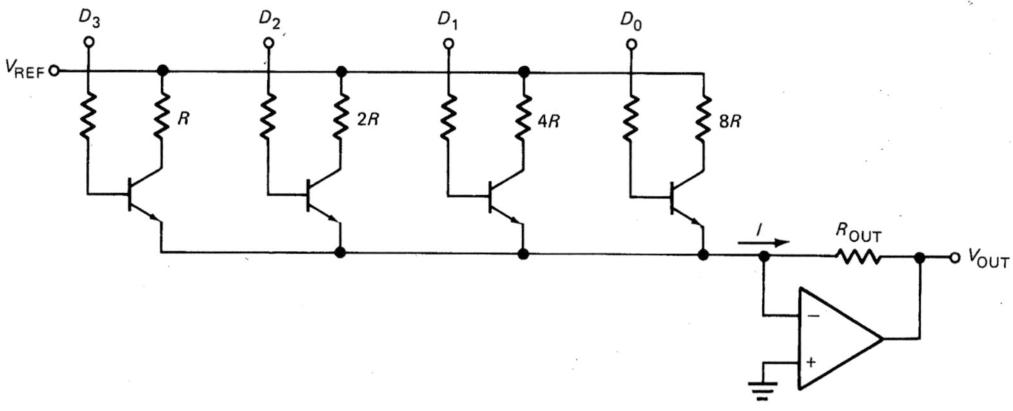

Fig. A1-5 Transistor switches for D/A converter.

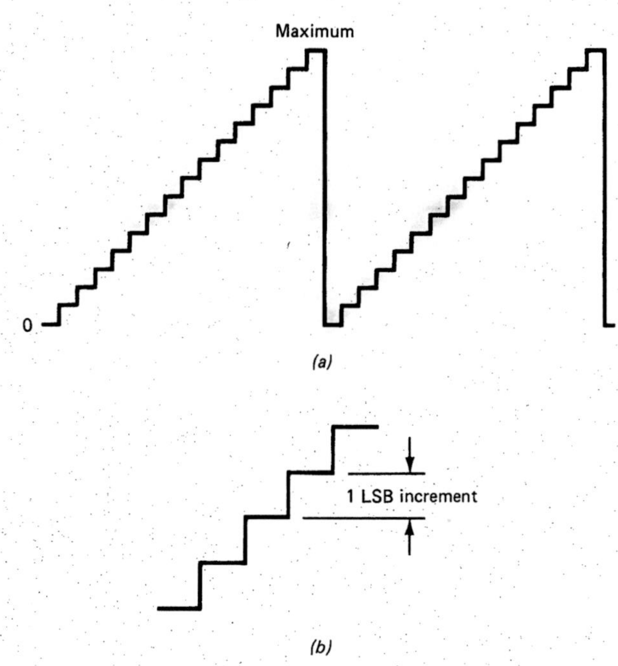

Fig. A1-6 (*a*) Staircase output current; (*b*) each step equals an LSB increment.

The first INR A produces accumulator contents of 00H. Subsequent INR executions produce 01H, 02H, . . . , 0FH, 10H, 11H, . . . , 1FH, 20H, 21H, . . . , FFH. As far as D_3 to D_0 are concerned, they see a nibble stream of 0000, 0001, 0010, 0011, . . . , 1111, 0000, 0001, and so on.

Figure A1-6*a* illustrates how the output current of the D/A converter appears. As each input nibble is latched into port 22H, the output current moves one step higher until reaching the maximum current. Then the cycle repeats. If all resistors are exact and all transistors matched, all steps are identical in size.

Resolution

In the perfect staircase of Fig. A1-6*b* a step is called an *LSB increment* because it is produced by a change in the LSB. One way to measure the quality of a D/A converter is its *resolution*, the ratio of the LSB increment to the maximum output. As a formula,

$$\text{Resolution} = \frac{1}{2^n - 1} \qquad \text{(A1-6)}$$

For instance, a 4-bit D/A converter has a resolution of

$$\text{Resolution} = \frac{1}{2^4 - 1} = \frac{1}{15}$$

This is sometimes read as 1 part in 15.

The number of different steps an *n*-bit converter produces is

$$\text{Steps} = 2^n - 1 \qquad \text{(A1-6}a\text{)}$$

Therefore, an alternative way to think of resolution is

$$\text{Resolution} = \frac{1}{\text{steps}} \qquad \text{(A1-6}b\text{)}$$

Percent resolution is given by

$$\text{Percent resolution} = \text{resolution} \times 100\% \qquad \text{(A1-7)}$$

If the resolution is 1 part in 15, then

$$\text{Percent resolution} = \tfrac{1}{15} \times 100\% = 6.67\%$$

The greater the number of bits, the better the resolution. With Eqs. A1-6 and A1-7 we can calculate the resolution and percent resolution for more bits. Table A1-2 is a summary of the resolution for converters with 4 to 18 bits.

Because the number of bits determines the resolution in Eq. A1-6, an indirect way to specify resolution is by stating the number of bits. For instance, an 8-bit converter has 8-bit resolution, a 10-bit converter has 10-bit resolution, and so on. This is a quick and easy way to pin down the resolution. When necessary, Eqs. A1-6, A1-6*a*, and A1-7 can give additional information.

Accuracy

In a D/A converter, *absolute accuracy* refers to how close each output current is to its ideal value. In Fig. A1-5 absolute accuracy depends on the reference voltage, resistor tolerance, transistor mismatch, and so forth. In a typical application, a trimmer adjustment is included to set the full-scale output at a preassigned value.

Relative accuracy refers to how close each output level is to its ideal fraction of full-scale output. With a 4-bit

TABLE A1-2. RESOLUTION

Bits	Resolution	Percent
4	1 part in 15	6.67
6	1 part in 63	1.59
8	1 part in 255	0.392
10	1 part in 1,023	0.0978
12	1 part in 4,095	0.0244
14	1 part in 16,383	0.0061
16	1 part in 65,535	0.00153
18	1 part in 262,143	0.000381

converter, the ideal output levels as a fraction of full-scale should be 0, $\frac{1}{15}$, $\frac{2}{15}$, $\frac{3}{15}$, and so on. Because data sheets specify relative accuracy rather than absolute accuracy, our subsequent discussions will emphasize relative accuracy.

Relative accuracy depends mainly on the tolerance of the weighted resistors in Fig. A1-5. If they are exactly R, $2R$, $4R$, and $8R$, all steps equal 1 LSB increment in Fig. A1-6*a*. When the resistors depart from ideal values, the steps may be larger or smaller than 1 LSB increment.

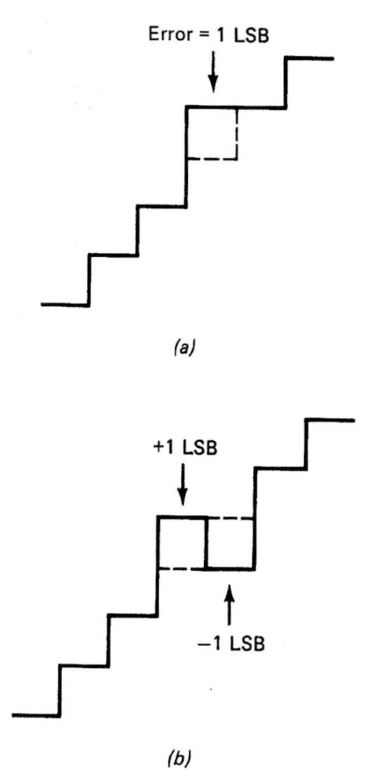

Fig. A1-7 Error specified in LSB increments.

Errors are specified in terms of LSB increments. For instance, Fig. A1-7*a* shows an error of 1 LSB; the actual output (solid line) differs from the ideal output (dashed line) by 1 LSB increment. If a negative error follows a positive error, the staircase can fall as shown in Fig. A1-7*b*. Here you see an error of + 1 LSB followed by an error of − 1 LSB.

Monotonicity

A *monotonic D/A converter* is one that produces an increase in output current for each successive digital input. The staircases of Fig. A1-7*a* and *b* are not monotonic because they do not produce an increase for each digital input. Figure A1-7*a* is almost monotonic, but Fig. A1-7*b* is far from monotonic. Monotonicity is the least we can expect from a D/A converter because it only makes sense; the output should increase when the input does.

For a D/A converter to be monotonic the error must be less than $\pm\frac{1}{2}$ LSB at each output level. Why? Because in the worst case, a $+\frac{1}{2}$-LSB error followed by a $-\frac{1}{2}$-LSB error produces the critical level where monotonicity is about to be lost. Figure A1-8 illustrates this critical case, an error of $+\frac{1}{2}$ LSB followed by an error of $-\frac{1}{2}$ LSB. If the error of a converter is less than $\pm\frac{1}{2}$ LSB for each output level, we are guaranteed a rising current for each successive digital input. Almost all commercially available D/A converters are monotonic because they have an accuracy of better than $\pm\frac{1}{2}$ LSB at each output level.

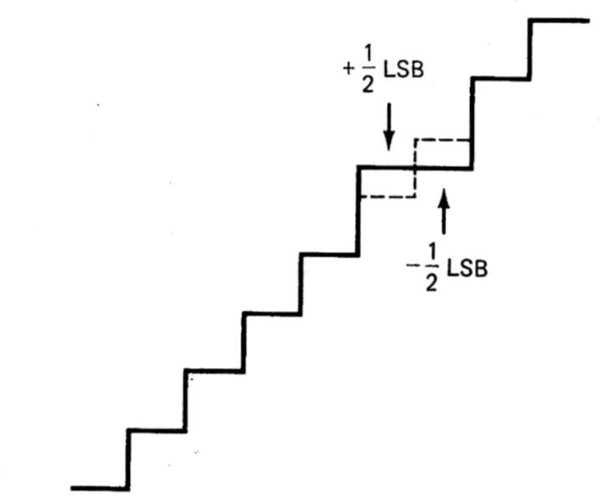

Fig. A1-8 Critical level for monotonicity.

Settling Time

After you apply a digital input, it takes a D/A converter anywhere from nanoseconds to microseconds to produce the correct output. *Settling time* is defined as the time it takes for the converter output to stabilize to within $\frac{1}{2}$ LSB of its final value. This time depends on the stray capacitance, saturation delay time, and other factors. Settling time is important because it places a limit on how fast you can change the digital inputs.

Disadvantages of Weighted Resistors

For a weighted-resistor circuit to be monotonic the tolerance of the resistors must be less than the percent resolution. For instance, if the resolution is $\frac{1}{15}$ (6.67 percent), resistors with a tolerance of less than ±6.67 percent will produce a monotonic staircase. If the resolution is $\frac{1}{255}$ (about 0.4 percent), the resistors need a tolerance of better than ±0.4 percent for a monotonic output. As you see, 4 bits are no problem, but 8 bits are.

Another difficulty arises with weighted resistors. As the number of bits increases, the range of resistance values gets awkward. For 8 bits, we need resistances of R, $2R$, $4R$, . . . , $128R$. The largest resistance is 128 times the smallest. For a 12-bit converter, the largest resistance needs to be 2,048 times the smallest. Because of the tolerance and range problems, mass production of weighted-resistor D/A converters is impractical.

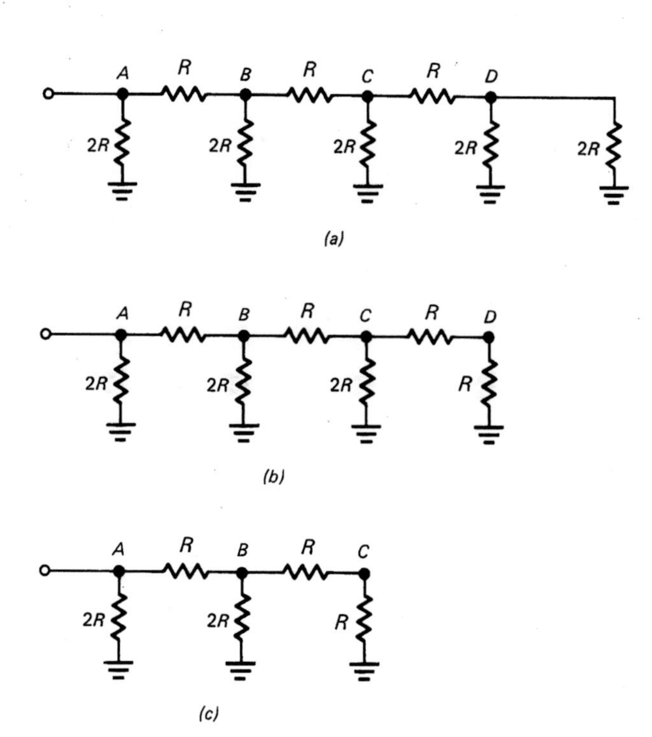

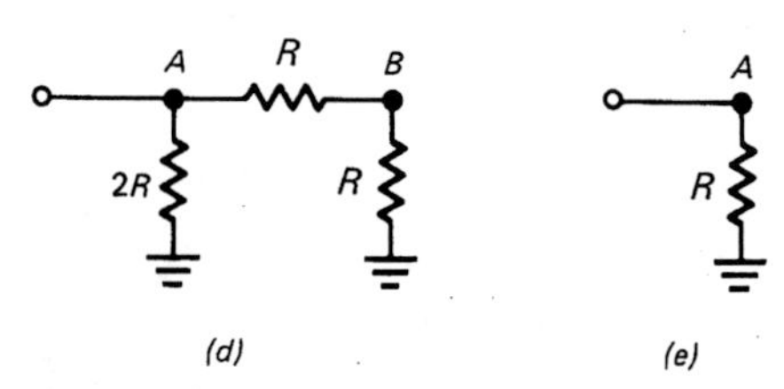

Fig. A1-9 *R*-2*R* ladder.

A1-3 THE LADDER METHOD

One way to get around the problems of a binary-weighted resistors is to use a *ladder* circuit. Figure A1-9*a* is an example of the *R*-2*R* ladder commonly used in integrated D/A converters. Only two resistance values are needed; this eliminates the range problem. Furthermore, since the resistors are on the same chip, they have almost identical characteristics; this minimizes the tolerance problem. In other words, as the number of bits increases, an integrated ladder can divide the current much more accurately than a binary-weighted circuit.

Ladder Properties

An *R*-2*R* ladder does something interesting to the impedance at different points in the circuit. To begin with, the two resistors at node D in Fig. A1-9*a* are in parallel and may be reduced to an equivalent resistance *R*, shown in Fig. A1-9*b*. Now, to the right of node C we have *R* in series with *R*, a total of 2*R*. Since node C has 2*R* is in parallel with 2*R*, the circuit reduces to Fig. A1-9*c*.

Looking into the left side of node B (Fig. A1-9*c*), we see 2*R* in parallel with 2*R*. Therefore, the circuit reduces to Fig. A1-9*d*. Again, 2*R* is in parallel with 2*R*, so the circuit reduces to the single *R* shown in Fig. A1-9*e*.

Figure A1-10 summarizes ladder impedances. Do you see the point? Looking into the left side of a node, we always see an equivalent resistance of *R*. Just to the right of each node, we always see a resistance of 2*R*. This impedance phenomenon is the key to analyzing modern D/A converters because they use the ladders instead of weighted resistors.

Binary Division of Current

Figure A1-11 shows how a ladder can divide the current into binary levels. The typical D/A converter has a reference current set by the user. In this example, the reference current is 2 mA. The bottom of each 2*R* resistor is grounded in either switch position. When a switch is to the right, the current through a 2*R* resistor flows to the upper ground. When a switch is to the left, the lower ground sinks the current. With all the switches to the right, as shown in Fig. A1-11, I_{OUT} is zero.

Here is how the ladder divides the 2 mA of reference current. Just to the right of node *A* we see an equivalent resistance of 2*R*. Therefore, the 2 mA of input current divides equally at node *A*. Similarly, at node *B* we see 2*R* in parallel with 2*R*; again, the current divides equally into 0.5-mA branch currents. This process continues through the ladder, so that we wind up with the upper grounds sinking 1, 0.5, 0.25, and 0.125 mA.

Other Switch Positions

When we move the switches, we do not change the way the current divides at the nodes. It still divides equally at each node. But when a switch is to the left, it steers the

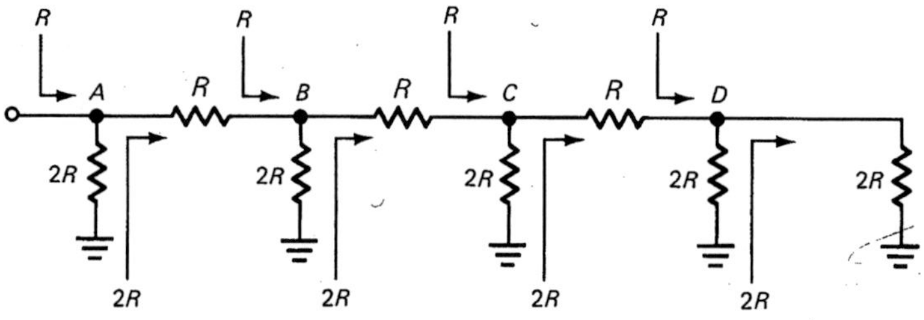

Fig. A1-10 Ladder impedances.

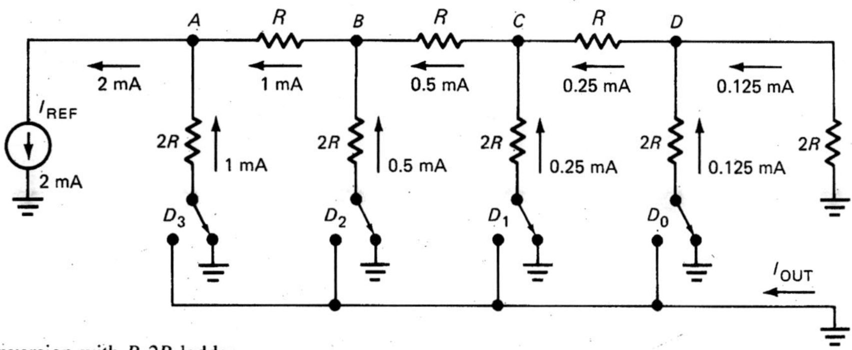

Fig. A1-11 D/A conversion with R-$2R$ ladder.

current into the lower ground. Bits D_3 to D_0 control the transistorized switches. From previous discussions, we can see that

$$I_{\text{OUT}} = (D_3 + 2^{-1}D_2 + 2^{-2}D_1 + 2^{-3}D_0)\frac{I_{\text{REF}}}{2} \quad \text{(A1-8)}$$

Therefore, the output current of a 4-bit ladder is from 0 to $\frac{15}{16}I_{\text{REF}}$.

More Bits

A similar analysis applies to longer ladders. The output current is

$$I_{\text{OUT}} = (D_{n-1} + 2^{-1}D_{n-2} + \cdots + 2^{1-n}D_0)\frac{I_{\text{REF}}}{2} \quad \text{(A1-9)}$$

For instance, an 8-bit ladder produces a maximum output current of $\frac{255}{256}I_{\text{REF}}$. The LSB increment is $\frac{1}{255}I_{\text{REF}}$.

Why Steer Current

Current steering may seem more complicated than necessary, but there is good reason for it. The currents throughout the ladder remain constant; all that changes are the ground points. Constant current implies constant voltage, which means that stray capacitance in the ladder has little effect. In other words, we do not get the usual exponential charge and discharge associated with a change in voltage. This reduces the settling time. For this reason, IC converters often use the current-steering approach shown in Fig. A1-11.

A1-4 THE COUNTER METHOD OF A/D CONVERSION

Figure A1-12 shows the simplest but least used method of A/D conversion. V_{IN} is the analog input voltage. D_7 to D_0 are the digital output. The digital output drives a D/A converter, which produces an analog output V_{OUT}. When *COUNT* is high, the counter counts upward. When *COUNT* is low, the counter stops. For convenience, an 8-bit D/A converter and 8-bit counter are used, but the idea applies to any number of bits.

Operation

The A/D conversion takes place as follows. First, the *START* pulse goes low, clearing the counter. When the

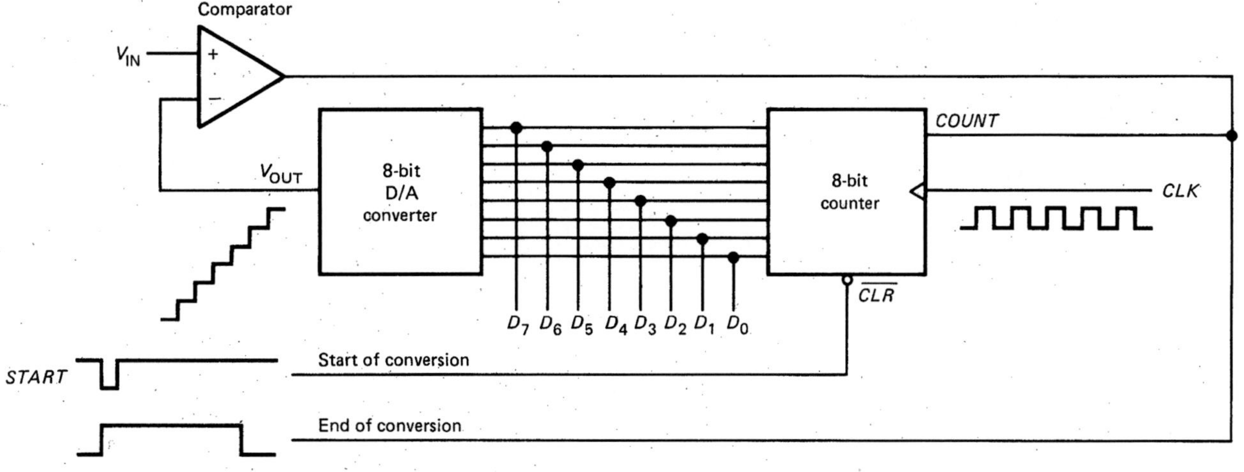

Fig. A1-12 A/D conversion with counter.

START pulse returns high, the counter is ready to go. Initially, V_{OUT} is zero; therefore, the op amp has a high output and *COUNT* is high. The counter starts counting upward from zero. Since the output of the counter drives a D/A converter, the converter output is a positive voltage staircase. As long as V_{IN} is greater than V_{OUT}, the op amp has a positive output, *COUNT* remains high, and the staircase voltage keeps rising.

At some point along the staircase, the next step makes V_{OUT} greater than V_{IN}. This forces *COUNT* to go low, and the counter stops. Now, the digital output D_7 to D_0 is the digital equivalent of the analog input. The negative-going edge of the *COUNT* signal is used as an *end-of-conversion* signal; this tells other circuits that the A/D conversion is finished.

If the analog input V_{IN} is changed, external circuits must send another *START* pulse to *start the conversion.* This clears the count and a new cycle begins. When the digital data is ready, the end-of-conversion signal has a falling edge.

Disadvantage

The main disadvantage of the *counter method* is its slow speed. In the worst case (maximum analog input) the counter has to reach the maximum count before the staircase voltage is greater than the analog input. For an 8-bit converter, this means a conversion time of 255 clock periods. For a 12-bit converter, the conversion time is 4,095 clock periods.

A1-5 SUCCESSIVE APPROXIMATION

The most widely used approach in A/D conversion is the *successive-approximation method* (see Fig. A1-13). As before, the output of a D/A converter drives the inverting input of an op-amp comparator. The difference, however, is in how the SAR register converges on the digital equivalent. (SAR stands for *successive-approximation register.*) When the conversion is finished, the digital equivalent is transferred to the output buffer register.

MSB First

When the start-of-conversion signal goes low, the SAR register is cleared and V_{OUT} drops to zero. When the start-of-conversion signal goes high, the conversion begins. Instead of counting up 1 bit at a time, the successive-approximation method starts by setting the MSB. In other words, during the first clock pulse the control circuit loads a high MSB into the SAR register, whose output then equals

1000 0000

As soon as this digital output appears, V_{OUT} jumps to $\frac{128}{255}$ times full-scale. If this is more than V_{IN}, the negative output of the comparator signals the control circuit to reset the MSB. On the other hand, if V_{OUT} is less than V_{IN}, the positive output of the comparator indicates that the MSB is to remain set. In some designs, setting and testing the MSB take place during the first clock pulse following the start of conversion. In other designs, several clock pulses may be needed to set the MSB, test it, and reset it if necessary.

Remaining Bits

Let us assume that the MSB was not reset. The SAR register contents are now 1000 0000. The next clock pulse will set

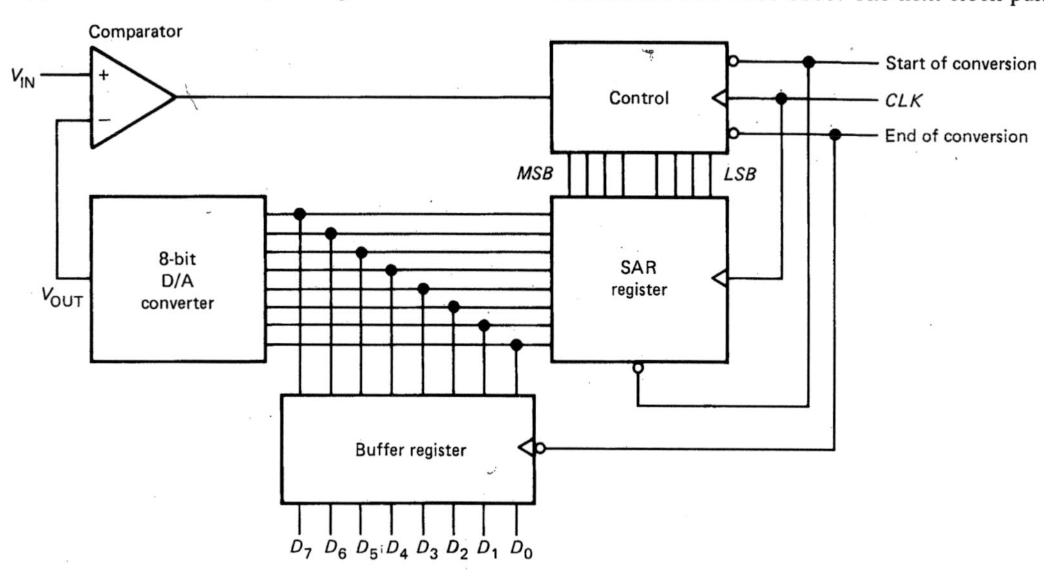

Fig. A1-13 A/D conversion by successive approximation.

D_6, giving a digital output of

1100 0000

V_{OUT} now steps to $\frac{192}{255}$ times full-scale. If V_{OUT} is greater than V_{IN}, the negative op-amp output causes D_6 to reset. If V_{OUT} is less than V_{IN}, D_6 remains set.

During the remaining clock pulses, successive bits are set and tested. Whenever a bit causes V_{OUT} to exceed V_{IN}, the bit is reset. In this way, all bits are set, tested, and reset if necessary. With the fastest circuits, the conversion is finished after eight clock pulses, and the D/A output is the analog equivalent of the register contents. Slower designs take longer because more clock pulses are needed to set, test, and possibly reset each bit.

Output Buffer

When the conversion is finished, the control circuit sends out a low end-of-conversion signal. The falling edge of this signal loads the digital equivalent into the buffer register. In this way, the digital output will remain even though we start a new conversion cycle.

Advantage

The main advantage of the successive-approximation method is speed. At best, it takes only n clock pulses to produce n-bit resolution of the analog signal. This is a big improvement over the counter method. Even with slower designs, the successive-approximation method is still considerably better than the counter method.

APPENDIX 2. BINARY-HEXADECIMAL-DECIMAL EQUIVALENTS

Binary	Hexadecimal	UB Decimal	LB Decimal
0000 0000	00	0	0
0000 0001	01	256	1
0000 0010	02	512	2
0000 0011	03	768	3
0000 0100	04	1,024	4
0000 0101	05	1,280	5
0000 0110	06	1,536	6
0000 0111	07	1,792	7
0000 1000	08	2,048	8
0000 1001	09	2,304	9
0000 1010	0A	2,560	10
0000 1011	0B	2,816	11
0000 1100	0C	3,072	12
0000 1101	0D	3,328	13
0000 1110	0E	3,584	14
0000 1111	0F	3,840	15
0001 0000	10	4,096	16
0001 0001	11	4,352	17
0001 0010	12	4,608	18
0001 0011	13	4,864	19
0001 0100	14	5,120	20
0001 0101	15	5,376	21
0001 0110	16	5,632	22
0001 0111	17	5,888	23
0001 1000	18	6,144	24
0001 1001	19	6,400	25
0001 1010	1A	6,656	26
0001 1011	1B	6,912	27
0001 1100	1C	7,168	28
0001 1101	1D	7,424	29
0001 1110	1E	7,680	30
0001 1111	1F	7,936	31
0010 0000	20	8,192	32
0010 0001	21	8,448	33
0010 0010	22	8,704	34
0010 0011	23	8,960	35
0010 0100	24	9,216	36
0010 0101	25	9,472	37
0010 0110	26	9,728	38
0010 0111	27	9,984	39
0010 1000	28	10,240	40
0010 1001	29	10,496	41
0010 1010	2A	10,752	42
0010 1011	2B	11,008	43
0010 1100	2C	11,264	44
0010 1101	2D	11,520	45
0010 1110	2E	11,776	46
0010 1111	2F	12,032	47
0011 0000	30	12,288	48
0011 0001	31	12,544	49
0011 0010	32	12,800	50
0011 0011	33	13,056	51
0011 0100	34	13,312	52
0011 0101	35	13,568	53
0011 0110	36	13,824	54
0011 0111	37	14,080	55
0011 1000	38	14,336	56
0011 1001	39	14,592	57
0011 1010	3A	14,848	58
0011 1011	3B	15,104	59
0011 1100	3C	15,360	60
0011 1101	3D	15,616	61
0011 1110	3E	15,872	62
0011 1111	3F	16,128	63
0100 0000	40	16,384	64
0100 0001	41	16,640	65
0100 0010	42	16,896	66
0100 0011	43	17,152	67
0100 0100	44	17,408	68
0100 0101	45	17,664	69
0100 0110	46	17,920	70
0100 0111	47	18,176	71
0100 1000	48	18,432	72
0100 1001	49	18,688	73
0100 1010	4A	18,944	74
0100 1011	4B	19,200	75
0100 1100	4C	19,456	76
0100 1101	4D	19,712	77
0100 1110	4E	19,968	78
0100 1111	4F	20,224	79
0101 0000	50	20,480	80
0101 0001	51	20,736	81
0101 0010	52	20,992	82
0101 0011	53	21,248	83
0101 0100	54	21,504	84
0101 0101	55	21,760	85
0101 0110	56	22,016	86
0101 0111	57	22,272	87
0101 1000	58	22,528	88
0101 1001	59	22,784	89
0101 1010	5A	23,040	90
0101 1011	5B	23,296	91
0101 1100	5C	23,552	92
0101 1101	5D	23,808	93
0101 1110	5E	24,064	94
0101 1111	5F	24,320	95

Binary	Hexadecimal	UB Decimal	LB Decimal
0110 0000	60	24,576	96
0110 0001	61	24,832	97
0110 0010	62	25,088	98
0110 0011	63	25,344	99
0110 0100	64	25,600	100
0110 0101	65	25,856	101
0110 0110	66	26,112	102
0110 0111	67	26,368	103
0110 1000	68	26,624	104
0110 1001	69	26,880	105
0110 1010	6A	27,136	106
0110 1011	6B	27,392	107
0110 1100	6C	27,648	108
0110 1101	6D	27,904	109
0110 1110	6E	28,160	110
0110 1111	6F	28,416	111
0111 0000	70	28,672	112
0111 0001	71	28,928	113
0111 0010	72	29,184	114
0111 0011	73	29,440	115
0111 0100	74	29,696	116
0111 0101	75	29,952	117
0111 0110	76	30,208	118
0111 0111	77	30,464	119
0111 1000	78	30,720	120
0111 1001	79	30,976	121
0111 1010	7A	31,232	122
0111 1011	7B	31,488	123
0111 1100	7C	31,744	124
0111 1101	7D	32,000	125
0111 1110	7E	32,256	126
0111 1111	7F	32,512	127
1000 0000	80	32,768	128
1000 0001	81	33,024	129
1000 0010	82	33,280	130
1000 0011	83	33,536	131
1000 0100	84	33,792	132
1000 0101	85	34,048	133
1000 0110	86	34,304	134
1000 0111	87	34,560	135
1000 1000	88	34,816	136
1000 1001	89	35,072	137
1000 1010	8A	35,328	138
1000 1011	8B	35,584	139
1000 1100	8C	35,840	140
1000 1101	8D	36,096	141
1000 1110	8E	36,352	142
1000 1111	8F	36,608	143
1001 0000	90	36,864	144
1001 0001	91	37,120	145
1001 0010	92	37,376	146
1001 0011	93	37,632	147
1001 0100	94	37,888	148
1001 0101	95	38,144	149
1001 0110	96	38,400	150
1001 0111	97	38,656	151
1001 1000	98	38,912	152
1001 1001	99	39,168	153
1001 1010	9A	39,424	154
1001 1011	9B	39,680	155
1001 1100	9C	39,936	156
1001 1101	9D	40,192	157
1001 1110	9E	40,448	158
1001 1111	9F	40,704	159
1010 0000	A0	40,960	160
1010 0001	A1	41,216	161
1010 0010	A2	41,472	162
1010 0011	A3	41,728	163
1010 0100	A4	41,984	164
1010 0101	A5	42,240	165
1010 0110	A6	42,496	166
1010 0111	A7	42,752	167
1010 1000	A8	43,008	168
1010 1001	A9	43,264	169
1010 1010	AA	43,520	170
1010 1011	AB	43,776	171
1010 1100	AC	44,032	172
1010 1101	AD	44,288	173
1010 1110	AE	44,544	174
1010 1111	AF	44,800	175
1011 0000	B0	45,056	176
1011 0001	B1	45,312	177
1011 0010	B2	45,568	178
1011 0011	B3	45,824	179
1011 0100	B4	46,080	180
1011 0101	B5	46,336	181
1011 0110	B6	46,592	182
1011 0111	B7	46,848	183
1011 1000	B8	47,104	184
1011 1001	B9	47,360	185
1011 1010	BA	47,616	186
1011 1011	BB	47,872	187
1011 1100	BC	48,128	188
1011 1101	BD	48,384	189
1011 1110	BE	48,640	190
1011 1111	BF	48,896	191
1100 0000	C0	49,152	192
1100 0001	C1	49,408	193
1100 0010	C2	49,664	194
1100 0011	C3	49,920	195

APPENDIX 2. BINARY-HEXADECIMAL-DECIMAL EQUIVALENTS (*Continued*)

Binary	Hexadecimal	UB Decimal	LB Decimal	Binary	Hexadecimal	UB Decimal	LB Decimal
1100 0100	C4	50,176	196	1110 0010	E2	57,856	226
1100 0101	C5	50,432	197	1110 0011	E3	58,112	227
1100 0110	C6	50,688	198	1110 0100	E4	58,368	228
1100 0111	C7	50,944	199	1110 0101	E5	58,624	229
1100 1000	C8	51,200	200	1110 0110	E6	58,880	230
1100 1001	C9	51,456	201	1110 0111	E7	59,136	231
1100 1010	CA	51,712	202	1110 1000	E8	59,392	232
1100 1011	CB	51,968	203	1110 1001	E9	59,648	233
1100 1100	CC	52,224	204	1110 1010	EA	59,904	234
1100 1101	CD	52,480	205	1110 1011	EB	60,160	235
1100 1110	CE	52,736	206	1110 1100	EC	60,416	236
1100 1111	CF	52,992	207	1110 1101	ED	60,672	237
1101 0000	D0	53,248	208	1110 1110	EE	60,928	238
1101 0001	D1	53,504	209	1110 1111	EF	61,184	239
1101 0010	D2	53,760	210	1111 0000	F0	61,440	240
1101 0011	D3	54,016	211	1111 0001	F1	61,696	241
1101 0100	D4	54,272	212	1111 0010	F2	61,952	242
1101 0101	D5	54,528	213	1111 0011	F3	62,208	243
1101 0110	D6	54,784	214	1111 0100	F4	62,464	244
1101 0111	D7	55,040	215	1111 0101	F5	62,720	245
1101 1000	D8	55,296	216	1111 0110	F6	62,976	246
1101 1001	D9	55,552	217	1111 0111	F7	63,232	247
1101 1010	DA	55,808	218	1111 1000	F8	63,488	248
1101 1011	DB	56,064	219	1111 1001	F9	63,744	249
1101 1100	DC	56,320	220	1111 1010	FA	64,000	250
1101 1101	DD	56,576	221	1111 1011	FB	64,256	251
1101 1110	DE	56,832	222	1111 1100	FC	64,512	252
1101 1111	DF	57,088	223	1111 1101	FD	64,768	253
1110 0000	E0	57,344	224	1111 1110	FE	65,024	254
1110 0001	E1	57,600	225	1111 1111	FF	65,280	255

APPENDIX 3. 7400 SERIES TTL

Number	Function
7400	Quad 2-input NAND gates
7401	Quad 2-input NAND gates (open collector)
7402	Quad 2-input NOR gates
7403	Quad 2-input NOR gates (open collector)
7404	Hex inverters
7405	Hex inverters (open collector)
7406	Hex inverter buffer-driver
7407	Hex buffer-drivers
7408	Quad 2-input AND gates
7409	Quad 2-input AND gates (open collector)
7410	Triple 3-input NAND gates
7411	Triple 3-input AND gates
7412	Triple 3-input NAND gates (open collector)
7413	Dual Schmitt triggers
7414	Hex Schmitt triggers
7416	Hex inverter buffer-drivers
7417	Hex buffer-drivers
7420	Dual 4-input NAND gates
7421	Dual 4-input AND gates
7422	Dual 4-input NAND gates (open collector)
7423	Expandable dual 4-input NOR gates
7425	Dual 4-input NOR gates
7226	Quad 2-input TTL-MOS interface NAND gates
7427	Triple 3-input NOR gates
7428	Quad 2-input NOR buffer
7430	8-input NAND gate
7432	Quad 2-input OR gates
7437	Quad 2-input NAND buffers
7438	Quad 2-input NAND buffers (open collector)
7439	Quad 2-input NAND buffers (open collector)
7440	Dual 4-input NAND buffers
7441	BCD-to-decimal decoder–Nixie driver
7442	BCD-to-decimal decoder
7443	Excess 3-to-decimal decoder
7444	Excess Gray-to-decimal
7445	BCD-to-decimal decoder-driver
7446	BCD-to-seven segment decoder-drivers (30-V output)
7447	BCD-to-seven segment decoder-drivers (15-V output)
7448	BCD-to-seven segment decoder-drivers
7450	Expandable dual 2-input 2-wide AND-OR-INVERT gates
7451	Dual 2-input 2-wide AND-OR-INVERT gates
7452	Expandable 2-input 4-wide AND-OR gates
7453	Expandable 2-input 4-wide AND-OR-INVERT gates
7454	2-input 4-wide AND-OR-INVERT gates
7455	Expandable 4-input 2-wide AND-OR-INVERT gates
7459	Dual 2-3 input 2-wide AND-OR-INVERT gates
7460	Dual 4-input expanders
7461	Triple 3-input expanders
7462	2-2-3-3 input 4-wide expanders
7464	2-2-3-4 input 4-wide AND-OR-INVERT gates
7465	4-wide AND-OR-INVERT gates (open collector)
7470	Edge-triggered *JK* flip-flop
7472	*JK* master-slave flip-flop
7473	Dual *JK* master-slave flip-flop
7474	Dual *D* flip-flop
7475	Quad latch
7476	Dual *JK* master-slave flip-flop
7480	Gates full adder
7482	2-bit binary full adder
7483	4-bit binary full adder
7485	4-bit magnitude comparator
7486	Quad EXCLUSIVE-OR gate
7489	64-bit random-access read-write memory
7490	Decade counter
7491	8-bit shift register
7492	Divide-by-12 counter
7493	4-bit binary counter
7494	4-bit shift register
7495	4-bit right-shift–left-shift register
7496	5-bit parallel-in–parallel-out shift register
74100	4-bit bistable latch
74104	*JK* master-slave flip-flop
74105	*JK* master-slave flip-flop
74107	Dual *JK* master-slave flip-flop
74109	Dual *JK* positive-edge-triggered flip-flop
74116	Dual 4-bit latches with clear
74121	Monostable multivibrator
74122	Monostable multivibrator with clear
74123	Monostable multivibrator
74125	Three-state quad bus buffer
74126	Three-state quad bus buffer
74132	Quad Schmitt trigger
74136	Quad 2-input EXCLUSIVE-OR gate
74141	BCD-to-decimal decoder-driver
74142	BCD counter-latch-driver
74145	BCD-to-decimal decoder-driver
74147	10/4 priority encoder
74148	Priority encoder
74150	16-line-to-1-line multiplexer
74151	8-channel digital multiplexer
74152	8-channel data selector-multiplexer

APPENDIX 3. 7400 SERIES TTL (*Continued*)

Number	Function	Number	Function
74153	Dual 4/1 multiplexer	74190	Up-down decade counter
74154	4-line–to–16-line decoder-demultiplexer	74191	Synchronous binary up-down counter
74155	Dual 2/4 demultiplexer	74192	Binary up-down counter
74156	Dual 2/4 demultiplexer	74193	Binary up-down counter
74157	Quad 2/1 data selector	74194	4-bit directional shift register
74160	Decade counter with asynchronous clear	74195	4-bit parallel-access shift register
74161	Synchronous 4-bit counter	74196	Presettable decade counter
74162	Synchronous 4-bit counter	74197	Presettable binary counter
74163	Synchronous 4-bit counter	74198	8-bit shift register
74164	8-bit serial shift register	74199	8-bit shift register
74165	Parallel-load 8-bit serial shift register	74221	Dual one-shot Schmitt trigger
74166	8-bit shift register	74251	Three-state 8-channel multiplexer
74173	4-bit three-state register	74259	8-bit addressable latch
74174	Hex *F* flip-flop with clear	74276	Quad *JK* flip-flop
74175	Quad *D* flip-flop with clear	74279	Quad debouncer
74176	35-MHz presettable decade counter	74283	4-bit binary full adder with fast carry
74177	35-MHz presettable binary counter	74284	Three-state 4-bit multiplexer
74179	4-bit parallel-access shift register	74285	Three-state 4-bit multiplexer
74180	8-bit odd-even parity generator-checker	74365	Three-state hex buffers
74181	Arithmetic-logic unit	74366	Three-state hex buffers
74182	Look-ahead carry generator	74367	Three-state hex buffers
74184	BCD-to-binary converter	74368	Three-state hex buffers
74185	Binary-to-BCD converter	74390	Individual clocks with flip-flops
74189	Three-state 64-bit random-access memory	74393	Dual 4-bit binary counter

APPENDIX 4. PINOUTS AND FUNCTION TABLES

74LS83

The 74LS83 is a 4-bit full adder; the binary output is

$$\mathbf{S} = \mathbf{A} + \mathbf{B}$$

74LS83
A3 1
S2 2
A2 3
B2 4
VCC 5
S1 6
B1 7
A1 8
16 B3
15 S3
14 CARRY OUT
13 CARRY IN
12 GND
11 B0
10 A0
9 S0

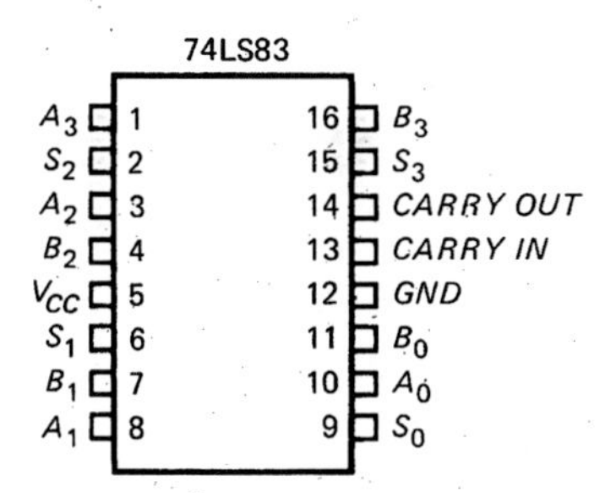

Fig. A4-1

In Fig. A4-1, pins 1, 3, 8, and 10 are the **A** input (A_3, A_2, A_1, A_0); pins 16, 4, 7, and 11 are the **B** input (B_3, B_2, B_1, B_0); and pins 15, 2, 6, and 9 are the **S** output (S_3, S_2, S_1, S_0). Pin 13 is the CARRY IN, and pin 14 is the CARRY OUT.

74LS157

This chip is a word multiplexer. Two words of 4 bits each are the inputs; one word of 4 bits is the output. The two input words are designated **L** (left) and **R** (right); the output word is **Y.** In Fig. A4-2, pin 1 (SELECT) and pin 15 (STROBE) are control inputs. The **L** word goes to pins 14, 11, 5, 2 (L_3, L_2, L_1, L_0), and the **R** word goes to pins 13, 10, 6, and 3 (R_3, R_2, R_1, R_0).

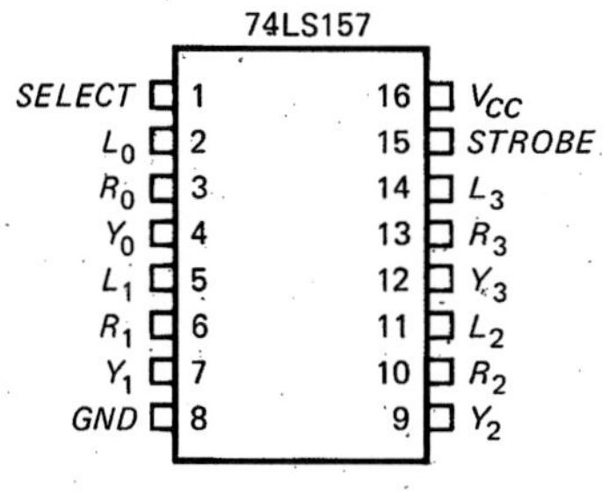

74LS157
SELECT 1
L0 2
R0 3
Y0 4
L1 5
R1 6
Y1 7
GND 8
16 VCC
15 STROBE
14 L3
13 R3
12 Y3
11 L2
10 R2
9 Y2

Fig. A4-2

TABLE A4-1. FUNCTION TABLE

STROBE	*SELECT*	**Y**	**Comment**
1	X	0	Output goes low
0	0	**L**	Output equals left word
0	1	**R**	Output equals right word

As indicated in Table A4-1, a high *STROBE* input produces a low output, no matter what the input words. When *STROBE* is low, the *SELECT* input controls the operation. A low *SELECT* will send the **L** word to the output; a high *SELECT* sends the **R** word to the output.

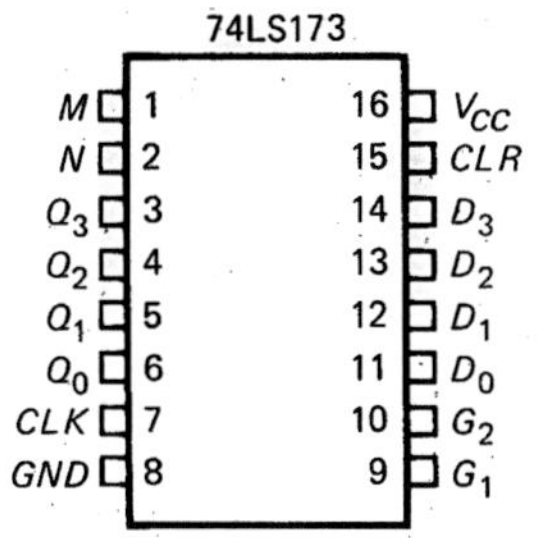

74LS173
M 1
N 2
Q3 3
Q2 4
Q1 5
Q0 6
CLK 7
GND 8
16 VCC
15 CLR
14 D3
13 D2
12 D1
11 D0
10 G2
9 G1

Fig. A4-3

74LS173

The 74LS173 is a 4-bit buffer register with three-state outputs. In Fig. A4-3, pins 14, 13, 12, and 11 are the data inputs (D_3, D_2, D_1, D_0). Pins 3, 4, 5, and 6 are the data outputs (Q_3, Q_2, Q_1, Q_0). Pins 9 and 10 (G_1 and G_2) are the input control. Pins 1 and 2 (M and N) are the output control.

As shown in Table A4-2, both M and N must be low to get a Q output. If either M or N (or both) is high, the output is three-stated (floating or high impedance).

When M and N are both low, Table A4-3 applies. As indicated, a high *CLEAR* will clear all Q bits to 0. When *CLEAR* is low, G_1 and G_2 control input loading. If either G_1 or G_2 (or both) are high, no change takes place in the Q bits. When both G_1 and G_2 are low, the next positive clock edge loads the input data.

TABLE A4-2. OUTPUT CONTROL

M	N	**Output**
0	0	Connected
0	1	Hi-Z
1	0	Hi-Z
1	1	Hi-Z

TABLE A4-3. FUNCTION TABLE FOR $M = 0$ AND $N = 0$

CLEAR	*CLOCK*	G_1	G_2	D_n	Q_n	**Comment**
1	X	X	X	X	0	Clear output
0	0	X	X	X	NC	No change
0	↑	1	X	X	NC	No change
0	↑	X	1	X	NC	No change
0	↑	0	0	0	0	Reset bit n
0	↑	0	0	1	1	Set bit n

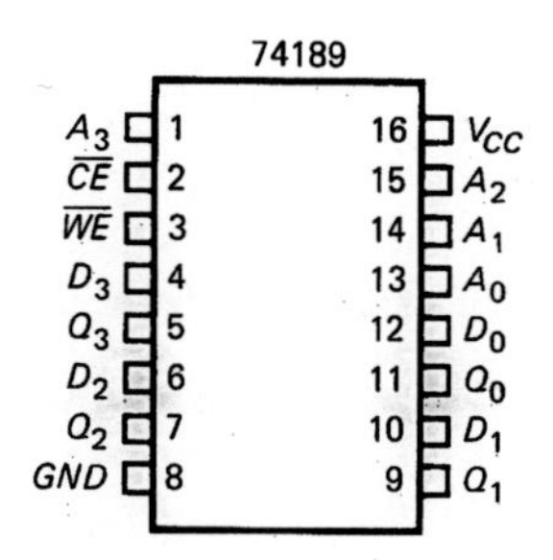

Fig. A4-4

74189

The 74189 is a 64-bit RAM organized as 16 words of 4 bits each. In Fig. A4-4 pins 1, 15, 14, and 13 are the address inputs (A_3, A_2, A_1, A_0). Pins 4, 6, 10, and 12 are the data inputs (D_3, D_2, D_1, D_0). Pins 5, 7, 9, and 11 are the data outputs (Q_3, Q_2, Q_1, Q_0).

TABLE A4-4. FUNCTION TABLE

$\overline{CE}$	$\overline{WE}$	Output	Comment
1	X	Hi-Z	Do nothing
0	0	Hi-Z	Write complement
0	1	Stored word	Read

Table A4-4 summarizes the operation of this read-write memory. When $\overline{CE}$ is high, the output is three-stated (high impedance). When $\overline{CE}$ is low and $\overline{WE}$ is low, the complement of the input data word is stored at the addressed memory location; during this write operation, the output is three-stated. When $\overline{CE}$ is low and $\overline{WE}$ is high, the stored word appears at the output.

APPENDIX 5. SAP-1 PARTS LIST

Chips

C1: 74LS107, dual *JK* master-slave flip-flop
C2: 74LS107
C3: 74LS126, quad three-state normally open switches
C4: 74LS173, buffer register, three-state outputs, 4 bits
C5: 74LS157, 2-to-1 nibble multiplexer
C6: 74189, 64-bit (16 × 4) static RAM, three-state outputs
C7: 74189
C8: 74LS173
C9: 74LS173
C10: 74LS173
C11: 74LS173
C12: 74LS126
C13: 74LS126
C14: 74LS86, quad 2-input EXCLUSIVE-OR gates
C15: 74LS86
C16: 74LS83, quad full adders
C17: 74LS83
C18: 74LS126
C19: 74LS126
C20: 74LS173
C21: 74LS173
C22: 74LS173
C23: 74LS173
C24: 7400, quad 2-input NAND gates
C25: 74LS10, triple 3-input NAND gates
C26: 74LS00
C27: 7404, hex inverter
C28: NE555, timer
C29: 74LS107
C30: LM340T-5, voltage regulator, 5 V
C31: 74LS04, hex inverter
C32: 74LS20, dual 4-input NAND gates
C33: 74LS20
C34: 74LS20
C35: 74LS04
C36: 74LS107
C37: 74LS107
C38: 74LS107
C39: 74LS00
C40: 74LS00
C41: 74LS00
C42: 74LS00
C43: 74LS00
C44: 74LS20
C45: 74LS10
C46: 74LS00
C47: 74LS04
C48: 74LS04

Diodes

D1: 1N4001, rectifier diode, 50 PIV, 1 A
D2: 1N4001
D3: 1N4001
D4: 1N4001

Switches

S1: SPST DIP switch, 4 bits
S2: DPST on-off
S3: SPST DIP, 8 bits
S4: SPST push button, momentary, normally open
S5: SPDT push button, momentary
S6: SPDT push button, momentary
S7: SPDT on-on switch

Miscellaneous

Resistors: eight 1-kΩ, fourteen 10-kΩ, one 18-kΩ, one 36-kΩ
Capacitors: 0.01-μF, 0.1-μF, 1000-μF (50 V)
Transformer: F-25X = 115 V primary, 12.6 V secondary CT, 1.5 A
Fuse: $\frac{3}{8}$-A slow blow

Totals

1N4001–4	74LS20–4
LM340T-5–1	74LS83–2
NE555–1	74LS86–2
7400–1	74LS107–6
74LS00–7	74LS126–5
7404–1	74LS157–1
74LS04–4	74LS173–9
74LS10–2	74189–2

APPENDIX 6. 8085 INSTRUCTIONS

Instruction	Op Code	*T* states	Flags	Main Effect
ACI byte	CE	7	All	A ← A + CY + byte
ADC A	8F	4	All	A ← A + A + CY
ADC B	88	4	All	A ← A + B + CY
ADC C	89	4	All	A ← A + C + CY
ADC D	8A	4	All	A ← A + D + CY
ADC E	8B	4	All	A ← A + E + CY
ADC H	8C	4	All	A ← A + H + CY
ADC L	8D	4	All	A ← A + L + CY
ADC M	8E	7	All	A ← A + M_{HL} + CY
ADD A	87	4	All	A ← A + A
ADD B	80	4	All	A ← A + B
ADD C	81	4	All	A ← A + C
ADD D	82	4	All	A ← A + D
ADD E	83	4	All	A ← A + E
ADD H	84	4	All	A ← A + H
ADD L	85	4	All	A ← A + L
ADD M	86	7	All	A ← A + M_{HL}
ADI byte	C6	7	All	A ← A + byte
ANA A	A7	4	All	A ← A AND A
ANA B	A0	4	All	A ← A AND B
ANA C	A1	4	All	A ← A AND C
ANA D	A2	4	All	A ← A AND D
ANA E	A3	4	All	A ← A AND E
ANA H	A4	4	All	A ← A AND H
ANA L	A5	4	All	A ← A AND L
ANA M	A6	7	All	A ← A AND M_{HL}
ANI byte	E6	7	All	A ← A AND byte
CALL address	CD	18	None	PC ← address
CC address	DC	18/9	None	PC ← address if $CY = 1$
CM address	FC	18/9	None	PC ← address if $S = 1$
CMA	2F	4	None	A ← $\overline{A}$
CMC	3F	4	CY	CY ← $\overline{CY}$
CMP A	BF	4	All	Z ← 1 if A = A
CMP B	B8	4	All	Z ← 1 if A = B
CMP C	B9	4	All	Z ← 1 if A = C
CMP D	BA	4	All	Z ← 1 if A = D
CMP E	BB	4	All	Z ← 1 if A = E
CMP H	BC	4	All	Z ← 1 if A = H
CMP L	BD	4	All	Z ← 1 if A = L
CMP M	BE	7	All	Z ← 1 if A = M_{HL}
CNC address	D4	18/9	None	PC ← address if $CY = 0$
CNZ address	C4	18/9	None	PC ← address if $Z = 0$
CP address	F4	18/9	None	PC ← address if $S = 0$
CPE address	EC	18/9	None	PC ← address if $P = 1$
CPI byte	FE	7	All	Z ← 1 if A = byte
CPO address	E4	18/9	None	PC ← address if $P = 0$
CZ address	CC	18/9	None	PC ← address if $Z = 1$
DAA	27	4	All	A ← BCD number
DAD B	09	10	CY	HL ← HL + BC
DAD D	19	10	CY	HL ← HL + DE
DAD H	29	10	CY	HL ← HL + HL

Instruction	Op Code	*T* states	Flags	Main Effect
DAD SP	39	10	CY	HL ← HL + SP
DCR A	3D	4	All but CY	A ← A − 1
DCR B	05	4	All but CY	B ← B − 1
DCR C	0D	4	All but CY	C ← C − 1
DCR D	15	4	All but CY	D ← D − 1
DCR E	1D	4	All but CY	E ← E − 1
DCR H	25	4	All but CY	H ← H − 1
DCR L	2D	4	All but CY	L ← L − 1
DCR M	35	10	All but CY	$M_{HL} \leftarrow M_{HL} - 1$
DCX B	0B	6	None	BC ← BC − 1
DCX D	1B	6	None	DE ← DE − 1
DCX H	2B	6	None	HL ← HL − 1
DCX SP	3B	6	None	SP ← SP − 1
DI	F3	4	None	Disable interrupts
EI	FB	4	None	Enable interrupts
HLT	76	5	None	Stop processing
IN byte	DB	10	None	A ← byte
INR A	3C	4	All but CY	A ← A + 1
INR B	04	4	All but CY	B ← B + 1
INR C	0C	4	All but CY	C ← C + 1
INR D	14	4	All but CY	D ← D + 1
INR E	1C	4	All but CY	E ← E + 1
INR H	24	4	All but CY	H ← H + 1
INR L	2C	4	All but CY	L ← L + 1
INR M	34	10	All but CY	$M_{HL} \leftarrow M_{HL} + 1$
INX B	03	6	None	BC ← BC + 1
INX D	13	6	None	DE ← DE + 1
INX H	23	6	None	HL ← HL + 1
INX SP	33	6	None	SP ← SP + 1
JC address	DA	10/7	None	PC ← address if *CY* = 1
JM address	FA	10/7	None	PC ← address if *S* = 1
JMP address	C3	10	None	PC ← address
JNC address	D2	10/7	None	PC ← address if *CY* = 0
JNZ address	C2	10/7	None	PC ← address if *Z* = 0
JP address	F2	10/7	None	PC ← address if *S* = 0
JPE address	EA	10/7	None	PC ← address if *P* = 1
JPO address	E2	10/7	None	PC ← address if *P* = 0
JZ address	CA	10/7	None	PC ← address if *Z* = 1
LDA address	3A	13	None	$A \leftarrow M_{adr}$
LDAX B	0A	7	None	$A \leftarrow M_{BC}$
LDAX D	1A	7	None	$A \leftarrow M_{DE}$
LHLD address	2A	16	None	$H \leftarrow M_{adr}$
LXI B, dble	01	10	None	BC ← dble
LXI D, dble	11	10	None	DE ← dble
LXI H, dble	21	10	None	HL ← dble
LXI SP, dble	31	10	None	SP ← dble
MOV A,A	7F	4	None	A ← A
MOV A,B	78	4	None	A ← B
MOV A,C	79	4	None	A ← C
MOV A,D	7A	4	None	A ← D
MOV A,E	7B	4	None	A ← E
MOV A,H	7C	4	None	A ← H

APPENDIX 6. 8085 INSTRUCTIONS (*Continued*)

Instruction	Op Code	*T* states	Flags	Main Effect
MOV A,L	7D	4	None	A ← L
MOV A,M	7E	7	None	A ← M_{HL}
MOV B,A	47	4	None	B ← A
MOV B,B	40	4	None	B ← B
MOV B,C	41	4	None	B ← C
MOV B,D	42	4	None	B ← D
MOV B,E	43	4	None	B ← E
MOV B,H	44	4	None	B ← H
MOV B,L	45	4	None	B ← L
MOV B,M	46	7	None	B ← M_{HL}
MOV C,A	4F	4	None	C ← A
MOV C,B	48	4	None	C ← B
MOV C,C	49	4	None	C ← C
MOV C,D	4A	4	None	C ← D
MOV C,E	4B	4	None	C ← E
MOV C,H	4C	4	None	C ← H
MOV C,L	4D	4	None	C ← L
MOV C,M	4E	7	None	C ← M_{HL}
MOV D,A	57	4	None	D ← A
MOV D,B	50	4	None	D ← B
MOV D,C	51	4	None	D ← C
MOV D,D	52	4	None	D ← D
MOV D,E	53	4	None	D ← E
MOV D,H	54	4	None	D ← H
MOV D,L	55	4	None	D ← L
MOV D,M	56	7	None	D ← M_{HL}
MOV E,A	5F	4	None	E ← A
MOV E,B	58	4	None	E ← B
MOV E,C	59	4	None	E ← C
MOV E,D	5A	4	None	E ← D
MOV E,E	5B	4	None	E ← E
MOV E,H	5C	4	None	E ← H
MOV E,L	5D	4	None	E ← L
MOV E,M	5E	7	None	E ← M_{HL}
MOV H,A	67	4	None	H ← A
MOV H,B	60	4	None	H ← B
MOV H,C	61	4	None	H ← C
MOV H,D	62	4	None	H ← D
MOV H,E	63	4	None	H ← E
MOV H,H	64	4	None	H ← H
MOV H,L	65	4	None	H ← L
MOV H,M	66	7	None	H ← M_{HL}
MOV L,A	6F	4	None	L ← A
MOV L,B	68	4	None	L ← B
MOV L,C	69	4	None	L ← C
MOV L,D	6A	4	None	L ← D
MOV L,E	6B	4	None	L ← E
MOV L,H	6C	4	None	L ← H
MOV L,L	6D	4	None	L ← L
MOV L,M	6E	7	None	L ← M_{HL}
MOV M,A	77	7	None	M_{HL} ← A

Instruction	Op Code	*T* states	Flags	Main Effect
MOV M,B	70	7	None	$M_{HL} \leftarrow B$
MOV M,C	71	7	None	$M_{HL} \leftarrow C$
MOV M,D	72	7	None	$M_{HL} \leftarrow D$
MOV M,E	73	7	None	$M_{HL} \leftarrow E$
MOV M,H	74	7	None	$M_{HL} \leftarrow H$
MOV M,L	75	7	None	$M_{HL} \leftarrow L$
MVI A,byte	3E	7	None	A ← byte
MVI B,byte	06	7	None	B ← byte
MVI C,byte	0E	7	None	C ← byte
MVI D,byte	16	7	None	D ← byte
MVI E,byte	1E	7	None	E ← byte
MVI H,byte	26	7	None	H ← byte
MVI L,byte	2E	7	None	L ← byte
MVI M,byte	36	10	None	M_{HL} ← byte
NOP	00	4	None	Delay
ORA A	B7	4	All	A ← A OR A
ORA B	B0	4	All	A ← A OR B
ORA C	B1	4	All	A ← A OR C
ORA D	B2	4	All	A ← A OR D
ORA E	B3	4	All	A ← A OR E
ORA H	B4	4	All	A ← A OR H
ORA L	B5	4	All	A ← A OR L
ORA M	B6	7	All	A ← A OR M_{HL}
ORI byte	F6	7	All	A ← A OR byte
OUT byte	D3	10	None	Port byte ← A
PCHL	E9	6	None	PC ← HL
POP B	C1	10	None	$B \leftarrow M_{stk}$
POP D	D1	10	None	$D \leftarrow M_{stk}$
POP H	E1	10	None	$H \leftarrow M_{stk}$
POP PSW	F1	10	None	$F \leftarrow M_{stk}$, $A \leftarrow M_{stk} - 1$
PUSH B	C5	12	None	$M_{stk} - 1 \leftarrow B$, $M_{stk} - 2 \leftarrow C$
PUSH D	D5	12	None	$M_{stk} - 1 \leftarrow D$, $M_{stk} - 2 \leftarrow E$
PUSH H	E5	12	None	$M_{stk} - 1 \leftarrow H$, $M_{stk} - 2 \leftarrow L$
PUSH PSW	F5	12	None	$M_{stk} - 1 \leftarrow A$, $M_{stk} - 2 \leftarrow F$
RAL	17	4	CY	Rotate all left
RAR	1F	4	CY	Rotate all right
RC	D8	12/6	None	PC ← return address if $CY = 1$
RET	C9	10	None	PC ← return address
RIM	20	4	None	A ← I
RLC	07	4	CY	Rotate left with carry
RM	F8	12/6	None	PC ← return address if $S = 1$
RNC	D0	12/6	None	PC ← return address if $CY = 0$
RNZ	C0	12/6	None	PC ← return address if $Z = 0$
RP	F0	12/6	None	PC ← return address if $S = 0$
RPE	E8	12/6	None	PC ← return address if $P = 1$
RPO	E0	12/6	None	PC ← return address if $P = 0$
RRC	0F	4	CY	Rotate right with carry
RST 0	C7	12	None	PC ← 0000H
RST 1	CF	12	None	PC ← 0008H
RST 2	D7	12	None	PC ← 0010H
RST 3	DF	12	None	PC ← 0018H
RST 4	E7	12	None	PC ← 0020H
RST 5	EF	12	None	PC ← 0028H

APPENDIX 6. 8085 INSTRUCTIONS (*Continued*)

Instruction	Op Code	*T* states	Flags	Main Effect
RST 6	F7	12	None	PC ← 0030H
RST 7	FF	12	None	PC ← 0038H
RZ	C8	12/6	None	PC ← return address if $Z = 1$
SBB A	9F	4	All	A ← A − A − CY
SBB B	98	4	All	A ← A − B − CY
SBB C	99	4	All	A ← A − C − CY
SBB D	9A	4	All	A ← A − D − CY
SBB E	9B	4	All	A ← A − E − CY
SBB H	9C	4	All	A ← A − H − CY
SBB L	9D	4	All	A ← A − L − CY
SBB M	9E	7	All	A ← A − M − CY
SBI byte	DE	7	All	A ← A − byte − CY
SHLD address	22	16	None	$M_{adr+1} \leftarrow H, M_{adr} \leftarrow L$
SIM	30	4	None	I ← A
SPHL	F9	6	None	SP ← HL
STA address	32	13	None	$M_{adr} \leftarrow A$
STAX B	02	7	None	$M_{BC} \leftarrow A$
STAX D	12	7	None	$M_{DE} \leftarrow A$
STC	37	4	CY	CY ← 1
SUB A	97	4	All	A ← A − A
SUB B	90	4	All	A ← A − B
SUB C	91	4	All	A ← A − C
SUB D	92	4	All	A ← A − D
SUB E	93	4	All	A ← A − E
SUB H	94	4	All	A ← A − H
SUB L	95	4	All	A ← A − L
SUB M	96	7	All	A ← A − M
SUI byte	D6	7	All	A ← A − byte
XCHG	EB	4	None	HL ↔ DE
XRA A	AF	4	All	A ← A XOR A
XRA B	A8	4	All	A ← A XOR B
XRA C	A9	4	All	A ← A XOR C
XRA D	AA	4	All	A ← A XOR D
XRA E	AB	4	All	A ← A XOR E
XRA H	AC	4	All	A ← A XOR H
XRA L	AD	4	All	A ← A XOR L
XRA M	AE	7	All	A ← A XOR M
XRI byte	EE	7	All	A ← A XOR byte
XTHL	E3	16	None	HL ↔ stack

APPENDIX 7. MEMORY LOCATIONS: POWERS OF 2

Address Bits	Hexadecimal	Decimal	Power of 2
0000 0000 0000 0001	0001H	1	0
0000 0000 0000 0010	0002H	2	1
0000 0000 0000 0100	0004H	4	2
0000 0000 0000 1000	0008H	8	3
0000 0000 0001 0000	0010H	16	4
0000 0000 0010 0000	0020H	32	5
0000 0000 0100 0000	0040H	64	6
0000 0000 1000 0000	0080H	128	7
0000 0001 0000 0000	0100H	256	8
0000 0010 0000 0000	0200H	512	9
0000 0100 0000 0000	0400H	1,024	10
0000 1000 0000 0000	0800H	2,048	11
0001 0000 0000 0000	1000H	4,096	12
0010 0000 0000 0000	2000H	8,192	13
0100 0000 0000 0000	4000H	16,384	14
1000 0000 0000 0000	8000H	32,768	15

APPENDIX 8. MEMORY LOCATIONS: 16K AND 8K INTERVALS

Address Bits	Hexadecimal	Decimal	Zone
Zone bits = $A_{15}A_{14}$			
0000 0000 0000 0000	0000H	0	0
0011 1111 1111 1111	3FFFH	16,383	
0100 0000 0000 0000	4000H	16,384	1
0111 1111 1111 1111	7FFFH	32,767	
1000 0000 0000 0000	8000H	32,768	2
1011 1111 1111 1111	BFFFH	49,151	
1100 0000 0000 0000	C000H	49,152	3
1111 1111 1111 1111	FFFFH	65,535	
Zone bits = $A_{15}A_{14}A_{13}$			
0000 0000 0000 0000	0000H	0	0
0001 1111 1111 1111	1FFFH	8,191	
0010 0000 0000 0000	2000H	8,192	1
0011 1111 1111 1111	3FFFH	16,383	
0100 0000 0000 0000	4000H	16,384	2
0101 1111 1111 1111	5FFFH	24,575	
0110 0000 0000 0000	6000H	24,576	3
0111 1111 1111 1111	7FFFH	32,767	
1000 0000 0000 0000	8000H	32,768	4
1001 1111 1111 1111	9FFFH	40,959	
1010 0000 0000 0000	A000H	40,960	5
1011 1111 1111 1111	BFFFH	49,151	
1100 0000 0000 0000	C000H	49,152	6
1101 1111 1111 1111	DFFFH	57,343	
1110 0000 0000 0000	E000H	57,344	7
1111 1111 1111 1111	FFFFH	65,535	

APPENDIX 9. MEMORY LOCATIONS: 4K INTERVALS

Address Bits	Hexadecimal	Decimal	Zone
Zone bits = $A_{15}A_{14}A_{13}A_{12}$			
0000 0000 0000 0000	0000H	0	0
0000 1111 1111 1111	0FFFH	4,095	
0001 0000 0000 0000	1000H	4,096	1
0001 1111 1111 1111	1FFFH	8,191	
0010 0000 0000 0000	2000H	8,192	2
0010 1111 1111 1111	2FFFH	12,287	
0011 0000 0000 0000	3000H	12,288	3
0011 1111 1111 1111	3FFFH	16,383	
0100 0000 0000 0000	4000H	16,384	4
0100 1111 1111 1111	4FFFH	20,479	
0101 0000 0000 0000	5000H	20,480	5
0101 1111 1111 1111	5FFFH	24,575	
0110 0000 0000 0000	6000H	24,576	6
0110 1111 1111 1111	6FFFH	28,671	
0111 0000 0000 0000	7000H	28,672	7
0111 1111 1111 1111	7FFFH	32,767	
1000 0000 0000 0000	8000H	32,768	8
1000 1111 1111 1111	8FFFH	36,863	
1001 0000 0000 0000	9000H	36,864	9
1001 1111 1111 1111	9FFFH	40,959	
1010 0000 0000 0000	A000H	40,960	10
1010 1111 1111 1111	AFFFH	45,055	
1011 0000 0000 0000	B000H	45,056	11
1011 1111 1111 1111	BFFFH	49,151	
1100 0000 0000 0000	C000H	49,152	12
1100 1111 1111 1111	CFFFH	53,247	
1101 0000 0000 0000	D000H	53,248	13
1101 1111 1111 1111	DFFFH	57,343	
1110 0000 0000 0000	E000H	57,344	14
1110 1111 1111 1111	EFFFH	61,439	
1111 0000 0000 0000	F000H	61,440	15
1111 1111 1111 1111	FFFFH	65,535	

APPENDIX 10. MEMORY LOCATIONS: 2K INTERVALS

Address Bits	Hexadecimal	Decimal	Zone	Address Bits	Hexadecimal	Decimal	Zone
			Zone bits = $A_{15}A_{14}A_{13}A_{12}A_{11}$				
0000 0000 0000 0000	0000H	0	0	1000 0000 0000 0000	8000H	32,768	16
0000 0111 1111 1111	07FFH	2,047		1000 0111 1111 1111	87FFH	34,815	
0000 1000 0000 0000	0800H	2,048	1	1000 1000 0000 0000	8800H	34,816	17
0000 1111 1111 1111	0FFFH	4,095		1000 1111 1111 1111	8FFFH	36,863	
0001 0000 0000 0000	1000H	4,096	2	1001 0000 0000 0000	9000H	36,864	18
0001 0111 1111 1111	17FFH	6,143		1001 0111 1111 1111	97FFH	38,911	
0001 1000 0000 0000	1800H	6,144	3	1001 1000 0000 0000	9800H	38,912	19
0001 1111 1111 1111	1FFFH	8,191		1001 1111 1111 1111	9FFFH	40,959	
0010 0000 0000 0000	2000H	8,192	4	1010 0000 0000 0000	A000H	40,960	20
0010 0111 1111 1111	27FFH	10,239		1010 0111 1111 1111	A7FFH	43,007	
0010 1000 0000 0000	2800H	10,240	5	1010 1000 0000 0000	A800H	43,008	21
0010 1111 1111 1111	2FFFH	12,287		1010 1111 1111 1111	AFFFH	45,055	
0011 0000 0000 0000	3000H	12,288	6	1011 0000 0000 0000	B000H	45,056	22
0011 0111 1111 1111	37FFH	14,335		1011 0111 1111 1111	B7FFH	47,103	
0011 1000 0000 0000	3800H	14,336	7	1011 1000 0000 0000	B800H	47,104	23
0011 1111 1111 1111	3FFFH	16,383		1011 1111 1111 1111	BFFFH	49,151	
0100 0000 0000 0000	4000H	16,384	8	1100 0000 0000 0000	C000H	49,152	24
0100 0111 1111 1111	47FFH	18,431		1100 0111 1111 1111	C7FFH	51,199	
0100 1000 0000 0000	4800H	18,432	9	1100 1000 0000 0000	C800H	51,200	25
0100 1111 1111 1111	4FFFH	20,479		1100 1111 1111 1111	CFFFH	53,247	
0101 0000 0000 0000	5000H	20,480	10	1101 0000 0000 0000	D000H	53,248	26
0101 0111 1111 1111	57FFH	22,527		1101 0111 1111 1111	D7FFH	55,295	
0101 1000 0000 0000	5800H	22,538	11	1101 1000 0000 0000	D800H	55,296	27
0101 1111 1111 1111	5FFFH	24,575		1101 1111 1111 1111	DFFFH	57,343	
0110 0000 0000 0000	6000H	24,576	12	1110 0000 0000 0000	E000H	57,344	28
0110 0111 1111 1111	67FFH	26,623		1110 0111 1111 1111	E7FFH	59,391	
0110 1000 0000 0000	6800H	26,624	13	1110 1000 0000 0000	E800H	59,392	29
0110 1111 1111 1111	6FFFH	28,671		1110 1111 1111 1111	EFFFH	61,439	
0111 0000 0000 0000	7000H	28,672	14	1111 0000 0000 0000	F000H	61,440	30
0111 0111 1111 1111	77FFH	30,719		1111 0111 1111 1111	F7FFH	63,487	
0111 1000 0000 0000	7800H	30,720	15	1111 1000 0000 0000	F800H	63,488	31
0111 1111 1111 1111	7FFFH	32,767		1111 1111 1111 1111	FFFFH	65,535	

APPENDIX 11. MEMORY LOCATIONS: 1K INTERVALS

Zone bits = $A_{15}A_{14}A_{13}A_{12}A_{11}A_{10}$

Address Bits	Hexadecimal	Decimal	Zone	Address Bits	Hexadecimal	Decimal	Zone
0000 0000 0000 0000 0000 0011 1111 1111	0000H 03FFH	0 1,023	0	0101 0000 0000 0000 0101 0011 1111 1111	5000H 53FFH	20,480 21,503	20
0000 0100 0000 0000 0000 0111 1111 1111	0400H 07FFH	1,024 2,047	1	0101 0100 0000 0000 0101 0111 1111 1111	5400H 57FFH	21,504 22,527	21
0000 1000 0000 0000 0000 1011 1111 1111	0800H 0BFFH	2,048 3,071	2	0101 1000 0000 0000 0101 1011 1111 1111	5800H 5BFFH	22,528 23,551	22
0000 1100 0000 0000 0000 1111 1111 1111	0C00H 0FFFH	3,072 4,095	3	0101 1100 0000 0000 0101 1111 1111 1111	5C00H 5FFFH	23,552 24,575	23
0001 0000 0000 0000 0001 0011 1111 1111	1000H 13FFH	4,096 5,119	4	0110 0000 0000 0000 0110 0011 1111 1111	6000H 63FFH	24,576 25,599	24
0001 0100 0000 0000 0001 0111 1111 1111	1400H 17FFH	5,120 6,143	5	0110 0100 0000 0000 0110 0111 1111 1111	6400H 67FFH	25,600 26,623	25
0001 1000 0000 0000 0001 1011 1111 1111	1800H 1BFFH	6,144 7,167	6	0110 1000 0000 0000 0110 1011 1111 1111	6800H 6BFFH	26,624 27,647	26
0001 1100 0000 0000 0001 1111 1111 1111	1C00H 1FFFH	7,168 8,191	7	0110 1100 0000 0000 0110 1111 1111 1111	6C00H 6FFFH	27,648 28,671	27
0010 0000 0000 0000 0010 0011 1111 1111	2000H 23FFH	8,192 9,215	8	0111 0000 0000 0000 0111 0011 1111 1111	7000H 73FFH	28,672 29,695	28
0010 0100 0000 0000 0010 0111 1111 1111	2400H 27FFH	9,216 10,239	9	0111 0100 0000 0000 0111 0111 1111 1111	7400H 77FFH	29,696 30,719	29
0010 1000 0000 0000 0010 1011 1111 1111	2800H 2BFFH	10,240 11,263	10	0111 1000 0000 0000 0111 1011 1111 1111	7800H 7BFFH	30,720 31,743	30
0010 1100 0000 0000 0010 1111 1111 1111	2C00H 2FFFH	11,264 12,287	11	0111 1100 0000 0000 0111 1111 1111 1111	7C00H 7FFFH	31,744 32,767	31
0011 0000 0000 0000 0011 0011 1111 1111	3000H 33FFH	12,288 13,311	12	1000 0000 0000 0000 1000 0011 1111 1111	8000H 83FFH	32,768 33,791	32
0011 0100 0000 0000 0011 0111 1111 1111	3400H 37FFH	13,312 14,335	13	1000 0100 0000 0000 1000 0111 1111 1111	8400H 87FFH	33,792 34,815	33
0011 1000 0000 0000 0011 1011 1111 1111	3800H 3BFFH	14,336 15,359	14	1000 1000 0000 0000 1000 1011 1111 1111	8800H 8BFFH	34,816 35,839	34
0011 1100 0000 0000 0011 1111 1111 1111	3C00H 3FFFH	15,360 16,383	15	1000 1100 0000 0000 1000 1111 1111 1111	8C00H 8FFFH	35,840 36,863	35
0100 0000 0000 0000 0100 0011 1111 1111	4000H 43FFH	16,384 17,407	16	1001 0000 0000 0000 1001 0011 1111 1111	9000H 93FFH	36,864 37,887	36
0100 0100 0000 0000 0100 0111 1111 1111	4400H 47FFH	17,408 18,431	17	1001 0100 0000 0000 1001 0111 1111 1111	9400H 97FFH	37,888 38,911	37
0100 1000 0000 0000 0100 1011 1111 1111	4800H 4BFFH	18,432 19,455	18	1001 1000 0000 0000 1001 1011 1111 1111	9800H 9BFFH	38,912 39,935	38
0100 1100 0000 0000 0100 1111 1111 1111	4C00H 4FFFH	19,456 20,479	19	1001 1100 0000 0000 1001 1111 1111 1111	9C00H 9FFFH	39,936 40,959	39

APPENDIX 11. MEMORY LOCATIONS: 1K INTERVALS (*Continued*)

Zone bits = $A_{15}A_{14}A_{13}A_{12}A_{11}A_{10}$

Address Bits	Hexadecimal	Decimal	Zone	Address Bits	Hexadecimal	Decimal	Zone
1010 0000 0000 0000 1010 0011 1111 1111	A000H A3FFH	40,960 41,983	40	1101 0000 0000 0000 1101 0011 1111 1111	D000H D3FFH	53,248 54,271	52
1010 0100 0000 0000 1010 0111 1111 1111	A400H A7FFH	41,984 43,007	41	1101 0100 0000 0000 1101 0111 1111 1111	D400H D7FFH	54,272 55,295	53
1010 1000 0000 0000 1010 1011 1111 1111	A800H ABFFH	43,008 44,031	42	1101 1000 0000 0000 1101 1011 1111 1111	D800H DBFFH	55,296 56,319	54
1010 1100 0000 0000 1010 1111 1111 1111	AC00H AFFFH	44,032 45,055	43	1101 1100 0000 0000 1101 1111 1111 1111	DC00H DFFFH	56,320 57,343	55
1011 0000 0000 0000 1011 0011 1111 1111	B000H B3FFH	45,056 46,079	44	1110 0000 0000 0000 1110 0011 1111 1111	E000H E3FFH	57,344 58,367	56
1011 0100 0000 0000 1011 0111 1111 1111	B400H B7FFH	46,080 47,103	45	1110 0100 0000 0000 1110 0111 1111 1111	E400H E7FFH	58,368 59,391	57
1011 1000 0000 0000 1011 1011 1111 1111	B800H BBFFH	47,104 48,127	46	1110 1000 0000 0000 1110 1011 1111 1111	E800H EBFFH	59,392 60,415	58
1011 1100 0000 0000 1011 1111 1111 1111	BC00H BFFFH	48,128 49,151	47	1110 1100 0000 0000 1110 1111 1111 1111	EC00H EFFFH	60,416 61,439	59
1100 0000 0000 0000 1100 0011 1111 1111	C000H C3FFH	49,152 50,175	48	1111 0000 0000 0000 1111 0011 1111 1111	F000H F3FFH	61,440 62,463	60
1100 0100 0000 0000 1100 0111 1111 1111	C400H C7FFH	50,176 51,199	49	1111 0100 0000 0000 1111 0111 1111 1111	F400H F7FFH	62,464 63,487	61
1100 1000 0000 0000 1100 1011 1111 1111	C800H CBFFH	51,200 52,223	50	1111 1000 0000 0000 1111 1011 1111 1111	F800H FBFFH	63,488 64,511	62
1100 1100 0000 0000 1100 1111 1111 1111	CC00H CFFFH	52,224 53,247	51	1111 1100 0000 0000 1111 1111 1111 1111	FC00H FFFFH	64,512 65,535	63

APPENDIX 12. PROGRAMMING MODELS

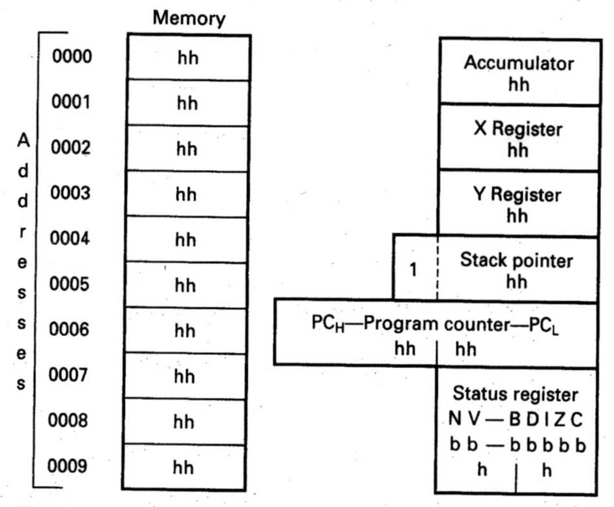

Fig. A12-1 6502 programming model.

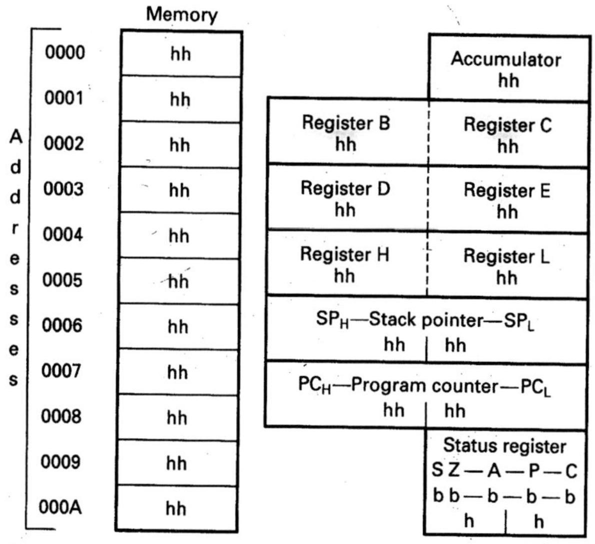

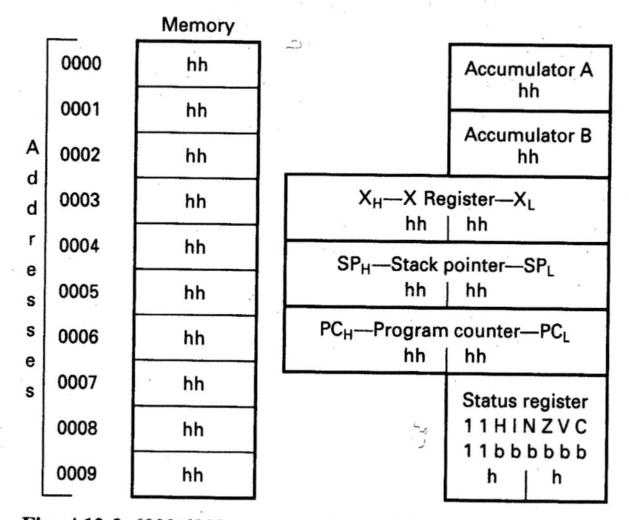

Fig. A12-2 6800/6808 programming model.

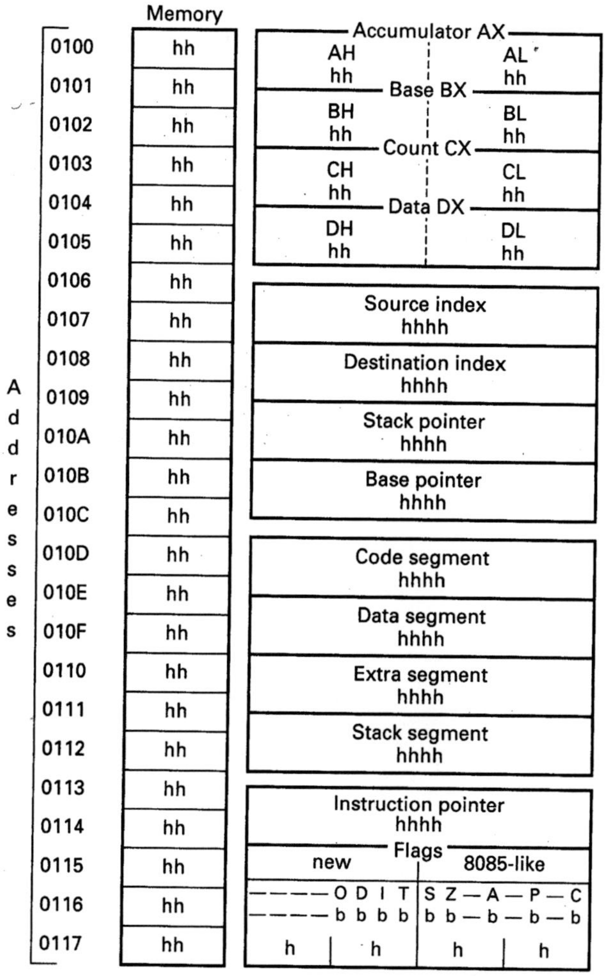

Memory

Addresses

0000 hh
0001 hh
0002 hh
0003 hh
0004 hh
0005 hh
0006 hh
0007 hh
0008 hh
0009 hh
000A hh

Accumulator hh

Register B hh | Register C hh

Register D hh | Register E hh

Register H hh | Register L hh

SP_H—Stack pointer—SP_L
hh | hh

PC_H—Program counter—PC_L
hh | hh

Status register
S Z — A — P — C
b b — b — b — b
h | h

Fig. A12-3 8085/Z80 (8085/8080 subset) programming model.

Memory

Addresses

0100 hh
0101 hh
0102 hh
0103 hh
0104 hh
0105 hh
0106 hh
0107 hh
0108 hh
0109 hh
010A hh
010B hh
010C hh
010D hh
010E hh
010F hh
0110 hh
0111 hh
0112 hh
0113 hh
0114 hh
0115 hh
0116 hh
0117 hh

Accumulator AX
AH hh | AL hh

Base BX
BH hh | BL hh

Count CX
CH hh | CL hh

Data DX
DH hh | DL hh

Source index hhhh

Destination index hhhh

Stack pointer hhhh

Base pointer hhhh

Code segment hhhh

Data segment hhhh

Extra segment hhhh

Stack segment hhhh

Instruction pointer hhhh

Flags
new | 8085-like
— — — — O D I T | S Z — A — P — C
— — — — b b b b | b b — b — b — b
h | h | h | h

Fig. A12-4 8088/8086 programming model.

Answers to Odd-Numbered Problems

CHAP. 1. 1-1. a. 1 b. 2 c. 2½ 1-3. a. 10 b. 2 c. 5 d. 16 1-5. 1,024, 4,096, 8K 1-7. 1010 1100, 172 1-9. 201 1-11. 1100 0111, 199 1-13. 111000 1-15. 1001 0110 1-17. F52B, F52C, F52D, F52E, F52F, F530 1-19. a. 1111 1111 b. 1010 1011 1100 c. 1100 1101 0100 0010 d. 1111 0011 0010 1001 1-21. 0011 1110, 0000 1110, 1101 0011, 0010 0000, 0111 0110 1-23. a. 4,095 b. 16,383 c. 32,740 d. 46,040 1-25. 16,384, 16K 1-27. 0000, FFFF 1-29. a. EE b. 1D7B c. 3BFF d. B8B5 1-31. a. 87 b. 9,043 c. 597,266 1-33. 100 1100, 100 1001, 101 0011, 101 0100

CHAP. 2. 2-1. One or more, one 2-3. Noninverter 2-5. 64, 000000 2-7. 3, 9, C, F 2-9. 128, 1111111 2-11. 0, 59 2-13. $Y = \overline{A + B}$, low 2-15. 8 2-17. 0, $Y = \bar{A} + \bar{B} + \bar{C}$, 000 to 110, 111 2-19. $Y = \overline{ABC}$, 0 2-21. $Y = \overline{AB + CD}$, 16, 0000, 0001, 0010, 0100, 0101, 0110, 1000, 1001, 1010 2-23. a. 0000 b. 0001 c. *JIM* d. *OPR* 2-25. a. Positive b. Negative c. Positive d. Negative

CHAP. 3. 3-1. High; low; inverter 3-3. None, Z_5, Z_6 3-5. Q is 1, $\overline{Q}$ is 0 3-7. Change the output NOR gate of Fig. 3-28*a* to a bubbled AND gate; all bubbles cancel leaving the simplified circuit of Fig. 3-28*b*. 3-9. 0, 1 3-11. 512 3-13. 16; 0, 1, 1, 0 3-15. 1, 0, inverter 3-17. a. None b. Z_7 c. Z_2 d. X_2 and Y_2 3-19. 0, 1 3-21. 512 3-23. Low, high 3-25. a. 0 b. 1 c. 1 d. 1 3-27. a. 11010b. 01001 c. 11111 d. 10010 3-29. Remove the inverter 3-31. a. $CARRY = 0$, $SUM = 0$ b. 0, 1 c. 0, 1 d. 1, 0 3-33. a. 0011 1100 b. 0101 0000 1100 c. 0001 1110 0101 1100 d. 1111 0000 1101 0010

CHAP. 4. 4-1. 1.075 mA, 1.387 mA 4-3. 5 4-5. All; b, c, f, g

CHAP. 5. 5-1. $\overline{AB}C\overline{D}$, $AB\overline{C}D$, $ABC\overline{D}$

5-3.

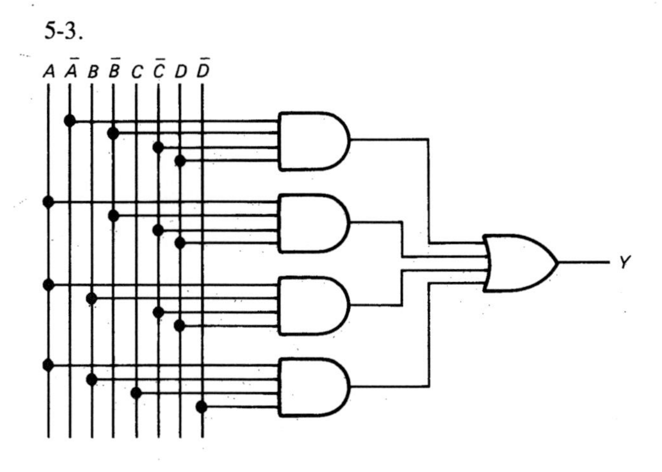

5-5.

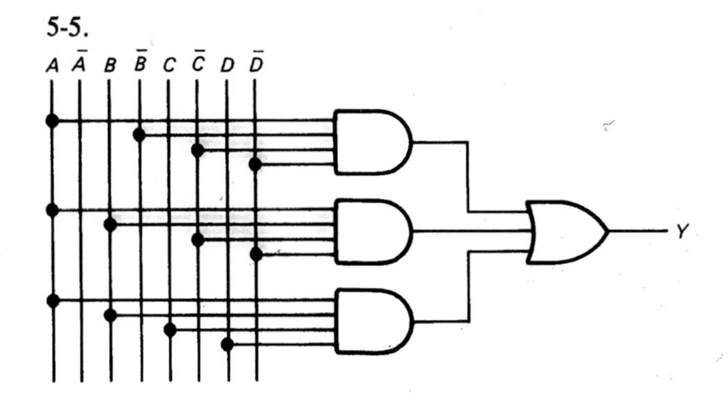

5-7.

	$\overline{C}\overline{D}$	$\overline{C}D$	CD	$C\overline{D}$
$\overline{A}\overline{B}$	0	0	0	0
$\overline{A}B$	0	0	0	0
AB	1	1	1	1
$A\overline{B}$	1	1	1	1

5-9.

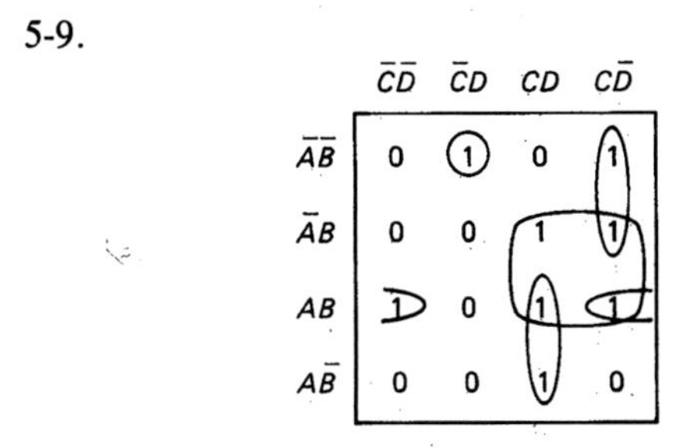

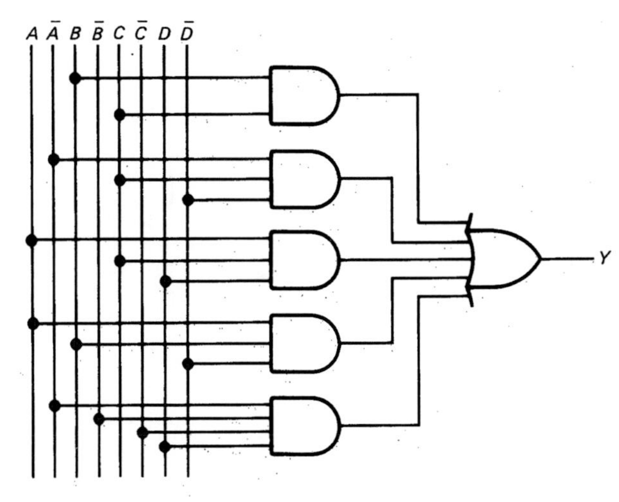

5-11.

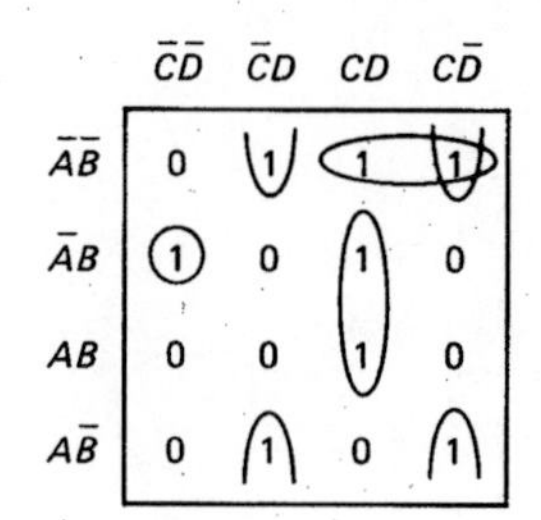

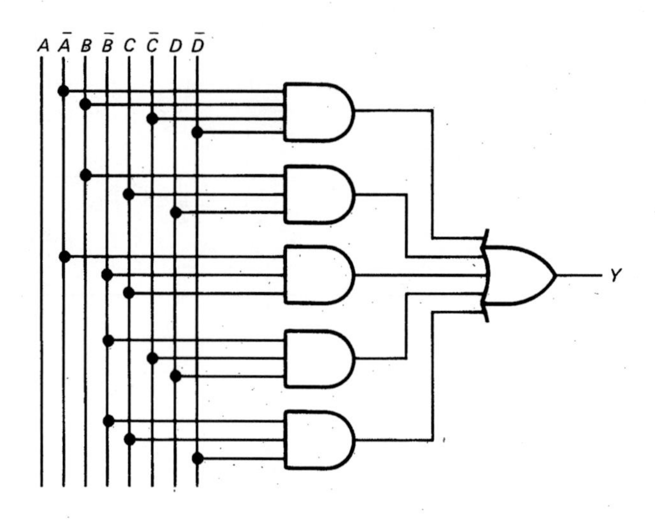

5-13.

	$\bar{C}\bar{D}$	$\bar{C}D$	CD	$C\bar{D}$
$\bar{A}\bar{B}$	0	1	0	1
$\bar{A}B$	0	0	1	1
AB	X	X	X	X
$A\bar{B}$	0	0	X	X

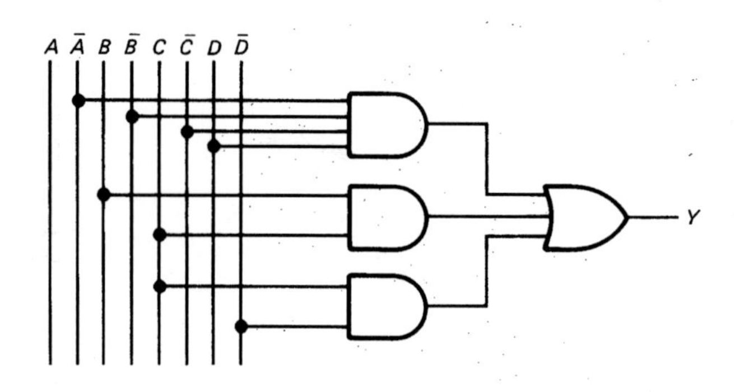

5-15.

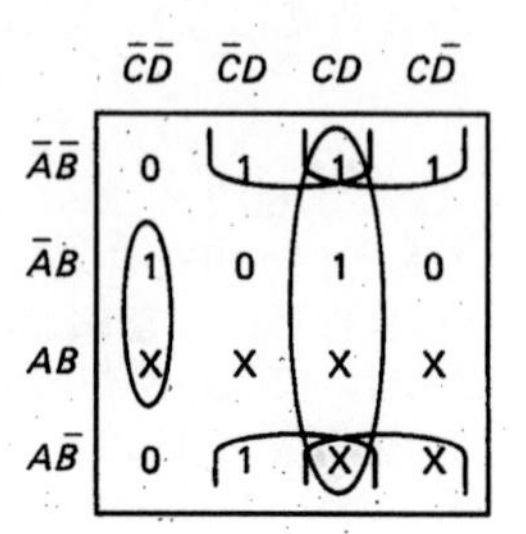

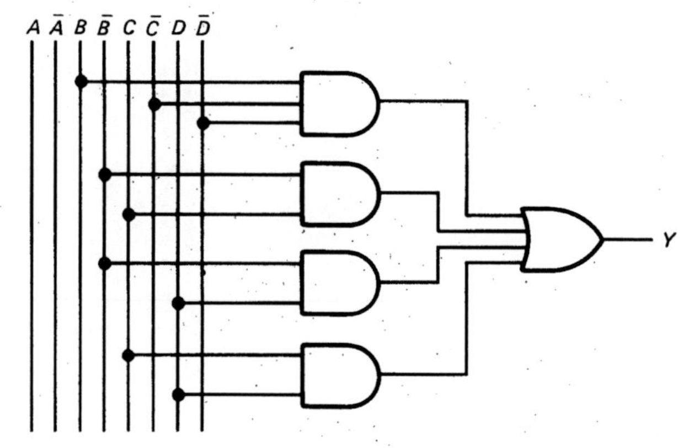

CHAP. 6. 6-1. a. 0001 1000, 18H b. 0010 0100, 24H c. 0010 1010, 2AH d. 0110 0011, 63H 6-3. a. 7BH b. 78H c. A8H d. D1H 6-5. a. +30 b. −7 c. −28 d. +49 6-7. a. F9H b. 01H c. 03H d. 1FH 6-9. a. 1110 1101, EDH b. 1101 0000, D0H c. 0010 0101, 25H d. 1101 1111, DFH 6-11. 9BH, DDH

CHAP. 7. 7-1. a. C b. G 7-3. a. 0000 b. 1001 7-5. 3 MHz; the output frequency is half the input frequency 7-7. $Q = 0, Y = 1; Q = 1, Y = CLK$

CHAP. 8. 8-1. a. 0001 0111 b. 1000 1101 8-3. 385 Ω 8-5. 4 μs 8-7. 6.4 μs 8-9. 65,535 8-11. 1 μs, 6 μs 8-13. 1.6 μs, 0.2 μs 8-15. Two answers: 7490 (divide by 10) and 7492 (divide by 6), or 7490 (divide by 5) and 7492 (divide by 12) 8-17. 136 8-19. a. 0, 1 b. 1, 1 c. 0

CHAP. 9. 9-1. 16,384 9-3. 12

9-5.

Address	Data
DDDD	UDDD UDDU
DDDU	DUUU UUDD
DDUD	DDUU DUUD
DDUU	DDUD DDUU
DUDD	DDDU DUUU
DUDU	DUDU UUUU
DUUD	UUUD UUDU
DUUU	UUUU UDDD

9-7. 63 9-9. BFFFH; 49,151 9-11. a. 47, 212, 207, 110, 83, 122 b. 36,357

CHAP. 10. 10-1.

Address	Mnemonic
0H	LDA DH
1H	ADD EH
2H	SUB FH
3H	OUT
4H	HLT
DH	05H
EH	04H
FH	06H

10-3.

Address	Mnemonic
0H	LDA BH
1H	ADD CH
2H	SUB DH
3H	ADD EH
4H	SUB FH
5H	HLT
BH	08H
CH	04H
DH	03H
EH	05H
FH	02H

10-5.

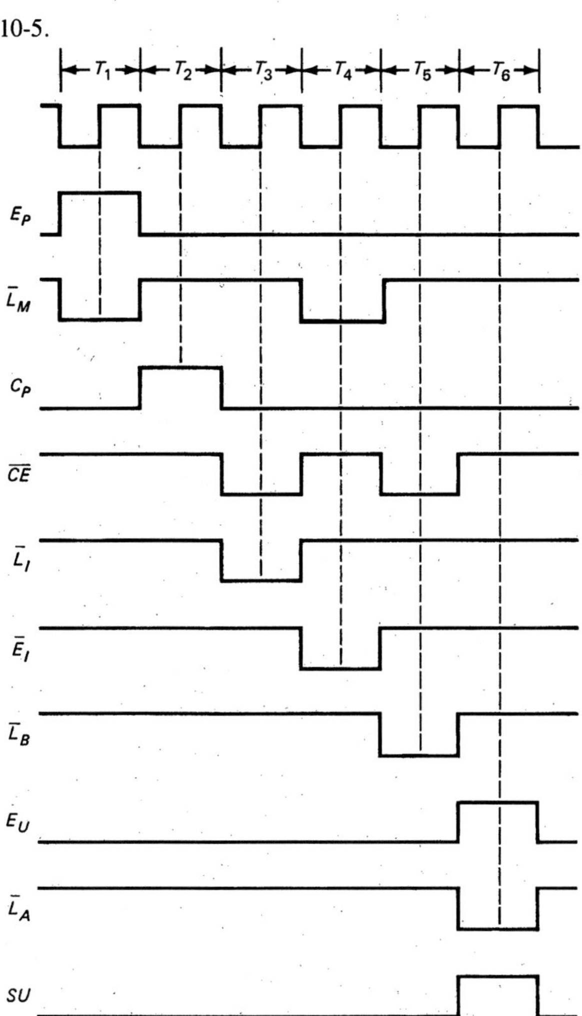

10-7. LDA: 1A3H or 0001 1010 0011, 2C3H or 0010 1100 0011, 3E3H or 0011 1110 0011; SUB: 1A3H or 0001 1010 0011, 2E1H or 0010 1110 0001, 3CFH or 0011 1100 1111 10-9. a. Negative edge; *CLK* is on its rising edge b. High c. Low d. High 10-11. a. Low b. Low c. High

CHAP. 11. 11-1.

Mnemonic
MVI A,64H
MVI B,96H
MVI C,C8H
HLT

11-3.

Mnemonic
MVI A,32H
STA 4000H
MVI A,33H
STA 4001H
MVI A,34H
STA 4002H
HLT

11-5.

Mnemonic
MVI A,44H
MVI B,22H
ADD B
STA 5000H
HLT

11-7. a. 120 b. 119 c. Change the first instruction to MVI C,D2H

11-9.

Mnemonic
MVI A,00H
MVI B,19H
MVI C,07H
CALL F006H
STA 2000H
HLT

11-11.

Label	Mnemonic
	IN 01H
	ANI 01H
	JNZ ODD
	MVI A,45H
	JMP DONE
ODD:	MVI A,4FH
DONE:	MVI C,08H
AGAIN:	OUT 04H
	RAR
	DCR C
	JNZ AGAIN
	HLT

11-13.

Address	Contents
2000H	DBH
2001H	02H
2002H	E6H
2003H	01H
2004H	CAH
2005H	00H
2006H	20H
2007H	DBH
2008H	01H
2009H	32H
200AH	00H
200BH	40H
200CH	76H

11-15.

Address	Contents
2000H	0EH
2001H	23H
2002H	0DH
2003H	C2H
2004H	02H
2005H	20H
2006H	C9H

11-17.

Label	Mnemonic
	MVI A,05H
LOOP:	CALL F020H
	DCR A
	JNZ LOOP
	RET

Address	Contents
E100H	3EH
E101H	05H
E102H	CDH
E103H	20H
E104H	F0H
E105H	3DH
E106H	C2H
E107H	02H
E108H	E1H
E109H	C9H

11-19.

Address	Contents
F080H	3EH
F081H	06H
F082H	32H
F083H	93H
F084H	F0H
F085H	CDH
F086H	60H
F087H	F0H
F088H	3AH
F089H	93H
F08AH	F0H
F08BH	3DH
F08CH	32H
F08DH	93H
F08EH	F0H
F08FH	C2H
F090H	85H
F091H	F0H
F092H	C9H

11-21.

Address	Contents
2000H	D3H
2001H	04H
2002H	0EH
2003H	42H
2004H	0DH
2005H	C2H
2006H	04H
2007H	20H
2008H	2FH
2009H	00H
200AH	C3H
200BH	00H
200CH	20H

CHAP. 12. 12-1.

Mnemonic
MVI A,00H
MVI B,01H
MVI C,59H
MVI D,02H
MVI E,F1H
ADD C
ADD E
MOV L,A
MVI A,00H
ADC B
ADD D
MOV H,A
HLT

An alternative solution is

Mnemonic
MVI A,F1H
ADI 59H
MOV L,A
MVI A,02H
ACI 01H
MOV H,A
HLT

12-3.

Label	Mnemonic
	LXI H,4FFFH
LOOP:	INX H
	MOV B,M
	MOV A,H

```
      ADI 40H
      MOV H,A
      MOV M,B
      SUI 40H
      MOV H,A
      CPI 53H
      JNZ LOOP
      MOV A,L
      CPI FFH
      JNZ LOOP
      HLT
```

12-5.

Label	Mnemonic
	LXI SP,E000H
	MVI A,00H
	MVI B,FFH
LOOP:	INR A
	OUT 22H
	CALL F010H
	DCR B
	JNZ LOOP
	HLT

12-7.

Label	Mnemonic
	LXI SP,E000H
	LXI H,5FFFH
LOOP:	INX H
	MOV A,M
	OUT 22H
	CALL F020H
	MOV A,H
	CPI 61H
	JNZ LOOP
	MOV A,L
	CPI FFH
	JNZ LOOP
	HLT

12-9.

Label	Mnemonic
	LXI SP,E000H
	LXI H,4FFFH
LOOP:	INX H
	MOV A,M
	MOV B,08H
AGAIN:	OUT 22H
	CALL F010H
	RAR
	DCR B
	JNZ AGAIN
	MOV A,L
	CPI FFH
	JNZ LOOP
	HLT

CHAP. 14. 14-1. How you would accomplish your task without a computer. 14-3. Branch. 14-5. The subroutine (part of the program) needs to be written only once but can then be used many times. 14-7. Formula translation. 14-9. Creating a language which would encourage programmers to write by using what are considered "correct" programming practices.

CHAP. 15. 15-1. By its address. 15-3. 1,048,576. 15-5. The accumulator. 15-7. Registers are faster. 15-9. The status register (or condition code register or flag register). 15-11. The carry flag. 15-13. No. 15-15. DE. 15-17. C581. 15-19. 8 bits. 15-21. 256 bytes. 15-23. 16 bits. 15-25. Nothing. They are always set. 15-27. None. 15-29. It is named AX and is 16 bits wide with an 8-bit upper half (called AH) and an 8-bit lower half (called AL). 15-31. The instruction pointer. 15-33. 65,536 bytes.

CHAP. 16. 16-1. Nothing. 16-3. The original number in the accumulator is still there. 16-5. 00. 16-7. It copies the contents of the Y register to the accumulator. 16-9. STY. 16-11. 01. 16-13. 16. 16-15. **CleaR** accumulator **A**.

16-17.

Addr	Obj	Assembler	Comment
0000	C6	LDAB #$89	Load the number immediately following the LDAB op code (C6) into accumulator B (89)
0001	89		
0002	17	TBA	Transfer (copy) the contents of B to A
0003	3E	WAI	Stop

16-19. 76. 16-21. It copies the contents of register C to register B. 16-23. STA aaaa *[LD (aaaa),A]*. 16-25. DEBUG. 16-27. Register or memory. 16-29. DL. 16-31. The contents of memory location 4456_{16}. 16-33. It stands for assemble and it translates 8088/8086 mnemonics into machine code. 16-35. It executes one instruction and then displays the current values of all registers and stops.

16-37.

```
-a
9522:0100 mov BL,89
9522:0102 mov CL,BL
9522:0104

-u 100 103
9522:0100 B389          MOV  BL,89
9522:0102 88D9          MOV  CL,BL

-r
AX=0000  BX=0000  CX=0000  DX=0000  SP=ADDE  BP=0000  SI=0000  DI=0000
DS=9522  ES=9522  SS=9522  CS=9522  IP=0100   NV UP EI PL NZ NA PO NC
9522:0100 B389          MOV     BL,89

-t
AX=0000  BX=0089  CX=0000  DX=0000  SP=ADDE  BP=0000  SI=0000  DI=0000
DS=9522  ES=9522  SS=9522  CS=9522  IP=0102   NV UP EI PL NZ NA PO NC
9522:0102 88D9          MOV     CL,BL

-t
AX=0000  BX=0089  CX=0089  DX=0000  SP=ADDE  BP=0000  SI=0000  DI=0000
DS=9522  ES=9522  SS=9522  CS=9522  IP=0104   NV UP EI PL NZ NA PO NC
```

Note: Answers to Chapters 18 to 23 are in the teacher's manual.

Index

Note: For entries marked with (#), refer also to specific families listed under "Microprocessor families."